VOLUME ONE HUNDRED AND THIRTY FOUR

INTERNATIONAL REVIEW OF NEUROBIOLOGY

Nonmotor Parkinson's: The Hidden Face

Management and the Hidden Face of Related Disorders

INTERNATIONAL REVIEW OF NEUROBIOLOGY

VOLUME 134

SERIES EDITOR

PATRICIA JANAK
Janak Lab, Dunning Hall
Baltimore, MD, USA

PETER JENNER
Division of Pharmacology and Therapeutics
GKT School of Biomedical Sciences
King's College, London, UK

VOLUME ONE HUNDRED AND THIRTY FOUR

INTERNATIONAL REVIEW OF NEUROBIOLOGY

Nonmotor Parkinson's: The Hidden Face

Management and the Hidden Face of Related Disorders

Edited by

K RAY CHAUDHURI

National Parkinson Foundation International Centre of Excellence, Kings College and Kings College Hospital; Maurice Wohl Clinical Neuroscience Institute, Kings College; National Institute for Health Research (NIHR) Mental Health Biomedical Research Centre (BRC), London, United Kingdom

NATALIYA TITOVA

Federal State Budgetary Educational Institution of Higher Education, "N.I. Pirogov Russian National Research Medical University" of the Ministry of Healthcare of the Russian Federation, Moscow, Russia

ELSEVIER

ACADEMIC PRESS

An imprint of Elsevier

Academic Press is an imprint of Elsevier
50 Hampshire Street, 5th Floor, Cambridge, MA 02139, United States
525 B Street, Suite 1800, San Diego, CA 92101-4495, United States
The Boulevard, Langford Lane, Kidlington, Oxford OX5 1GB, United Kingdom
125 London Wall, London, EC2Y 5AS, United Kingdom

First edition 2017

ISBN: 978-0-12-812603-5
ISSN: 0074-7742

Publisher: Zoe Kruze
Acquisition Editor: Kirsten Shankland
Editorial Project Manager: Andrea Gallego Ortiz
Production Project Manager: Surya Narayanan Jayachandran
Cover Designer: Greg Harris

Typeset by SPi Global, India

CONTENTS

Section 3
Therapeutics and Nonmotor Symptoms in Parkinson's Disease

Section 5
Nonmotor Parkinson's and Future Directions

CONTRIBUTORS

Ammar Al-Chalabi
Maurice Wohl Clinical Neuroscience Institute, King's College London, London, United Kingdom

Ľubomíra Anderková
School of Medicine and Movement Disorders Center, St. Anne's University Hospital; Applied Neuroscience Research Group, Central European Institute of Technology, CEITEC MU, Masaryk University, Brno, Czech Republic

Angelo Antonini
University of Padua and Hospital San Camillo IRCCS, Venice, Italy

Richard A. Armstrong
Vision Sciences, Aston University, Birmingham, United Kingdom

Olga V. Babkina
Russian Medical Academy of Professional Continuous Education, Centre of Extrapyramidal Disorders, Moscow, Russia

Kailash P. Bhatia
Institute of Neurology, London, United Kingdom

Kalyan B. Bhattacharyya
RG Kar Medical College & Hospital, Kolkata, India

Samar Biswas
Bangur Institute of Neurosciences, Institute of Post graduate Medical Education and Research, Kolkata, India

Hugo M. Briceño
Movement Disorders Unit, Westmead Hospital, Sydney, NSW, Australia

Francisco Cardoso
UFMG, Belo Horizonte, Brazil

K Ray Chaudhuri
National Parkinson Foundation International Centre of Excellence, Kings College and Kings College Hospital; Maurice Wohl Clinical Neuroscience Institute, Kings College; National Institute for Health Research (NIHR) Mental Health Biomedical Research Centre (BRC), London, United Kingdom

Achcha Sh. Chimagomedova
Russian Medical Academy of Professional Continuous Education, Centre of Extrapyramidal Disorders, Moscow, Russia

Shyamal K. Das
Bangur Institute of Neurosciences, Institute of Post graduate Medical Education and Research, Kolkata, India

Ton Fang
Maurice Wohl Clinical Neuroscience Institute, King's College London, London, United Kingdom

Alessandro F. Fois
Movement Disorders Unit, Westmead Hospital, Sydney, NSW, Australia

Christiana Franke
Charité—Universitätsmedizin Berlin, Berlin, Germany

Steven Frucht
Icahn School of Medicine at Mount Sinai, New York, NY, United States

Victor S.C. Fung
Movement Disorders Unit, Westmead Hospital; Sydney Medical School, University of Sydney, Sydney, NSW, Australia

Marialuisa Gandolfi
Neuromotor and Cognitive Rehabilitation Research Center (CRRNC), University of Verona; Neurorehabilitation Unit, Azienda Ospedaliera Universitaria Integrata, Verona, Italy

Christos Ganos
University Medical Center Hamburg-Eppendorf, Hamburg, Germany

Christian Geroin
Neuromotor and Cognitive Rehabilitation Research Center (CRRNC), University of Verona, Verona, Italy

Sarah Graner
Cedars-Sinai Medical Center, Los Angeles, CA, United States

Elliot Hogg
Cedars-Sinai Medical Center, Los Angeles, CA, United States

Beomseok Jeon
Seoul National University, College of Medicine, Seoul, South Korea

Julia A. Johnson
Kings College Hospital NHS Foundation Trust, London, United Kingdom

Wolfgang H. Jost
Parkinson-Klinik Ortenau, Wolfach; University of Freiburg, Freiburg/Breisgau, Germany

Felix Jozsa
Maurice Wohl Clinical Neuroscience Institute, King's College London, London, United Kingdom

Ryul Kim
Seoul National University, College of Medicine, Seoul, South Korea

Lisa Klingelhoefer
Technical University Dresden, Dresden, Germany

Kimberly Kwei
Icahn School of Medicine at Mount Sinai, New York, NY, United States

Oleg S. Levin
Russian Medical Academy of Professional Continuous Education, Centre of Extrapyramidal Disorders, Moscow, Russia

Anna Lewis
Geriatric Medicine (Movement Disorders), Lincoln County Hospital; University of Lincoln, Lincoln, United Kingdom

Elena A. Lyashenko
Russian Medical Academy of Professional Continuous Education, Centre of Extrapyramidal Disorders, Moscow, Russia

Mariusz Madaliński
The Pennine Acute Hospitals NHS Trust, Manchester, United Kingdom

Davide Martino
University of Calgary, Calgary, AB, Canada

Vinod Metta
Imperial College Hospitals NHS; University College London, London, United Kingdom

Laurie K. Mischley
Bastyr University Research Institute, Kenmore, WA, United States

Ulrika Mundt-Petersen
Lund University, Faculty of Medicine, Lund, Sweden

Per Odin
Lund University, Faculty of Medicine, Lund, Sweden

Chandrasekhara Padmakumar
John Hunter Hospital, Newcastle, NSW, Australia

Neelanjana Paul
ICARE Institute of Medical Sciences and Research, Haldia, India

Tamara M. Pringsheim
University of Calgary, Calgary, AB, Canada

Heinz Reichmann
Technical University Dresden, Dresden, Germany

Irena Rektorová
School of Medicine and Movement Disorders Center, St. Anne's University Hospital; Applied Neuroscience Research Group, Central European Institute of Technology, CEITEC MU, Masaryk University, Brno, Czech Republic

Miguel Rosa-Grilo
National Parkinson Foundation (NPF) International Center of Excellence at King's College Hospital; The Maurice Wohl Clinical Neuroscience Institute; National Institute for Health Research (NIHR) Mental Health Biomedical Research Centre, Institute of Psychology, Psychiatry and Neurosciences, King's College London, London, United Kingdom

Raymond L. Rosales
University of Santo Tomas Hospital, Manila, Philippines

Tamara C. Sanchez
Hospital Universitari Germans Trias i Pujol, Barcelona, Spain

Jarosław Sławek
Medical University of Gdańsk; St. Adalbert Hospital, "Copernicus" Ltd., Gdańsk, Poland

Jagdish C. Sharma
Geriatric Medicine (Movement Disorders), Lincoln County Hospital; University of Lincoln, Lincoln, United Kingdom

Janine A. Simons
University of Lübeck, Lübeck, Germany

Natalia A. Skripkina
Russian Medical Academy of Professional Continuous Education, Centre of Extrapyramidal Disorders, Moscow, Russia

Nicola Smania
Neuromotor and Cognitive Rehabilitation Research Center (CRRNC), University of Verona; Neurorehabilitation Unit, Azienda Ospedaliera Universitaria Integrata, Verona, Italy

Maria Stamelou
HYGEIA Hospital; University of Athens, Athens, Greece; Neurology Clinic, Philipps University Marburg, Marburg, Germany

Fabrizio Stocchi
University and Institute for Research and Medical Care, IRCCS San Raffaele, Rome, Italy

Alexander Storch
University of Rostock; German Centre for Neurodegenerative Diseases (DZNE) Rostock, Rostock, Germany

Indu Subramanian
UCLA/West LA VA, Los Angeles, CA, United States

Michele Tagliati
Cedars-Sinai Medical Center, Los Angeles, CA, United States

Michele Tinazzi
Neurology Unit, Movement Disorders Division, University of Verona, Verona, Italy

Nataliya Titova
Federal State Budgetary Educational Institution of Higher Education "N.I. Pirogov Russian National Research Medical University" of the Ministry of Healthcare of the Russian Federation, Moscow, Russia

Julie Ann Kristy L. Torres
University of Santo Tomas Hospital, Manila, Philippines

Margherita Torti
University and Institute for Research and Medical Care, IRCCS San Raffaele, Rome, Italy

Jeffrey Wertheimer
Cedars-Sinai Medical Center, Los Angeles, CA, United States

PREFACES

MESSAGE FROM PROFESSOR GÜNTHER DEUSCHL, EUROPOEAN ACADEMY OF NEUROLOGY

Parkinson's disease is one of several important degenerative diseases, but it has become the model for research and clinical progress for neurodegenerative conditions in general. Many aspects of neurodegeneration have been studied and successfully understood for the first time in this disease. It was the first disease for which a clearly identified pathology was discovered, the first with a known transmitter deficit which could be replaced, the first in which cell transplantation for the brain was used, and the first for which deep brain stimulation was discovered as a knowledge-based therapy.

While most of this significant progress has been made during the past 5 decades, the study of the clinical symptoms lagged somewhat behind, and nowadays it is hard to understand why the clinicians' views were so much focused on the motor symptoms alone. This is nicely documented by looking at the Webster Scale of 1968 which contained Seborrhea as the only nonmotor symptom. The Hoehn and Yahr Scale of 1967 which we still currently use is only focused on the motor symptoms and does not even mention dementia in stage 5 of the condition. Certainly psychiatric issues were also dealt with in these days, but it was considered a concomitant disease—mostly Alzheimer's—rather than an integral part of Parkinson's disease. Urinary problems were certainly also recognized and treated, but it was not considered an important feature of the disease process. Only a handful papers were discussing such nonmotor features as a manifestation of the disease itself in the past century, and the systematic study of nonmotor symptoms started not before the early 21st century.

Beyond a number of review articles the current book is for the first time summarizing the current knowledge on nonmotor conditions, their pathologic, biochemical, and neurophysiological basis on the one hand and the many clinical manifestations on the other hand. Particular care has been taken to cover the therapeutic options for nonmotor symptoms which we are just starting to understand. K Ray Chaudhuri, and the Non-Motor Study Group of the Movement Disorder Society (particularly the trainee section represented by Nataliya Titova) which he leads have done an excellent job in systematically discover the many ramifications of clinical symptoms. Together with the new concepts of progression of disease pathology and the subsequent involvement of many different neuronal circuits we now understand these symptoms as part and parcel of the "whole-brain" disease—Parkinson's disease. This book is a wonderful and concise summary of the progress which has been made so far. The authors and the editors owe our thanks for undertaking this huge task.

I wish this book a broad readership and hope that it will further help to disseminate the new concept of Parkinson's disease, and that it will help to better understand the disease, our patients, and their needs.

Günther Deuschl
President of the European Association of Neurology (EAN)

MESSAGE FROM PROFESSOR CHRISTOPHER G. GOETZ, INTERNATIONAL PARKINSON'S AND MOVEMENT DISORDERS SOCIETY

Coinciding with the 200th anniversary of the celebrated *Essay on the Shaking Palsy* by James Parkinson, this remarkably complete work focuses on a burgeoning area of research and clinical challenge: the nonmotor elements of Parkinson's disease. The importance of nonmotor problems in a disease that is diagnosed by criteria strictly related to the motor system may seem counterintuitive, but the gamut of autonomic, cognitive, and behavior elements of Parkinson's disease justifies this pointed focus. The editors have recruited a wide array of international authors, some the most senior in the field and some more junior colleagues with high credentials, to compose a large, two-volume reference source with chapters covering historical, clinical, genetic, neurochemical, and physiological underpinnings of nonmotor elements of Parkinson's disease and related disorders. Such topics have been summarized briefly in prior texts, but basic scientists and the neurological, psychiatric, and medical professionals who confront Parkinson's disease on a daily basis have not heretofore had a comprehensive reference source of this dimension. An excellent introductory chapter by B. Hurwitz on Parkinson himself and his appreciation of nonmotor symptoms in *paralysis agitans* documents the few words Parkinson devoted to these problems and places his seminal study in historical context. Genetics issues, so important to 21st century science, are covered in two particularly well-developed chapters by E.-K. Tan and C. Klein. More elusive topics of pivotal concern to clinicians are discussed with sophistication, two examples being the excellent chapter on Apathy by J. Pagonabarraga and J. Kulisevsky, and a highly up-to-date discussion of Fatigue by B. Kluger. In an era of increasing focus on ancillary interventions and resources, Nutrition, Exercise, and the concept of Personalized Care are included into the Management section. Thanks to the editors' particular attention, each chapter is a separate and autonomous discussion, but works well with the entire volume, ensuring a natural flow from one topic to another.

Setting this large and impressive new work next to James Parkinson's tiny 1817 pamphlet is a fitting visual image of the progress made over 200 years. This current contribution elegantly expands on the observations by Parkinson, but the editors and authors are duly cognizant that the *Essay* remains the inspirational source for this modern work. *Nonmotor Symptoms:*

The Hidden Face of Parkinson's beautifully and humbly complements the original work by Parkinson, honoring 200 years of scientific progress and bearing no intent to replace Parkinson's original work. In my view, the tenor taken by the author team and the editors echoes the clarity and integrity for which James Parkinson himself was known. In this light, I cite the words of Jean-Martin Charcot, the most celebrated clinical neurologist of the 19th century whose 1888 comments on Parkinson's essay resonate with regard to the quality of this new scientific contribution:

> Read the entire book and it will provide you with the satisfaction and knowledge that one always gleans from a direct clinical description made by an honest and careful observer.

Christopher G. Goetz, MD
Professor of Neurological Sciences, Professor of Pharmacology
Rush University Medical Center, Chicago, IL, United States
President of the International Parkinson and Movement Disorder Society 2017–2019

Citation: Charcot, J-. M. (1887–1888). Leçon du 12 juin 1888. *Leçons du mardi à la Salpêtrière* (pp. 526–541). Paris: Bureau du progrès. In English: Goetz, C. G. (1987). Lesson of June 12 1888. *Charcot, the clinician: The Tuesday lessons* (pp. 123–140). New York: Raven Press.

THE EDITORS WORDS

James Parkinson the famous London physician who described Parkinson's disease (PD) 200 years back was an astute clinician and observer. He was said to have given several Ale-House (currently known as a pub) sermons where he spoke on medicine and geology. His medical foresight is reflected in the fact that he described not just the core motor features but also several non-motor features we now regard as integral to the phenotype of PD. These include dribbling of saliva, dysphagia, constipation, sleep dysfunction, delirium, and pain. In addition, he also mentioned presymptomatic "brachial" pain. This was the first suggestion that nonmotor symptoms (NMS) may appear in the prodromal stage of PD.

Yet the scientific community was distracted by the dramatic response of motor symptoms of PD to levodopa, a seminal discovery in the 1960s. Unfortunately this meant that the critical role of NMS in PD was somewhat ignored. While individual NMS were recognized, no measures addressed the holistic concept of NMS till the early 2000. Relevant holistic tools were developed and validated in early 2000 around the time when Bill Langston described the NMS of PD as the submerged part of the iceberg that is PD.

This book is therefore dedicated to celebrate the 200 years since James Parkinson described the eponymous condition. The book is meant to celebrate the advances we have made in the nonmotor aspects of PD in the last 100 years. The concept of NMS in PD ranges from neuropathological heterogeneity highlighted by work from Braak as well as Jellinger to a multitude of neurotransmitter deficiency and clinical phenotypes supported by varying biomarkers. We have tried to highlight these issues in the chapters included in two volumes of this book, possibly the most extensive book on the topic ever written.

The book reflects a global reach and we are indeed grateful to key opinion leaders around the world who have agreed to contribute to the book in spite of working through a holiday period in December–January 2016–2017. The range of topics also reflects the diversity of NMS in PD, but we have also for the first time attempted to address NMS in parkinsonism as well as related movement disorders. These include essential tremor, cerebrovascular disease, chorea, Wilson's disease, Huntington's disease, amyotrophic lateral sclerosis, tics, and dystonia. An unique feature is a carer-focused chapter contributed to by a real-time caregiver for PD as well as contributions from allied health specialists. There are also chapters focused on palliative care, exercise, and complimentary therapies and their impact on NMS.

There are several other points related to this book that is worth mentioning. The concept and the idea for this book was developed between Nataliya Titova (N.T.) and K Ray Chaudhuri (K.R.C.) based on the fact that many general neurologists did not appear to have a grasp of nonmotor issues in PD. This unmet need brought together a relatively established clinical scientist (K.R.C.) coediting this book with an emerging neurologist and scientist (N.T.) representing the trainee as well as the junior subsection of the movement disorders community. The trainee and junior involvement is also evident in several chapters where the trainees have either taken the lead or coauthored chapters. The engagement of trainees as well as juniors is indeed a key strategy for the MDS as well as the EAN, and we are privileged that the presidents of both societies have written forewords for this book.

It is sad that in spite of the foresight of Parkinson, nonmotor aspects of PD remain a key unmet need in PD. Although awareness has grown and learned societies such as the MDS have now dedicated nonmotor study groups, yet for much of the world PD remains a pure motor disorder. The overwhelming evidence that NMS burden drives quality of life of the patient and carers thus has remained unexplored and is the hidden face of PD. The book cover we chose reflects this fact and comes from the Ukrainian artist Oleg Shupliak. Oleg Shupliak is known for illusory or camouflage painting which are presented as if not one, but two, or even three pictures. The viewers can discover for themselves the different images that are apparent or hidden. We felt this style of art is analogous to our message of NMS of PD being the hidden but a real face of PD. The cover hides the face of the famous Dutch painter Van Gogh. A museum celebrating the life and works of Van Gogh is located in Amsterdam, and it so happens that the EAN 2017 congress was held in Amsterdam. This was the reason we aimed for the book launch in this city.

From a personal angle both K.R.C. and N.T. acknowledge the enormous effort and work that went into the planning of this book over the Christmas period in 2016 and the gratitude we feel toward all authors, most of whom we approached over this period. Their contributions, within a strict deadline we enforced, have ensured that the book is not just destined for the shelf but is credible, relevant, and hopefully will be widely cited.

Finally the book will not be complete without the support of Elsevier Press and in particular Andrea Gallego Oritz who has worked ceaselessly to support us and other authors. N.T. would also like to acknowledge the support from the Department of Neurology, Neurosurgery and Medical Genetics of N.I. Pirogov Russian National Research Medical University. In particular, she would like to acknowledge the help, support, and advice from Professor Elena Katunina. In addition, she would also acknowledge Professor G.N. Avakyan for entry to the world of neurology and movement disorders and Professor E.I. Gusev, the head of the department, for the opportunity to develop her career in movement disorders. We also acknowledge the media interest in the book in countries such as India and hope that general neurologists, physicians, as well as trainees from all walks of neuroscience and related specialities will benefit from the chapters. We also hope allied health specialists as well as expert PD patients and carers will find the book of interest. If this is achieved, we will feel we have gone someway toward unraveling the hidden face of PD.

K Ray Chaudhuri and Nataliya Titova

SECTION 2B

Specific Nonmotor Symptoms in Parkinson's Disease, Part 2

CHAPTER TWENTY-FIVE

Autonomic Dysfunction in Parkinson's Disease: Cardiovascular Symptoms, Thermoregulation, and Urogenital Symptoms

Wolfgang H. Jost[*,†,1]
*Parkinson-Klinik Ortenau, Wolfach, Germany
†University of Freiburg, Freiburg/Breisgau, Germany
[1]Corresponding author: e-mail address: w.jost@parkinson-klinik.de

Contents

Abstract

Parkinson's disease is characterized by motor, autonomic, and neuropsychiatric symptoms.

These occur in varying degrees in all stages of the disease. Among the autonomic disorders, cardiovascular, urogenital, gastrointestinal, and thermoregulatory disorders are particularly relevant. Because of the significant impact on the quality of life, appropriate diagnostics and therapy should be carried out at all stages of the disease. Within cardiovascular disorders drop of blood pressure after orthostasis and nondipper behavior are very important, but also the influence of cardiovascular medication. Nocturnal

International Review of Neurobiology, Volume 134
ISSN 0074-7742
http://dx.doi.org/10.1016/bs.irn.2017.04.003

sweating is the most common problem of thermoregulation disturbances, for which there is no satisfactory therapy. Urgency, nocturia, and incontinence are of particular note within the urological problems.

1. INTRODUCTION

Autonomic regulatory disorders were long considered to be but late symptoms in Parkinson's disease (PD), but this has since proven erroneous. They can in fact present prior to diagnosis as well as in all subsequent phases of the disease (Jost, 1999; Jost & Augustis, 2015; Korchounov, Kessler, Yakhno, Damulin, & Schipper, 2005). Parkinson (1817) already identified such symptoms in almost all areas of autonomic functioning in his patients. The undeniable success of levodopa (L-dopa) treatment was certainly one cause for our focus of attention to become so limited as only to monitor motor symptoms and their therapy.

Consensus has in the meantime been attained taking PD neither as a purely nigrostriatal degeneration nor as beginning in the substantia nigra. There is still debate on whether the disease initiates in the periphery and ascends centrally (Braak, Rüb, Gai, & Del Tredici, 2003; Pan-Montojo et al., 2010), or whether peripheral and central disease processes evolve parallel to each other. There is clinical confirmation rather for the involvement of the entire autonomic nervous system. The influence of autonomic disturbances on quality of life is tremendous in all stages of the disease (Barone et al., 2009; Duncan et al., 2014).

The cause of the symptoms is an involvement both of the central and the peripheral, postganglionic autonomic nervous system due to the underlying disease, which has been well recognized for over 50 years now (Den Hartog Jager & Bethlem, 1960). Morphological studies have demonstrated Lewy bodies at cerebral and spinal levels, in sympathetic trunk ganglia, in cardiac nerve cells, in the Gl. submandibularis and the intestinal plexus, and further areas of the peripheral, autonomic nervous system (Adler et al., 2016; Braak, de Vos, Bohl, & Del Tredici, 2006; Den Hartog Jager & Bethlem, 1960; Fasano, Visanji, Liu, Lang, & Pfeiffer, 2015; Jost, 1999).

^{123}I-meta-iodobenzylguanidin (MIBG) scintigraphy has already proven useful in differential diagnostics, for example, for the indirect demonstration of peripheral, postganglionic degeneration. And more recent work has shown that such a finding correlates with the stage of the disease (Fujishiro et al., 2008), although this has not yet been secured reliably for stage I according to Hoehn and Yahr (Jost, Del Tredici, Landvogt, & Braune, 2010).

2. CARDIOVASCULAR SYMPTOMS

Considering the various autonomic complaints, cardiovascular disturbances have evoked the most intensive clinical and scientific interest, on the one hand because they frequently induce serious symptoms and, on the other because different well standardized and noninvasive methods investigation are readily available. There is no one singular test for research of all cardiovascular disturbances (Jost, 1999; Oka et al., 2007), but the lying-to-standing orthostatic tests (the modified Schellong Test) are viewed as an effective screening tool. The most important method of differential diagnostics is MIBG scintigraphy (Jost et al., 2010). Further work will be needed, however, to determine whether specific ECG analyses are reliable enough for early diagnoses.

Cardiovascular symptoms present in all phases of the disease, and thus, even in de novo patients (Strano et al., 2016). Approximately half of the patients complain of nonspecific symptoms such as transient visual impairment, nausea, lightheadedness, and dizziness. Loss of consciousness is rare, although serious injury can easily ensue in such cases. A tendency toward hypotension is often reinforced by ingestion (Jost, 1999; Korczyn, 1990).

The most frequent symptom detailed is orthostatic hypotension, occurring in over half the patients (Allcock, Ullyart, Kenny, & Burn, 2004). Postprandial hypotension has been reported (Umehara, Nakahara, Matsuno, Toyoda, & Oka, 2016). Furthermore, all other cardiovascular regulatory mechanisms have been described as dysfunctioning. As a case in point, in an investigation of five noninvasive cardiovascular functional tests in 100 Parkinson patients, only three had a completely inconspicuous finding (Jost, 1999), while 72 patients demonstrated pathological results in at least two tests. Altogether, pathological or conspicuous findings were found for 84% of the patients.

Based on a large number of publications, resting blood pressure can reliably be described as tendentially lower in Parkinson patients, without being significantly different from that of the normal population. But it is striking that Parkinson patients less often present with hypertonia (Jost, 1999).

It must be kept in mind that Parkinson patients are so-called nondippers (Oh et al., 2013; Sommer, Aral-Becher, & Jost, 2011): that is, they do not show nocturnal lowering of blood pressure. This has immediate clinical relevance and should be considered carefully when blood pressure is being adjusted, which means that long-term blood pressure measurement should definitely be conducted before any attempt at adjustment. Cardiovascular disturbances correlate neither with the dopaminergic deficit nor with motor fluctuations (Oh et al., 2013; Storch et al., 2013).

Several studies (Chaudhuri et al., 1997; Thomaides et al., 2005; Umehara et al., 2016) have clearly shown that ingestion in patients with an idiopathic Parkinson syndrome leads to both (1) a worsening in cardiovascular adaptation and the occurrence of an otherwise subclinical orthostatic dysregulation and (2) motor symptoms (particularly in the advanced stages of the disease). Admittedly, there are also studies which failed to find any relevant influence of ingestion on resting blood pressure (Jost, 1999).

2.1 Blood Pressure and Pulse Reaction in Orthostasis

A distinct fall in blood pressure is found in many PD patients with active *orthostasis* and frequently becomes symptomatic. Such a pathological drop in pressure has been documented in all stages of the disease (Jost & Augustis, 2015). It is striking to note that the drop is not only distinct but also that it does not normalize within the first 10min (Jost, 1999).

There is some debate in the literature as to the relationship of *orthostasis* to the severity and duration of PD. Orthostatic hypotension can present simultaneously with or prior to the emergence of motor symptoms (Goldstein, 2006; Jost & Augustis, 2015; Korchounov et al., 2005). A recent publication demonstrated pathological hypotension in many de novo patients and found furthermore that there was no significant deterioration over the course of the disease (Jost & Augustis, 2015).

The cause of this orthostatic hypotension is most probably multifactorial. A deficient adaptation in blood pressure occurs under strain. Frequently, blame is placed on a lack of activity, that is, insufficient circulatory training. An essential factor is also seen in fact in the drug treatments themselves. But basically the hypotension in PD should be considered an intrinsic symptom of the primary disease itself (Jost, 1999), a symptom which is then reinforced by additional factors. The cardiovascular symptoms are most probably independent of the striatal deficiency in dopamine (Oh et al., 2013).

In PD, degeneration is of course found in the nigrostriatal system, but has also been found in the central and the peripheral autonomic nervous system (Braak et al., 2002; Jost, 1999). And the consequent influence on hormonal regulation can be viewed as an essential cause of the autonomic functional disturbances (Jost, 1999). There is thus room for discussing whether damage also to the postganglionic sympathetic efferences is of significance (Kaye, Gage, Kimber, Storey, & Trend, 2006). As a result of the extensive degeneration in the noradrenergic system in PD patients, reduced norepinephrine concentrations can be seen in the corresponding brain regions. In the case of active orthostasis, there is a distinct reduction in the rise of serum noradrenaline (Bellon, Jost, Schimrigk, König, & Zeppezauer, 1996). It is now generally held that a systemic sympathetic denervation exists with varying degrees of severity and distribution (Goldstein, Holmes, Dendi, Bruce, & Li, 2002).

2.2 Influence of the Parkinson Therapy on Cardiovascular Regulation

One explanation regularly advanced for the very frequent cardiovascular disturbances in PD entails the influence of the standard drug treatments themselves. A distinct tendency toward hypotension has been especially found for treatment with L-dopa and/or dopamine agonists (Jost, 1999). This, however, disregards the observations that (1) hypotension tendencies in PD patients with orthostasis were described prior to the introduction of drug treatments, in particular L-dopa (in the 1960s), and that (2) the same holds true for as yet untreated patients (Jost, 1999; Jost & Augustis, 2015; Parkinson, 1817). Furthermore, in early descriptions, in which a marked hypotensive effect of L-dopa was described, there was either no additional application of a decarboxylase inhibitor or else a very high dose was administered up to a gram or more (De Deurwaerdèrea, Di Giovanni, & Milland, 2017). Based on current information, the hypotonic effect manifests itself especially in the initial phase of L-dopa therapy (Jost, 1999). After longer treatment, blood pressure values regress to previous levels (Jost, 1999).

In addition to a possible reinforcement of hypotension under L-dopa treatment, a similar effect has been seen with dopamine agonists. Particularly when using ergoline substances a negative influence can be expected on norepinephrine release, although the differences between the different substances can be substantial. Whether the weaker influence of nonergoline agonists (Jost, 1999) on norepinephrine release can be confirmed will depend on further work. An essential influence of COMT inhibitors is improbable (Lyytinen, Sovijarvi, Kaakkola, Gordin, & Teräväinen, 2001).

For therapy with anticholinergics and other Parkinson medications, a negative influence on blood pressure and cardiovascular functional tests has been described, which could be relevant as well (Jost, 1999).

A good many patients presenting with high blood pressure require in the course of the disease less blood pressure medication. This is probably due to the primary degenerative process.

2.3 Therapeutic Possibilities for Cardiovascular Problems

When orthostatic problems occur in Parkinson patients, conservative therapeutic measures are primarily recommended (Low & Singer, 2008). These involve an elevated position for the head during sleep periods (so as to reduce sodium release while in a horizontal position) as well as sufficient fluid intake (2–3 l/d) and in some few cases a sodium-enriched diet (which may only be partially helpful). The use of support stockings is advisable here, although the effect is only limited. Regular balneological or physiotherapeutic treatment for training the circulation system is also useful. If the patients complain particularly of postprandial hypotension, several smaller meals can be recommended. The quality, however, of such nondrug treatments is limited (Low & Singer, 2008; Schoffer, Henderson, O'Maley, & O'Sullivan, 2007).

Altering the Parkinson-specific medication has to considered with special care as a reduction of dosage can clearly worsen the symptoms and thus the general status of the condition. Nonetheless, medications which can be assumed to have a substantial influence on blood pressure are preferentially the ones to be altered. The positive influence of domperidone (10 mg tid up to 20 mg tid) must be viewed as only slight (Schoffer et al., 2007). Only for the exceptional case (when the patient suffers from substantial hypotension and the measures described earlier fail to give improvement) and only then can administering fludrocortisone or midodrine (an alpha receptor agonist) be discussed (Low & Singer, 2008; Schoffer et al., 2007). Yohimbine is now seen as obsolete due to central nervous effects, and dihydroergotamine is insufficiently effective. Pyridostigmine is under discussion (Low & Singer, 2008).

Midodrine can be considered the most effective drug here (Izcovich, González Malla, Manzotti, Catalano, & Guyatt, 2014; Low & Singer, 2008); e.g., 2.5 mg in the morning, up to 2.5 mg tid, in severe cases up to 5 mg tid).

Fludrocortisone (0.05–0.2 mg/die) is the most potent synthetic mineralocorticoid and primarily functions through increasing sodium reabsorption by raising the intravasal volume and the cardiac output as well as secondarily through influencing the electrolyte household, meaning that it compensates for hyperaldosteronism.

Using droxidopa (L-Threo-DOPS) has proven advantageous. The substance, however, has only been approved in certain countries, although it can be obtained through the international pharmacy (Northera®, 100/200/300 mg). But conclusive data are still lacking (Hauser, Isaacson, Lisk, Hewitt, & Rowse, 2015).

It is essential while working with Parkinson patients that any tendency toward hypotension is kept in mind and hypotension-inducing drugs must be avoided.

There are repeated descriptions of deep brain stimulation having a positive influence on autonomic regulations disturbances. This is also the case for orthostatic dysregulation (Stemper et al., 2006).

3. THERMOREGULATION SYMPTOMS (PERSPIRATION-RELATED SYMPTOMS)

One-third to a half of Parkinson patients can be assumed to have a disturbance in their perspiration regulation (Jost, 1999). Defective thermoregulation seems to be more severe in MSA patients, but is very common in PD patients as well (Augustis, Saferis, & Jost, 2017). Both an increase (Fischer, Gemende, Marsch, & Fischer, 2001; Jost, 1999; Oka et al., 2007) and a decrease (Aminoff & Wilcox, 1971; Jost, 1999) in perspiration have been described. Most often the patients complain of profuse perspiration attacks.

Episodic and strong attacks of perspiration without a stimulus (such as physical strain, increased external temperature, and fever) occur in as much as half of the patients (Jost, 1999; Oppenheim, 1902). Sweating in these cases involves prominently the face and the upper body. These attacks occur most often at night and the patients complain not only of sweating but also of a feeling of heat, anxiety, and fatigue. In many cases, perspiration is so pronounced that a change of clothing is required (Jost, 1999; Oppenheim, 1902; Taly & Muthane, 1992). Interestingly, two-thirds of the cases of sweating attacks occur simultaneously with a severe akinesia or an off-state (Pursiainen, Haapaniemi, Korpelainen, Sotaniemi, & Myllylä, 2007; Taly & Muthane, 1992; Witjas et al., 2002). On the other hand perspiration is often independent of the motor state (Ossig et al., 2016).

A decrease in perspiration tendency or in fact a lack of perspiration secretion has also been reported on in different research groups (Aminoff & Wilcox, 1971; Jost, 1999).

The cause of this distinct disturbance in perspiration could be a participation of the hypothalamus, that is, the center of thermoregulation.

Due to the disturbance in thermoregulatory perspiration, many Parkinson patients feel unwell, especially in warm temperatures. In addition to the disturbance in perspiration a lower body temperature has also been described, in particular for nighttime and early mornings (Martignoni, Pacchetti, Godi, Micieli, & Nappi, 1995).

3.1 Therapy for Disturbances in Perspiration

When patients present with distinct disturbances in perspiration, it can be of advantage to give anticholinergics, for example, glycopyrrolat (1–3 mg/die) or pirenzepine 25–100 mg (because of the cholinergic transmission) and possibly also beta receptor blockers. Especially when episodic and profuse perspiration is the case, administering propranolol has been reported as being of help. In the case of focal hyperhidrosis, which is however seldom, the use of topic externa on an aluminum base or injections of botulinum toxin can be considered. Conservative therapy with sage teas or extracts (3–4 × 100 mg) is but rarely successful.

Of course, attention has to be given first and foremost to attaining sufficient dopaminergic stimulation, as a nonmotor fluctuation just might be possible here (nonmotor-off/NMOS) (Ossig et al., 2016).

4. URINARY SYMPTOMS

The earliest reports of PD already mentioned urogenital disturbances (Lewy, 1913; Parkinson, 1817). Urogenital and bronchopulmonary infections are the most frequent causes of death in syndrome.

Patients complain particularly of imperative urinary urgency, pollakisuria, nycturia, and urinary incontinence (Jost, 2013). In a good many cases, the bladder problems are not even observed by the patients themselves in the early phase. In addition to micturition problems and urinary incontinence disturbances of sexuality are also prominent.

4.1 Disturbances of Micturition

The causes for bladder disturbances have not yet been explained to satisfaction. One explanation entails the degeneration of the dopaminergic system, relying on the fact that the basal ganglia and specifically dopamine have an inhibitory effect on micturition. Damage to the dopaminergic system results in hyperactivity in the detrusor muscle of the bladder (Sakakibara, Uchiyama, Yamanishi, & Kishi, 2010) and possibly in a deficient regulation of M. sphincter urethrae externus. There are a number of further alternative causes in consideration, for example, an impairment in prefrontal inhibition (Jost, 2013). Detrusor contractions can be viewed as an indication of extra-pyramidal participation, and peripheral neurogenic damage can be seen as a participation of corticospinal pathways (Andersen, Hebjørn, Frimodt-Møller, Walter, & Worm-Petersen, 1976).

Problems with the bladder present in 37%–93% of PD patients examined (Aminoff & Wilcox, 1971; Jost, 1999; Martignoni et al., 1995). The majority of urodynamic studies describe detrusor hyperactivity in these patients (Andersen et al., 1976; Jost, 1999; Raz, 1976). They also repeatedly detailed a combination of an increase in detrusor pressure and incomplete micturition.

But there are also studies describing detrusor hypoactivity in Parkinson patients (Aminoff & Wilcox, 1971; Jost, 2013). The relevant question here is whether the bladder is secondarily weakened by an obstruction, for example, prostate hypertrophy. Furthermore, the possible influence of anticholinergic medication must always be kept in mind.

Unfortunately, the different studies are limited in their degree of comparability to one another since the basic diagnostic work is often dissatisfactory and does not reliably differentiate between the different forms of Parkinson's syndromes. Importantly, all forms of micturition dysfunction can present in Parkinson syndromes, although detrusor hyperactivity is most prominent.

Bladder disturbances within the Parkinson syndromes can on the one hand occur completely independently of the primary disease but also, on the other, as a result of the medication administered or the primary disease itself (Raz, 1976; Schwab & England, 1958). It is basically difficult to distinguish these two disease factors from one another or to place them in a proper chronological order (Raz, 1976). The influence of Parkinson medication varies from patient to patient and is thus not in any way predictable (Winge, Werdelin, Nielsen, & Stimpel, 2004). A correlation with the motor disturbances has not been established (Korchounov et al., 2005; Lemack, Dewey, Roehrborn, O'Suilleabhain, & Zimmern, 2000) or is only partial (Ossig et al., 2016; Storch et al., 2013). It has been claimed that deep brain stimulation has a positive effect on bladder problems (Mock et al., 2016).

An effect of anticholinergics on bladder function is undeniable. It is furthermore well known that L-dopa (alpha- and beta-sympathomimetic) influences the tone of the smooth muscles of the urethra (Raz, 1976), but it also has central nervous control of bladder function. In contrast, however, Parkinson medications themselves can have a direct positive influence on bladder disturbance (see the relationship of D1–D2). Thus, an improvement has been found under L-dopa treatment since the detrusor activity and functional bladder obstruction are reduced (Raz, 1976). At present, there is discussion as to whether L-dopa initially has negative effects but then positive ones under long-term therapy (Brusa et al., 2007). Anticholinergics are attributed with similar effects. In the case of the rare hypoactive detrusor, however, anticholinergics further worsen the micturition condition.

In addition to classifying bladder disturbances according to hypo- and hyperactive bladders, corresponding to the clinical complaints and the findings from urography and cystography, a participation of the voluntary sphincter has also been found as well. Electromyographical analyses in Parkinson's patients have also revealed peripheral neurogenic damage in the striated bladder sphincter (Andersen et al., 1976; Galloway, 1983; Jost, 1999; Raz, 1976).

Electromyographical studies have shown not only peripheral neurogenic damage in the sphincter but also in individual patients, involuntary burst-like contractions of the sphincter urethrae (Galloway, 1983; Jost, 1999). This phenomenon can be termed urethrismus, in analogy to anismus (Jost, Merkle, & Müller-Lobeck, 1998).

As well as for cerebral and peripheral disturbances, there are also descriptions of central, infrapontine lesions leading to detrusor sphincter dyssynergia.

Summarizing, micturition disturbances in PD can be classified as:

- disturbances independent of the PD, for example, benign prostate hypertrophy
- pharmacological effects of the Parkinson medication (either direct or indirect)
- micturition as a result of the primary disease
- the rare detrusor hypoactivity
- rarely, detrusor sphincter *dyssynergia*

All these disturbances induce problems in micturition or urinary incontinence.

4.2 Urinary Incontinence

Parkinson himself already reported finding incontinentia urinae to be a typical symptom of PD (Parkinson, 1817). Approximately one-third of all patients present with urinary incontinence. Most of these cases involve urge incontinence, fewer have stress incontinence. This argues for disinhibited detrusor contractions, which has been described earlier as typical for PD.

There is no conclusive data as to the extent that stress incontinence should be interpreted as an indication of peripheral damage to the sphincter. In particular, the fact that severe cases of incontinence are found only in women supports viewing this incontinence as that found in an age-correlated population. In addition to micturition disturbances and incontinence, recurrent urethral infections have often been described in PD patients (Andersen et al., 1976).

An interesting aspect was recently described in a study on falls in PD patients: The only relevant relation found for the falls was urinary incontinence (Balash et al., 2005). Thus, the disturbance in bladder function was not only directly but also indirectly a relevant prognostic factor, as falls frequently lead to fractures.

4.3 Therapy of Bladder Disturbances

The goal of diagnostics and therapy is a controlled micturition, an end to involuntary urinary loss, and a complete micturition.

In micturition, disturbance and urinary incontinence therapy is preferentially conducted by urologists. Neurologists should, however, become aware of the possibilities and also the possible adverse effects of a urological therapy on neurological symptoms and therapy. In addition, in PD, a considerable number of medications are administered which can induce bladder disturbance, namely, among others anticholinergics, but in the long run almost all other Parkinson drugs.

In micturition disturbance, the following therapeutic strategies are recommended, corresponding to the different basic kinds of disturbance:

- Intermitting self-catheterization
- Antimuscarinergic drugs (anticholinergics)
- Spasticity-reducing medication (especially baclofen)
- Alpha receptor blockers
- Tricyclical antidepressants
- Sympathomimetics (for example, midodrine)
- Behavioral therapy (bladder training, biofeedback)
- Botulinum toxin (for detrusor hyperactivity, spasticity, and urethrism)

Intermittent self-catheterization earns special consideration here. The spectrum of drug treatments appears broader than it is in reality. Efficacy is concentrated mainly on anticholinergic, sympatholytic, and antispastic effects.

Detrusor hyperactivity is the most common disorder. The drugs most frequently administered for treatment of detrusor hyperactivity are trospium chloride and M3-receptor antagonists (darifenacin, solifenacin). Before any drug therapy, a residual urine must always be excluded first.

It is paramount to remember that all anticholinergic substances can also have central nervous effects, that is, they can essentially impair cognition (Sakakibara et al., 2010). And therefore, the non-CNS agents and the M3 receptor antagonists should be used. Because of the possible side effects, medications such as oxybutynin should be avoided. The available data are limited (Zesiewicz et al., 2010). The effectiveness of mirabegron is at the moment not yet resolved.

In the meantime, botulinum toxin is also approved for the treatment of neurogenic and idiopathic detrusor hyperactivity and should be considered for therapy-resistant cases.

In benign prostatic hypertrophy drugs, like tamsulosin, are recommended.

As is frequently the case, unfortunately, urinary incontinence receives very little attention in PD patients. Here, toilet training and especially assistive equipment are made use of. But therapeutic medication can in some cases also be considered. At present only one drug has been approved for the therapy of stress incontinence (duloxetine), although data on its efficiency in PD are lacking.

5. CONCLUSION

Autonomic disorders occur very frequently in PD. It is important to recognize and treat them. In many cases, an interdisciplinary cooperation must take place.

Possible influences on the autonomic disorders caused by the therapy and effects of the therapy on PD must be considered.

Among the cardiovascular disorders, orthostatic hypotension and lack of blood pressure drop during the night are particularly noteworthy. This must also be taken into account when treating the blood pressure.

Disturbances of the thermoregulation are very frequent and reduce the quality of life. Night sweats are particularly frequent. The therapeutic possibilities are limited.

Detrusor hyperactivity is the most common urological symptomatology. Residual urine is to be excluded. Therapy of the choice is the use of antimuscarinergics.

REFERENCES

Adler, C. H., Dugger, B. N., Hentz, J. G., Hinni, M. L., Lott, D. G., et al. (2016). Peripheral synucleinopathy in early Parkinson's disease: Submandibular gland needle biopsy findings. *Movement Disorders*, *31*(2), 250–256.

Allcock, L. M., Ullyart, K., Kenny, R. A., & Burn, D. J. (2004). Frequency of orthostatic hypotension in a community based cohort of patients with Parkinson's disease. *Journal of Neurology, Neurosurgery, and Psychiatry*, *75*(10), 1470–1471.

Aminoff, M. J., & Wilcox, C. S. (1971). Assessment of autonomic function in patients with a parkinsonian syndrome. *British Medical Journal*, *4*(5779), 80–84.

Andersen, J. T., Hebjørn, S., Frimodt-Møller, C., Walter, S., & Worm-Petersen, J. (1976). Disturbances of micturition in Parkinson's disease. *Acta Neurologica Scandinavica*, *53*(3), 161–170.

Augustis, S., Saferis, V., & Jost, W. H. (2017). Autonomic disturbances including impaired hand thermoregulation in multiple system atrophy and Parkinson's disease. *Journal of Neural Transmission*. http://dx.doi.org/10.1007/s00702-016-1665-8.

Balash, Y., Peretz, C., Leibovich, G., Herman, T., Hausdorff, J. M., & Giladi, N. (2005). Falls in outpatients with Parkinson's disease. *Journal of Neurology*, *252*(11), 1310–1315.

Barone, P., Antonini, A., Colosimo, C., Marconi, R., Morgante, L., Avarello, T. P., et al. (2009). The PRIAMO study: A multicenter assessment of nonmotor symptoms and their impact on quality of life in Parkinson's disease. *Movement Disorders*, *24*(11), 1641–1649.

Bellon, A. K., Jost, W. H., Schimrigk, K., König, J., & Zeppezauer, M. (1996). Hormonelle regulation beim Parkinson syndrome. *Deutsche medizinische Wochenschrift*, *121*(36), 1077–1083.

Braak, H., de Vos, R. A., Bohl, J., & Del Tredici, K. (2006). Gastric a-synuclein immunoreactive inclusions in Meissner's and Auerbach's plexuses in cases staged for Parkinson's disease related brain pathology. *Neuroscience Letters*, *396*(1), 67–72.

Braak, H., Del Tredici, K., Bratzke, H., Hamm-Clement, J., Sandmann-Keil, D., & Rüb, U. (2002). Staging of the intracerebral inclusion body pathology associated with idiopathic Parkinson's disease. *Journal of Neurology*, *249*(S3), 1–5.

Braak, H., Rüb, U., Gai, W. P., & Del Tredici, K. (2003). Idiopathic Parkinson's disease. Possible routes by which vulnerable neuronal types may be subject to neuroinvasion by an unknown pathogen. *Journal of Neural Transmission*, *110*(5), 517–536.

Brusa, L., Petta, F., Pisani, A., Moschella, V., Iani, C., Stanzione, P., et al. (2007). Acute vs chronic effects of L-dopa on bladder function in patients with mild Parkinson disease. *Neurology*, *68*(18), 1455–1459.

Chaudhuri, K. R., Ellis, C., Love-Jones, S., Thomaides, T., Clift, S., et al. (1997). Postprandial hypotension and parkinsonian state in Parkinson's disease. *Movement Disorders*, *12*(6), 877–884.

De Deurwaerdèrea, P., Di Giovanni, G., & Milland, M. J. (2017). Expanding the repertoire of L-DOPA's actions: A comprehensive review of its functional neurochemistry. *Progress in Neurobiology*, *151*, 57–110.

Den Hartog Jager, W. A., & Bethlem, J. (1960). The distribution of Lewy bodies in the central and autonomic nervous system in idiopathic paralysis agitans. *Journal of Neurology, Neurosurgery, and Psychiatry*, *23*, 283–290.

Duncan, G. W., Khoo, T. K., Yarnall, A. J., O'Brien, J. T., Coleman, S. Y., et al. (2014). Health-related quality of life in early Parkinson's disease: The impact of nonmotor symptoms. *Movement Disorders*, *29*(2), 195–202.

Fasano, A., Visanji, N. P., Liu, L. W., Lang, A. E., & Pfeiffer, R. F. (2015). Gastrointestinal dysfunction in Parkinson's disease. *Lancet Neurology*, *14*(6), 625–639.

Fischer, M., Gemende, I., Marsch, W. C., & Fischer, P. A. (2001). Skin function and skin disorders in Parkinson's disease. *Journal of Neural Transmission*, *108*(2), 205–213.

Fujishiro, H., Frigerio, R., Burnett, M., Klos, K. J., Josephs, K. A., et al. (2008). Cardiac sympathetic denervation correlates with clinical and pathologic stages of Parkinson's disease. *Movement Disorders*, *23*(8), 1085–1092.

Galloway, N. T. (1983). Urethral sphincter abnormalities in parkinsonism. *British Journal of Urology*, *55*(6), 691–693.

Goldstein, D. S. (2006). Orthostatic hypotension as an early finding in Parkinson's disease. *Clinical Autonomic Research: Official Journal of the Clinical Autonomic Research Society*, *16*(1), 46–54.

Goldstein, D. S., Holmes, C. S., Dendi, R., Bruce, S. R., & Li, S. T. (2002). Orthostatic hypotension from sympathetic denervation in Parkinson's disease. *Neurology*, *58*(8), 1247–1255.

Hauser, R. A., Isaacson, S., Lisk, J. P., Hewitt, L. A., & Rowse, G. (2015). Droxidopa for the short-term treatment of symptomatic neurogenic orthostatic hypotension in Parkinson's disease (nOH306B). *Movement Disorders*, *30*(5), 646–654.

Izcovich, A., González Malla, C., Manzotti, M., Catalano, H. N., & Guyatt, G. (2014). Midodrine for orthostatic hypotension and recurrent reflex syncope. A systemic review. *Neurology*, *83*(13), 1170–1177.

Jost, W. H. (1999). *Autonome Regulationsstörungen beim Parkinson-Syndrom*. Aachen: Shaker-Verlag.

Jost, W. H. (2013). Urological problems in Parkinson's disease: Clinical aspects. *Journal of Neural Transmission*, *120*(4), 587–591.

Jost, W. H., & Augustis, S. (2015). Severity of orthostatic hypotension in the course of Parkinson's disease: No correlation with the duration of the disease. *Parkinsonism & Related Disorders*, *21*(3), 314–316.

Jost, W. H., Del Tredici, K., Landvogt, C., & Braune, S. (2010). Importance of 123I-metaiodobenzylguanidin (MIBG) scintigraphy/SPECT for diagnosis and differential diagnostics of Parkinson syndromes. *Neurodegenerative Diseases*, 7(5), 341–347.

Jost, W. H., Merkle, W., & Müller-Lobeck, H. (1998). Urethrismus accounting for voiding disorder. *Urology*, *52*(2), 352.

Kaye, J., Gage, H., Kimber, A., Storey, L., & Trend, P. (2006). Excess burden of constipation in Parkinson's disease: A pilot study. *Movement Disorders*, *21*(8), 1270–1273.

Korchounov, A., Kessler, K. R., Yakhno, N. N., Damulin, I. V., & Schipper, H. I. (2005). Determinants of autonomic dysfunction in idiopathic Parkinson's disease. *Journal of Neurology*, *252*(12), 1530–1536.

Korczyn, A. D. (1990). Autonomic nervous system disturbances in Parkinson's disease. *Advances in Neurology*, *53*, 463–468.

Lemack, G. E., Dewey, R. B., Roehrborn, C. R., O'Suilleabhain, P. E., & Zimmern, P. E. (2000). Questionnaire-based assessment of bladder dysfunction in patients with mild to moderate Parkinson's disease. *Urology*, *56*(2), 250–254.

Lewy, F. H. (1913). Zur pathologischen Anatomie der Paralysis agitans. *Deutsche Zeitschrift für Nervenheilkunde*, *50*, 50–55.

Low, P. A., & Singer, W. (2008). Management of neurogenic orthostatic hypotension: An update. *Lancet Neurology*, 7(5), 451–458.

Lyytinen, J., Sovijarvi, A., Kaakkola, S., Gordin, A., & Teräväinen, H. (2001). The effect of catechol-O-methyltransferase inhibition with entacapone on cardiovascular autonomic responses in L-dopa-treated patients with Parkinson's disease. *Clinical Neuropharmacology*, *24*(1), 50–57.

Martignoni, E., Pacchetti, C., Godi, L., Micieli, G., & Nappi, G. (1995). Autonomic disorders in Parkinson's disease. *Journal of Neural Transmission. Supplementum*, *45*, 11–19.

Mock, S., Osborn, D. J., Brown, E. T., Stuart Reynolds, W., Turchan, M., et al. (2016). The impact of pallidal and subthalamic deep brain stimulation on urologic function in Parkinson's disease. *Neuromodulation*, *19*(7), 717–723.

Oh, Y. S., Kim, J. S., Chung, Y. A., You, IR., Yang, D. W., et al. (2013). Orthostatic hypotension, non-dipping and striatal dopamine in Parkinson disease. *Neurological Sciences*, *34*(4), 557–560.

Oka, H., Yoshioka, M., Onouchi, K., Morita, M., Mochio, S., et al. (2007). Characteristics of orthostatic hypotension in Parkinson's disease. *Brain, 130*(Pt.9), 2425–2432.

Oppenheim, H. (1902). Kleine Beiträge zur Neuropathologie. *Journal für Psychologie und Neurologie, 1*, 129–146.

Ossig, C., Sippel, D., Fauser, M., Gandor, F., Jost, W. H., et al. (2016). Assessment of nonmotor fluctuations using a diary in advanced Parkinson's disease. *Journal of Parkinson's Disease, 6*(3), 597–607.

Pan-Montojo, F., Anichtchik, O., Dening, Y., Knels, L., Pursche, S., et al. (2010). Progression of Parkinson's disease pathology is reproduced by intragastric administration of rotenone in mice. *PloS One, 5*, e8762. http://dx.doi.org/10.1371/journal.pone.0008762.

Parkinson, J. (1817). *An essay on the shaking palsy*. London: Sherwood, Neely, and Jones.

Pursiainen, V., Haapaniemi, T. H., Korpelainen, J. T., Sotaniemi, K. A., & Myllylä, V. V. (2007). Sweating in parkinsonian patients with wearing-off. *Movement Disorders, 22*(6), 828–832.

Raz, S. (1976). Parkinsonism and neurogenic bladder. *Urological Research, 4*(3), 133–138.

Sakakibara, R., Uchiyama, T., Yamanishi, T., & Kishi, M. (2010). Genitourinary dysfunction in Parkinson's disease. *Movement Disorders, 25*(1), 2–12.

Schoffer, K. L., Henderson, R. D., O'Maley, K., & O'Sullivan, J. D. (2007). Nonpharmocological treamtent, fludrocortisone, and domperidone for orthostatic hypotension in Parkinson's disease. *Movement Disorders, 22*(11), 1543–1549.

Schwab, R. S., & England, A. C. (1958). Parkinson's disease. *Journal of Chronic Diseases, 8*(4), 488–509.

Sommer, S., Aral-Becher, B., & Jost, W. (2011). Nondipping in Parkinson's disease. *Parkinson's Disease, 2011*, 897586. http://dx.doi.org/10.4061/2011/897586.

Stemper, B., Beric, A., Welsch, G., Haendl, T., Sterio, D., & Hilz, M. J. (2006). Deep brain stimulation improves orthostatic regulation of patients with Parkinson disease. *Neurology, 67*(10), 1781–1785.

Storch, A., Schneider, C. B., Wolz, M., Stürwald, Y., Nebe, A., et al. (2013). Nonmotor fluctuations in Parkinson disease: Severity and correlation with motor complications. *Neurology, 80*(9), 800–809.

Strano, S., Fanciulli, A., Rizzo, M., Marinelli, P., Palange, P., et al. (2016). Cardiovascular dysfunction in untreated Parkinson's disease: A multi-modality assessment. *Journal of the Neurological Sciences, 370*, 251–255.

Taly, A. B., & Muthane, U. B. (1992). Involvement of peripheral nervous system in juvenile Parkinson's disease. *Acta Neurologica Scandinavica, 85*(4), 272–275.

Thomaides, T., Karapanayiotides, T., Zoukos, Y., Haeropoulos, C., Kerezoudi, E., et al. (2005). Gastric emptying after semi-solid food in multiple system atrophy and Parkinson's disease. *Journal of Neurology, 252*(9), 1055–1059.

Umehara, T., Nakahara, A., Matsuno, H., Toyoda, C., & Oka, H. (2016). Predictors of postprandial hypotension in elderly patients with de novo Parkinson's disease. *Journal of Neural Transmission, 123*(11), 1331–1339.

Winge, K., Werdelin, L. M., Nielsen, K. K., & Stimpel, H. (2004). Effects of dopaminergic treatment on bladder function in Parkinson's disease. *Neurourology and Urodynamics, 23*(7), 689–696.

Witjas, T., Kaphan, E., Azulay, J. P., Blin, O., Ceccaldi, M., et al. (2002). Nonmotor fluctuations in Parkinson's disease: Frequent and disabling. *Neurology, 59*(3), 408–413.

Zesiewicz, T. A., Sullivan, K. L., Arnulf, I., Chaudhuri, K. R., Morgan, J. C., et al. (2010). Practice parameter: Treatment of nonmotor symptoms of Parkinson disease: Report of the quality standards subcommittee of the American Academy of Neurology. *Neurology, 74*(11), 924–931.

CHAPTER TWENTY-SIX

The Gut and Nonmotor Symptoms in Parkinson's Disease

Lisa Klingelhoefer, Heinz Reichmann[1]
Technical University Dresden, Dresden, Germany
[1]Corresponding author: e-mail address: Heinz.Reichmann@uniklinikum-dresden.de

Contents

Abstract

Gastrointestinal (GI) symptoms are one of the most common nonmotor symptoms (NMS) in patients with Parkinson's disease (PD) involving the whole GI tract (GIT) and being evident throughout the whole course of the disease. Furthermore, constipation serves as a risk factor for PD as well as an early prodromal NMS of PD. The gut as gateway to the environment with its enteric nervous system (ENS) plays a crucial role in the neurodegenerative process that leads to PD. Alpha-synucleinopathy as the pathological hallmark of PD could be found within the whole GIT in a rostrocaudal gradient interacting with the ENS, the gut microbiome, and enteric glial cells. Bidirectional interactions between the ENS and the central nervous system (CNS) via a brain–gut–enteric microbiota axis have been reported. As well as there is evidence out of animal, autopsy, and epidemiological studies that α-synuclein spreads via rostrocranial transmission by transsynaptic cell-to-cell transfer via the sympathetic and parasympathetic nervous system to the CNS causing the typical neuropathological changes of PD. Recognition of GI NMS as prodromal markers of PD as well as a better understanding of the brain–gut connection offers new insights in the pathophysiology of PD and might provide the opportunity of PD diagnosis before CNS involvement. Hereby the opportunity for development of neuroprotective and disease-modifying therapeutics, respectively, seem to be promising. This chapter covers the variety of GI NMS and its consequences in PD as well as the important role of the gut as part of the pathological process in PD.

International Review of Neurobiology, Volume 134
ISSN 0074-7742
http://dx.doi.org/10.1016/bs.irn.2017.05.027

1. INTRODUCTION

Gastrointestinal (GI) nonmotor symptoms (NMS) are common in patients with Parkinson's disease (PD). They are various, involve the whole GI tract (GIT), and are evident in all stages of PD.

GI NMS have a negative influence on reliable therapy with oral antiparkinsonian medication ranging from dysphagia with swallowing problems of the tablets till malabsorption caused by delayed gastric emptying and/or prolonged transport times, which can lead to loss of medication effect, accumulation of doses, or even motor fluctuations (Djaldetti, Baron, Ziv, & Melamed, 1996; Muller et al., 2006). Next to their negative impact on motor control, GI NMS can lead to stigmatization due to sialorrhoea with drooling of saliva, weight loss due to malabsorption, abdominal pain due to constipation, and even life-threatening complications such as aspiration with pneumonia in dysphagia or perforation by obstruction in the intestinal tract; and therefore generally result in major discomfort with reduced health-related quality of life in PD patients (Barone et al., 2009; Martinez-Martin, Rodriguez-Blazquez, Kurtis, & Chaudhuri, 2011; Schrag, Jahanshahi, & Quinn, 2000). Regular assessment of GI NMS and a specific treatment is of major importance in the long-term care of PD patients.

Next to being bothersome NMS, constipation serves as an early premotor symptom of PD and is valued as a risk factor for developing PD (Abbott et al., 2001; Berg et al., 2015; Postuma et al., 2015). Corresponding to this clinical observation, the Braak staging system postulates that PD is a neurodegenerative disorder of both the brain and the gut as the neuropathological process of PD seems to start in the enteric nervous system (ENS) of the gut in some extent. From there Parkinson pathology spreads via rostrocranial transmission by transsynaptic cell-to-cell transfer via the sympathetic and parasympathetic nervous system to the central nervous system (CNS) causing the typical neuropathological changes of PD with selective degeneration of dopaminergic neurons of the substantia nigra pars compacta, the formation of Lewy bodies, a reactive gliosis and a progressive central neurodegeneration (Braak, Del Tredici, et al., 2003; Dickson, Uchikado, Fujishiro, & Tsuboi, 2010; Goedert, Spillantini, Del Tredici, & Braak, 2013; Halliday, McCann, & Shepherd, 2012; Jellinger, 2003; Klingelhoefer & Reichmann, 2015). This establishes the possibility that the gut serves as a gateway to the environment, and hereby environmental substances can trigger PD pathogenesis. Therefore, not a single causative factor, but rather a multifactorial process, resulting from interactions between genetic defects, environmental factors, and an individual susceptibility are important for PD development affecting different anatomical structures and neurotransmitters (Jellinger, 2009; Reichmann, 2011; Titova, Padmakumar, Lewis, & Chaudhuri, 2016).

2. GI SYMPTOMS IN PD

GI symptoms are commonly observed at all stages of PD (Pfeiffer, 2003) with a higher percentage of complaints being observed in advanced disease stages (Chaudhuri, Healy, & Schapira, 2006; Goetze et al., 2006). Up to 30% of PD patients report any GI symptom (Martinez-Martin, 2011; Rodriguez-Violante, Cervantes-Arriaga, Villar-Velarde, & Corona, 2010). The wide range of GI NMS is summarized in Table 1 providing the anatomic site of presentation, possible causes, and consequences of the most important GI NMS in PD.

Constipation is one of the most common NMS of PD with a prevalence of 28%–80% (Barone et al., 2009; Chaudhuri, Martinez-Martin, et al., 2006; Jost, 1997; Martinez-Martin et al., 2007; Pfeiffer, 2011) and up to sixfold more common among PD patients than among age- and sex-matched controls (Cersosimo et al., 2013; Chaudhuri, Martinez-Martin, et al., 2006; Edwards, Pfeiffer, Quigley, Hofman, & Balluff, 1991; Korczyn, 1990). Furthermore, constipation is one of the earliest NMS in PD, being present up to 15 years before motor PD manifestation (Abbott et al., 2001; Adams-Carr et al., 2015; Ashraf, Pfeiffer, Park, Lof, & Quigley, 1997; Cersosimo et al., 2013; Postuma, Gagnon, Pelletier, & Montplaisir, 2013; Savica et al., 2009). In the Honolulu Heart Program 6790 men without PD at baseline have been monitored for incidence of PD over 24 years (Abbott et al., 2001). 96 participants developed PD after an average time of 12 years. After adjustment for various factors that might influence constipation, such as age, nicotine, and coffee consumption, men who experienced fewer bowel movements than one per day had a 2.7-fold, 4.1-fold, and 4.5-fold greater risk of PD than men who experienced one, two, or more than two bowel movements per day, respectively. These findings demonstrate that infrequent bowel movements are associated with an elevated risk of developing PD. The underlying mechanisms of constipation in PD are unknown, but neurodegeneration of autonomic centers of the ENS and the CNS are likely (Cersosimo & Benarroch, 2008; Lebouvier et al., 2008). Therefore, constipation is a risk factor for PD as well as an early prodromal NMS of PD.

Table 1 Gastrointestinal Nonmotor Symptoms With Their Anatomic Site of Presentation, Possible Causes, Resulting Consequences, and the Corresponding Reference in PD

Anatomic Site	Symptom	Cause	Consequence	References
Mouth	Dental problems	• Impaired fine motor skills interfering with teeth brushing • Reduced opening of the jaw • Change of salivary pH due to sialorrhea or dry mouth • Change in microbiota	• Poor oral hygiene • Caries • Periodontal disease • Loss of teeth • Bad breath	Bakke, Larsen, Lautrup, and Karlsborg (2011), Hanaoka and Kashihara (2009), and Muller, Palluch, and Jackowski (2011)
	Sialorrhea with drooling of saliva	• Impaired salivary clearance caused by dysphagia or an inability to maintain saliva within the oral cavity (due to hypomimia with mouth opening, dropped head) • Overproduction of saliva	• Stigmatization and social embarrassment • Dental problems • Difficulty speaking • Increased risk of aspiration with respiratory infection	Kalf et al. (2011), Rana, Yousuf, Awan, and Fattah (2012), and Srivanitchapoom, Pandey, and Hallett (2014)

	Impaired taste	• PD progression with involvement of the frontal operculum or orbitofrontal cortex as impaired taste is associated with impaired cognitive function and advanced stage of PD • Dysfunction seems to rely on a specific tongue region • Chemosensory interaction with olfaction and decreased trigeminal function	• None as the majority of PD patients does not notice any change in taste perception in everyday life	Cecchini, Fasano, Boschi, Osculati, and Tinazzi (2015), Cecchini et al. (2014), Deeb et al. (2010), Doty et al. (2015), Hummel et al. (1996), Kim et al. (2011), Landis et al. (2010), and Lang et al. (2006)
Pharynx/ esophagus	Dysphagia	• PD motor symptoms (bradykinesia and rigidity) • Central basal ganglia dysfunction with hypometabolism in the supplementary motor area and the anterior cingulate cortex • Peripheral dysfunction of motor and sensory pharyngeal nerves	• Increased risk of aspiration with respiratory infection • Weight loss and malnutrition	Kikuchi et al. (2013), Mu et al. (2012), and Pfeiffer (2011)

Continued

Table 1 Gastrointestinal Nonmotor Symptoms With Their Anatomic Site of Presentation, Possible Causes, Resulting Consequences, and the Corresponding Reference in PD—cont'd

Anatomic Site	Symptom	Cause	Consequence	References
Stomach	Nausea and vomiting, early satiety, excessive fullness, abdominal distension due to delayed gastric emptying and gastroparesis	• Central nervous system disturbance • Enteric nervous system disturbance • Disturbance of local neuropeptide function (e.g., ghrelin) • More likely with solid meals	• Weight loss and malnutrition	Braak, de Vos, Bohl, and Del Tredici (2006) and Cheung and Wu (2013)
Large intestine	Slow-transit constipation	• Impairment of colonic motility by enteric and central nervous system disturbance	• Abdominal pain and discomfort • Perforation • Ileus	Anderson et al. (2007), Jost and Schrank (1998), and Pfeiffer (2011)
Rectum	Outlet constipation and defecatory difficulties with unsatisfactory voiding of bowel and anismus	• Paradoxical anal sphincter muscle contraction during defecation (kind of focal dystonia) • Involuntary and uncoordinated contractions of the pelvic floor • Reduced anorectal sensitivity due to dysfunction of afferent pathways	• Abdominal pain and discomfort • Ileus	Ashraf, Pfeiffer, and Quigley (1994), Bassotti et al. (2000), Edwards, Quigley, Harned, Hofman, and Pfeiffer (1994), and Jost, Schrank, Herold, and Leiss (1999)

Clinically constipation associated with PD can be classified as either "slow-transit" constipation, caused by prolonged intestinal passage (Edwards et al., 1991; Jost & Schrank, 1998; Sakakibara et al., 2003) or "outlet" constipation, caused by anorectal dysfunction; patients might experience one or both (Cersosimo & Benarroch, 2012; Sakakibara et al., 2011; Wang, Sung, Wang, & Tsai, 2013). Furthermore, increased intestinal transit times have been recorded in treated and untreated PD patients as well as in PD patients without symptomatic constipation (Sakakibara et al., 2003).

3. THE GUT AS PART OF THE PATHOLOGICAL PROCESS IN PD

Pathological investigations have indicated that the neuropathological changes in PD follow a specific chronological and regional pattern (Braak et al., 1996; Gaspar & Gray, 1984). Braak and colleagues developed a pathological staging system for PD, which correlates well with the clinical presentation at the different stages of the disease (Braak, Del Tredici, et al., 2003; Braak, Ghebremedhin, Rub, Bratzke, & Del Tredici, 2004; Del Tredici, Rub, De Vos, Bohl, & Braak, 2002; Wolters & Braak, 2006). It is proposed that via a dual-hit mechanism the neurodegenerative process starts in the olfactory bulb following inhalation and in the ENS of the gut following ingestion of a neurotropic pathogen (Hawkes, Del Tredici, & Braak, 2007; Reichmann, 2011). Thereafter, pathology progresses anterograde from the olfactory system into the temporal lobe and retrograde from the gut via sympathetic and parasympathetic nerves to the dorsal motor nucleus of the vagus (DMNV) or the intermediolateral nucleus (IML) of the spinal cord, respectively, and further to the brain stem (Bloch, Probst, Bissig, Adams, & Tolnay, 2006; Braak & Del Tredici, 2009; Braak, Del Tredici, et al., 2003; Braak, Rub, Gai, & Del Tredici, 2003; Braak, Sastre, Bohl, de Vos, & Del Tredici, 2007; Del Tredici & Braak, 2012; Fig. 1). This results in a progressive degeneration of the dopaminergic nigrostriatal system and a widespread of pathology affecting different extranigral anatomical structures and neurotransmitters (Jellinger, 2003, 2015). Therefore, it is nowadays thought of PD being a syndrome rather than a disease (Titova et al., 2016).

These hypotheses of pathological development and progression of PD are supported as well as challenged by findings in pathological studies, animal studies, and clinical observational studies. In transgenic mice models, overexpressing normal human α-synuclein in neurons, insoluble α-synuclein

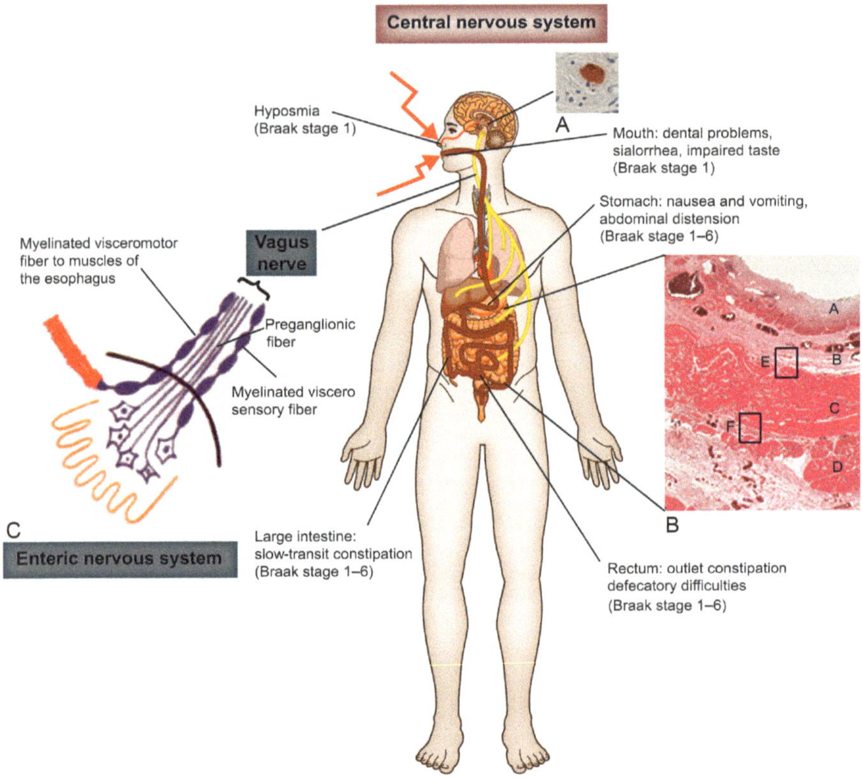

Fig. 1 Scheme of the brain and the gastrointestinal system presenting a dual hit (*red flash*) of the olfactory and the gastrointestinal system. From there pathology spreads via the olfactory nerve to the brain stem (*red line*) and/or after inhalation/ingestion (*red line*) from the gut via sympathetic and parasympathetic nervous system (*yellow line*) in rostrocranial direction to the central nervous system. In the text boxes gastrointestinal NMS and corresponding Braak stages are summarized. Histological figures of: (A) alpha-synuclein inclusion similar to Lewy body (B) intestinal section with A: mucosa, B: submucosa, C: rings muscles, D: longitudinal muscles and the enteric nervous system with E: Meissner's plexus, and F: Auerbach's plexus. (C) Drawing of the vagus nerve as the connector between enteric and central nervous system.

aggregates in enteric neurons associated with colonic dysmotility were found preceding the onset of central changes (Hallett, McLean, Kartunen, Langston, & Isacson, 2012; Wang et al., 2012). In another transgenic mice model prolonged colonic transit times due to early synuclein-immunoreactive aggregates in the ENS preceding CNS changes were found as well (Kuo et al., 2010). Further studies showed that pathological changes occur in the ENS before CNS involvement in PD (Grathwohl, Steiner,

Britschgi, & Brundin, 2013; Lebouvier, Neunlist, et al., 2010). Lewy bodies and Lewy neurites were found in the ENS of the gut, the olfactory bulb, in sympathetic and parasympathetic ganglia, the IML of the spinal cord and in the CNS (Beach et al., 2009; Braak et al., 2006, 2007; Del Tredici, Hawkes, Ghebremedhin, & Braak, 2010; Minguez-Castellanos et al., 2007; Orimo et al., 2005). Alpha-synucleinopathy was detected in the Auerbach plexus, the Meissner plexus, and the olfactory bulb in people with a diagnosis of PD, as well as in random people who did not yet present with typical motor symptoms or CNS pathology of PD (Braak et al., 2006, 2007; Kupsky, Grimes, Sweeting, Bertsch, & Cote, 1987; Lebouvier et al., 2008; Qualman, Haupt, Yang, & Hamilton, 1984; Wakabayashi, Takahashi, Ohama, & Ikuta, 1990; Wakabayashi, Takahashi, Takeda, Ohama, & Ikuta, 1988). Studies focusing on Lewy pathology in the ENS of patients with Lewy body diseases have found α-synuclein in the ENS of nearly every patient with PD (Beach et al., 2010; Gelpi et al., 2014). Furthermore, a substantial number of PD patients present with Lewy pathology in the ENS up to 20 years before CNS manifestation and the percentage of phosphorylated α-synuclein is comparable in prodromal and manifest PD stage (Schneider et al., 2016; Stokholm, Danielsen, Hamilton-Dutoit, & Borghammer, 2016). Hereby the prognostic value of endoscopic biopsies with detection of Lewy neurites in the submucosa in PD patients is discussed (Hilton et al., 2014; Lebouvier, Coron, et al., 2010; Shannon et al., 2012), but aggregated α-synuclein is not specific for PD (Visanji et al., 2015). Furthermore, there is a rostrocaudal gradient of α-synucleinopathy within the GIT with submandibular gland and lower esophagus having the highest frequency of Lewy bodies followed by stomach, small intestine, colon, and rectum (Beach et al., 2010; Wakabayashi et al., 1988). This rostrocaudal distribution of Lewy pathology coincides with the vagal innervation from the DMNV as well as the distribution of dopaminergic neurons in the ENS (Cersosimo & Benarroch, 2012; Li, Pham, Tamir, Chen, & Gershon, 2004). Autopsies have shown that late-life constipation in people with no clinical signs of parkinsonism or dementia before death is associated with incidental Lewy bodies in the substantia nigra and the locus coeruleus and with low neuron density in the substantia nigra (Abbott et al., 2007; Petrovitch et al., 2009). Cross-sectional studies suggest that incidental Lewy bodies are present in ~8% of elderly people (Jellinger, 2004; Parkkinen, Soininen, Laakso, & Alafuzoff, 2001; Saito et al., 2004), and incidental Lewy Body disease (ILBD) was described in almost 30% of aged brains without any neuropsychiatric disorder or extrapyramidal symptom (Jellinger, 2004). The relevance of ILBD and the clinical impact of Lewy body-related α-synucleinopathy in normal aging are still not known (Dickson et al., 2008; Halliday et al., 2012). New data suggest that the phosphorylation of α-synuclein and the accumulation of phosphorylated α-synuclein might be a process of aging as phosphorylated α-synuclein was mainly found in the ENS of older aged healthy controls and not or only in small amounts in younger aged healthy controls (Bottner et al., 2012; Muntane, Ferrer, & Martinez-Vicente, 2012; Visanji et al., 2015). Taking together that the efficacy of cellular repair mechanism, e.g., to degrade misfolded proteins is reduced with increasing age, aging may be critical for PD and PD may be a local expression of aging on cell populations which, by their characteristics, are highly vulnerable to agents promoting aging (Rodriguez, Rodriguez-Sabate, Morales, Sanchez, & Sabate, 2015; Vilchez, Saez, & Dillin, 2014).

The gut microbiome comprises the collective genome of all the microorganisms residing in the GIT. Its composition is not static throughout the host's life span, and further changes occur dependent of diet, environment, and disease state (Ghaisas, Maher, & Kanthasamy, 2016; Voreades, Kozil, & Weir, 2014). Gut microbes play important roles in aiding digestion, production, and absorption of essential metabolites; development and regulation of the immune system; and development of the nervous system (Littman & Pamer, 2011; Walker & Lawley, 2013). Dysbiosis, defined as a shift in the balance of the gut microbiota composition, can result in disturbance of the host's health and has been implicated in different diseases such as inflammatory bowel disease, irritable bowel syndrome, diabetes type 2, metabolic syndrome, autism, and Alzheimer's disease (Mayer, Tillisch, & Gupta, 2015; Walker & Lawley, 2013). Furthermore, a brain–gut–enteric microbiota axis for bidirectional interactions between the enteric microbiota and the CNS has been recognized (Grenham, Clarke, Cryan, & Dinan, 2011; Mayer et al., 2015; Rhee, Pothoulakis, & Mayer, 2009). In an animal model, using mice that overexpress α-synuclein, the presence of gut microbiota was required for the development of motor deficits, the activation of microglia, and the development of α-synucleinopathology (Sampson et al., 2016). Antibiotic treatment ameliorated, while microbial recolonization promoted pathophysiology in this mice model. Furthermore, colonization of these α-synuclein-overexpressing mice with microbiota from PD patients enhanced motor impairments compared to microbiota transplants from healthy human donors (Sampson et al., 2016). These findings suggest that alterations in the human intestinal microbiome represent a risk factor for PD. This hypothesis is consistent with findings of altered intestinal microbiomes in PD patients and an influence of gut microbiota on the activity of enteric neurons (Cryan & Dinan, 2012; Forsythe & Kunze, 2013; Scheperjans, Aho, et al., 2015). It has been shown that the abundance of specific bacterial families together with the severity of constipation identifies PD patients with 67% sensitivity and 90% specificity (Scheperjans, Aho, et al., 2015). Like in the α-synuclein—overexpressing mice model (Sampson et al., 2016), an association of gut microbiome composition and PD motor phenotype was found (Scheperjans, Aho, et al., 2015). Furthermore, other groups could show that dependent of motor phenotype (nontremor dominant vs tremor dominant) the extent of α-synuclein pathology in the ENS varies with nontremor dominant PD being associated with a more extensive α-synuclein pathology in the ENS (Lebouvier, Neunlist, et al., 2010; Marras & Lang, 2013; van Rooden et al., 2011).

Next to the gut microbiome, the enteric glial cells as counterpart for the brain astrocytes seem to be involved in GI dysfunction as well as in PD initiation and development (Clairembault, Leclair-Visonneau, Neunlist, & Derkinderen, 2015). In PD patients neurochemical changes, proinflammatory cytokines, and modulation of intestinal epithelial barrier integrity have been observed as indirect sign for dysregulation and loss of enteric glial cells (Bassotti et al., 2006; Clairembault et al., 2014, 2015; De Giorgio et al., 2012; Devos et al., 2013). Furthermore, a positive correlation between gut permeability, which is regulated by enteric glial cells, and intestinal levels of α-synuclein was observed in untreated patients with early PD (Forsyth et al., 2011). Alpha-synuclein plays a crucial role in neuroinflammation by triggering and/or potentiating glial cell activation (Gao et al., 2011; Sanchez-Guajardo, Tentillier, & Romero-Ramos, 2015) which in turn contributes to the initiation of α-synuclein misfolding (Devos et al., 2013).

The involvement of the ENS, enteric glial cells, and the gut microbiome is in line with the observation that smoking cigarettes and ingesting caffeine are associated with a reduced risk of PD (Derkinderen, Shannon, & Brundin, 2014; Scheperjans, Pekkonen, Kaakkola, & Auvinen, 2015) as both, smoking and caffeine, change the composition of the microbiome in a way that mitigates intestinal inflammation (Biedermann et al., 2014; Derkinderen et al., 2014). Less inflammation leads to less aggregation of α-synuclein in the ENS, thereby reducing the risk of PD by minimizing propagation of α-synuclein to the CNS (Derkinderen et al., 2014). Furthermore, it has been shown that peripherally induced inflammation can induce central inflammation and degeneration of dopaminergic neurons (Bodea et al., 2014; Villaran et al., 2010). Again the influence of aging might play a role as the microbiome diversity diminishes with aging (Dinan & Cryan, 2017).

4. THE GUT–BRAIN CONNECTION

Pan-Montojo et al. (2010) used chronic intragastric administration via a gastric tube of low doses of rotenone to initiate neuropathological changes typical of PD in wild-type mice (the Dresden Parkinson Model). Only local effects of rotenone on the ENS and no systemic effects were induced which could be proven by either high-performance liquid chromatographic analysis or measurements of mitochondrial complex I activity of blood, brain and brain stem, and muscle samples from the animals. In this rotenone-treated mice model next to the development of PD-associated motor symptoms in the mice, a spatiotemporal development of pathological manifestations corresponding well to the Braak staging of the neurodegenerative process and timings of clinical motor symptom presentation in PD patients could be observed. After a short time of rotenone treatment, typical pathological changes of PD, such as α-synuclein aggregation, were observed in the ENS, IML, and DMNV without any motor dysfunction of the mice. After a longer time of rotenone treatment, α-synuclein aggregation and loss of dopaminergic neurons were observed in the substantia nigra pars compacta as well as motor dysfunction of the mice appeared. Several other studies from different groups supported these findings and provided further evidence that administration of rotenone leads to typical pathological changes of PD, such as α-synuclein accumulation and selective nigrostriatal dopaminergic neurodegeneration (Betarbet et al., 2000; Chorfa et al., 2014; Inden et al., 2007; Sherer, Kim, Betarbet, & Greenamyre, 2003; Silva, Einarsdottir, Fink, & Uversky, 2013; Yuan et al., 2015). However, there have been inconsistent findings concerning degeneration of nondopaminergic neurons, activation of glial cells, changes of GI motility, and neuropathological changes in the ENS (Inden et al., 2007; Tasselli et al., 2013).

In vitro and in vivo studies documented that α-synuclein can be transported between neurons (Angot et al., 2012; Danzer et al., 2012; Desplats et al., 2009). Autopsy studies of PD patients who had received fetal mesencephalic transplants showed that α-synuclein spreads from host to graft cells where it accumulated in the grafted neurons (Kordower & Brundin, 2009; Kordower, Chu, Hauser, Freeman, & Olanow, 2008; Li et al., 2008). In the rotenone-treated mice model the progression of PD pathology was stopped if the connection between the ENS and sympathetic or parasympathetic neurons to the CNS was interrupted (Pan-Montojo et al., 2012). Hemivagotomy or partial sympathectomy before gastric rotenone treatment delayed the development of motor symptoms and accumulation of α-synuclein in neurons was significantly less in the DMNV ipsilateral to the hemivagotomy than contralateral. In contrast there was no delay of GI symptoms in rotenone-treated and -operated mice relative to control mice that were rotenone-treated but not operated on (Pan-Montojo et al., 2012). Holmqvist et al. showed that α-synuclein after being injected into the intestinal wall of rats is actively transported via the vagal nerve in a time-dependent manner (Holmqvist et al., 2014). These findings are supported by a large epidemiological study which observed that patients who underwent truncal vagotomy, but not superselective vagotomy, are at lower risk of PD than the general population (Svensson et al., 2015). The incidence of PD among people with a follow-up period >20 years after vagotomy was 0.65 per 1000 person-years, compared with 1.28 per 1000 person-years in matched controls and adjusted for possible confounders such as smoking and comorbidities, who did not undergo vagotomy (Svensson et al., 2015). Unfortunately the strength of these findings have been questioned by another group analyzing the same dataset with an expanded time frame and using different statistical analysis (Tysnes et al., 2015).

5. CONCLUSION

GI NMS are exceptional in PD as on the one hand they need to be monitored and treated throughout the course of PD as they are serious to take and quality of life reducing NMS, and on the other hand constipation serves as an early prodromal marker and risk factor for PD. Furthermore, the neuropathological process that leads to PD seems to start in the gut with detection of α-synucleinopathy throughout the whole GIT in a rostrocaudal gradient. Next to the ENS, the gut microbiome and enteric glial cells become more important in the hypothesis of PD development. Taking together that constipation serves as a prodromal NMS in PD, different studies reported aggregated α-synuclein in the ENS in the presence and absence of CNS α-synucleinopathy, the role of gut microbiota, and the active transport of α-synuclein via the vagal nerve to the CNS, all these observations offer a range of targets for disease modifying or even neuroprotective therapies. Actively balancing the gut microbiome might even have the potential of a primary preventive medicine against PD.

ACKNOWLEDGMENTS

We would like to thank Dr. J. Schultz-Schaeffer, Head of Prion and Dementia Research, National Reference Centre for Spongiform Encephalopathies, University Hospital Göttingen, Germany, for the histological sections (Fig. 1A and B).

REFERENCES

Abbott, R. D., Petrovitch, H., White, L. R., Masaki, K. H., Tanner, C. M., Curb, J. D., et al. (2001). Frequency of bowel movements and the future risk of Parkinson's disease. *Neurology*, *57*(3), 456–462.

Abbott, R. D., Ross, G. W., Petrovitch, H., Tanner, C. M., Davis, D. G., Masaki, K. H., et al. (2007). Bowel movement frequency in late-life and incidental Lewy bodies. *Movement Disorders*, *22*(11), 1581–1586. http://dx.doi.org/10.1002/mds.21560.

Adams-Carr, K. L., Bestwick, J. P., Shribman, S., Lees, A., Schrag, A., & Noyce, A. J. (2016). Constipation preceding Parkinson's disease: A systematic review and meta-analysis. *Journal of Neurology, Neurosurgery, and Psychiatry*, *87*, 710–716. http://dx.doi.org/10.1136/jnnp-2015-311680.

Anderson, G., Noorian, A. R., Taylor, G., Anitha, M., Bernhard, D., Srinivasan, S., et al. (2007). Loss of enteric dopaminergic neurons and associated changes in colon motility in an MPTP mouse model of Parkinson's disease. *Experimental Neurology*, *207*(1), 4–12. http://dx.doi.org/10.1016/j.expneurol.2007.05.010.

Angot, E., Steiner, J. A., Lema Tome, C. M., Ekstrom, P., Mattsson, B., Bjorklund, A., et al. (2012). Alpha-synuclein cell-to-cell transfer and seeding in grafted dopaminergic neurons in vivo. *PLoS One*, *7*(6), e39465. http://dx.doi.org/10.1371/journal.pone.0039465.

Ashraf, W., Pfeiffer, R. F., Park, F., Lof, J., & Quigley, E. M. (1997). Constipation in Parkinson's disease: Objective assessment and response to psyllium. *Movement Disorders*, *12*(6), 946–951. http://dx.doi.org/10.1002/mds.870120617.

Ashraf, W., Pfeiffer, R. F., & Quigley, E. M. (1994). Anorectal manometry in the assessment of anorectal function in Parkinson's disease: A comparison with chronic idiopathic constipation. *Movement Disorders*, *9*(6), 655–663. http://dx.doi.org/10.1002/mds.870090612.

Bakke, M., Larsen, S. L., Lautrup, C., & Karlsborg, M. (2011). Orofacial function and oral health in patients with Parkinson's disease. *European Journal of Oral Sciences*, *119*(1), 27–32. http://dx.doi.org/10.1111/j.1600-0722.2010.00802.x.

Barone, P., Antonini, A., Colosimo, C., Marconi, R., Morgante, L., Avarello, T. P., et al. (2009). The PRIAMO study: A multicenter assessment of nonmotor symptoms and their impact on quality of life in Parkinson's disease. *Movement Disorders*, *24*(11), 1641–1649. http://dx.doi.org/10.1002/mds.22643.

Bassotti, G., Maggio, D., Battaglia, E., Giulietti, O., Spinozzi, F., Reboldi, G., et al. (2000). Manometric investigation of anorectal function in early and late stage Parkinson's disease. *Journal of Neurology, Neurosurgery, and Psychiatry*, *68*(6), 768–770.

Bassotti, G., Villanacci, V., Maurer, C. A., Fisogni, S., Di Fabio, F., Cadei, M., et al. (2006). The role of glial cells and apoptosis of enteric neurones in the neuropathology of intractable slow transit constipation. *Gut*, *55*(1), 41–46. http://dx.doi.org/10.1136/gut.2005.073197.

Beach, T. G., Adler, C. H., Sue, L. I., Vedders, L., Lue, L., White Iii, C. L., et al. (2010). Multi-organ distribution of phosphorylated alpha-synuclein histopathology in subjects with Lewy body disorders. *Acta Neuropathologica*, *119*(6), 689–702. http://dx.doi.org/10.1007/s00401-010-0664-3.

Beach, T. G., White, C. L., 3rd, Hladik, C. L., Sabbagh, M. N., Connor, D. J., Shill, H. A., et al. (2009). Olfactory bulb alpha-synucleinopathy has high specificity and sensitivity for Lewy body disorders. *Acta Neuropathologica*, *117*(2), 169–174. http://dx.doi.org/10.1007/s00401-008-0450-7.

Berg, D., Postuma, R. B., Adler, C. H., Bloem, B. R., Chan, P., Dubois, B., et al. (2015). MDS research criteria for prodromal Parkinson's disease. *Movement Disorders*, *30*(12), 1600–1611. http://dx.doi.org/10.1002/mds.26431.

Betarbet, R., Sherer, T. B., MacKenzie, G., Garcia-Osuna, M., Panov, A. V., & Greenamyre, J. T. (2000). Chronic systemic pesticide exposure reproduces features of Parkinson's disease. *Nature Neuroscience*, *3*(12), 1301–1306. http://dx.doi.org/10.1038/81834.

Biedermann, L., Brulisauer, K., Zeitz, J., Frei, P., Scharl, M., Vavricka, S. R., et al. (2014). Smoking cessation alters intestinal microbiota: Insights from quantitative investigations on human fecal samples using FISH. *Inflammatory Bowel Diseases*, *20*(9), 1496–1501. http://dx.doi.org/10.1097/MIB.0000000000000129.

Bloch, A., Probst, A., Bissig, H., Adams, H., & Tolnay, M. (2006). Alpha-synuclein pathology of the spinal and peripheral autonomic nervous system in neurologically unimpaired elderly subjects. *Neuropathology and Applied Neurobiology*, *32*(3), 284–295. http://dx.doi.org/10.1111/j.1365-2990.2006.00727.x.

Bodea, L. G., Wang, Y., Linnartz-Gerlach, B., Kopatz, J., Sinkkonen, L., Musgrove, R., et al. (2014). Neurodegeneration by activation of the microglial complement-phagosome pathway. *The Journal of Neuroscience*, *34*(25), 8546–8556. http://dx.doi.org/10.1523/JNEUROSCI.5002-13.2014.

Bottner, M., Zorenkov, D., Hellwig, I., Barrenschee, M., Harde, J., Fricke, T., et al. (2012). Expression pattern and localization of alpha-synuclein in the human enteric nervous system. *Neurobiology of Disease*, *48*(3), 474–480. http://dx.doi.org/10.1016/j.nbd.2012.07.018.

Braak, H., Braak, E., Yilmazer, D., de Vos, R. A., Jansen, E. N., & Bohl, J. (1996). Pattern of brain destruction in Parkinson's and Alzheimer's diseases. *Journal of Neural Transmission*, *103*(4), 455–490.

Braak, H., de Vos, R. A., Bohl, J., & Del Tredici, K. (2006). Gastric alpha-synuclein immunoreactive inclusions in Meissner's and Auerbach's plexuses in cases staged for Parkinson's disease-related brain pathology. *Neuroscience Letters*, *396*(1), 67–72. http://dx.doi.org/10.1016/j.neulet.2005.11.012.

Braak, H., & Del Tredici, K. (2009). Neuroanatomy and pathology of sporadic Parkinson's disease. *Advances in Anatomy, Embryology, and Cell Biology*, *201*, 1–119.

Braak, H., Del Tredici, K., Rub, U., de Vos, R. A., Jansen Steur, E. N., & Braak, E. (2003a). Staging of brain pathology related to sporadic Parkinson's disease. *Neurobiology of Aging*, *24*(2), 197–211.

Braak, H., Ghebremedhin, E., Rub, U., Bratzke, H., & Del Tredici, K. (2004). Stages in the development of Parkinson's disease-related pathology. *Cell and Tissue Research*, *318*(1), 121–134. http://dx.doi.org/10.1007/s00441-004-0956-9.

Braak, H., Rub, U., Gai, W. P., & Del Tredici, K. (2003b). Idiopathic Parkinson's disease: Possible routes by which vulnerable neuronal types may be subject to neuroinvasion by an unknown pathogen. *Journal of Neural Transmission*, *110*(5), 517–536. http://dx.doi.org/10.1007/s00702-002-0808-2.

Braak, H., Sastre, M., Bohl, J. R., de Vos, R. A., & Del Tredici, K. (2007). Parkinson's disease: Lesions in dorsal horn layer I, involvement of parasympathetic and sympathetic pre- and postganglionic neurons. *Acta Neuropathologica*, *113*(4), 421–429. http://dx.doi.org/10.1007/s00401-007-0193-x.

Cecchini, M. P., Fasano, A., Boschi, F., Osculati, F., & Tinazzi, M. (2015). Taste in Parkinson's disease. *Journal of Neurology*, *262*(4), 806–813. http://dx.doi.org/10.1007/s00415-014-7518-1.

Cecchini, M. P., Osculati, F., Ottaviani, S., Boschi, F., Fasano, A., & Tinazzi, M. (2014). Taste performance in Parkinson's disease. *Journal of Neural Transmission (Vienna, Austria: 1996)*, *121*(2), 119–122. http://dx.doi.org/10.1007/s00702-013-1089-7.

Cersosimo, M. G., & Benarroch, E. E. (2008). Neural control of the gastrointestinal tract: Implications for Parkinson disease. *Movement Disorders*, *23*(8), 1065–1075. http://dx.doi.org/10.1002/mds.22051.

Cersosimo, M. G., & Benarroch, E. E. (2012). Pathological correlates of gastrointestinal dysfunction in Parkinson's disease. *Neurobiology of Disease*, *46*(3), 559–564. http://dx.doi.org/10.1016/j.nbd.2011.10.014.

Cersosimo, M. G., Raina, G. B., Pecci, C., Pellene, A., Calandra, C. R., Gutierrez, C., et al. (2013). Gastrointestinal manifestations in Parkinson's disease: Prevalence and occurrence before motor symptoms. *Journal of Neurology*, *260*(5), 1332–1338. http://dx.doi.org/10.1007/s00415-012-6801-2.

Chaudhuri, K. R., Healy, D. G., & Schapira, A. H. (2006b). Non-motor symptoms of Parkinson's disease: Diagnosis and management. *Lancet Neurology*, *5*(3), 235–245. http://dx.doi.org/10.1016/S1474-4422(06)70373-8.

Chaudhuri, K. R., Martinez-Martin, P., Schapira, A. H., Stocchi, F., Sethi, K., Odin, P., et al. (2006a). International multicenter pilot study of the first comprehensive self-completed nonmotor symptoms questionnaire for Parkinson's disease: The NMSQuest study. *Movement Disorders*, *21*(7), 916–923. http://dx.doi.org/10.1002/mds.20844.

Cheung, C. K., & Wu, J. C. (2013). Role of ghrelin in the pathophysiology of gastrointestinal disease. *Gut and Liver*, 7(5), 505–512. http://dx.doi.org/10.5009/gnl.2013.7.5.505.

Chorfa, A., Lazizzera, C., Betemps, D., Morignat, E., Dussurgey, S., Andrieu, T., et al. (2014). A variety of pesticides trigger in vitro alpha-synuclein accumulation, a key event in Parkinson's disease. *Archives of Toxicology*, http://dx.doi.org/10.1007/s00204-014-1388-2.

Clairembault, T., Kamphuis, W., Leclair-Visonneau, L., Rolli-Derkinderen, M., Coron, E., Neunlist, M., et al. (2014). Enteric GFAP expression and phosphorylation in Parkinson's disease. *Journal of Neurochemistry*, *130*(6), 805–815. http://dx.doi.org/10.1111/jnc.12742.

Clairembault, T., Leclair-Visonneau, L., Neunlist, M., & Derkinderen, P. (2015). Enteric glial cells: New players in Parkinson's disease? *Movement Disorders*, *30*(4), 494–498. http://dx.doi.org/10.1002/mds.25979.

Cryan, J. F., & Dinan, T. G. (2012). Mind-altering microorganisms: The impact of the gut microbiota on brain and behaviour. *Nature Reviews. Neuroscience*, *13*(10), 701–712. http://dx.doi.org/10.1038/nrn3346.

Danzer, K. M., Kranich, L. R., Ruf, W. P., Cagsal-Getkin, O., Winslow, A. R., Zhu, L., et al. (2012). Exosomal cell-to-cell transmission of alpha synuclein oligomers. *Molecular Neurodegeneration*, 7, 42. http://dx.doi.org/10.1186/1750-1326-7-42.

De Giorgio, R., Giancola, F., Boschetti, E., Abdo, H., Lardeux, B., & Neunlist, M. (2012). Enteric glia and neuroprotection: Basic and clinical aspects. *American Journal of Physiology. Gastrointestinal and Liver Physiology*, *303*(8), G887–G893. http://dx.doi.org/10.1152/ajpgi.00096.2012.

Deeb, J., Shah, M., Muhammed, N., Gunasekera, R., Gannon, K., Findley, L. J., et al. (2010). A basic smell test is as sensitive as a dopamine transporter scan: Comparison of olfaction, taste and DaTSCAN in the diagnosis of Parkinson's disease. *QJM*, *103*(12), 941–952. http://dx.doi.org/10.1093/qjmed/hcq142.

Del Tredici, K., & Braak, H. (2012). Spinal cord lesions in sporadic Parkinson's disease. *Acta Neuropathologica*, *124*(5), 643–664. http://dx.doi.org/10.1007/s00401-012-1028-y.

Del Tredici, K., Hawkes, C. H., Ghebremedhin, E., & Braak, H. (2010). Lewy pathology in the submandibular gland of individuals with incidental Lewy body disease and sporadic Parkinson's disease. *Acta Neuropathologica*, *119*(6), 703–713. http://dx.doi.org/10.1007/s00401-010-0665-2.

Del Tredici, K., Rub, U., De Vos, R. A., Bohl, J. R., & Braak, H. (2002). Where does Parkinson disease pathology begin in the brain? *Journal of Neuropathology and Experimental Neurology*, *61*(5), 413–426.

Derkinderen, P., Shannon, K. M., & Brundin, P. (2014). Gut feelings about smoking and coffee in Parkinson's disease. *Movement Disorders*, *29*(8), 976–979. http://dx.doi.org/10.1002/mds.25882.

Desplats, P., Lee, H. J., Bae, E. J., Patrick, C., Rockenstein, E., Crews, L., et al. (2009). Inclusion formation and neuronal cell death through neuron-to-neuron transmission of alpha-synuclein. *Proceedings of the National Academy of Sciences of the United States of America*, *106*(31), 13010–13015. http://dx.doi.org/10.1073/pnas.0903691106.

Devos, D., Lebouvier, T., Lardeux, B., Biraud, M., Rouaud, T., Pouclet, H., et al. (2013). Colonic inflammation in Parkinson's disease. *Neurobiology of Disease*, *50*, 42–48. http://dx.doi.org/10.1016/j.nbd.2012.09.007.

Dickson, D. W., Fujishiro, H., DelleDonne, A., Menke, J., Ahmed, Z., Klos, K. J., et al. (2008). Evidence that incidental Lewy body disease is pre-symptomatic Parkinson's disease. *Acta Neuropathologica*, *115*(4), 437–444. http://dx.doi.org/10.1007/s00401-008-0345-7.

Dickson, D. W., Uchikado, H., Fujishiro, H., & Tsuboi, Y. (2010). Evidence in favor of Braak staging of Parkinson's disease. *Movement Disorders*, *25*(Suppl. 1), S78–S82. http://dx.doi.org/10.1002/mds.22637.

Dinan, T. G., & Cryan, J. F. (2017). Gut instincts: Microbiota as a key regulator of brain development, ageing and neurodegeneration. *The Journal of Physiology*, *595*(2), 489–503. http://dx.doi.org/10.1113/JP273106.

Djaldetti, R., Baron, J., Ziv, I., & Melamed, E. (1996). Gastric emptying in Parkinson's disease: Patients with and without response fluctuations. *Neurology*, *46*(4), 1051–1054.

Doty, R. L., Nsoesie, M. T., Chung, I., Osman, A., Pawasarat, I., Caulfield, J., et al. (2015). Taste function in early stage treated and untreated Parkinson's disease. *Journal of Neurology*, *262*(3), 547–557. http://dx.doi.org/10.1007/s00415-014-7589-z.

Edwards, L. L., Pfeiffer, R. F., Quigley, E. M., Hofman, R., & Balluff, M. (1991). Gastrointestinal symptoms in Parkinson's disease. *Movement Disorders*, *6*(2), 151–156. http://dx.doi.org/10.1002/mds.870060211.

Edwards, L. L., Quigley, E. M., Harned, R. K., Hofman, R., & Pfeiffer, R. F. (1994). Characterization of swallowing and defecation in Parkinson's disease. *The American Journal of Gastroenterology*, *89*(1), 15–25.

Forsyth, C. B., Shannon, K. M., Kordower, J. H., Voigt, R. M., Shaikh, M., Jaglin, J. A., et al. (2011). Increased intestinal permeability correlates with sigmoid mucosa alpha-synuclein staining and endotoxin exposure markers in early Parkinson's disease. *PLoS One*, *6*(12), e28032. http://dx.doi.org/10.1371/journal.pone.0028032.

Forsythe, P., & Kunze, W. A. (2013). Voices from within: Gut microbes and the CNS. *Cellular and Molecular Life Sciences*, *70*(1), 55–69. http://dx.doi.org/10.1007/s00018-012-1028-z.

Gao, H. M., Zhang, F., Zhou, H., Kam, W., Wilson, B., & Hong, J. S. (2011). Neuroinflammation and alpha-synuclein dysfunction potentiate each other, driving chronic progression of neurodegeneration in a mouse model of Parkinson's disease. *Environmental Health Perspectives*, *119*(6), 807–814. http://dx.doi.org/10.1289/ehp.1003013.

Gaspar, P., & Gray, F. (1984). Dementia in idiopathic Parkinson's disease. A neuropathological study of 32 cases. *Acta Neuropathologica*, *64*(1), 43–52.

Gelpi, E., Navarro-Otano, J., Tolosa, E., Gaig, C., Compta, Y., Rey, M. J., et al. (2014). Multiple organ involvement by alpha-synuclein pathology in Lewy body disorders. *Movement Disorders*, *29*(8), 1010–1018. http://dx.doi.org/10.1002/mds.25776.

Ghaisas, S., Maher, J., & Kanthasamy, A. (2016). Gut microbiome in health and disease: Linking the microbiome-gut-brain axis and environmental factors in the pathogenesis of systemic and neurodegenerative diseases. *Pharmacology & Therapeutics*, *158*, 52–62. http://dx.doi.org/10.1016/j.pharmthera.2015.11.012.

Goedert, M., Spillantini, M. G., Del Tredici, K., & Braak, H. (2013). 100 Years of Lewy pathology. *Nature Reviews. Neurology*, *9*(1), 13–24. http://dx.doi.org/10.1038/nrneurol.2012.242.

Goetze, O., Nikodem, A. B., Wiezcorek, J., Banasch, M., Przuntek, H., Mueller, T., et al. (2006). Predictors of gastric emptying in Parkinson's disease. *Neurogastroenterology and Motility*, *18*(5), 369–375. http://dx.doi.org/10.1111/j.1365-2982.2006.00780.x.

Grathwohl, S. A., Steiner, J. A., Britschgi, M., & Brundin, P. (2013). Mind the gut: Secretion of alpha-synuclein by enteric neurons. *Journal of Neurochemistry*, *125*(4), 487–490. http://dx.doi.org/10.1111/jnc.12191.

Grenham, S., Clarke, G., Cryan, J. F., & Dinan, T. G. (2011). Brain-gut-microbe communication in health and disease. *Frontiers in Physiology*, *2*, 94. http://dx.doi.org/10.3389/fphys.2011.00094.

Hallett, P. J., McLean, J. R., Kartunen, A., Langston, J. W., & Isacson, O. (2012). Alpha-synuclein overexpressing transgenic mice show internal organ pathology and autonomic deficits. *Neurobiology of Disease*, *47*(2), 258–267. http://dx.doi.org/10.1016/j.nbd.2012.04.009.

Halliday, G., McCann, H., & Shepherd, C. (2012). Evaluation of the Braak hypothesis: How far can it explain the pathogenesis of Parkinson's disease? *Expert Review of Neurotherapeutics*, *12*(6), 673–686. http://dx.doi.org/10.1586/ern.12.47.

Hanaoka, A., & Kashihara, K. (2009). Increased frequencies of caries, periodontal disease and tooth loss in patients with Parkinson's disease. *Journal of Clinical Neuroscience*, *16*(10), 1279–1282. http://dx.doi.org/10.1016/j.jocn.2008.12.027.

Hawkes, C. H., Del Tredici, K., & Braak, H. (2007). Parkinson's disease: A dual-hit hypothesis. *Neuropathology and Applied Neurobiology*, *33*(6), 599–614. http://dx.doi.org/10.1111/j.1365-2990.2007.00874.x.

Hilton, D., Stephens, M., Kirk, L., Edwards, P., Potter, R., Zajicek, J., et al. (2014). Accumulation of alpha-synuclein in the bowel of patients in the pre-clinical phase of Parkinson's disease. *Acta Neuropathologica*, *127*(2), 235–241. http://dx.doi.org/10.1007/s00401-013-1214-6.

Holmqvist, S., Chutna, O., Bousset, L., Aldrin-Kirk, P., Li, W., Bjorklund, T., et al. (2014). Direct evidence of Parkinson pathology spread from the gastrointestinal tract to the brain in rats. *Acta Neuropathologica*, *128*(6), 805–820. http://dx.doi.org/10.1007/s00401-014-1343-6.

Hummel, T., Barz, S., Lotsch, J., Roscher, S., Kettenmann, B., & Kobal, G. (1996). Loss of olfactory function leads to a decrease of trigeminal sensitivity. *Chemical Senses*, *21*(1), 75–79.

Inden, M., Kitamura, Y., Takeuchi, H., Yanagida, T., Takata, K., Kobayashi, Y., et al. (2007). Neurodegeneration of mouse nigrostriatal dopaminergic system induced by repeated oral administration of rotenone is prevented by 4-phenylbutyrate, a chemical chaperone. *Journal of Neurochemistry*, *101*(6), 1491–1504. http://dx.doi.org/10.1111/j.1471-4159.2006.04440.x.

Jellinger, K. A. (2003). Alpha-synuclein pathology in Parkinson's and Alzheimer's disease brain: Incidence and topographic distribution—A pilot study. *Acta Neuropathologica*, *106*(3), 191–201. http://dx.doi.org/10.1007/s00401-003-0725-y.

Jellinger, K. A. (2004). Lewy body-related alpha-synucleinopathy in the aged human brain. *Journal of Neural Transmission*, *111*(10 – 11), 1219–1235. http://dx.doi.org/10.1007/s00702-004-0138-7.

Jellinger, K. A. (2009). Recent advances in our understanding of neurodegeneration. *Journal of Neural Transmission (Vienna, Austria: 1996)*, *116*(9), 1111–1162. http://dx.doi.org/10.1007/s00702-009-0240-y.

Jellinger, K. A. (2015). Neuropathobiology of non-motor symptoms in Parkinson disease. *Journal of Neural Transmission (Vienna, Austria: 1996)*, *122*(10), 1429–1440. http://dx.doi.org/10.1007/s00702-015-1405-5.

Jost, W. H. (1997). Gastrointestinal motility problems in patients with Parkinson's disease. Effects of antiparkinsonian treatment and guidelines for management. *Drugs & Aging*, *10*(4), 249–258.

Jost, W. H., & Schrank, B. (1998). Defecatory disorders in de novo parkinsonians—Colonic transit and electromyogram of the external anal sphincter. *Wiener Klinische Wochenschrift*, *110*(15), 535–537.

Jost, W. H., Schrank, B., Herold, A., & Leiss, O. (1999). Functional outlet obstruction: Anismus, spastic pelvic floor syndrome, and dyscoordination of the voluntary sphincter muscles. Definition, diagnosis, and treatment from the neurologic point of view. *Scandinavian Journal of Gastroenterology*, *34*(5), 449–453.

Kalf, J. G., Munneke, M., van den Engel-Hoek, L., de Swart, B. J., Borm, G. F., Bloem, B. R., et al. (2011). Pathophysiology of diurnal drooling in Parkinson's disease. *Movement Disorders*, *26*(9), 1670–1676. http://dx.doi.org/10.1002/mds.23720.

Kikuchi, A., Baba, T., Hasegawa, T., Kobayashi, M., Sugeno, N., Konno, M., et al. (2013). Hypometabolism in the supplementary and anterior cingulate cortices is related to dysphagia in Parkinson's disease: A cross-sectional and 3-year longitudinal cohort study. *BMJ Open*, *3*(3). http://dx.doi.org/10.1136/bmjopen-2012-002249.

Kim, H. J., Jeon, B. S., Lee, J. Y., Cho, Y. J., Hong, K. S., & Cho, J. Y. (2011). Taste function in patients with Parkinson disease. *Journal of Neurology*, *258*(6), 1076–1079. http://dx.doi.org/10.1007/s00415-010-5884-x.

Klingelhoefer, L., & Reichmann, H. (2015). Pathogenesis of Parkinson disease—The gut-brain axis and environmental factors. *Nature Reviews. Neurology*, *11*(11), 625–636. http://dx.doi.org/10.1038/nrneurol.2015.197.

Korczyn, A. D. (1990). Autonomic nervous system disturbances in Parkinson's disease. *Advances in Neurology*, *53*, 463–468.

Kordower, J. H., & Brundin, P. (2009). Lewy body pathology in long-term fetal nigral transplants: Is Parkinson's disease transmitted from one neural system to another? *Neuropsychopharmacology*, *34*(1), 254. http://dx.doi.org/10.1038/npp.2008.161.

Kordower, J. H., Chu, Y., Hauser, R. A., Freeman, T. B., & Olanow, C. W. (2008). Lewy body-like pathology in long-term embryonic nigral transplants in Parkinson's disease. *Nature Medicine*, *14*(5), 504–506. http://dx.doi.org/10.1038/nm1747.

Kuo, Y. M., Li, Z., Jiao, Y., Gaborit, N., Pani, A. K., Orrison, B. M., et al. (2010). Extensive enteric nervous system abnormalities in mice transgenic for artificial chromosomes containing Parkinson disease-associated alpha-synuclein gene mutations precede central nervous system changes. *Human Molecular Genetics*, *19*(9), 1633–1650. http://dx.doi.org/10.1093/hmg/ddq038.

Kupsky, W. J., Grimes, M. M., Sweeting, J., Bertsch, R., & Cote, L. J. (1987). Parkinson's disease and megacolon: Concentric hyaline inclusions (Lewy bodies) in enteric ganglion cells. *Neurology*, *37*(7), 1253–1255.

Landis, B. N., Scheibe, M., Weber, C., Berger, R., Bramerson, A., Bende, M., et al. (2010). Chemosensory interaction: Acquired olfactory impairment is associated with decreased taste function. *Journal of Neurology*, *257*(8), 1303–1308. http://dx.doi.org/10.1007/s00415-010-5513-8.

Lang, C. J., Leuschner, T., Ulrich, K., Stossel, C., Heckmann, J. G., & Hummel, T. (2006). Taste in dementing diseases and parkinsonism. *Journal of the Neurological Sciences*, *248*(1–2), 177–184. http://dx.doi.org/10.1016/j.jns.2006.05.020.

Lebouvier, T., Chaumette, T., Damier, P., Coron, E., Touchefeu, Y., Vrignaud, S., et al. (2008). Pathological lesions in colonic biopsies during Parkinson's disease. *Gut*, *57*(12), 1741–1743. http://dx.doi.org/10.1136/gut.2008.162503.

Lebouvier, T., Coron, E., Chaumette, T., Paillusson, S., Bruley des Varannes, S., Neunlist, M., et al. (2010a). Routine colonic biopsies as a new tool to study the enteric nervous system in living patients. *Neurogastroenterology and Motility*, *22*(1), e11–e14. http://dx.doi.org/10.1111/j.1365-2982.2009.01368.x.

Lebouvier, T., Neunlist, M., Bruley des Varannes, S., Coron, E., Drouard, A., N'Guyen, J. M., et al. (2010b). Colonic biopsies to assess the neuropathology of Parkinson's disease and its relationship with symptoms. *PLoS One*, *5*(9), e12728. http://dx.doi.org/10.1371/journal.pone.0012728.

Li, J. Y., Englund, E., Holton, J. L., Soulet, D., Hagell, P., Lees, A. J., et al. (2008). Lewy bodies in grafted neurons in subjects with Parkinson's disease suggest host-to-graft disease propagation. *Nature Medicine*, *14*(5), 501–503. http://dx.doi.org/10.1038/nm1746.

Li, Z. S., Pham, T. D., Tamir, H., Chen, J. J., & Gershon, M. D. (2004). Enteric dopaminergic neurons: Definition, developmental lineage, and effects of extrinsic denervation. *The Journal of Neuroscience*, *24*(6), 1330–1339. http://dx.doi.org/10.1523/JNEUROSCI.3982-03.2004.

Littman, D. R., & Pamer, E. G. (2011). Role of the commensal microbiota in normal and pathogenic host immune responses. *Cell Host & Microbe*, *10*(4), 311–323. http://dx.doi.org/10.1016/j.chom.2011.10.004.

Marras, C., & Lang, A. (2013). Parkinson's disease subtypes: Lost in translation? *Journal of Neurology, Neurosurgery, and Psychiatry*, *84*(4), 409–415. http://dx.doi.org/10.1136/jnnp-2012-303455.

Martinez-Martin, P. (2011). The importance of non-motor disturbances to quality of life in Parkinson's disease. *Journal of the Neurological Sciences*, *310*(1–2), 12–16. http://dx.doi.org/10.1016/j.jns.2011.05.006.

Martinez-Martin, P., Rodriguez-Blazquez, C., Kurtis, M. M., & Chaudhuri, K. R. (2011). The impact of non-motor symptoms on health-related quality of life of patients with Parkinson's disease. *Movement Disorders*, *26*(3), 399–406. http://dx.doi.org/10.1002/mds.23462.

Martinez-Martin, P., Schapira, A. H., Stocchi, F., Sethi, K., Odin, P., MacPhee, G., et al. (2007). Prevalence of nonmotor symptoms in Parkinson's disease in an international setting; study using nonmotor symptoms questionnaire in 545 patients. *Movement Disorders*, *22*(11), 1623–1629. http://dx.doi.org/10.1002/mds.21586.

Mayer, E. A., Tillisch, K., & Gupta, A. (2015). Gut/brain axis and the microbiota. *The Journal of Clinical Investigation*, *125*(3), 926–938. http://dx.doi.org/10.1172/JCI76304.

Minguez-Castellanos, A., Chamorro, C. E., Escamilla-Sevilla, F., Ortega-Moreno, A., Rebollo, A. C., Gomez-Rio, M., et al. (2007). Do alpha-synuclein aggregates in autonomic plexuses predate Lewy body disorders?: A cohort study. *Neurology*, *68*(23), 2012–2018. http://dx.doi.org/10.1212/01.wnl.0000264429.59379.d9.

Mu, L., Sobotka, S., Chen, J., Su, H., Sanders, I., Adler, C. H., et al. (2012). Altered pharyngeal muscles in Parkinson disease. *Journal of Neuropathology and Experimental Neurology*, *71*(6), 520–530. http://dx.doi.org/10.1097/NEN.0b013e318258381b.

Muller, T., Erdmann, C., Bremen, D., Schmidt, W. E., Muhlack, S., Woitalla, D., et al. (2006). Impact of gastric emptying on levodopa pharmacokinetics in Parkinson disease patients. *Clinical Neuropharmacology*, *29*(2), 61–67.

Muller, T., Palluch, R., & Jackowski, J. (2011). Caries and periodontal disease in patients with Parkinson's disease. *Special Care in Dentistry*, *31*(5), 178–181. http://dx.doi.org/10.1111/j.1754-4505.2011.00205.x.

Muntane, G., Ferrer, I., & Martinez-Vicente, M. (2012). Alpha-synuclein phosphorylation and truncation are normal events in the adult human brain. *Neuroscience*, *200*, 106–119. http://dx.doi.org/10.1016/j.neuroscience.2011.10.042.

Orimo, S., Amino, T., Itoh, Y., Takahashi, A., Kojo, T., Uchihara, T., et al. (2005). Cardiac sympathetic denervation precedes neuronal loss in the sympathetic ganglia in Lewy body disease. *Acta Neuropathologica*, *109*(6), 583–588. http://dx.doi.org/10.1007/s00401-005-0995-7.

Pan-Montojo, F., Anichtchik, O., Dening, Y., Knels, L., Pursche, S., Jung, R., et al. (2010). Progression of Parkinson's disease pathology is reproduced by intragastric administration of rotenone in mice. *PLoS One*, *5*(1), e8762. http://dx.doi.org/10.1371/journal.pone.0008762.

Pan-Montojo, F., Schwarz, M., Winkler, C., Arnhold, M., O'Sullivan, G. A., Pal, A., et al. (2012). Environmental toxins trigger PD-like progression via increased alpha-synuclein release from enteric neurons in mice. *Scientific Reports*, *2*, 898. http://dx.doi.org/10.1038/srep00898.

Parkkinen, L., Soininen, H., Laakso, M., & Alafuzoff, I. (2001). Alpha-synuclein pathology is highly dependent on the case selection. *Neuropathology and Applied Neurobiology*, *27*(4), 314–325.

Petrovitch, H., Abbott, R. D., Ross, G. W., Nelson, J., Masaki, K. H., Tanner, C. M., et al. (2009). Bowel movement frequency in late-life and substantia nigra neuron density at death. *Movement Disorders*, *24*(3), 371–376. http://dx.doi.org/10.1002/mds.22360.

Pfeiffer, R. F. (2003). Gastrointestinal dysfunction in Parkinson's disease. *Lancet Neurology*, *2*(2), 107–116.

Pfeiffer, R. F. (2011). Gastrointestinal dysfunction in Parkinson's disease. *Parkinsonism & Related Disorders*, *17*(1), 10–15. http://dx.doi.org/10.1016/j.parkreldis.2010.08.003.

Postuma, R. B., Berg, D., Stern, M., Poewe, W., Olanow, C. W., Oertel, W., et al. (2015). MDS clinical diagnostic criteria for Parkinson's disease. *Movement Disorders*, *30*(12), 1591–1601. http://dx.doi.org/10.1002/mds.26424.

Postuma, R. B., Gagnon, J. F., Pelletier, A., & Montplaisir, J. (2013). Prodromal autonomic symptoms and signs in Parkinson's disease and dementia with Lewy bodies. *Movement Disorders*, *28*(5), 597–604. http://dx.doi.org/10.1002/mds.25445.

Qualman, S. J., Haupt, H. M., Yang, P., & Hamilton, S. R. (1984). Esophageal Lewy bodies associated with ganglion cell loss in achalasia. Similarity to Parkinson's disease. *Gastroenterology*, *87*(4), 848–856.

Rana, A. Q., Yousuf, M. S., Awan, N., & Fattah, A. (2012). Impact of progression of Parkinson's disease on drooling in various ethnic groups. *European Neurology*, *67*(5), 312–314. http://dx.doi.org/10.1159/000336054.

Reichmann, H. (2011). View point: Etiology in Parkinson's disease. Dual hit or spreading intoxication. *Journal of the Neurological Sciences*, *310*(1–2), 9–11. http://dx.doi.org/10.1016/j.jns.2011.04.016.

Rhee, S. H., Pothoulakis, C., & Mayer, E. A. (2009). Principles and clinical implications of the brain-gut-enteric microbiota axis. *Nature Reviews. Gastroenterology & Hepatology*, *6*(5), 306–314. http://dx.doi.org/10.1038/nrgastro.2009.35.

Rodriguez, M., Rodriguez-Sabate, C., Morales, I., Sanchez, A., & Sabate, M. (2015). Parkinson's disease as a result of aging. *Aging Cell*, *14*(3), 293–308. http://dx.doi.org/10.1111/acel.12312.

Rodriguez-Violante, M., Cervantes-Arriaga, A., Villar-Velarde, A., & Corona, T. (2010). Prevalence of non-motor dysfunction among Parkinson's disease patients from a tertiary

referral center in Mexico City. *Clinical Neurology and Neurosurgery*, *112*(10), 883–885. http://dx.doi.org/10.1016/j.clineuro.2010.07.021.
Saito, Y., Ruberu, N. N., Sawabe, M., Arai, T., Kazama, H., Hosoi, T., et al. (2004). Lewy body-related alpha-synucleinopathy in aging. *Journal of Neuropathology and Experimental Neurology*, *63*(7), 742–749.
Sakakibara, R., Kishi, M., Ogawa, E., Tateno, F., Uchiyama, T., Yamamoto, T., et al. (2011). Bladder, bowel, and sexual dysfunction in Parkinson's disease. *Parkinson's Disease*, *2011*, 924605. http://dx.doi.org/10.4061/2011/924605.
Sakakibara, R., Odaka, T., Uchiyama, T., Asahina, M., Yamaguchi, K., Yamaguchi, T., et al. (2003). Colonic transit time and rectoanal videomanometry in Parkinson's disease. *Journal of Neurology, Neurosurgery, and Psychiatry*, *74*(2), 268–272.
Sampson, T. R., Debelius, J. W., Thron, T., Janssen, S., Shastri, G. G., Ilhan, Z. E., et al. (2016). Gut microbiota regulate motor deficits and neuroinflammation in a model of Parkinson's disease. *Cell*, *167*(6), 1469–1480, e1412. http://dx.doi.org/10.1016/j.cell.2016.11.018.
Sanchez-Guajardo, V., Tentillier, N., & Romero-Ramos, M. (2015). The relation between alpha-synuclein and microglia in Parkinson's disease: Recent developments. *Neuroscience*, *302*, 47–58. http://dx.doi.org/10.1016/j.neuroscience.2015.02.008.
Savica, R., Carlin, J. M., Grossardt, B. R., Bower, J. H., Ahlskog, J. E., Maraganore, D. M., et al. (2009). Medical records documentation of constipation preceding Parkinson disease: A case-control study. *Neurology*, *73*(21), 1752–1758. http://dx.doi.org/10.1212/WNL.0b013e3181c34af5.
Scheperjans, F., Aho, V., Pereira, P. A., Koskinen, K., Paulin, L., Pekkonen, E., et al. (2015a). Gut microbiota are related to Parkinson's disease and clinical phenotype. *Movement Disorders*, *30*(3), 350–358. http://dx.doi.org/10.1002/mds.26069.
Scheperjans, F., Pekkonen, E., Kaakkola, S., & Auvinen, P. (2015b). Linking smoking, coffee, urate, and Parkinson's disease—A role for gut microbiota? *Journal of Parkinson's Disease*, *5*, 255–262. http://dx.doi.org/10.3233/JPD-150557.
Schneider, S. A., Boettner, M., Alexoudi, A., Zorenkov, D., Deuschl, G., & Wedel, T. (2016). Can we use peripheral tissue biopsies to diagnose Parkinson's disease? A review of the literature. *European Journal of Neurology*, *23*(2), 247–261. http://dx.doi.org/10.1111/ene.12753.
Schrag, A., Jahanshahi, M., & Quinn, N. (2000). How does Parkinson's disease affect quality of life? A comparison with quality of life in the general population. *Movement Disorders*, *15*(6), 1112–1118.
Shannon, K. M., Keshavarzian, A., Mutlu, E., Dodiya, H. B., Daian, D., Jaglin, J. A., et al. (2012). Alpha-synuclein in colonic submucosa in early untreated Parkinson's disease. *Movement Disorders*, *27*(6), 709–715. http://dx.doi.org/10.1002/mds.23838.
Sherer, T. B., Kim, J. H., Betarbet, R., & Greenamyre, J. T. (2003). Subcutaneous rotenone exposure causes highly selective dopaminergic degeneration and alpha-synuclein aggregation. *Experimental Neurology*, *179*(1), 9–16.
Silva, B. A., Einarsdottir, O., Fink, A. L., & Uversky, V. N. (2013). Biophysical characterization of alpha-synuclein and rotenone interaction. *Biomolecules*, *3*(3), 703–732. http://dx.doi.org/10.3390/biom3030703.
Srivanitchapoom, P., Pandey, S., & Hallett, M. (2014). Drooling in Parkinson's disease: A review. *Parkinsonism & Related Disorders*, *20*(11), 1109–1118. http://dx.doi.org/10.1016/j.parkreldis.2014.08.013.
Stokholm, M. G., Danielsen, E. H., Hamilton-Dutoit, S. J., & Borghammer, P. (2016). Pathological alpha-synuclein in gastrointestinal tissues from prodromal Parkinson disease patients. *Annals of Neurology*, *79*(6), 940–949. http://dx.doi.org/10.1002/ana.24648.

Svensson, E., Horvath-Puho, E., Thomsen, R. W., Djurhuus, J. C., Pedersen, L., Borghammer, P., et al. (2015). Vagotomy and subsequent risk of Parkinson's disease. *Annals of Neurology*, *78*, 522–529. http://dx.doi.org/10.1002/ana.24448.

Tasselli, M., Chaumette, T., Paillusson, S., Monnet, Y., Lafoux, A., Huchet-Cadiou, C., et al. (2013). Effects of oral administration of rotenone on gastrointestinal functions in mice. *Neurogastroenterology and Motility*, *25*(3), e183–e193. http://dx.doi.org/10.1111/nmo.12070.

Titova, N., Padmakumar, C., Lewis, S. J., & Chaudhuri, K. R. (2016). Parkinson's: A syndrome rather than a disease? *Journal of Neural Transmission (Vienna, Austria: 1996)*. http://dx.doi.org/10.1007/s00702-016-1667-6. [Epub ahead of print].

Tysnes, O. B., Kenborg, L., Herlofson, K., Steding-Jessen, M., Horn, A., Olsen, J. H., et al. (2015). Does vagotomy reduce the risk of Parkinson's disease? *Annals of Neurology*, *78*(6), 1011–1012.

van Rooden, S. M., Colas, F., Martinez-Martin, P., Visser, M., Verbaan, D., Marinus, J., et al. (2011). Clinical subtypes of Parkinson's disease. *Movement Disorders*, *26*(1), 51–58. http://dx.doi.org/10.1002/mds.23346.

Vilchez, D., Saez, I., & Dillin, A. (2014). The role of protein clearance mechanisms in organismal ageing and age-related diseases. *Nature Communications*, *5*, 5659. http://dx.doi.org/10.1038/ncomms6659.

Villaran, R. F., Espinosa-Oliva, A. M., Sarmiento, M., De Pablos, R. M., Arguelles, S., Delgado-Cortes, M. J., et al. (2010). Ulcerative colitis exacerbates lipopolysaccharide-induced damage to the nigral dopaminergic system: Potential risk factor in Parkinson's disease. *Journal of Neurochemistry*, *114*(6), 1687–1700. http://dx.doi.org/10.1111/j.1471-4159.2010.06879.x.

Visanji, N. P., Marras, C., Kern, D. S., Al Dakheel, A., Gao, A., Liu, L. W., et al. (2015). Colonic mucosal a-synuclein lacks specificity as a biomarker for Parkinson disease. *Neurology*, *84*(6), 609–616. http://dx.doi.org/10.1212/WNL.0000000000001240.

Voreades, N., Kozil, A., & Weir, T. L. (2014). Diet and the development of the human intestinal microbiome. *Frontiers in Microbiology*, *5*, 494. http://dx.doi.org/10.3389/fmicb.2014.00494.

Wakabayashi, K., Takahashi, H., Ohama, E., & Ikuta, F. (1990). Parkinson's disease: An immunohistochemical study of Lewy body-containing neurons in the enteric nervous system. *Acta Neuropathologica*, *79*(6), 581–583.

Wakabayashi, K., Takahashi, H., Takeda, S., Ohama, E., & Ikuta, F. (1988). Parkinson's disease: The presence of Lewy bodies in Auerbach's and Meissner's plexuses. *Acta Neuropathologica*, *76*(3), 217–221.

Walker, A. W., & Lawley, T. D. (2013). Therapeutic modulation of intestinal dysbiosis. *Pharmacological Research*, *69*(1), 75–86. http://dx.doi.org/10.1016/j.phrs.2012.09.008.

Wang, L., Magen, I., Yuan, P. Q., Subramaniam, S. R., Richter, F., Chesselet, M. F., et al. (2012). Mice overexpressing wild-type human alpha-synuclein display alterations in colonic myenteric ganglia and defecation. *Neurogastroenterology and Motility*, *24*(9), e425–e436. http://dx.doi.org/10.1111/j.1365-2982.2012.01974.x.

Wang, C. P., Sung, W. H., Wang, C. C., & Tsai, P. Y. (2013). Early recognition of pelvic floor dyssynergia and colorectal assessment in Parkinson's disease associated with bowel dysfunction. *Colorectal Disease*, *15*(3), e130–e137. http://dx.doi.org/10.1111/codi.12105.

Wolters, E. C., & Braak, H. (2006). Parkinson's disease: Premotor clinico-pathological correlations. *Journal of Neural Transmission. Supplementum*, (70), 309–319.

Yuan, Y. H., Yan, W. F., Sun, J. D., Huang, J. Y., Mu, Z., & Chen, N. H. (2015). The molecular mechanism of rotenone-induced alpha-synuclein aggregation: Emphasizing the role of the calcium/GSK3beta pathway. *Toxicology Letters*, *233*(2), 163–171. http://dx.doi.org/10.1016/j.toxlet.2014.11.029.

CHAPTER TWENTY-SEVEN

Constipation in Parkinson's Disease

Fabrizio Stocchi[1], Margherita Torti[1]
University and Institute for Research and Medical Care, IRCCS San Raffaele, Rome, Italy
[1]Corresponding authors: e-mail address: fabrizio.stocchi@fastwebnet.it; margherita.torti@sanraffaele.it

Contents

Abstract

Constipation is one of the main and disabling nonmotor symptoms in Parkinson's disease (PD), with a prevalence ranging from 24.6% to 63% according to the different diagnostic criteria used to define chronic constipation. Constipation is currently recognized as a risk factor of PD in relation to the number of evacuation per week and its severity. Moreover, several studies have demonstrated that constipation may precede the occurrence of motor symptoms underlying an earlier involvement of the enteric nervous system and the dorsal motor nucleus of the vagus in the α-synuclein pathology. In PD, constipation is mainly due to slower colonic transit or puborectalis dyssynergia, but the concomitant use of antiparkinsonian, pain, and antidepressant medications may worsen it. An accurate diagnosis and an adequate treatment of constipation it is pivotal to prevent complications such as intestinal occlusion and to ensure an optimal clinical response to levodopa.

International Review of Neurobiology, Volume 134
ISSN 0074-7742
http://dx.doi.org/10.1016/bs.irn.2017.06.003

1. INTRODUCTION

Parkinson's disease (PD) is a neurodegenerative disorder characterized by motor and nonmotor features. Several studies suggest that gastrointestinal dysfunctions are one of the most common and disabling nonmotor symptoms in PD patients playing an important role in determining quality of life (Barone et al., 2009; Cersosimo et al., 2013; Martinez-Martin et al., 2007). Moreover, they have been associated with potentially life-threatening complications such as pulmonary aspiration, intestinal obstruction, and perforation ranking among the first causes of emergency admission in PD population. Gastrointestinal symptoms are also recognized to have a role in motor fluctuations due to an erratic absorption of levodopa. Within gastrointestinal symptoms, constipation is the most common and one of the most distressing one.

2. CONSTIPATION: A PRODROMAL SIGN AND A RISK FACTOR FOR PD

Constipation is a recognized symptom of the disease and it is demonstrated to even precede motor features. Indeed, number of bowel movements per week (Abbott et al., 2001; Gao, Chen, Schwarzschild, & Ascherio, 2011) and severity of constipation (Lin, Lin, Liu, & Chang Ch, 2014) were found to be associated with an increased risk of developing PD.

The role of constipation as a risk factor for PD seems to confirm the Braak's model for the pathodynamics of Lewy pathology in PD, with an early involvement of the enteric nervous system and dorsal motor nucleus of the vagus (DMV) even before degeneration of substantia nigra begins. An autopsy survey published by the Arizona PD Consortium has shown the occurrence of Lewy pathology in the peripheral nervous system, especially in the enteric nervous system, in almost all patients with PD (Beach et al., 2009). α-Synuclein staining has been reported in biopsies of the submucosa of the ascending colon taken during routine screening of patients with mild-to-moderate PD. A study performed on 10 de novo PD patients has demonstrated staining for α-synuclein in the lamina propria of the colonic submucosa. 3-Nitrotyrosine staining, a marker of oxidative stress, was also seen in the majority of subjects even though it was not specific for PD (Shannon et al., 2012).

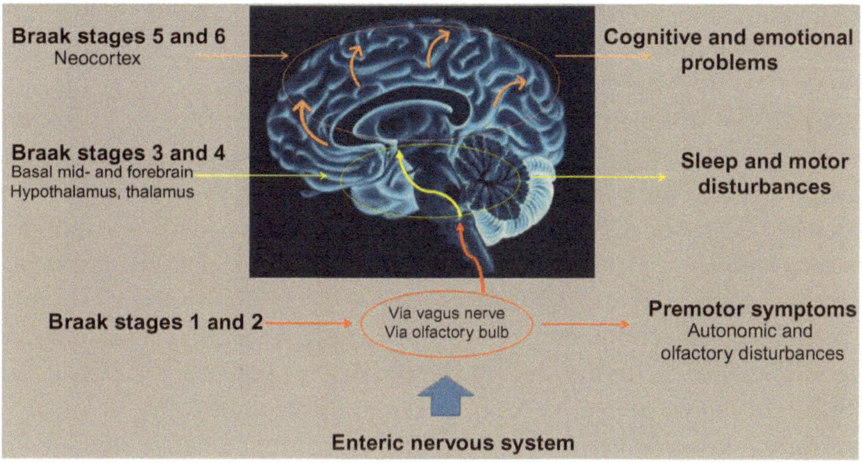

Fig. 1 Braak staging: progression of the Lewy body pathology.

The extension of Lewy body pathology in the enteric nervous system correlates with the severity of LB deposition in substantia nigra (Braak, de Vos, Bohl, & Del Tredici, 2006). According to Braak's staging system (Fig. 1; Braak et al., 2002), the α-synuclein pathology spreads from enteric nervous system and olfactory bulbs to substantia nigra and other nuclear grays of the midbrain defining a nonrandom, rostrocaudal progression confirmed in postmortem studies (Beach et al., 2009; Shannon et al., 2012). If it was first hypothesized that an unidentified agent with neurotrophic properties could enter the brain through the ENS, there is now growing evidence that α-synuclein itself could act as a pathogenic factor spreading from one cell to another resembling a prion-like transmission (Visanji, Brooks, Hazrati, & Lang, 2013). The DMV is affected at the earliest stage of Lewy body pathology in PD (Gai, Blumbergs, Geffen, & Blessing, 1992), and its involvement is the first step for the spreading of LB pathology in the brain. Indeed, studies performed in toxic models of parkinsonism have demonstrated that hemivagotomy was able to stop progression of symptoms (Klingelhoefer & Reichmann, 2015), and subjects underwent to truncal vagotomy were found to be at lower risk of developing PD (Svensson et al., 2015).

3. MICROBIOTA AND SIBO

The human microbiota consists of 10–100 trillion symbiotic microbial cells harbored by each person, primarily bacteria in the gut. Microbiota brain–gut axis interactions are significantly modulated by microbiota via immunological, neuroendocrine, and direct neural mechanisms (Mulak & Bonaz, 2015). Dysregulation of the brain–gut microbiota axis in PD may be associated with gastrointestinal dysfunctions frequently preceding motor symptoms, as well as with the pathogenesis of PD itself, supporting the hypothesis that the pathological process spreads from the gut to the brain (Braak et al., 2006). Scheperjans et al. reported alterations in the composition of microbiota in PD and its association with clinical phenotypes of the disease (Scheperjans et al., 2015).

An important component of microbiota is represented by small intestinal bacteria and recent studies showed an overgrowth of small intestinal bacterial (SIBO) in PD patients (Tan et al., 2014), probably induced by an impaired gastrointestinal motility and the resulting altered imbalance of microbiota.

Gut dysbiosis and SIBO can increase intestinal permeability causing excessive stimulation of the innate immune system and systemic inflammation, mechanism involved in the initiation of α-synuclein deposits (de Vos & de Vos, 2012; Klingelhoefer & Reichmann, 2015; Visanji et al., 2013).

Quantitative and qualitative alterations of microbiota could influence disease progression and severity, through a negative effect on pharmacological response. It has indeed reported that SIBO can affect levodopa absorption and participate to the development of motor fluctuations. PD patients with SIBO were found to suffer from more severe motor fluctuations which significantly improved after antibiotic treatment (Fasano et al., 2013).

4. DEFINITION AND DIAGNOSTIC CRITERIA

The American College of Gastroenterology (2005) defines constipation as an "unsatisfactory defecation characterized by infrequent stools, difficult stool passage or both. Difficult stool passage includes straining, a sense of difficulty passing stool, incomplete evacuation, hard/lumpy stool, prolonged time to pass stool. Chronic constipation is defined as the presence of these symptoms for at least three months."

Functional chronic constipation was defined according to the Rome III Diagnostic Criteria (Table 1; Longstreth et al., 2006).

5. EPIDEMIOLOGY

The prevalence of constipation in PD ranges from 24.6% to 63% according to the different diagnostic criteria used to define chronic

Table 1 Rome III Diagnostic Criteria for Functional Chronic Constipation

1. Chronic constipation must include two of the following conditions[a]: • Straining • Lumpy or hard stools • Sensation of incomplete evacuation • Sensation of anorectal obstruction/blockage • Manual maneuvers to facilitate defecation • Less than three defecations per week
2. Loose stools are rarely present without the use of laxatives
3. Insufficient criteria for irritable bowel syndrome
4. No structural abnormalities

[a]In at least 25% of defecation in the last 3 months with symptoms onset at least 6 months prior to diagnosis.

constipation, the type of scales used and to study population enrolled (early or advanced, selected, or nonselected). The above-mentioned Rome criteria have indeed not properly validated in PD and therefore not widely used to assess constipation in this particular population. A modified version of Rome III criteria was used to analyze constipation, finding that this symptom was almost three times more prevalent in PD than in controls (59% vs 21%) (Drossman & Dumitrascu, 2006; Evatt et al., 2009; Kaye, Gage, Kimber, Storey, & Trend, 2006). Other scales used to evaluate constipation were the Scales for Outcomes in Parkinson's disease—Autonomic (SCOPA-AUT) that analyze the frequency of bowel movement, straining and fecal incontinence (Damian et al., 2012; Forjaz et al., 2010), and the Nonmotor Symptoms Scale (NMSS) that investigates frequency and severity (Chaudhuri et al., 2007).

Considering the most commonly used criterion of frequency, (i.e., <3 bowel movements per week), the median prevalence of constipation in PD was found to be 50% with a range varying from 20 to 63. Prevalence of straining generally exceeded prevalence of decreased frequency of bowel movements, showing also variation across various studies (33%–83%) (Knudsen, Krogh, Østergaard, & Borghammer, 2017).

Different prevalence rates were also found between patients with akinetic-rigid form and tremor-dominant phenotype: 64% vs 49% in advanced parkinsonian patients (Khedr, Fetoh El, Khalifa, Ahmed, & Beh El, 2013) and 45% vs 21% in a study on de novo PD patients (Pont-Sunyer et al., 2015). A large number of studies have finally demonstrated a clear increase in constipation prevalence related to disease progression/Hoehn and Yahr stages, regardless of definition or evaluation scales used.

6. PATHOPHYSIOLOGY

Constipation in PD can result from several causes. In parkinsonism, constipation is due to slower colonic transit or to outlet-type dysfunction or both. Many authors have showed that mean colon transit time is almost doubled in PD patients (Edwards, Quigley, Harned, Hofman, & Pfeiffer, 1994). Early involvement of enteric neurons may explain the finding that constipation is a very early sign of PD (Abbott et al., 2001). There is accumulation of Lewy bodies in VIPergic neurons of the ENS (Wakabayashi, Takahashi, Ohama, Takeda, & Ikuta, 1993), suggesting that slow intestinal transit in PD may result primarily from impaired reflex relaxation of the distal smooth muscle due to loss of inhibitory motor neurons. However, there is also age-related loss of excitatory cholinergic neurons in the colon (Phillips & Powley, 2005; Wade & Hornby, 2005), and impaired cholinergic function likely contributes to slow colonic transit time in PD. The 5-HT4 agonists mosapride (Liu et al., 2005) and tegaserod (Morgan & Sethi, 2007) improve constipation in patients with PD, presumably by promoting ACh release from myenteric neurons. Loss of enteric dopamine neurons has also been reported in PD (Singaram et al., 1995), but the functional significance of this is uncertain. Dopamine cells constitute only 1%–2% of the total population of ENS neurons in humans and dopamine, acting via presynaptic D2 receptors, inhibiting ACh release and intestinal motility. There is selective loss of enteric dopaminergic neurons in the MPTP (1-methyl 4-phenyl 1,2,3,6-tetrahydropyridine) mouse model of PD. Electrophysiological recording of neural-mediated muscle contraction in isolated colon from MPTP-treated animals showed a defect in relaxation. These animals exhibit transient increase in colon motility, but no changes in gastric emptying or small intestine transit (Anderson et al., 2007). Other authors (Wakabayashi, Takahachi, Ohama, & Ikuta, 1989; Wakabayashi, Takahachi, Ohama, & Ikuta, 1990) have shown most Lewy bodies in the gut to be in VIP-immunoreactive neurons, whereas in the sympathetic ganglia, all Lewy body-containing neurons were immunoreactive for tyrosine hydroxylase. These findings suggest that slow colonic transit may be related to a direct involvement of the colonic myenteric plexus by the PD process.

Outlet obstruction has been attributed to a focal dystonia of the pelvic floor sustained by a failure of the puborectalis muscle to relax during efforts to defecate, or by its paradoxical contraction (Mathers, Kempster, Swash, & Lees, 1988). It is characterized by excessive recruitment of synergistic and antagonistic muscle groups during voluntary activity and lack of reciprocal inhibition. Mathers et al. (1989) showed that the selective pattern of muscle contraction and inhibition that accompanies defecation straining is disturbed in some parkinsonian patients suffering from constipation. Paradoxical activation of the puborectalis and EAS muscles occurred, and some patients also showed a tendency to recruit gluteal muscles during simulated defecation straining.

During defecation, the puborectalis muscle normally relaxes, opening the anorectal angle to promote outward passage of stool. When the puborectalis fails to relax or contracts during defecation, as seen in some PD patients, the forward passage of stool is impaired and defecation obstructed. Sometimes paradoxical anal sphincter muscle contraction resembling anismus-type pelvic outlet obstruction may also occur (Mathers et al., 1988). Anorectal manometry has revealed several abnormalities in parkinsonian patients, including low basal and impaired squeeze pressures, prominent phasic fluctuations during squeeze, and a hypercontractile response to rectosphincteric reflex (Pierantozzi et al., 2006). Stocchi et al. (2000) showed that resting and squeezing anal pressures had an inverse direct relationship with disease duration. Maximal anal contraction pressure was higher in patients with a lower degree of disability than in the more severe patients. A significant relationship between anal pressure and age of the patients and disease duration was found. Altered voluntary anal contraction and paradoxical anal contraction during straining have been repeatedly reported in PD patients (Edwards et al., 1994).

7. DIAGNOSIS

A review of the dietary habits should be performed in order to implement fiber and water intake and to modify sedentary habits.

A complete medical history is essential and should include information on any recent introduction or changes in dosage of any prescription medications, herbal agents, and over-the-counter medications (Table 2). Opiates are often prescribed to treat drug-resistant pain that is very common in PD patients and can induce a rapid worsening of constipation symptoms.

Table 2 Drugs Inducing Constipation

Analgesic
FANS and opiates
Anticholinergic
Atropine, antidepressant, and antiparkinsonian
Antipsychotics
Antihistaminics
Antiepileptics
Antihypertensive
Calcium-channel blockers, clonidine, and methyldopa
Diuretics
Metals
Aluminum (antiacids), calcium, and iron supplement

A detailed medical history should be performed to identify the other causes of patient's constipation and to rule out any other significant disease. A reduction in the caliber of the stool or the presence of blood in the stool could suggest a possible mass lesion, as the alternans of diarrhea and constipation could indicate irritable bowel syndrome. Family history for colorectal cancer should be investigated.

8. MOTOR FLUCTUATIONS AND CONSTIPATION

Constipation may contribute to the pathogenesis of motor fluctuations: the excessive distension of the intestinal wall may delay gastric emptying through a cologastric reflex (Kellow, Gill, & Wingate, 1987), determining a delay or a failure in the absorption of a levodopa dose. Astarloa, Mena, Sanchez, de la Vega, and de Yébenes (1992) observed that a significant improvement of constipation in PD patients determined an improvement in extrapyramidal function, which was probably related to an increase in the levodopa absorption and bioavailability due to the augmentation in gastric emptying and in gastrointestinal motility. This could explain the beneficial effect of treatments improving constipation on extrapyramidal symptoms.

9. TREATMENT (TABLE 3)

General approach to constipation may include recommending increased fluid intake and a dietary regimen based on high levels of dietary fiber. Flatulence, bloating, and cramps are common side effects encountered when bran fiber is first introduced. Astarloa et al. (1992) reported that a diet rich in insoluble fibers produced a significant improvement in constipation, as indicated by an increase in stool frequency and an improvement in stool consistency. Synthetic bulk agents (e.g., psyllium) are showed to require an adequate amount of fluid intake; otherwise, they may worsen constipation. Ashraf, Pfeiffer, Park, Lof, and Quigley (1997) reported that psyllium increased stool frequency when administered for 4 weeks without altering colonic transit or anorectal function in PD patients with confirmed constipation. Since fibers can delay gastric emptying, they should be avoided in subjects suffering from gastroparesis.

Osmotic laxatives are widely used in general population and can be useful in parkinsonian patients. In a double-blind controlled trial, Zangaglia et al. (2007) showed that isosmotic macrogol electrolyte solution improved constipation in PD patients. The drug was well tolerated and did not affect PD motor symptomatology. Osmotic laxatives should be used with caution in case of renal and cardiac dysfunction and in older patients because of their tendency to cause electrolyte abnormalities and volume overload.

9.1 Newer Agents

Prucalopride, a selective high-affinity 5-HT4 receptor agonist, has been marketed in the last years for chronic constipation. It has greater selectivity for KCHN-2 channels improving gastric mobility, colonic transit, and increasing bowel movements (Bouras et al., 2001). Prucalopride has never been tested in PD. The efficacy of mosapride, another 5-HT4 receptor agonist, has been evaluated in a small Japanese trial and has been found to improve bowel movements and defecation (Liu et al., 2005). Very recently a trial with relamorelin has been performed in parkinsonian subject suffering from chronic constipation, but unfortunately the

Table 3 Treatment Options for Constipation
Products Used in the Symptomatic Treatment of Constipation

Medication	Recommended Dose	Onset of Action	Side Effects
Bulk agents			
Psyllium	20 g	12–72 h	Bloating and flatus
Methylcellulose	20 g		
Polycarbophil	20 g		
Osmotic laxatives			
Magnesium	30–45 mL QD	1–6 h	Caution in patient with kidney failure
Sodium phosphate	20–45 mL QD		Hyperphosphatemia
Nonabsorbable sugars			
Lactulose	15–30 mL QD or bid	28–48 h	Bloating
Sorbitol	15–30 mL QD or bid	28–48 h	Bloating
PEG	17 g bid or tid	12–24 h	Bloating and cramping
Stimulant laxatives			
Senna	8–34 mg QD	6–12 h	
Bisacodyl	5–10 mg QD		
5-HT4 receptor agonists			
Mosapride	15 mg QD		Diarrhea, dry mouth, headache
Prucalopride	1 mg QD to bid		Nausea, diarrhea, bloating
Synthetic ghrelin agonists			
Relamorelin[a]	100 mcg QD		Headache
Chloride channel activators			
Lubiprostone	24 mcg bid		Nausea, diarrhea, headache

[a]Subcutaneous injection, not marketed yet.

study failed due to poor enrollment (The Parkinson Study Group, 2017). Relamorelin is a synthetic ghrelin agonist that has been demonstrated to promote gastric emptying, colon transit, and propagated colonic contractions.

Lubiprostone is a chloride channel activator approved by FDA for treatment of chronic idiopathic constipation in adults. It is also demonstrated to improve abdominal bloating, pain, and straining. Common side effects are headache, nausea, and diarrhea. A trial with lubiprostone has been performed on 54 PD patients at a dosage of 24 μg once and twice a day over a period of 4 weeks, showing improvements in constipation and subjective pain (Barboza, Okun, & Moshiree, 2015; Ondo et al., 2012).

9.2 Dyssinergic Defecation

The difficulties in relaxing the pelvic floor and paradoxical puborectalis contraction generally respond dramatically to dopaminergic drugs. Paradoxical EMG activity in anal sphincter muscles tends to persist even in the "on" phase. However, a definite improvement in pelvic outlet obstruction has been demonstrated with proctography after apomorphine administration (Mathers et al., 1989). This evidence extends the concept of an "off"-period pelvic floor dystonia in PD, advanced by Mathers et al. (1988).

Patients should therefore try to pass the feces when "on" or take a quick acting drug (soluble levodopa or apomorphine) before starting defecation (Mathers et al., 1989; Stocchi et al., 1996). Unfortunately, some patients still experience difficulty defecating when motor disability has otherwise been reversed by dopaminergic drugs. In these patients, the injection of botulinum toxin in the puborectalis muscle and/or in the external anal sphincter may be useful. Botulinum toxin injection may be used also in those patients with painful anismus-type dystonia (Albanese, Brisinda, Bentivoglio, & Maria, 2003).

Patients should also be advised to assume a proper position to evacuate (Fig. 2). Indeed, in the sitting position, defecation is more difficult due to the contraction of the puborectalis muscle, which chokes the rectum maintaining fecal continence (Fig. 3A). To facilitate the evacuation, it is therefore indicated to assume a "squatting position" (Fig. 3B): in this position, the puborectalis muscle is relaxed allowing an easier and more complete feces elimination.

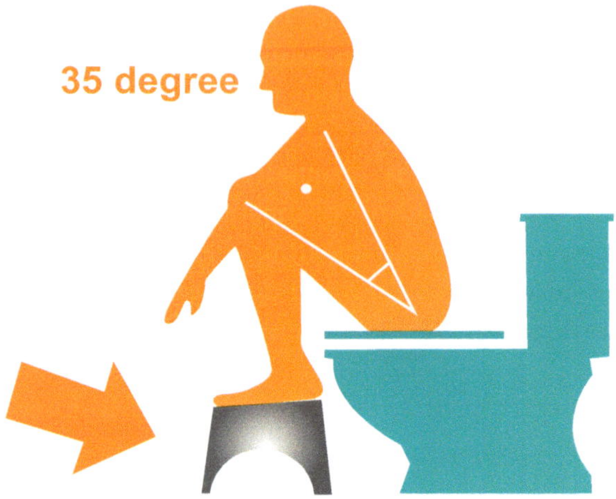

Fig. 2 Squatting position.

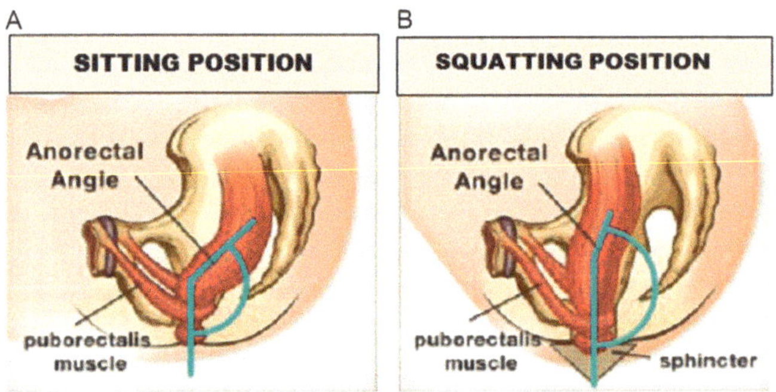

Fig. 3 (A) In the sitting position the puborectalis muscle "chokes" the rectum maintaining continence. (B) Squatting allows to relax the puborectalis muscle straightening the anorectal angle.

10. ACUTE EMERGENCY ADMISSION

Special attention needs to be paid to the acute onset of symptoms like pain, fever, nausea, and vomiting that could indicate a total or a partial occlusion.

A retrospective study performed in 2015 to analyze outcomes of colorectal surgery in PD population showed that 54.3% of parkinsonian patients receive surgery in the emergency department compared to non-PD population (38.6%). The high percentage of GI symptoms during the course of the disease could induce the clinician to underestimate or not recognize symptoms of an intestinal obstruction or an underlying disease like colorectal cancer. Among all emergency surgeries performed in PD patients more than 20% were due to intestinal obstruction and were treated with open approaches. The morbidity and the mortality rates were higher in PD compared to the general population and were mainly related to aspiration pneumonia and sepsis, highlighting the importance of an early diagnosis and treatment of occlusive symptoms (Hwang et al., 2015).

Early recognition and treatment of constipation in PD patients may also prevent complications like sigmoid volvulus that remains an under recognized complication with mortality rates up to 50%. Few case reports have been published (Blackley, Maguire, & Daniels, 2016; Caplan, Jacobson, Rubinstein, & Rotman, 1965; Rosenthal & Marshall, 1987) associating PD and volvulus, but the real incidence of this complication is unknown. Endoscopic detorsion has lower complication rates and should always be attempted as first approach even if it has high recurrence rates. Signs of possible necrosis or perforation require instead of immediate surgical intervention.

REFERENCES

Abbott, R. D., Petrovitch, H., White, L. R., Masaki, K. H., Tanner, C. M., Curb, J. D., et al. (2001). Frequency of bowel movements and the future risk of Parkinson's disease. *Neurology*, *57*, 456–462.

Albanese, A., Brisinda, G., Bentivoglio, A. R., & Maria, G. (2003). Treatment of outlet obstruction constipation in Parkinson's disease with botulinum neurotoxin A. *American Journal of Gastroenterology*, *98*, 1439–1440.

American College of Gastroenterology Chronic Constipation Task Force. (2005). An evidence-based approach to the management of chronic constipation in North America. *The American Journal of Gastroenterology*, *100*(S1), 1–4.

Anderson, G., Noorian, A. R., Taylor, G., Anitha, M., Bernhard, D., Srinivasan, S., et al. (2007). Loss of enteric dopaminergic neurons and associated changes in colon motility in an MPTP mouse model of Parkinson's disease. *Experimental Neurology*, *207*, 4–12.

Ashraf, W., Pfeiffer, R. P., Park, F., Lof, J., & Quigley, E. M. (1997). Constipation in Parkinson's disease: Objective assessment and response to psyllium. *Movement Disorders*, *6*, 946–951.

Astarloa, R., Mena, M. A., Sanchez, V., de la Vega, L., & de Yébenes, J. G. (1992). Clinical and pharmacological effects of a diet rich in insoluble fiber on Parkinson's disease. *Clinical Neuropharmacology*, *15*, 375–380.

Barboza, J. L., Okun, M. S., & Moshiree, B. (2015). The treatment of gastroparesis, constipation and small intestinal bacterial overgrowth syndrome in patients with Parkinson's disease. *Expert Opinion on Pharmacotherapy*, *16*(16), 2449–2464.

Barone, P., Antonini, A., Colosimo, C., Marconi, R., Morgante, L., Avarello, T. P., et al. PRIAMO study group. (2009). The PRIAMO study: A multicenter assessment of nonmotor symptoms and their impact on quality of life in Parkinson's disease. *Movement Disorders*, *24*, 1641–1649.

Beach, T. G., Adler, C. H., Sue, L. I., Vedders, L., Lue, L., White Iii, C. L., et al.Arizona Parkinson's Disease Consortium, (2009). Multi-organ distribution of phosphorylated alpha-synuclein histopathology in subjects with Lewy body disorders. *Acta Neuropathologica*, *119*, 689–702.

Blackley, S., Maguire, C., & Daniels, T. (2016). Seven cases of sigmoid volvulus in Parkinson's disease. *The Journal of the Royal College of Physicians of Edinburgh*, *46*(3), 157–159.

Bouras, E. P., Camilleri, M., Burton, D. D., Thomforde, G., McKinzie, S., & Zinsmeister, A. R. (2001). Prucalopride accelerates gastrointestinal and colonic transit in patients with constipation without a rectal evacuation disorder. *Gastroenterology*, *120*, 354–360.

Braak, H., de Vos, R. A., Bohl, J., & Del Tredici, K. (2006). Gastric alpha-synuclein immunoreactive inclusions in Meissner's and Auerbach's plexuses in cases staged for Parkinson's disease-related brain pathology. *Neuroscience Letters*, *396*, 67–72.

Braak, H., Del Tredici, K., Braztke, H., Hamm-Clement, J., Sandmann-Keil, D., & Rub, U. (2002). Staging the intracerebral inclusion body pathology associated with idiopathic Parkinson's disease (preclinical and clinical stages). *Journal of Neurology*, *249*(Suppl. 3). III/1-5.

Caplan, L. H., Jacobson, H. G., Rubinstein, B. M., & Rotman, M. Z. (1965). Megacolon and volvulus in Parkinson's disease. *Radiology*, *85*, 73–79.

Cersosimo, M. G., Raina, G. B., Pecci, C., Pellene, A., Calandra, C. R., Gutiérrez, C., et al. (2013). Gastrointestinal manifestations in Parkinson's disease: Prevalence and occurrence before motor symptoms. *Journal of Neurology*, *260*, 1332–1338.

Chaudhuri, K. R., Martinez-Martin, P., Brown, R. G., Sethi, K., Stocchi, F., Odin, P., et al. (2007). The metric properties of a novel non-motor symptoms scale for Parkinson's disease: Results from an international pilot study. *Movement Disorders*, *22*(13), 1901–1911.

Damian, A., Adler, C. H., Hentz, J. G., Shill, H. A., Caviness, J. N., Sabbagh, M. N., et al. (2012). Autonomic function, as self-reported on the SCOPA-autonomic questionnaire, is normal in essential tremor but not in Parkinson's disease. *Parkinsonism & Related Disorders*, *18*(10), 1089–1093.

de Vos, W. M., & de Vos, E. A. (2012). Role of the intestinal microbiome in health and disease: From correlation to causation. *Nutrition Reviews*, *70*(Suppl 1), S45–S56.

Drossman, D., & Dumitrascu, D. L. (2006). Rome III: New standard for functional gastrointestinal disorders. *Journal of Gastrointestinal and Liver Diseases*, *15*(3), 237–241.

Edwards, L. L., Quigley, E. M. M., Harned, R. K., Hofman, R., & Pfeiffer, R. F. (1994). Characterization of swallowing and defecation in Parkinson's disease. *American Journal of Gastroenterology*, *89*, 15–25.

Evatt, M. L., Chaudhuri, K. R., Chou, K. L., Cubo, E., Hinson, V., Kompoliti, K., et al. (2009). Dysautonomia rating scales in Parkinson's disease: Sialorrhea, dysphagia, and constipation—Critique and recommendations by movement disorders task force on rating scales for Parkinson's disease. *Movement Disorders*, *24*(5), 635–646.

Fasano, A., Bove, F., Gabrielli, M., Petracca, M., Zocco, M. A., Ragazzoni, E., et al. (2013). The role of small intestinal bacterial overgrowth in Parkinson's disease. *Movement Disorders*, *28*, 1241–1249.

Forjaz, M. J., Ayala, A., Rodriguez-Blazquez, C., Frades-Payo, B., Martinez-Martin, P., & Longitudinal Parkinson's Disease Patient Study, Estudio longitudinal de pacientes con enferedad de Parkinson—ELEP Group. (2010). Assessing autonomic symptoms of Parkinson's disease with the SCOPA-AUT: A new perspective from Rasch analysis. *European Journal of Neurology*, *17*(2), 273–279.

Gai, W. P., Blumbergs, P. C., Geffen, L. B., & Blessing, W. W. (1992). Age-related loss of dorsal vagal neurons in Parkinson's disease. *Neurology*, *42*, 2106–2111.

Gao, X., Chen, H., Schwarzschild, M. A., & Ascherio, A. (2011). A prospective study of bowel movement frequency and risk of Parkinson's disease. *American Journal of Epidemiology*, *174*, 546–551.

Hwang, G. S., Young, M. T., Hanna, M. H., Menon, G., Phelan, M., Carmichael, J. C., et al. (2015). Colorectal surgery in Parkinson's disease—Outcomes and predictor of mortality. *International Journal of Colorectal Disease*, *30*, 1051–1058.

Kaye, J., Gage, H., Kimber, A., Storey, L., & Trend, P. (2006). Excess burden of constipation in Parkinson's disease: A pilot study. *Movement Disorders*, *21*(8), 1270–1273.

Kellow, J. E., Gill, R. C., & Wingate, D. L. (1987). Modulation of human upper gastrointestinal motility by rectal distension. *Gut*, *28*, 864–868.

Khedr, E. M., Fetoh El, N. A., Khalifa, H., Ahmed, M. A., & Beh El, K. M. (2013). Prevalence of non-motor features in a cohort of Parkinson's disease patients. *Clinical Neurology and Neurosurgery*, *115*, 673–677.

Klingelhoefer, L., & Reichmann, H. (2015). Pathogenesis of Parkinson disease—The gut-brain axis and environmental factors. *Nature Reviews Neurology*, *11*(11), 625–636.

Knudsen, K., Krogh, K., Østergaard, K., & Borghammer, P. (2017). Constipation in Parkinson's disease: Subjective symptoms, objective markers, and new perspectives. *Movement Disorders*, *32*(1), 94–105.

Lin, C. H., Lin, J. W., Liu, Y. C., & Chang Ch, W. R. M. (2014). Risk of Parkinson's disease following severe constipation: A nationwide, population-based cohort study. *Parkinsonism & Related Disorders*, *20*, 1371–1375.

Liu, Z., Sakakibara, R., Odaka, T., Uchiyama, T., Uchiyama, T., Yamamoto, T., et al. (2005). Mosapride citrate, a novel 5-HT4 agonist and partial 5-HT3 antagonist, ameliorates constipation in parkinsonian patients. *Movement Disorders*, *20*(6), 680–686.

Longstreth, G. F., Thompson, W. G., Chey, W. D., Houghton, L. A., Mearin, F., & Spiller, R. C. (2006). Functional bowel disorders. *Gastroenterology*, *130*, 1480–1491.

Martinez-Martin, P., Schapira, A. H., Stocchi, F., Sethi, K., Odin, P., MacPhee, G., et al. (2007). Prevalence of nonmotor symptoms in Parkinson's disease in an international setting; study using nonmotor symptoms questionnaire in 545 patients. *Movement Disorders*, *22*, 1623–1629.

Mathers, S. E., Kempster, P. A., Law, P. J., Frankel, J. P., Bartram, C. I., Lees, A. J., et al. (1989). Anal sphincter dysfunction in Parkinson's disease. *Archives of Neurology*, *46*, 1061–1064.

Mathers, S. E., Kempster, P. A., Swash, M., & Lees, A. J. (1988). Constipation and paradoxical puborectalis contraction in anismus and Parkinson's disease: A dystonic phenomenon? *Journal of Neurology, Neurosurgery, and Psychiatry*, *51*, 1503–1507.

Morgan, J. C., & Sethi, K. D. (2007). Tegaserod in constipation associated with Parkinson disease. *Clinical Neuropharmacology*, *30*, 52–54.

Mulak, A., & Bonaz, B. (2015). Brain-gut-microbiota axis in Parkinson's disease. *World Journal of Gastroenterology*, *21*(37), 10609–10620.

Ondo, W. G., Kenney, C., Sullivan, K., Davidson, A., Hunter, C., Jahan, I., et al. (2012). Placebo-controlled trial of lubiprostone for constipation associated with Parkinson disease. *Neurology*, *78*(21), 1650–1654.

Phillips, R. J., & Powley, T. L. (2005). Innervation of the gastrointestinal tract: Patterns of aging. *Autonomic Neuroscience*, *2007*(136), 1–19.

Pierantozzi, M., Pietroiusti, A., Brusa, L., Galati, S., Stefani, A., Lunardi, G., et al. (2006). Helicobacter pylori eradication and L-dopa absorption in patients with PD and motor fluctuations. *Neurology*, *66*, 1824–1829.

Pont-Sunyer, C., Hotter, A., Gaig, C., Seppi, K., Compta, Y., Katzenschlager, R., et al. (2015). The onset of nonmotor symptoms in Parkinson's disease (the ONSET PD study). *Movement Disorders*, *30*, 229–237.

Rosenthal, M. J., & Marshall, C. E. (1987). Sigmoid volvulus in association with parkinsonism. Report of four cases. *Journal of the American Geriatrics Society*, *35*, 683–684.

Scheperjans, F., Aho, V., Pereira, P. A., Koskinen, K., Paulin, L., Pekkonen, E., et al. (2015). Gut microbiota are related to Parkinson's disease and clinical phenotype. *Movement Disorders*, *30*(3), 350–358.

Shannon, K. M., Kesharvarzian, A., Mutlu, E., Dodiya, H. B., Daian, D., Jaglin, J. A., et al. (2012). Alpha-synuclein in colonic submucosa in early untreated Parkinson's disease. *Movement Disorders*, *6*, 709–715.

Singaram, C., Ashraf, W., Gaumnitz, E. A., Torbey, C., Sengupta, A., Pfeiffer, R., et al. (1995). Dopaminergic defect of enteric nervous system in Parkinson's disease patients with chronic constipation. *The Lancet*, *346*, 861–864.

Stocchi, F., Badiali, D., Vacca, L., D'Alba, L., Bracci, F., Ruggieri, S., et al. (2000). Anorectal function in multiple system atrophy and Parkinson's disease. *Movement Disorders*, *15*(1), 71–76.

Stocchi, F., Barbato, L., Bramante, L., Nordera, G., Vacca, L., & Ruggieri, S. (1996). Fluctuating parkinsonism: A pilot study of a single afternoon dose of levodopa methyl ester. *Journal of Neurology*, *243*, 377–380.

Svensson, E., Horváth-Puhó, E., Thomsen, R. W., Djurhuus, J. C., Pedersen, L., Borghammer, P., et al. (2015). Vagotomy and subsequent risk of Parkinson's disease. *Annals of Neurology*, *78*(4), 522–529.

Tan, A. H., Mahadeva, S., Thalha, A. M., Gibson, P. R., Kiew, C. K., Yeat, C. M., et al. (2014). Small intestinal bacterial overgrowth in Parkinson's disease. *Parkinsonism & Related Disorders*, *20*, 535–540.

The Parkinson Study Group. (2017). A randomized trial of relamorelin for constipation in Parkinson's disease (MOVE-PD): Trial results and lessons learned. *Parkinsonism & Related Disorders*, *37*, 101–105.

Visanji, N. P., Brooks, P. L., Hazrati, L. N., & Lang, A. E. (2013). The prion hypothesis in Parkinson's disease: Braak to the future. *Acta Neuropathologica Communications*, *1*, 2.

Wade, P. R., & Hornby, P. J. (2005). Age-related neurodegenerative changes and how they affect the gut. *Science of Aging Knowledge Environment*, *2005*(12), pe8.

Wakabayashi, K., Takahachi, H., Ohama, E., & Ikuta, F. (1989). Tyrosine hydroxylase-immunoreactive intrinsic neurons in the Auerbach's and Meissner's plexuses of humans. *Neuroscience Letters*, *96*, 259–263.

Wakabayashi, K., Takahachi, H., Ohama, E., & Ikuta, F. (1990). Parkinson's disease: An immunohistochemical study of Lewy body-containing neurons in the enteric nervous system. *Acta Neuropathologica*, *79*, 581–583.

Wakabayashi, K., Takahashi, H., Ohama, E., Takeda, S., & Ikuta, F. (1993). Lewy bodies in the visceral autonomic nervous system in Parkinson's disease. *Advances in Neurology*, *60*, 609–612.

Zangaglia, R., Martignoni, E., Glorioso, M., Ossola, M., Riboldazzi, G., Calandrella, D., et al. (2007). Macrogol for the treatment of constipation in Parkinson's disease. A randomized placebo-controlled study. *Movement Disorders*, *22*, 1239–1244.

FURTHER READING

Kalf, J. G., de Swart, B. J., Borm, G. F., Bloem, B. R., & Munneke, M. (2009). Prevalence and definition of drooling in Parkinson's disease: A systematic review. *Journal of Neurology*, *256*(9), 1391–1396.

Sullivan, K. L., Staffetti, J. F., Hauser, R. A., Dunne, P. B., & Zesiewicz, T. A. (2006). Tegaserod (Zelnorm) for the treatment of constipation in Parkinson' disease. *Movement Disorders*, *21*, 115–116.

Tjeerdsma, H. C., Smout, A. J. P. M., & Akkermans, L. M. A. (1993). Voluntary suppression of defecation delays gastric emptying. *Digestive Diseases and Sciences*, *38*, 832–836.

CHAPTER TWENTY-EIGHT

Understanding and Treating Pain Syndromes in Parkinson's Disease

Marialuisa Gandolfi[*,†], Christian Geroin[*], Angelo Antonini[‡], Nicola Smania[*,†,2], Michele Tinazzi[§,1,2]

[*]Neuromotor and Cognitive Rehabilitation Research Center (CRRNC), University of Verona, Verona, Italy
[†]Neurorehabilitation Unit, Azienda Ospedaliera Universitaria Integrata, Verona, Italy
[‡]University of Padua and Hospital San Camillo IRCCS, Venice, Italy
[§]Neurology Unit, Movement Disorders Division, University of Verona, Verona, Italy
[1]Corresponding author: e-mail address: michele.tinazzi@univr.it

Contents

Abstract

Pain affects many people with Parkinson's disease (PD) and diminishes their quality of life. Different types of pain have been described, but their related pathophysiological mechanisms remain unclear. The aim of this chapter is to provide movement disorders specialists an update about the pathophysiology of pain and a practical guide for the management of pain syndromes in clinical practice. This chapter reviews current knowledge on the pathophysiological mechanisms of sensory changes and pain in PD, as well as assessment and treatment procedures to manage these symptoms. In summary, changes in peripheral and central pain processing have been demonstrated in PD patients. A decrease in pain threshold and tolerance to several stimuli, a reduced nociceptive withdrawal reflex, a reduced pain threshold, and abnormal pain-induced activation in cortical pain-related areas have been reported. There is no direct association between improvement of motor symptoms and sensory/pain changes, suggesting that motor and nonmotor symptoms do not inevitably share the same mechanisms. Special care in pain assessment in PD is warranted by the specific pathophysiological aspects

[2] NS and MT contributed equally.

International Review of Neurobiology, Volume 134
ISSN 0074-7742
http://dx.doi.org/10.1016/bs.irn.2017.05.013

and the complexity of motor and nonmotor symptoms associated with pain symptoms. Rehabilitation may represent a valid option to manage pain syndromes in PD. However, further research in this field is needed. An integrated approach to pain involving a multidisciplinary team of medical specialists and rehabilitation experts should allow a comprehensive approach to pain in PD.

1. BACKGROUND

Pain is one of the most common cause of disability with significant impact on quality of life of patients and their families resulting in major clinical and socioeconomic burden. Because millions of people suffer from pain-related conditions (Dzau & Pizzo, 2014), it is also the most common reason for physician office visits, loss of work time, and productivity (Stewart, Ricci, Chee, Morganstein, & Lipton, 2003).

Several taxonomies of pain have been described in the medical literature. The International Association for the Study of Pain (IASP) Subcommittee on Taxonomy defines pain as "an unpleasant sensory and emotional experience associated with actual or potential tissue damage, or described in terms of such damage," whereby other domains such as emotion and cognition are implicated in the origin, maintenance of pain, and its management (Williams & Craig, 2016).

From a pathophysiological perspective, pain can be generally classified into "nociceptive" and "neuropathic." Nociceptive pain is described as pain arising from actual or potential damage to nonneural tissue consequent to nociceptors activation (i.e., dystonia-related pain occurring in the off periods, or due to skeletal deformation, abnormal posture, and muscle rigidity). These conditions can overstimulate peripheral nociceptors. Conversely, neuropathic pain is defined as pain due to a lesion of the central of peripheral somatosensory system.

Pain represents one of the most common nonmotor symptoms affecting people with Parkinson's disease (PD). It was first described in PD by James Parkinson in 1817 in his work "An Essay on the Shaking Palsy" (Parkinson, 2002) and lately reappraised because it represents a common symptom that leads to disability and results in diminished quality of life (Corallo et al., 2017; Ozturk, Gundogdu, Kocer, Comoglu, & Cakci, 2016). A widely accepted classification of pain in PD based on its etiology is the Ford criteria (Ford, 2010). A further classification distinguishing PD-related pain and PD-unrelated pain has also been proposed (Nègre-Pagès et al., 2008; see Table 1).

The prevalence of pain ranges from 40% to 85% (Broen, Braaksma, Patijn, & Weber, 2012; Nègre-Pagès et al., 2008). Musculoskeletal pain is the most frequent (up to 58.5%), followed by radicular-peripheral

Table 1 Definitions and Types of Pain in Parkinson's Disease

Type of Pain	Clinical Features
According to Ford's definition (Ford, 2010)	
Dystonic	Pain involving body parts affected by dystonia that can be observed as forceful and painful muscular contractions and/or sustained twisting movements and abnormal postures. Different parts of the body can be involved, including upper and lower extremities and pharyngeal and facial musculature. This pain can change in relation to the level of medication (i.e., off dystonia, early morning dystonia, beginning-of-dose and end-of-dose dystonia, peak-dose dystonia)
Musculoskeletal	Aching, cramping, arthralgic, myalgic sensations in joints and muscles (i.e., skeletal deformities, reduced joint mobility, postural abnormalities, antalgic gait). It can be provoked by parkinsonian rigidity, increased stiffness, and reduced mobility. Musculoskeletal pain can improve with levodopa medication
Radicular-peripherical neuropathic	Pain in a nerve territory or root correlated with motor or sensory signs of nerve or root entrapment
Akathisia	Subjective sense of restlessness that can be associated with an urge to move
Central neuropathic	Neuropathic sensations such as burning, tingling, formication not confined to the root, or nerve territory; this pain cannot be explained by an internal lesion or musculoskeletal lesions, dystonia, and rigidity
Other definitions related to chronic pain (Nègre-Pagès et al., 2008)	
Non-PD pain	A type of pain not related to PD but due to other causes
PD pain	A type of pain caused or exacerbated by PD
PD pain direct	A pain directly related to PD and not related to other causes according to medical record (clinical examination, medical history, laboratory tests, imaging findings)
PD pain indirect	A type of pain related to other diseases such as osteoarthritis and PD worsens in intensity due to abnormal postures, movements, or rigidity

neuropathic (38%), dystonic (33.5%), and central neuropathic pain (8.5%). Chronic pain affects 61.8% of PD patients: 60.1% of pain syndromes are PD-related and the remaining are not PD-related but due to another origin (i.e., osteoarthritis) (Nègre-Pagès et al., 2008).

Pain affects all body parts, including the shoulder, upper and lower limbs, and trunk. Unusual pain syndromes, which affect approximately 5.7% of PD patients, may involve the head, face, pharynx, abdomen, epigastrium, pelvis, rectum, and genitalia (Barone et al., 2009).

The literature is scant on pain predictors in PD. A greater risk of developing pain was reported in PD patients with motor complications, female gender, worse parkinsonian symptoms, depression, associated medical conditions (e.g., diabetes, osteoporosis, rheumatic disease, arthritis), and back pain (Broetz, Eichner, Gasser, Weller, & Steinbach, 2007; Defazio et al., 2008; Ehrt, Larsen, & Aarsland, 2009; McNamara, Stavitsky, Harris, Szent-Imrey, & Durso, 2010; Nègre-Pagès et al., 2008; Tinazzi et al., 2006; Zambito Marsala et al., 2011).

Genetic factors have also been correlated with an increased risk for developing pain in PD (Greenbaum et al., 2012; Li, Chen, Yin, & Zhang, 2014). In advanced stages of PD, pain has been correlated with a significant reduction in quality of life (Valkovic et al., 2015).

Diagnosis of different pain syndromes is an essential phase of clinical visits, especially in personalized symptomatic treatment. Types of pain differ remarkably from one another and may be experienced singularly or coexist in the same patient.

Here, we review recent research investigating pain in PD. The aim of this chapter is to provide movement disorders specialists an update on the pathophysiology of pain and a practical guide for the management of pain syndromes in clinical practice.

2. PAIN PATHOPHYSIOLOGY IN PEOPLE WITH PD

In 1968, Melzack and Casey proposed three dimensions of pain that could influence each other (Melzack & Casey, 1968). First, the sensory-discriminative dimension that refers to the sensation of pain. It involves the stimulation of nociceptors, the transmission of stimuli in the dorsal horn neurons of the spinal cord and then to the higher center brain neurons through the medial and lateral pain systems. Descending pathways originating in the brain stem and cerebral structures (i.e., periaqueductal areas) play a major role in the integration and modulation of the sensitivity of the dorsal horn neurons through serotoninergic, noradrenergic, and dopaminergic networks (Millan, 1999). Second, the motivational–affective dimension denotes how the pain worries the patient and the degree to which it is experienced as unpleasant. Finally, the cognitive–evaluative dimension is based on the patient's beliefs and previous experience of pain. The latter may affect the outcome, interfering with both the sensory-discriminative and the motivational–affective dimensions.

The neural underpinning of pain is complex. It has been theorized as a "pain matrix," a "fluid system" comprising several interacting networks that are jointly activated by painful stimuli (Garcia-Larrea & Peyron, 2013). It is based on a hierarchical involvement of different neural centers. The first-order network is the "nociceptive matrix" that receives spinothalamic projections and senses the bodily specificity of pain in terms of location, quality, and intensity. Therefore, damage to this matrix results in selective pain deficits. The transition from cortical nociception to conscious pain relies on a second-order network that is not nociceptive specific (i.e., posterior parietal lobe, prefrontal and anterior insular areas). Joint activation of these second-order regions does not evoke pain but rather its conscious perception, attentional modulation, and control of vegetative reactions. Finally, third-order areas include the orbitofrontal, perigenual, and limbic networks involved in the pain experience, emotions, and expectations. To sum up, pain results from continuous interactions between these three subsystems, and substantial changes in the pain experience can be achieved by acting on any one of them (Garcia-Larrea & Peyron, 2013).

Converging preclinical and clinical evidence suggests a central role for dopamine neurotransmission in modulating pain perception and analgesia (Jarcho, Mayer, Jiang, Feier, & London, 2012). The experience of pain, in fact, can be modulated by dysregulation in dopamine signaling both directly and indirectly. Dopamine dysregulation can directly enhance or diminish the propagation of nociceptive signals and, indirectly, influence affective and cognitive processes (Jarcho et al., 2012). The latter are involved in the experience and interpretation of nociceptive signals. Note that disorders linked to excessive dopamine neurotransmission are characterized by hyposensitivity to pain (i.e., schizophrenia). In contrast, disorders associated with deficits in dopamine, such as mood disorders and PD, are characterized by hypersensitivity to pain and high rates of comorbid chronic pain.

PD patients typically complain of primary symptoms of their disorder. While pain in PD can be caused by primary motor symptoms (Beiske, Loge, Rønningen, & Svensson, 2009), it may also arise independently (Lee, Walker, Hildreth, & Prentice, 2006). The progressive neurodegeneration in PD affects pain processing at multiple levels (Fil et al., 2013) in either its peripheral transmission and sensory-discriminative processing or higher-order centers involved in its reception and interpretation (Conte, Khan, Defazio, Rothwell, & Berardelli, 2013; Fil et al., 2013;

Geroin, Gandolfi, Bruno, Smania, & Tinazzi, 2016). The former involves peripheral nervous system abnormalities such as reduced unmyelinated nerve fiber density (21%) (Kanda, Tsukagoshi, Oda, Miyamoto, & Tanabe, 1996), cutaneous denervation with loss of free and encapsulated nerve endings, as demonstrated by skin biopsies (Nolano et al., 2008), and nociceptor neurodegeneration (Reichling & Levine, 2011). Changes in nociceptors, and the resulting peripheral deafferentation, could play a role in the pathogenesis of sensory dysfunction in PD (Nolano et al., 2008). However, recent evidence has reported that peripheral mechanisms may be not as important as central mechanisms to pain processing in patients with PD (Zambito-Marsala et al., 2017). The latter involves changes in the spinal cord and several brain areas (Braak et al., 2003; Braak, Rüb, Jansen Steur, Del Tredici, & de Vos, 2005). Braak et al. (2003, 2005) divided the course of PD into six stages during which pathological changes in the anatomical structures involved in pain take place. In the first two stages (premotor period), the changes affect the olfactory bulb and progress toward the medulla oblongata and pontine tegmentum, then the nucleus raphe magnus and gigantocellular reticular nucleus and the descending antinociceptive pathways (Gebhart, 2004). In addition, the presence of Lewy bodies in the ceruleus/subceruleus area can influence the autonomous, motivational–emotional, and cognitive dimensions of pain (Braak et al., 2003, 2005; Brefel-Courbon et al., 2005; Chudler & Dong, 1995; Gebhart, 2004; Kuraishi, Fukui, Shiomi, Akaike, & Takagi, 1978; Millan, 2002; Nolano et al., 2008; Pertovaara, 2006; Scherder, Wolters, Polman, Sergeant, & Swaab, 2005; Voisin, Guy, Chalus, & Dallel, 2005; Willis & Westlund, 1997).

Disturbances in this pain-inhibiting region can increase the sensation of pain (Scherder et al., 2005). In the following four symptomatic stages, pathological changes involve the substantia nigra (stage 3), the mesocortex (stage 4), and the neocortex (stages 5 and 6). These changes may play a relevant role in the pathogenesis of higher-order dysfunctions related to pain. For a full discussion, please refer to Fil et al. (2013).

Pain perception in PD may not be directly related with striatal dopaminergic dysfunction but could perhaps reflect extrastriatal dopaminergic dysfunction, with an imbalance between the sensory and the affective cerebral nociceptive pathways. Indeed, significant thinning in several cortical regions have been reported in PD with persistent pain. The contribution of frontal, prefrontal, and insular areas in nociceptive modulation and accumbens–hippocampus disconnection have been highlighted (Polli et al., 2016).

PD patients with pain exhibit a variety of symptoms that differentiate them from healthy controls and from patients without pain. PD patients are noted to have, independent of pain symptoms, a low pain threshold to electrical stimulation (Gerdelat-Mas et al., 2007; Mylius et al., 2009; Perrotta et al., 2011; Zambito Marsala et al., 2011) and low pain tolerance for acute noxious stimulation (Djaldetti et al., 2004; Lim et al., 2008) compared to healthy controls. The same finding can be extended to the pain threshold to cold water. Pain-free PD patients and those with primary central pain reported a lower pain threshold to cold water than healthy subjects. No differences were found between PD patients with pain and pain-free PD patients (Brefel-Courbon, Ory-Magne, Thalamas, Payoux, & Rascol, 2013; Brefel-Courbon et al., 2005). In PD the pain threshold to electrical and cold water stimuli is reduced and the presence of pain does not seem to influence the outcome.

The evaluation of pain threshold to heated thermode has yielded different results regardless of the presence of pain and depending on the type of pain. Pain-free patients and PD patients with primary central pain have a lower pain threshold to heated thermode than healthy controls (Djaldetti et al., 2004; Schestatsky et al., 2007), whereas PD patients complaining of musculoskeletal pain have a threshold similar to healthy controls (Mylius et al., 2009). Zambito et al. confirmed previous findings and reported a low pain tolerance that decreases as PD progresses (Zambito Marsala et al., 2011). A possible correlation between sensory thresholds and demographic/clinical features of PD patients has been speculated. Female gender, dyskinesia, medical conditions associated with painful symptoms, and postural abnormalities secondary to rigidity/bradykinesia may contribute to the onset of spontaneous pain in predisposed subjects. Yet, pain threshold and pain tolerance are decreased irrespective of the presence of pain. In patients with pain, no relationship was found between the pain threshold and the intensity/type of pain (Zambito Marsala et al., 2011). The administration of L-dopa can increase the threshold for acute noxious stimulation by increasing the levels of dopamine (Schestatsky et al., 2007).

Electrical stimulation combined with electromyography allows evaluation of the nociceptive flexion reflex that characterizes pain processing at the spinal cord level. Increased subjective pain sensitivity and increased spinal nociception appear to be reversible by dopaminergic treatment owing to reduced descending pain inhibition. In 2007, Gerdelat et al. provided evidence of dopaminergic modulation of an objective pain threshold in PD patients. Moreover, the decrease in the nociceptive flexion reflex threshold in the "off" state, compared with controls, confirms the existence of an objective pain perception disturbance in PD (Gerdelat-Mas et al., 2007). Perrotta et al. (2011) used the temporal summation threshold of the nociceptive withdrawal reflex and the related pain sensation to evaluate facilitation in pain processing at the spinal level. Seven PD patients with unilateral signs (Hoehn and Yahr [H&Y] stage 1) and eight patients in a more advanced stage of the disease with bilateral parkinsonian signs (H&Y stages 2 and 2.5) were evaluated in both the "on" and the "off" state. The results showed a reduced temporal summation threshold and increased painful sensation in the PD patients as compared with the healthy subjects, suggesting significant facilitation in the temporal summation of pain. This facilitation was more evident in the patients with bilateral signs and in the more affected side in the patients with unilateral signs. Although no significant changes were measured after L-dopa administration, a slight improvement was detected. These findings suggest that the increased gain in pain processing at the spinal level may be related to degenerative phenomena that involve the supraspinal projections. Possibly, these supraspinal projections are implicated in the modulation of pain processing and increased risk of developing pain conditions in PD patients. Mylius et al. confirmed previous findings and hypothesized that reduced activation of the diffuse noxious inhibitory control system was apparently not associated with increased pain sensitivity, suggesting that diffuse noxious inhibitory control-like mechanisms do not significantly contribute to clinical pain in PD (Mylius et al., 2011).

The functional status of cerebral structures responding to nociceptive inputs has been investigated using CO_2 laser-evoked potentials (LEP), a noninvasive exploration consisting of laser stimulation delivered over the hairy skin that gives rise to an N2/P2 complex at the vertex generated by inputs conveyed by Aδ fibers (Valeriani, Rambaud, & Mauguière, 1996). In healthy subjects, the N2/P2 complex is generated by the anterior cingulate cortex with a possible contribution from the insular regions bilaterally (Garcia-Larrea, Frot, & Valeriani, 2003). Moreover, it is preceded by an earlier lateralized negative component (N1), probably originating from the opercular (SII) cortex (Garcia-Larrea et al., 2003). Tinazzi et al. were the first to investigate whether central processing of nociceptive inputs is abnormal in PD (Tinazzi et al., 2008, 2009). Eighteen pain-free PD patients with the unilateral bradykinetic-rigid syndrome were studied during the "on" and the "off" state. The data showed specific abnormalities in PD independent of pain and affected side. Compared to the healthy subjects, PD patients showed comparable N2 and P2 latencies (Tinazzi et al., 2008, 2009). However, the N2/P2 peak-to-peak amplitude was significantly lower in the PD patients than in the controls, and no significant changes were measured between the "on" and the "off" state. In a subsequent paper by the same authors, 11 PD patients complaining of shoulder pain (musculoskeletal pain ipsilateral to motor symptoms) were evaluated with the same research protocol (Tinazzi et al., 2010). Intriguingly, the N2/P2 amplitude obtained following stimulation of the painful shoulder was significantly lower than that obtained in response to nonpainful shoulder stimulation and as compared with the pain-free PD patients. No significant correlation was observed between muscular pain intensity and N2/P2 amplitude abnormalities. These results suggest that pain-free PD patients have abnormal nociceptive input processing that appears to be independent of motor signs and dopaminergic stimulation. These alterations are more evident in the presence of musculoskeletal pain (Tinazzi et al., 2008, 2009, 2010).

Zambito-Marsala et al. (2017) found comparable N1, N2, and P2 latencies in PD patients and normal subjects. The pain-free PD patients showed a significantly lower N2/P2 amplitude than the healthy controls in the clinically affected side, while the N1/P1 amplitude did not differ.

Schestatsky et al. investigated nine PD patients with primary central pain, defined as the absence of any identified cause of pain. In the "off" condition, the patients with primary central pain had lower heat pain and laser pinprick thresholds, higher LEP amplitudes, and less habituation of laser-induced sudomotor skin responses as compared to the pain-free PD patients and the healthy subjects. In the "on" condition, these differences were lessened or vanished. These findings showed that conduction along peripheral and central pain pathways is abnormal in PD patients with or without primary central pain, and an evident abnormal control of the effects of pain inputs on autonomic centers. These abnormalities were attenuated by L-dopa, suggesting that in PD patients with primary central pain the dysfunction may occur in dopamine-dependent centers involved in regulating both autonomic function and inhibitory modulation of pain inputs (Schestatsky et al., 2007).

Evidence from neuroimaging studies (i.e., FMRI and PET) has indicated that in PD an abnormal pain-induced activation in the sensory-discriminative pain processing mediated via the insula, and in the affective motivational processing mediated by the anterior cingulate and prefrontal cortex (Brefel-Courbon et al., 2005).

In summary, changes in peripheral and central pain processing have been demonstrated in PD patients. A decrease in pain threshold and tolerance to several stimuli, a reduced nociceptive withdrawal reflex, a reduced pain threshold, and abnormal pain-induced activation in cortical pain-related areas have been reported. These abnormalities may depend on increased activity in both the ascending lateral and medial pain pathways (Scherder et al., 2005) and on decreased basal ganglia dopamine levels. Changes in pain processing depend on the neurodegeneration of dopaminergic pathways and on other nondopaminergic pain-mediating structures such as the diencephalon, limbic system, brain stem, and spinal cord (Braak et al., 2003). Finally, the presence of chronic pain conditions that typically affect elderly frail people (i.e., osteoporosis, rheumatic diseases, and arthritis) may contribute to sustain abnormal pain processing (Defazio, Gigante, Mancino, & Tinazzi, 2013).

The flaw of the existing literature might inspire future studies on this topic. First, only patients under chronic dopaminergic treatment have been investigated. It might be interesting investigate naïve patients in future studies. Second, the population tested were heterogeneous regarding the presence of pain, type of pain, and type of motor symptoms (i.e., unilateral vs bilateral). Finally, heterogeneity of results might depend on the heterogeneity of other methodological issues (i.e., research protocols, stimulation parameters).

3. EVALUATION OF PAIN SYNDROMES IN PEOPLE WITH PD

To date, no standardized assessment protocols have been developed to evaluate pain in PD (Geroin et al., 2016). Electrodiagnostic tests (i.e., electromyography and electroneurography), quantitative sensory testing, LEP, skin biopsy, autonomic tests, microneurography, and functional neuroimaging (Magrinelli, Zanette, & Tamburin, 2013) have been used to investigate the integrity of peripheral and central pain pathways. Transcutaneous electrical stimulators are of interest because they permit assessment of the thresholds of different types of afferent fibers (i.e., Aβ, Aδ, and C) via different sinusoidal frequencies and electrical intensities (Chen et al., 2015). The functional status of cerebral structures responding to nociceptive inputs can be noninvasively explored with CO_2 LEP. These procedures are mainly used in research settings essentially for two reasons. On the one hand, they require specific instrumental tools and expertise that are not usually available in clinical contexts. Most recently, the King's Parkinson's Disease Pain Scale (KPPS) was developed and validated in 178 PD patients vs 83 age- and sex-matched controls showing higher KPPS total score mostly in the domains of musculoskeletal, chronic, fluctuation-related, nocturnal, orofacial, and radicular pain (Chaudhuri, Rizos, Trenkwalder, et al., 2015).

Therefore, clinical assessment together with dedicated scales remains the mainstay in clinical practice. The evaluation of negative and positive signs of neuropathic pain is a simple bedside test to distinguish between different types of pain and to plan specific interventions (Magrinelli et al., 2013). Both unidimensional and multidimensional outcome measures are used to further characterize pain. The former includes visual analogue and numerical rating scales to evaluate pain intensity. Among the multidimensional scales, those commonly used in clinical practice are Douleur Neuropathique 4 Questionnaires, the Pain Catastrophizing Scale, the PainDETECT, the Neuropathic Pain Symptom Inventory, and the Leeds Assessment of Neuropathic Symptoms and Signs. All these outcome measures are used to investigate whether neuropathic symptoms and signs are present. The McGill Pain Questionnaire (as well as its short form), the Brief Pain Inventory, and the Gracely Box Scale evaluate different qualities of pain, including the sensorial and emotional experience. Among the nonspecific measures, the Medical Outcomes Study 36-Item Short Form, the palliative care assessment, the Nottingham Health Profile, and the Unified Parkinson's Disease Rating Scale are the most widely used clinical rating scales to measure PD severity (Goetz et al., 2008). The other scales listed in the table are considered nonspecific measures of pain. They assess multiple domains that may be affected in PD including pain, but they are not specifically focused on pain evaluation.

Many drawbacks to applying the existing outcome measures in PD patients should be acknowledged. First, they do not consider cognitive deficits that may influence the patient's performance (Cury et al., 2016). For instance, studies evaluating pain thresholds may be affected by deficits in attention, motivation-based processes, and depression (Chaudhuri, Healy, Schapira, & National Institute for Clinical Excellence, 2006; Czernecki et al., 2002). Note that depression is associated with higher pain sensitivity in PD patients (Ehrt et al., 2009). Furthermore, primary motor symptoms (i.e., rigidity, tremor, and akinesia) may sustain continuous sensory stimuli that can conflict with superimposed sensory stimuli such as a mechanical liminal stimulus or thermal probe (Cury et al., 2016). Such disorders need to be evaluated in parallel with pain.

The observation that PD population received more analgesic prescription than the general population (Brefel-Courbon et al., 2009) might allow clinicians including the total consumption of acute medication as an indirect measure of pain.

In conclusion, special care in pain assessment in PD is warranted by the specific pathophysiological aspects and the complexity of motor and nonmotor symptoms associated with pain symptoms. Pain management in PD should involve a specific decision-making algorithm that includes both uni- and multidimensional clinical rating scales along with instrumental assessments, when necessary. This would help to improve the management of pain in PD from diagnosis to treatment. In parallel, the tool could improve our understanding of the mechanisms underlying pain in PD.

4. PROCEDURES FOR TREATING PAIN IN PEOPLE WITH PD

Treatment of pain in PD should be driven by multidisciplinary clinical pathways that include integrated pharmacological and nonpharmacological (i.e., rehabilitation) approaches to treatment (Geroin et al., 2016). Implementing multidisciplinary approaches is particularly urgent in PD because, as demonstrated by Beiske et al. in a cross-sectional survey, pain symptoms are often overlooked by clinicians: 50% of patients receive neither pharmacotherapy nor physiotherapy (Beiske et al., 2009). Scant reporting on their use may be one explanation for the lack of acceptance of multidisciplinary procedures. The body of evidence for pharmacologic treatment, surgical, and rehabilitative intervention is similarly scarce (Perez-Lloret et al., 2012). Our aim here is to summarize the pharmacological, surgical, and rehabilitative procedures commonly reported in the literature. Since surgical interventions are mostly focused on deep brain stimulation (DBS) of the subthalamic nucleus (STN), whilst rehabilitation studies are scant, potential targets of new therapy options (i.e., physical therapy) will be also discussed regarding what is known about the pathophysiological mechanisms of pain in PD.

5. PHARMACOLOGICAL THERAPY

The effects of levodopa and dopamine agonists on pain syndromes can be justified, at least in part, by the involvement of basal ganglia in pain processing. This assumption is reinforced by clinical observations that patients exhibit pain especially during the off state (Juri, Rodriguez-Oroz, & Obeso, 2010) and that levodopa intake normalizes pain perception abnormalities (Gerdelat-Mas et al., 2007). Consequently, optimization of treatment with levodopa and other antiparkinsonian medications should be the first step in the management of PD-related pain. Patients report non-motor symptoms more frequently during the wearing-off than the on-state (Nebe & Ebersbach, 2009; Stacy, 2010).

Barone et al., in a double-blind, placebo-controlled trial, evaluated the effects of pramipexole on depression and other outcomes including pain. The results showed no significant differences between pramipexole and placebo on pain (Barone et al., 2010). A case report and a case series suggested that apomorphine produced some beneficial effect in treating dystonic and pelvic pain (Factor, Brown, & Molho, 2000; Frankel, Lees, Kempster, & Stern, 1990). Finally, rotigotine reported improved pain, though some of its effects were attributable to benefits in motor function and sleep disturbances (RECOVER study) (Ghys, Surmann, Whitesides, & Boroojerdi, 2011; Kassubek et al., 2014; Trenkwalder, Kies, Rudzinska, et al., 2011). Recently, a double-blind placebo-controlled trail has been published on the effects of a dopamine agonist on PD-associated pain (DOLORES study) (Rascol et al., 2016). A numerical improvement in pain was observed in favor of rotigotine, and proportion of responders was higher in the rotigotine (60%) vs placebo (47%). Rotigotine treatment effects were associated with a higher improvement in KPPS domain "fluctuation-related pain" and PDQ-8 scores than the placebo. Future large-scale confirmatory study is needed because this study was underpowered to observe statistical differences.

In people with PD, there is no evidence to encourage the use of a specific analgesic medication. Acetaminophen is generally recommended, as reported by clinical experience with other neurological diseases (Perez-Lloret et al., 2012). Tramadol and oxycodone are a complementary therapy that may be considered if pain syndromes are not relieved with first-line analgesics. Codeine and morphine may also be used with caution owing to their psychotropic effects on PD patients (Ghoche, 2012). Duloxetine was evaluated for the treatment of several types of painful symptoms in PD. Pregabalin and gabapentin may have a benefit in patients, especially in those with radicular pain, but no studies to date have addressed patients with PD (Finnerup et al., 2015).

The efficacy and safety of opiate analgesics has been recently studied in PD patients experiencing chronic pain. A fixed combination of oxycodone with the peripheral opiate antagonist naloxone (which minimizes the risk of constipation) has offered the possibility of using this drug to treat pain in PD patients. One small observational study showed significant pain relief as assessed by reductions in numeric rating scales and in BPI scores (Madeo et al., 2015). Another large double-blind placebo-controlled randomized study (PANDA) recruited patients with at least one type of severe pain, and an average 24-h pain score of at least 6 out of 10. Although the primary endpoint (24-h pain score at week 16) was not significant, assessments of 24-h pain at other time points during the study and other secondary endpoints (e.g., responder rates for 24-h pain scores) favored treatment with oxycodone–naloxone (Trenkwalder et al., 2015).

Patients who do not respond to dopaminergic treatment adjustments or experience poor tolerance or adverse effects with the use of systemic analgesics may benefit from localized treatments. This approach may be more useful for treating musculoskeletal pain, but it is not exclusive.

Botulinum toxin (BTX) injections, in addition to their neuromuscular action in movement disorders, may relieve pain through analgesic mechanisms (Sim, 2011). A retrospective analysis investigating the potential usefulness of BTX in PD patients suggested that BTX treatment is safe and useful in the treatment of pain in this population (Bruno, Fox, Mancini, & Miyasaki, 2016). Likewise, a double-blind, placebo-controlled, crossover pilot study is ongoing to study the effects of BTX type A to reduce pain in people with advanced stage PD (ClinicalTrials.gov Identifier: NCT02472210). Cordivari et al. found moderate benefit in treating clenched fist with BTX injections in patients with PD (Cordivari, Misra, Catania, & Lees, 2001). BTX has been used to treat abnormal postures such as Pisa syndrome with positive effect on pain (Bonanni, Thomas, Varanese, Scorrano, & Onofrj, 2007; Tassorelli et al., 2014).

6. SURGICAL APPROACHES

Implanted spinal cord stimulation (SCS), pallidotomy, and DBS are the main surgical procedures to reduce pain in PD. Most studies are focused on DBS, in which the globus pallidus internus (GPi) (Loher, Burgunder, Weber, Sommerhalder, & Krauss, 2002) and the STN with unilateral or bilateral stimulation are the targets of choice for the treatment of motor symptoms in advanced PD. Chronic DBS of these targets produces beneficial effects in motor function and disability (DBS Study Group, 2001).

Implanted SCS of the dorsal column is an alternative therapeutic option for treating neuropathic chronic pain conditions in PD (Fénelon et al., 2012; Hassan, Amer, Alwaki, & Elborno, 2013; Oakley & Prager, 2002). The mechanisms behind its action are supported by a complex relationship between multiple structures at several levels of the nervous system (Oakley & Prager, 2002) and are, at least in part, explained by the gate control theory (Melzack & Wall, 1965). Chronic, sympathetic-mediated, and neuropathic pain respond well to SCS, with no detectable side effects noted in the nervous system (Oakley & Prager, 2002).

Pallidotomy is used to achieve short- and long-term effects on dystonia and muscle pain. Laitinen, Bergenheim, and Hariz (1992) demonstrated that the percentage of patients reporting dystonia and/or pain is decreased after Leksell's posteroventral pallidotomy. It has long-lasting effects on other parkinsonian symptoms such as tremor, rigidity, and bradykinesia. Honey et al. found a long-term effect on pain reduction in 21 patients with PD-related

Table 2 Studies on Pain Management With Pallidotomy and Spinal Cord Stimulation in Parkinson's Disease

Author	Pain Type	Body Localization	Type of Treatment	Main Results
Hassan et al. (2013)	Chronic neuropathic pain	Neck and upper extremities	Bilateral spinal cord stimulation at C2 level	Pain reduction from 8 to 2 on VAS score
Fénelon et al. (2012)	Neuropathic	Lower limbs	Spinal cord stimulation at T9–T10 level	Pain symptoms improved with stimulation on
Honey et al. (1999)	Pain associated with dystonic movements, dysesthetic pain, somatic pain exacerbated by PD and pain of musculoskeletal origin	Lower and upper limbs	Unilateral pallidotomy	Pain reduction at 6 weeks and after 1 year following pallidotomy

PD, Parkinson's disease; *VAS*, visual analogue scale.

pain at 6 weeks and 1 year after surgery (Honey, Stoessl, Tsui, Schulzer, & Calne, 1999; see Table 2).

The deep brain stimulation of the subthalamic nucleus (STN-DBS) has been found to reduce pain in 87% of PD patients, remarkably in the off medication (Cury et al., 2014; Kim et al., 2008; Sürücü, Baumann-Vogel, Uhl, Imbach, & Baumann, 2013; Witjas et al., 2007). Dystonic pain is the most responsive (Jung et al., 2015) approaching the 100% rate of improvement, followed by central (92%), neuritic/radicular (63%), and musculoskeletal pain (61%). The pain involves the head, neck, upper and lower extremities, and trunk. Pain in the trunk, identified as low back pain, was the least responsive (14%) (Kim et al., 2008). A reduction in pain intensity correlates with improved quality of life after surgery (Cury et al., 2014). STN-DBS is superior to dopaminergic treatment for reducing pain in PD (Sürücü et al., 2013). The response of pain symptoms to DBS can be predicted by L-dopa challenge tests assessing pain severity. This diagnostic procedure can provide evidence on whether a patient with severe pain should undergo DBS for potential pain relief (Sürücü et al., 2013).

Pain detection threshold and tolerance tested with mechanical stimuli can be increased with STN-DBS, whereas the effects induced by thermal stimuli were not significant or unclear (Marques et al., 2013). STN-DBS has no effect on the detection of large fiber-mediated sensations (Aβ) such as the mechanical and vibration detection thresholds (Ciampi de Andrade et al., 2012). However, it does appear to modulate the small fiber-dependent (Aδ and C) sensory threshold, increasing the mechanical pain threshold and heat pain thresholds but reducing the cold pain threshold. It also increases sensitivity to innocuous thermal sensations by reducing the warm detection threshold and increasing the cold detection threshold (Ciampi de Andrade et al., 2012). The thermal sensory limen and the cold and warm detection thresholds can be improved with STN-DBS (Gierthmühlen et al., 2010; Maruo et al., 2011). No statistically significant changes in warm and cold detection thresholds or heat and cold pain thresholds were found with on-stimulation (Spielberger, Wolf, Kress, Seppi, & Poewe, 2011). The subjective heat pain threshold can be increased with STN-DBS. It reduces pain-induced cerebral activity in the somatosensory cortex, but has no effect in pain-free PD patients (Dellapina et al., 2012). Finally, Belasen et al. (2017) explored whether altering the frequency of STN-DBS can change pain perception as measured by quantitative sensory testing and found that low-frequency stimulation can be useful in treating chronic pain in PD and can modulate the mechanical and thermal detection thresholds to a greater extent than high-frequency stimulation.

STN-DBS can have short- and long-term effects on reducing pain in PD. Pellaprat et al. (2014) reported a decrease in pain symptoms 12 months after surgery. Oshima et al. (2012) found a significant decrease of 69% in VAS pain score at 6 months and a decrease of 80% at 12 months after surgery in people with PD with predominantly musculoskeletal pain. Kim et al. (Jung et al., 2015; Kim, Jeon, Lee, Paek, & Kim, 2012) showed additional improvement at 24 months and 8 years, despite novel central and musculoskeletal pain symptoms at follow-up.

Although STN-DBS produces beneficial effects on reducing pain syndromes, a direct correlation between pain relief and motor control has not been found (Marques et al., 2013; Wolz et al., 2012). This indicates that not all pain syndromes are the consequence of reduced mobility and augmented muscle tone. Studies showing the psychophysical and pain modulation effect of DBS are reported in Table 3.

In summary, three STN-DBS pain-relief mechanisms have been distinguished. First, STN-DBS may increase pain tolerance and decrease pain perception. Pain tolerance refers to the psychological aspects of pain, a multifaceted interaction between cognitive and affective functions, while pain perception denotes the sensory-discriminative component of pain

Table 3 Studies on Psychophysical Modulations and Pain Management With STN or GPi DBS in Parkinson's Disease

Author	Pain Type	Body Localization	Type of DBS	Main Results
Belasen et al. (2017)	Chronic pain	Hand, low back, shoulder, ankle, thumb, epigastric region, chest	Unilateral and bilateral STN-DBS	In PD with chronic pain, STN LFS significantly reduced WDT, as compared with HFS or in the off state. LFS significantly increased MDT and VDT, as compared with thresholds following HFS
Jung et al. (2015)	Central, neuritic/radicular, musculoskeletal, and dystonic	Upper and lower extremities, neck, trunk, and head	Unilateral and bilateral STN-DBS	83% of patients reported improvement in pain symptoms over a long-term follow-up period of 8 years
Pellaprat et al. (2014)	n.a.	Upper and lower limbs, head, neck, and trunk	Bilateral STN-DBS	Stim-on decreased pain symptoms: 19% of patients were pain free after 12 months
Cury et al. (2014)	Central, radicular/neuropathic, musculoskeletal, and dystonic	Upper and lower limbs, head, neck, and back (especially low back)	STN-DBS	DBS reduced pain after surgery with improvements of symptoms in 69% of patients. Dystonic pain was the most responsive to the therapy, followed by musculoskeletal pain. Neuropathic and central pain were not influenced by intervention
Marques et al. (2013)	Central pain	Upper limb, hand	Bilateral STN-DBS	MPT and MPTo increased with on-stim and on-levodopa, as compared with off state

Continued

Table 3 Studies on Psychophysical Modulations and Pain Management With STN or GPi DBS in Parkinson's Disease—cont'd

Author	Pain Type	Body Localization	Type of DBS	Main Results
Sürücü et al. (2013)	Central, neuritic/radicular, dystonic, and musculoskeletal	Multifocal, lumbar spine, neck, arm, leg, and abdominal/visceral	STN-DBS vs levodopa	Eight patients with on-medical showed improvement in pain. Greater improvement observed with on-stim as compared with levodopa, with long-lasting effects of 41 months
Oshima et al. (2012)	Somatic pain exacerbated by PD, central, musculoskeletal, dystonic, neuritic–radicular	Back, face, abdomen, neck, upper and lower limbs	Bilateral STN-DBS	Postsurgery improvements in pain intensity: 75% reduction in VAS score at 2 weeks, 69% at 6 months, and 80% at 12 months with on-stim
Kim et al. (2012)	Central, dystonic, radiculoneuritic, and musculoskeletal	Upper and lower extremities, head, and neck	Unilateral and bilateral STN-DBS	Improvement in pain symptoms after 3 and 24 months
Wolz et al. (2012)	n.a.	n.a.	Bilateral STN-DBS	DBS-on, no changes in pain
Dellapina et al. (2012)	Nociceptive, neuropathic	Trunk, upper and lower limbs	Bilateral STN-DBS	Stim had no effect in pain-free patients. Significant increase in HPT. Significant reduction in cerebral activity in the cerebellum and somatosensory cortex in patients with pain with on-stim
Ciampi de Andrade et al. (2012)	Musculoskeletal, dystonic	n.a.	Bilateral STN-DBS	With on-stim, increased MPT, HPT but decreased CPT. CDT increased but WDT decreased. VAS score in SuH and InC reduced with on-stim. No changes in MDT, VDT with on-stim

Maruo et al. (2011)	—	—	Bilateral STN-DBS	No differences in CPT and HPT. With stim-on, a lower CDT and WDT
Spielberger et al. (2011)	n.a.	n.a.	Bilateral STN-DBS and levodopa	No significant changes in WDT, CDT, HPT, or CPT with on-stim
Gierthmühlen et al. (2010)	Nociceptive, neuropathic	Neck, back, upper (hands) and lower extremities	Bilateral STN-DBS and levodopa	No effect of STN-DBS on thermal and mechanical pain thresholds. Improvement of pain symptoms and intensity, CDT, WDT, and TSL with on-stim
Kim et al. (2008)	Central, dystonic, musculoskeletal, neuritic–radicular	Upper and lower extremities, head, neck, and trunk	Unilateral and bilateral STN-DBS	Pain symptoms improved after 3 months
Witjas et al. (2007)	n.a.	n.a.	Bilateral STN-DBS	Pain symptoms improved after 12 months
Loher et al. (2002)	n.a.	Neck, trunk, upper and lower extremities	Unilateral and bilateral GPi-DBS	Pain and dysesthesia improved 3–5 days after surgery and maintained at follow-up of 12 months

CDT, cold detection threshold; *CP*, chronic pain; *CPT*, cold pain threshold; *DBS*, deep brain stimulation; *EP*, experimental-induced pain; *GPi*, globus pallidus internus; *HFS*, high-frequency stimulation; *HPT*, heat pain threshold; *InC*, infrathreshold cold stimulation; *LFS*, low-frequency stimulation; *Medical off*, without PD medication; *Medical on*, with PD medication; *MPT*, mechanical pain threshold; *MPTo*, mechanical pain tolerance; *n.a.*, not available; *off-stim*, DBS turned off; *on-stim*, DBS turned on; *Stim*, DBS stimulation; *STN*, subthalamic nucleus; *SuH*, suprathreshold heat stimulation; *TSL*, thermal sensory limen; *WDT*, warm detection threshold.

(Maruo et al., 2011). STN-DBS modulates the sensory-discriminatory portion of pain, which is thought to result from interactions between the descending pain inhibitory system and STN (Marques et al., 2013). The sensory-discriminative component of pain is then integrated at the somatosensory cortices, while the related affective components are related to activity of the nucleus accumbens, insular cortex, and anterior cingulate gyrus (Cury et al., 2016; Ogino et al., 2007). In the limbic component of STN, DBS may modulate the nucleus accumbens, which is involved in the modulation of emotional and motivational–affective status and represents part of a potent pain-suppression system (Altier & Stewart, 1999; Gear, Aley, & Levine, 1999).

Second, a reduction in pain leads to a decrease in muscle tone due to musculoskeletal or diatonic pain, resulting in additional pain relief (Bonanni et al., 2007; Tassorelli et al., 2014). This mechanism has been advanced as a valid explanation for the effects of pallidotomy and dopaminergic medication. Finally, a further possible mechanism may be related to the consequences of the improved motor function induced by STN-DBS. The model suggests that pain is coupled with motor behavior adaptations that involve the reorganization of activity between and within muscles as well as biomechanical changes (Hodges, 2011). These adaptations may involve multiple levels of the motor system and work as "protective mechanisms" to avoid further injury and related pain. The adaptations can be complementary, additive, or competitive, with short-term benefits but can have potential long-term consequences that limit motor performance such as decreased movement, increased load, and decreased movement variability (Hodges, 2011).

7. REHABILITATION

Rehabilitation is perhaps the most commonly used procedure as an adjunct to drug therapy to treat movement disorders in people with PD (Abbruzzese, Marchese, Avanzino, & Pelosin, 2016). Recent studies demonstrated that rehabilitation could induce clinically important benefits, particularly for gait and balance (Abbruzzese et al., 2016; Smania et al., 2010; Tomlinson et al., 2012). The effects of intensive exercise in promoting cell proliferation and neuronal differentiation in animal models have been reported in a large cohort of studies, and these neuroplastic effects are probably related to increased expression of a variety of neurotrophic factors (Abbruzzese et al., 2016; Petzinger et al., 2013). However, the effect of physical exercise on neurodegenerative processes of PD and pain is not fully understood. A metaepidemiological study of randomized trials on the effects of a "physiotherapy" intervention on self-reported pain in adults reported an overall moderate effect of physiotherapy on pain (Ginnerup-Nielsen, Christensen, Thorborg, Tarp, & Henriksen, 2016). Some evidence suggests that exercise may influence the experience of pain in people with PD. The effects of physiotherapy were evaluated in two studies in which pain was not the primary outcome measure. Reuter et al., in a randomized controlled trial, reported the effects of flexibility and relaxation exercises, walking or Nordic walking in 90 people with PD. After 6 months of intervention, all patients reported a reduction in painful syndromes in the different parts of the body (i.e., neck, hip, iliosacral joint). When compared with the relaxation and flexibility group, the walking and Nordic walking group showed a greater decrease in intensity of back pain (Reuter et al., 2011). Rodrigues de Paula et al., in a single-group, uncontrolled study, investigated the effects of a 12-week exercise program in 20 people with PD and reported a slight (non-significant) improvement in pain after treatment (Rodrigues de Paula, Teixeira-Salmela, Coelho de Morais Faria, Rocha de Brito, & Cardoso, 2006).

Several theories have been proposed to explain the mechanisms of pain relief for rehabilitation. These include activation of gait control mechanisms, acting as a counterirritant and activation of endogenous opioids (Sluka, 2016). Moreover, exercise can promote restoration of function removing peripheral irritant and potentially modulate pain through dopaminergic and non-dopaminergic pain inhibitory pathways (Allen, Moloney, van Vliet, & Canning, 2015; Sluka, 2016). Exercise can promote also changes in the anatomical, physiological, functional mechanisms of the brain in response to changes in a person's behavior or surroundings (Mattson, 2014; Petzinger et al., 2013).

The role of alternative and complementary medicine requires further investigation. Acupuncture, massage therapy, vitamins, and herbs are frequently used as alternative and complementary therapies in PD (Lee & Lim, 2017; Rajendran, Thompson, & Reich, 2001). No evidence is available about their effectiveness in providing pain relief, suggesting their complementary role. Shulman et al. (2002) showed that about 85% of patients reported subjective improvement in PD symptoms (including pain) after acupuncture. Donoyama and Ohkoshi (2012) investigated the effects of traditional Japanese massage therapy on PD symptoms in 10 patients and reported that the patients experienced a significant improvement in pain symptoms after massage therapy.

A preliminary report showed a link between electromagnetic therapy combined with physical therapy and pain relief in PD. Electromagnetic therapy (i.e., transcranial magnetic stimulation, repetitive transcranial magnetic stimulation, high-frequency transcranial magnetic stimulation, and pulsed electromagnetic field therapy) is a noninvasive and safe form of transcranial magnetic stimulation used in the clinical setting to treat motor and nonmotor symptoms in people with PD (Vadalà et al., 2015). Kodama et al. (2011) described in a case report a patient with severely painful off-period dystonia in the unilateral lower limb. Delivery of low-frequency (0.9 Hz subthreshold) repetitive transcranial magnetic stimulation over the inhibitory (contralateral) primary motor area combined with physical therapy reportedly relived the patient's pain. As pointed out by the authors, this treatment should be further verified in such patients. Table 4 summarizes study on pain management with nonpharmachological and surgical interventions. Table 5 summarizes potential rehabilitative interventions for pain management in PD.

Table 4 Studies on Pain Management With Nonpharmacological/Surgical Interventions in Parkinson's Disease

Author	Pain Type	Body Localization	Type of Treatment	Main Results
Rodrigues de Paula et al. (2006)	n.a.	n.a.	Exercise program	Painful symptoms improved after 12 weeks (trend toward significance)
Reuter et al. (2011)	n.a.	Toes, feet, knees, back, iliosacral joint, hip, hands, arms, neck	Flexibility exercises and Nordic walking	Painful symptoms improved after 6-week training program
Kodama et al. (2011)	Dystonic pain	Lower limb	Repetitive transcranial magnetic stimulation over the inhibitory primary motor area + physical therapy	Reduced painful dystonia and walking disturbances

VAS, visual analogue scale.

Table 5 Potential Rehabilitative Interventions for Pain Management in Parkinson's Disease

Type of Treatment	References for Neurophysiological Rationale	Pain Syndromes That Might Be Potentially Treated
Transcranial magnetic stimulation	Moisset, de Andrade, and Bouhassira (2016) and Klein et al. (2015)	CN
Transcranial direct current stimulation	Lefaucheur et al. (2017) and Knotkova, Nitsche, and Cruciani (2013)	CN, RN, M
Transcutaneous neuromuscular electrical stimulation	Schuhfried, Crevenna, Fialka-Moser, and Paternostro-Sluga (2012)	M
Graded motor imagery	Priganc and Stralka (2011)	CN
Mirror therapy	Lamont, Chin, and Kogan (2011) and McCabe (2011)	CN
Manual therapy (mobilization, manipulation, and myofascial techniques)	Lascurain-Aguirrebeña, Newham, and Critchley (2016), Haavik and Murphy (2012), Vigotsky and Bruhns (2015), and Simmonds, Miller, and Gemmell (2012)	M, RN
Back school	Moffett and McLean (2006)	M
McKenzi	Hefford (2008)	M
Exercises	Berdishevsky et al. (2016)	M
Massage	Vigotsky and Bruhns (2015)	M
Dry needling	Dunning et al. (2014)	M
Ultrasound	Baker, Robertson, and Duck (2001)	M
Kinesio taping	Wu, Hong, and Chou (2015)	M
Diathermy	Cacolice, Scibek, and Martin (2013)	M
Low-level laser therapy	van Middelkoop et al. (2011)	M
Hydrotherapy, balneotherapy, and spa treatment	Bender et al. (2005)	M

CN, central neuropathic; *D*, dystonic; *M*, musculoskeletal; *RN*, radicular-peripheral neuropathic.

8. CONCLUSIONS

This chapter highlights the many procedures that have been implemented to manage painful syndromes in PD. Selecting the most appropriate treatment intervention based on the type of pain, stage of disease, disability, and the biopsychosocial model of pain is essential. Pharmacological and nonpharmacological interventions should be recommended. However, future studies are needed to evaluate the integration of pharmacologic with nonpharmacological approaches. An integrated approach to pain involving a multidisciplinary team of medical specialists and rehabilitation experts should allow a comprehensive approach to pain in PD.

REFERENCES

Abbruzzese, G., Marchese, R., Avanzino, L., & Pelosin, E. (2016). Rehabilitation for Parkinson's disease: Current outlook and future challenges. *Parkinsonism & Related Disorders*, *22*, S60–S64.

Allen, N. E., Moloney, N., van Vliet, V., & Canning, C. G. (2015). The rationale for exercise in the management of pain in Parkinson's disease. *Journal of Parkinson's Disease*, *5*, 229–239.

Altier, N., & Stewart, J. (1999). The role of dopamine in the nucleus accumbens in analgesia. *Life Sciences*, *65*, 2269–2287.

Baker, K. G., Robertson, V. J., & Duck, F. A. (2001). A review of therapeutic ultrasound: Biophysical effects. *Physical Theraphy*, *81*, 1351–1358.

Barone, P., Antonini, A., Colosimo, C., Marconi, R., Morgante, L., Avarello, T. P., et al. (2009). The PRIAMO study: A multicenter assessment of nonmotor symptoms and their impact on quality of life in Parkinson's disease. *Movement Disorders*, *24*, 1641–1649.

Barone, P., Pöewe, W., Albrecht, S., Debieuvre, C., Massey, D., Rascol, O., et al. (2010). Pramipexole for the treatment of depressive symptoms in patients with Parkinson's disease: A randomised, double-blind, placebo-controlled trial. *Lancet Neurology*, *9*, 573–580.

Beiske, A. G., Loge, J. H., Rønningen, A., & Svensson, E. (2009). Pain in Parkinson's disease: Prevalence and characteristics. *Pain*, *141*, 173–177.

Belasen, A., Rizvi, K., Gee, L. E., Yeung, P., Prusik, J., Ramirez-Zamora, A., et al. (2017). Effect of low-frequency deep brain stimulation on sensory thresholds in Parkinson's disease. *Journal of Neurosurgery*, *126*, 397–403.

Bender, T., Karagülle, Z., Bálint, G. P., Gutenbrunner, C., Bálint, P. V., & Sukenik, S. (2005). Hydrotherapy, balneotherapy, and spa treatment in pain management. *Rheumatology International*, *25*, 220–224.

Berdishevsky, H., Lebel, V. A., Bettany-Saltikov, J., Rigo, M., Lebel, A., Hennes, A., et al. (2016). Physiotherapy scoliosis-specific exercises—A comprehensive review of seven major schools. *Scoliosis and Spinal Disorders*, *11*, 20.

Bonanni, L., Thomas, A., Varanese, S., Scorrano, V., & Onofrj, M. (2007). Botulinum toxin treatment of lateral axial dystonia in Parkinsonism. *Movement Disorders*, *22*(14), 2097–2103.

Braak, H., Del Tredici, K., Rüb, U., de Vos, R. A., Jansen Steur, E. N., & Braak, E. (2003). Staging of brain pathology related to sporadic Parkinson's disease. *Neurobiology of Aging*, *24*, 197–211.

Braak, H., Rüb, U., Jansen Steur, E. N., Del Tredici, K., & de Vos, R. A. (2005). Cognitive status correlates with neuropathologic stage in Parkinson disease. *Neurology, 64*, 1404–1410.

Brefel-Courbon, C., Grolleau, S., Thalamas, C., Bourrel, R., Allaria-Lapierre, V., Loï, R., et al. (2009). Comparison of chronic analgesic drugs prevalence in Parkinson's disease, other chronic diseases and the general population. *Pain, 141*, 14–18.

Brefel-Courbon, C., Ory-Magne, F., Thalamas, C., Payoux, P., & Rascol, O. (2013). Nociceptive brain activation in patients with neuropathic pain related to Parkinson's disease. *Parkinsonism & Related Disorders, 19*, 548–552.

Brefel-Courbon, C., Payoux, P., Thalamas, C., Ory, F., Quelven, I., Chollet, F., et al. (2005). Effect of levodopa on pain threshold in Parkinson's disease: A clinical and positron emission tomography study. *Movement Disorders, 20*, 1557–1563.

Broen, M. P., Braaksma, M. M., Patijn, J., & Weber, W. E. (2012). Prevalence of pain in Parkinson's disease: A systematic review using the modified QUADAS tool. *Movement Disorders, 27*(4), 480.

Broetz, D., Eichner, M., Gasser, T., Weller, M., & Steinbach, J. P. (2007). Radicular and nonradicular back pain in Parkinson's disease: A controlled study. *Movement Disorders, 22*(6), 853.

Bruno, V. A., Fox, S. H., Mancini, D., & Miyasaki, J. M. (2016). Botulinum toxin use in refractory pain and other symptoms in Parkinsonism. *The Canadian Journal of Neurological Sciences, 43*, 697–702.

Cacolice, P. A., Scibek, J. S., & Martin, R. (2013). Diathermy: A literature review of current research and practices. *Orthopaedic Physical Therapy Practice, 25*(3), 155–161.

Chaudhuri, K. R., Healy, D. G., Schapira, A. H., & National Institute for Clinical Excellence. (2006). Non-motor symptoms of Parkinson's disease: Diagnosis and management. *Lancet Neurology, 5*, 235–245.

Chaudhuri, K. R., Rizos, A., Trenkwalder, C., et al. (2015). King's Parkinson's disease pain scale, the first scale for pain in PD: An international validation. *Movement Disorders, 30*(12), 1623–1631.

Chen, Y., Mao, C. J., Li, S. J., Wang, F., Chen, J., Zhang, H. J., et al. (2015). Quantitative and fiber-selective evaluation of pain and sensory dysfunction in patients with Parkinson's disease. *Parkinsonism & Related Disorders, 21*, 361–365.

Chudler, E. H., & Dong, W. K. (1995). The role of the basal ganglia in nociception and pain. *Pain, 60*, 3–38.

Ciampi de Andrade, D., Lefaucheur, J. P., Galhardoni, R., Ferreira, K. S., Brandão Paiva, A. R., Bor-Seng-Shu, E., et al. (2012). Subthalamic deep brain stimulation modulates small fiber-dependent sensory thresholds in Parkinson's disease. *Pain, 153*, 1107–1113.

Conte, A., Khan, N., Defazio, G., Rothwell, J. C., & Berardelli, A. (2013). Pathophysiology of somatosensory abnormalities in Parkinson disease. *Nature Reviews. Neurology, 9*, 687–697.

Corallo, F., De Cola, M. C., Buono, V. L., Di Lorenzo, G., Bramanti, P., & Marino, S. (2017). Observational study of quality of life of Parkinson's patients and their caregivers. *Psychogeriatrics, 17*, 97–102.

Cordivari, C., Misra, V. P., Catania, S., & Lees, A. J. (2001). Treatment of dystonic clenched fist with botulinum toxin. *Movement Disorders, 16*, 907–913.

Cury, R. G., Galhardoni, R., Fonoff, E. T., Dos Santos Ghilardi, M. G., Fonoff, F., Arnaut, D., et al. (2014). Effects of deep brain stimulation on pain and other nonmotor symptoms in Parkinson disease. *Neurology, 83*, 1403–1409.

Cury, R. G., Galhardoni, R., Fonoff, E. T., Perez Lloret, S., Dos Santos Ghilardi, M. G., Barbosa, E. R., et al. (2016). Sensory abnormalities and pain in Parkinson disease and its modulation by treatment of motor symptoms. *European Journal of Pain, 20*, 151–165.

Czernecki, V., Pillon, B., Houeto, J. L., Pochon, J. B., Levy, R., & Dubois, B. (2002). Motivation, reward, and Parkinson's disease: Influence of dopatherapy. *Neuropsychologia, 40*, 2257–2267.

Deep-Brain Stimulation for Parkinson's Disease Study Group. (2001). Deep-brain stimulation of the subthalamic nucleus or the pars interna of the globus pallidus in Parkinson's disease. *The New England Journal of Medicine, 345*, 956–963.

Defazio, G., Berardelli, A., Fabbrini, G., Martino, D., Fincati, E., Fiaschi, A., et al. (2008). Pain as a nonmotor symptom of Parkinson disease: Evidence from a case-control study. *Archives of Neurology, 65*, 1191–1194.

Defazio, G., Gigante, A., Mancino, P., & Tinazzi, M. (2013). The epidemiology of pain in Parkinson's disease. *Journal of Neural Transmission), 120*, 583–586.

Dellapina, E., Ory-Magne, F., Regragui, W., Thalamas, C., Lazorthes, Y., Rascol, O., et al. (2012). Effect of subthalamic deep brain stimulation on pain in Parkinson's disease. *Pain, 153*, 2267–2273.

Djaldetti, R., Shifrin, A., Rogowski, Z., Sprecher, E., Melamed, E., & Yarnitsky, D. (2004). Quantitative measurement of pain sensation in patients with Parkinson disease. *Neurology, 62*, 2171–2175.

Donoyama, N., & Ohkoshi, N. (2012). Effects of traditional Japanese massage therapy on various symptoms in patients with Parkinson's disease: A case-series study. *Journal of Alternative and Complementary Medicine, 18*, 294–299.

Dunning, J., Butts, R., Mourad, F., Young, I., Flannagan, S., & Perreault, T. (2014). Dry needling: A literature review with implications for clinical practice guidelines. *The Physical Therapy Review, 19*, 252–265.

Dzau, V. J., & Pizzo, P. A. (2014). Relieving pain in America: Insights from an Institute of Medicine committee. *JAMA, 312*, 1507–1508.

Ehrt, U., Larsen, J. P., & Aarsland, D. (2009). Pain and its relationship to depression in Parkinson disease. *The American Journal of Geriatric Psychiatry, 17*, 269–275.

Factor, S. A., Brown, D. L., & Molho, E. S. (2000). Subcutaneous apomorphine injections as a treatment for intractable pain in Parkinson's disease. *Movement Disorders, 15*, 167–169.

Fénelon, G., Goujon, C., Gurruchaga, J. M., Cesaro, P., Jarraya, B., Palfi, S., et al. (2012). Spinal cord stimulation for chronic pain improved motor function in a patient with Parkinson's disease. *Parkinsonism & Related Disorders, 18*, 213–214.

Fil, A., Cano-de-la-Cuerda, R., Muñoz-Hellín, E., Vela, L., Ramiro-González, M., & Fernández-de-Las-Peñas, C. (2013). Pain in Parkinson disease: A review of the literature. *Parkinsonism & Related Disorders, 19*, 285–294.

Finnerup, N. B., Attal, N., Haroutounian, S., McNicol, E., Baron, R., Dworkin, R. H., et al. (2015). Pharmacotherapy for neuropathic pain in adults: A systematic review and meta-analysis. *Lancet Neurology, 14*, 162–173.

Ford, B. (2010). Pain in Parkinson's disease. *Movement Disorders, 25*(Suppl. 1), S98–103.

Frankel, J. P., Lees, A. J., Kempster, P. A., & Stern, G. M. (1990). Subcutaneous apomorphine in the treatment of Parkinson's disease. *Journal of Neurology, Neurosurgery, and Psychiatry, 53*, 96–101.

Garcia-Larrea, L., Frot, M., & Valeriani, M. (2003). Brain generators of laser-evoked potentials: From dipoles to functional significance. *Neurophysiologie Clinique, 33*, 279–292.

Garcia-Larrea, L., & Peyron, R. (2013). Pain matrices and neuropathic pain matrices: A review. *Pain, 154*, S29–S43.

Gear, R. W., Aley, K. O., & Levine, J. D. (1999). Pain-induced analgesia mediated by mesolimbic reward circuits. *The Journal of Neuroscience, 19*(16), 7175–7181.

Gebhart, G. F. (2004). Descending modulation of pain. *Neuroscience and Biobehavioral Reviews, 27*, 729–737.

Gerdelat-Mas, A., Simonetta-Moreau, M., Thalamas, C., Ory-Magne, F., Slaoui, T., Rascol, O., et al. (2007). Levodopa raises objective pain threshold in Parkinson's disease: A RIII reflex study. *Journal of Neurology, Neurosurgery, and Psychiatry, 78*, 1140–1142.

Geroin, C., Gandolfi, M., Bruno, V., Smania, N., & Tinazzi, M. (2016). Integrated approach for pain management in Parkinson disease. *Current Neurology and Neuroscience Reports, 16*, 28.

Ghoche, R. (2012). The conceptual framework of palliative care applied to advanced Parkinson's disease. *Parkinsonism & Related Disorders, 18*, S2–S5.

Ghys, L., Surmann, E., Whitesides, J., & Boroojerdi, B. (2011). Effect of rotigotine on sleep and quality of life in Parkinson's disease patients: Post hoc analysis of RECOVER patients who were symptomatic at baseline. *Expert Opinion on Pharmacotherapy, 12*, 1985–1998.

Gierthmühlen, J., Arning, P., Binder, A., Herzog, J., Deuschl, G., Wasner, G., et al. (2010). Influence of deep brain stimulation and levodopa on sensory signs in Parkinson's disease. *Movement Disorders, 25*, 1195–1202.

Ginnerup-Nielsen, E., Christensen, R., Thorborg, K., Tarp, S., & Henriksen, M. (2016). Physiotherapy for pain: A meta-epidemiological study of randomised trials. *British Journal of Sports Medicine, 50*, 965–971.

Goetz, C. G., Tilley, B. C., Shaftman, S. R., Stebbins, G. T., Fahn, S., Martinez-Martin, P., et al. (2008). Movement Disorder Society-sponsored revision of the Unified Parkinson's Disease Rating Scale (MDS-UPDRS): Scale presentation and clinimetric testing results. *Movement Disorders, 23*, 2129–2170.

Greenbaum, L., Tegeder, I., Barhum, Y., Melamed, E., Roditi, Y., & Djaldetti, R. (2012). Contribution of genetic variants to pain susceptibility in Parkinson disease. *European Journal of Pain, 16*, 1243–1250.

Haavik, H., & Murphy, B. (2012). The role of spinal manipulation in addressing disordered sensorimotor integration and altered motor control. *Journal of Electromyography and Kinesiology, 22*, 768–776.

Hassan, S., Amer, S., Alwaki, A., & Elborno, A. (2013). A patient with Parkinson's disease benefits from spinal cord stimulation. *Journal of Clinical Neuroscience, 20*, 1155–1156.

Hefford, C. (2008). McKenzie classification of mechanical spinal pain: Profile of syndromes and directions of preference. *Manual Therapy, 13*, 75–81.

Hodges, P. W. (2011). Pain and motor control: From the laboratory to rehabilitation. *Journal of Electromyography and Kinesiology, 21*, 220–228.

Honey, C. R., Stoessl, A. J., Tsui, J. K., Schulzer, M., & Calne, D. B. (1999). Unilateral pallidotomy for reduction of parkinsonian pain. *Journal of Neurosurgery, 91*, 198–201.

Jarcho, J. M., Mayer, E. A., Jiang, Z. K., Feier, N. A., & London, E. D. (2012). Pain, affective symptoms, and cognitive deficits in patients with cerebral dopamine dysfunction. *Pain, 153*, 744–754.

Jung, Y. J., Kim, H. J., Jeon, B. S., Park, H., Lee, W. W., & Paek, S. H. (2015). An 8-year follow-up on the effect of subthalamic nucleus deep brain stimulation on pain in Parkinson disease. *JAMA Neurology, 72*, 504–510.

Juri, C., Rodriguez-Oroz, M., & Obeso, J. A. (2010). The pathophysiological basis of sensory disturbances in Parkinson's disease. *Journal of the Neurological Sciences, 289*, 60–65.

Kanda, T., Tsukagoshi, H., Oda, M., Miyamoto, K., & Tanabe, H. (1996). Changes of unmyelinated nerve fibers in sural nerve in amyotrophic lateral sclerosis, Parkinson's disease and multiple system atrophy. *Acta Neuropathologica, 91*(2), 145–154.

Kassubek, J., Chaudhuri, K. R., Zesiewicz, T., Surmann, E., Boroojerdi, B., Moran, K., et al. (2014). Rotigotine transdermal system and evaluation of pain in patients with Parkinson's disease: A post hoc analysis of the RECOVER study. *BMC Neurology, 14*, 42.

Kim, H. J., Jeon, B. S., Lee, J. Y., Paek, S. H., & Kim, D. G. (2012). The benefit of subthalamic deep brain stimulation for pain in Parkinson disease: A 2-year follow-up study. *Neurosurgery*, *70*, 18–23.

Kim, H. J., Paek, S. H., Kim, J. Y., Lee, J. Y., Lim, Y. H., Kim, M. R., et al. (2008). Chronic subthalamic deep brain stimulation improves pain in Parkinson disease. *Journal of Neurology*, *255*, 1889–1894.

Klein, M. M., Treister, R., Raij, T., Pascual-Leone, A., Park, L., Nurmikko, T., et al. (2015). Transcranial magnetic stimulation of the brain: Guidelines for pain treatment research. *Pain*, *156*, 1601–1614.

Knotkova, H., Nitsche, M. A., & Cruciani, R. A. (2013). Putative physiological mechanisms underlying tDCS analgesic effects. *Frontiers in Human Neuroscience*, *7*, 628.

Kodama, M., Kasahara, T., Hyodo, M., Aono, K., Sugaya, M., Koyama, Y., et al. (2011). Effect of low-frequency repetitive transcranial magnetic stimulation combined with physical therapy on L-dopa-induced painful off-period dystonia in Parkinson's disease. *American Journal of Physical Medicine & Rehabilitation*, *90*, 150–155.

Kuraishi, Y., Fukui, K., Shiomi, H., Akaike, A., & Takagi, H. (1978). Microinjection of opioids into the nucleus reticularis gigantocellularis of the rat: Analgesia and increase in the normetanephrine level in the spinal cord. *Biochemical Pharmacology*, *27*, 2756–2758.

Laitinen, L. V., Bergenheim, A. T., & Hariz, M. I. (1992). Leksell's posteroventral pallidotomy in the treatment of Parkinson's disease. *Journal of Neurosurgery*, *76*, 53–61.

Lamont, K., Chin, M., & Kogan, M. (2011). Mirror box therapy: Seeing is believing. *Explore (New York, N.Y.)*, *7*, 369–372.

Lascurain-Aguirrebeña, I., Newham, D., & Critchley, D. J. (2016). Mechanism of action of spinal mobilizations: A systematic review. *Spine (Phila Pa 1976)*, *41*, 159–172.

Lee, S. H., & Lim, S. (2017). Clinical effectiveness of acupuncture on Parkinson disease: A PRISMA-compliant systematic review and meta-analysis. *Medicine*, *96*(3), e5836.

Lee, M. A., Walker, R. W., Hildreth, T. J., & Prentice, W. M. (2006). A survey of pain in idiopathic Parkinson's disease. *Journal of Pain and Symptom Management*, *32*, 462–469.

Lefaucheur, J. P., Antal, A., Ayache, S. S., Benninger, D. H., Brunelin, J., Cogiamanian, F., et al. (2017). Evidence-based guidelines on the therapeutic use of transcranial direct current stimulation (tDCS). *Clinical Neurophysiology*, *128*, 56–92.

Li, W., Chen, Y., Yin, B., & Zhang, L. (2014). Pain in Parkinson's disease associated with COMT gene polymorphisms. *Behavioural Neurology*, *2014*, 304203.

Lim, S. Y., Farrell, M. J., Gibson, S. J., Helme, R. D., Lang, A. E., & Evans, A. H. (2008). Do dyskinesia and pain share common pathophysiological mechanisms in Parkinson's disease? *Movement Disorders*, *23*, 1689–1695.

Loher, T. J., Burgunder, J. M., Weber, S., Sommerhalder, R., & Krauss, J. K. (2002). Effect of chronic pallidal deep brain stimulation on off period dystonia and sensory symptoms in advanced Parkinson's disease. *Journal of Neurology, Neurosurgery, and Psychiatry*, *73*, 395–399.

Madeo, G., Schirinzi, T., Natoli, S., Pierantozzi, M., Stefani, A., Dauri, M., et al. (2015). Efficacy and safety profile of prolonged release oxycodone in combination with naloxone (OXN PR) in Parkinson's disease patients with chronic pain. *Journal of Neurology*, *262*, 2164–2170.

Magrinelli, F., Zanette, G., & Tamburin, S. (2013). Neuropathic pain: Diagnosis and treatment. *Practical Neurology*, *13*, 292–307.

Marques, A., Chassin, O., Morand, D., Pereira, B., Debilly, B., Derost, P., et al. (2013). Central pain modulation after subthalamic nucleus stimulation: A crossover randomized trial. *Neurology*, *81*, 633–640.

Maruo, T., Saitoh, Y., Hosomi, K., Kishima, H., Shimokawa, T., Hirata, M., et al. (2011). Deep brain stimulation of the subthalamic nucleus improves temperature sensation in patients with Parkinson's disease. *Pain*, *152*, 860–865.

Mattson, M. P. (2014). Interventions that improve body and brain bioenergetics for Parkinson's disease risk reduction and therapy. *Journal of Parkinson's Disease, 4*, 1–13.
McCabe, C. (2011). Mirror visual feedback therapy. A practical approach. *Journal of Hand Therapy, 24*, 170–178.
McNamara, P., Stavitsky, K., Harris, E., Szent-Imrey, O., & Durso, R. (2010). Mood, side of motor symptom onset and pain complaints in Parkinson's disease. *International Journal of Geriatric Psychiatry, 25*, 519–524.
Melzack, R., & Casey, K. L. (1968). In *The skin senses Proceedings of the first international symposium on the skin senses.*
Melzack, R., & Wall, P. D. (1965). Pain mechanisms: A new theory. *Science, 150*(3699), 971–979.
Millan, M. J. (1999). The induction of pain: An integrative review. *Progress in Neurobiology, 57*(1), 1–164.
Millan, M. J. (2002). Descending control of pain. *Progress in Neurobiology, 66*, 355–474.
Moffett, J., & McLean, S. (2006). The role of physiotherapy in the management of non-specific back pain and neck pain. *Rheumatology (Oxford, England), 45*, 371–378.
Moisset, X., de Andrade, D. C., & Bouhassira, D. (2016). From pulses to pain relief: An update on the mechanisms of rTMS induced analgesic effects. *European Journal of Pain, 20*, 689–700.
Mylius, V., Brebbermann, J., Dohmann, H., Engau, I., Oertel, W. H., & Möller, J. C. (2011). Pain sensitivity and clinical progression in Parkinson's disease. *Movement Disorders, 26*, 2220–2225.
Mylius, V., Engau, I., Teepker, M., Stiasny-Kolster, K., Schepelmann, K., Oertel, W. H., et al. (2009). Pain sensitivity and descending inhibition of pain in Parkinson's disease. *Journal of Neurology, Neurosurgery, and Psychiatry, 80*, 24–28.
Nebe, A., & Ebersbach, G. (2009). Pain intensity on and off levodopa in patients with Parkinson's disease. *Movement Disorders, 24*, 1233–1237.
Nègre-Pagès, L., Regragui, W., Bouhassira, D., Grandjean, H., Rascol, O., & DoPaMiP Study Group. (2008). Chronic pain in Parkinson's disease: The cross-sectional French DoPaMiP survey. *Movement Disorders, 23*, 1361–1369.
Nolano, M., Provitera, V., Estraneo, A., Selim, M. M., Caporaso, G., Stancanelli, A., et al. (2008). Sensory deficit in Parkinson's disease: Evidence of a cutaneous denervation. *Brain, 131*, 1903–1911.
Oakley, J. C., & Prager, J. P. (2002). Spinal cord stimulation: Mechanisms of action. *Spine (Phila Pa 1976), 27*, 2574–2583.
Ogino, Y., Nemoto, H., Inui, K., Saito, S., Kakigi, R., & Goto, F. (2007). Inner experience of pain: Imagination of pain while viewing images showing painful events forms subjective pain representation in human brain. *Cerebral Cortex, 17*, 1139–1146.
Oshima, H., Katayama, Y., Morishita, T., Sumi, K., Otaka, T., Kobayashi, K., et al. (2012). Subthalamic nucleus stimulation for attenuation of pain related to Parkinson disease. *Journal of Neurosurgery, 116*, 99–106.
Ozturk, E. A., Gundogdu, I., Kocer, B., Comoglu, S., & Cakci, A. (2016). Chronic pain in Parkinson's disease: Frequency, characteristics, independent factors, and relationship with health-related quality of life. *Journal of Back and Musculoskeletal Rehabilitation*. Epub ahead of print.
Parkinson, J. (2002). An essay on the shaking palsy. 1817. *The Journal of Neuropsychiatry and Clinical Neurosciences, 14*, 223–236.
Pellaprat, J., Ory-Magne, F., Canivet, C., Simonetta-Moreau, M., Lotterie, J. A., Radji, F., et al. (2014). Deep brain stimulation of the subthalamic nucleus improves pain in Parkinson's disease. *Parkinsonism & Related Disorders, 20*, 662–664.
Perez-Lloret, S., Rey, M. V., Dellapina, E., Pellaprat, J., Brefel-Courbon, C., & Rascol, O. (2012). Emerging analgesic drugs for Parkinson's disease. *Expert Opinion on Emerging Drugs, 17*, 157–171.

Perrotta, A., Sandrini, G., Serrao, M., Buscone, S., Tassorelli, C., Tinazzi, M., et al. (2011). Facilitated temporal summation of pain at spinal level in Parkinson's disease. *Movement Disorders*, *26*, 442–448.

Pertovaara, A. (2006). Noradrenergic pain modulation. *Progress in Neurobiology*, *80*, 53–83.

Petzinger, G. M., Fisher, B. E., McEwen, S., Beeler, J. A., Walsh, J. P., & Jakowec, M. W. (2013). Exercise-enhanced neuroplasticity targeting motor and cognitive circuitry in Parkinson's disease. *Lancet Neurology*, *12*(7), 716–726.

Polli, A., Weis, L., Biundo, R., Thacker, M., Turolla, A., Koutsikos, K., et al. (2016). Anatomical and functional correlates of persistent pain in Parkinson's disease. *Movement Disorders*, *31*, 1854–1864.

Priganc, V. W., & Stralka, S. W. (2011). Graded motor imagery. *Journal of Hand Therapy*, *24*(2), 164–168.

Rajendran, P. R., Thompson, R. E., & Reich, S. G. (2001). The use of alternative therapies by patients with Parkinson's disease. *Neurology*, *57*, 790–794.

Rascol, O., Zesiewicz, T., Chaudhuri, K. R., Asgharnejad, M., Surmann, E., Dohin, E., et al. (2016). A randomized controlled exploratory pilot study to evaluate the effect of rotigotine transdermal patch on Parkinson's disease-associated chronic pain. *Journal of Clinical Pharmacology*, *56*, 852–861.

Reichling, D. B., & Levine, J. D. (2011). Pain and death: Neurodegenerative disease mechanisms in the nociceptor. *Annals of Neurology*, *69*, 13–21.

Reuter, I., Mehnert, S., Leone, P., Kaps, M., Oechsner, M., & Engelhardt, M. (2011). Effects of a flexibility and relaxation programme, walking, and nordic walking on Parkinson's disease. *Journal of Aging Research*, *2011*, 232473.

Rodrigues de Paula, F., Teixeira-Salmela, L. F., Coelho de Morais Faria, C. D., Rocha de Brito, P., & Cardoso, F. (2006). Impact of an exercise program on physical, emotional, and social aspects of quality of life of individuals with Parkinson's disease. *Movement Disorders*, *21*, 1073–1077.

Scherder, E., Wolters, E., Polman, C., Sergeant, J., & Swaab, D. (2005). Pain in Parkinson's disease and multiple sclerosis: Its relation to the medial and lateral pain systems. *Neuroscience and Biobehavioral Reviews*, *29*, 1047–1056.

Schestatsky, P., Kumru, H., Valls-Solé, J., Valldeoriola, F., Marti, M. J., Tolosa, E., et al. (2007). Neurophysiologic study of central pain in patients with Parkinson disease. *Neurology*, *69*, 2162–2169.

Schuhfried, O., Crevenna, R., Fialka-Moser, V., & Paternostro-Sluga, T. (2012). Non-invasive neuromuscular electrical stimulation in patients with central nervous system lesions: An educational review. *Journal of Rehabilitation Medicine*, *44*, 99–105.

Shulman, L. M., Wen, X., Weiner, W. J., Bateman, D., Minagar, A., Duncan, R., et al. (2002). Acupuncture therapy for the symptoms of Parkinson's disease. *Movement Disorders*, *17*, 799–802.

Sim, W. S. (2011). Application of botulinum toxin in pain management. *The Korean Journal of Pain*, *24*, 1–6.

Simmonds, N., Miller, P., & Gemmell, H. (2012). A theoretical framework for the role of fascia in manual therapy. *Journal of Bodywork and Movement Therapies*, *16*, 83–93.

Sluka, K. A. (2016). *Mechanisms and management of pain for the physical therapist.* Seattle, USA: IASP Press.

Smania, N., Corato, E., Tinazzi, M., Stanzani, C., Fiaschi, A., Girardi, P., et al. (2010). Effect of balance training on postural instability in patients with idiopathic Parkinson's disease. *Neurorehabilitation and Neural Repair*, *24*, 826–834.

Spielberger, S., Wolf, E., Kress, M., Seppi, K., & Poewe, W. (2011). The influence of deep brain stimulation on pain perception in Parkinson's disease. *Movement Disorders*, *26*, 1367–1368.

Stacy, M. (2010). The wearing-off phenomenon and the use of questionnaires to facilitate its recognition in Parkinson's disease. *Journal of Neural Transmission*, *117*, 837–846.

Stewart, W. F., Ricci, J. A., Chee, E., Morganstein, D., & Lipton, R. (2003). Lost productive time and cost due to common pain conditions in the US workforce. *JAMA*, *290*, 2443–2454.

Sürücü, O., Baumann-Vogel, H., Uhl, M., Imbach, L. L., & Baumann, C. R. (2013). Subthalamic deep brain stimulation versus best medical therapy for L-dopa responsive pain in Parkinson's disease. *Pain*, *154*, 1477–1479.

Tassorelli, C., De Icco, R., Alfonsi, E., Bartolo, M., Serrao, M., Avenali, M., et al. (2014). Botulinum toxin type A potentiates the effect of neuromotor rehabilitation of Pisa syndrome in Parkinson disease: A placebo controlled study. *Parkinsonism & Related Disorders*, *20*(11), 1140–1144.

Tinazzi, M., Del Vesco, C., Defazio, G., Fincati, E., Smania, N., Moretto, G., et al. (2008). Abnormal processing of the nociceptive input in Parkinson's disease: A study with CO2 laser evoked potentials. *Pain*, *136*, 117–124.

Tinazzi, M., Del Vesco, C., Fincati, E., Ottaviani, S., Smania, N., Moretto, G., et al. (2006). Pain and motor complications in Parkinson's disease. *Journal of Neurology, Neurosurgery, and Psychiatry*, *77*, 822–825.

Tinazzi, M., Recchia, S., Simonetto, S., Defazio, G., Tamburin, S., Moretto, G., et al. (2009). Hyperalgesia and laser evoked potentials alterations in hemiparkinson: Evidence for an abnormal nociceptive processing. *Journal of the Neurological Sciences*, *276*, 153–158.

Tinazzi, M., Recchia, S., Simonetto, S., Tamburin, S., Defazio, G., Fiaschi, A., et al. (2010). Muscular pain in Parkinson's disease and nociceptive processing assessed with CO2 laser-evoked potentials. *Movement Disorders*, *25*(2), 213–220.

Tomlinson, C. L., Patel, S., Meek, C., Herd, C. P., Clarke, C. E., Stowe, R., et al. (2012). Physiotherapy versus placebo or no intervention in Parkinson's disease. *The Cochrane Database of Systematic Reviews*, *8*, CD002817.

Trenkwalder, C., Chaudhuri, K. R., Martinez-Martin, P., Rascol, O., Ehret, R., Vališ, M., et al. (2015). Prolonged-release oxycodone-naloxone for treatment of severe pain in patients with Parkinson's disease (PANDA): A double-blind, randomised, placebo-controlled trial. *Lancet Neurology*, *14*, 1161–1170.

Trenkwalder, C., Kies, B., Rudzinska, M., et al. (2011). Rotigotine effects on early morning motor function and sleep in Parkinson's disease: A double-blind, randomized, placebo-controlled study (RECOVER). *Movement Disorders*, *26*, 90–99.

Vadalà, M., Vallelunga, A., Palmieri, L., Palmieri, B., Morales-Medina, J. C., & Iannitti, T. (2015). Mechanisms and therapeutic applications of electromagnetic therapy in Parkinson's disease. *Behavioral and Brain Functions*, *11*, 26.

Valeriani, M., Rambaud, L., & Mauguière, F. (1996). Scalp topography and dipolar source modelling of potentials evoked by CO2 laser stimulation of the hand. *Electroencephalography and Clinical Neurophysiology*, *100*, 343–353.

Valkovic, P., Minar, M., Singliarova, H., Harsany, J., Hanakova, M., Martinkova, J., et al. (2015). Pain in Parkinson's disease: A cross-sectional study of its prevalence, types, and relationship to depression and quality of life. *PLoS One*, *10*, e0136541.

van Middelkoop, M., Rubinstein, S. M., Kuijpers, T., Verhagen, A. P., Ostelo, R., Koes, B. W., et al. (2011). A systematic review on the effectiveness of physical and rehabilitation interventions for chronic non-specific low back pain. *European Spine Journal*, *20*, 19–39.

Vigotsky, A. D., & Bruhns, R. P. (2015). The role of descending modulation in manual therapy and its analgesic implications: A narrative review. *Pain Research Treatment*, *2015*, 292805.

Voisin, D. L., Guy, N., Chalus, M., & Dallel, R. (2005). Nociceptive stimulation activates locus coeruleus neurones projecting to the somatosensory thalamus in the rat. *The Journal of Physiology*, *566*, 929–937.

Williams, A. C., & Craig, K. D. (2016). Updating the definition of pain. *Pain*, *157*, 2420–2423.

Willis, W. D., & Westlund, K. N. (1997). Neuroanatomy of the pain system and of the pathways that modulate pain. *Journal of Clinical Neurophysiology*, *14*, 2–31.

Witjas, T., Kaphan, E., Régis, J., Jouve, E., Chérif, A. A., Péragut, J. C., et al. (2007). Effects of chronic subthalamic stimulation on nonmotor fluctuations in Parkinson's disease. *Movement Disorders*, *22*, 1729–1734.

Wolz, M., Hauschild, J., Koy, J., Fauser, M., Klingelhöfer, L., Schackert, G., et al. (2012). Immediate effects of deep brain stimulation of the subthalamic nucleus on nonmotor symptoms in Parkinson's disease. *Parkinsonism & Related Disorders*, *18*, 994–997.

Wu, W. T., Hong, C. Z., & Chou, L. W. (2015). The Kinesio taping method for myofascial pain control. *Evidence-Based Complementary and Alternative Medicine*, *2015*, 950519.

Zambito-Marsala, S., Erro, R., Bacchin, R., Fornasier, A., Fabris, F., Lo Cascio, C., et al. (2017). Abnormal nociceptive processing occurs centrally and not peripherally in pain-free Parkinson disease patients: A study with laser-evoked potentials. *Parkinsonism & Related Disorders*, *34*, 43–48.

Zambito Marsala, S., Tinazzi, M., Vitaliani, R., Recchia, S., Fabris, F., Marchini, C., et al. (2011). Spontaneous pain, pain threshold, and pain tolerance in Parkinson's disease. *Journal of Neurology*, *258*, 627–633.

FURTHER READING

Gatchel, R. J., McGeary, D. D., McGeary, C. A., & Lippe, B. (2014). Interdisciplinary chronic pain management: Past, present, and future. *The American Psychologist*, *69*, 119–130.

CHAPTER TWENTY-NINE

Sexual Dysfunctions in Parkinson's Disease: An Underrated Problem in a Much Discussed Disorder

Kalyan B. Bhattacharyya[*,1], **Miguel Rosa-Grilo**[†,‡,§,1]
[*]RG Kar Medical College & Hospital, Kolkata, India
[†]National Parkinson Foundation (NPF) International Center of Excellence at King's College Hospital, London, United Kingdom
[‡]The Maurice Wohl Clinical Neuroscience Institute, King's College London, London, United Kingdom
[§]National Institute for Health Research (NIHR) Mental Health Biomedical Research Centre, Institute of Psychology, Psychiatry and Neurosciences, King's College London, London, United Kingdom
[1]Corresponding authors: e-mail address: kalyanbrb@gmail.com; miguel.grilo@kcl.ac.uk

Contents

Abstract

Sexual dysfunctions (SDs) are one of the most neglected nonmotor symptoms in Parkinson's disease (PD). A number of reasons including social and cultural factors might explain, at least partially, why SD is still one of the most underrecognized aspects of the condition after 200 years since the very first description by James Parkinson. SD has not been extensively investigated, however, a number of studies have shown a high prevalence of decreased libido, orgasmic dysfunction in both men and women with PD, and erectile dysfunction in male subjects. Moreover, SD in PD also comprises the increasingly recognized hypersexuality that is often associated with PD treatment. Taken together, SD in PD includes a remarkable range of symptoms and conditions that often require a multidisciplinary approach regarding assessment, investigation, and treatment.

Parkinson's disease (PD) is the most often discussed movement disorder. Ever since James Parkinson published his classic treatise, *An Essay on the Shaking Palsy* in 1817, numerous papers, articles, and monographs have appeared in the last two centuries. Most of the early accounts seemed to send

International Review of Neurobiology, Volume 134
ISSN 0074-7742
http://dx.doi.org/10.1016/bs.irn.2017.05.019

one singular message that PD is a disease of the motor system, where tremor, rigidity, bradykinesia, postural instability, loss of arm swing, stooped and short shuffling gait, hypomimia, and hypophonia are all that constitute the entity. Incidentally, James Parkinson drew his conclusions from only six cases and it is likely that the nonmotor symptoms (NMSs) were submerged within the overwhelmingly florid and visible motor signs and he felt that ... *the senses and intellect being unimpaired.* However, he alluded to some NMS, albeit in a passing manner, and concluded the clinical account with the statement that .... *and at last, constant sleepiness, with slight delirium and other marks of exhaustion, announce the wished-for relief* (Bhattacharyya, 2011; Parkinson, 1817). Later works, particularly in the last 50 years, have incontrovertibly proved that the NMS are frequent, troublesome, and ubiquitous, and some authors are now beginning to question whether the disease should be renamed as *Parkinson's disease syndrome* or *Parkinson's disease complex* (Titova, Padmakumar, Lewis, & Chaudhuri, 2016).

Sexual dysfunctions (SDs) are one of the most neglected, underreported, and underrecognized aspects of PD. The reasons are manifold. The majority of the patients feel discomfited and embarrassed to broach the issue before the neurologist during consultation. Moreover, many feel that the declining sexual urge and performance is a concomitant inevitable of old age, while others feel it is inconsequential in the face of the distressing motor and other NMS and tend to ignore the fact that healthy sexual relationship is one of the key aspects for a happy and contented life and normal social dealings (Koller et al., 1990; Lipe, Longstreth, Bird, & Linde, 1990; McInnes, 2003). It is equally true that neurologists also seem to be somewhat insensitive and less attentive toward NMS. In a landmark article in *Lancet Neurology* in 2006, Chaudhuri, Healy and Schapira observed that in a prospective study involving 101 patients, neurologists did not discuss these symptoms with the patients in more than 50% of the cases and they attributed the reason to limited consultation time in their busy clinic, attitude of the patients, and their caregivers, or even the lack of awareness on their part that these symptoms can be the most distressing ones, ahead of the more obvious and manifest motor problems (Chaudhuri, Healy, & Schapira, 2006). It is known that sexual urge, activity, and frequency wane with age, illnesses, and psychosocial factors. One study reported that decline in sexual desire was observed in about 35% of normal women aged between 18 and 70 years, while erectile dysfunction was a feature in 52% of male subjects aged between 40 and 70 years. Furthermore, premature ejaculation was observed in 30% of subjects in the age range of 18–59 years (Fugl-Meyer & Fugl-Meyer, 1999;

Laumann, Paik, & Rosen, 1999; Lindau et al., 2007; Nusbaum, Hamilton, & Lenahan, 2003). However, impaired sexual function in PD seems to be more prevalent in male and female subjects with PD; a range of SDs including decreased libido and erectile dysfunction can occur in up to 80% of male PD patients, and in women with PD, decreased libido and difficulties in reaching orgasm can reach 84% and 75%, respectively (Bronner, Royter, Korczyn, & Giladi, 2004; Kummer, Cardoso, & Teixeira, 2009; Sakakibara et al., 2001; Waters & Smolowitz, 2013). In Table 1, different types of SD and their prevalence are shown in more detail.

The important causes of SD in PD include motor problems like rigidity, bradykinesia, slowness of movement, difficulty in turning in bed, emotional instability, cognitive deficits, sleep problems, and side effects of antiparkinsonian drugs (Bronner et al., 2004; Schrag, Jahanshahi, & Quinn, 2000; Welsh, Hung, & Waters, 1997). Features like drooling, excessive sweating and seborrhea, and hypomimia often make patients feel self-conscious, and sleep disturbances, such as rapid eye movement disorder, periodic leg movement of sleep and restless leg syndrome, may force the couple to sleep in separate beds or even rooms (Basson, 2001; Wermuth & Stenager, 1992). Depression in PD is considered to be another contributing factor (Giladi et al., 2000). Bronner et al. reported that patients refrain from approaching their partners sexually due to the fear of rejection, while their spouse feel that they are being ignored (Bronner, Royter, Korczyn, & Giladi, 2003). Problems of sexual arousal, behavior, and orgasm were common in young males with greater duration of the disease and some workers reported reduced frequency of intercourse, erectile dysfunction, and difficulty in reaching orgasm in men with PD (Jacobs, Vieregge, & Vieregge, 2000; Lucon et al., 2001; Yu, Roane, Miner, Fleming, & Rogers, 2004). In an important retrospective study over 14 years on 32,616 men with erectile dysfunction aged between 40 and 70 years, Gao et al. reported that there was a 3.8 times higher incidence of developing PD in later years (Gao et al., 2007). Berman et al. reported that among women the most frequently observed SDs encountered were decline in desire, arousal, vaginal dryness resulting in dyspareunia, vaginismus, and diminished libido (Berman, Adhikari, & Goldstein, 2000). It has been seen that women fail to reach climax following abrupt decline in desire and these problems increase with the progression of the disease. Furthermore, Fischer et al. reported that PD in the male subjects adversely affected desire, arousal, and orgasm in the female partner and improvement in their sexual function was associated with better sexual gratification in the female partner (Basson, 1996;

Table 1 Sexual Dysfunction in Parkinson's Disease

	Symptoms	Definition	Prevalence in PD	Studies
Women	Decreased libido; deficiency or absence of desire for sexual activity	Female hypoactive sexual desire dysfunction[a]	46.9%–84%	Bronner et al. (2004), Sakakibara et al. (2001), and Waters and Smolowitz (2013)
	Difficulties in reaching orgasm	Female orgasmic dysfunction[a]	75%	Bronner et al. (2004)
	Increased libido and repetitive behaviors toward sexual gratification outside the accepted social and personal bounds	Hypersexuality	0.5%	Weintraub et al. (2010)
Men	Decreased libido; deficiency or absence of desire for sexual activity	Male hypoactive sexual desire disorder[a]	83%	Sakakibara et al. (2001)
	Decrease in penile erection; inability to attain or maintain penile erection sufficient for sexual intercourse	Erectile dysfunction[a]	42.6%–79%	Bronner et al. (2004), Kummer et al. (2009), and Sakakibara et al. (2001)
	Significant decrease in latency time to ejaculation	Premature ejaculation[a]	40.6%	Bronner et al. (2004)
	Decrease in ejaculation; absence of normal ejaculation	Anejaculation[a]	79%	Sakakibara et al. (2001)
	Decrease in orgasm; inability to reach orgasm	Anorgasmia[a]	87%	Sakakibara et al. (2001)
	Increased libido and repetitive behaviors toward sexual gratification outside the accepted social and personal bounds	Hypersexuality	5.2%	Weintraub et al. (2010)

[a]Consensus Statement from the Fourth International Consultation on Sexual Medicine (McCabe et al., 2016).

Fisher, Rosen, Eardley, Sand, & Goldstein, 2005). The Derogatis Interview for Sexual Functioning Self Report scale showed reduction in five domains of sexual gratification in the female partners of PD patients namely, arousal, behavior, orgasm, fantasy, and drive, while Koller et al. used a semistructured questionnaire and reported autonomic dysfunction as an important cause of erectile dysfunction in PD (Brown, Jahanshahi, Quinn, & Marsden, 1990; Koller et al., 1990).

1. PARKINSON'S DISEASE AND HYPERSEXUALITY

The central dopaminergic pathways play an important role in sexual functions in men, including the augmentation or the decline in libido and erectile activity and levodopa can alter the latency of ejaculation and orgasm (Ahlenius, Engel, Larsson, & Svensson, 1982; Atagün et al., 2013). In two controlled studies, it has been shown that among the dopaminergic agents, bromocriptine and pergolide decrease latency of ejaculation and induce spontaneous ejaculation, respectively, while apomorphine leads to penile erection in the rat model (Ahlenius et al., 1982; Mizusawa et al., 2002). Antidepressants are frequently prescribed in PD patients since depression is a common NMS, and it has been observed that selective serotonin reuptake inhibitors (SSRIs) can lead to SD, while tricyclic antidepressants like, amitriptyline or nortriptyline and SSRIs delay ejaculation with high dosages. Atypical neuroleptics like, risperidone and clozapine, block dopamine and serotonin receptors and also cause problems like delayed ejaculation (Csoka, Bahrick, & Mehtonen, 2008; Montejo, Llorca, Izquierdo, & Rico-Villademoros, 2001).

Since the mid-1990s, a curious complication following dopaminergic drug therapy has been described. This has been referred as *dopamine dysregulation syndrome* (DDS), *dopamine drug addiction*, *compulsive dopaminergic drug use*, *repetitive behavior* and *hedonistic homeostatic dysregulation* (Cantello, 2000; Evans, Lawrence, Potts, Appel, & Lees, 2005; Giovannoni, O'Sullivan, Turner, Manson, & Lees, 2000; Kurlan, 2004; McKeon et al., 2007; Serrano-Dueñas, 2002; Weintraub, 2008). It is to be noted that Quinn et al. and Vogel et al. alluded to this problem as early as in 1983, about 15 years after the introduction of levodopa in clinical practice (Quinn, Toone, Lang, Marsden, & Parkes, 1983; Vogel & Schiffter, 1983). Its incidence is in the range of 3%–5% and it is characterized by compulsive behavioral patterns like pathological and internet gambling, constant fiddling with objects and dismantling and reorganizing them, albeit mostly unsuccessfully, drug hoarding, and hypersexuality (HS), often collectively known as *punding*, and a number of terms like *sexual addiction*, *sexual compulsivity*, *sexual impulsivity*, and *increased interest in sex* are assigned to it (Gallagher, 2010). Weintraub et al. named this entire gamut of behavioral abnormalities as *impulse control disorder* (ICD) and the risk factors include male sex, younger age at onset, personal and family history of drug abuse, bipolar disorders, traits of gambling, and impulsive personality, while Lawrence et al. and Djamshidian et al. observed similar personality traits like, impulsive behavior, addictions, tendency toward jumping into conclusions, and impaired decision making as the premorbid features of this condition (Djamshidian et al., 2012; Lawrence, Evans, & Lees, 2003; O'Sullivan, Evans, & Lees, 2009; Weintraub & Nirenberg, 2013). Solla et al. found that higher incidence of motor complications was an important contributory factor for this disorder and suggested a possible direct effect of levodopa therapy (Solla et al., 2011). Codling et al. suggested that ICD may be treated as an example of disinhibitory psychopathology where there is a failure of executive function to inhibit reinforcing responses (Codling, Shaw, & David, 2015). In a further detailed multicentric study on 3096 treated patients, involving 46 movement disorders centers all over Europe and the United States, Weintraub et al. applied the Minnesota Impulsive Disorders Interview score for compulsive sexual behavior and buying and identified that 3.5% of such subjects suffered from HS and related problems and there was a 2–3.5-fold increase in the development of ICDs, particularly in those who were treated with dopaminergic agents. Concurrent use of levodopa, younger age, being unmarried, history of cigarette smoking, and a family history of gambling habit were additional risk factors (Weintraub et al., 2010). Features like, increased frequency of erection, higher libido, compulsive masturbation, and sexually demanding behavior characterize the syndrome and HS has been defined as an ICD, along with problems like gambling, compulsive shopping, compulsive eating, and compulsive hobbyism and punding or repetitive goal-less task (Uitti et al., 1989). Some patients may develop intense feeling of jealousy and suicidal ideation, while paraphilia, or repetitive and intense sexual urge in socially inappropriate settings, calling for legal interventions, may occasionally be observed (Bronner et al., 2003; Harvey, 1988; Riley, 2002). Aberrations in the form of qualitative alterations in sexuality like, transvestism, or gender identity disorders have also been reported (Courty, Durif, Zenut, Courty, & Lavarenne, 1997; Odiyoor, Kobylecki, Hackett, Silverdale, & Kellett, 2009; Sohtaoglu, Demiray, Kenangil, Ozekmekci, & Erginoz, 2010). Interestingly, it is seen that though most of the components of ICD disappear with discontinuation of dopaminergic therapy, HS tend to persist for a longer time (Cools, 2006; Evans & Lees, 2004). In a study on 129 PD patients in Kolkata, India, using the Nonmotor Symptom Assessment Scale for Parkinson's disease, Bhattacharyya et al. observed that SD was rare and those were mostly of the nature of hyposexuality and issues such as loss of libido and declining interest in sex were the predominant complaints. Only one patient had history of surreptitious intake of levodopa and the resultant development of HS. The reasons for lesser incidence of reported hyposexuality can be attributed to a general feeling that old age has robbed them of sexual conditioning and performance as a natural inevitability, cultural inhibitions as prevalent in most of the developing countries, like an utmost inhibition for broaching such intensely private issues in spite of the fact that they are victims of deprivation of a natural and pleasurable physiological phenomenon, and a certain feeling of resignation to fate. The lower incidence of HS may could be attributed to the low dose of levodopa or dopamine agonists prescribed, since PD patients in India seem to be intolerant to high dose (<600 mg daily being the usually highest prescribed dosage in general), since orthostatic hypotension, and early development of fluctuations and dyskinesia are the most troublesome complications reported in this country (Bhattacharyya et al., in press).

ICD and HS are considered as manifestations of frontal lobe inhibition, and disruption in the reciprocal loops between prefrontal cortex and the striatum following dopamine deficiency is implicated in the genesis of this condition (Harvey, 1988; Klos, Bower, Josephs, Matsumoto, & Ahlskog, 2005; Lawrence et al., 2003; Uitti et al., 1989; Weintraub & Nirenberg, 2013). It is postulated that primary degeneration of the reward system and secondary changes following pulsatile stimulation of the dopaminergic system by levodopa and other related drugs, particularly administered with early onset of the disease, are responsible for the development of this syndrome, while denervation supersensitivity leading to an augmented excitatory activity of levodopa and dopaminergic drugs has been proposed as an alternative hypothesis (Langstrom & Hanson, 2006; Shapiro, Chang, Munson, Okun, & Fernandez, 2006; Simonet, Fernandez, Cerdan, & Duarte, 2016). Otherwise, compulsive use of levodopa may be due to progressive neuroadaptations in the dopaminergic projections in the circuitry involving the nucleus accumbens which lead to excessive release of dopamine from the ventral tier or the pars compactus in the substantia nigra, described by Baxter and Olszewski in 1954 (Lawrence et al., 2003; Szarfman, Doraiswamy,

Tonning, & Levine, 2006; Vitale et al., 2013). This sort of habituation following long-term usage of levodopa or the related agents and the consequent feeling of pleasure or hedonism forces the individual to increase the dose further. Bearn et al. in a study on 10 PD patients taking levodopa in the prescribed dose and an equal number of patients who had been consuming extra amount observed that in the latter group the majority of subjects had been experiencing unpleasant off period with dysphoria and consequently, were forced to revert to the higher dose of the drug, in spite of experiencing troublesome dyskinesia (Bearn, Evans, Kelleher, Turner, & Lees, 2004). The authors concluded that some sort of maladaptive dependence on levodopa, with involvement of the dopaminergic pathways was the reason behind such craving for levodopa. This promotes damage to the dopamine receptors in the striatum, which impairs goal-directed mental functions and enhancement of sensitization to dopaminergic therapy (Evans et al., 2004; Merims & Giladi, 2008; Solla, Bortolato, Cannas, Mulas, & Marrosu, 2015; Vitale et al., 2011).

The precise mechanism of HS in PD is not clearly known (Codling et al., 2015). The aberrant cognitive profile is attributed primarily to a deficit in impulse control with inability to learn from mistakes and failures, and poor performance on memory test (Poletti & Bonuccelli, 2013). Recent works implicate the issue of immediate gratifying behavior, in spite of the knowledge of potential adverse consequences. Bechara et al. have developed a system to assess the preference of choosing cards from a pack which have the apparent potential of providing immediate financial benefit, and another set which might incur occasional loss but bring about intermittent small gains, though the results have been nonconvincing (Bechara, Damasio, Tranel, & Damasio, 1997). Using 11C–raclopride PET scan in 18 patients with ICD and HS, O′Sullivan et al. documented greater reduction in raclopride-binding potential in the ventral striatum of substantia nigra following levodopa challenge test and attributed this finding to higher response of the striatal reward circuitry (O'Sullivan et al., 2011).

2. NEUROBIOLOGY OF HS IN PD

There is an ongoing debate whether HS in PD is a behavioral problem, an example of ICD, or a variant of obsessive–compulsive disorder (Harvey, 1988; Kor, Fogel, Reid, & Potenza, 2013). Animal studies have clearly shown that the concentration of dopamine in the nucleus accumbens increases at the time of orgasm and dwindles rapidly during the refractory phase, which mimics the euphoric state that follows administration of opiates after a period of abstinence and this strongly suggests an addictive nature of HS. Features like, craving, incessant desire for finding source of fun and enjoyment with scant attention to adverse consequences, diminished self-control over ones desire and compulsive performance strongly mimic addictive behavior (Blackburn, Pfaus, & Phillips, 1992; Kor et al., 2013; Leeman & Potenza, 2012; Weintraub & Nirenberg, 2013). The importance of mesolimbic reward system, striatum and ventral medial prefrontal cortex following indulgence in sex, addiction, and pathological gambling has been established using functional brain imaging studies and the ratio of the D3 and other dopaminergic receptors in increased in the mesolimbic system in HS following levodopa therapy (Bechara et al., 1997; Georgiadis, Kringelbach, & Pfaus, 2012; Probst & van Eimeren, 2013). As a matter of fact, it is of interest to note the one of the distinguishing features between dopaminergic drugs and levodopa is the affinity for D3 receptors and this perhaps explains to a large extent, the higher risk of HS and ICD following treatment with agonists (Harvey, 1988; Voon, Mehta, & Hallett, 2011).

The alternative hypothesis concerning the mechanism of HS centers on the concept of loss of inhibition. Bancroft et al. and Fiorino et al. advanced a novel theory of duality of sexual excitation and inhibition. It presupposes that in the face of other challenging tasks, the inhibitory mechanism takes over and leads to waning of excitation, while breakdown of this inbuilt inhibition leads to reckless escapade and prurient sexual misadventures (Bancroft, Graham, Janssen, & Sanders, 2009; Fiorino & Phillips, 1999). Some workers have suggested that HS in general population is related to general impulsiveness and since impulsivity is centered around the noradrenergic connections in the right inferior frontal gyrus, prefrontal lobe lesions often lead to hypersexual behavior along with a medley of executive dysfunction like, set shifting, planning, and loss of emotional control (Rees, Fowler, & Maas, 2007; Rogers et al., 1999).

3. ASSESSMENT AND INVESTIGATION OF SDs IN PARKINSON'S DISEASE

Addressing SD should be made taking into consideration cultural and social aspects of the group or community the patient belong to. Moreover, involvement of the patient's partner should be considered, and expectations of the couple toward their sexual life investigated. In PD, no validated instrument exists which can incontrovertibly diagnose SD or ICD in PD (Weintraub et al., 2010). Minnesota Impulsive Disorders Interview measures some of the items, including sexual behavior, but is not a full-proof test (Weintraub et al., 2010). Once SD is diagnosed, the first important step is to recognize whether they belong to the category of lack of arousal, diminished libido, erectile dysfunction, vaginismus, or the socially unacceptable HS or ICD, and plan investigations accordingly. The onset and progression of the problem should be identified, the nature of the illness in terms of type and time of sexual activity should be assessed and sexual history before the onset of PD ascertained. It is always important to consider comorbid illnesses like cardiovascular disease, diabetes mellitus, hyperlipidemia, hypertension, genitourinary disorders, depression, anxiety, etc., and personal habits such as smoking, intake of tea or coffee, and alcohol or any substance abuse. Diagnostic tests range from routine blood tests with glucose and lipid profile, and hormonal profile, to more extensive neurophysiologic and vascular investigations. Further testing should be made after multidisciplinary discussion and can include electromyography, nerve conduction studies, thermal and vibratory threshold, bulbocavernous latency, penile plethysmography, Doppler ultrasound, penile arteriography, nocturnal penile tumescence study, pharmacocavernosography, and among others.

4. MANAGEMENT OF SDs IN PARKINSON'S DISEASE

Obviously, a multidimensional approach to SD must be sought in PD, however, this not mean that the neurologist should not be actively involved in identifying risk factors and discussing therapeutic plans. Once appropriate investigations have been carried out, the treatment must be individualized. In general, erectile dysfunction responds well to five cyclic guanosine monophosphate phosphodiesterase inhibitors (PDE5 inhibitors) like sildenafil, or the related drugs tadalafil or vardenafil. The latter have been studied in patients with SD secondary to other neurological conditions, however sildenafil has been investigated in PD and has shown significant improvements (Giuliano et al., 2006; Hussain, Brady, Swinn, Mathias, & Fowler, 2001; Raffaele et al., 2002). Monitoring of orthostatic hypotension is recommended when prescribing PDE5 inhibitors. These drugs augment the nitric oxide-mediated relaxation pathway in the corpora cavernosum and spongiosum and thereby increases the availability of cyclic guanosine monophosphate in the related tissues. Thus, these agents do not increase the arousal; they only help to maintain the erection (Hussain et al., 2001;

Montorsi et al., 2004; Raffaele et al., 2002; Voon, Sohr, et al., 2011; Zesiewicz, Helal, & Hauser, 2000). Sublingual apomorphine, intracavernosal injection of papaverine or prostaglandin E1, intraurethral application of vasodilators like, alprostadil, vacuum constrictor pump, and penile prosthesis have been suggested when oral drugs fail (Bronner & Vodusek, 2011). Antidepressants like SSRIs delay ejaculation, while psychotherapy and counseling may be required for some patients. For example, increasing open sexual communication, practicing comfortable positions, training in intimacy, emphasis on foreplay, etc., are worth trying in refractory cases (Bronner & Vodusek, 2011).

The management of HS in PD is rather complicated. The first step is to reduce the dose of levodopa or dopamine agonists and it is seen that even substituting the dopamine agonist with another agent may be of help in some cases (Driver-Dunckley, Samanta, & Stacy, 2003; *Non-motor symptoms of Parkinson's disease*, 2009). Alternatively, considering dopamine agonists with a different dopamine receptor profile might be entertained as well. However, a number of studies have indicated that these patients may develop troublesome dopamine agonist withdrawal syndrome. The condition can include a myriad of symptoms, such as severe anxiety, panic, social phobia, agoraphobia, dysphoria, depression, pain, diaphoresis, orthostatic hypotension, drug cravings, and suicidal ideation and these do not abate with replacement of levodopa or dopamine agonists (Cunnington, White, & Hood, 2012; Pondal et al., 2013; Rabinak & Nirenberg, 2010). The relationship between deep brains stimulation and ICD is complex and though the problem wanes in intensity in some patients after subthalamic nucleus stimulation, presumably owing to consumption of lesser amount of dopaminergic drugs, it has also been shown by others that ICD and HS may worsen or may actually begin following the procedure and some workers attributed this to the placement of the lead in the medial and lower part of the subthalamic nucleus of Luys and the resultant stimulation of the nearby hypothalamus (Ardouin et al., 2006; Mamikonyan et al., 2008; Smeding et al., 2007). One further work has shown that the subjects had more difficulty in decision making and higher impulsivity when the stimulator was turned on (Frank, Samanta, Moustafa, & Sherman, 2007). The usefulness of newer atypical antipsychotics and antidepressants like, SSRIs, quetiapine, olanzapine, or risperidone is not completely stabilized in this context, and its use might depend on previous personal clinical experience (Grant et al., 2006; Samuel et al., 2015; Sevincok, Akoglu, & Akyol, 2007).

5. CONCLUDING REMARKS

The issue of PD and SD is rather complex, and SD is an under-recognized problem in PD, be it of the nature of erectile and ejaculatory dysfunctions, problems in arousal, drive or orgasm, or of the socially more embarrassing, ICD. It is now established that dopaminergic therapy in high dose for a long time or DBS surgery is associated with higher incidence of errant sexual misadventures and the consequent social and legal fallout. Attention to the details of the patient and the partner in terms of hyposexuality, counseling, sex therapy, behavioral therapy, sex education, and change in attitude shall go a long way toward improving the problem and drugs which help to maintain erection and delay ejaculation are always worth a trial. For treating ICD and HS, the neurologist should be careful that the therapeutic dosage of dopamimetic agents be kept at as minimal as possible and this is the first important step toward preventing this complication of disastrous import. Dopaminergic drugs must be reduced and stopped if necessary, at the first indication of ICD with HS, carefully weighing the pros and the cons, like the degree of parkinsonism and the severity of DDS, though the issue of dopamine agonist withdrawal syndrome should be kept in mind. Management of sexual problems in PD is a challenging issue for the neurologist and the caregiver alike.

REFERENCES

Ahlenius, S., Engel, J., Larsson, K., & Svensson, L. (1982). Effects of pergolide and bromocriptine on male rat sexual behavior. *Journal of Neural Transmission*, *54*(3–4), 165–170.

Ardouin, C., Voon, V., Worbe, Y., Abouazar, N., Czernecki, V., Hosseini, H., et al. (2006). Pathological gambling in Parkinson's disease improves on chronic subthalamic nucleus stimulation. *Movement Disorders*, *21*(11), 1941–1946. http://dx.doi.org/10.1002/mds.21098.

Atagün, M., Atmaca, B., Özer, F., Özben, S., Tıraş, R., Çetin, S., et al. (2013). Determinants of sexual dysfunction in male and female patients with Parkinson's disease. *The Journal of Psychiatry and Neurological Sciences*, *26*, 28–35.

Bancroft, J., Graham, C. A., Janssen, E., & Sanders, S. A. (2009). The dual control model: Current status and future directions. *Journal of Sex Research*, *46*(2–3), 121–142. http://dx.doi.org/10.1080/00224490902747222.

Basson, R. (1996). Sexuality and Parkinson's disease. *Parkinsonism & Related Disorders*, *2*(4), 177–185.

Basson, R. (2001). Sex and idiopathic Parkinson's disease. *Advances in Neurology*, *86*, 295–300.

Bearn, J., Evans, A., Kelleher, M., Turner, K., & Lees, A. (2004). Recognition of a dopamine replacement therapy dependence syndrome in Parkinson's disease: A pilot study. *Drug and Alcohol Dependence*, *76*(3), 305–310. http://dx.doi.org/10.1016/j.drugalcdep.2004.06.005.

Bechara, A., Damasio, H., Tranel, D., & Damasio, A. R. (1997). Deciding advantageously before knowing the advantageous strategy. *Science*, *275*(5304), 1293–1295.

Berman, J. R., Adhikari, S. P., & Goldstein, I. (2000). Anatomy and physiology of female sexual function and dysfunction: Classification, evaluation and treatment options. *European Urology*, *38*(1), 20–29. doi:20247.

Bhattacharyya, K., Das, D., Bose, P., Datta, A., Mandal, B., & P., B. (in press). Sexual behaviour in a cohort of 129 cases of Parkinson's disease in a Tertiary Care Hospital in Kolkata, India. Annals of Indian Academy of Neurology.

Bhattacharyya, K. B. (2011). *Eminent neuroscientists: Their lives and works* (1st ed.). Kolkata: Academic Publishers.

Blackburn, J. R., Pfaus, J. G., & Phillips, A. G. (1992). Dopamine functions in appetitive and defensive behaviours. *Progress in Neurobiology*, *39*(3), 247–279.

Bronner, G., Royter, V., Korczyn, A. D., & Giladi, N. (2004). Sexual dysfunction in Parkinson's disease. *Journal of Sex & Marital Therapy*, *30*(2), 95–105. http://dx.doi.org/10.1080/00926230490258893.

Bronner, G., Royter, V., Korczyn, A. D., & Giladi, N. (Eds.), (2003). *Sexuality and Parkinson's disease.* Totowa, NJ: Humana Press.

Bronner, G., & Vodusek, D. B. (2011). Management of sexual dysfunction in Parkinson's disease. *Therapeutic Advances in Neurological Disorders*, *4*(6), 375–383. http://dx.doi.org/10.1177/1756285611411504.

Brown, R. G., Jahanshahi, M., Quinn, N., & Marsden, C. D. (1990). Sexual function in patients with Parkinson's disease and their partners. *Journal of Neurology, Neurosurgery, and Psychiatry*, *53*(6), 480–486.

Cantello, R. (2000). Hedonistic homeostatic dysregulation in patients with Parkinson's disease on dopamine replacement therapies. *Journal of Neurology, Neurosurgery, and Psychiatry*, *69*(4), 566–567.

Chaudhuri, K. R., Healy, D. G., & Schapira, A. H. (2006). Non-motor symptoms of Parkinson's disease: Diagnosis and management. *Lancet Neurology*, *5*(3), 235–245. http://dx.doi.org/10.1016/s1474-4422(06)70373-8.

Codling, D., Shaw, P., & David, A. S. (2015). Hypersexuality in Parkinson's disease: Systematic review and report of 7 new cases. *Movement Disorders Clinical Practice*, *2*(2), 116–126. http://dx.doi.org/10.1002/mdc3.12155.

Cools, R. (2006). Dopaminergic modulation of cognitive function-implications for L-DOPA treatment in Parkinson's disease. *Neuroscience and Biobehavioral Reviews*, *30*(1), 1–23. http://dx.doi.org/10.1016/j.neubiorev.2005.03.024.

Courty, E., Durif, F., Zenut, M., Courty, P., & Lavarenne, J. (1997). Psychiatric and sexual disorders induced by apomorphine in Parkinson's disease. *Clinical Neuropharmacology*, *20*(2), 140–147.

Csoka, A. B., Bahrick, A., & Mehtonen, O. P. (2008). Persistent sexual dysfunction after discontinuation of selective serotonin reuptake inhibitors. *The Journal of Sexual Medicine*, *5*(1), 227–233. http://dx.doi.org/10.1111/j.1743-6109.2007.00630.x.

Cunnington, A. L., White, L., & Hood, K. (2012). Identification of possible risk factors for the development of dopamine agonist withdrawal syndrome in Parkinson's disease. *Parkinsonism & Related Disorders*, *18*(9), 1051–1052. http://dx.doi.org/10.1016/j.parkreldis.2012.05.012.

Djamshidian, A., O'Sullivan, S. S., Sanotsky, Y., Sharman, S., Matviyenko, Y., Foltynie, T., et al. (2012). Decision making, impulsivity, and addictions: Do Parkinson's disease patients jump to conclusions? *Movement Disorders*, *27*(9), 1137–1145. http://dx.doi.org/10.1002/mds.25105.

Driver-Dunckley, E., Samanta, J., & Stacy, M. (2003). Pathological gambling associated with dopamine agonist therapy in Parkinson's disease. *Neurology*, *61*(3), 422–423.

Evans, A. H., Katzenschlager, R., Paviour, D., O'Sullivan, J. D., Appel, S., Lawrence, A. D., et al. (2004). Punding in Parkinson's disease: Its relation to the dopamine dysregulation syndrome. *Movement Disorders*, *19*(4), 397–405. http://dx.doi.org/10.1002/mds.20045.

Evans, A. H., Lawrence, A. D., Potts, J., Appel, S., & Lees, A. J. (2005). Factors influencing susceptibility to compulsive dopaminergic drug use in Parkinson disease. *Neurology, 65*(10), 1570–1574. http://dx.doi.org/10.1212/01.wnl.0000184487.72289.f0.

Evans, A. H., & Lees, A. J. (2004). Dopamine dysregulation syndrome in Parkinson's disease. *Current Opinion in Neurology, 17*(4), 393–398.

Fiorino, D. F., & Phillips, A. G. (1999). Facilitation of sexual behavior in male rats following D-amphetamine-induced behavioral sensitization. *Psychopharmacology, 142*(2), 200–208.

Fisher, W. A., Rosen, R. C., Eardley, I., Sand, M., & Goldstein, I. (2005). Sexual experience of female partners of men with erectile dysfunction: The female experience of men's attitudes to life events and sexuality (FEMALES) study. *The Journal of Sexual Medicine, 2*(5), 675–684. http://dx.doi.org/10.1111/j.1743-6109.2005.00118.x.

Frank, M. J., Samanta, J., Moustafa, A. A., & Sherman, S. J. (2007). Hold your horses: Impulsivity, deep brain stimulation, and medication in parkinsonism. *Science, 318*(5854), 1309–1312. http://dx.doi.org/10.1126/science.1146157.

Fugl-Meyer, A. R., & Fugl-Meyer, K. S. (1999). Sexual disabilities, problems and satisfaction in 18–74 year old swedes. *Scandinavian Journal of Sexology, 2*(2), 79–105.

Gallagher, S. (2010). Treating Parkinson's disease. Dopamine dysregulation syndrome and impulse control. *British Journal of Neuroscience Nursing, 6*(1), 548–552.

Gao, X., Chen, H., Schwarzschild, M. A., Glasser, D. B., Logroscino, G., Rimm, E. B., et al. (2007). Erectile function and risk of Parkinson's disease. *American Journal of Epidemiology, 166*(12), 1446–1450. http://dx.doi.org/10.1093/aje/kwm246.

Georgiadis, J. R., Kringelbach, M. L., & Pfaus, J. G. (2012). Sex for fun: A synthesis of human and animal neurobiology. *Nature Reviews Urology, 9*(9), 486–498. http://dx.doi.org/10.1038/nrurol.2012.151.

Giladi, N., Treves, T. A., Paleacu, D., Shabtai, H., Orlov, Y., Kandinov, B., et al. (2000). Risk factors for dementia, depression and psychosis in long-standing Parkinson's disease. *Journal of Neural Transmission (Vienna, Austria: 1996), 107*(1), 59–71. http://dx.doi.org/10.1007/s007020050005.

Giovannoni, G., O'Sullivan, J. D., Turner, K., Manson, A. J., & Lees, A. J. (2000). Hedonistic homeostatic dysregulation in patients with Parkinson's disease on dopamine replacement therapies. *Journal of Neurology, Neurosurgery, and Psychiatry, 68*(4), 423–428.

Giuliano, F., Rubio-Aurioles, E., Kennelly, M., Montorsi, F., Kim, E. D., Finkbeiner, A. E., et al. (2006). Efficacy and safety of vardenafil in men with erectile dysfunction caused by spinal cord injury. *Neurology, 66*(2), 210–216. http://dx.doi.org/10.1212/01.wnl.0000194260.43583.32.

Grant, J. E., Potenza, M. N., Hollander, E., Cunningham-Williams, R., Nurminen, T., Smits, G., et al. (2006). Multicenter investigation of the opioid antagonist nalmefene in the treatment of pathological gambling. *The American Journal of Psychiatry, 163*(2), 303–312. http://dx.doi.org/10.1176/appi.ajp.163.2.303.

Harvey, N. S. (1988). Serial cognitive profiles in levodopa-induced hypersexuality. *The British Journal of Psychiatry, 153*, 833–836.

Hussain, I. F., Brady, C. M., Swinn, M. J., Mathias, C. J., & Fowler, C. J. (2001). Treatment of erectile dysfunction with sildenafil citrate (Viagra) in parkinsonism due to Parkinson's disease or multiple system atrophy with observations on orthostatic hypotension. *Journal of Neurology, Neurosurgery, and Psychiatry, 71*(3), 371–374.

Jacobs, H., Vieregge, A., & Vieregge, P. (2000). Sexuality in young patients with Parkinson's disease: A population based comparison with healthy controls. *Journal of Neurology, Neurosurgery, and Psychiatry, 69*(4), 550–552.

Klos, K. J., Bower, J. H., Josephs, K. A., Matsumoto, J. Y., & Ahlskog, J. E. (2005). Pathological hypersexuality predominantly linked to adjuvant dopamine agonist therapy in

Parkinson's disease and multiple system atrophy. *Parkinsonism & Related Disorders*, *11*(6), 381–386. http://dx.doi.org/10.1016/j.parkreldis.2005.06.005.
Koller, W. C., Vetere-Overfield, B., Williamson, A., Busenbark, K., Nash, J., & Parrish, D. (1990). Sexual dysfunction in Parkinson's disease. *Clinical Neuropharmacology*, *13*(5), 461–463.
Kor, A., Fogel, Y., Reid, R. C., & Potenza, M. N. (2013). Should hypersexual disorder be classified as an addiction? *Sexual Addiction & Compulsivity*, *20*(1–2). http://dx.doi.org/10.1080/10720162.2013.768132.
Kummer, A., Cardoso, F., & Teixeira, A. L. (2009). Loss of libido in Parkinson's disease. *The Journal of Sexual Medicine*, *6*(4), 1024–1031. http://dx.doi.org/10.1111/j.1743-6109.2008.01083.x.
Kurlan, R. (2004). Disabling repetitive behaviors in Parkinson's disease. *Movement Disorders*, *19*(4), 433–437. http://dx.doi.org/10.1002/mds.10625.
Langstrom, N., & Hanson, R. K. (2006). High rates of sexual behavior in the general population: Correlates and predictors. *Archives of Sexual Behavior*, *35*(1), 37–52. http://dx.doi.org/10.1007/s10508-006-8993-y.
Laumann, E. O., Paik, A., & Rosen, R. C. (1999). Sexual dysfunction in the United States: Prevalence and predictors. *JAMA*, *281*(6), 537–544.
Lawrence, A. D., Evans, A. H., & Lees, A. J. (2003). Compulsive use of dopamine replacement therapy in Parkinson's disease: Reward systems gone awry? *Lancet Neurology*, *2*(10), 595–604.
Leeman, R. F., & Potenza, M. N. (2012). Similarities and differences between pathological gambling and substance use disorders: A focus on impulsivity and compulsivity. *Psychopharmacology*, *219*(2), 469–490. http://dx.doi.org/10.1007/s00213-011-2550-7.
Lindau, S. T., Schumm, L. P., Laumann, E. O., Levinson, W., O'Muircheartaigh, C. A., & Waite, L. J. (2007). A study of sexuality and health among older adults in the United States. *The New England Journal of Medicine*, *357*(8), 762–774. http://dx.doi.org/10.1056/NEJMoa067423.
Lipe, H., Longstreth, W. T., Bird, T. D., & Linde, M. (1990). Sexual function in married men with Parkinson's disease compared to married men with arthritis. *Neurology*, *40*(9), 1347–1349. Retrieved from https://www.ncbi.nlm.nih.gov/pubmed/2392215.
Lucon, M., Pinto, A. S., Simm, R. F., Haddad, M. S., Arap, S., Lucon, A. M., et al. (2001). Assessment of erectile dysfunction in patients with Parkinson's disease. *Arquivos de Neuro-Psiquiatria*, *59*(3-a), 559–562.
Mamikonyan, E., Siderowf, A. D., Duda, J. E., Potenza, M. N., Horn, S., Stern, M. B., et al. (2008). Long-term follow-up of impulse control disorders in Parkinson's disease. *Movement Disorders*, *23*(1), 75–80. http://dx.doi.org/10.1002/mds.21770.
McCabe, M. P., Sharlip, I. D., Atalla, E., Balon, R., Fisher, A. D., Laumann, E., et al. (2016). Definitions of sexual dysfunctions in women and men: A consensus statement from the fourth international consultation on sexual medicine 2015. *The Journal of Sexual Medicine*, *13*(2), 135–143. http://dx.doi.org/10.1016/j.jsxm.2015.12.019.
McInnes, R. A. (2003). Chronic illness and sexuality. *The Medical Journal of Australia*, *179*(5), 263–266. Retrieved from https://www.ncbi.nlm.nih.gov/pubmed/12924976. https://www.mja.com.au/system/files/issues/179_05_010903/mci10673_fm.pdf.
McKeon, A., Josephs, K. A., Klos, K. J., Hecksel, K., Bower, J. H., Michael Bostwick, J., et al. (2007). Unusual compulsive behaviors primarily related to dopamine agonist therapy in Parkinson's disease and multiple system atrophy. *Parkinsonism & Related Disorders*, *13*(8), 516–519. http://dx.doi.org/10.1016/j.parkreldis.2007.04.004.
Merims, D., & Giladi, N. (2008). Dopamine dysregulation syndrome, addiction and behavioral changes in Parkinson's disease. *Parkinsonism & Related Disorders*, *14*(4), 273–280. http://dx.doi.org/10.1016/j.parkreldis.2007.09.007.

Mizusawa, H., Hedlund, P., Sjunnesson, J., Brioni, J. D., Sullivan, J. P., & Andersson, K. E. (2002). Enhancement of apomorphine-induced penile erection in the rat by a selective alpha(1D)-adrenoceptor antagonist. *British Journal of Pharmacology, 136*(5), 701–708. http://dx.doi.org/10.1038/sj.bjp.0704773.

Montejo, A. L., Llorca, G., Izquierdo, J. A., & Rico-Villademoros, F. (2001). Incidence of sexual dysfunction associated with antidepressant agents: A prospective multicenter study of 1022 outpatients. Spanish working group for the study of psychotropic-related sexual dysfunction. *The Journal of Clinical Psychiatry, 62*(Suppl. 3), 10–21.

Montorsi, F., Padma-Nathan, H., Buvat, J., Schwaibold, H., Beneke, M., Ulbrich, E., et al. (2004). Earliest time to onset of action leading to successful intercourse with vardenafil determined in an at-home setting: A randomized, double-blind, placebo-controlled trial. *The Journal of Sexual Medicine, 1*(2), 168–178. http://dx.doi.org/10.1111/j.1743-6109.2004.04025.x.

Non-motor symptoms of Parkinson's disease (Vol. II). (2009). New York: Oxford University Press.

Nusbaum, M. R., Hamilton, C., & Lenahan, P. (2003). Chronic illness and sexual functioning. *American Family Physician, 67*(2), 347–354.

O'Sullivan, S. S., Evans, A. H., & Lees, A. J. (2009). Dopamine dysregulation syndrome: An overview of its epidemiology, mechanisms and management. *CNS Drugs, 23*(2), 157–170. http://dx.doi.org/10.2165/00023210-200923020-00005.

O'Sullivan, S. S., Wu, K., Politis, M., Lawrence, A. D., Evans, A. H., Bose, S. K., et al. (2011). Cue-induced striatal dopamine release in Parkinson's disease-associated impulsive–compulsive behaviours. *Brain, 134*(Pt. 4), 969–978. http://dx.doi.org/10.1093/brain/awr003.

Odiyoor, M., Kobylecki, C., Hackett, R. J., Silverdale, M. A., & Kellett, M. W. (2009). Pramipexole and gender identity disorder: Expanding the phenotype of hypersexuality in Parkinson's disease. *Movement Disorders, 24*(16), 2434–2435. http://dx.doi.org/10.1002/mds.22848.

Parkinson, J. (1817). *An essay on the shaking palsy*. London: Sherwood, Neely & Jones.

Poletti, M., & Bonuccelli, U. (2013). Acute and chronic cognitive effects of levodopa and dopamine agonists on patients with Parkinson's disease: A review. *Therapeutic Advances in Psychopharmacology, 3*(2), 101–113. http://dx.doi.org/10.1177/2045125312470130.

Pondal, M., Marras, C., Miyasaki, J., Moro, E., Armstrong, M. J., Strafella, A. P., et al. (2013). Clinical features of dopamine agonist withdrawal syndrome in a movement disorders clinic. *Journal of Neurology, Neurosurgery, and Psychiatry, 84*(2), 130–135. http://dx.doi.org/10.1136/jnnp-2012-302684.

Probst, C. C., & van Eimeren, T. (2013). The functional anatomy of impulse control disorders. *Current Neurology and Neuroscience Reports, 13*(10), 386. http://dx.doi.org/10.1007/s11910-013-0386-8.

Quinn, N. P., Toone, B., Lang, A. E., Marsden, C. D., & Parkes, J. D. (1983). Dopa dose-dependent sexual deviation. *The British Journal of Psychiatry, 142*, 296–298.

Rabinak, C. A., & Nirenberg, M. J. (2010). Dopamine agonist withdrawal syndrome in Parkinson disease. *Archives of Neurology, 67*(1), 58–63. http://dx.doi.org/10.1001/archneurol.2009.294.

Raffaele, R., Vecchio, I., Giammusso, B., Morgia, G., Brunetto, M. B., & Rampello, L. (2002). Efficacy and safety of fixed-dose oral sildenafil in the treatment of sexual dysfunction in depressed patients with idiopathic Parkinson's disease. *European Urology, 41*(4), 382–386.

Rees, P. M., Fowler, C. J., & Maas, C. P. (2007). Sexual function in men and women with neurological disorders. *Lancet, 369*(9560), 512–525. http://dx.doi.org/10.1016/s0140-6736(07)60238-4.

Riley, D. E. (2002). Reversible transvestic fetishism in a man with Parkinson's disease treated with selegiline. *Clinical Neuropharmacology*, *25*(4), 234–237.

Rogers, R. D., Everitt, B. J., Baldacchino, A., Blackshaw, A. J., Swainson, R., Wynne, K., et al. (1999). Dissociable deficits in the decision-making cognition of chronic amphetamine abusers, opiate abusers, patients with focal damage to prefrontal cortex, and tryptophan-depleted normal volunteers: Evidence for monoaminergic mechanisms. *Neuropsychopharmacology*, *20*(4), 322–339. http://dx.doi.org/10.1016/s0893-133x(98)00091-8.

Sakakibara, R., Shinotoh, H., Uchiyama, T., Sakuma, M., Kashiwado, M., Yoshiyama, M., et al. (2001). Questionnaire-based assessment of pelvic organ dysfunction in Parkinson's disease. *Autonomic Neuroscience*, *92*(1–2), 76–85. http://dx.doi.org/10.1016/s1566-0702(01)00295-8.

Samuel, M., Rodriguez-Oroz, M., Antonini, A., Brotchie, J. M., Ray Chaudhuri, K., Brown, R. G., et al. (2015). Management of impulse control disorders in Parkinson's disease: Controversies and future approaches. *Movement Disorders*, *30*(2), 150–159. http://dx.doi.org/10.1002/mds.26099.

Schrag, A., Jahanshahi, M., & Quinn, N. (2000). How does Parkinson's disease affect quality of life? A comparison with quality of life in the general population. *Movement Disorders*, *15*(6), 1112–1118.

Serrano-Dueñas, M. (2002). Chronic dopamimetic drug addiction and pathological gambling in patients with Parkinson's disease—Presentation of four cases. *German Journal of Psychiatry*, *2*(5), 62–66.

Sevincok, L., Akoglu, A., & Akyol, A. (2007). Quetiapine in a case with Parkinson disease and pathological gambling. *Journal of Clinical Psychopharmacology*, *27*(1), 107–108. http://dx.doi.org/10.1097/JCP.0b013e31802e98c3.

Shapiro, M. A., Chang, Y. L., Munson, S. K., Okun, M. S., & Fernandez, H. H. (2006). Hypersexuality and paraphilia induced by selegiline in Parkinson's disease: Report of 2 cases. *Parkinsonism & Related Disorders*, *12*(6), 392–395. http://dx.doi.org/10.1016/j.parkreldis.2006.01.010.

Simonet, C., Fernandez, B., Cerdan, D. M., & Duarte, J. (2016). Hypersexuality induced by rasagiline in monotherapy in Parkinson's disease. *Neurological Sciences*, *37*(11), 1889–1890. http://dx.doi.org/10.1007/s10072-016-2668-9.

Smeding, H. M., Goudriaan, A. E., Foncke, E. M., Schuurman, P. R., Speelman, J. D., & Schmand, B. (2007). Pathological gambling after bilateral subthalamic nucleus stimulation in Parkinson disease. *Journal of Neurology, Neurosurgery, and Psychiatry*, *78*(5), 517–519. http://dx.doi.org/10.1136/jnnp.2006.102061.

Sohtaoglu, M., Demiray, D. Y., Kenangil, G., Ozekmekci, S., & Erginoz, E. (2010). Long term follow-up of Parkinson's disease patients with impulse control disorders. *Parkinsonism & Related Disorders*, *16*(5), 334–337. http://dx.doi.org/10.1016/j.parkreldis.2010.02.006.

Solla, P., Bortolato, M., Cannas, A., Mulas, C. S., & Marrosu, F. (2015). Paraphilias and paraphilic disorders in Parkinson's disease: A systematic review of the literature. *Movement Disorders*, *30*(5), 604–613. http://dx.doi.org/10.1002/mds.26157.

Solla, P., Cannas, A., Floris, G. L., Orofino, G., Costantino, E., Boi, A., et al. (2011). Behavioral, neuropsychiatric and cognitive disorders in Parkinson's disease patients with and without motor complications. *Progress in Neuro-Psychopharmacology & Biological Psychiatry*, *35*(4), 1009–1013. http://dx.doi.org/10.1016/j.pnpbp.2011.02.002.

Szarfman, A., Doraiswamy, P. M., Tonning, J. M., & Levine, J. G. (2006). Association between pathologic gambling and parkinsonian therapy as detected in the food and drug administration adverse event database. *Archives of Neurology*, *63*(2), 299–300. author reply 300. 10.1001/archneur.63.2.299-b.

Titova, N., Padmakumar, C., Lewis, S. J., & Chaudhuri, K. R. (2016). Parkinson's: A syndrome rather than a disease? *Journal of Neural Transmission (Vienna, Austria: 1996)*. http://dx.doi.org/10.1007/s00702-016-1667-6.

Uitti, R. J., Tanner, C. M., Rajput, A. H., Goetz, C. G., Klawans, H. L., & Thiessen, B. (1989). Hypersexuality with antiparkinsonian therapy. *Clinical Neuropharmacology*, *12*(5), 375–383.

Vitale, C., Santangelo, G., Erro, R., Errico, D., Manganelli, F., Improta, I., et al. (2013). Impulse control disorders induced by rasagiline as adjunctive therapy for Parkinson's disease: Report of 2 cases. *Parkinsonism & Related Disorders*, *19*(4), 483–484. http://dx.doi.org/10.1016/j.parkreldis.2012.11.008.

Vitale, C., Santangelo, G., Trojano, L., Verde, F., Rocco, M., Grossi, D., et al. (2011). Comparative neuropsychological profile of pathological gambling, hypersexuality, and compulsive eating in Parkinson's disease. *Movement Disorders*, *26*(5), 830–836. http://dx.doi.org/10.1002/mds.23567.

Vogel, H. P., & Schiffter, R. (1983). Hypersexuality—A complication of dopaminergic therapy in Parkinson's disease. *Pharmacopsychiatria*, *16*(4), 107–110. http://dx.doi.org/10.1055/s-2007-1017459.

Voon, V., Mehta, A. R., & Hallett, M. (2011). Impulse control disorders in Parkinson's disease: Recent advances. *Current Opinion in Neurology*, *24*(4), 324–330. http://dx.doi.org/10.1097/WCO.0b013e3283489687.

Voon, V., Sohr, M., Lang, A. E., Potenza, M. N., Siderowf, A. D., Whetteckey, J., et al. (2011). Impulse control disorders in Parkinson disease: A multicenter case—Control study. *Annals of Neurology*, *69*(6), 986–996. http://dx.doi.org/10.1002/ana.22356.

Waters, C., & Smolowitz, J. (2013). Impaired sexual function. In R. F. Pfeiffer & I. Bodis-Wollner (Eds.), *Parkinson's disease and nonmotor dysfunction* (pp. 173–186). Totowa, NJ: Humana Press.

Weintraub, D. (2008). Dopamine and impulse control disorders in Parkinson's disease. *Annals of Neurology*, *64*(Suppl. 2), S93–100. http://dx.doi.org/10.1002/ana.21454.

Weintraub, D., Koester, J., Potenza, M. N., Siderowf, A. D., Stacy, M., Voon, V., et al. (2010). Impulse control disorders in Parkinson disease: A cross-sectional study of 3090 patients. *Archives of Neurology*, *67*(5), 589–595. http://dx.doi.org/10.1001/archneurol.2010.65.

Weintraub, D., & Nirenberg, M. J. (2013). Impulse control and related disorders in Parkinson's disease. *Neurodegenerative Diseases*, *11*(2), 63–71. http://dx.doi.org/10.1159/000341996.

Welsh, M., Hung, L., & Waters, C. H. (1997). Sexuality in women with Parkinson's disease. *Movement Disorders*, *12*(6), 923–927. http://dx.doi.org/10.1002/mds.870120614.

Wermuth, L., & Stenager, E. (1992). Sexual aspects of Parkinson's disease. *Seminars in Neurology*, *12*(2), 125–127. http://dx.doi.org/10.1055/s-2008-1041166.

Yu, M., Roane, D. M., Miner, C. R., Fleming, M., & Rogers, J. D. (2004). Dimensions of sexual dysfunction in Parkinson disease. *The American Journal of Geriatric Psychiatry*, *12*(2), 221–226.

Zesiewicz, T. A., Helal, M., & Hauser, R. A. (2000). Sildenafil citrate (Viagra) for the treatment of erectile dysfunction in men with Parkinson's disease. *Movement Disorders*, *15*(2), 305–308.

CHAPTER THIRTY

Osteoporosis: A Hidden Nonmotor Face of Parkinson's Disease

Vinod Metta[*,†,1], **Tamara C. Sanchez**[‡], **Chandrasekhara Padmakumar**[§]
[*]Imperial College Hospitals NHS, London, United Kingdom
[†]University College London, London, United Kingdom
[‡]Hospital Universitari Germans Trias i Pujol, Barcelona, Spain
[§]John Hunter Hospital, Newcastle, NSW, Australia
[1]Corresponding author: e-mail address: vinod.metta@nhs.net

Contents

Abstract

Osteoporosis is a "hidden nonmotor face" of Parkinson's disease and a cause of considerable morbidity in the older general population and in Parkinson's disease patients. Some regard this as a "hidden epidemic." Women are overrepresented and have considerable problems related to osteoporosis. In general osteoporosis leads to reduced mobility aggravating the motor syndrome of PD. The nonmotor aspects and impact of osteoporosis in PD have remained unexplored. Possible nonmotor consequences include a range of pain syndromes related to local pain, fractures, falls, and injuries as well as pathological fractures and radiculopathy. In addition depression, sleep dysfunction, dementia, as well as fear of falling also complicate the clinical picture. Quality of life deteriorates both for the patient and career. Pathways of care do not always include assessments for osteoporosis and needs to become obligatory particularly in older female PD patients. Active management strategies then need to be undertaken for osteoporosis in PD. Related motor and nonmotor consequences also highlight the importance of multidisciplinary treatment in PD particularly when dealing with osteoporosis.

International Review of Neurobiology, Volume 134
ISSN 0074-7742
http://dx.doi.org/10.1016/bs.irn.2017.05.034

1. INTRODUCTION

In Parkinson's disease (PD), osteoporosis represents a hidden nonmotor symptom with motor and nonmotor consequences. Wood and Walker (2005) have referred to this as a "hidden epidemic." It is known that PD patients run a higher risk of falls because of the progression of PD (postural instability). In many such cases fractures complicate the falls and may be related to underlying osteoporosis (Wood & Walker, 2005). PD patients have lower bone mineral density (BMD) values in comparison to both age- and sex-matched individuals who do not have PD (Invernizzi, Carda, Viscontini, & Cisari, 2009; Malochet-Guinamand, Durif, & Thomas, 2015; van den Bos et al., 2013) conferring susceptibility to osteoporosis. In addition various comorbidities complicate the clinical scenario and the skeletal makeup in PD which include loss of bone mass (BM) and osteoporosis (Invernizzi et al., 2009; van den Bos et al., 2013; Wood & Walker, 2005). Falls, fractures, and also pathological fractures related to osteoporosis lead to a range of secondary nonmotor symptoms (NMS) such as pain of various types as well as depression, anxiety, and sleep disorders. These will be discussed in this chapter.

Osteoporosis is a bone disease and is defined as a bone density of 2.5 standard deviations (SD) below that of a young adult, measured by dual-energy X-ray absorptiometry (DEXA) at the hip (Table 1). A value on DEXA below the 2.5 SD is considered as osteopenia. This definition is based on World Health Organization (WHO) classification as shown in Table 1.

Vitamin D deficiency plays a key role in the pathogenesis of osteoporosis, and the role of vitamin D status is particularly important in PD because it reduces BMD (Wood & Walker, 2005). Supporting this concept is the fact that treatment with bisphosphonates, vitamin D, and calcium can increase BMD and reduce fractures in PD patients. Hypovitaminosis D has been correlated with an increased risk of falls and fractures (Broe et al., 2007). In a study by Evatt et al. (2008) vitamin D insufficiency was noted to be significantly more frequent in PD patients than in age-matched healthy controls and patients with Alzheimer's disease (AD). Vitamin D insufficiency was reported in 55% of PD, 36% of age-matched healthy controls, and 41% AD patients. The greatest decrease of BMD in patients is found in the lumbar spine as well as the femoral neck with advanced stages of the PD affliction. This is aggravated by low calcium intake and sunlight exposure (Invernizzi et al., 2009; Malochet-Guinamand et al., 2015; van den Bos

Table 1 The WHO Definition of Osteoporosis
WHO Definition of Osteoporosis Based on BMD

Classification	BMD	T-Score
Normal	Within 1 SD of the mean level for a young-adult reference population	T-score at −1.0 and above
Low bone mass (Osteopenia)	Between 1.0 and 2.5 SD below that of the mean level for a young-adult reference population	T-score between −1.0 and −2.5
Osteoporosis	2.5 SD or more below that of the mean level for a young-adult reference population	T-score at or below −2.5
Severe or established Osteoporosis	2.5 SD or more below that of the mean level for a young-adult reference population	T-score at or below −2.5 with one or more fractures

BMD, bone mineral density. T-score = measurement unit that compares a person's bone density with what is normally expected in a healthy young adult of the same sex, using standard deviations.

et al., 2013). Gender differences in reporting NMS of PD have also been shown (Wood & Walker, 2005), and studies indicate that the female gender is an independent predictor of osteoporosis in PD.

Vitamin D has been shown to have effects on mesencephalic neurons and protects these from insults causing neurodegeneration in vitro and reduces the effects of 6-OH-dopamine (Raglione, Sorbi, & Nacmias, 2011). In addition, vitamin D is involved in genetic regions closely associated with PD such as the promoter region of glial cell line-derived neurotrophic factor (GDNF) and neurturina regulating the expression of GDNF and tyrosine hydroxylase. An association between VDR gene polymorphism, PD, and vitamin D-binding protein has been recently identified and has been suggested as a potential biomarker (Raglione et al., 2011).

In PD patients there is a higher prevalence of insufficiency or reduction of 25(OH)D compared to AD patients or healthy controls (Evatt et al., 2008). Inadequate sun exposure and inadequate dietary intake could explain this condition, but it is not clearly understood whether hypovitaminosis D is a factor influencing the genesis of Parkinsonism or if PD itself leads to hypovitaminosis D.

Studies suggest numbers of mechanisms could contribute to loss of BD occurring in PD patients (Malochet-Guinamand et al., 2015). There is a correlation with BM and the weight and body mass index (BMI). One study

(Malochet-Guinamand et al., 2015) reported after age adjustment of the women in their study of osteoporotic Fractures revealed a 7.3% lower BMD than other women. Analysis via a multivariate methodology revealed the PD women had a markedly lower BMD with 2.1% than the other women in the study. This was after the weight adjustment among other criteria. Also body weight explained 72% of the BMD difference between the PD group and the control group.

Osteoporosis and osteopenia related to PD is thought to be the direct and indirect result of several factors as listed below (Invernizzi et al., 2009):

- ➢ Female gender
- ➢ Vitamin D deficiency
- ➢ Reduced exposure to sunlight
- ➢ Low body weight
- ➢ Nutritional status
- ➢ Vitamin B12 and folate deficiency
- ➢ Hyperhomocysteinemia (due to reduced vitamin B12 and folate or L-Dopa treatment)
- ➢ PD duration and severity
- ➢ Immobility and reduced BM
- ➢ Decreased muscle strength
- ➢ Other neuroendocrinal status (increased serum concentration of undercarboxylated osteocalcin)

1.1 Osteoporosis and Other NMS of PD

Osteoporosis and osteopenia are common in PD particularly the older population. Increasing life expectancy in PD and prolonged survival underpin the increased prevalence of osteoporosis in older PD. The prevalence figures of osteoporosis rises from 0.6% in those aged 65–69 years to approximately 3% in those older than 80 (Bezza, Ouzzif, Naji, et al., 2008). Broadly women are more susceptible and figures suggest a rate of 91% in women and 61% in men (Invernizzi et al., 2009). Little data exist in relation to secondary NMS that may be related to osteoporosis in PD. It is known that long-term chronic conditions such as osteoarthritis and osteoporosis may act synergistically with aging-related nonmotor issues such as progressive loss of vision as well as hearing (Melton and Riggs (1985). With superimposition of impaired balance and reflexes the overall risk factor for injury following seemingly minor traumatic mechanisms in the older patient becomes more evident. This factor may explain the high rates of falls and fractures in older PD. Many of these patients experience a rage of NMS as a result, either acutely in the hospitalized state or chronically at home (e.g., depression). Wood and Walker (2005) reported a statistically significant link with increasing age and depression as assessed by the geriatric depression score. These patients also had a higher Hoehn and Yahr score, and gait and balance abnormality. A summary of studies reporting falls and fractures is shown in Table 2 (adapted from Invernizzi et al., 2009)

A range of NMS occur in PD as a direct or indirect consequence of osteoporosis (Table 3). These depend on the severity of osteoporosis as well as the mobility status of the patient. For instance, fractures can cause local or radicular pain; however, a severe fall in an elderly PD may lead to hospitalization. Many such patients would suffer from the trauma of the fall as well as acute hospitalization-related problems which often are nonmotor in nature (neuropsychiatric and sleep dysfunction). Respiratory and urinary infections related to underlying immobility could in turn also trigger neuropsychiatric as well as sleep problems (Table 3). An additional aspect is that in female PD high doses of levodopa secondary to intrajejunal infusion therapy may precipitate nutritional deficiencies which could cause hyperhomocysteinemia. The latter is considered a risk factor for osteoporosis.

1.2 Why PD Patients Are More at Risk of Osteoporosis?

A review by Malochet-Guinamand et al. (2015) revealed that Parkinson's patients experience a measurable reduction of BMD compared to the age-matched controls. Several variables operate to lead to reduced BMD and bone strength in PD (Fig. 1). These include metabolic and pharmacological issues such as levodopa treatment (high dose in particular), nutritional status which affects body weight and homocysteine levels and vitamin D status (exposure to sunlight is a key issue). Other variables that contribute to reduced BMD in PD are mechanical problems such as muscle strength, low physical exercise, decreased mobility and immobility in some cases. These factors are shown in Fig. 1 (taken from van den Bos et al., 2013).

Bone densities have been shown to be different between white Caucasian south Asian and black African subjects (Zengin et al., 2016). However, there are no such studies in PD particularly related to ethnic differences and bone density, antiosteoporotic medications are mainly biphospohonates.

Table 2 Studies Reporting Falls, Bone Mineral Density, and Fractures in a Series of Reports in PD

Authors	Type	Number of Subjects and Sex	Outcome Measures	Results
Stalenhoef, Diederiks, Knottnerus, Kester, and Crebolder (2002)	PCS	311, age > 70 CD	Falls, falls injuries	33% fell; 45% fall-related injuries
Dargent-Molina et al. (1996)	PCS	7575 W, age ≥ 75	BMD fem neck DEXA, self-reported physical capacity, neuromuscular function, mobility, visual function, medications	154 first hip fracture
Fink et al. (2005)	PCS	5995 CD M, age 65, 0.9% PD	BMD hip and spine DEXA and QCT	Risk of multiple future falls OR = 2.91, CI [95%] = 1.55–5.46
Taylor et al. (2004)	PCS	6.787 CD W, age 66	BMD hip DEXA number of hip fractures	602 (8.9%) hip fracture
Woodford and Walker (2005)	PCS	246 PD patients	Emergency hospital admissions of PD patients over a 4-year period	Falls and fractures were, respectively, the first and eighth cause of hospital admission
Koller, Glatt, Vetere-Overfield, and Hassanein (1989)	RCS	100 PD patients (61 M, 39 W)	Frequency, circumstances, consequences of falling	38%—falls; 17 patients—fall-related fractures (4 hip fractures)
Wielinski, Erickson-Davis, Wichmann, Walde-Douglas, and Parashos (2005)	RCS	1092 PD patients	Number of falls within the past 2 years, related injuries	55.9%—1 fall in the past 2 years; 65.0%—injury; 33.0%—fracture
Johnell, Melton, Atkinson, O'Fallon, and Kurland (1992)	RCS	138 PD patients, 138 ASMC	Record of hip fractures	59% PD vs 44% AMSC ($P < 0.002$)

Genever, Downes, and Medcalf (2005)	RCS 200	39 PD patients, 200 AMSC	Record of fracture at any body site	PD 15% vs AMSC 7.5%; $P=0.007$
Lorefalt, Toss, and Granerus (2007)	PCS	26 PD, 26 AMSC	BMD hip and total body (including head) DEXA	BMD of total body, hip, neck is decreased in PD patients ($P<0.05$)
Taggart and Crawford (1995)	PCS	51 PD, 51 AMSC	BMD spine, hip DEXA	PD hip BMD < 10% AMSC ($P=0.014$)
Kao, Chen, Wang, Chia, and Yeh (1994)	PCS	22 PD patients, 22 AMSC	Spine BMD DEXA	BMD PD < AMSC
Sato, Kaji, Tsuru, and Oizumi (2001)	PCS	115 PD patients	BMD 2nd Met CXD	Low BMD alone is an important risk factor for hip fracture in PD patients
Yamada, Kachi, and Ando (1995)	PCS	82 PD patients, 99 ASMC	Spine BMD DEXA	Spine BMD is lower in PD than in control
Ishizaki, Harada, Katayama, Abe, and Nakamura (1993)	PCS	70 PD patients, 46 ASMC	BMD MSXP	Osteopenia 26 (59%) of the 44 PD F, 5 (19%) of the 26 PD M, 6 (24%) of the 25 AMSC F, 2 (9%) of the 21 AMSC M
Wood and Walker (2005)	X-sect	105 PD patients	Hip and spine BMD DEXA	63% W osteoporotic; 28% W osteopenic; 20% M osteoporotic; 41% M osteopenic
Bezza et al. (2008)	PCS	52 PD patients, 52 ASMC	BMD hip, spine, Dexa X-ray, osteocalcin, CTX, PTH	9 PD (17.3%) osteoporotic; 28 PD (53.8%) osteopenic
Kamanli, Ardicoglu, Ozgocmen, and Yoldas (2008)	PCS	24 PD patients, 31 ASMC	BMD spine, fem neck, Ward, troch, bilat hands DXA	PD F BMD bilat hands and fem neck < BMD AMSC ($P<0.05$)

Continued

Table 2 Studies Reporting Falls, Bone Mineral Density, and Fractures in a Series of Reports in PD—cont'd

Authors	Type	Number of Subjects and Sex	Outcome Measures	Results
Di Monaco, Vallero, Di Monaco, Tappero, and Cavanna (2006)	CCS	56 hip fracture, 28 PD, 28 ASMC	BMD tot fem, fem neck, troch, intertroch, Ward (unfractured femur) DXA	T-score did not differ significantly between PD patients and AMSC
Di Monaco, Vallero, Di Monaco, Tappero, and Cavanna (2008)	PCS	38 PD hip fracture, 38 AMSC hip fracture	BMD tot fem, fem neck, troch, intertroch DEXA	BMD lower in PD troch fractures than cervical fractures ($P=0.028$)
Schneider et al. (2008)	PCS	8105 CD W, 73 PD	Fem neck, tot fem BMD DEXA, number of fractures, BMI	Mean tot fem BMD was 7.3% lower in PD group. PD patients had a 2.6-fold higher age-adjusted risk for incident hip fracture
Yamanashi et al. (2005)	PCS	714 fractured patients	Second hip fracture, age, gender, fracture type, cognitive impairment, comorbidities, medical conditions	45 second hip fractures; no significant difference in the incidence of second hip fracture as to gender or age; increased risk of a second hip fracture was associated with senile dementia and PD

2nd Met, 2nd metacarpal bone; *ASMC*, age–sex-matched controls; *bilat hands*, bilateral hands; *BMD*, bone mineral density; *BMI*, body mass index; *CCS*, case–control study; *CD*, community-dwelling; *CTX*, C-telopeptide; *DEXA*, dual energy X-ray absorptiometry; *F*, female; *fem neck*, femoral neck; *intertroch*, intertrochanteric; *M*, male; *MSXP*, multiple scanning X-ray photodensitometry; *PCS*, prospective cohort study; *PD*, Parkinson's disease; *PTH*, parathyroid hormone; *RCS*, retrospective cohort study; *tot fem*, total femur; *troch*, trochanter; *W*, women; *X-sect*, cross-sectional.

Adapted from Invernizzi, M., Carda, S., Viscontini, G. S., & Cisari, C. (2009). Osteoporosis in Parkinson's disease. *Parkinsonism and Related Disorders, 15*(5), 339–346. doi:10.1016/j.parkreldis.2009.02.009.

Table 3 Bone Status, Osteoporosis and PD, and Consequent Direct or Indirect Nonmotor Symptoms

PD and Osteoporosis	Nonmotor Symptoms
Decreased BMD	Local and radicular pain, posture-related pain
Fractures	Local pain
Fractures with immobility (hospitalized patient)	Secondary infection (respiratory and urinary) Neuropsychiatric complications (hallucinations, delirium) Sleep dysfunction (insomnia and confusional states)
Osteoporosis in older subjects Severe pain related to osteoporosis	Depression, anxiety
Female patient on intrajejunal levodopa infusion and osteoporosis	Nutritional deficiency triggering hyperhomocysteinemia

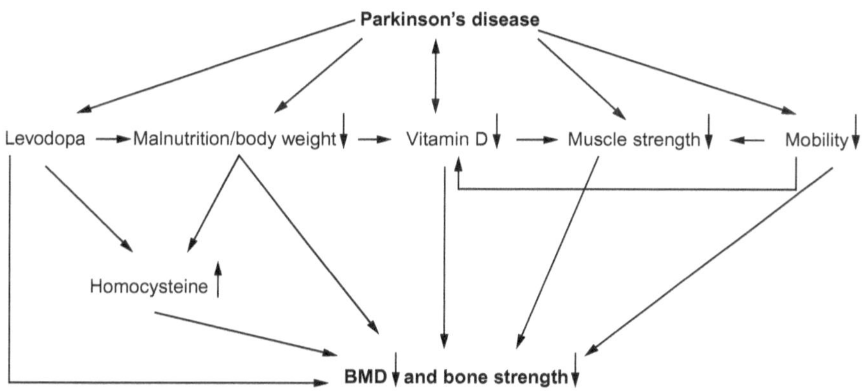

Fig. 1 Bone and Parkinson's disease. *BMD*, bone mineral density.

2. DIAGNOSIS AND TREATMENT

Screening for osteoporosis in PD should be a quality standard in clinic which would enable appropriate therapeutic interventions to be initiated (van den Bos et al., 2013). All patients with PD should be evaluated for clinical risk factors like thyroid disease, gonadal hormone deficiency, risk of falls, immobilization, inadequate activity, smoking, and alcohol intake (Fig. 2). Screening for vitamins D, B12, folic acid, calcium, thyroid function, and

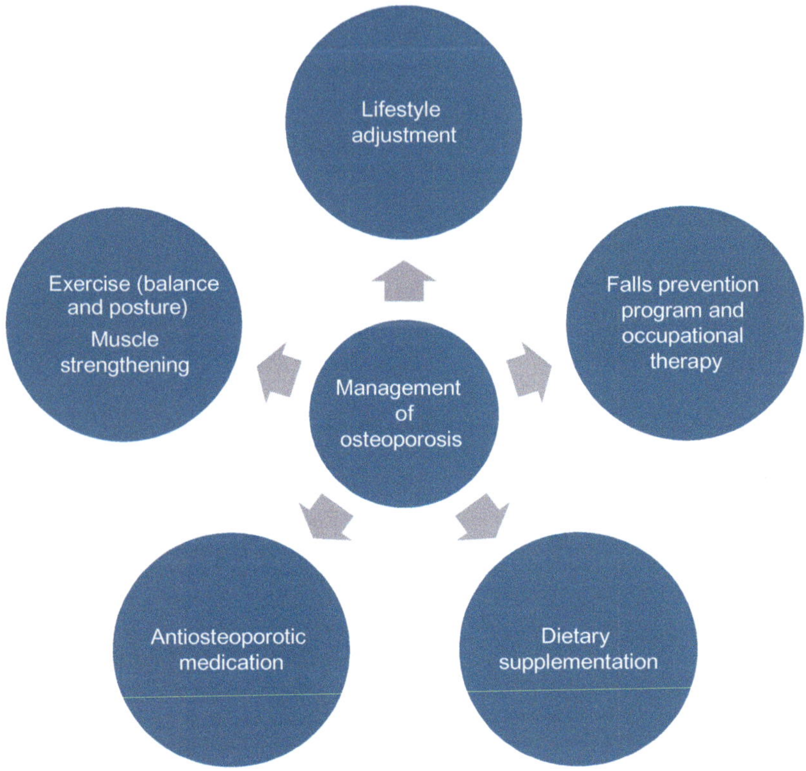

Fig. 2 A summary of the various enablers of a successful osteoporosis management programme for Parkinson's disease.

parathormone levels should be done on a regular basis (Invernizzi et al., 2009; Malochet-Guinamand et al., 2015; van den Bos et al., 2013).

Radiological diagnosis of osteoporosis is based on BMD test. BMD test measures bone density at the spine and at the hip for the PD patient. Another tool called FRAX has been recently developed to evaluate fracture risk of patients (Kanis, Johnell, Oden, Johansson, & McCloskey, 2008). FRAX is based on individual patient models that integrate the risks associated with clinical risk factors as well as BMD at the femoral neck. Whether this will be effective in PD is unknown.

There are no specific guidelines related to management of osteoporosis and PD. Broadly treatment needs to be started with a combination of biphosphonates combined with vitamin D and ensuring a regular intake of calcium containing food or supplements. Fig. 2 shows the other important aspects for a holistic management of osteoporosis in PD.

Treatment with medications is mainly based on DEXA (T score < −2.5) and FRAX scores (10-year risk for fracture >3% hip, >20% of any major osteoporotic fracture). Mainly two types of pharmacological treatments are available to treat patients with osteoporosis. Antiresorptive agents which prevent bone loss and preserve architecture, improve quality of bone, reduce the risk of vertebral fractures. Biphosponates have been studied for osteoporosis in PD and risedronate in men with PD was studied in 121patients in a 2-year, randomized, double-blind, placebo-controlled trial. Risedronate (2.5 mg) and ergocalciferol (1000 IU) daily were compared with ergocalciferol (1000 IU) alone as well as placebo (Sato, Honda, & Iwamoto, 2007). BMD improved in the risedronate group and overall reduced the relative risk of hip fracture by 0.33 (95% confidence interval: 0.09–1.2). In another study, benefit with 17.5 mg risedronate and ergocalciferol over ergocalciferol and placebo in elderly PD patients was reported by the same authors. Another biphosphonate, aledronate was reported to reduce the relative risk of hip fractures in a study in PD by 0.29 (95% confidence interval: 0.10–0.85) (Sato, Iwamoto, Kanoko, & Satoh, 2006). Although some have criticized these studies for the absence of use of DEXA at hip for BMD measurement, the data indicate that biphosphonates are useful for osteoporosis in PD and should be used. Zoledronic acid and denosumab (monoclonal antibody) have also been reported to reduce the risk of nonvertebral and hip fractures. Anabolic agents have also been tried, and rhPTH [1–34] (teriparatide) increases bone density and size, improves quality of bone, and reduces the risk of vertebral and nonvertebral fractures especially for advanced PD patients. When swallowing is a issue parenteral treatment options like monoclonal antibody (denosumab, 60 mg subcutaneous injection) every 6 months and the anabolic teriparatide daily subcutaneous injections have been shown to be beneficial (Neer et al., 2001).

Dietary and nutritional manipulation is important, and vitamin D, B12, and folate status need to be addressed and corrected if required. One study has shown that folate supplementation and vitamin B12 in levodopa-treated PD patients may prevent bone loss and hyperhomocysteinemia (Malochet-Guinamand et al., 2015), the latter a risk factor for osteoporosis. Vitamin K is another nutritional factor that may cause carboxylation of gamma-carboxyglutamic acid-rich residue related to types of bone proteins such as osteocalcin. Vitamin K supplementation improved metacarpal BMD of PD patients aged 65 and older in a small controlled study with a mean 5-year follow-up (Malochet-Guinamand et al., 2015).

Lifestyle-related management approach is related to addressing risk factors such as smoking and excessive alcohol consumption. Smoking needs to be stopped while alcohol intake needs moderating. Exercise is to be encouraged as it is believed this helps strengthening specific affected area of the body as well as muscle strength and balance improvement which reduced risk of falls (Fig. 2). However, this concept is not strictly evidence based. Falls prevention program is an important aspect of management of PD patients with osteoporosis, and these fracture risk reduction programs are aimed at reducing the number and severity of falls. In the elderly, falls prevention program consists of correcting decreased visual acuity, reducing alertness, and balance altering medications. Occupation therapy input is required for improvement of home environment (slippery floors, creating space and familiarization with furniture, insufficient lighting, handrails, bathroom, and kitchen adaptations). Behavioral adaptations include avoiding multitasking and taking time while attempting to stand (orthostatic hypotension may complicate).

3. CONCLUSIONS

Osteoporosis is a relatively underexplored nonmotor aspect of PD but with major functional consequence. The pathophysiology is a little unclear, and patients with PD have a lower BMD than an age-matched control population. Frequent falls complicate PD and reduced BM leads to increased fracture risk with consequent morbidity and mortality. The bone loss itself is multifactorial and is contributed to by a combination of immobility, vitamin D deficiency, hyperhomocysteinemia, poor nutritional state, and muscle weakness. Treatment with bisphosphonates, calcium, and a close input from occupational therapist is important to manage osteoporosis in PD as is an effective falls prevention program.

REFERENCES

Bezza, A., Ouzzif, Z., Naji, H., et al. (2008). Prevalence and risk factors of osteoporosis in patients with Parkinson's disease. *Rheumatology International*, *28*, 1205–1209. http://dx.doi.org/10.1007/s00296-008-0632-6.

Broe, K. E., Chen, T. C., Weinberg, J., Bischoff-Ferrari, H. A., Holick, M. F., & Kiel, D. P. (2007). A higher dose of vitamin D reduces the risk of falls in nursing home residents: A randomized, multiple-dose study. *Journal of the American Geriatrics Society*, *55*, 234–239. http://dx.doi.org/10.1111/j.1532-5415.2007.01048.x.

Dargent-Molina, P., Favier, F., Grandjean, H., Baudoin, C., Schott, A. M., Hausherr, E., et al. (1996). Fall-related factors and risk of hip fracture: The EPIDOS prospective study. *Lancet*, *348*(9021), 145–149. http://dx.doi.org/10.1016/S0140-6736(96)01440-7.

Di Monaco, M., Vallero, F., Di Monaco, R., Tappero, R., & Cavanna, A. (2006). Bone mineral density in hip-fracture patients with Parkinson's disease: A case-control study.

Archives of Physical Medicine and Rehabilitation, *87*(11), 1459–1462. http://dx.doi.org/10.1016/j.apmr.2006.07.265.

Di Monaco, M., Vallero, F., Di Monaco, R., Tappero, R., & Cavanna, A. (2008). Type of hip fracture in patients with Parkinson disease is associated with femoral bone mineral density. *Archives of Physical Medicine and Rehabilitation*, *89*(12), 2297–2301. http://dx.doi.org/10.1016/j.apmr.2008.06.016.

Evatt, M. L., Delong, M. R., Khazai, N., Rosen, A., Triche, S., & Tangpricha, V. (2008). Prevalence of Vitamin D insufficiency in patients with Parkinson's disease and Alzheimer disease. *Archives of Neurology*, *65*(10), 1348–1352. http://dx.doi.org/10.1001/archneur.65.10.1348.

Fink, H. A., Kuskowski, M. A., Orwoll, E. S., Cauley, J. A., Ensrud, K. E., & Osteoporotic Fractures in Men (MrOS) Study Group (2005). Association between Parkinson's disease and low bone density and falls in older men: The osteoporotic fractures in men study. *Journal of the American Geriatrics Society*, *53*(9), 1559–1564. http://dx.doi.org/10.1111/j.1532-5415.2005.53464.x.

Genever, R. W., Downes, T. W., & Medcalf, P. (2005). Fracture rates in Parkinson's disease compared with age- and gender-matched controls: A retrospective cohort study. *Age and Ageing*, *34*(1), 21–24. http://dx.doi.org/10.1093/ageing/afh203.

Invernizzi, M., Carda, S., Viscontini, G. S., & Cisari, C. (2009). Osteoporosis in Parkinson's disease. *Parkinsonism & Related Disorders*, *15*(5), 339–346. http://dx.doi.org/10.1016/j.parkreldis.2009.02.009.

Ishizaki, F., Harada, T., Katayama, S., Abe, H., & Nakamura, S. (1993). Bone changes in Parkinson's disease. *Nō to Shinkei*, *45*(8), 719–724.

Johnell, O., Melton, L. J., 3rd., Atkinson, E. J., O'Fallon, M. W., & Kurland, L. T. (1992). Fracture risk in patients with parkinsonism: A population-based study in Olmsted country, Minnesota. *Age and Ageing*, *21*(1), 32–38.

Kamanli, A., Ardicoglu, O., Ozgocmen, S., & Yoldas, T. K. (2008). Bone mineral density in patients with Parkinson's disease. *Aging Clinical and Experimental Research*, *20*, 277–279. http://dx.doi.org/10.1007/BF03324774.

Kanis, J. A., Johnell, O., Oden, A., Johansson, H., & McCloskey, E. (2008). FRAX™ and the assessment of fracture probability in men and women from the UK. *Osteoporosis International*, *19*(4), 385–397. http://dx.doi.org/10.1007/s00198-007-0543-5.

Kao, C. H., Chen, C. C., Wang, S. J., Chia, L. G., & Yeh, S. H. (1994). Bone mineral density in patients with Parkinson's disease measured by dual photon absorptiometry. *Nuclear Medicine Communications*, *15*(3), 173–177.

Koller, W. C., Glatt, S., Vetere-Overfield, B., & Hassanein, R. (1989). Falls and Parkinson's disease. *Clinical Neuropharmacology*, *12*(2), 98–105.

Lorefalt, B., Toss, G., & Granerus, A. K. (2007). Bone mass in elderly patients with Parkinson's disease. *Acta Neurologica Scandinavica*, *116*(4), 248–254. http://dx.doi.org/10.1111/j.1600-0404.2007.00875.x.

Malochet-Guinamand, S., Durif, F., & Thomas, T. (2015). Parkinson's disease: A risk factor for osteoporosis. *Joint, Bone, Spine*, *82*(6), 406–410. http://dx.doi.org/10.1016/j.jbspin.2015.03.009.

Melton, L. J., 3rd., & Riggs, B. L. (1985). Risk factors for injury after a fall. *Clinics in Geriatric Medicine*, *1*(3), 525–539.

Neer, R. M., Arnaud, C. D., Zanchetta, J. R., Prince, R., Gaich, G. A., Reginster, J. Y., et al. (2001). Effect of parathyroid hormone (1-34) on fractures and bone mineral density in postmenopausal women with osteoporosis. *The New England Journal of Medicine*, *344*, 1434–1441. http://dx.doi.org/10.1056/NEJM200105103441904.

Raglione, L. M., Sorbi, S., & Nacmias, B. (2011). Osteoporosis and Parkinson's disease. *Clinical Cases in Mineral and Bone Metabolism*, *8*(3), 16–18.

Sato, Y., Honda, Y., & Iwamoto, J. (2007). Risedronate and ergocalciferol prevent hip fracture in elderly men with Parkinson disease. *Neurology*, *68*, 911–915. http://dx.doi.org/10.1212/01.wnl.0000257089.50476.92.

Sato, Y., Iwamoto, J., Kanoko, T., & Satoh, K. (2006). Alendronate and vitamin D2 for prevention of hip fracture in Parkinson's disease: A randomized controlled trial. *Movement Disorders*, *21*, 924–929. http://dx.doi.org/10.1002/mds.20825.

Sato, Y., Kaji, M., Tsuru, T., & Oizumi, K. (2001). Risk factors for hip fracture among elderly patients with Parkinson's disease. *Journal of the Neurological Sciences*, *182*(2), 89–93. http://dx.doi.org/10.1016/S0022-510X(00)00458-5.

Schneider, J. L., Fink, H. A., Ewing, S. K., Ensrud, K. E., Cummings, S. R., & Study of Osteoporotic Fractures (SOF) Research Group. (2008). The association of Parkinson's disease with bone mineral density and fracture in older women. *Osteoporosis International*, *19*(7), 1093–1097. http://dx.doi.org/10.1007/s00198-008-0583-5.

Stalenhoef, P. A., Diederiks, J. P. M., Knottnerus, J. A., Kester, A. D. M., & Crebolder, H. F. J. M. (2002). A risk model for the prediction of recurrent falls in community-dwelling elderly: A prospective cohort study. *Journal of Clinical Epidemiology*, *55*(11), 1088–1094. http://dx.doi.org/10.1016/S0895-4356(02)00502-4.

Taggart, H., & Crawford, V. (1995). Reduced bone density of the hip in elderly patients with Parkinson's disease. *Age and Ageing*, *24*(4), 326–328.

Taylor, B. C., Schreiner, P. J., Stone, K. L., Fink, H. A., Cummings, S. R., Nevitt, M. C., et al. (2004). Long-term prediction of incident hip fracture risk in elderly white women: Study of osteoporotic fractures. *Journal of the American Geriatrics Society*, *52*(9), 1479–1486. http://dx.doi.org/10.1111/j.1532-5415.2004.52410.x.

van den Bos, F., Speelman, A. D., Samson, M., Munneke, M., Bloem, B. R., & Verhaar, H. J. (2013). Parkinson's disease and osteoporosis. *Age and Ageing*, *42*(2), 156–162. http://dx.doi.org/10.1093/ageing/afs161.

Wielinski, C. L., Erickson-Davis, C., Wichmann, R., Walde-Douglas, M., & Parashos, S. A. (2005). Falls and injuries resulting from falls among patients with Parkinson's disease and other parkinsonian syndromes. *Movement Disorders*, *20*(4), 410–415. http://dx.doi.org/10.1002/mds.20347.

Wood, B., & Walker, R. (2005). Osteoporosis in Parkinson's disease. *Movement Disorders*, *20*, 1636–1640. http://dx.doi.org/10.1002/mds.20643.

Woodford, H., & Walker, R. (2005). Emergency hospital admissions in idiopathic Parkinson's disease. *Movement Disorders*, *20*(9), 1104–1108. http://dx.doi.org/10.1002/mds.20485.

Yamada, T., Kachi, T., & Ando, K. (1995). Osteoporosis and fractures in Parkinson's disease. *Nippon Ronen Igakkai Zasshi*, *32*(10), 637–640. http://dx.doi.org/10.3143/geriatrics.32.637.

Yamanashi, A., Yamazaki, K., Kanamori, M., Mochizuki, K., Okamoto, S., Koide, Y., et al. (2005). Assessment of risk factors for second hip fractures in Japanese elderly. *Osteoporosis International*, *16*(10), 1239–1246. http://dx.doi.org/10.1007/s00198-005-1835-2.

Zengin, A., Pye, S. R., Cook, M. J., Adams, J. E., Wu, F. C. W., O'Neill, T. W., et al. (2016). Ethnic differences in bone geometry between White, Black and South Asian men in the UK. *Bone*, *9*, 180–185. http://dx.doi.org/10.1016/j.bone.2016.07.018.

FURTHER READING

Healthit.gov. (2014). *The Government & Health IT. Retrieved from* https://www.healthit.gov/patients-families/government-health-it.

Melamed, E., & Djaldetti, R. (2006). Camptocormia in Parkinson's disease. *Journal of Neurology*, *253*(Suppl. 7), VII14–VII16. http://dx.doi.org/10.1007/s00415-006-7004-5.

Simon, H. (2017). *Parkinson disease–Possible complications. March 28, 2017.* New York Times.com.

Upadhyaya, C. D., Starr, P. A., & Mummaneni, P. V. (2010). Spinal deformity and Parkinson disease: A treatment algorithm–Pisa syndrome. *Neurosurgical Focus*, *28*(3), E5. http://dx.doi.org/10.3171/2010.1.FOCUS09288.

CHAPTER THIRTY-ONE

Weight in Parkinson's Disease: Phenotypical Significance

Jagdish C. Sharma[*,†,1], **Anna Lewis**[*,†]
[*]Geriatric Medicine (Movement Disorders), Lincoln County Hospital, Lincoln, United Kingdom
[†]University of Lincoln, Lincoln, United Kingdom
[1]Corresponding author: e-mail address: jagdish.sharma@ulh.nhs.uk

Contents

Abstract

Body weight in Parkinson's disease (PD) is a significant nonmotor feature. Weight homeostasis is a complex physiological process and gets deranged in PD patients leading to changes in weight. While both the low and high body weight have been reported as risk factors for PD, the majority of PD patients have a lower weight and a subset of patients lose weight during the course of the disease, while a small proportion gain weight. A number of clinical parameters such as older age, impaired cognition, severity of disease, and an imbalance of food intake determined by satiety and hunger

International Review of Neurobiology, Volume 134
ISSN 0074-7742
http://dx.doi.org/10.1016/bs.irn.2017.04.011

hormones have been reported to be associated with but not the cause of weight change. Low body weight and weight loss have a negative impact on disease severity, dyskinesia quality of life, and mortality indicative of disease progression. An early assessment of olfactory impairment seems to identify patients at risk of weight loss, the patients with more severe olfactory loss—anosmic group, lose weight as compared to the patients with some preservation of olfaction, the hyposmic group. Higher levodopa dose per kilogram body weight increases the risk of dyskinesia, higher body weight seems to be protective against this complication. The identification of PD patients according to the nonmotor phenotype of "Park-olfaction-weight-phenotype" and the "olfaction-weight-dyskinesia" triad should help to develop strategies to prevent weight reduction and improve general health and complications of PD patients. The phenotype seems to reflect a differential prodromal pathology and influence clinical disease. Higher body weight patients would benefit from life style changes to achieve a healthy profile. Weight monitoring and weight orientated approach to management of PD patients should help to improve their outcome. Body weight change might be a surrogate to disease progression and may be used to investigate neuroprotection strategies.

1. INTRODUCTION

"The power of conveying the food to the mouth is at length so much impeded that the patient is obliged to consent to be fed by others. The bowels ... demand stimulating medicines of very considerable power... Food is conveyed to the mouth, food is with difficulty retained in the mouth until masticated and then difficultly swallowed."

James Parkinson (1817).

James Parkinson described a number of gastrointestinal features in his description of the malady in 1817, which we now recognize as Parkinson's disease. He, however, did not make any comments about any observations of body weight in this condition.

The neurodegenerative and other long-term conditions such as chronic pulmonary disease and cardiac failure have an impact on body weight during the course of the disease and quite often determine their prognosis. Parkinson's disease (PD) is a progressive neurodegenerative disease primarily involving the nigrostriatum and the patients are prone to weight changes, particularly weight loss, similar to conditions like Alzheimer's disease (Sergi, De Rui, Coin, Inelman, & Manzato, 2013). Alzheimer's disease is a significant predictor of weight loss; and weight loss is associated with more severe disease and higher mortality (White, Pieper, Schmader, &

Fillenbaum, 1996; White, Pieper, & Schmader, 1998). Similarly the patients with chronic pulmonary disease lose weight due to muscular apoptosis (Alvar et al., 2002) and muscle wasting (Woulters, Creutzberg, & Schols, 2002); while the patients with chronic heart failure frequently suffer from irreversible weight loss (cardiac cachexia) associated with impaired survival (Anker et al., 2003). Over the last several years, authors have investigated body weight status and weight changes in PD and its significance in disease management. The purpose of this chapter is to describe these and how body weight might have implications on managing patients with PD.

2. BODY WEIGHT HOMEOSTASIS AND CHANGES IN PD (FIG. 1)

The human beings are dependent on achieving and maintaining a body weight on the basis of ingestion of nutrition, its assimilation and

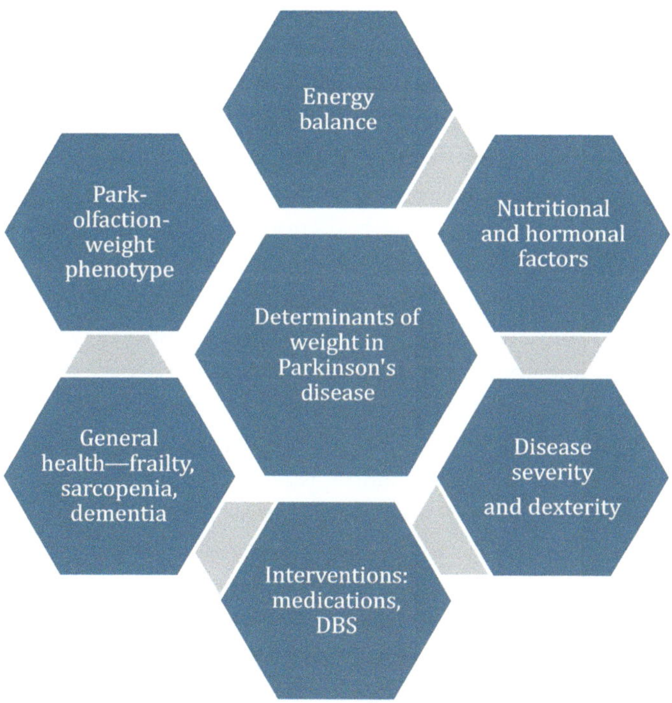

Fig. 1 Factors for weight determination in Parkinson's disease.

expenditure. A change in their balance would lead to weight imbalance. Body weight is controlled by a balance between central (orexins and anorexigenic hormones and peripheral signals of satiety pathways, ghrelin (the hunger hormone), and leptin). The intricate balance between multiple orexigenic and anorexigenic signals in the hypothalamus and the dysregulation of ghrelin would influence body weight by controlling hunger, food intake, and energy consumption. Leptin, produced by adipose tissue, produces an anorexigenic influence leading to satiety and thus interrupts food intake, whereas ghrelin, a peptide synthesized by gastric mucosa during fasting, promotes hunger and increased food intake leading to weight gain and stimulates growth hormone secretion. Presence of appetite boosting and satiety-related hormones would determine the amount of food ingestion; changes in body weight and food intake in neurodegenerative conditions have been the subject of a number of studies (Aziz et al., 2008; de Boer, Ter Horst, & Lorist, 2013). The previously held concept of satiety and hunger centers has now been replaced by a more complex network of neurons of homeostatic and hedonic systems which ultimately control body weight (Benarroch, 2010; Meister, 2007; Saper, Chou, & Elmquist, 2002).

There are a number of neuroendocrine abnormalities in PD including disruption of melatonin, disturbance of glucose, insulin resistance and bone metabolism, and body weight changes (De Pablo-Fernández et al., 2017). Some of these abnormalities may determine and influence body weight and weight changes in PD.

Dopamine and serotonin are both implicated in food intake and thus weight variation: dopamine is involved in feeding behavior by modulating reward property of food and motivation for food intake (Narayanan, Guarnieri, & DiLeone, 2010; Palmiter, 2007; Volkow, Wang, & Baler, 2011). Orexin and melanin-concentration hormone neurons are reduced in PD and would contribute to weight changes (Fiszer et al., 2010; Fronczek et al., 2007; Thannickal, Lai, & Siegel, 2008). Dopamine and dopamine receptors D2 play a central role in feeding behavior, the system is however very complex. It seems that dopamine is not required for liking of food but is necessary for motivational process of wanting (Berridge, Ho, Richard, & Di Felice Antonio, 2010). Dopamine release in dorsal striatum is considered to be essential for normal feeding in humans (Volkow et al., 2002). Dopaminergic system has a complex influence on food behavior. Excessive dopaminergic state has an impulsive behavior leading to binge eating and nocturnal eating behavior (Weintraub et al., 2010), whereas the hypodopaminergic state may lead to hyperphagic state of snack eating, both have different effect on final body weight.

The serotonergic system similarly has a potential to alter body weight in PD. A positive correlation has been reported between serotonin transporter availability and body mass index (BMI). A PET scan study reports increased serotonin transporter binding in patients who experienced weight gain in PD (Politis & Loane, 2011). Degeneration of serotonergic system with low levels of serotonin in PD may explain the preference of sweets and chocolates in PD (Aden et al., 2011; Fox, Chuang, & Brotchie, 2009; Woltz et al., 2009), this might be a compensatory phenomenon since the high carbohydrate diet increases brain dopamine levels and thus provide a feel-good factor. A balance between energy intake and expenditure, dependent on intricate factors as described underneath, would influence weight.

A number of factors of nonmotor nature may influence weight in PD patients such as the impairment of homeostatic regulation, gastrointestinal dysfunction, feeling of gastric fullness due to delayed gastric emptying, severe constipation, impaired olfaction, taste, mood, apathy, motivation, insomnia, and fatigue.

3. WEIGHT MEASUREMENT

Weight measurements methodology in PD has not been standardized; methods have included a pragmatic method of out-patients weighing scales (Sharma, Macnamara, Hasoon, Vassallo, & Ross, 2006; Sharma, Bachmann, & Linzasoro, 2010), standing or sitting posture electronic scales or vertical telescopic stadiometer (Barichella, Cereda, et al., 2016), measuring weight in light clothes with shoes on or off, use of hospital medical records, historical records from primary care notes (Uc et al., 2006), even a recall by the patients which is considered to be highly inaccurate (Dahl & Reynolds, 2013). In some studies, the exact method for weight measurement is not reported (Lee et al., 2016). Thus methodological variations of weight measurements and demographical differences between studies may partly account for inconsistent results. One method of weight measurement being promoted is the BOD-POD technology using air displacement plethysmography, although there are no reports of its use in PD patients. BOD-POD has a good test and retest reliability (Anderson, 2007; Tucker, Lecheminant, & Bailey, 2014).

4. BODY WEIGHT STATUS AND CHANGE IN PD

4.1 Weight as a Risk Factor for PD

High and low weight and BMI have both been reported to be a risk for developing PD. One study has shown high BMI as risk factor for developing PD with a hazard ratio of 2.03 for BMI $>$ 30 compared to BMI $<$ 23 (Hu et al., 2006). Other studies have also reported a reduction in body weight prior to the clinical diagnosis (Chen, Zhang, Hernan, Willett, & Ascherio, 2003; Cheshire & Wszolek, 2005; Logroscino, Sesso, Paffenbarger, & Lee, 2007). Another study showed a low BMI at diagnosis, however on retrospective analysis, it would appear that in this group there had been a very recent reduction in BMI, suggestive that these people actually had early presymptomatic disease and their loss of weight was related to the onset of PD (Cheshire & Wszolek, 2005). Higher adiposity (measured by triceps skinfold thickness) increased incidence of PD threefold (3.7 vs 11.1/10,000 person years) (Abbott et al., 2002). The relationship between BMI and risk of PD remains controversial. Since PD pathology commences several years prior to clinical presentation, there are likely to be variations in body weight in different patients and probably populations. There is a need for large well-designed studies with uniform weight measurements.

4.2 Weight Profile in PD

Body weight changes in PD have been extensively studied. Longitudinal observational studies have reported how body weight varies in PD patients at the time of diagnosis and during the previous or subsequent years. It has been recognized that advanced stage PD patients have lower body weight and BMI than the general population (Bachmann, Zapf, Brunner, & Trenkwalder, 2009; van der Marck et al., 2012). However in this cohort of 166 advanced PD patients, only 4.2% patients were underweight. Additionally, the patients with more advanced disease have lower BMI (van der Marck et al., 2012). This large meta-analysis involving pooled data of 871 patients and 736 controls revealed that PD patients had a significantly lower BMI than controls—there was a significant difference of 1.73 in BMI between the PD patients (lower BMI) and the controls.

This is however not always the case as some studies have revealed a higher body weight in PD patients (Barichella, Marczewska, Vairo,

Canes, & Pezzoli, 2003). The explanation was the higher BMI in general population including the PD persons. These authors have also reported from a case control study of 600 patients vs 600 healthy controls that the PD patients had a lower BMI as compared to the controls and the PD patients reported a higher food intake (Barichella, Cereda, et al., 2016). In this cohort, there were more underweight patients than the controls (4.35% vs 0.3% and fewer obese—20.5% vs 30.8%). Advanced PD patients may develop "sarcopenic obesity" with excessive adipose tissue and depletion of lean body mass (Petroni et al., 2003). The mechanisms of weight fluctuations are very complex (Kistner, Lhommee, & Krack, 2014).

5. CHANGES IN BODY WEIGHT IN PD (FIG. 1)

PD patients may over the years gain weight, remain stable or lose weight, there being weight losers, and nonweight losers (Fig. 2).

6. WEIGHT GAIN AND OBESITY IN PD

PD is a progressive disease primarily involving the nigrostriatal dopaminergic system. Dopamine is known to be the key agent for food reward

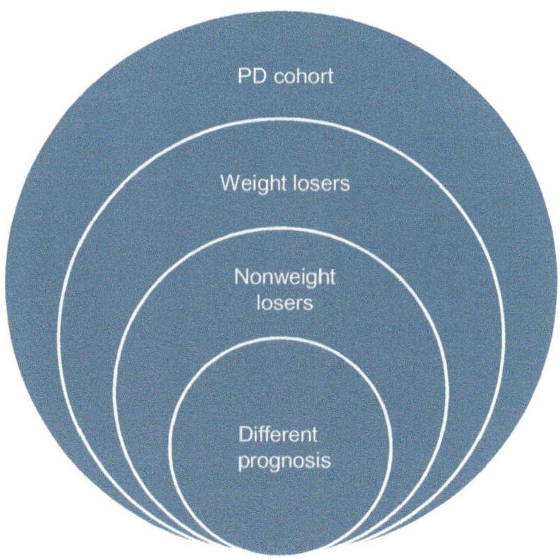

Fig. 2 Changes in weight profile in Parkinson's disease.

and control of food intake. A variable proportion of PD patients have been reported to be overweight—60% in Italy (Barichella, Marczewska, Vairo, et al., 2003), 50% in Germany (Bachmann et al., 2009), a large proportion of patients going for deep brain stimulation (DBS) are noticed to be overweight. A reduction in the dopaminergic striatal response has been shown to be a contributor for weight gain and obesity.

6.1 Dopaminergic System

While a large subset of PD patients are of low weight or lose weight during disease progression, a significant proportion are of higher weight and BMI or gain weight (Barichella, Marczewska, Vairo, et al., 2003; Hu et al., 2006; Sharma & Turton, 2012). One explanation for weight gain is the lower D2 receptor availability in the striatum, possibly resulting in obesity and eating disorders (Wang et al., 2001). Higher body weight or central obesity has been reported as risk factors for PD (Chen et al., 2004; Hu et al., 2006), and some PD patients have a greater midlife skinfold thickness (Abbott et al., 2002).

Some treatments can reverse this problem of weight loss and actually lead to an increase in weight. Pramipexole caused significant weight gain after 3 months of therapy, which may be due to the effect on limbic D3 receptors involved in the control of feeding (Kumru, Santamaria, Valldeoriola, Marti, & Tolosa, 2006) or a reduced energy expenditure due to reduction in muscular rigidity. Weight gain however is not with all the dopamine agonists (Sharma & Hasoon, 2009) (see later) possibly due to differential D-receptor stimulation; D3 stimulation is noticed to lead to compulsive behavior including binge eating (Kumru et al., 2006; Lawrence, Evans, & Lees, 2003; Nirenberg & Waters, 2006).

6.2 Atypical Neuroleptics

The use of atypical neuroleptics (risperidone, olanzapine, chlorpromazine), the drugs used in some patients with PD-related psychosis, is known to increase weight the mechanism being of H1-receptor blockade or hyperprolactenemia (Kapur & Marques, 2016; Kroeze et al., 2003; Wetterling & Mussigbrodt, 1999). A differential baseline striatal activity might determine why some patients gain weight on antipsychotics in PD and some of the dopaminergic drugs may have this effect (Kapur & Marques, 2016).

6.3 Deep Brain Stimulation

DBS also has effects far beyond symptomatic improvement and reduction in motor complications. One study showed 5.9 kg weight gain 3 months postprocedure (Moghaddasi & Boshtam, 2010) and another showed 2.9 kg (4.65%) weight gain after 21.3 months (Strowd et al., 2016). The effect on weight is not limited to PD patients with an increase in BMI of 1.6 kg/m^2 at 7 months and 2.4 kg/m^2 at 72 months demonstrated in a group of patients treated for primary dystonia (Wolf et al., 2016) and weight gain also being observed in patients given DBS for essential tremor (Strowd et al., 2010). The major factor in weight gain with DBS to globus pallidus internus (GPI) in PD is likely to be improvement in motor symptoms as treatment to this region is more selective for motor symptoms than DBS to the subthalamic nucleus (STN). One study reports significantly greater weight gain at 6 months after DBS to STN compared to GPI (increase in BMI 8.4% vs 3.3%), which would suggest that there are other mechanisms involved in weight gain than just reduction in motor complications (Sauleau et al., 2016, 2009). Another study however found this difference was only significant in a group of patients treated for dystonia and there was no difference in weight gain between the two groups in PD patients (Mills, Scherzer, Starr, & Ostrem, 2012).

6.4 Mechanism of Weight Gain After DBS

One mechanism of weight gain after DBS seems to be the normalization of energy metabolism (Montaurier et al., 2007), other explanations are the decrease in dopaminergic medications and improvement in dyskinesia, there is however a poor correlation between weight gain and objective dyskinesia score (Barichella, Marczewska, Mariani, et al., 2003; Macia et al., 2004; Ondo et al., 2000; Vitek et al., 2000). A reduced growth hormone secretion after DBS with resultant decreased lipolysis has been proposed to lead to weight gain (Friess et al., 2001; Nam & Marcus, 2000). Reduction in free living energy expenditure and an increase in body fat is another explanation for weight gain following DBS (Jorgensen, Werdelin, Lokkegaard, Westertrep, & Simonsen, 2012).

Patients should be advised about strategies for weight management (Barichella, Marczewska, Mariani, et al., 2003; Tuite et al., 2005). Because levodopa reduces glucose oxidation, reducing the dose given after DBS could help reduce any effect of weight gain on glycemia (Perlemoine et al., 2005).

6.5 Other Potential Mechanisms for Weight Gain

- The mechanisms of weight gain therefore remain unexplained but are likely to be multifactorial, also contributed to by metabolic and homeostatic mechanisms.
- Patients with early PD have been noted to gain weight despite an increase in physical activity and without an increase in energy intake (Vikdahl, Carlsson, Linder, Forsgren, & Haglin, 2014).
- A hypothesis of decreased energy expenditure has been put forward (Toth, Fishman, & Poehlman, 1997) but not supported by other studies (Chen et al., 2003; Levi, Cox, Lugron, Hodkinson, & Tomkins, 1990; Markus, Cox, & Tomkins, 1992; Markus, Tomkins, & Stern, 1993).
- There is an increase in fat mass (Bannier et al., 2009) thought to be due to a combination of reduced energy expenditure without a reduction in energy intake (Macia et al., 2004), reduced motor complications, and changes in energy metabolism (Rieu et al., 2011).
- Both lipid and protein oxidation are reduced, leading to weight gain, while glucose oxidation increases (Perlemoine et al., 2005).
- There is a potential role of hormones with increased levels of ghrelin and neuropeptide Y found following treatment (Markaki et al., 2012) with probable dysregulation of hypothalamic secretion (Novakova et al., 2007).
- One explanation for weight gain in a subset of PD patients might be the difference in olfaction and weight-related phenotype as described later (Sharma & Turton, 2012).

7. WEIGHT LOSS IN PD

Approximately 30%–70% patients lose weight, whereas some of the patients might remain stable or gain weight. That PD patients lose weight as compared to the healthy controls is long established (Beyer, Palarino, Michalek, Busenbark, & Koller, 1995; Jankovic, Wooten, Van der Linden, & Jansson, 1992; Markus et al., 1993). The exact timing and stage of weight loss is not quite clear since some patients tend to gain weight in early stage of PD, possibly due to a reduced physical activity (Vikdahl et al., 2014). A longitudinal study over 10 years reveals weight loss in PD patients both in the early and advanced stages and there was a significant, >5%, weight loss in more than half the patients as compared to the healthy controls (Chen et al., 2003). This prospective epidemiology study of 468 individuals suggested that PD patients start to lose weight several years before the diagnosis. A 10-year study of 49 patients showed that a group of PD patients lost 7.7% of their body weight vs only 0.2% body weight loss in controls (Uc et al., 2006). With greater than 5% body weight loss being considered significant this affected 55.6% of the PD group but only 20.5% of the control group. There was a steady decline in weight during the course of 10 years, including the early stage but more rapid and pronounced in the late phase. Another smaller study demonstrated 73% of patients with PD lost weight over a 12-month period, despite an increase in energy intake (Lorefalt et al., 2004). A more recent study reports two types of patients in PD, weight losers and weight stable or weight gainers, based on the weight profile changes over a 5-year duration (Sharma & Turton, 2012) as described later. Thus a significant proportion, but not all, patients with PD lose weight during the course of the disease.

7.1 Why Do Some PD Patients Lose Weight?

Why do PD patients lose weight has been a matter of interest and research. Several factors have been reported which seem to be the associations of weight loss rather than the categorically proven cause. There are reports of no significant differences in weight losers and nonweight losers as regards the factors of age of onset, disease severity, gender, cognition, motor features, bulbar involvement, depression, anxiety, and cognition are concerned (Jankovic et al., 1992; Sharma & Hasoon, 2009). It has been suggested that weight loss might be due to hyposmia (Chen et al., 2003), abnormality of autonomic function, malabsorption, disease progression, increased metabolism, increased energy expenditure, impaired intestinal motility, dyskinesia, depression, dysphagia, chewing difficulty, hand–mouth in-coordination, increased energy requirement, cognitive impairment, and increasing age (Bachmann & Trenkwalder, 2006; Barichella, Cereda, et al., 2016; Lorefalt, Granerus, & Unosson, 2006).

7.1.1 Nutritional Intake—Dysphagia and Motor Dexterity

BMI was found to be significantly lower in a dysphagic group of PD patients compared with a nondysphagic group (19.1 vs 21.6 kg/m^2, $P < 0.005$) (Nozaki, Saito, Matsumura, Miyai, & Kang, 1999). Bulbar dysfunction was however not reported to be a cause of weight loss in a study of 281 PD patients as compared to 86 patients with progressive supranuclear palsy (Jankovic et al., 1992). In another study of 304 PD patients of 4.3 years follow-up, there was no difference in swallowing difficulties between weight losers (140 patients) and nonweight losers (164 patients) (Sharma & Hasoon, 2009). Although impaired hand to mouth dexterity seems to be a plausible cause of low food intake in PD patients, this was not noticed to be the case in a study where all underweight patients could eat independently (Bachmann et al., 2009). Dysphagia is usually a feature in advanced disease and weight loss at this stage is more likely to be due to disease severity rather than dysphagia.

7.1.2 Hormonal Factors

Weight loss is associated with lower levels of leptin, the hormone that regulates appetite thus reducing food intake (Fiszer et al., 2010; Lorefalt, Toss, & Granerus, 2009), although the differences were not significant as compared to the weight stable patients in another study (Evidente, Caviness, Adler, Gwinn-Hardy, & Pratley, 2001). Hormones may play a role in weight loss with dopamine dysfunction producing anorexigenic signals in the hypothalamus and the dysregulation of ghrelin (the "hunger hormone"), with lower levels found in patients with low BMI and a further reduction in levels following the usual postprandial fall (De Pablo-Fernández et al., 2017).

7.1.3 Energy Balance—Intake and Expenditure and Nutrition

Energy intake and expenditure have been investigated in PD patients in several studies albeit in small number of patients. In early stage young age patients the investigation reveals no difference in energy intake as compared to the healthy controls (Vikdahl et al., 2014). Resting energy expenditure is higher in more advanced disease and less expenditure in on-state (Capecci et al., 2013).

A study comparing a weight loss group and weight stable group showed no difference in energy intake, daily and resting energy expenditure or physical activity (Delikanaki-Skaribas, Trail, Wong, & Lai, 2009). This was also the case in another similar study reporting no difference in daily energy expenditure in PD patients and controls (Toth et al., 1997). A study reporting on 3 days food diary revealed no difference in energy intake between weight losers and nonweight losers, 1637 ± 467 vs 1672 ± 532 calories, $P = 0.74$ (Sharma & Turton, 2012). Weight loss is not caused by reduced energy intake in early or advanced stage (Chen et al., 2003; Uc et al., 2006). An anthropometric study however revealed increased prevalence of under-nutrition in PD (Markus et al., 1993).

On the contrary, there is evidence of weight loss in PD persons despite an increased energy intake (Chen et al., 2003; Lorefalt et al., 2004) possibly due to increased rigidity or dyskinesia (Markus et al., 1992) at different stages of PD. These were the studies of small number of PD patients. Thus, there is no convincing evidence of energy intake and expenditure-related factors to explain weight loss in PD patients.

7.1.4 Disease Severity

Weight loss has been reported to be more prevalent in more severe PD patients (Lorefalt et al., 2004). There is a steady fall in weight as the disease progresses despite an increase or no reduction in energy intake (Chen et al., 2003; Sharma & Turton, 2012; Uc et al., 2006). Patients who lose weight have a more rapid increase in UPDRS score than those whose weight remains stable or increases (Wills et al., 2016) and those with Hoehn–Yahr stage 3 disease had lower BMI than those with stage 2 disease (Barichella, Cereda, et al., 2016). It has been shown on scanning people with PD that a low BMI is a predictor of low DAT activity in the putamen, ventral striatum, caudate nucleus, and total striatum, suggesting a lower density of dopaminergic neurons (Lee et al., 2016). Thus there is an association between clinical disease severity, increase dopaminergic degeneration, and weight loss. This could explain why loss of weight and BMI $<18.5\,kg/m^2$ are associated with increased disease severity (Umehara, Nakahara, Matsuno, Toyoda, & Oka, 2016). Does this indicate that maintaining body weight might be a surrogate for less neurodegeneration?

7.1.5 Gender and Age

Female PD patients seem to be at a higher risk of weight loss in a logistic regression analysis (Sharma & Turton, 2012). One set of data showed 37% of females lost >10 kg body weight compared to only 7% of men (Sakajiri & Takamori, 1997), another showed 8.5% weight loss in females but only 4.3% loss in males (Durrieu et al., 1992). BMI similarly differed between females with PD vs controls but with no significant difference between the male groups. Several studies have reported older age (Jaafar, Gray, Porter, Turnbull, & Walker, 2010; Lorefalt et al., 2004; Uc et al., 2006) and multiple comorbidities (Akbar et al., 2015) as risk factors for weight loss.

7.1.6 Olfaction

Chen commented on the potential role of impaired olfaction as a cause of weight loss in PD (Chen et al., 2003). This has been investigated recently (Sharma & Turton, 2012). Olfaction was tested with University of Pennsylvania smell identification test (UPSIT) (Doty, Sharman, & Dann, 1984). The patients with UPSIT value below the sample median, classified as the anosmic group, had much higher frequency (55%) of weight loss than those with UPSIT value above the median (the hyposmic group—16%), $P = 0.001$. In addition the anosmic patients had lost weight over the previous 5 years, whereas the hyposmic patients had maintained or gained weight. The anosmic group finally weighed significantly less as compared to the hyposmic group, 75 vs 83 kg, $P = 0.01$. There was a steady decline in olfactory scores and body weight in increasing Hoehn and Yahr stage (UPSIT scores H&Y 1–26 vs H&Y 4–15, $P = 0.02$; weight H&Y 1–84 vs H&Y 4–76 kg, $P = 0.04$) (Fig. 3).

It has been hypothesized on the basis of a study that there are two phenotypes of PD patients: phenotype A—severe loss of olfaction (anosmia), associated with lower initial body weight, weight loss, and higher risk of dyskinesia with progression of pre- and postmotor disease; Phenotype B—with less loss of olfaction (hyposmia), higher initial body weight, stable or

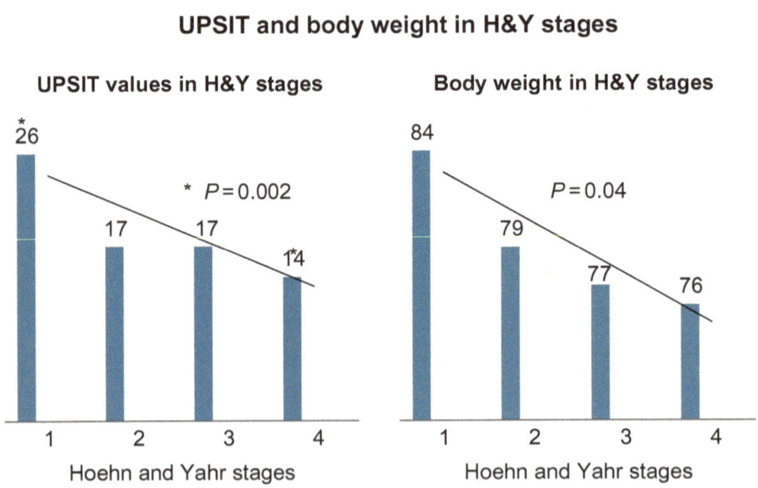

Fig. 3 Decline in olfaction and weight with disease progression in Hoehn and Yahr stages. *Data from Sharma, J. C., & Turton, J. (2012). Olfaction, dyskinesia and profile of weight change in Parkinson's disease: Identifying neurodegenerative phenotypes.* Parkinsonism & Related Disorders, 18*(8), 964–970.*

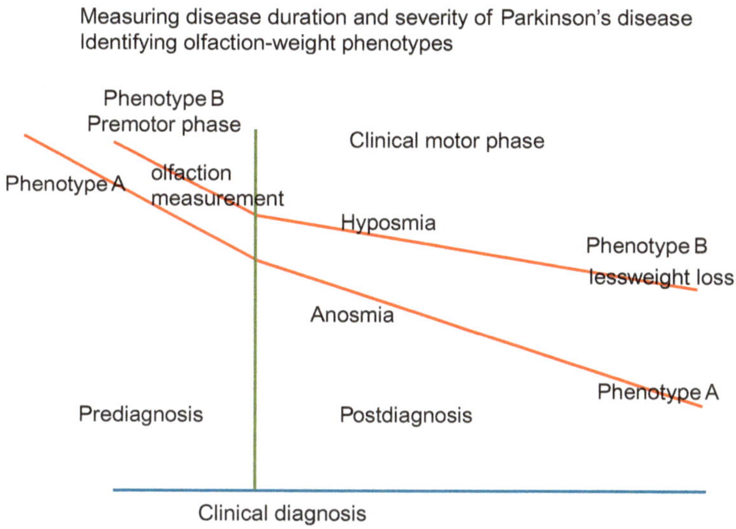

Fig. 4 Disease progression in olfaction-weight phenotypes. Phenotype A—longer preclinical duration more rapid progression with weight loss. Phenotype B—shorter preclinical duration, slower disease progression—no/less weight loss.

increasing weight, and lower risk of dyskinesia with disease progression (Sharma & Turton, 2012; Sharma & Vassallo, 2014) (Fig. 4).

The reason for their weight loss was hypothesized to be due to the potential differences in the severity of neurodegenerative process since the energy intake in the two groups was similar; anosmic group -1605 ± 474 vs 1740 ± 532 calories in the hyposmic group, $P = 0.19$. The degree of olfactory loss was the most significant variable as compared to the confounding factors to predict weight loss in this cohort of PD patients, the other significant factors were the Hoehn and Yahr stage and female gender. The mechanism for weight loss may simply be the wasting effect of neurodegenerative process and the onset of sarcopenia.

Poor olfaction has also been linked to a decrease in MMSE (Vikdahl, Domellöf, Forsgren, & Håglin, 2015) and it is possible that those patients with anosmia have more rapidly progressive neurodegeneration or that they have had a longer duration of premotor prodromal disease (Postuma et al., 2012).

7.1.7 Dementia/Depression/Hallucination

Impaired cognition and hallucinations have been noted to be associated with weight loss (Chen et al., 2003; Lorefalt et al., 2004). This may be similar to weight loss in Alzheimer's disease (Poehlman & Dvorak, 2000) where patients lose weight without a hypermetabolic state. In one study, the factors associated with weight loss in PD were the age at diagnosis, visual hallucinations, and possibly dementia (Uc et al., 2006). These may not be the cause of weight loss but the associated features in weight losers related to advanced disease. The authors concluded that the neurodegeneration involving both motor and nonmotor systems may be associated with weight loss (Uc et al., 2006). There no reports about depression and weight loss in PD.

7.1.8 Structural Changes and Sarcopenia

As PD progresses, there is likely to be the onset of sarcopenia (loss of muscle mass with advancing age) since both are age related. Sarcopenia would be a contributory factor for low body weigh in PD. A small study measuring gait and hand grip did not reveal an increased prevalence of sarcopenia in PD (Loureiro, Yamaguchi, Villegas, Ferreira, & Israel, 2016), a larger study also reported sarcopenia an infrequent condition (only 6.6%) in PD as assessed on European Working Group on Sarcopenia in older people (EWGSOP) criteria (Barichella, Pinelli, et al., 2016). The prevalence of sarcopenia ranges from 6.6% (Barichella, Pinelli, et al., 2016) to 62% prevalence (Abou-Raya, Abou-Raya, El Khadrawy, & Helmii, 2014). Vetrano et al. found prevalence of 40.2% in males and 22.4% in females using EWGSOP criteria, whereas using other criteria in the same group, the prevalence was much lower (Vetrano et al., 2015). There is an indication of a significant association between disease severity and early stage sarcopenia (Drey et al., 2017). PD patients have a reduction in mid-upper arm muscle circumference and handgrip strength as disease progresses (Lindskov, Sjöberg, Hagell, & Westergren, 2016) and their mid-arm muscle circumference was below the 10th percentile in 23% of patients (Petroni et al., 2003). It would appear that disease severity and not sarcopenia is a significant independent factor to explain weight loss in a cohort of PD, sarcopenia may be contributing albeit to only a small extent.

7.1.9 Dopaminergic Medication

There are very limited reports on the possible association of dopaminergic medications and body weight. Cabergoline and pergolide are reported to lead to weight loss, ropinirole was neutral, whereas pramipexole patients have a higher body weight and tend to gain weight (Kumru et al., 2006; Nirenberg & Waters, 2006; Sharma & Hasoon, 2009; Sharma & Ross, 1999). Levodopa therapy has been implicated as a factor in weight loss in

PD patients (Bachmann & Trenkwalder, 2006; Palhagen et al., 2005; Vardi et al., 1976). One investigation reveals no relationship between body weight and weight change as compared to the levodopa equivalent dose of medications (Vikdahl et al., 2014).

It is likely that several factors jointly lead to weight change, whether loss or gain, in an individual patient and no one factor may apply to all weight changes.

8. SIGNIFICANCE OF LOW BODY WEIGHT AND WEIGHT LOSS IN PD

8.1 Nutritional Status, Frailty, and General Health

PD patients are at risk of under-nutrition, more so if there is dyskinesia, in the advanced disease (Vikdahl et al., 2014). Low body weight and weight loss would increase the risk of under-nutrition and malnutrition and predispose to negative health impact such as risk of falls, fatigue, osteoporosis, and fractures in addition to the consequences of malnutrition (Bachmann & Trenkwalder, 2006). There has been malnutrition in up to 24% and an element of malnutrition in up to 3%–60% patients (Sheard, Ash, Silburn, & Kerr, 2001). There is a change in body composition in PD with a predominant loss of fat mass, indicative of malnutrition (Beyer et al., 1995; Lorefalt et al., 2009). Malnutrition also puts patients at increased risk of infection, reduced independence, and increasing need for long-term care and more frequent hospitalization (Holden, 2001; Kashihara, 2006). PD patients seem to have a reduced protein and a higher carbohydrate intake. This seems to relate to the degree of olfactory loss, whether this is a cause or effect relationship is not clear (Aden et al., 2011). The underweight PD patients may also suffer from reduced intake of micronutrients but their relevance is not clear.

Due to low weight being a marker of general frailty phenotype (Cesari, Gambassi, Van Kan, & Vellars, 2014), the PD patients, particularly women are at higher risk of frailty (Roland, Jakobi, & Jones, 2011; Roland, Cornett, Theou, Jakobi, & Jones, 2012) and hence the frailty-related risks of low physical reserves and a propensity to higher impact of any external stressors (Clegg, Young, Lliffe, Rikkert, & Rockwood, 2013) such as increased risk of diseases and mortality. Weight loss is associated with increasing fatigue (Okuma, 2012).

Those patients who lose weight show a more rapid cognitive decline over a 36-month period (Kim et al., 2012) and low BMI is also associated with more severe dysautonomia and postural hypotension (Nakamura, Suzuki, Ueda, Hirayama, & Katsuno, 2017; Umehara et al., 2016).

8.1.1 Body Weight and Dyskinesia

A series of studies have highlighted the relationship between body weight and levodopa-induced dyskinesia (LID). A study of 220 PD (Sharma et al., 2006) patients with the Hoehn and Yahr stage of 3 at the initial assessment, excluding patients with initial dyskinesia, revealed that the patients who developed dyskinesia during follow-up had lower initial body weight than those who remained free of dyskinesia, 72 ± 17 vs 75 ± 15 kg; the patients developing dyskinesia had lost weight prior to developing dyskinesia and the weight at the end of the study follow-up was significantly lower in dyskinetic patients than the nondyskinetic patients, 66 ± 17 vs 76 ± 15 kg, $P = 0.002$. In this study, the dyskinetic patients had received a higher levodopa dose per kilogram body weight, 9 ± 4 vs 6 ± 4 mg/kg, $P = 0.001$. The levodopa dose per kilogram body weight was more significant for the risk of dyskinesia than absolute dose of levodopa. In addition, the weight losers were reported to be several times, 21% vs 6%, more likely to develop dyskinesia than nonweight losers. The higher the dose of levodopa per kilogram, higher the risk of dyskinesia (2.7% at the dose of <2.88 mg/kg as compared to 29% at the dose of >8.8 mg/kg). The authors proposed a concept of cascade of levodopa dose and weight-related dyskinesia in PD (LD-WD-PD). Thus the patients with low body weight would receive a higher levodopa dose per kilogram causing LID, which further perpetuates weight loss and further dyskinesia (Sharma et al., 2006).

In a similar analysis of the pooled patients from the 056 and REAL-PET studies of levodopa vs ropinirole (Sharma, Ross, Rascol, & Brooks, 2008), analyzing 263 patients who had received levodopa out of the total 427, the patients who developed dyskinesia had received a significantly higher levodopa dose per kilogram as compared to those who remained free of dyskinesia, 9.2 ± 4.3 vs 6.9 ± 5.2 mg/kg, $P = 0.001$; the female patients developed dyskinesia at a lower mg/kg dose—10 ± 4 mg (males) vs 8 ± 3 mg (females), $P = 0.03$. The dyskinetic patients were lighter in initial body weight. Interestingly in both these studies of early and well-established PD patients (Sharma et al., 2006, 2008), the dyskinetic patients had received a dose of 9–10 mg/kg body weight as compared to the nondyskinetic patients receiving approximately 7 mg/kg body weight. In both these studies, the patients developing dyskinesia were lighter in weight at the initial stage.

In another (Olanow et al., 2013) study of a secondary analysis of the STRIDE-PD data of comparison of levodopa + carbidopa vs levodopa + carbidopa + entacapone in early PD patients, the patients receiving more than 4 mg/kg body weight were at a higher risk of dyskinesia at 208 weeks of follow-up. In this study, body weight was a significant factor for dyskinesia and there was a linear correlation between the rate of development of dyskinesia and L-dopa dose, the risk of dyskinesia escalated above the levodopa dose of 6 mg/kg weight.

In a review of the classification of risk factors for dyskinesia, current body weight and the history of weight loss were identified as the significant modifiable factors which can be adjusted according to the patients' physical habitus to reduce the risk of dyskinesia (Sharma et al., 2010). The factors of age, disease duration, disease severity, and genetic influence are the nonmodifiable factors. Pharmacokinetics study reveals that low body weight is associated with higher levodopa levels under the curve, i.e., there is an inverse relationship between body weight and levodopa levels (Muller, Woitalla, Saft, & Kuhn, 2006; Zappia et al., 2002). This would be the obvious explanation for the increased risk of dyskinesia in low body weight patients for a similar levodopa dose. The lesson is that the patients with a lower body weight are at a higher risk of LID for a similar dose of levodopa, the higher body weight thus being protective against LID.

8.1.2 Weight-Related Gender Differences for Dyskinesia

Not only do female patients develop dyskinesia at a lower levodopa dose per kilogram body weight (Sharma et al., 2008), they are also prone to develop dyskinesia earlier (Hassin-Baer et al., 2011) than male patients, the median time for onset of dyskinesia being 4 years in females and 6 years for males, $P=0.004$. This risk may be explained on the basis of their lower body weight and hence reflect the increased levodopa concentrations in nigrostriatal tissue and body. For a similar weight loss, females have higher risk of dyskinesia than males (Sharma et al., 2006), perhaps due to a difference in pharmacokinetics in men and women determined by their body weight.

8.1.3 Body Weight and Mortality and Quality of Life

Whatever limited information is available indicates an increased mortality in weight-loser PD patients, the risk is higher in men than in women according to a retrospective study (Walker, Davidson, & Gray, 2012). Steady annual decline in body weight was associated with an increased mortality and reduced quality of life (Akbar et al., 2015). Higher mortality in weight losers may simply be a reflection of advancing disease but may be a component of the weight loss-frailty phenotype. Another study reported weight loss to be independently associated with poor quality of life, measured on PDQ8 (significant variables for quality of life—UPDRS score—0.001, dyskinesia—0.003, weight loss—0.04), revealing that weight loss has a significant negative impact on quality of life (Sharma, 2013). Quality of life has been shown to be reduced in weight losers with an increase in PDQ-39 (Parkinson's Disease Questionnaire) score, particularly in the mobility and ADL components (Bachmann & Trenkwalder, 2006). Some of the impact on quality of life may be secondary to increasing frailty and dyskinesia in these patients.

9. IDENTIFYING WEIGHT-LOSER PD PATIENTS—THE PARK-OLFACTION-WEIGHT PHENOTYPE?

There are no differences in clinical parameters of weight losers and nonweight losers to enable early indication for weight loss in subsequent progression of PD. It may however be possible to identify potential weight losers by an early assessment of olfactory impairment in PD patients, perhaps soon after the clinical diagnosis and reassessed during follow-up. As detailed earlier (Sharma & Turton, 2012), the weight losers were noticed to have suffered more severe loss of olfaction (anosmic group); thus, a low score on olfaction measurement would indicate a higher risk of weight loss in the subsequent years and thus the risk of LID as compared to the patients with less severe loss of olfaction (hyposmic group) due to receiving a higher levodopa dose per kilogram thus entering the "levodopa dose-weight-dyskinesia" cascade. Dyskinetic patients in this study had steadily lost weight prior to developing dyskinesia, i.e., weight loss preceded dyskinesia. Thus the weight losers belong to the phenotype of anosmic patients, i.e., the *Park-olfaction-weight-phenotype* (Fig. 4), patients in the anosmic range being at risk of weight loss and dyskinesia—the triad of "olfaction-weight-dyskinesia" (Fig. 5). The phenotypical characteristics support the notion of heterogeneity of PD which may be demonstrable at an early stage of clinical presentation (Erro et al., 2013) and describe one of the nonmotor predominant subtype of PD (Sauerbier, Jenner, Todorova, & Chaudhuri,

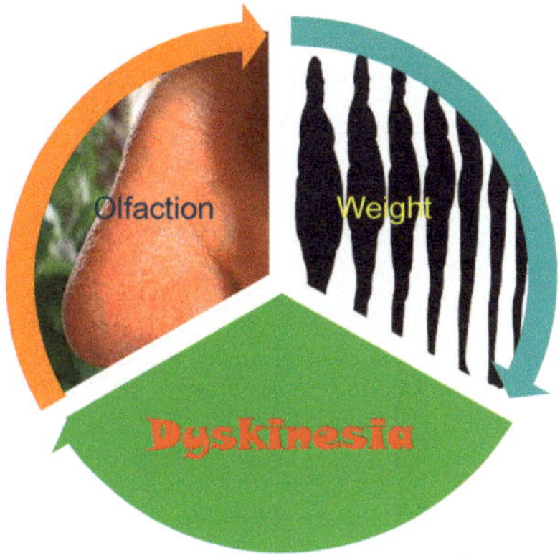

Fig. 5 The triad of olfaction-weight-dyskinesia.

2016). It seems that the characteristics and pathology of prodromal phase determine and influence clinical disease and its progression.

10. CONCLUSIONS

- Body weight is a significant nonmotor feature in PD.
- Body weight profile changes during the disease progression.
- Some patients lose weight, while the others remain stable or gain weight.
- There are complex factors leading to change in body weight.
- Weight loss is not a benign phenomenon, it has prognostic implications and is associated with negative impact on general health, poor quality of life, and higher mortality.
- Low body weight at an early stage or during the course of disease predisposes patients to the levodopa-related dyskinesia.
- Patients at risk of weight loss may be identified by their severe loss of olfaction (anosmia vs hyposmia) at an early stage.
- The degree of olfactory loss (anosmia vs hyposmia) may be an index of severity and/or duration of preclinical neurodegenerative process (Fig. 4) and more severe clinical disease.
- The prodromal phase seems to determine the clinical subtype of PD.
- An identification of Park-olfaction-weight phenotype may provide measures for management strategies in individual patients for prevention and amelioration of weight loss and dyskinesia.
- A rule of the clinical triad of "olfaction-weight-dyskinesia" may be of some value in patient management.

11. RECOMMENDATIONS AND IMPLICATIONS ON DISEASE MANAGEMENT

- PD patients should be actively monitored for initial weight and subsequent body weight changes.
- Strategies need to be developed to prevent the onset of dyskinesia in PD since dyskinesia has a negative impact on quality of life.
- Early identification of potential weight losers (anosmic phenotype A of Park-olfaction-weight-phenotype) should lead to strategies for measures against weight loss. Weight gain may not be possible but should be aimed for. A high carbohydrate diet where possible may be of value.
- An assessment of the triad of "olfaction-weight-dyskinesia" should be a reminder in identifying and managing patients.
- Levodopa dose should be adjusted according to the body weight and any subsequent weight loss should alert to the risk of dyskinesia. Non-levodopa medications should be used as appropriate to achieve best motor symptom control.
- Patients should be advised for nutritional intake to prevent weight loss and the consequences of malnutrition.
- Overweight PD patients should be advised for general health matters and life style changes to prevent negative health impact including cardiovascular disease.
- Overweight PD patients undergoing DBS should be advised of the potential for weight gain as the consequence of surgical treatment and counseled accordingly.
- Further research into prodromal characteristics and disease progression is warranted, using weight as a surrogate measure of disease progression in neuroprotection trials.

REFERENCES

Abbott, R. D., Ross, G. W., White, L. R., Nelson, J. S., Masaki, K. H., Tanner, C. M., et al. (2002). Midlife adiposity and the future risk of Parkinson's disease. *Neurology*, *59*(7), 1051–1057.

Abou-Raya, S., Abou-Raya, A., El Khadrawy, T., & Helmii, M. (2014). Prevalence of sarcopenia and osteopenia/osteoporosis in older adults with parkinson's disease: A cross-sectional analysis. *Annals of the Rheumatic Diseases*, *71*(Suppl. 3), 600.

Aden, E., Carlsson, M., Poortvilet, E., Stenlund, H., Linder, I., Edstrom, M., et al. (2011). Dietary intake and olfactory function in newly diagnosed Parkinson's disease: A case control study. *Nutritional Neuroscience*, *14*(1), 25–31.

Akbar, U., He, Y., Dai, Y., Hack, N., Malaty, I., McFarland, N. R., et al. (2015). Weight loss and impact on quality of life in Parkinson's disease. *PloS One*, *10*(5), e0124541.

Alvar, G. N., Sauleda, J., Miralles, C., Gomez, C., Togores, B., Sala, E., et al. (2002). Skeletal muscle apoptosis and weight loss in chronic obstructive pulmonary disease. *American Journal of Respiratory and Critical Care Medicine*, *166*(4), 485–489.

Anderson, D. E. (2007). Reliability of air displacement plethysmography. *Journal of Strength and Conditioning Research*, *21*(1), 169–172.

Anker, S., Negassa, A., Coats, A. J. S., Afzal, R., Poole-Wilson, P. A., Cohn, J. N., et al. (2003). Prognostic importance of weight loss in chronic heart failure and the effect of treatment with angiotensin-converting-enzyme inhibitors: An observational study. *The Lancet*, *361*, 1077–1083.

Aziz, N. A., van der Marck, M. A., Pijl, H., Olde Rikkert, M. G., Bloem, B. R., & Roos, R. A. (2008). Weight loss in neurodegenerative disorders. *Journal of Neurology*, *255*, 1872–1880.

Bachmann, C. G., & Trenkwalder, C. (2006). Body weight in patients with Parkinson's disease. *Movement Disorders*, *21*(11), 1824–1830.

Bachmann, C. G., Zapf, A., Brunner, E., & Trenkwalder, C. (2009). Dopaminergic treatment is associated with decreased body weight in patients with Parkinson's disease and dyskinesia. *European Journal of Neurology*, *16*(8), 895–901.

Bannier, S., Montaurier, C., Derost, P. P., Ulla, M., Lemaire, J. J., Boirie, Y., et al. (2009). Overweight after deep brain stimulation of the subthalamic nucleus in Parkinson's disease: Long term follow-up. *Journal of Neurology, Neurosurgery, and Psychiatry*, *80*(5), 484–488.

Barichella, M., Cereda, E., Cassani, E., Pinelli, G., Iorio, L., Ferri, V., et al. (2016). Dietary habits and neurological features of Parkinson's disease patients: Implications for practice. *Clinical Nutrition.* pii: S0261-5614(16)30155-8.

Barichella, M., Pinelli, G., Iorio, L., Cassani, E., Valentino, A., Pusani, C., et al. (2016). Sarcopenia and dynapenia in patients with Parkinsonism. *Journal of the American Medical Directors Association*, *17*, 640–646.

Barichella, M., Marczewska, A. M., Mariani, C., Landi, A., Vairo, A., & Pezzoli, G. (2003). Body weight gain rate in patients with Parkinson's disease after deep brain stimulation. *Movement Disorders*, *18*(11), 1337–1340.

Barichella, M., Marczewska, A., Vairo, A., Canes, M., & Pezzoli, G. (2003). Is under weightedness still a major problem in Parkinson's disease patients? *European Journal of Clinical Nutrition*, *57*(4), 543–547.

Benarroch, E. E. (2010). Neural control of feeding behaviour: Overview and clinical correlations. *Neurology*, *74*(20), 1643–1650.

Berridge, K. C., Ho, C. Y., Richard, J. M., & Di Felice Antonio, A. G. (2010). The tempted brain eats: Pleasure and desire circuits in obesity and eating disorders. *Brain Research*, *1350*, 43–46.

Beyer, P. L., Palarino, M. Y., Michalek, D., Busenbark, K., & Koller, W. C. (1995). Weight changes and body composition in patients with Parkinson's disease. *Journal of the American Dietetic Association*, *95*(9), 879–883.

Capecci, M., Petrelli, M., Emanuelli, B., Millevolte, M., Nicolai, A., Provinciali, L., et al. (2013). Rest energy expenditure in Parkinson's disease: Role of disease progression and dopaminergic therapy. *Parkinsonism & Related Disorders*, *19*(2), 238–241.

Cesari, M., Gambassi, G., Van Kan, G. A., & Vellars, B. (2014). The frailty phenotype and the frailty index: Different instruments for different purposes. *Age and Ageing*, *43*(1), 10–12.

Chen, H., Zhang, S. M., Hernan, M. A., Willett, W. C., & Ascherio, A. (2003). Weight loss in Parkinson's disease. *Annals of Neurology*, *53*(5), 676–679.

Chen, H., Zhang, S. M., Schwarzschild, M. A., Hernan, M. A., Willett, W. C., & Ascherio, A. (2004). Obesity and the risk of Parkinson's disease. *American Journal of Epidemiology*, *199*, 547–555.

Cheshire, W. P., & Wszolek, Z. K. (2005). Body mass index is reduced early in Parkinson's disease. *Parkinsonism & Related Disorders*, *11*(1), 35–38.

Clegg, A., Young, J., Lliffe, S., Rikkert, M. O., & Rockwood, K. (2013). Frailty in elderly people. *Lancet*, *381*(9868), 752–762.

Dahl, A. K., & Reynolds, C. A. (2013). Accuracy of recalled body weight. A study with 20 years follow-up. *Obesity*, *21*(6), 1293–1298.

de Boer, A., Ter Horst, G. I., & Lorist, M. M. (2013). Physiological and psychological age-related changes associated with reduced food intake in older persons. *Ageing Research Reviews*, *12*, 316–328.

De Pablo-Fernández, E., Breen, D. P., Bouloux, P. M., Barker, R. A., Foltynie, T., & Warner, T. T. (2017). Neuroendocrine abnormalities in Parkinson's disease. *Journal of Neurology, Neurosurgery, and Psychiatry*, *88*(2), 176–185.

Delikanaki-Skaribas, E., Trail, M., Wong, W. W., & Lai, E. C. (2009). Daily energy expenditure, physical activity, and weight loss in Parkinson's disease patients. *Movement Disorders*, *24*(5), 667–671.

Doty, R. L., Sharman, P., & Dann, M. (1984). Development of University of Pennsylvania smell identification test: A microencapsulated test for olfactory function. *Physiology & Behavior*, *32*, 489–502.

Drey, M., Hasman, S. E., Krenovsky, J., Hobert, M. A., Straub, S., & Elshehabi, M. (2017). Association between early markers of Parkinson's disease and sarcopenia. *Frontiers in Ageing Neuroscience*, *9*, 53. http://dx.doi.org/10.3389/fnagi.2017.00053.

Durrieu, G., Llau, M. E., Rascol, O., Senard, J. M., Rascol, A., & Montastruc, J. L. (1992). Parkinson's disease and weight loss: A study with anthropometric and nutritional assessment. *Clinical Autonomic Research*, *2*(3), 153–157.

Erro, R., Vitale, C., Amboni, M., Picillo, M., Moccia, M., Longo, K., et al. (2013). Heterogeneity of early Parkinson's disease: A cluster analysis on newly diagnosed untreated patients. *PloS One*, *8*(8), e70244.

Evidente, V. G., Caviness, J. N., Adler, C. H., Gwinn-Hardy, K. A., & Pratley, R. E. (2001). Serum leptin concentrations and satiety in Parkinson's disease patients with and without weight loss. *Movement Disorders*, *16*(5), 924–927.

Fiszer, U., Michalowska, M., Baranowska, B., Wolinska-Witort, E., Jeske, W., Jethon, M., et al. (2010). Leptin and ghrelin concentrations and weight loss in Parkinson's disease. *Acta Neurologica Scandinavica*, *121*(4), 230–236.

Fox, S. H., Chuang, R., & Brotchie, J. M. (2009). Serotonin and Parkinson's disease: On movement, mood and madness. *Movement Disorders*, *24*, 1255–1260.

Friess, E., Kuempfel, T., Winkelmann, J., Schmid, D., Uhr, M., Rupprecht, R., et al. (2001). Increased growth hormone response to apomorphine in Parkinson's disease compared with multiple system atrophy. *Archives of Neurology*, *58*, 241–246.

Fronczek, R., Overeem, S., Lee, S. Y., Hegmann, I. M., van Pelt, J., van Duinen, S. G., et al. (2007). Hypocretin (orexin) loss in Parkinson's disease. *Brain*, *130*, 1577–1585.

Hassin-Baer, S., Molchadski, I., Cohen, O. S., Nitzan, Z., Efrati, L., Tunkel, O., et al. (2011). Gender effect on time to levodopa-induced dyskinesias. *Journal of Neurology*, *258*(11), 2048–2053.

Holden, K. E. (2001). Unintentional weight loss and its management in patients with Parkinson's disease. *Nutrition in Clinical Care*, *4*(3), 131–139.

Hu, G., Jousilahti, P., Nissinen, A., Antikainen, R., Kivipelto, M., & Tuomilehto, J. (2006). Body mass index and the risk of Parkinson disease. *Neurology*, *67*(11), 1955–1959.

Jaafar, A. F., Gray, W. K., Porter, B., Turnbull, E. J., & Walker, R. W. (2010). A cross-sectional study of the nutritional status of community-dwelling people with idiopathic Parkinson's disease. *BMC Neurology*, *10*, 124.

Jankovic, J., Wooten, M., Van der Linden, C., & Jansson, B. (1992). Low body weight in Parkinson's disease. *Southern Medical Journal*, *85*(4), 351–354.

Jorgensen, H. U., Werdelin, L., Lokkegaard, A., Westertrep, K. R., & Simonsen, L. (2012). Free living energy expenditure after deep brain stimulation for Parkinson's disease. *Clinical Physiology and Functional Imaging*, *32*, 214–220.

Kapur, S., & Marques, T. R. (2016). Dopamine, striatum, antipsychotics and questions about weight gain. *JAMA Neurology*, *73*, 107–108.

Kashihara, K. (2006). Weight loss in Parkinson's disease. *Journal of Neurology*, *253*(Suppl. 7), vii38–vii41.

Kistner, A., Lhommee, E., & Krack, P. (2014). Mechanisms of body weight fluctuations in Parkinson's disease. *Frontiers in Neurology*, *5*, 1–15.

Kim, H. J., Oh, E. S., Lee, J. H., Moon, J. S., Oh, J. E., Shin, J. W., et al. (2012). Relationship between changes of body mass index (BMI) and cognitive decline in Parkinson's disease (PD). *Archives of Gerontology and Geriatrics*, *55*(1), 70–72.

Kroeze, W. K., Hufeisen, S. J., Popadak, B. A., Renock, S. M., Steinberg, S., Ernsberger, P., et al. (2003). H1-histamine receptor affinity predicts short-term weight gain for typical and atypical antipsychotic drugs. *Neuropsychopharmacology*, *28*, 519–526.

Kumru, H., Santamaria, J., Valldeoriola, F., Marti, M. J., & Tolosa, E. (2006). Increase in body weight after pramipexole treatment in parkinson's disease. *Movement Disorders*, *21*(11), 1972–1974.

Lawrence, A. D., Evans, A. H., & Lees, A. J. (2003). Compulsive use of dopamine replacement therapy in Parkinson's disease: Reward systems gone away? *Lancet Neurology*, *2*, 595–604.

Lee, J. J., Oh, J. S., Ham, J. H., Lee, D. H., Lee, I., Sohn, Y. H., et al. (2016). Association of body mass index and the depletion of nigrostriatal dopamine in Parkinson's disease. *Neurobiology of Aging*, *38*, 197–204.

Levi, S., Cox, M., Lugron, M., Hodkinson, M., & Tomkins, A. (1990). Increased energy expenditure in Parkinson's disease. *BMJ*, *301*, 1256–1257.

Lindskov, S., Sjöberg, K., Hagell, P., & Westergren, A. (2016). Weight stability in Parkinson's disease. *Nutritional Neuroscience*, *19*(1), 11–20.

Logroscino, G., Sesso, H. D., Paffenbarger, R. S., & Lee, J. M. (2007). Body mass index and risk of Parkinson's disease: A prospective cohort study. *American Journal of Epidemiology*, *166*, 1186–1190.

Lorefalt, B., Ganowiak, W., Palhagen, S., Toss, G., Unosson, M., & Granerus, A. K. (2004). Factors of importance for weight loss in elderly patients with Parkinson's disease. *Acta Neurologica Scandinavica*, *110*(3), 180–187.

Lorefalt, B., Granerus, A., & Unosson, M. (2006). Avoidance of solid food in weight losing older patients with Parkinson's disease. *Journal of Clinical Nursing*, *15*(11), 1404–1412.

Lorefalt, B., Toss, G., & Granerus, A. K. (2009). Weight loss, body fat mass and leptin in Parkinson's disease. *Movement Disorders*, *24*(6), 885–890.

Loureiro, A. P. C., Yamaguchi, I. L. P., Villegas, M. P., Ferreira, V. L., & Israel, V. L. (2016). Predicting risk of sarcopenia in people with Parkinson's disease. *Movement Disorders*, *31*(Suppl. 2), S578, Abstract 1752.

Macia, F., Perlemoine, C., Coman, I., Guehl, D., Burbaud, P., Cuny, E., et al. (2004). Parkinson's disease patients with bilateral deep brain stimulation gain weight. *Movement Disorders*, *19*(2), 206–212.

Markaki, E., Ellul, J., Kefalopoulou, Z., Trachani, E., Theodoropoulou, A., Kyriazopoulou, V., et al. (2012). The role of ghrelin, neuropeptide Y and leptin peptides in weight gain after deep brain stimulation for Parkinson's disease. *Stereotactic and Functional Neurosurgery*, *90*(2), 104–112.

Markus, H. S., Cox, M., & Tomkins, A. M. (1992). Raised energy expenditure in Parkinson's disease and its relationship to muscle rigidity. *Clinical Science (London)*, *83*(2), 199–204.

Markus, H. S., Tomkins, A. M., & Stern, G. M. (1993). Increased prevalence of undernutrition in Parkinson's disease and its relationship to clinical disease parameters. *Journal of Neural Transmission. Parkinson's Disease and Dementia Section*, *5*(2), 117–125.
Meister, B. (2007). Neurotransmitters in key neurons of the hypothalamus that regulate feeding behavior and body weight. *Physiology & Behaviour*, *92*(1-2), 263–271.
Mills, K. A., Scherzer, R., Starr, P. A., & Ostrem, J. L. (2012). Weight change after globus pallidus internus or subthalamic nucleus deep brain stimulation in Parkinson's disease and dystonia. *Stereotactic and Functional Neurosurgery*, *90*(6), 386–393.
Moghaddasi, M., & Boshtam, M. (2010). Weight changes in Parkinson's disease patients after subthalamic nucleus deep brain stimulation surgery. *Acta Neurologica Belgica*, *110*(4), 311–313.
Montaurier, C., Morio, B., Tasker, R., Derost, P., Arnaud, P., Brandolini-Bunlon, M., et al. (2007). Mechanisms of body weight gain in patients with Parkinson's disease after subthalamic stimulation. *Brain*, *130*(Pt. 7), 1808–1818.
Muller, T., Woitalla, D., Saft, C., & Kuhn, W. (2006). Levodopa in plasma correlates with body weight of parkinsonian patients. *Parkinsonism & Related Disorders*, *6*(3), 171–173.
Nakamura, T., Suzuki, M., Ueda, M., Hirayama, M., & Katsuno, M. (2017). Lower body mass index is associated with orthostatic hypotension in Parkinson's disease. *Journal of the Neurological Sciences*, *372*, 14–18.
Nam, S. Y., & Marcus, C. (2000). Growth hormone and adipocyte function in obesity. *Hormone Research*, *53*(Suppl. 1), 87–97.
Narayanan, N. S., Guarnieri, D. J. & DiLeone, R., J. (2010). Metabolic hormones, dopamine circuits and feeding. Frontiers in Neuroendocrinology 31, 104–12.
Nirenberg, M. J., & Waters, C. (2006). Compulsive eating and weight gain related to dopamine agonist use. *Movement Disorders*, *21*(4), 524–529.
Novakova, L., Ruzicka, E., Jech, R., Serranova, T., Dusek, P., & Urgosik, D. (2007). Increase in body weight is a non-motor side effect of deep brain stimulation of the subthalamic nucleus in Parkinson's disease. *Neuro Endocrinology Letters*, *28*(1), 21–25.
Nozaki, S., Saito, T., Matsumura, T., Miyai, I., & Kang, J. (1999). Relationship between weight loss and dysphagia in patients with Parkinson's disease. *Rinshō Shinkeigaku*, *39*(10), 1010–1014.
Okuma, Y. (2012). Fatigue and weight loss in Parkinson's disease. *Brain and Nerve*, *64*(4), 384–393.
Olanow, C. W., Kieburtz, K., Rascol, O., Poewe, W., Schapira, A. H., Emre, M., et al. (2013). Factors predictive of the development of levodopa induced dyskinesia and wearing off in Parkinson's disease. *Movement Disorders*, *28*(8), 1064–1071.
Ondo, W. G., Ben-Aire, L., Jankovic, J., Lai, E., Contant, C., & Grossman, R. (2000). Weight gain following unilateral pallidotomy in parkinson's disease. *Acta Neurologica Scandinavica*, *101*(2), 79–84.
Palhagen, S., Lorefait, B., Carrison, M., Ganowiak, W., Toss, G., Unosson, M., et al. (2005). Does L-dopa treatment contribute to reduction in body weight in elderly patients with Parkinson's disease. *Acta Neurologica Scandinavica*, *111*(1), 12–20.
Palmiter, R. D. (2007). Is dopamine a physiologically relevant mediator of feeding behaviour? *Trends in Neurosciences*, *30*(8), 375–381.
Perlemoine, C., Macia, F., Tison, F., Coman, I., Guehl, D., Burbaud, P., et al. (2005). Effects of subthalamic nucleus deep brain stimulation and levodopa on energy production rate and substrate oxidation in Parkinson's disease. *The British Journal of Nutrition*, *93*(2), 191–198.
Petroni, M. L., Albani, G., Bicchiega, V., Baudo, S., Vinci, C., Montesano, A., et al. (2003). Body composition in advanced-stage Parkinson's disease. *Acta Diabetologica*, *40*(Suppl. 1), S187–S190.

Politis, M., & Loane, C. (2011). Serotonergic dysfunction in Parkinsons' disease and its relevance to disability. *The Scientific World Journal, 11*, 1724–1736.

Postuma, R. B., Aarsland, D., Barone, P., Burn, D. J., Hawkes, C. H., Oertel, W., et al. (2012). Identifying prodromal Parkinson's disease: Pre-motor disorders in Parkinson's disease. *Movement Disorders, 27*(5), 617–626.

Poehlman, E. T., & Dvorak, R. V. (2000). Energy expenditure, energy intake and weight loss in Alzheimer's disease. *The American Journal of Clinical Nutrition, 71*(2), 650S–655S.

Rieu, I., Derost, P., Ulla, M., Marques, A., Debilly, B., De Chazeron, I., et al. (2011). Body weight gain and deep brain stimulation. *Journal of Neurological Science, 310*(1-2), 267–270.

Roland, K. P., Jakobi, J. M., & Jones, G. R. (2011) Contributing factors to frailty status in persons with Parkinson's disease. 15th International congress of Parkinson's disease and movement disorders, Vol. 26, S296.

Roland, K. P., Cornett, K. M., Theou, O., Jakobi, J. M., & Jones, G. R. (2012). Concurrence of frailty and Parkinson's disease. *Journal of Frailty & Aging, 1*(3), 123–127.

Sakajiri, K., & Takamori, M. (1997). Body fat loss in patients with Parkinson's disease. *Rinshō Shinkeigaku, 37*(7), 611–614.

Saper, C. B., Chou, T. C., & Elmquist, J. K. (2002). The need to feed: Homeostatic and hedonic control of eating. *Neuron, 36*(2), 199–211.

Sauerbier, A., Jenner, P., Todorova, A., & Chaudhuri, K. R. (2016). Non motor subtypes and Parkinson's disease. *Parkinsonism & Related Disorders, 22*(Suppl. 1), S41–S46.

Sauleau, P., Drapier, S., Duprez, J., Houvenaghel, J. F., Dondaine, T., Haegelen, C., et al. (2016). Weight gain following Pallidal deep brain stimulation: A PET study. *PloS One, 11*(4), e0153438.

Sauleau, P., Leray, E., Rouaud, T., Drapier, S., Drapier, D., Blanchard, S., et al. (2009). Comparison of weight gain and energy intake after subthalamic versus pallidal stimulation in Parkinson's disease. *Movement Disorders, 24*(14), 2149–2155.

Sergi, D., De Rui, M., Coin, A., Inelman, E. M., & Manzato, E. (2013). Weight loss and Alzheimer's disease: Temporal and aetiology connections. *The Proceedings of the Nutrition Society, 72*(1), 160–165.

Sharma, J. C., & Ross, I. N. (1999). Long term role of pergolide as an adjunct therapy in Parkinson's disease: Influence on disability, blood pressure and levodopa syndrome. *Parkinsonism & Related Disorders, 5*, 111–114.

Sharma, J. C., Macnamara, L., Hasoon, M., Vassallo, M., & Ross, I. (2006). Cascade of levodopa dose and weight related dyskinesia in Parkinson's disease (LD-WD-PD cascade). *Parkinsonism & Related Disorders, 12*(8), 499–505.

Sharma, J. C., Ross, I. N., Rascol, O., & Brooks, D. (2008). Relationship between weight, levodopa and dyskinesia: The significance of levodopa dose per kilogram body weight. *European Journal of Neurology, 15*(5), 493–496.

Sharma, J. C., & Hasoon, M. (2009). Clinical characteristics of weight losers in Parkinson's disease. *Movement Disorders, 24*(Suppl. 1), S538.

Sharma, J. C., Bachmann, C. G., & Linzasoro, G. (2010). Classifying risk factors for dyskinesia in Parkinson's disease. *Parkinsonism & Related Disorders, 16*(8), 490–497.

Sharma, J. C., & Turton, J. (2012). Olfaction, dyskinesia and profile of weight change in Parkinson's disease: Identifying neurodegenerative phenotypes. *Parkinsonism & Related Disorders, 18*(8), 964–970.

Sharma, J. C. (2013). In *Weight changes in Parkinson's disease and its relationship with long term prognosis and dyskinesia. 20th Parkinson's disease and related disorders conference. Geneva.* Abstract.

Sharma, J. C., & Vassallo, M. (2014). Prognostic significance of weight changes in Parkinson's disease: The Park-weight phenotype. *Neurodegenerative Disease Management, 4*(4), 309–316.

Sheard, J. M., Ash, S., Silburn, P. A., & Kerr, G. K. (2001). Prevalence of malnutrition in Parkinson's disease: A systematic review. *Nutrition Reviews, 69*(9), 520–532.
Strowd, R. E., Cartwright, M. S., Passmore, L. V., Ellis, T. L., Tatter, S. B., & Siddiqui, M. S. (2010). Weight change following deep brain stimulation for movement disorders. *Journal of Neurology, 257*(8), 1293–1297.
Strowd, R. E., Herco, M., Passmore-Griffin, L., Avery, B., Haq, I., Tatter, S. B., et al. (2016). Association between subthalamic nucleus deep brain stimulation and weight gain: Results of a case-control study. *Clinical Neurology and Neurosurgery, 140*, 38–42.
Thannickal, T. C., Lai, Y. Y., & Siegel, J. M. (2008). Hypocretin (orexin) and melanin concentrating hormone loss and the symptoms of Parkinson's disease. *Brain, 131*, e87.
Toth, M. J., Fishman, P. S., & Poehlman, E. T. (1997). Free living daily energy expenditure in Parkinson's disease. *Neurology, 48*(1), 88–91.
Tucker, L. A., Lecheminant, J. D., & Bailey, B. W. (2014). Test-retest reliability of the bod pod: The effect of multiple assessments. *Perceptual and Motor Skills, 118*(2), 563–570.
Tuite, P. J., Maxwell, R. E., Ikramuddin, S., Kotz, C. M., Billington, C. J., Laseski, M. A., et al. (2005). Weight and body mass index in Parkinson's disease patients after deep brain stimulation surgery. *Parkinsonism & Related Disorders, 11*(4), 247–252.
Uc, E. Y., Struck, L. K., Rodnitzky, R. L., Zimmerman, B., Dobson, J., & Evans, W. J. (2006). Predictors of weight loss in Parkinson's disease. *Movement Disorders, 21*(7), 930–936.
Umehara, T., Nakahara, A., Matsuno, H., Toyoda, C., & Oka, H. (2016). Body weight and dysautonomia in early Parkinson's disease. *Acta Neurologica Scandinavica, 135*(5), 560–567. http://dx.doi.org/10.1111/ane.12633..
van der Marck, M. A., Dicke, H. C., Uc, E. Y., Kentin, Z. H., Borm, G. F., Bloem, B. R., et al. (2012). Body mass index in Parkinson's disease: A meta-analysis. *Parkinsonism & Related Disorders, 18*(3), 263–267.
Vardi, J., Oberman, Z., Rabey, I., Streifler, M., Ayalon, D., & Herzberg, M. (1976). Weight loss in patients treated long term with levodopa. *Journal of the Neurological Sciences, 30*(1), 33–40.
Vetrano, D. L., Pisciotta, M. S., Brandi, V., Lo Monaco, M. R., Laudisio, A., Onder, G., et al. (2015). Prevalence of sarcopenia in Parkinson's disease. European Geriatric. *Medicine, 6*(1), S93.
Vikdahl, M., Carlsson, M., Linder, J., Forsgren, L., & Haglin, L. (2014). Weight gain and increased central obesity in early phase of Parkinson's disease. *Clinical Nutrition, 33*(6), 1132–1139.
Vikdahl, M., Domellöf, M. E., Forsgren, L., & Håglin, L. (2015). Olfactory function, eating ability, and visceral obesity associated with MMSE three years after Parkinson's disease diagnosis. *Journal of Nutrition, Health & Aging, 19*(9), 894–900.
Vitek, J. L., Bakay, R. A., Freeman, A., Evatt, M., Green, J., McDonald, W., et al. (2000). Randomised trial of pallidotomy versus medical therapy for Parkinsons' disease. *Annals of Neurology, 53*(5), 558–569.
Volkow, N. D., Wang, G. J., Fowler, J. S., Logan, J., Jayne, M., Franceschi, D., et al. (2002). "Non-hedonic" food motivation in humans involves dopamine striatum and methylphenidate amplifies this effect. *Synapse, 44*, 175–180.
Volkow, N. D., Wang, G. I., & Baler, R. D. (2011). Reward, dopamine and the control of food intake. *Trends in Cognitive Sciences, 15*, 37–46.
Walker, R., Davidson, M., & Gray, W. (2012). Gender differences in 1-year survival rates after weight loss in idiopathic Parkinson's disease. *International Journal of Palliative Nursing, 18*(1), 35–39.
Wang, G. J., Volkow, N. D., Logan, J., Pappas, N. R., Wong, C. T., Zhu, W., et al. (2001). Brain dopamine and obesity. *Lancet, 357*, 354–357.

Wetterling, T., & Mussigbrodt, H. E. (1999). Weight gain: Side effects of atypical neuroleptics? *Journal of Clinical Psychopharmacology*, *19*, 316–321.

White, H., Pieper, C., Schmader, K., & Fillenbaum, G. (1996). Weight change in Alzheimer's disease. *Journal of the American Geriatrics Society*, *44*, 265–272.

White, H., Pieper, C., & Schmader, K. (1998). The association of weight change in Alzheimer's disease with severity and mortality: A longitudinal analysis. *Journal of the American Geriatrics Society*, *46*, 1223–1227.

Weintraub, D., Koester, J., Potenza, M. N., Siderowf, A. D., Stacy, M., Voon, V., et al. (2010). Impulse control disorders in Parkinson's disease: A cross sectional study of 3090 patients. *Archives of Neurology*, *67*, 589–595.

Wills, A. M., Pérez, A., Wang, J., Su, X., Morgan, J., Rajan, S. S., et al. (2016). Association between change in body mass index, unified Parkinson's disease rating scale scores, and survival among persons with Parkinson disease: Secondary analysis of longitudinal data from NINDS exploratory trials in Parkinson disease long-term study 1. *JAMA Neurology*, *73*(3), 321–328.

Wolf, M. E., Capelle, H. H., Lütjens, G., Ebert, A. D., Hennerici, M. G., Krauss, J. K., et al. (2016). Body weight gain in patients with bilateral deep brain stimulation for dystonia. *Journal of Neural Transmission (Vienna)*, *123*(3), 261–267.

Woltz, M., Kaminsky, A., Lohle, M., Koch, R., Storch, A., & Reichmann, H. (2009). Chocolate consumption is increased in Parkinson's disease. Results from a self-questionnaire study. *Journal of Neurology*, *256*, 488–492.

Woulters, E. F. M., Creutzberg, E. C., & Schols, A. M. J. W. (2002). Systemic effects in COPD. *Chest*, *121*, 127S–130S.

Zappia, M., Crescibene, L., Arabia, G., Nicoletti, G., Bagala, A., Bastone, L., et al. (2002). Body weight influences pharmacokinetics of levodopa in Parkinson's disease. *Clinical Neuropharmacology*, *25*(2), 79–82.

FURTHER READING

Barichella, M., Cereda, E., & Pezzoli, G. (2009). Major nutritional issues in the management of Parkinson's disease. *Movement Disorders*, *24*(13), 1881–1892.

Bernhardt, D., Müller, H. P., Ludolph, A. C., Dupuis, L., & Kassubek, J. (2016). Body fat distribution in Parkinson's disease: An MRI-based body fat quantification study. *Parkinsonism & Related Disorders*, *33*, 84–89.

Jellinger, K. A. (2012). Neuropathology of sporadic Parkinson's disease: Evaluation and changes of concept. *Movement Disorders*, *27*(1), 8–30.

Thenganatt, M. A., & Jankovic, J. (2014). Parkinson's disease subtypes. *JAMA Neurology*, *71*(4), 499–504.

CHAPTER THIRTY-TWO

Visual Dysfunction in Parkinson's Disease

Richard A. Armstrong[1]
Vision Sciences, Aston University, Birmingham, United Kingdom
[1]Corresponding author: e-mail address: r.a.armstrong@aston.ac.uk

Contents

Abstract

This chapter describes the visual problems likely to be encountered in Parkinson's disease (PD) and whether such signs are useful in differentiating the parkinsonian syndromes. Visual dysfunction in PD may involve visual acuity, contrast sensitivity, color discrimination, pupil reactivity, saccadic and pursuit eye movements, motion perception, visual fields, and visual processing speeds. In addition, disturbance of visuospatial

International Review of Neurobiology, Volume 134
ISSN 0074-7742
http://dx.doi.org/10.1016/bs.irn.2017.04.007

orientation, facial recognition problems, rapid eye movement (REM) sleep behavior disorder, and chronic visual hallucinations may be present. Problems affecting pupil reactivity, stereopsis, pursuit eye movement, and visuomotor adaptation, when accompanied by REM sleep behavior disorder, could be early features of PD. Dementia associated with PD is associated with enhanced eye movement problems, visuospatial deficits, and visual hallucinations. Visual dysfunction may be a useful diagnostic feature in differentiating PD from other parkinsonian symptoms, visual hallucinations, visuospatial dysfunction, and variation in saccadic eye movement problems being particularly useful discriminating features.

1. INTRODUCTION

Parkinson's disease (PD) patients exhibit a variety of nonmotor symptoms including loss of will, decreased activity, and poverty of behavior (Antal, Bandini, Keri, & Bodis-Wollner, 1998) as well as depression, sleep problems, cognitive impairment, associated dementia, and autonomic, gastrointestinal, and sensory dysfunction (Armstrong, 2008). Sensory problems may include visual dysfunction, loss of smell, auditory problems, and "restless legs" syndrome (Earley, 2003). A variety of visual signs and symptoms have been reported in PD including defects in primary vision such as visual acuity (VA), color vision, and eye movement, and deficits in more complex visual tasks involving the ability to judge distance or the shape of an object (Armstrong, 2008, 2016a; Weil et al., 2016). Visual deficits in PD are important in influencing overall motor function (Diederich, Raman, Leurgans, & Goetz, 2002) and are a risk factor for developing hallucinations (Matsui et al., 2006), and unmet visual needs can have a significant influence on the overall quality of life of the patient (Armstrong, 2008).

PD is the most common member of the "parkinsonian syndromes," a group of closely related motor disorders which also includes progressive supranuclear palsy (PSP) (Armstrong, 2011), dementia with Lewy bodies (DLB) (Armstrong, 2012), multiple system atrophy (MSA) (Armstrong, 2014), and corticobasal degeneration (CBD) (Armstrong, 2016b). There is considerable overlap among members of this group, and differential diagnosis can be challenging, especially early in the disease. Nevertheless, where visual dysfunction is present, it may be a useful additional diagnostic feature (Armstrong, 2008). Hence, this chapter describes: (1) the dopamine pathways of eye and brain and the neuropathology of PD, (2) the visual problems likely to be encountered in PD, (3) whether visual symptoms are useful in the differential diagnosis of parkinsonian syndromes, and (4) the visual care and quality of life of the patient.

2. DOPAMINE PATHWAYS IN EYE AND BRAIN

A number of neurotransmitter deficits have been observed in PD, but dysfunction of the dopamine neurotransmitter system is most characteristic of the disorder and may be responsible for a number of visual deficits. Dopamine is an important neurotransmitter in the retina being present in amacrine cells along the inner border of the inner nuclear layer (INL) (Dowling, Ehinger, & Hedden, 1976) and is also accumulated by interplexiform cells (Frederick, Rayborn, Laties, Lam, & Hollyfield, 1982). Two types of amacrine cells could be involved: (1) type 1 cells which send ascending processes to synapse with γ-aminobutyric acid interplexiform cells in stratum 1 of the INL and (2) type 2 cells which have dendrites stratifying above those of the type 1 cells of the inner plexiform layer (IPL). Hence, dopamine may be involved in the organization of the ganglion cell and bipolar cell receptive fields, and may influence the activity of the photoreceptors (Masson, Mestre, & Blin, 1993). Dopamine is also involved in the coupling of the horizontal and amacrine lateral system (Djamgoz, Hankins, Hirano, & Archer, 1997). Thinning of the retinal nerve fiber layer has been recorded in PD (Moreno-Ramos, Benito-Leon, Villarejo, & Bermejo-Pareja, 2013) affecting the INL in parafoveal regions (Fig. 1) and especially in those patients exhibiting visual hallucinations but without overt signs of Parkinson's disease dementia (PDD) (Lee et al., 2014). In addition, pathological changes have been observed in retinal blood vessels in PD, especially affecting the veins (Kromer et al., 2016).

A number of dopamine pathways are present in human brain (Fig. 2). First, the striatonigral pathway originates in the substantia nigra (cell group A9) and terminates in the striatum (caudate nucleus and putamen). Second, two pathways originate in the ventral tegmentum (cell groups A8 and A10) and project to the nucleus accumbens and limbic system (mesolimbic pathway), and frontal cortex (mesocortical pathway). There are also dopamine pathways within the hypothalamus, which transmit dopamine to the pituitary gland, and four pathways connecting the ventral tegmentum to the amygdala, hippocampus, cingulate gyrus, and olfactory bulb. Within the cerebral cortex, significant dopamine activity is largely limited to frontal and limbic areas, and there is significantly less activity in visual cortex (Nguyen-Legros, Harnois, DiPaolo, & Simon, 1993). Nevertheless, cerebral metabolic rates for glucose are reduced by up to 23% in primary visual cortex (area V1) of PD patients (Eberling, Richardson, Reed, Wolfe, & Jagust, 1994). Reductions in dopamine levels in basal ganglia and frontal cortex

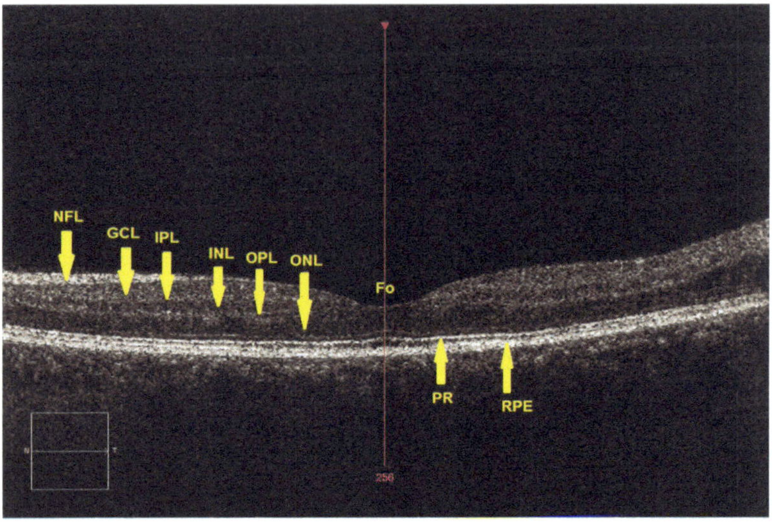

Fig. 1 Retinal image of a normal control subject using optical coherence tomography (OCT) showing the various layers: *Fo*, location of fovea; *GCL*, ganglion cell layer; *INL*, inner nuclear layer; *IPL*, inner plexiform layer; *N*, nasal direction; *NFL*, nerve fiber layer; *ONL*, outer nuclear layer; *OPL*, outer plexiform layer; *PR*, photoreceptor layer; *RPE*, retinal pigment epithelium; *T*, temporal direction. A specific thinning of INL in the parafoveal region has been recorded in Parkinson's disease (PD). *(Image courtesy Dr. R Heitmar, Aston University)*

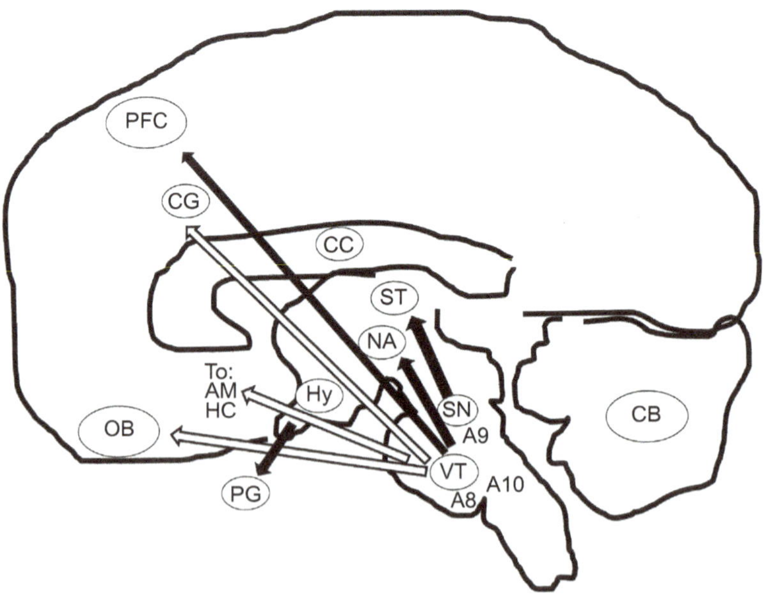

Fig. 2 The main dopamine pathways of the human brain (*AM*, amygdala; *CB*, cerebellum; *CC*, corpus callosum; *CG*, cingulate gyrus; *HC*, hippocampus; *Hy*, hypothalamus; *NA*, nucleus accumbens; *OB*, olfactory bulb; *PFC*, prefrontal cortex; *PG*, pituitary gland; *SN*, substantia nigra; *ST*, striatum; *VT*, ventral tegmentum).

may also deplete levels in the superior colliculus and therefore could be a factor in eye movement dysfunction (Crawford, Goodrich, Henderson, & Kennard, 1989). Dopamine also has a peripheral role in sympathetic ganglia, visceral ganglia, and in artery walls. Hence, reductions in dopamine in some of these regions could be a factor contributing to eye movement problems and defects in pupil reactivity in PD.

There is also increasing interest in the clinical effects of cholinergic deficits affecting the basal forebrain and tegmental pedunculopontine complex projection in PD (Bohnen et al., 2014; Muller and Bohnen, 2013; Muller et al., 2015). Whereas dopamine denervation is frequent in PD with minimal or no cognitive change, significant cortical cholinergic denervation is a major degenerative process associated with cognitive decline and PDD (Muller and Bohnen, 2013). Extranigral pathway and cholinergic deficits are also particularly common in PD patients exhibiting "freezing of gait" (Bohnen et al., 2014).

3. NEUROPATHOLOGICAL CHANGES

Neurons in the substantia nigra in PD often contain cytoplasmic "inclusions" called Lewy bodies (LB) which result from the degeneration of cytoskeletal filaments. LB may differ from other types of neurofibrillary pathology, such as the neurofibrillary tangles commonly found in Alzheimer's disease (AD), in that they contain abnormal aggregates of the protein α-synuclein (Spillantini et al., 1998). α-Synuclein is a small presynaptic protein, and the entire molecule undergoes a conformational change in PD resulting in the insoluble protein that forms a major component of the LB.

In PD, LB are confined largely to subcortical regions such as the substantia nigra and striatum, but in PDD, the pathology may spread to significantly affect the cerebral cortex including areas of visual cortex. In addition, studies have demonstrated a significantly more complex and widespread synucleinopathy especially in PDD affecting cortical and limbic regions (Armstrong et al., 2014; Saito et al., 2003) (Fig. 3). Hence, as well as LB, PDD cases also exhibit Lewy neurites (LN) which appear as abnormal threadlike structures, and Lewy grains (LG) resembling the argyrophilic dots and grains (AG) commonly observed in other neurodegenerative disorders such as argyrophilic grain disease (AGD) (Braak & Braak, 1998; Tolnay, Monsch, Staehelin, & Probst, 1999; Zhukareva et al., 2002), AD (Sabbagh et al., 2009), and in elderly cognitively normal brains

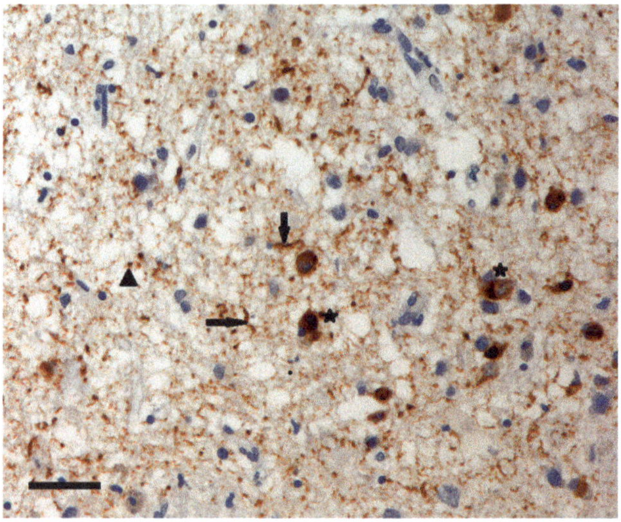

Fig. 3 Neuropathology of the cerebral cortex in a case of Parkinson's disease dementia (PDD) showing the presence of Lewy bodies (LB) (*stars*), Lewy neurites (LN) (*arrows*), and Lewy grains (LG) (*arrowheads*) (α-synuclein immunohistochemistry, hematoxylin stain, magnification bar = 25 μm).

(Ding et al., 2006; Josephs et al., 2008). Hence, spread of this pathology from subcortical to cortical regions could be an important factor in the development of PDD (Armstrong et al., 2014).

4. VISUAL DYSFUNCTION IN PD

The visual problems reported in PD are summarized in Table 1. Aspects of visual function have been divided into various categories based largely on those used to characterize vision in the dementias (Armstrong & Kergoat, 2015). Although these aspects of vision are described individually, patients can exhibit multiple visual defects (Armstrong, 2008).

4.1 Visual Acuity

Patients with PD frequently complain of poor vision especially as the disease progresses (Archibald, Clarke, Mosimann, & Burn, 2011; Repka et al., 1996), low contrast VA being especially affected (Jones & Donaldson, 1995; Jones et al., 1992; Lin et al., 2015). Impaired VA may also be a risk factor for the development of the chronic hallucinations most frequently associated with PDD (Matsui et al., 2006).

Table 1 Summary of Visual Signs and Symptoms of Parkinson's Disease (PD)

Ocular Aspect	Change in PD	References
Visual acuity	Poor, especially at low contrast	Jones, Donaldson, and Timmings (1992), Repka, Claro, Loupe, and Reich (1996), and Lin et al. (2015)
Color vision	Vision blurred for colored stimuli	Price, Feldman, Adelberg, and Kayne (1992)
	Shortened color fusion time	Buttner et al. (1993)
	Progressive deterioration	Diederich et al. (2002)
Visual fields	Increase in glaucomatous visual field defects	Bayer, Keller, Ferrari, and Maag (2002)
	Side effect of surgery	Biousse et al. (1998)
Saccadic eye movement	Reaction time and velocity of horizontal gaze slower	Shibasaki, Tsuji, and Kuroiwa (1979)
	Hypometria	Crawford, Goodrich, Henderson, and Kennard (1989) and Shaunak et al. (1999)
	Amplitude increased after cued saccades	Winograd-Gurvich, Georgiou-Karistianis, Fitzgerald, Millist, and White (2006)
Smooth pursuit movement	May be affected early in disease	Bares et al. (2003)
	Superimposed saccades	Shibasaki et al., (1979)
Stereopsis	Impaired, especially if poor VA	Cronin-Golomb et al. (1991) and Kiyosawa et al. (1989)
Optokinetic nystagmus	Abnormal in some patients	Shibasaki et al. (1979)
Convergence	Impaired, associated with large exophoria, diplopia	Corin, Bender, and Elizan (1971) and Lepore (2006)
Blink frequency	Reduced, abnormal tear film Dry eye and reduced vision	Biousse et al. (2004)
Blink reflex	Habituation not observed	Garland (1952)

Continued

Table 1 Summary of Visual Signs and Symptoms of Parkinson's Disease (PD)—cont'd

Ocular Aspect	Change in PD	References
Pupil reactivity	Larger after light adaptation with anisocoria	Micieli et al. (1991)
	Longer reflex latency and constriction time	Biousse et al. (2004)
Contrast sensitivity	Abnormal in some cases; intermediate to high frequencies affected	Hutton, Morris, and Elias (1993) and Bulens, Meerwaldt, Van der Wildt, and Van Deursen (1987)
EEG	Theta-GFS increased in PD when eyes closed	Pugnetti et al. (2010)
ERG	Reduced amplitude of "b" wave	Gottlob, Schneider, Heider, and Skrandies (1987)
PERG	Reduced amplitudes	Gottlob et al. (1987)
	Delayed P50	Peppe et al. (1995), Tagliati, Bodis-Wollner, Kovanecz, and Stanzione (1994), and Tagliati, Bodis-Wollner, and Yahr (1996)
Cortical VEP	Delayed P100 to checkerboard stimulus	Bodis-Wollner and Yahr (1978) and Bodis-Wollner, Yahr, Mylin, and Thornton (1982)
Chromatic VEP	Increased latency and reduced Amplitude (esp. blue-yellow)	Sartucci et al. (2006)
ERP	Delayed P300	Wang et al. (2000) and Li et al. (2003)
Visuospatial orientation	Difficulty in judging verticals, position of body parts, and in route-walking tasks	Davidsdottir, Cronin-Golomb, and Lee (2005)
Visuomotor adaptation	May be affected early in PD	Sato, Onitsuka, Ninomiya, Nakamura, and Kanba (2014)
Orientation and motion discrimination	Impaired	Trick, Kaski, and Steinman (1994)
Reading ability	Commonly affected in PD	Yu et al. (2016)

Table 1 Summary of Visual Signs and Symptoms of Parkinson's Disease (PD)—cont'd

Ocular Aspect	Change in PD	References
Facial perception	Impaired ability to perceive and imagine faces	Lang (1979) and Sprengelmeyer et al. (2003)
REM sleep disorder	A possible early feature of PD	Oertel et al. (2014)
Visual hallucinations	Frequent in PD in association with poor VA and cognition	Diederich, Goetz, and Stebbins (2005) and Fenelon, Mahieux, Huon, and Ziegler (2000)

EEG, electroencephalogram; *ERG*, electroretinogram; *ERP*, event-related potentials; *PERG*, pattern electroretinogram; *REM*, rapid eye movement; *SF*, spatial frequency; *VA*, visual acuity; *VEP*, visual evoked potentials.

4.2 Color Vision

Vision is often blurred in PD when colored stimuli are used (Price et al., 1992) with reduced perception of monochromatic contours especially when dark-green, light-blue, and dark-red stimuli are used (Buttner et al., 1993). Nevertheless, using the Farnsworth-Munsell 100-hue test, color visual discrimination was not consistently impaired in the early stages of PD, elevation of error scores exceeding the upper limit of normality in only three cases (Vesela et al., 2001). Color vision deficits in PD may be associated with significant loss of cells in the ganglion cell layer (GCL) especially at the macula (Polo et al., 2016).

4.3 Visual Fields

Relatively few studies have examined the visual fields of PD patients. Retrospective analysis of ophthalmic charts from PD patients using a cup-to-disc ratio of 0.8 or greater to define glaucoma revealed glaucomatous visual field defects in approximately 24% of patients, suggesting there may be an increased rate of glaucoma in PD (Bayer et al., 2002). In addition, intraocular pressure was slightly higher in PD patients with glaucoma compared with glaucoma patients without PD. Of eight PD patients diagnosed with glaucoma, five were considered to have low-tension glaucoma. Visual fields have also been investigated in PD patients undergoing posterior pallidotomy, a surgical procedure which can damage the optic tract (Biousse et al., 1998). Of 40 such patients, three had visual field defects which could have been attributable to the surgery, viz., contralateral superior quadrantanopias, associated in two patients with small paracentral scotomas. Although less common than in other parkinsonian syndromes such as CBD, lateralization of the pathology may occur in PD which could also result in visual field problems (Armstrong, 2016b).

4.4 Eye Movement

Assessment of oculomotor function is usually carried out using electrooculography (EOG) and on patients in which cognitive impairment is not serious enough to compromise the tests. Hence, EOG responses are often normal in PD if the eyes are in the primary position or when resting. However, convergence and divergence were studied in 18 PD patients and 18 controls using infrared VOG (Hanuska et al., 2015). The latency of both movements was increased in PD, divergence being slow and exhibiting hypometria, i.e., "under reaching of task," whereas other convergence measures were similar to controls. Convergence can be associated with a relatively large outward deviation of the eye (exophoria), and the result is often double vision (diplopia) (Corin et al., 1971; Lepore, 2006).

Abnormal saccadic and pursuit eye movements have been reported in about three-quarters of PD patients, reaction time, and maximum saccadic velocity of horizontal gaze being slower in PD than controls (Shibasaki et al., 1979). Saccades may also exhibit hypometria (Crawford et al., 1989), while smooth pursuit movements may be interrupted by small superimposed saccades (Shibasaki et al., 1979). The amplitude of saccadic eye movements is often increased in normal subjects when there is a change from externally cued saccades to self-paced saccades, and this effect is often greater in PD (Winograd-Gurvich et al., 2006). In addition, when the delay of remembered (imagined) saccades was gradually increased in untreated PD patients, there was a marked hypometria of saccadic gain at all delays, suggesting dysfunction of the striatocollicular inhibitory pathway (Shaunak et al., 1999). Spatial working memory has also been studied in relation to eye movements in which a sequence of four targets was memorized by the patient, and then the eyes were instructed to move to fixate the targets in their correct order. In PD, several discrete saccadic eye movements of reduced amplitude were necessary before reaching the final eye position, and the patients also exhibited an increase in errors while remembering the target sequence (Hodgson, Dittrich, & Kennard, 1999). The results suggest that problems with memory representation could be an early symptom of PD.

EOG recordings have been made before and after apomorphine treatment in early-stage patients confirming that pursuit movements may also be affected (Bares et al., 2003). In addition, patients with PD often have difficulty in sustaining repetitive actions, and hence, pursuits may exhibit a reduction in response amplitude and a progressive decline in response with stimulus repetition.

4.5 Stereopsis

Binocular depth perception (stereopsis) is mediated by various neural pathways involving the thalamus and posterior parietal lobe, regions likely to be affected in PD. Hence, PD patients can exhibit impairment of stereopsis especially in those with reduced VA and color perception (Cronin-Golomb et al., 1991; Kiyosawa et al., 1989), which may be attributable to pathology in extrastriate cortical areas (Sun et al., 2014).

4.6 Nystagmus

Abnormal optokinetic nystagmus (train nystagmus), an abnormal rhythmical movement of the eyes induced when a patient looks at a moving object (Shibasaki et al., 1979), has been reported in PD.

4.7 Blink Reflex

A "staring" appearance resulting from reduced rates of blinking is commonly observed in PD patients (Biousse et al., 2004). Reduced blink rates can also lead to an abnormal tear film, dry eye, and reduced vision. A characteristic ocular sign of PD is the blink reflex, elicited by a light tap above the bridge of the nose, successive taps in normal individuals producing less and less response as the reflex habituates (Garland, 1952). In PD, however, the blink reflex may not disappear on repeated tapping, and its duration may be increased as a result of loss of dopamine neurons (Peshori, Schicatano, Gopalaswamy, Sahay, & Evinger, 2001).

4.8 Pupil Reactivity

Larger pupil diameters and unequal pupil sizes (anisocoria) have been observed in PD in light adapted eyes, less difference being apparent after dark adaptation (Micieli et al., 1991). Longer light reflex latencies and constriction times have also been observed while contraction amplitudes may be reduced (Biousse et al., 2004), suggesting early involvement of the parasympathetic system in PD (Postuma, Gagnon, Pelletier, & Montplaisir, 2013). In addition, the maximum contraction ability of the iris muscle measured in vitro is greater in PD than in controls, suggesting that the muscle may have acquired adaptive sensitivity changes (Patil & Mauger, 1992).

4.9 Contrast Sensitivity

Contrast sensitivity (CS) provides a measure of visual performance across a wide range of spatial frequencies and contrasts. CS is reduced in some PD patients, performance at high or intermediate frequencies being most commonly affected (Bulens et al., 1987; Hutton et al., 1993). In addition, a decrease in CS may occur with disease progression which could contribute to poor VA. Abnormalities in CS could be related to significant loss of cells in the GCL (Polo et al., 2016) and dopamine dysfunction in the retina (Ture, Inci, & Gedizlioglu, 2007), but are often orientation specific, which also suggests cortical involvement (Rodnitzky, 2005).

4.10 Electroencephalography

Electroencephalography (EEG) in response to eye opening and to 12-Hz photic stimulation has been studied in PD using global field synchronization (GFS), a technique which assesses correlation among waves of different frequency (Pugnetti et al., 2010). When eyes were closed, theta-GFS was increased, and using 12-Hz intermittent photic stimulation, reactivity of posterior electrodes was decreased in PD, consistent with disruption of posterior anatomical pathways (Pugnetti et al., 2010).

4.11 Electroretinogram

The amplitude of the electroretinogram (ERG) "b" wave, an indicator of INL function (Fig. 1), is reduced in PD under a variety of light conditions (Gottlob et al., 1987). The amplitude of the pattern response ERG (PERG) to a checkerboard stimulus is also decreased in PD (Gottlob et al., 1987) and the latency of the P50 component delayed (Peppe et al., 1995), suggesting that retinal dopamine depletion may result in reduced ERG responses to peak stimuli (Bodis-Wollner & Tzelepi, 1998). In addition, steady-state PERG to a sinusoidal grating was affected over a range of spatial frequencies, the pattern of age-related loss being different in PD (Tagliati et al., 1996). In PD, a specific deficit at medium spatial frequencies was present accompanied by a distorted PERG spatial frequency response function. PERG is also sensitive to dopamine manipulation in the monkey retina (Tagliati et al., 1994). Using a selective D2 antagonist, treatment affected the PERG to a sinusoidal vertical grating presented at four spatial frequencies (Tagliati et al., 1994). Hence, dopamine may be involved in retinal processing, and the D2 receptor may be necessary for spatial–temporal tuning of pattern vision. Subsequently it was shown that the two dopamine receptors D1 and D2 play different roles in the retina and therefore could affect vision differently in PD (Stanzione et al., 1999).

4.12 Visual Evoked Potentials

Visual evoked responses (VEP) to colored stimuli are affected in PD supporting the hypothesis that dopamine is involved in the retinal color system (Barbata, Rinalduzzi, Laurenti, Ruggieri, & Accornero, 1994). In some patients, amplitude is decreased and latency increased for all chromatic stimuli and especially for those using blue-yellow horizontal gratings (Sartucci et al., 2006). Increased latency of the P100 peak to a checkerboard stimulus has also been reported in PD, suggesting a delay in visual processing (Bodis-Wollner & Yahr, 1978; Bodis-Wollner et al., 1982; Onofrj & Bodis-Wollner, 1982).

4.13 Event-Related Potentials

Studies have also suggested that event-related potentials (ERP) may be useful in the study of PD. Hence, lower-level complex information processing has been studied in PD using the ERP and a typical "oddball" experimental design in which a rare stimulus is interspersed among a common stimulus. Abnormal ERP responses in PD are often associated with declining cognitive and motor dysfunctional scale scores (Li et al., 2003). In a study of the P300 component of the ERP, believed to reflect orientation, attention, stimulus evaluation, and memory (Koga, Nagata, & Hirata, 1997), reductions in reaction time were actually less in PD than in other parkinsonian syndromes such as PSP and CBD (Wang et al., 2000).

4.14 Complex Visual Functions

There are prominent deficits in PD involving neuropsychological tests which require self-motivation and a demanding response from the patient (Antal et al., 1998). Hence, PD patients may exhibit a variety of deficits in visuospatial orientation (Davidsdottir et al., 2005; Levin et al., 1991) including difficulty in judging verticals and the position of body parts, and in carrying out a route-walking task. Visuospatial working memory may also be impaired early in the disease process as a result of degeneration of the basal ganglia, the dorsal visual stream, and the frontal–prefrontal cortex (Antal et al., 1998). In a problem solving task involving arranging colored balls in pockets on a computer screen, PD patients made more errors than controls and also did not show dissociation in the amount of time fixating the two halves of the display (Hodgson, Tiesman, Owen, & Kennard, 2002). The results suggested difficulties in encoding and/or maintaining an objective during a problem solving task in PD.

Visuomotor adaptation may also be affected early in PD (Sato et al., 2014). Hence, in two tasks carried out over 3 days involving mirror reading and reversed vision using a prism, mirror reading time was increased after 1 day, and there was significantly more slowing on the reversed vision task in early-stage PD (Sato et al., 2014). In addition, motor adaptation to imposed visual rotation was significantly enhanced when tested days later after a sleep-dependent memory consolidation task. PD patients and controls exhibited similar movements and adaptation to rotation but after a few days, PD patients did not show this consolidation (Marinelli et al., 2009). There may also be motor perception deficits in PD especially when attempting to perceive biological motion (Jaywant, Shiffrar, Roy, & Cronin-Golomb, 2016). Hence, in 26 PD patients and 24 controls presented with videos showing human walking, PD patients were less sensitive to motion than controls, a deficit likely to be attributable to higher-level processing rather than primary vision.

PD patients show impairments of orientation and motion discrimination (Trick et al., 1994). In addition, impairments in the ability to perceive and imagine faces have been reported in PD (Lang, 1979). Treated and untreated patients exhibit facial recognition problems, but these deficits were most frequently present in the untreated group (Sprengelmeyer et al., 2003).

Reading problems are common in PD but are not well reported. Hence, a 40-year old with a 6-year history of PD treated with dopamine replacement therapy had good VA and normal eye movements (Yu et al., 2016). However, the patient exhibited slower eye movements during reading together with an increasing number of progressive saccades, smaller saccade amplitudes, increasing regressive saccades, and longer fixation times. Hence, these deficits are likely to be attributable to dysfunction in higher-order motor planning and cognitive ability than a primary oculomotor problem.

4.15 Rapid Eye Movement Sleep Behavior Disorder

Rapid eye movement (REM) sleep behavior disorder is characterized by a loss of REM sleep-associated atonia, motor activation during dreaming, and a disturbing dream content and is a significant feature of PD. In addition, a high proportion of patients with REM sleep behavior disorder develop either an α-synuclein related parkinsonian disorder, i.e., PD, DLB, or MSA, within 10–20 years (Oertel et al., 2014), or PD itself (Chen et al., 2014). Hence, REM sleep behavioral disorder could be an important early sign of a parkinsonian syndrome. Cases with confirmed REM sleep behavioral disorder may also exhibit postural instability.

4.16 Visual Hallucinations

Visual hallucinations are frequently reported in PD (Fenelon et al., 2000; Diederich et al., 2005) especially in those developing dementia and in patients treated with dopamine replacement therapy. In a large study of PD, hallucinations occurred in the previous 3 months in 40% of patients examined, being visual in 22% and auditory in 10% of patients (Fenelon et al., 2000). A number of factors may predict hallucinations in PD including a severe cognitive defect, daytime somnolence, longer duration of disease, poor VA, and reduced activity in primary visual cortex (Eberling et al., 1994). Hallucinations in PD are often complex with flickering lights, and illusionary misconceptions often preceding the most common manifestation, viz., stereotypical colorful images (Diederich et al., 2005). Recently, multimodal tactile and visual hallucinations ("string" hallucinations) have been observed in small numbers of PD patients characterized by the appearance of "threads" emerging from the hands (Perez-Perez, Pagonabarraga, Fernandez-Bobadilla, & Kulisevsky, 2016).

5. DISCUSSION AND CONCLUSIONS

5.1 Visual Features of PD

Visual dysfunction in PD may involve VA, CS, color discrimination, pupil reactivity, eye movements, motion perception, visual field defects, and visual processing speeds. In addition, disturbance of visuospatial orientation, facial recognition problems, REM sleep behavior disorder, and chronic visual hallucinations may be present. A number of visual features may be characteristic of the early stages of PD including problems affecting pupil reactivity (Postuma et al., 2013), smooth pursuit movements dysfunction

(Bares et al., 2003), visuomotor adaptation (Sato et al., 2014), and the presence of REM sleep behavior disorder. The presence of any of these features in undiagnosed individuals may raise a concern of possible PD. Since early diagnosis may enable intervention and possible neuroprotection (Goldman & Postuma, 2014), further visual studies of early-stage PD are needed.

Many of the visual features present in early and moderate stage PD are significantly enhanced in PDD. However, some features may be particularly exacerbated in the presence of dementia including color vision problems (Postuma, Gagnon, Vendette, Desjardins, & Montplaisir, 2011), changes in papillary function (Granholm et al., 2003), and changes in the EEG (Pugnetti et al., 2010). There may also be visual features particularly indicative of PDD such as prominent visual hallucinations (Archibald et al., 2011; Hely et al., 1996; Ransmayer, 2000; Williams, Warren, & Lees, 2008), increasingly severe eye movement problems (Debruin, Lees, & Daniel, 1992; Mosimann et al., 2004), and defects in visuospatial orientation (Ala, Hughes, Kyrouac, Ghobrial, & Elble, 2001).

5.2 Differential Diagnosis

Many of the visual features of PD overlap with those of other parkinsonian syndromes resulting in problems in differential diagnosis. Nevertheless, in patients with an indeterminate parkinsonian syndrome, the presence of visual hallucinations is an important symptom which may indicate underlying α-synuclein pathology, i.e., supporting a diagnosis of PD or DLB rather than PSP or CBD which have a tau-based pathology (Williams et al., 2008). Distinguishing PD from PSP can be an especially difficult early in the disease process. Nevertheless, atypical features of PSP include slowing of upward saccades, moderate slowing of downward saccades, the presence of a full range of voluntary vertical eye movements, a curved trajectory of oblique saccades, and the absence of square-wave jerks (Averbach-Heller, Paulson, Davoff, & Leigh, 1999). Useful in separating PSP from PD is the presence in the former of vertical supranuclear gaze palsy, fixation instability, lid retraction, blepharospasm, and apraxia of eyelid opening and closing (Friedman, Jankovic, & McCrary, 1992). Downgaze palsy is probably the single most useful diagnostic clinical sign of PSP (Grandas & Esteban, 1994).

Distinguishing PD from DLB is necessary because patients with visual hallucinations may be treated with antipsychotic drugs, a hazardous treatment in DLB (Kaufer, 2004). DLB and PD generally exhibit similarities on saccadic eye movement tasks. Nevertheless, a newly developed portable saccadometer has been used to compare parkinsonian syndromes (Antoniades, Bak, Carpenter, Hodges, & Barker, 2007) and suggested that a combination of saccadic parameters may enable PD and DLB to be distinguished. In addition, deficits in orientation, "trail-making," and reading the names of colors (Stroop test) which may indicate DLB rather than PD (Mondon et al., 2007).

MSA is a complex disorder whose symptoms include parkinsonism, ataxia, and autonomic dysfunction (Lantos, 1994). A number of visual symptoms may suggestive MSA including excessive square-wave jerks, mild to moderate hypometria of saccades, impaired vestibuloocular reflex, and the presence of nystagmus (Anderson et al., 2008). Visual hallucinations, unrelated to medication, are also rare in MSA compared with PD (Kumbler & Kornhuber, 2002). Moreover, the color VEP is affected in PD but not in MSA (Sartucci et al., 2006). An additional feature which may be useful in separating PD and MSA is an ability to fixate an object, which is abnormal in a significant proportion of patients with MSA but less so in PD (Rascol et al., 1995). In addition, eye movements recorded during sinusoidal tracking by VOG show that in MSA, saccades correct for position error (catch-up saccades), while in PD, saccades are directed toward future target positions (anticipatory saccades) (Pinkhardt et al., 2009).

CBD is a particularly complex disorder and distinguishing it from other parkinsonian syndromes can be difficult (Schneider, Watts, Gearing, Brewer, & Mirra, 1997; Wenning et al., 1998). However, a combination of increased latency of saccadic eye movements (Revaud-Pechoux et al., 2000; Massman, Kreiter, Jankovic, & Doody, 1996; Pierrot-Deselligny & Rivauld-Pechoux, 2003), impaired smooth pursuit movements (Rinne, Lee, Thompson, & Marsden, 1994), and visuospatial dysfunction especially involving object-based tasks (Mori et al., 2000) may be more typical of CBD, while vertical gaze palsy, visual hallucinations, sleep disturbance, and an impaired ERG are less typical. It is also possible that visuospatial deficits are more prominent in CBD than in PDD.

5.3 Visual Care of the Patient

PD results in both motor and nonmotor symptoms which can affect many aspects of the quality of life of the patient, and a multidisciplinary approach is often necessary to management and care including pharmacological intervention and rehabilitation (Ekker, Janssen, Nonnekes, Bloem, & de Vries, 2016). Visual deficits in the parkinsonian syndromes, where present, are often important in influencing overall motor function and the multiplicity of visual deficits often markedly debilitating (Rajagopal, Bateman, & Van Stavern, 2012). Hence, identifying and correcting visual problems as far as possible may reduce the chances that a patient of a serious fall and improve quality of life. Eye-care practitioners can work in collaboration with the patient and health care providers to identify and manage the specific visual deficits of the patient. First, a decline in VA is fairly typical of the parkinsonian syndromes in general and should be investigated in any suspected case of PD. Second, visual fields should be tested as the pathology may be lateralized which could result in a hemianopia (Formaglio et al., 2009; Rey et al., 1995). Third, the external eye should be monitored as blepharospasm and possibly dry eye may be present, these features being an essentially treatable visual feature of parkinsonian syndromes (Piccione, Mancini, Tonin, & Bizzarini, 1997; Rana, Kabir, Dogu, Patel, & Khondker, 2012; Tolosa & Compta, 2005). Artificial tears can be used to help dry eyes, while blepharospasm has been treated successfully with botulinum toxin A injected at the junction of the preseptal and pretarsal parts of the levator palpebrae and orbicularis oculi muscles (Piccione et al., 1997). Fourth, patients with difficulties in opening their eyelids may have "lid crutches" fitted to spectacle frames that can hold the lids open. Fifth, PD patients may have a particular difficulty with saccadic eye movements and deficits in gaze control may influence "stepping behavior" which may increase the risk of trips or falls (Di Fabrio, Zampieri, & Tuite, 2008). In addition, impaired postural control in PD is common and a major contributor to falls (Hill, Stuart, Lord, Del Din, & Rochester, 2016). Balance training in combination with eye movement and visual awareness exercises has been shown to improve gaze control in PSP and may also help in PD. Sixth, PD patients may have difficult in locating objects in visual space (Bak, Caine, Nearn, & Hodges, 2006) and strategies to improve object recognition in the home may be beneficial. In addition, to maximize VA, cataract surgery and appropriate refractive correction can improve overall motor performance and associated therapies. Seventh, although it is unclear whether driving accidents are more frequent in PD, there may be a potential problem involving driving because of the low contrast VA deficits in PD (Lin et al., 2015). In addition, some treatments for PD are likely to produce moderate effects on fitness to drive (Alvarez, 2016).

In conclusion, visual dysfunction in PD can involve VA, CS, color discrimination, pupil reactivity, saccadic and pursuit eye movements, motion perception, visual fields, and visual processing speeds. In addition, disturbance of visuospatial orientation, facial recognition problems, REM sleep behavior disorder, and chronic visual hallucinations may be present. Visual dysfunction in PD can be a useful diagnostic indicator in differentiating the parkinsonian symptoms, the presence of visual hallucinations, visuospatial problems, and variation in saccadic eye movement dysfunction being useful discriminating features.

REFERENCES

Ala, T. A., Hughes, L. F., Kyrouac, G. A., Ghobrial, M. W., & Elble, R. J. (2001). Pentagon copying is more impaired in dementia with Lewy bodies than in Alzheimer's disease. *Journal of Neurology, Neurosurgery, and Psychiatry*, *70*, 483–488.

Alvarez, F. J. (2016). Parkinson's disease, antiparkinson medicines, and driving. *Expert Review of Neurotherapeutics*, *16*, 1023–1032.

Anderson, T., Luxon, L., Quinn, N., Daniel, S., Marsden, C. D., & Bronstein, A. (2008). Oculomotor function in multiple system atrophy: Clinical and laboratory features in 30 patients. *Movement Disorders*, *23*, 977–984.

Antal, A., Bandini, P., Keri, S., & Bodis-Wollner, I. (1998). Visuo cognitive dysfunctions in Parkinson's disease. *Clinical Neuroscience*, *5*, 147–152.

Antoniades, C. A., Bak, T. H., Carpenter, R. H. S., Hodges, J. R., & Barker, R. A. (2007). Diagnostic potential of saccadometry in progressive supranuclear palsy. *Biomarkers in Medicine*, *1*, 487–490.

Archibald, N. K., Clarke, M. P., Mosimann, U. P., & Burn, D. J. (2011). Visual symptoms in Parkinson's disease and Parkinson's disease dementia. *Movement Disorders*, *26*, 2387–2395.

Armstrong, R. A. (2008). Visual signs and symptoms of Parkinson's disease. *Clinical & Experimental Optometry*, *91*, 129–138.

Armstrong, R. A. (2011). Visual signs and symptoms of progressive supranuclear palsy. *Clinical & Experimental Optometry*, *95*, 150–160.

Armstrong, R. A. (2012). Visual signs and symptoms of dementia with Lewy bodies. *Clinical & Experimental Optometry*, *94*, 621–630.

Armstrong, R. A. (2014). Visual signs and symptoms of multiple system atrophy. *Clinical & Experimental Optometry*, *97*, 483–491.

Armstrong, R. A. (2016a). Oculo-visual dysfunction in Parkinson's disease. *Journal of Parkinsons' Disease*, *5*, 715–726.

Armstrong, R. A. (2016b). Visual signs and symptoms of corticobasal degeneration. *Clinical & Experimental Optometry*, *99*, 498–506.

Armstrong, R. A., & Kergoat, H. (2015). Oculo-visual changes and clinical considerations affecting older patients with dementia. *Ophthalmic and Physiological Optics*, *35*, 352–376.

Armstrong, R. A., Kotzbauer, P. T., Perlmutter, J. S., Campbell, M. C., Hurth, K. M., Schmidt, R. E., et al. (2014). A quantitative study of α-synuclein pathology in fifteen cases of dementia associated with Parkinson's disease. *Journal of Neural Transmission*, *121*, 171–181.

Averbach-Heller, L., Paulson, G. W., Davoff, R. B., & Leigh, R. J. (1999). Whipple's disease mimicking progressive supranuclear palsy, the diagnostic value of eye movement recording. *Journal of Neurology, Neurosurgery, and Psychiatry*, *66*, 532–535.

Bak, T. H., Caine, D., Nearn, V. C., & Hodges, J. R. (2006). Visuo-spatial functions in atypical parkinsonian syndromes. *Journal of Neurology, Neurosurgery, and Psychiatry, 77*, 454–456.

Barbata, L., Rinalduzzi, S., Laurenti, M., Ruggieri, S., & Accornero, N. (1994). Color visual evoked potentials in Parkinson's disease. *Electroencephalograpy and Clinical Neurophysiology, 92*, 169–172.

Bares, M., Brazdil, M., Kanovsky, P., Jurak, P., Daniel, P., Kukleta, M., et al. (2003). The effect of apomorphine administration on smooth pursuit ocular movements in early Parkinson's disease. *Parkinsonism & Related Disorders, 9*, 139–144.

Bayer, A. U., Keller, O. N., Ferrari, F., & Maag, K. P. (2002). Association of glaucoma with neurodegenerative diseases with apoptotic cell death: Alzheimer's disease and Parkinson's disease. *American Journal of Ophthalmology, 133*, 135–137.

Biousse, V., Newman, N. J., Carroll, C., Mewes, K., Vitek, J. L., Bakay, R. A. E., et al. (1998). Visual fields in patients with posterior GPi pallidotomy. *Neurology, 50*, 258–265.

Biousse, V., Skibell, B. C., Watts, R. L., Loupe, D. N., Drews-Botsch, C., & Newman, N. J. (2004). Ophthalmologic features of Parkinson's disease. *Neurology, 62*, 177–180.

Bodis-Wollner, I., & Tzelepi, A. (1998). The push-pull action of dopamine on spatial tuning of the monkey retina: The effects of dopaminergic deficiency and selective D-1 and D-2 receptor ligands on the pattern electroretinogram. *Vision Research, 38*, 1479–1487.

Bodis-Wollner, I., & Yahr, M. (1978). Measurement of visual evoked potentials in Parkinson's disease. *Brain, 101*, 661–671.

Bodis-Wollner, I., Yahr, M. D., Mylin, L., & Thornton, J. (1982). Dopaminergic deficiency and delayed visual evoked-potentials in humans. *Annals of Neurology, 11*, 478–483.

Bohnen, N. I., Frey, K. A., Studenski, S., Kotagal, V., Koeppe, R. A., Constantine, G. M., et al. (2014). Extra-nigral pathological conditions are common in Parkinson's disease with freezing of gait: An in vivo positron emission tomographic study. *Movement Disorders, 29*, 1118–1124.

Braak, H., & Braak, E. (1998). Argyrophilic grain disease: Frequency of occurrence in different age categories and neuropathologic diagnostic criteria. *Journal of Neural Transmission, 105*, 801–819.

Bulens, C., Meerwaldt, J. D., Van der Wildt, G. J., & Van Deursen, J. B. P. (1987). Effect of levodopa treatment on contrast sensitivity in Parkinson's disease. *Annals of Neurology, 22*, 365–369.

Buttner, T., Kuhn, W., Klotz, P., Steinberg, R., Voss, L., Bulgaru, P., et al. (1993). Disturbance of colour perception in Parkinson's disease. *Journal of Neural Transmission, 6*, 11–15.

Chen, T. Z., Xu, G. J., Zhou, G. A., Wang, J. R., Chan, P., & Du, Y. F. (2014). Postural sway in idiopathic rapid eye movement sleep behaviour disorder: A potential marker of prodromal Parkinson's disease. *Brain Research, 1559*, 26–32.

Corin, M. S., Bender, M. B., & Elizan, T. S. (1971). Oculomotor function in patients with Parkinson's disease. *Journal of the Neurological Sciences, 15*, 251–265.

Crawford, T. J., Goodrich, S., Henderson, L., & Kennard, C. (1989). Predictive responses in Parkinson's disease: Manual keypresses and saccadic eye movements to regular stimulus events. *Journal of Neurology, Neurosurgery, and Psychiatry, 52*, 1033–1042.

Cronin-Golomb, A., Corkin, S., Rizzo, J. F., Cohen, J., Growdon, J. H., & Banks, J. H. (1991). Visual dysfunction in Alzheimer disease: Relation to normal aging. *Annals of Neurology, 29*, 41–52.

Davidsdottir, S., Cronin-Golomb, A., & Lee, A. (2005). Visual and spatial symptoms in Parkinson's disease. *Vision Research, 45*, 1285–1296.

Debruin, V. M., Lees, A. J., & Daniel, S. E. (1992). Diffuse Lewy body disease presenting with supranuclear gaze palsy, parkinsonism and dementia: A case report. *Movement Disorders, 7*, 355–358.

Di Fabrio, R. P., Zampieri, C., & Tuite, P. (2008). Gaze control and foot kinematics during stair climbing: Characteristics leading to fall risk in progressive supranuclear palsy. *Physical Therapy*, *88*, 240–250.

Diederich, N. J., Goetz, C. G., & Stebbins, G. T. (2005). Repeated visual hallucinations in Parkinsons's disease as disturbed external/internal perceptions: Focussed review and a new integrative model. *Movement Disorders*, *20*, 130–140.

Diederich, N. J., Raman, R., Leurgans, S., & Goetz, C. G. (2002). Progressive worsening of spatial and chromatic processing deficits in Parkinson's disease. *Archives of Neurology*, *59*, 1249–1252.

Ding, Z. T., Wang, Y., Jiang, Y. P., Hashizume, Y., Yoshida, M., Mimuro, M., et al. (2006). Characteristics of alpha-synucleinopathy in centenarians. *Acta Neuropathologica*, *111*, 450–458.

Djamgoz, M. B. A., Hankins, M. W., Hirano, J., & Archer, S. N. (1997). Neurobiology or retinal dopamine in relation to degenerative states of the tissue. *Vision Research*, *37*, 3509–3529.

Dowling, J. E., Ehinger, B., & Hedden, W. L. (1976). Interplexiform cell: New type of retinal neuron. *Investigative Ophthalmology & Vision Science*, *15*, 916–926.

Earley, C. J. (2003). Restless legs syndrome. *New England Journal of Medicine*, *348*, 2103–2109.

Eberling, J. L., Richardson, B. C., Reed, B. R., Wolfe, N., & Jagust, W. J. (1994). Cortical glucose metabolism in Parkinson's disease without dementia. *Neurobiology of Aging*, *15*, 329–335.

Ekker, M. S., Janssen, S., Nonnekes, J., Bloem, B. R., & de Vries, N. M. (2016). Neurorehabilitation for Parkinson's disease: Future perspectives for behavioural adaptation. *Parkinsonism & Related Disorders*, *22*, S73–S77.

Fenelon, G., Mahieux, F., Huon, R., & Ziegler, M. (2000). Hallucinations in Parkinson's disease: Prevalence, phenomenology and risk factors. *Brain*, *123*, 733–745.

Formaglio, M., Krolek-Salmon, P., Tillikete, C., Bernard, M., Croisile, B., & Vighetto, A. (2009). Homonymous hemianopia and posterior cortical atrophy. *Revue Neurologique*, *2009*(165), 256–262.

Frederick, J. M., Rayborn, M. E., Laties, A. M., Lam, D. M. K., & Hollyfield, J. G. (1982). Dopaminergic neurones in the human retina. *Journal of Comparative Neurology*, *210*, 65–79.

Friedman, D. I., Jankovic, J., & McCrary, J. A. (1992). Neuro-ophthalmic findings in progressive supranuclear palsy. *Journal of Clinical Neuro-Ophthalmology*, *12*, 104–109.

Garland, H. G. (1952). Refresher course for general practitioners: Parkinsonism. *British Medical Journal*, *1*, 153–155.

Goldman, J. G., & Postuma, R. (2014). Premotor and non-motor features of Parkinson's disease. *Current Opinion in Neurology*, *27*, 434–441.

Gottlob, I., Schneider, E., Heider, W., & Skrandies, W. (1987). Alteration of visual evoked potentials and electroretinograms in Parkinson's disease. *Electroencephalograpy and Clinical Neurophysiology*, *66*, 349–357.

Grandas, F., & Esteban, A. (1994). Eyelid motor abnormalities in progressive supranuclear palsy. *Journal of Neural Transmission (Supp)*, *42*, 33–41.

Granholm, E., Morris, S., Galasko, D., Shults, C., Rogers, E., & Vukov, B. (2003). Tropicamide effects on pupil size and papillary light reflexes in Alzheimer's and Parkinson's disease. *International Journal of Psychophysiology*, *47*, 95–115.

Hanuska, J., Bonnet, C., Rusz, J., Sieger, T., Jech, R., Rivaud-Pechoux, S., et al. (2015). Fast vergence eye movements are disrupted in Parkinson's disease: A video-oculography study. *Parkinsonism & Related Disorders*, *21*, 797–799.

Hely, M. A., Reid, W. G., Halliday, G. M., McRitchie, D. A., Leicester, J., Joffe, R., et al. (1996). Diffuse Lewy body disease: Clinical features in nine cases without coexistant Alzheimer's disease. *Journal of Neurology, Neurosurgery, and Psychiatry*, *60*, 531–538.

Hill, E., Stuart, S., Lord, S., Del Din, S., & Rochester, L. (2016). Vision, visuo-cognition and postural control in Parkinson's disease: An associative pilot study. *Gait and Posture, 48*, 74–76.

Hodgson, T. L., Dittrich, W. H., Henderson, L., & Kennard, C. (1999). Eye movements and spatial working memory in Parkinson's disease. *Neuropsychologia, 37*, 927–938.

Hodgson, T. L., Tiesman, B., Owen, A. M., & Kennard, C. (2002). Abnormal gaze strategies during problem solving in Parkinson's disease. *Neuropsychologia, 40*, 411–422.

Hutton, J. T., Morris, J. L., & Elias, J. W. (1993). Levodopa improves spatial contrast sensitivity in Parkinson's disease. *Archives of Neurology, 50*, 721–724.

Jaywant, A., Shiffrar, M., Roy, S., & Cronin-Golomb, A. (2016). Impaired perception of biological motion in Parkinson's disease. *Neuropsychology, 30*, 720–730.

Jones, R. D., & Donaldson, I. M. (1995). Fractionation of visuo-perceptual dysfunction in Parkinson's disease. *Journal of the Neurological Sciences, 131*, 43–50.

Jones, R. D., Donaldson, I. M., & Timmings, P. L. (1992). Impairment of high-contrast visual activity in Parkinson's disease. *Movement Disorders, 7*, 232–238.

Josephs, K. A., Whitwell, J. L., Parisi, J. E., Knopman, D. S., Boeve, B. F., Geda, Y. E., et al. (2008). Argyrophilic grains: A distinct disease or an additive pathology? *Neurobiology of Aging, 29*, 566–573.

Kaufer, D. I. (2004). Pharmacologic treatment expectations in the management of dementia with Lewy bodies. *Dementia and Geriatric Cognitive Disorders, 17*, 32–39.

Kiyosawa, M., Bosley, T. M., Chawluk, J., Jamieson, D., Scatz, N. J., Savino, P. J., et al. (1989). Alzheimer's disease with prominent visual symptoms; clinical and metabolic evaluation. *Ophthalmology, 96*, 1077–1085.

Koga, Y., Nagata, K., & Hirata, K. (1997). Hypothesis on the P300 generators based on visual P300 results in neurological disorders. *Brain Topography Today: International Congress Series, 1147*, 354–357.

Kromer, R., Buhmann, C., Hidding, U., Keseru, M., Keseru, D., Hassenstein, A., et al. (2016). Evaluation of retinal vessel morphology in patients with Parkinson's disease using optical coherence tomography. *PloS One, 11*. e0161136.

Kumbler, E., & Kornhuber, M. (2002). Delusional parasitosis in multiple system atrophy. *Nervenarzt, 73*, 380–383.

Lang, P. J. (1979). A bio-informational theory of emotional imagery. *Psychophysiology, 16*, 495–512.

Lantos, P. L. (1994). (1994) Cellular pathology of multiple system atrophy: A review. *Journal of Neurology, Neurosurgery, and Psychiatry, 57*, 129–133.

Lee, J. Y., Kim, J. M., Ahn, J., Kim, H. J., Jeon, B. S., & Kim, T. W. (2014). Retinal nerve fibre layer thickness and visual hallucinations in Parkinson's disease. *Movement Disorders, 29*, 61–67.

Lepore, F. E. (2006). Parkinson's disease and diplopia. *Neuro-opthalmology, 30*, 37–40.

Levin, B. E., Llabre, M. M., Reisman, S., Weiner, W. J., Sanchez-Ramos, J., Singer, C., et al. (1991). Visuospatial impairments in Parkinson's disease. *Neurology, 41*, 365–369.

Li, M., Kuroiwa, Y., Wang, L. H., Kamitani, T., Takahashi, T., Suzuki, Y., et al. (2003). Early sensory information processes are enhanced on visual oddball and S1-S2 tasks in Parkinson's disease: A visual event-related potentials study. *Parkinsonism & Related Disorders, 9*, 329–340.

Lin, T. P., Rigby, H., Adler, J. S., Hentz, J. G., Balcer, L. J., Galetta, S. L., et al. (2015). Abnormal visual contrast acuity in Parkinson's disease. *Journal of Parkinson's Disease, 5*, 125–130.

Marinelli, L., Crupi, D., Di Rocco, A., Eidelberg, D., Abbruzzese, G., & Ghilardi, M. F. (2009). Learning and consolidation of visuo-motor adaptation in Parkinson's disease. *Parkinsonism & Related Disorders, 15*, 6–11.

Massman, P. J., Kreiter, K. T., Jankovic, J., & Doody, R. S. (1996). Neuropsychological functioning in cortical-basal ganglionic degeneration: Differentiation from Alzheimer's disease. *Neurology, 46*, 720–726.

Masson, G., Mestre, D., & Blin, O. (1993). Dopaminergic modulation of visual sensitivity in man. *Fundamental & Clinical Pharmacy*, 7, 449–463.

Matsui, H., Udakam, F., Tamuram, A., Oda, M., Kubori, T., Nishinaka, K., et al. (2006). Impaired visual acuity as a risk factor for visual hallucinations in Parkinsons's disease. *Journal of Geriatric Psychiatric and Neurology, 19*, 36–40.

Micieli, G., Tassorelli, C., Martignoni, E., Pachetti, C., Bruggi, P., Magri, M., et al. (1991). Disordered pupil reactivity in Parkinson's disease. *Clinical Autonomic Research, 1*, 55–58.

Mondon, K., Gochard, A., Margue, A., Armand, A., Beauchamp, D., Prunier, C., et al. (2007). Visual recognition memory differentiates dementia with Lewy bodies and Parkinson's disease dementia. *Journal of Neurology, Neurosurgery, and Psychiatry, 78*, 738–741.

Moreno-Ramos, T., Benito-Leon, J., Villarejo, A., & Bermejo-Pareja, F. (2013). Retinal nerve fibre layer thinning in dementia associated with Parkinson's disease, dementia with Lewy bodies, and Alzheimer's disease. *Journal of Alzheimer Disease, 34*, 659–664.

Mori, E., Shimomura, T., Fujimori, M., Hirono, N., Imamura, T., Hashimoto, M., et al. (2000). Visuoperceptual impairment in dementia with Lewy bodies. *Archives of Neurology, 57*, 489–493.

Mosimann, U. P., Mather, G., Wesnes, K. A., O'Brien, J. T., Burn, D. J., & McKeith, I. G. (2004). Visual perception in Parkinson's disease dementia and dementia with Lewy bodies. *Neurology, 63*, 2091–2096.

Muller, M. L. T. M., & Bohnen, N. I. (2013). Cholinergic dysfunction in Parkinson's disease. *Current Neurology and Neuroscience Reports, 13*. Article Number 377.

Muller, M. L. T. M., Bohnen, N. I., Kotagal, V., Scott, P. J. H., Koeppe, R. A., Frey, K. A., et al. (2015). Clinical markers for identifying cholinergic deficits in Parkinson's disease. *Movement Disorders, 30*, 269–273.

Nguyen-Legros, J., Harnois, C., DiPaolo, T., & Simon, A. (1993). The retinal dopamine system in Parkinson's disease. *Clinical Vision Science, 8*, 1–12.

Oertel, W. H., Depboylu, C., Krenzer, M., Vadasz, D., Ries, V., Sixel-Doring, F., et al. (2014). REM sleep behaviour disorder as a prodromal stage of alpha-synucleinopathies: Symptoms, epidemiology, pathophysiology, diagnosis and therapy. *Nervenarzt, 85*, 19–25.

Onofrj, M., & Bodis-Wollner, I. (1982). Dopaminergic deficiency causes delayed visual evoked-potentials in rats. *Annals of Neurology, 11*, 484–490.

Patil, P. N., & Mauger, T. F. (1992). Cholinergic sensitivity of irides from donors with various pathological conditions and lens implants. *N-S Archives of Pharmacology, 346*, 620–628.

Peppe, A., Stanzione, P., Pierelli, F., De Angelis, D., Pierantozzi, M., & Bernardi, G. (1995). Visual alterations in de novo Parkinson's disease: Pattern electroretinogram latencies are more delayed and more reversible by levodopa than are visual evoked potentials. *Neurology, 45*, 1144–1148.

Perez-Perez, J., Pagonabarraga, J., Fernandez-Bobadilla, R., & Kulisevsky, J. (2016). "string hallucinations": Multimodal tactile and visual hallucinations in Parkinson's disease. *Movement Disorders Clinical Practice, 3*, 180–183.

Peshori, K. R., Schicatano, E. J., Gopalaswamy, R., Sahay, E., & Evinger, C. (2001). Aging of the trigeminal blink system. *Experimental Brain Research, 136*, 351–363.

Piccione, F., Mancini, E., Tonin, P., & Bizzarini, M. (1997). Botulinum toxin treatment of apraxia of eyelid opening in progressive supranuclear palsy: Report of two cases. *Archives of Physical Medical and Rehabilitation, 78*, 525–529.

Pierrot-Deselligny, C., & Rivauld-Pechoux, S. (2003). Contribution of oculomotor exploitation for the etiological diagnosis of parkinsonian syndromes. *Revue Neurologique, 159*, S75–S81.
Pinkhardt, E. H., Kassubek, J., Sussmuth, S., Ludolph, A. C., Becker, W., & Jurgens, R. (2009). Comparison of smooth pursuit eye movement deficits in multiple system atrophy and Parkinson's disease. *Journal of Neurology, 256*, 1438–1446.
Polo, V., Satue, M., Rodrigo, M. J., Otin, S., Alarcia, R., Bambo, M. P., et al. (2016). Visual dysfunction and its correlation with retinal changes in patients with Parkinson's disease: An observational cross-sectional study. *BMJ Open, 6*. e009658.
Postuma, R. B., Gagnon, J. F., Pelletier, A., & Montplaisir, J. (2013). Prodromal autonomic symptoms and signs in Parkinson's disease and dementia with Lewy bodies. *Movement Disorders, 28*, 597–604.
Postuma, R. B., Gagnon, J. F., Vendette, M., Desjardins, C., & Montplaisir, J. Y. (2011). Olfaction and color vision identify impending neurodegeneration in rapid eye movement sleep disorder. *Annals of Neurology, 69*, 811–818.
Price, M. J., Feldman, R. G., Adelberg, D., & Kayne, H. (1992). Abnormalities in colour vision and contrast sensitivity in Parkinson's disease. *Neurology, 42*, 887–890.
Pugnetti, L., Baglio, F., Farina, E., Alberoni, M., Calabrese, E., Gambini, A., et al. (2010). EEG evidence of posterior cortical disconnection in Parkinson's disease and related dementias. *International Journal of Neuroscience, 120*, 88–98.
Rajagopal, R., Bateman, R., & Van Stavern, G. P. (2012). Visual involvement in corticobasal syndrome. *Journal of Neuro-Ophthalmology, 32*, 338–340.
Rana, A. O., Kabir, A., Dogu, O., Patel, A., & Khondker, S. (2012). Prevalence of blepharospasm and apraxia of eye opening in patients with parkinsonism, cervical dystonia, and essential tremor. *European Neurology, 68*, 318–321.
Ransmayer, G. (2000). Dementia with Lewy bodies: Prevalence, clinical spectrum and natural history. In P. Riederer, D. B. Calne, R. Horowski, Y. Mizuno, C. W. Olanow, W. Poewe, et al. (Eds.), *Advances in Research on Neurodegeneration: Vol. 8* (pp. 303–314).
Rascol, O., Sabatini, U., Fabre, N., Senard, J. M., Simonett-Amoreau, M., Montastruc, J. L., et al. (1995). Abnormal vestibulo-ocular cancellation in multiple system atrophy and progressive supranuclear palsy but not in Parkinson's disease. *Movement Disorders, 10*, 163–170.
Repka, M. X., Claro, M. C., Loupe, D. N., & Reich, S. G. (1996). Ocular motility in Parkinson's disease. *Journal of Pediatric Ophthalmology and Strabismus, 33*, 144–147.
Revaud-Pechoux, S., Vidailhet, M., Gallouedec, G., Litvan, I., Gaymard, B., & Pierrot-Deseilligny, C. (2000). Longitudinal ocular motor study in corticobasal degeneration and progressive supranuclear palsy. *Neurology, 54*, 1029–1032.
Rey, G. J., Tomer, R., Levin, B. E., Sanchez-Ramos, J., Bowen, B., & Bruce, J. H. (1995). Psychiatric symptoms, atypical dementia, and left visual-field inattention in corticobasal ganglionic degeneration. *Movement Disorders, 10*, 106–110.
Rinne, J. O., Lee, M. S., Thompson, P. D., & Marsden, C. D. (1994). Corticobasal degeneration: A clinical study of 36 cases. *Brain, 117*, 1183–1196.
Rodnitzky, R. L. (2005). Visual dysfunction in Parkinson's disease. *Clinical Neuroscience, 5*, 102–106.
Sabbagh, M. N., Sandhu, S. S., Farlow, M. R., Veddeis, L., Shell, H. A., Caviness, J. N., et al. (2009). Correlation of clinical features with argyrophilic grains at autopsy. *Alzheimer Disease and Associated Disorders, 23*, 229–233.
Saito, Y., Kawashimam, A., Ruberum, N. N., Fujiwara, H., Koyama, S., Sawaki, M., et al. (2003). Accumulation of phosphorylated α-synuclein in aging human brain. *Journal of Neuropathology and Experimental Neurology, 62*, 644–654.
Sartucci, F., Orlandi, G., Bonuccelli, U., Borghetti, D., Murri, L., Orsini, C., et al. (2006). Chromatic pattern-reversal electroretinogram (ChPERG) are spared in multiple system atrophy compared with Parkinson's disease. *Neurological Science, 26*, 395–401.

Sato, E., Onitsuka, T., Ninomiya, H., Nakamura, I., & Kanba, S. (2014). Prism adaptation and perceptual skill learning deficits in early-stage Parkinson's disease. *Neuropsychobiology, 70*, 165–172.

Schneider, J. A., Watts, R. L., Gearing, M., Brewer, R. P., & Mirra, S. S. (1997). Corticobasal degeneration: Neuropathological and clinical heterogeneity. *Neurology, 48*, 959–969.

Shaunak, S., O'Sullivan, E., Blunt, S., Lawden, M., Crawford, T., Henderson, L., et al. (1999). Remembered saccades with variable delay in Parkinson's disease. *Movement Disorders, 14*, 80–86.

Shibasaki, H., Tsuji, S., & Kuroiwa, Y. (1979). Oculomotor abnormalities in Parkinson's disease. *Archives of Neurology, 36*, 360–364.

Spillantini, M. G., Crowther, R. A., Jakes, R., Cairns, N. J., Lantos, P., & Goedert, M. (1998). Filamentous alpha-synuclein inclusions link multiple system atrophy with Parkinson's disease and dementia with Lewy bodies. *Neuroscience Letters, 251*, 205–208.

Sprengelmeyer, R., Young, A. W., Mahn, K., Schroeder, U., Woitalla, D., Buttner, T., et al. (2003). Facial expression recognition in people with medicated and unmedicated Parkinson's disease. *Neuropsychology, 41*, 1047–1057.

Stanzione, P., Bodis-Wollner, I., Pierantozzi, M., Semprini, R., Tagliati, M., Peppe, A., et al. (1999). A mixed D1 and D2 antagonist does not replay pattern electroretinogram alterations observed with a selective D2 antagonist in normal humans: Relationship with Parkinson's disease pattern electroretinogram alterations. *Clinical Neurophysiology, 110*, 82–85.

Sun, L., Zhang, H., Gu, Z. Q., Cao, M., Li, D. W., & Chan, P. (2014). Stereopsis impairment is associated with decreased color perception and worse motor performance in Parkinson's disease. *European Journal of Medical Research*. Article 19.

Tagliati, M., Bodis-Wollner, I., Kovanecz, I., & Stanzione, P. (1994). Spatial-frequency tuning of the monkey pattern ERG depends on D2 receptor-linked action of dopamine. *Vision Research, 34*, 2051–2057.

Tagliati, M., Bodis-Wollner, I., & Yahr, M. D. (1996). The pattern electroretinogram in Parkinson's disease reveals lack of retinal spatial tuning. *Electroencephalograpy and Clinical Neurophysiology, 100*, 1–11.

Tolnay, M., Monsch, A. U., Staehelin, H. B., & Probst, A. (1999). Argyrophilic grain disease: A disorder distinct from Alzheimer's disease. *Pathologe, 20*, 159–168.

Tolosa, E., & Compta, Y. (2005). Early and prominent blepharospasm in untreated patients with parkinsonism should raise suspicion of progressive supranuclear palsy, multiple system atrophy, or corticobasal degeneration. *Journal of Neurology, 253*, 7–13.

Trick, G. L., Kaski, B., & Steinman, S. B. (1994). Visual impairment in Parkinson's disease: Deficits in orientation and motion discrimination. *Optometry and Vision Science, 71*, 242–245.

Ture, S., Inci, I., & Gedizlioglu, M. (2007). Abnormalities of contrast sensitivity, visual fields and visual evoked potentials in Parkinson's disease and effect of dopaminergic treatment. *Journal of Neurology, 254*, 93.

Vesela, O., Ruzicka, E., Jech, R., Roth, J., Stepenkova, K., Mecir, P., et al. (2001). Colour discrimination is not a reliable early marker of Parkinson's disease. *Journal of Neurology, 248*, 975–978.

Wang, L. H., Kuroiwa, Y., Kamitani, T., Li, M., Takahashi, T., Suzuk, Y., et al. (2000). Visual event-related potentials in progressive supranuclear palsy, corticobasal degeneration, and Parkinsons' disease. *Journal of Neurology, 247*, 356–363.

Weil, R. S., Schrag, A. E., Warren, J. D., Crutch, S. J., Lees, A. J., & Morris, H. R. (2016). Visual dysfunction in Parkinson's disease. *Brain, 139*, 2827–2843.

Wenning, G. K., Litvan, I., Jankovic, J., Granata, R., Mangone, C. A., McKee, A., et al. (1998). Natural history and survival of 14 patients with corticobasal degeneration confirmed at postmortem examination. *Journal of Neurology, Neurosurgery, and Psychiatry, 64*, 184–189.

Williams, D. R., Warren, J. D., & Lees, A. J. (2008). Using the presence of visual hallucinations to differentiate Parkinson's disease from atypical parkinsonism. *Journal of Neurology, Neurosurgery, and Psychiatry*, *79*, 652–655.

Winograd-Gurvich, C., Georgiou-Karistianis, N., Fitzgerald, P. B., Millist, L., & White, O. B. (2006). Self-paced saccades and saccades to oddball targets in Parkinson's disease. *Brain Research*, *1106*, 134–141.

Yu, C. Y., Lee, T., Shariati, A., Santini, V., Poston, K., & Liao, Y. J. (2016). Abnormal eye movement behaviour during reading in Parkinson's disease. *Parkinsonism & Related Disorders*, *32*, 130–132.

Zhukareva, V., Shah, K., Uryu, K., Braak, H., del Tredici, K., Sundarraj, S., et al. (2002). Biochemical analysis of tau proteins in argyrophilic grain disease, Alzheimer's disease and Pick's disease: A comparative study. *American Journal of Pathology*, *161*, 1135–1141.

CHAPTER THIRTY-THREE

Nonmotor Fluctuations in Parkinson's Disease

Christiana Franke*, Alexander Storch†,‡,1
*Charité—Universitätsmedizin Berlin, Berlin, Germany
†University of Rostock, Rostock, Germany
‡German Centre for Neurodegenerative Diseases (DZNE) Rostock, Rostock, Germany
[1]Corresponding author: e-mail address: alexander.storch@med.uni-rostock.de

Contents

Abstract

The advanced stage of Parkinson's disease (PD) is characterized by motor complications such as motor fluctuations and dyskinesias induced by long-term levodopa treatment. Recent clinical research provides growing evidence that various nonmotor symptoms such as neuropsychiatric, autonomic, and sensory symptoms (particularly pain) also show fluctuations in patients with motor fluctuations (called nonmotor fluctuations or NMF). However, NMF have not yet been adequately considered in routine care of advanced PD patients and only few therapeutic studies are available. Since the pathophysiology of NMF remains largely unknown, innovative therapeutic concepts are largely missing. The close connection of NMF and motor fluctuations, however, strongly suggests that the strategies used to treat motor complications—namely continuous dopaminergic stimulation—also apply for the therapy of NMF. Future controlled clinical trials specifically addressing NMF are urgently warranted.

International Review of Neurobiology, Volume 134
ISSN 0074-7742
http://dx.doi.org/10.1016/bs.irn.2017.05.021

In 1817 James Parkinson described in his historical book "An essay on the shaking palsy," the major motor signs of the disease that would later bear his name: tremor, bradykinesia or hypokinesia, and gait disturbances (Parkinson, 1817). However, he also reported other features of the disease such as sleep disturbances, constipation, bladder dysfunction, and delirium (Parkinson, 1817). Consistently, Parkinson's disease (PD) is nowadays regarded as a progressive neurodegenerative multisystem disorder expressing motor symptoms such as tremor, bradykinesia, and muscle rigidity accompanied by various neuropsychiatric, autonomous, and sensory nonmotor symptoms (NMS). During the last years, the relevance of NMS for the management of PD and their pivotal impact on the quality of life (QoL) of the patients have acquired increasing attention to both the treating physicians and PD researchers (Chaudhuri, Healy, & Schapira, 2006; Martinez-Martin, 2011, 2014). NMS are indeed highly prevalent and can be present in 60% up to 100% of patients suffering from PD depending on the disease stage and assessment method (Kim et al., 2013; Krishnan, Sarma, Sarma, & Kishore, 2011; Martinez-Martin et al., 2007).

Levodopa is still the most effective symptomatic treatment of PD and is thus considered as the "gold standard" therapy, but its long-term use is associated with motor complications including wearing-Off and levodopa-induced dyskinesia, which occur in 40%–60% of patients after 5 years of treatment (Blanchet, Allard, Gregoire, Tardif, & Bedard, 1996; Rascol et al., 2000) and can substantially impair their QoL (Hechtner et al., 2014). Shortly after the introduction of levodopa to the therapeutic options in PD, Marsden and Parkes in 1976 added nonmotor fluctuations (NMF) to the spectrum of levodopa-induced complications by describing "mood swings" associated with the motor wearing-Off phenomenon: "... the patient is mobile, active, and happy but shows florid dyskinesias (...) and then (...) the patient reverts to profound akinesia often with fear, sweating, flushing and confusion" (Marsden & Parkes, 1976). Subsequent investigations confirmed these pivotal observation (Cantello, Gilli, Riccio, & Bergamasco, 1986; Hardie, Lees, & Stern, 1984) and stimulated further clinical research resulting in the concept of NMF: In temporal connection to motor Off periods, the patient experiences a deterioration of NMS (herein called nonmotor Off states), mainly involving neuropsychiatric, autonomic, and sensory symptoms (Riley & Lang, 1993; Seki et al., 2013; Storch et al., 2015, 2013). It became clear from these studies that NMS do not represent nonmotor reactions to motor Off states, but can be conceptually separated from motor fluctuations.

1. EPIDEMIOLOGY AND RISK FACTORS OF NMF

The epidemiology of NMF is still largely unclear and previous investigations revealed heterogeneous data on the frequency of NMF. A first large study reporting a population-based evaluation of a general PD cohort including more than 300 patients with and without motor fluctuations (mean $\pm$ standard deviation age: 66 ± 10 years; mean disease duration of 10 ± 7 years) found that approx. 20% of all patients, but 33% of patient with motor fluctuations suffered from NMF (Brun et al., 2014). NMF occurred on average 10 ± 8 years after disease onset (Brun et al., 2014). This is in line with earlier studies in cohorts of PD patients with motor fluctuations reporting prevalences of NMF ranging from 17% up to 100% of patients (Kim et al., 2013; Krishnan et al., 2011; Raudino, 2001; Seki et al., 2013; Storch et al., 2015, 2013; Witjas et al., 2002). The reasons for this wide range of frequencies are most likely differences of the patient cohorts and the use of various evaluation tools with different sensitivities and specificities in detecting NMF due to the lack of established rating or assessment scales for NMF.

Interestingly, only 86% of patients with NMF also report motor fluctuations (Brun et al., 2014), indicating that NMF may occur independent of motor fluctuations in a relevant proportion of patients. In agreement, in a small cohort of patients assessed for NMF using a NMS diary, Ossig and coworkers detected more switches between NMS On- and Off-state periods (as a measure of NMF) in PD patients without motor fluctuations compared to healthy controls for depressed mood, concentration/attention, fatigue, and dizziness indicating again the occurrence of NMF in nonfluctuating PD patients (Ossig et al., 2016). However, for most neuropsychiatric symptoms, PD patients with motor fluctuations reported even more NMS state switches (a measure of NMF) as compared to patients without motor fluctuations (Ossig et al., 2016).

The search for risk or predictive factors associated with NMF is still at the very beginning, and prospective studies are only rarely available. For motor fluctuations, prospective studies suggest that higher motor scores and levodopa dosage, younger age at onset, and female gender are major risk factors for the development of motor fluctuations (Stocchi et al., 2010; Warren Olanow et al., 2013). A recent cross-sectional study suggests that longer disease duration is also risk factor for the development of motor fluctuations, but whether the duration of levodopa therapy is also predictive for the occurrence of motor fluctuations is still under debate (Cilia et al., 2014).

In general, the studies available suggest that the predictive factors for NMF are very similar to those for motor fluctuations, but additionally include the presence of motor fluctuations. A prospective study involving de novo PD patients who had been followed-up for 4 years revealed female gender as a major risk factor for the development of NMF (Picillo et al., 2016). In a large cross-sectional study on both fluctuating and nonfluctuating PD patients (Brun et al., 2014), disease duration, levodopa treatment duration, and levodopa equivalent dose were significantly higher in the group of patients with NMF compared to that without NMF. Interestingly, dysautonomia and motor fluctuations were also more frequent in the group of patients with NMF. These data are in large agreement with previous studies showing that disease duration, motor fluctuations, duration of levodopa therapy, levodopa equivalent dose are factors associated with NMF (Gunal, Nurichalichi, Tuncer, Bekiroglu, & Aktan, 2002; Seki et al., 2013; Witjas et al., 2002). Gunal and colleagues showed additional factors associated with NMF—particularly for sensory and autonomic NMF—such as age at disease onset (Gunal et al., 2002). From a practical point of view, the data show that it is of pivotal importance for the disease management to investigate or even screen all PD patients with motor fluctuations for the occurrence of NMF.

2. CLINICAL PRESENTATION OF NMF

In large cross-sectional studies, 60%–100% of PD patients report of at least one NMS being present indicating their importance in PD management patients (Kim et al., 2013; Krishnan et al., 2011; Raudino, 2001; Seki et al., 2013; Storch et al., 2015, 2013; Witjas et al., 2002). The commonly applied subdivision of NMS into neuropsychiatric, autonomic, and sensory NMS does not reflect the whole complexity of NMS, and an appropriate assignment can be challenging. Needless to say, most of the NMS are either of neuropsychiatric or autonomic origin, but some symptoms including dizziness, inner restlessness, and fatigue are difficult to categorize in one of these groups. Abnormalities of sensation and pain as well as sleep disturbances are additionally enlarging the spectrum. The differences largely result from differences between the patient cohorts, in particular with respect to the disease stage and the use of various evaluation tools with different sensitivities and specificities in detecting NMS. Comparison between studies is almost impossible, since even within circumscribed categories no exact definition of the investigated subitems is yet agreed on.

In general, most if not all NMS from all major NMS domains express fluctuations in both frequency and severity in conjunction to motor fluctuations in advanced PD patients (see Table 1 for details), but perceptual problems and hallucinations seem to fluctuate less compared to other NMS (Hillen & Sage, 1996; Riley & Lang, 1993; Seki et al., 2013; Storch et al., 2015, 2013). In several studies, patients rated fatigue as their most disruptive and predominantly fluctuating symptom (Storch et al., 2015, 2013), but others reported sensory abnormalities and pain to be the most disabling NMF followed by dysautonomic NMF (Witjas et al., 2002). The clinical impression is that NMS fluctuate in clusters (Storch et al., 2015, 2013); however, no detailed analysis has been performed to support this observation. In Sections 2.1–2.4, the detailed description of the clinical presentations of NMF follows the classification/categorization of NMS as used in the NMS-Scale with its nine domains (Chaudhuri et al., 2007; Martinez-Martin et al., 2009), but strongly focuses on neuropsychiatric, autonomic, and sensory NMS.

2.1 Neuropsychiatric Symptoms

Neuropsychiatric fluctuations (NMF of domains D3 [mood/apathy], D4 [perception/hallucination], and D5 [attention/memory] of the NMS-Scale; Chaudhuri et al., 2007; Martinez-Martin et al., 2009) are reported to be the most frequent and disabling NMF with 32%–100% of fluctuating PD patients experiencing neuropsychiatric NMF (Seki et al., 2013; Storch et al., 2015, 2013; Witjas et al., 2002). NMF of mood symptoms such as depression and anxiety indeed have the greatest negative impact on QoL (Storch et al., 2013). In the context of NMF, it is useful to divide neuropsychiatric NMS into "plus" (presumably occurring in hyperdopaminergic state and thus in conjunction with motor On or dyskinetic state) and "minus" symptoms (presumably generated by low dopaminergic stimulation and thus associated with motor Off state)—a discrimination of major significance on the fluctuating properties of this NMS category.

The spectrum of neuropsychiatric "minus" symptoms involves anxiety, depression, apathy, mutism, and cognitive impairment. Particularly these minus symptoms have an outrageous effect on the QoL of the patients (Martinez-Martin, Rodriguez-Blazquez, Kurtis, Chaudhuri, & NMSS Validation Group, 2011; Schrag, 2006) and seem to subsist mainly during motor Off state (Seki et al., 2013; Storch et al., 2015, 2013). Anxiety and depression account for 42% of a decreased QoL in PD patients

Table 1 Nonmotor Symptoms Reported to Fluctuate in Conjunction With Motor Fluctuations

Item/Domain	**Frequency[a]**	**Frequency × Severity[b]**
1. Lightheadedness	Off > On	Off > On
2. Fainting	Off = On	Off = On
D1. Cardiovascular	*Off > On*	*Off > On*
3. Daytime sleepiness	Off = On	Off > On
4. Fatigue	Off > On	Off > On
5. Difficulty falling asleep	Off = On	Off > On
6. Restless legs	Off = On	Off = On
D2. Sleep/fatigue	*Off > On*	*Off > On*
7. Lost interests in surroundings	Off > On	Off > On
8. Lack motivation	Off > On	Off > On
9. Feel nervous	Off > On	Off > On
10. Seem sad	Off > On	Off > On
11. Flat mood	Off > On	Off > On
12. Difficulty experiencing pleasure	Off > On	Off > On
D3. Mood/apathy	*Off > On*	*Off > On*
13. Hallucinations	Off = On	Off = On
14. Delusions	Off = On	Off = On
15. Double vision	Off = On	Off = On
D4. Perceptual problems/hallucinations	*Off = On*	*Off = On*
16. Concentration	Off > On	Off > On
17. Forget things or events	Off > On	Off > On
18. Forget to do things	Off = On	Off > On
D5. Attention/memory	*Off > On*	*Off > On*
19. Saliva	Off > On	Off > On
20. Swallowing	Off > On	Off > On
21. Constipation	Off = On	Off = On
D6. Gastrointestinal	*Off = On*	*Off > On*

Table 1 Nonmotor Symptoms Reported to Fluctuate in Conjunction With Motor Fluctuations—cont'd

Item/Domain	Frequency	Frequency × Severity
22. Urgency	Off > On	Off > On
23. Frequency	Off = On	Off > On
24. Nocturia	Off = On	Off > On
D7. Urinary	*Off > On*	*Off > On*
25. Interest in sex	Off = On	Off > On
26. Problems having sex	Off = On	Off > On
D8. Sexual dysfunction	*Off = On*	*Off > On*
27. Pain	Off > On	Off > On
28. Taste or smell	Off = On	Off = On
29. Weight change	Off = On	Off = On
30. Excessive sweating	Off > On	Off > On
D9. Miscellaneous	*Off > On*	*Off > On*
NMSS total	Off = On	Off > On

[a]Data are for the frequencies of symptoms of the NMS-Scale domain or item in motor On and Off state.
[b]Data are for the compound score of frequency × severity of symptoms of the NMS-Scale domain or item in motor On and Off state.
Data are adapted from Storch, A., Schneider, C. B., Klingelhofer, L., Odin, P., Fuchs, G., Jost, W. H., et al. (2015). Quantitative assessment of non-motor fluctuations in Parkinson's disease using the Non-Motor Symptoms Scale (NMSS). *Journal of Neural Transmission (Vienna, Austria: 1996), 122*(12), 1673–1684.

(Quelhas & Costa, 2009). Anxiety and depression show a particularly intense interdependence to motor Off symptoms (Ossig et al., 2016, 2017; Seki et al., 2013; Storch et al., 2015, 2013) since it is related to postural instability and freezing of gait as well as other features of gait impairment (Pontone et al., 2009). Indeed, the timing of mood fluctuations is tightly linked to that of motor fluctuations with respect to the onset of the Off periods, but mood Off periods show a longer duration as compared to motor Off periods (Ossig et al., 2016, 2017).

Apathy has a high prevalence rate of 20%–36% in de novo PD patients. The correlation between apathy, anxiety, depression, and the severity of motor symptoms has not finally been solved (den Brok et al., 2015). In treatment-naïve patients, apathy has been associated with severely pronounced motor symptoms and in 50% associated with depression and cognitive impairment (Dujardin et al., 2014). Apathy—similar to other mood symptoms—is fluctuating being more often and more severe in motor Off state as compared to motor On state (Storch et al., 2015, 2013). The domain including impairment of attention and concentration problems, forgetfulness, slowness of thinking and mental emptiness (D5 of the NMS-Scale) also shows clear fluctuating properties in conjunction with motor fluctuations (Menza, Sage, Marshall, Cody, & Duvoisin, 1990; Seki et al., 2013; Stefanova et al., 2015; Storch et al., 2015, 2013), but with a slower time course and longer Off-state period (Ossig et al., 2016, 2017).

Neuropsychiatric "plus" symptoms such as hallucinations, euphoria, and hyperactivity seem to be predominantly present during motor On phase (Storch et al., 2015; Witjas et al., 2002). Notably, using the NMS-Scale, not significant fluctuation of perceptual problems and hallucinations was observed (Storch et al., 2015). A careful distinction should be made about hallucinations linked to dementia and therefore occurring independent of motor fluctuations and symptoms such as excessive motivation and impulse control disorders, which are presumably related to side effects of dopaminergic medication and thus more likely fluctuating in conjunction with motor fluctuations. The challenge of differentiating these various symptoms in detail might be the reason for the absence of NMF in this NMS domain and might lead to misinterpretations of NMF particularly in motor On state in routine clinical settings.

Usually neuropsychiatric NMF have been attributed to develop mainly in parallel to motor fluctuations (Friedenberg & Cummings, 1989; Nissenbaum et al., 1987). Minus symptoms as well as motor Off symptoms might be clinical parameters of reduced mesolimbic and/or mesocortical dopaminergic stimulation. Activation of these circuits results in elevated mood and hyperactivity as rather "socially acceptable" neuropsychiatric plus symptoms along with motor On state. Finally, an overstimulation may result in pathologic "plus" symptoms such as hallucinations and aggressive behavior (Martinez-Fernandez, Schmitt, Martinez-Martin, & Krack, 2016; Rizos et al., 2014; Storch et al., 2013). Presumably, dopamine stimulation acts as a key factor for the modulation of neuropsychiatric NMF. A recent study by Ossig and coworkers showed a correlation of "minus" symptoms with motor Off states supporting this hypothesis (Ossig et al., 2016). In a 24-h analysis of the same data, a parallel course of anxiety and motor Off symptoms with peaks during morning, early afternoon, and evening was constituted (Ossig et al., 2017).

Cognitive fluctuations respond in a rather complex manner to an increase of levodopa or dopaminergic stimulation, where attention deficits improve, but executive functions deteriorate (Nieoullon, 2002). The more stable the levodopa availability, the less cognitive fluctuations emerge, assuming that the underlying mechanisms are eventually regulated by dopamine (Cools, 2006). Of note, using a visual analogue scales (VAS) as well as the NMS-Scale, concentration, and attention (domain D5) show fluctuations of both frequency and severity with more frequent and more severe symptoms in motor Off state (Storch et al., 2015, 2013).

2.2 Autonomic Symptoms

Fluctuations of autonomic NMS (NMS of domains D1 [cardiovascular], D6 [gastrointestinal], D7 [urinary], and D8 [sexual dysfunction] of the NMS-Scale; Chaudhuri et al., 2007; Martinez-Martin et al., 2009) are reported to be less frequent as fluctuations of neuropsychiatric NMS with 16%–90% of fluctuating PD patients experiencing autonomic NMF (Seki et al., 2013; Storch et al., 2015, 2013; Witjas et al., 2002). Autonomic NMS are usually assigned to the particular organ system involved—cardiovascular, gastrointestinal, and genitourinary symptoms. In general, autonomic NMS are frequent with a prevalence rate of over 80% (Jost, 2003). Extraordinary caution has to been paid if the presented autonomic feature is due to a NMS in the classical meaning or due to side effects of dopaminergic therapy or concomitant diseases.

Witjas and colleagues correlated dyspnoea, stridor, urinary urgency, and constipation with motor Off state (Witjas et al., 2002). It is, however, unclear how constipation is defined as a fluctuating NMS, and Storch and coworkers did not observe fluctuations of constipation (Storch et al., 2013). Symptoms of urinary and sexual dysfunction (except bladder urgency) display—in contrast to most other NMS—no fluctuation with respect to the frequency of the symptoms, but show more severe symptoms during motor Off periods (Storch et al., 2013). Bladder urgency is reported to be clearly associated with hypodopaminergic motor Off periods (Ossig et al., 2016, 2017; Seki et al., 2013; Storch et al., 2013; Witjas et al., 2002).

2.3 Sensory Symptoms and Pain

Fluctuations of sensory symptoms and pain (included in domain D9 [miscellaneous] of the NMS-Scale; Chaudhuri et al., 2007; Martinez-Martin et al.,

2009) occur with similar frequencies as those of autonomic NMS with 30%–94% of fluctuating PD patients experiencing NMF of sensory symptoms and/or pain (Seki et al., 2013; Storch et al., 2015, 2013; Witjas et al., 2002). There is, however, no clear indication that sensory symptoms such as olfactory dysfunction, which is present in up to 90% of PD patients (Cavaco et al., 2015) or vision problems, including diplopia, reduced contrast sensitivity, convergence insufficiency, and impaired color discrimination (Nowacka, Lubinski, Honczarenko, Potemkowski, & Safranow, 2014) fluctuate in conjunction with motor fluctuations (Storch et al., 2015, 2013; Witjas et al., 2002). On the other hand, pain and related symptoms are clearly fluctuating in conjunction with motor fluctuations. Nebe and coworkers initially show higher pain intensity during motor Off periods (Nebe & Ebersbach, 2009). Subsequent studies confirmed this result by showing more frequent and more severe pain symptoms during motor Off compared to motor On periods (Storch et al., 2015, 2013). During motor Off periods pain is more frequently noted on the limbs and the trunk (Ford, 2010). Pain in PD can be subdivided in musculoskeletal, neuropathic, and dystonia-related pain and discomfort due to akathisia (Defazio et al., 2008). Since central dopamine depletion and the disrupted projection of the striatum to the parietal somatosensory cortex are one rationale, peripherally dystonia and rigidity may be reason enough to cause pain (Chudler & Dong, 1995). These pain types should thus responsive to dopaminergic stimulation and particularly sensitive for fluctuations. Indeed, it is postulated that pain associated with motor Off is responsive to dopamine and therefore improves after dopaminergic medication is applied (Factor, Brown, & Molho, 2000). However, it still remains enigmatic which type of pain is fluctuating and how pain fluctuations are temporarily linked to motor fluctuations (Ossig et al., 2017).

2.4 Sleep and Fatigue

Sleep disorders and fatigue are included in domain D2 [sleep/fatigue] of the NMS-Scale (Chaudhuri et al., 2007; Martinez-Martin et al., 2009), but data on fluctuations of these symptoms are rare. The frequency of sleep symptoms/disorders in PD patients is up to 90% of patients in all disease stages. Sleep fragmentation with frequent nocturnal awakenings, restless legs syndrome, and REM sleep behavior disorder covers the spectrum of sleep disturbances. Repetitive nocturnal awakenings might be due to reduced dopaminergic stimulation leading to motor Off and the inability of the patient to turn around in bed. Dopaminergic stimulation could therefore ameliorate sleep quality (Trenkwalder et al., 2011). Insomnia and excessive daytime sleepiness might be the logical consequences of the reduced nighttime sleep, but also count as autonomous clinical pattern. Sleep symptoms show a heterogeneous picture with respect to their fluctuations: For daytime sleepiness and difficulties falling asleep, no differences were observed with respect to the occurrence rate, but the symptoms were more severe in motor Off states, but restless legs symptoms do not fluctuate (Storch et al., 2015, 2013).

Fatigue is common in PD patients and occurs throughout every stage of PD. The exact distinction of fatigue and fatigability and further classification in physical and cognitive fatigue and fatigability may be difficult to differentiate in clinical practice (Lou, 2015). Fatigue has also been described to occur concomitant to depression and has been associated with other NMS such as cognitive dysfunction, apathy, and sleep disorders (Dujardin et al., 2014; Skorvanek et al., 2015). In many studies fatigue is shown to have the highest negative impact on QoL of all NMS (Barone et al., 2009; Martinez-Martin et al., 2011). Nondopaminergic circuits, such as the noradrenergic, cholinergic, and serotoninergic systems, play a significant role in fatigue and fatigability. However, the progression of fatigue is reduced in PD patients receiving dopaminergic medication (Schifitto et al., 2008). Consistently, Storch and coworkers found a clear association of fatigue and motor states with more frequent and more severe fatigue during motor Off periods (Storch et al., 2015, 2013). Interestingly, patients experiencing fatigue only during motor On state have worse QoL compared to patients with fatigue in motor Off only or in motor On and Off states (Storch et al., 2013) underpinning the pivotal importance of the symptom fatigue for QoL particularly in good motor conditions.

3. NMF AND QOL

Aside from motor dysfunction, QoL in PD is strongly associated with the presence of NMS (Balestrino & Martinez-Martin, 2017; Li et al., 2010; Martinez-Martin, 2011, 2014; Martinez-Martin et al., 2011). The available data on NMF and QoL clearly show that NMF impair patient's QoL, most likely to a greater extent than motor fluctuations (Rieu et al., 2016; Stocchi et al., 2014; Storch et al., 2015, 2013; Witjas et al., 2002). With respect to specific NMS, mood symptoms and pain seem to have the greatest negative impact on QoL (Storch et al., 2013). The high variability of NMF with respect to both occurrence and severity between different motor Off-state periods within the individual patients (Fauser et al., 2015) leading to a large unpredictability of the respective burden of a given Off state might be one factor contributing to the large impact of NMF on QoL.

4. PATHOPHYSIOLOGY OF NMF

The pathophysiology of NMF is most likely heterogeneous and complex, and the pathophysiological knowledge on NMF is limited. On the one hand, the tight temporal link of some NMF (mainly mood, cognitive dysfunction, bladder urgency, and pain) and motor fluctuations (Gunal et al., 2002; Ossig et al., 2016, 2017) strongly implicates that dopaminergic factors—either via direct dopaminergic neurotransmission or via indirect mechanisms—are the key factors for the development of these NMF (Caceres-Redondo et al., 2014; Maricle, Nutt, Valentine, & Carter, 1995; Maricle, Valentine, Carter, & Nutt, 1998; Zhu et al., 2016). This view is underpinned by the observation that the risk factors for the development of both motor and NMF are very similar (see earlier for details). On the other hand, the different kinetics of these NMF, the variability in their severity, and the only weak correlation with motor impairment also suggest that other factors (neurotransmitter systems) contribute to the pathophysiology of these NMF. It is rather likely that the combination of the oscillation of dopaminergic stimulation and nondopaminergic mechanisms, which specifically mediate the respective fluctuating NMS, is responsible for the occurrence of NMF. Indeed, dopaminergic modulation of other neurotransmitter systems (serotoninergic, adrenergic, glutamatergic) is an attractive concept for understanding the pathophysiology and subsequent development of novel therapeutic strategies for NMF. In addition, the threshold and kinetics for dopaminergic stimulation for the improvement of motor functions seem—at least in some cases—different from that for the improvement of NMS (Evans, Farrell, Gibson, Helme, & Lim, 2012; Maricle et al., 1995, 1998).

On the other hand, some NMS (autonomic symptoms, fatigue) fluctuate independent of motor oscillations with respect to timing and severity (Gunal et al., 2002; Ossig et al., 2016, 2017) suggesting pathophysiological mechanisms independent or less dependent on dopaminergic stimulation. These mechanisms remain enigmatic.

5. MANAGEMENT OF NMF

Awareness and knowledge of NMF are inevitable in the management of PD patients throughout the course of the disease particularly in the advanced disease stage. Patients and caregivers may not be aware of NMS and their potential fluctuations or may attribute these symptoms to a different cause. Educational information should thus be provided by the treating physicians. These aspects together with the diversity of NMS as well as the heterogeneous patterns of NMF make the proper diagnosis time consuming and challenging, even for experienced movement disorder specialists. Recent data indeed show that motor and NMF (wearing Off) are already frequent in relatively early stages of PD and that they are underestimated by routine neurological evaluation (Stocchi et al., 2014). It is thus strongly recommended to use validated assessment tools (rating scales) for the regular screening of motor and NMF in PD patients already from early disease stages (~1–3 years after disease onset). The chosen treatment concepts need to bear in mind the diversity of NMS themselves as well as drug-induced and other secondary NMS developing throughout the disease. Fig. 1 displays a synopsis of the management of NMF proposing a five-step treatment approach.

5.1 Assessment Tools for NMS and NMF

The International Movement Disorder Society lists various rating scales for the assessment of NMS (http://www.movementdisorders.org/MDS/Education/Rating-Scales) including the NMS-Quest (Chaudhuri, Martinez-Martin, et al., 2006) and the NMS-Scale (Chaudhuri et al., 2007; Martinez-Martin et al., 2009). The NMS-Quest is a 30-item questionnaire designed as screening tool in a Yes/No answering format, allowing a specificity of approx. 89% for all NMS (Romenets et al., 2012). The NMS-Scale assesses the frequency and severity of NMS with high accuracy (Chaudhuri et al., 2007; Martinez-Martin et al., 2009). There are in addition numerous rating scales for individual NMS in PD, which are helpful in both daily routine and clinical trials: As examples, the Beck Depression Inventory (BDI), the Montgomery–Asberg Depression Rating Scale (MADRS), and the Hamilton Depression Scale (Ham-D) are commonly used to assess depression in PD and the response to antidepressant treatment (Schrag et al., 2007). Tools for measuring cognitive impairment comprise the

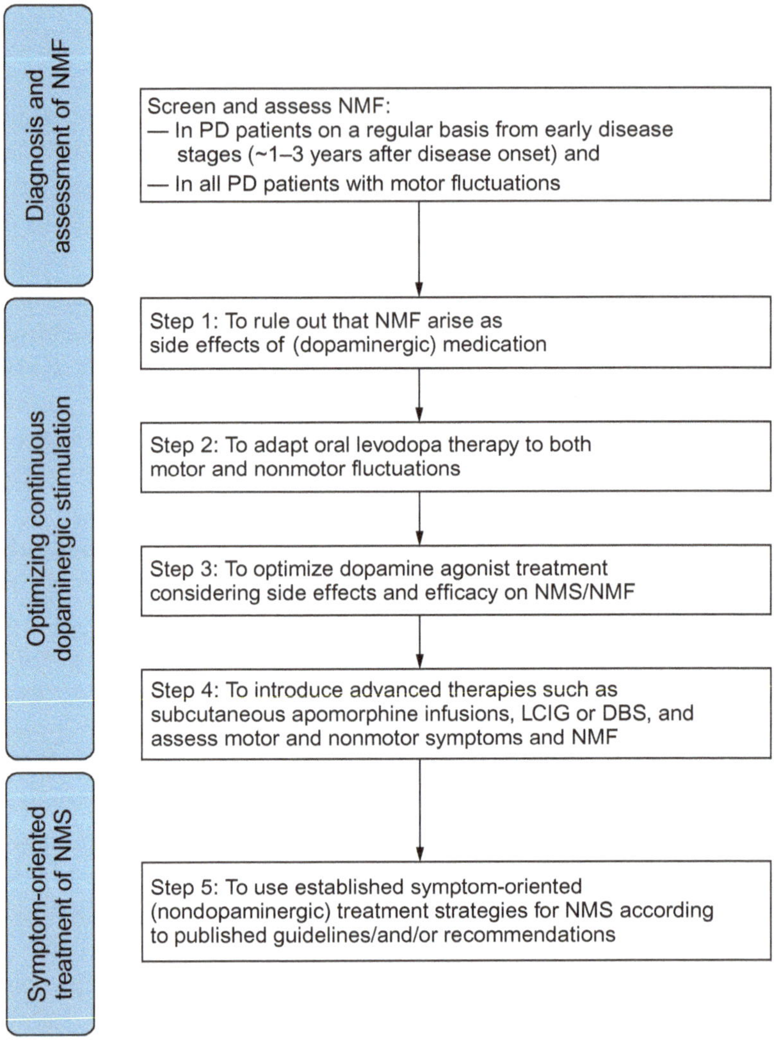

Fig. 1 Synopsis of management approach to nonmotor fluctuations (NMF). *DBS*, deep brain stimulation; *LCIG*, levodopa/carbidopa intestinal gel infusion; *NMS*, nonmotor symptoms; *PD*, Parkinson's disease.

Parkinson neuropsychometric dementia assessment (PANDA) (Kalbe et al., 2008) and the SCOPA-cognitive subscale (SCOPA-cog) (Marinus et al., 2003), but also the frequently used Montreal Cognitive Assessment Scale (MoCa) is well suited for detecting dementia in PD patients (Dalrymple-Alford et al., 2010). Pain can be examined by recently introduced King's Parkinson's disease pain scale (Chaudhuri et al., 2015).

On the other hand, there are no commonly accepted tools or rating instruments for the assessment of NMF and most rating scales are not validated or used in confirmatory studies. The initial studies used unstructured or structured interviews, in some cases combined with VAS to assess symptom severity (Hillen & Sage, 1996; Nebe & Ebersbach, 2009; Storch et al., 2013; Witjas et al., 2002). The group of Stacy and coworkers developed the wearing-Off questionnaire (WOQ) assessing both motor and NMS and their fluctuations in a qualitative manner (Antonini et al., 2011; Stacy et al., 2005, 2008; Stacy & Hauser, 2007). The 19-item version of this self-questionnaire (WOQ-19) is recommended by the Movement Disorder Society (MDS) Task Force on rating scales for PD as an effective and reliable tool to detect both motor and nonmotor wearing-Off phenomena in clinical practice.

Recently, Kleiner-Fisman and coworkers introduced the nonmotor fluctuation assessment instrument (NoMoFa) as a patient-based questionnaire using a semiquantitative assessment method (Kleiner-Fisman, Martine, Lang, & Stern, 2011). This questionnaire is, however, still not validated. The NMS-Scale quantitatively evaluates NMS by assessing the frequency and severity of 30 different NMS organized in nine domains (Chaudhuri et al., 2007; Martinez-Martin et al., 2009). Storch and coworkers modified the application of the NMS-Scale to quantitatively assess NMF (Table 1; Storch et al., 2015). The clinimetric attributes of this modified application in fluctuating PD patients including scoring range coverage, mean–median difference, and internal consistency measures are similar to those for the original utilization of the NMS-Scale independent of the motor state (Chaudhuri et al., 2007; Martinez-Martin et al., 2009). There are no test–retest data available. However, the NMS-Scale seems suitable for the quantitative assessment of NMF as needed in future clinical trials.

To assess motor fluctuations in clinical care as well as clinical trial, the PD home diary as introduced more than 10 years ago by Hauser, Deckers, and Lehert (2004) and Hauser et al. (2000, 2006). There is no such approach in NMF, but recently Ossig and coworkers introduced a NMS diary based on the questions of the NMS-Quest (Chaudhuri, Martinez-Martin, et al., 2006; Ossig et al., 2016). Although detailed data on the validity and reliability of the NMS diary are missing (as for the PD motor diary), the comparison of the diary data from fluctuating patients with healthy controls and nonfluctuators and the convergent validity data suggests that it measures NMF specifically in PD patients with motor fluctuations. However, it is noted that the diary misses pain as a major sensory NMF.

5.2 Treatment of NMF

Several guidelines and reviews have been published on the treatment of NMS: While the recommendations of the Movement Disorder Society (MDS; www.movementdisorders.org) and the American Academy of Neurology (AAN; www.aan.com) are strictly evidence based, most reviews also present expert opinions derived from clinical operating experience (Schrag, Sauerbier, & Chaudhuri, 2015; Seppi et al., 2011; Zesiewicz et al., 2010). However, these guidelines do not address the treatment of NMF. Although there is increasing evidence for the treatment of NMS experience (Chaudhuri, Healy, et al., 2006; Chaudhuri & Schapira, 2009; Schrag et al., 2015; Seppi et al., 2011; Zesiewicz et al., 2010), no such evidence is available for the treatment of NMF. Indeed, controlled trials with specific evaluation of therapies for NMF are lacking.

The principles for the treatment of NMF are in general very similar to those established for motor fluctuations (see Fig. 1). However, the earlier discussed pathophysiological considerations imply some additional strategies: (i) First, the combination of dopaminergic and nondopaminergic mechanisms with the latter mechanisms probably specific for the respective NMS suggests that it might be useful to combine the strategy of continuous dopaminergic stimulation (similar to that applied for motor fluctuations) with nondopaminergic therapeutic standards for the specific NMS according to published treatment guidelines (Schrag et al., 2015; Seppi et al., 2011; Zesiewicz et al., 2010). (ii) Second, the different dopaminergic thresholds for the improvement of motor and NMS (Evans et al., 2012; Maricle et al., 1995, 1998) require the adaption of dopaminergic medication/stimulation not only to the motor but also to the NMF.

The initial approach in the management of NMF is to optimize continuous dopaminergic stimulation according to the principles known for motor fluctuations (Chaudhuri, Rizos, & Sethi, 2013; Timpka, Mundt-Petersen, & Odin, 2016) as recommended in several national guidelines (e.g., Germany: https://www.dgn.org/leitlinien; UK: https://www.nice.org.uk; US: https://www.aan.com/Guidelines), but with clear attention to the fluctuation of NMS with respect to their occurrence and severity. This includes—as the first step—the initial rule-out that NMF arise as side effects of (dopaminergic) medication, mainly observed for neuropsychiatric symptoms and dizziness/orthostatic hypotension during motor On-state periods (see earlier).

The second step is to adapt oral levodopa therapy to both motor and NMF using levodopa dose fragmentation, prolonged-release levodopa formulations (particularly for bedtime use), and the introduction of dopamine-enhancing therapies, such as catechol-O-methyltransferase (COMT) inhibitors or monoamine oxidase-B (MAO-B) inhibitors. For the assessment of NMF, the WOQ-19 is recommended, but it does not allow the quantitative assessment of NMF. Since there are no controlled data available, there is currently no clear indication that one of the available compounds (entacapone, opicapone, and tolcapone as COMT inhibitors; selegiline, rasagiline, and safinamide as MAO-B inhibitors) is superior in the treatment of NMF. It is thus recommended to use the same strategies as those recommended for the treatment of motor fluctuations. However, the novel MAO-B inhibitor safinamide (with additional antiglutamatergic actions) was shown to improve not only motor complications in advanced PD but also ameliorate depression in a controlled clinical trial (Borgohain et al., 2014). The central acting COMT inhibitor tolcapone (Ceravolo et al., 2002; Gerlach, Ukai, Ozawa, & Riederer, 2002) is reported to be effective in reducing the motor wearing-Off phenomenon (Baas et al., 1997; Kurth et al., 1997; Rajput, Martin, Saint-Hilaire, Dorflinger, & Pedder, 1997) but also provides positive effects in several NMS, particularly antidepressive effects (Antonini et al., 2008; Ebersbach, Hahn, Lorrain, & Storch, 2010; Muller & TANIMOS Study Investigators, 2014). No data on NMF are available.

The third step (might be done in parallel to the second step) is to optimize dopamine agonist treatment considering side effects and efficacy on NMS. Several studies provide evidence that dopamine D3 agonists improve mood symptoms, particularly depression and apathy (Barone et al., 2010; Ray Chaudhuri et al., 2013; Thobois et al., 2013). Whether these effects are also present for fluctuating NMS is yet unknown. The post hoc analysis of a double-blind trial of transdermal dopamine agonist rotigotine vs placebo in PD patients with motor fluctuations suggests an improvement of pain as well as the NMS-Scale domains D2 sleep/fatigue and D3 mood/apathy during the early morning Off period (Kassubek et al., 2014; Ray Chaudhuri et al., 2013; Trenkwalder et al., 2011). The usefulness of apomorphine injections for the treatment of NMF is still under debate: Although apomorphine injections lead to a prompt improvement of Off-related motor and NMS (Pahwa, Koller, Trosch, Sherry, & APO303 Study Investigators, 2007; Trenkwalder et al., 2015), their intermittent, rapid, and short-acting effect might lead to compulsive use and aggravation of neuropsychiatric NMS.

The fourth step is to introduce advanced therapies such as subcutaneous apomorphine infusions, levodopa/carbidopa intestinal gel infusion (LCIG), or deep brain stimulation of the subthalamic nucleus (DBS-STN). These therapies are highly effective in reducing motor fluctuations (Katzenschlager et al., 2005; Krack et al., 2003; Martinez-Martin et al., 2015; Odin et al., 2015; Olanow et al., 2014; Trenkwalder et al., 2015), and all three treatment strategies have shown efficacy in NMS (Deuschl et al., 2006; Martinez-Fernandez et al., 2016; Martinez-Martin et al., 2015; Schuepbach et al., 2013; Wolz et al., 2012). Although there are no controlled data on NMF for DBS-STN, one small cohort study reported a reduction in NMF after surgery of approx. 60% with pain fluctuations showing the greatest (84%) and neuropsychiatric fluctuations the smallest reduction of 30% (Witjas et al., 2007). It is worth to note that the reduction of antiparkinsonian medication after DBS-STN surgery might provoke neuropsychiatric side effects such as apathy, depression, and anxiety, but the readministration of dopaminergic medication usually reverses these symptomsis condition (Thobois et al., 2010, 2013). Since there are no controlled data on NMS or NMF for these therapeutic strategies, the presence of NMS or NMF is not an indication for these invasive therapies per se (Odin et al., 2015). However, the presence and severity of both NMS and NMF should be seriously considered for the indication of an advanced PD treatment approach.

The fifth step in the approach to target NMF is to use established symptom-oriented (nondopaminergic) treatment strategies of NMS such as antidepressant drugs for depressive mood fluctuations or rivastigmine for fluctuations of cognitive dysfunction (Schrag et al., 2015; Seppi et al., 2011; Zesiewicz et al., 2010). Although the clinical experience shows that many NMF do not sufficiently respond to such treatment strategies, the introduction of such therapies should be considered in patients with NMF already optimized but not sufficiently controlled by continuous dopaminergic stimulation or DBS. This holds true in particular for those patients experiencing the respective NMS independent of their fluctuations. The treatment guidelines for the specific NMS are recommended for this approach (Schrag et al., 2015; Seppi et al., 2011; Zesiewicz et al., 2010).

Together, fluctuations of NMS have not yet been adequately considered in the management of advanced PD patients and only few therapeutic studies are available. Indeed, controlled trials on NMF as the primary endpoint—similar to those for motor fluctuations—are not yet reported. One major technical roadblock for such trials is the lack of a validated and accepted clinical scale for the specific assessment of NMF. Moreover, the pathophysiology remains largely unknown complicating the development of novel and innovative therapeutic concepts for NMF. However, the close connection of NMF and motor fluctuations strongly suggests that the strategies applied to treat motor complications—namely continuous dopaminergic stimulation—also apply for the therapy of NMF. Future controlled clinical trials specifically addressing NMF are urgently warranted.

ACKNOWLEDGMENTS

Both authors participated in drafting the manuscript and its critical revision. Both authors approved the manuscript prior to its submission.

REFERENCES

Antonini, A., Abbruzzese, G., Barone, P., Bonuccelli, U., Lopiano, L., Onofrj, M., et al. (2008). COMT inhibition with tolcapone in the treatment algorithm of patients with Parkinson's disease (PD): Relevance for motor and non-motor features. *Neuropsychiatric Disease and Treatment*, *4*(1), 1–9.

Antonini, A., Martinez-Martin, P., Chaudhuri, R. K., Merello, M., Hauser, R., Katzenschlager, R., et al. (2011). Wearing-off scales in Parkinson's disease: Critique and recommendations. *Movement Disorders*, *26*(12), 2169–2175.

Baas, H., Beiske, A. G., Ghika, J., Jackson, M., Oertel, W. H., Poewe, W., et al. (1997). Catechol-O-methyltransferase inhibition with tolcapone reduces the "wearing off" phenomenon and levodopa requirements in fluctuating parkinsonian patients. *Journal of Neurology, Neurosurgery, and Psychiatry*, *63*(4), 421–428.

Balestrino, R., & Martinez-Martin, P. (2017). Neuropsychiatric symptoms, behavioural disorders, and quality of life in Parkinson's disease. *Journal of the Neurological Sciences*, *373*, 173–178.

Barone, P., Antonini, A., Colosimo, C., Marconi, R., Morgante, L., Avarello, T. P., et al. (2009). The PRIAMO study: A multicenter assessment of nonmotor symptoms and their impact on quality of life in Parkinson's disease. *Movement Disorders*, *24*(11), 1641–1649.

Barone, P., Poewe, W., Albrecht, S., Debieuvre, C., Massey, D., Rascol, O., et al. (2010). Pramipexole for the treatment of depressive symptoms in patients with Parkinson's disease: A randomised, double-blind, placebo-controlled trial. *Lancet Neurology*, *9*(6), 573–580.

Blanchet, P. J., Allard, P., Gregoire, L., Tardif, F., & Bedard, P. J. (1996). Risk factors for peak dose dyskinesia in 100 levodopa-treated parkinsonian patients. *The Canadian Journal of Neurological Sciences*, *23*(3), 189–193.

Borgohain, R., Szasz, J., Stanzione, P., Meshram, C., Bhatt, M. H., Chirilineau, D., et al. (2014). Two-year, randomized, controlled study of safinamide as add-on to levodopa in mid to late Parkinson's disease. *Movement Disorders*, *29*(10), 1273–1280.

Brun, L., Lefaucheur, R., Fetter, D., Derrey, S., Borden, A., Wallon, D., et al. (2014). Non-motor fluctuations in Parkinson's disease: Prevalence, characteristics and management in a large cohort of parkinsonian outpatients. *Clinical Neurology and Neurosurgery*, *127*, 93–96.

Caceres-Redondo, M. T., Carrillo, F., Lama, M. J., Huertas-Fernandez, I., Vargas-Gonzalez, L., Carballo, M., et al. (2014). Long-term levodopa/carbidopa intestinal gel in advanced Parkinson's disease. *Journal of Neurology*, *261*(3), 561–569.

Cantello, R., Gilli, M., Riccio, A., & Bergamasco, B. (1986). Mood changes associated with "end-of-dose deterioration" in Parkinson's disease: A controlled study. *Journal of Neurology, Neurosurgery, and Psychiatry*, *49*(10), 1182–1190.

Cavaco, S., Goncalves, A., Mendes, A., Vila-Cha, N., Moreira, I., Fernandes, J., et al. (2015). Abnormal olfaction in Parkinson's disease is related to faster disease progression. *Behavioural Neurology*, *2015*, 976589.

Ceravolo, R., Piccini, P., Bailey, D. L., Jorga, K. M., Bryson, H., & Brooks, D. J. (2002). 18F-dopa PET evidence that tolcapone acts as a central COMT inhibitor in Parkinson's disease. *Synapse*, *43*(3), 201–207.

Chaudhuri, K. R., Healy, D. G., & Schapira, A. H. (2006). Non-motor symptoms of Parkinson's disease: Diagnosis and management. *Lancet Neurology*, *5*(3), 235–245.

Chaudhuri, K. R., Martinez-Martin, P., Brown, R. G., Sethi, K., Stocchi, F., Odin, P., et al. (2007). The metric properties of a novel non-motor symptoms scale for Parkinson's disease: Results from an international pilot study. *Movement Disorders*, *22*(13), 1901–1911.

Chaudhuri, K. R., Martinez-Martin, P., Schapira, A. H., Stocchi, F., Sethi, K., Odin, P., et al. (2006). International multicenter pilot study of the first comprehensive self-completed nonmotor symptoms questionnaire for Parkinson's disease: The NMSQuest study. *Movement Disorders*, *21*(7), 916–923.

Chaudhuri, K. R., Rizos, A., & Sethi, K. D. (2013). Motor and nonmotor complications in Parkinson's disease: An argument for continuous drug delivery? *Journal of Neural Transmission (Vienna, Austria: 1996)*, *120*, 1305–1320.

Chaudhuri, K. R., Rizos, A., Trenkwalder, C., Rascol, O., Pal, S., Martino, D., et al. (2015). King's Parkinson's disease pain scale, the first scale for pain in PD: An international validation. *Movement Disorders*, *30*(12), 1623–1631.

Chaudhuri, K. R., & Schapira, A. H. (2009). Non-motor symptoms of Parkinson's disease: Dopaminergic pathophysiology and treatment. *Lancet Neurology*, *8*(5), 464–474.

Chudler, E. H., & Dong, W. K. (1995). The role of the basal ganglia in nociception and pain. *Pain*, *60*(1), 3–38.

Cilia, R., Akpalu, A., Sarfo, F. S., Cham, M., Amboni, M., Cereda, E., et al. (2014). The modern pre-levodopa era of Parkinson's disease: Insights into motor complications from sub-Saharan Africa. *Brain*, *137*(Pt. 10), 2731–2742.

Cools, R. (2006). Dopaminergic modulation of cognitive function-implications for L-DOPA treatment in Parkinson's disease. *Neuroscience and Biobehavioral Reviews*, *30*(1), 1–23.

Dalrymple-Alford, J. C., MacAskill, M. R., Nakas, C. T., Livingston, L., Graham, C., Crucian, G. P., et al. (2010). The MoCA: Well-suited screen for cognitive impairment in Parkinson disease. *Neurology*, *75*(19), 1717–1725.

Defazio, G., Berardelli, A., Fabbrini, G., Martino, D., Fincati, E., Fiaschi, A., et al. (2008). Pain as a nonmotor symptom of Parkinson disease: Evidence from a case-control study. *Archives of Neurology*, *65*(9), 1191–1194.

den Brok, M. G., van Dalen, J. W., van Gool, W. A., Moll van Charante, E. P., de Bie, R. M., & Richard, E. (2015). Apathy in Parkinson's disease: A systematic review and meta-analysis. *Movement Disorders*, *30*(6), 759–769.

Deuschl, G., Schade-Brittinger, C., Krack, P., Volkmann, J., Schafer, H., Botzel, K., et al. (2006). A randomized trial of deep-brain stimulation for Parkinson's disease. *The New England Journal of Medicine*, *355*(9), 896–908.

Dujardin, K., Langlois, C., Plomhause, L., Carette, A. S., Delliaux, M., Duhamel, A., et al. (2014). Apathy in untreated early-stage Parkinson disease: Relationship with other non-motor symptoms. *Movement Disorders*, *29*(14), 1796–1801.

Ebersbach, G., Hahn, K., Lorrain, M., & Storch, A. (2010). Tolcapone improves sleep in patients with advanced Parkinson's disease (PD). *Archives of Gerontology and Geriatrics*, *51*(3), e125–e128.

Evans, A. H., Farrell, M. J., Gibson, S. J., Helme, R. D., & Lim, S. Y. (2012). Dyskinetic patients show rebound worsening of affect after an acute L-dopa challenge. *Parkinsonism & Related Disorders*, *18*(5), 514–519.

Factor, S. A., Brown, D. L., & Molho, E. S. (2000). Subcutaneous apomorphine injections as a treatment for intractable pain in Parkinson's disease. *Movement Disorders*, *15*(1), 167–169.

Fauser, M., Lohle, M., Ebersbach, G., Odin, P., Fuchs, G., Jost, W. H., et al. (2015). Intraindividual variability of nonmotor fluctuations in advanced Parkinson's disease. *Journal of Parkinson's Disease, 5*(4), 737–741.

Ford, B. (2010). Pain in Parkinson's disease. *Movement Disorders, 25*(Suppl. 1), S98–103.

Friedenberg, D. L., & Cummings, J. L. (1989). Parkinson's disease, depression, and the on-off phenomenon. *Psychosomatics, 30*(1), 94–99.

Gerlach, M., Ukai, W., Ozawa, H., & Riederer, P. (2002). Different modes of action of catecholamine-O-methyltransferase inhibitors entacapone and tolcapone on adenylyl cyclase activity in vitro. *Journal of Neural Transmission (Vienna, Austria: 1996), 109*(5–6), 789–795.

Gunal, D. I., Nurichalichi, K., Tuncer, N., Bekiroglu, N., & Aktan, S. (2002). The clinical profile of nonmotor fluctuations in Parkinson's disease patients. *The Canadian Journal of Neurological Sciences, 29*(1), 61–64.

Hardie, R. J., Lees, A. J., & Stern, G. M. (1984). On-off fluctuations in Parkinson's disease. A clinical and neuropharmacological study. *Brain, 107*(Pt. 2), 487–506.

Hauser, R. A., Deckers, F., & Lehert, P. (2004). Parkinson's disease home diary: Further validation and implications for clinical trials. *Movement Disorders, 19*(12), 1409–1413.

Hauser, R. A., Friedlander, J., Zesiewicz, T. A., Adler, C. H., Seeberger, L. C., O'Brien, C. F., et al. (2000). A home diary to assess functional status in patients with Parkinson's disease with motor fluctuations and dyskinesia. *Clinical Neuropharmacology, 23*(2), 75–81.

Hauser, R. A., Russ, H., Haeger, D. A., Bruguiere-Fontenille, M., Muller, T., & Wenning, G. K. (2006). Patient evaluation of a home diary to assess duration and severity of dyskinesia in Parkinson disease. *Clinical Neuropharmacology, 29*(6), 322–330.

Hechtner, M. C., Vogt, T., Zollner, Y., Schroder, S., Sauer, J. B., Binder, H., et al. (2014). Quality of life in Parkinson's disease patients with motor fluctuations and dyskinesias in five European countries. *Parkinsonism & Related Disorders, 20*(9), 969–974.

Hillen, M. E., & Sage, J. I. (1996). Nonmotor fluctuations in patients with Parkinson's disease. *Neurology, 47*(5), 1180–1183.

Jost, W. H. (2003). Autonomic dysfunctions in idiopathic Parkinson's disease. *Journal of Neurology, 250*(Suppl. 1), I28–I30.

Kalbe, E., Calabrese, P., Kohn, N., Hilker, R., Riedel, O., Wittchen, H. U., et al. (2008). Screening for cognitive deficits in Parkinson's disease with the Parkinson neuropsychometric dementia assessment (PANDA) instrument. *Parkinsonism & Related Disorders, 14*(2), 93–101.

Kassubek, J., Chaudhuri, K. R., Zesiewicz, T., Surmann, E., Boroojerdi, B., Moran, K., et al. (2014). Rotigotine transdermal system and evaluation of pain in patients with Parkinson's disease: A post hoc analysis of the RECOVER study. *BMC Neurology, 14*, 42.

Katzenschlager, R., Hughes, A., Evans, A., Manson, A. J., Hoffman, M., Swinn, L., et al. (2005). Continuous subcutaneous apomorphine therapy improves dyskinesias in Parkinson's disease: A prospective study using single-dose challenges. *Movement Disorders, 20*(2), 151–157.

Kim, H. S., Cheon, S. M., Seo, J. W., Ryu, H. J., Park, K. W., & Kim, J. W. (2013). Nonmotor symptoms more closely related to Parkinson's disease: Comparison with normal elderly. *Journal of the Neurological Sciences, 324*(1–2), 70–73.

Kleiner-Fisman, G., Martine, R., Lang, A. E., & Stern, M. B. (2011). Development of a nonmotor fluctuation assessment instrument for Parkinson disease. *Parkinson's Disease, 2011*. 292719.

Krack, P., Batir, A., Van Blercom, N., Chabardes, S., Fraix, V., Ardouin, C., et al. (2003). Five-year follow-up of bilateral stimulation of the subthalamic nucleus in advanced Parkinson's disease. *The New England Journal of Medicine, 349*(20), 1925–1934.

Krishnan, S., Sarma, G., Sarma, S., & Kishore, A. (2011). Do nonmotor symptoms in Parkinson's disease differ from normal aging? *Movement Disorders, 26*(11), 2110–2113.

Kurth, M. C., Adler, C. H., Hilaire, M. S., Singer, C., Waters, C., LeWitt, P., et al. (1997). Tolcapone improves motor function and reduces levodopa requirement in patients with Parkinson's disease experiencing motor fluctuations: A multicenter, double-blind, randomized, placebo-controlled trial. Tolcapone Fluctuator Study Group I. *Neurology, 48*(1), 81–87.

Li, H., Zhang, M., Chen, L., Zhang, J., Pei, Z., Hu, A., et al. (2010). Nonmotor symptoms are independently associated with impaired health-related quality of life in Chinese patients with Parkinson's disease. *Movement Disorders, 25*(16), 2740–2746.

Lou, J. S. (2015). Fatigue in Parkinson's disease and potential interventions. *NeuroRehabilitation, 37*(1), 25–34.

Maricle, R. A., Nutt, J. G., Valentine, R. J., & Carter, J. H. (1995). Dose-response relationship of levodopa with mood and anxiety in fluctuating Parkinson's disease: A double-blind, placebo-controlled study. *Neurology, 45*(9), 1757–1760.

Maricle, R. A., Valentine, R. J., Carter, J., & Nutt, J. G. (1998). Mood response to levodopa infusion in early Parkinson's disease. *Neurology, 50*(6), 1890–1892.

Marinus, J., Visser, M., Verwey, N. A., Verhey, F. R., Middelkoop, H. A., Stiggelbout, A. M., et al. (2003). Assessment of cognition in Parkinson's disease. *Neurology, 61*(9), 1222–1228.

Marsden, C. D., & Parkes, J. D. (1976). "On-off" effects in patients with Parkinson's disease on chronic levodopa therapy. *Lancet, 1*(7954), 292–296.

Martinez-Fernandez, R., Schmitt, E., Martinez-Martin, P., & Krack, P. (2016). The hidden sister of motor fluctuations in Parkinson's disease: A review on nonmotor fluctuations. *Movement Disorders, 31*(8), 1080–1094.

Martinez-Martin, P. (2011). The importance of non-motor disturbances to quality of life in Parkinson's disease. *Journal of the Neurological Sciences, 310*(1–2), 12–16.

Martinez-Martin, P. (2014). Nonmotor symptoms and health-related quality of life in early Parkinson's disease. *Movement Disorders, 29*(2), 166–168.

Martinez-Martin, P., Reddy, P., Katzenschlager, R., Antonini, A., Todorova, A., Odin, P., et al. (2015). EuroInf: A multicenter comparative observational study of apomorphine and levodopa infusion in Parkinson's disease. *Movement Disorders, 30*(4), 510–516.

Martinez-Martin, P., Rodriguez-Blazquez, C., Abe, K., Bhattacharyya, K. B., Bloem, B. R., Carod-Artal, F. J., et al. (2009). International study on the psychometric attributes of the non-motor symptoms scale in Parkinson disease. *Neurology, 73*(19), 1584–1591.

Martinez-Martin, P., Rodriguez-Blazquez, C., Kurtis, M. M., Chaudhuri, K. R., & NMSS Validation Group. (2011). The impact of non-motor symptoms on health-related quality of life of patients with Parkinson's disease. *Movement Disorders, 26*(3), 399–406.

Martinez-Martin, P., Schapira, A. H., Stocchi, F., Sethi, K., Odin, P., MacPhee, G., et al. (2007). Prevalence of nonmotor symptoms in Parkinson's disease in an international setting; study using nonmotor symptoms questionnaire in 545 patients. *Movement Disorders, 22*(11), 1623–1629.

Menza, M. A., Sage, J., Marshall, E., Cody, R., & Duvoisin, R. (1990). Mood changes and "on-off" phenomena in Parkinson's disease. *Movement Disorders, 5*(2), 148–151.

Muller, T., & TANIMOS Study Investigators. (2014). Tolcapone addition improves Parkinson's disease associated nonmotor symptoms. *Therapeutic Advances in Neurological Disorders*, 7(2), 77–82.

Nebe, A., & Ebersbach, G. (2009). Pain intensity on and off levodopa in patients with Parkinson's disease. *Movement Disorders, 24*(8), 1233–1237.

Nieoullon, A. (2002). Dopamine and the regulation of cognition and attention. *Progress in Neurobiology, 67*(1), 53–83.

Nissenbaum, H., Quinn, N. P., Brown, R. G., Toone, B., Gotham, A. M., & Marsden, C. D. (1987). Mood swings associated with the "on-off" phenomenon in Parkinson's disease. *Psychological Medicine, 17*(4), 899–904.

Nowacka, B., Lubinski, W., Honczarenko, K., Potemkowski, A., & Safranow, K. (2014). Ophthalmological features of Parkinson disease. *Medical Science Monitor, 20*, 2243–2249.
Odin, P., Ray Chaudhuri, K., Slevin, J. T., Volkmann, J., Dietrichs, E., Martinez-Martin, P., et al. (2015). Collective physician perspectives on non-oral medication approaches for the management of clinically relevant unresolved issues in Parkinson's disease: Consensus from an international survey and discussion program. *Parkinsonism & Related Disorders, 21*(10), 1133–1144.
Olanow, C. W., Kieburtz, K., Odin, P., Espay, A. J., Standaert, D. G., Fernandez, H. H., et al. (2014). Continuous intrajejunal infusion of levodopa-carbidopa intestinal gel for patients with advanced Parkinson's disease: A randomised, controlled, double-blind, double-dummy study. *Lancet Neurology, 13*(2), 141–149.
Ossig, C., Sippel, D., Fauser, M., Gandor, F., Jost, W. H., Ebersbach, G., et al. (2016). Assessment of nonmotor fluctuations using a diary in advanced Parkinson's disease. *Journal of Parkinson's Disease, 6*(3), 597–607.
Ossig, C., Sippel, D., Fauser, M., Gandor, F., Jost, W. H., Ebersbach, G., et al. (2017). Timing and kinetics of nonmotor fluctuations in advanced Parkinson's disease. *Journal of Parkinson's Disease, 7*, 325–330.
Pahwa, R., Koller, W. C., Trosch, R. M., Sherry, J. H., & APO303 Study Investigators. (2007). Subcutaneous apomorphine in patients with advanced Parkinson's disease: A dose-escalation study with randomized, double-blind, placebo-controlled crossover evaluation of a single dose. *Journal of the Neurological Sciences, 258*(1–2), 137–143.
Parkinson, J. (1817). *An essay on the shaking palsy*. London: Sherwood Neely & Jones.
Picillo, M., Palladino, R., Moccia, M., Erro, R., Amboni, M., Vitale, C., et al. (2016). Gender and non motor fluctuations in Parkinson's disease: A prospective study. *Parkinsonism & Related Disorders, 27*, 89–92.
Pontone, G. M., Williams, J. R., Anderson, K. E., Chase, G., Goldstein, S. A., Grill, S., et al. (2009). Prevalence of anxiety disorders and anxiety subtypes in patients with Parkinson's disease. *Movement Disorders, 24*(9), 1333–1338.
Quelhas, R., & Costa, M. (2009). Anxiety, depression, and quality of life in Parkinson's disease. *The Journal of Neuropsychiatry and Clinical Neurosciences, 21*(4), 413–419.
Rajput, A. H., Martin, W., Saint-Hilaire, M. H., Dorflinger, E., & Pedder, S. (1997). Tolcapone improves motor function in parkinsonian patients with the "wearing-off" phenomenon: A double-blind, placebo-controlled, multicenter trial. *Neurology, 49*(4), 1066–1071.
Rascol, O., Brooks, D. J., Korczyn, A. D., De Deyn, P. P., Clarke, C. E., & Lang, A. E. (2000). A five-year study of the incidence of dyskinesia in patients with early Parkinson's disease who were treated with ropinirole or levodopa. 056 Study Group. *The New England Journal of Medicine, 342*(20), 1484–1491.
Raudino, F. (2001). Non motor off in Parkinson's disease. *Acta Neurologica Scandinavica, 104*(5), 312–315.
Ray Chaudhuri, K., Martinez-Martin, P., Antonini, A., Brown, R. G., Friedman, J. H., Onofrj, M., et al. (2013). Rotigotine and specific non-motor symptoms of Parkinson's disease: Post hoc analysis of RECOVER. *Parkinsonism & Related Disorders, 19*(7), 660–665.
Rieu, I., Houeto, J. L., Pereira, B., De Chazeron, I., Bichon, A., Chereau, I., et al. (2016). Impact of mood and behavioral disorders on quality of life in Parkinson's disease. *Journal of Parkinson's Disease, 6*(1), 267–277.
Riley, D. E., & Lang, A. E. (1993). The spectrum of levodopa-related fluctuations in Parkinson's disease. *Neurology, 43*(8), 1459–1464.
Rizos, A., Martinez-Martin, P., Odin, P., Antonini, A., Kessel, B., Kozul, T. K., et al. (2014). Characterizing motor and non-motor aspects of early-morning off periods in Parkinson's disease: An international multicenter study. *Parkinsonism & Related Disorders, 20*(11), 1231–1235.

Romenets, S. R., Wolfson, C., Galatas, C., Pelletier, A., Altman, R., Wadup, L., et al. (2012). Validation of the non-motor symptoms questionnaire (NMS-Quest). *Parkinsonism & Related Disorders*, *18*(1), 54–58.
Schifitto, G., Friedman, J. H., Oakes, D., Shulman, L., Comella, C. L., Marek, K., et al. (2008). Fatigue in levodopa-naive subjects with Parkinson disease. *Neurology*, *71*(7), 481–485.
Schrag, A. (2006). Quality of life and depression in Parkinson's disease. *Journal of the Neurological Sciences*, *248*(1–2), 151–157.
Schrag, A., Barone, P., Brown, R. G., Leentjens, A. F., McDonald, W. M., Starkstein, S., et al. (2007). Depression rating scales in Parkinson's disease: Critique and recommendations. *Movement Disorders*, *22*(8), 1077–1092.
Schrag, A., Sauerbier, A., & Chaudhuri, K. R. (2015). New clinical trials for nonmotor manifestations of Parkinson's disease. *Movement Disorders*, *30*(11), 1490–1504.
Schuepbach, W. M., Rau, J., Knudsen, K., Volkmann, J., Krack, P., Timmermann, L., et al. (2013). Neurostimulation for Parkinson's disease with early motor complications. *The New England Journal of Medicine*, *368*(7), 610–622.
Seki, M., Takahashi, K., Uematsu, D., Mihara, B., Morita, Y., Isozumi, K., et al. (2013). Clinical features and varieties of non-motor fluctuations in Parkinson's disease: A Japanese multicenter study. *Parkinsonism & Related Disorders*, *19*(1), 104–108.
Seppi, K., Weintraub, D., Coelho, M., Perez-Lloret, S., Fox, S. H., Katzenschlager, R., et al. (2011). The movement disorder society evidence-based medicine review update: Treatments for the non-motor symptoms of Parkinson's disease. *Movement Disorders*, *26*(Suppl. 3), S42–S80.
Skorvanek, M., Gdovinova, Z., Rosenberger, J., Saeedian, R. G., Nagyova, I., Groothoff, J. W., et al. (2015). The associations between fatigue, apathy, and depression in Parkinson's disease. *Acta Neurologica Scandinavica*, *131*(2), 80–87.
Stacy, M., Bowron, A., Guttman, M., Hauser, R., Hughes, K., Larsen, J. P., et al. (2005). Identification of motor and nonmotor wearing-off in Parkinson's disease: Comparison of a patient questionnaire versus a clinician assessment. *Movement Disorders*, *20*(6), 726–733.
Stacy, M., & Hauser, R. (2007). Development of a patient questionnaire to facilitate recognition of motor and non-motor wearing-off in Parkinson's disease. *Journal of Neural Transmission (Vienna, Austria: 1996)*, *114*(2), 211–217.
Stacy, M. A., Murphy, J. M., Greeley, D. R., Stewart, R. M., Murck, H., & Meng, X. (2008). The sensitivity and specificity of the 9-item wearing-off questionnaire. *Parkinsonism & Related Disorders*, *14*(3), 205–212.
Stefanova, E., Ziropadja, L., Stojkovic, T., Stankovic, I., Tomic, A., Jecmenica-Lukic, M., et al. (2015). Mild cognitive impairment in early Parkinson's disease using the movement disorder society task force criteria: Cross-sectional study in Hoehn and Yahr stage 1. *Dementia and Geriatric Cognitive Disorders*, *40*(3–4), 199–209.
Stocchi, F., Antonini, A., Barone, P., Tinazzi, M., Zappia, M., Onofrj, M., et al. (2014). Early DEtection of wEaring off in Parkinson disease: The DEEP study. *Parkinsonism & Related Disorders*, *20*(2), 204–211.
Stocchi, F., Rascol, O., Kieburtz, K., Poewe, W., Jankovic, J., Tolosa, E., et al. (2010). Initiating levodopa/carbidopa therapy with and without entacapone in early Parkinson disease: The STRIDE-PD study. *Annals of Neurology*, *68*(1), 18–27.
Storch, A., Schneider, C. B., Klingelhofer, L., Odin, P., Fuchs, G., Jost, W. H., et al. (2015). Quantitative assessment of non-motor fluctuations in Parkinson's disease using the Non-Motor Symptoms Scale (NMSS). *Journal of Neural Transmission (Vienna, Austria: 1996)*, *122*(12), 1673–1684.
Storch, A., Schneider, C. B., Wolz, M., Sturwald, Y., Nebe, A., Odin, P., et al. (2013). Nonmotor fluctuations in Parkinson disease: Severity and correlation with motor complications. *Neurology*, *80*(9), 800–809.

Thobois, S., Ardouin, C., Lhommee, E., Klinger, H., Lagrange, C., Xie, J., et al. (2010). Non-motor dopamine withdrawal syndrome after surgery for Parkinson's disease: Predictors and underlying mesolimbic denervation. *Brain*, *133*(Pt. 4), 1111–1127.

Thobois, S., Lhommee, E., Klinger, H., Ardouin, C., Schmitt, E., Bichon, A., et al. (2013). Parkinsonian apathy responds to dopaminergic stimulation of D2/D3 receptors with piribedil. *Brain*, *136*(Pt. 5), 1568–1577.

Timpka, J., Mundt-Petersen, U., & Odin, P. (2016). Continuous dopaminergic stimulation therapy for Parkinson's disease—Recent advances. *Current Opinion in Neurology*, *29*(4), 474–479.

Trenkwalder, C., Chaudhuri, K. R., Garcia Ruiz, P. J., LeWitt, P., Katzenschlager, R., Sixel-Doring, F., et al. (2015). Expert consensus group report on the use of apomorphine in the treatment of Parkinson's disease—Clinical practice recommendations. *Parkinsonism & Related Disorders*, *21*(9), 1023–1030.

Trenkwalder, C., Kies, B., Rudzinska, M., Fine, J., Nikl, J., Honczarenko, K., et al. (2011). Rotigotine effects on early morning motor function and sleep in Parkinson's disease: A double-blind, randomized, placebo-controlled study (RECOVER). *Movement Disorders*, *26*(1), 90–99.

Warren Olanow, C., Kieburtz, K., Rascol, O., Poewe, W., Schapira, A. H., Emre, M., et al. (2013). Factors predictive of the development of levodopa-induced dyskinesia and wearing-off in Parkinson's disease. *Movement Disorders*, *28*(8), 1064–1071.

Witjas, T., Kaphan, E., Azulay, J. P., Blin, O., Ceccaldi, M., Pouget, J., et al. (2002). Nonmotor fluctuations in Parkinson's disease: Frequent and disabling. *Neurology*, *59*(3), 408–413.

Witjas, T., Kaphan, E., Regis, J., Jouve, E., Cherif, A. A., Peragut, J. C., et al. (2007). Effects of chronic subthalamic stimulation on nonmotor fluctuations in Parkinson's disease. *Movement Disorders*, *22*(12), 1729–1734.

Wolz, M., Hauschild, J., Fauser, M., Klingelhofer, L., Reichmann, H., & Storch, A. (2012). Immediate effects of deep brain stimulation of the subthalamic nucleus on nonmotor symptoms in Parkinson's disease. *Parkinsonism & Related Disorders*, *18*, 994–997.

Zesiewicz, T. A., Sullivan, K. L., Arnulf, I., Chaudhuri, K. R., Morgan, J. C., Gronseth, G. S., et al. (2010). Practice parameter: Treatment of nonmotor symptoms of Parkinson disease: Report of the Quality Standards Subcommittee of the American Academy of Neurology. *Neurology*, *74*(11), 924–931.

Zhu, M., Li, M., Ye, D., Jiang, W., Lei, T., & Shu, K. (2016). Sensory symptoms in Parkinson's disease: Clinical features, pathophysiology, and treatment. *Journal of Neuroscience Research*, *94*(8), 685–692.

CHAPTER THIRTY-FOUR

Acute Presentation of Nonmotor Symptoms in Parkinson's Disease

Kimberly Kwei, Steven Frucht[1]
Icahn School of Medicine at Mount Sinai, New York, NY, United States
[1]Corresponding author: e-mail address: steven.frucht@mssm.edu

Contents

Abstract

There are a few syndromes involving the nonmotor symptoms of Parkinson's disease and other movement disorders that can quickly lead to severe morbidity and mortality, and, as such, need rapid identification and management. Among these are neuroleptic malignant syndrome, serotonin syndrome, dopamine agonist withdrawal syndrome, and dystonic storm. It is important to maintain a high index of suspicion for these disorders as lack of identification can lead to death. Many of these acutely occurring nonmotor syndromes are primarily the result of imbalances in dopaminergic and serotonergic systems due to changes in pharmacologic management of psychiatric disorders or Parkinson's disease. We discuss these acutely occurring nonmotor symptoms in order to raise awareness and also to highlight how these extremes in symptoms may uniquely shed light on the pathophysiology of Parkinson's disease.

1. INTRODUCTION

Most of the motor and nonmotor symptoms of movement disorders are managed in the outpatient setting, but there are a few syndromes that can rapidly lead to severe morbidity and mortality, requiring rapid identification and management. It is important to maintain a high index of suspicion for these disorders as lack of identification can lead to death. Among these are

International Review of Neurobiology, Volume 134
ISSN 0074-7742
http://dx.doi.org/10.1016/bs.irn.2017.06.002

neuroleptic malignant syndrome, serotonin syndrome, dopamine agonist withdrawal syndrome, and dystonic storm. The first three "emergencies" are primarily the result of imbalances in dopaminergic and serotonergic systems due to changes in pharmacologic management of psychiatric disorders or Parkinson's disease. The derangements they result in can be devastating, and have shed light on the function of these various neurotransmitters in the central nervous system.

2. NEUROLEPTIC MALIGNANT SYNDROME

Neuroleptic malignant syndrome (NMS) is characterized by a triad of fever, rigidity, and altered mental status in the setting of neuroleptic use. Autonomic dysfunction also frequently occurs (Shalev & Munitz, 1986). It is fairly rare with an incidence of about 0.2%–0.5% (Caroff & Mann, 1993; Kurlan, Hamill, & Shoulson, 1984; Shalev & Munitz, 1986), and failure to promptly diagnose the disorder can be fatal. Mortality is estimated to be about 20%. NMS has been described in a wide range of ages, from early childhood to old age. It is more likely to occur in young and middle-aged men (Kipps, Fung, Grattan-Smith, de Moore, & Morris, 2005). As first described in 1968 by Delay and Deniker, it was felt to be a syndrome characterized by fever and pallor, movement disorder, and pulmonary symptoms. The inclusion of pulmonary symptoms in the NMS likely reflected the complications of severe muscle rigidity, leading to rhabdomyolysis causing end-organ damage, and aspiration pneumonia, which contribute to the high mortality rate. NMS can occur very quickly after the onset of use, but can also occur up to 65 days after the initiation of the offending agent, with the average time of onset of NMS being 48 h after initiation of the drug (Shalev & Munitz, 1986).

A study of 52 patients found that the majority of patients who developed NMS were being treated with neuroleptics for psychiatric illness, but a few patients had only been given neuroleptics in the setting of preinduction anesthesia or to treat inpatient agitation (Kurlan et al., 1984). This study found that patients who develop NMS are usually on a standard dose of a neuroleptic, and the majority of the patients were on one or a combination of three neuroleptics: haloperidol, fluphenazine, or chlorpromazine, with haloperidol being the most frequently associated with NMS (Kurlan et al., 1984). Another early study looking at 150 case reports of NMS found that it occurs more frequently when a new neuroleptic is introduced, or when the dose of an existing one is increased (Shalev & Munitz, 1986).

While early studies mostly found NMS cases due to typical antipsychotics, it has also been described in patients taking atypical antipsychotics (Hasan & Buckley, 1998). NMS was not felt to be related to any particular underlying illness but specifically to neuroleptic use. After the disease has resolved, it is frequently impossible in patients with psychiatric disorders to stop antipsychotic treatment altogether. While most case reports suggest that restarting a neuroleptic usually does not lead to recurrence of NMS, recurrences may occur rarely and even result in death (Shalev & Munitz, 1986).

Because it is critical to identify NMS rapidly given its high mortality, efforts have been made to develop diagnostic criteria. An early study suggested a high likelihood of NMS when there was recent neuroleptic use coupled with the presence of either all three major criteria (fever, rigidity, and elevated creatine kinase levels) or two major and four minor criteria (tachycardia, abnormal blood pressure, tachypnea, altered consciousness, diaphoresis, and leukocytosis) (Levenson, 1985; Robottom, Weiner, & Factor, 2011). More recently, a panel of experts came to the consensus that the following criteria, in the setting of a negative workup for other causes and recent dopamine agonist exposure or dopamine agonist withdrawal, were strongly suggestive of NMS: hyperthermia, rigidity, change in mental status, an elevated creatine kinase, autonomic instability, tachycardia, and tachypnea (Gurrera et al., 2011).

NMS may be indistinguishable from the entity lethal catatonia, which is dangerous because the former requires a withdrawal of neuroleptics, while the latter is often treated with neuroleptics (Castillo, Rubin, & Holsboer-Trachsler, 1989). Both syndromes are characterized by fever, alteration of consciousness, and rigidity; in the case of NMS, the rigidity is due to severely increased muscle tone, while in lethal catatonia the patient is rigid because of overexcitement. In lethal catatonia, the rigidity seen is usually preceded by, and likely the result of, weeks to months of extreme anxiety and restlessness. In NMS, the rigidity does not have this behavioral prodrome and is also accompanied by tremors and dyskinesias. Yet NMS can sometimes be both preceded and followed by symptoms of catatonia, and catatonia can at times be worsened by antipsychotics. It has been postulated that NMS and lethal catatonia are diseases along the same spectrum (Luchini et al., 2013).

Although mortality is high, there are relatively few lasting effects of NMS once it has resolved. Treatment for NMS begins with withdrawal of all neuroleptics and then supportive care to try to mitigate the frequently fatal secondary effects. The dopamine agonists levodopa and bromocriptine are frequently used to treat NMS, but other dopamine agonists such as lisuride

(Rodriguez et al., 1990) and apomorphine (Wang & Hsieh, 2001) have also been used either when those more common dopamine agonists have failed or as initial therapy. Lisuride and apomorphine can be given parenterally, which is useful as patients may be too obtunded to safely swallow. When all else fails, it has also been found that electroconvulsive therapy (ECT) can be effective in treating NMS (Addonizio & Susman, 1987). ECT is also useful when it is difficult to differentiate between NMS and catatonia. Once neuroleptics are withdrawn, the symptoms of NMS gradually abate in 1–2 weeks.

Muscle relaxants such as dantrolene seem to help with the symptomatology of NMS even if they do not treat the cause of NMS. The use of dantrolene suggests that NMS and malignant hyperthermia share similar mechanism. However, the time course and precipitating agents differ between NMS and malignant hyperthermia. While the halogenated inhaled anesthetics that precipitate malignant hyperthermia work at the muscles themselves, neuroleptics exert a dopamine-blocking effect in the CNS which precipitate hyperthermia possibly due to effects in the hypothalamus and rigidity due to dopamine agonist effects in the striatum (Kurlan et al., 1984; Shalev & Munitz, 1986).

Patients on dopaminergic medications can develop an NMS-like syndrome when their dopaminergic medications are abruptly withdrawn, known as the parkinsonism hyperpyrexia syndrome (PHS). One of the first cases described a man with dyskinesias on levodopa, who developed a "fatal hyperpyrexia" with symptoms similar to NMS when his levodopa was reduced in an attempt to mitigate his dyskinesias (Sechi, Tanda, & Mutani, 1984). Given that the mechanism of NMS is thought to be neuroleptic-induced dopamine receptor blockade, it makes sense that tapering patients off their dopaminergic medications, as in the case of Parkinson's patients, might trigger an NMS-like syndrome. PHS was often observed when it was more common practice to put Parkinson's patients on "drug holidays" (Friedman, Feinberg, & Feldman, 1985). PHS has even been seen in some patients with very severe on–off phenomenon during "off" periods (Pfeiffer & Sucha, 1989). More recently, it has been observed that Parkinson's patients being treated with deep brain stimulation can undergo PHS as their dopaminergic drugs are being tapered (Factor, 2007), and conversely that a patient who had his DBS electrodes turned off developed PHS (Kadowaki, Hashimoto, Suzuki, Watanabe, & Hirata, 2011). Interestingly, it has also been observed in Parkinson's patients on dopaminergic medications during the summer months. It is thought that this seasonal dependence is caused by the warm weather causing dehydration that reduces the efficacy of dopaminergic drugs (Kuno, Mizuta, & Yamasaki, 1997). Treatment of PHS requires its rapid identification, as like NMS it can quickly lead to fatality. Levodopa is often used to treat PHS (Frucht, 2014), but the only treatment for PHS supported by a double blind, placebo-controlled study is methylprednisolone. Methylprednisolone pulse therapy was found to decrease the duration of PHS and also helped to mitigate its symptoms (Sato, Asoh, Metoki, & Satoh, 2003).

3. SEROTONIN SYNDROME

Serotonin syndrome was first described in patients taking monoamine oxidase inhibitors (MAOIs) who were given various doses of tryptophan (Oates & Sjoerdsma, 1960). Strikingly, several of these patients became altered as though they had undergone ethanol intoxication. The clinical hallmarks of serotonin syndrome are altered mental status, fever, and myoclonus. Serotonin syndrome shares features similar to those of NMS, but with the addition of myoclonus, hyperreflexia, seizures, and mood alterations (Radomski, Dursun, Reveley, & Kutcher, 2000; Robottom, Weiner, et al., 2011; Sternbach, 1991). While serotonin syndrome and NMS have similar symptoms, there are key clinical features that differentiate the two. The onset of serotonin syndrome occurs within minutes to hours, while NMS develops over days. Once the causative agents are withdrawn the symptoms of NMS abate slowly, while those of serotonin syndrome often resolve in less than 24 h.

The evidence that serotonin syndrome is due to overstimulation of serotonin receptors comes from animal models. Stimulation with serotonin receptor agonists can cause serotonin syndrome, while pretreatment with an agent that blocks 5-HT synthesis can prevent serotonin syndrome (Sternbach, 1991). Studies also suggest that both dopamine and serotonin are involved in the development of serotonin syndrome (Graham & Ilett, 1988). A study of patients on SSRIs showed that about 15% of patients who overdose on SSRIs develop serotonin syndrome (Watson et al., 2005), but the exact incidence is unclear. Serotonin syndrome can occur in all ages.

Although serotonin syndrome is frequently seen in patients on an MAOI with concomitant L-tryptophan, it has also been observed in patients being transitioned from the selective serotonin reuptake inhibitor (SSRI) fluoxetine to an MAOI. Patients taking concurrent lithium seem to be more prone to serotonin syndrome, and this has been supported by animal studies (Sternbach, 1991). MAOIs in combination with tricyclic antidepressants have also been found to cause serotonin syndrome.

Although serotonin syndrome is seen primarily in psychiatric patients on serotonergic medications, an early case report described a patient who developed serotonin syndrome while being transitioned from bromocriptine to levodopa (Sandyk, 1986). This patient's symptoms abated with treatment with a serotonin antagonist, methysergide. The Parkinson Study Group showed in 1993 that treatment with selegiline delayed the need for levodopa therapy in patients who had been diagnosed with Parkinson's disease within the previous 5 years (Parkinson Study Group, 1993). This study led to the widespread use of MAOIs in early Parkinson's disease. As depression is a well-known nonmotor symptom of Parkinson's disease, many patients began to be treated concomitantly with SSRIs and MAOIs, resulting in an increase in reports of Parkinson's patients with serotonin syndrome (Hébant et al., 2016; Weiss, 1995). A recent retrospective study of patients taking rasagiline with other antidepressants found no cases of serotonin syndrome (Panisset, Chen, Rhyee, Conner, & Mathena, 2014), which suggests that serotonin syndrome in patients taking MAO-B inhibitors and other antidepressants is relatively rare.

It is thought that depression in Parkinson's disease is at least partially due to dopamine deficiency in limbic pathways (Chaudhuri & Schapira, 2009). Thus an MAO-B selective agent, such as selegiline or rasagiline, may be better suited to treat depression in Parkinson's disease. A study of Parkinson's patients on rasagiline in combination with other antidepressants demonstrated that rasagiline caused a reduction in the "nonmotor symptoms" of PD, including depression, cognition, and daytime sleepiness when compared with patients on placebo (Smith, Eyal, & Weintraub, 2015).

Levodopa is metabolized into dopamine in serotonergic neurons, thus causing an increase in striatal serotonin. This increase in striatal serotonin is thought to contribute to the myoclonus seen in patients on levodopa therapy (Sandyk, 1986). Lewy body deposition in the substantia nigra causes dopamine depletion in the dorsal striatum, resulting in the classic Parkinson's symptomatology of rigidity, tremor, and akinesia. However, the deposition of Lewy bodies in the cortex and ventral striatum results in excessive cortical serotonin and dopamine, which is thought to be the cause of the psychosis seen in Parkinson's disease (Smith et al., 2015).

The frequent use of dopamine and serotonin in the treatment of Parkinson's disease was demonstrated in a case report of a patient with

PD being treated with DBS, amantadine, levodopa, clozapine, and rasagiline. The patient developed serotonin syndrome while being treated with ziprasidone for hallucinations and having recently been prescribed tramadol. Ziprasidone has high antagonistic activity toward the serotonin receptors 1D and 2C, but it is also a potent 5-HT1A agonist. Tramadol can act as a weak serotonin receptor inhibitor. DBS in the subthalamic nucleus decreases serotonin production from the serotonergic projections of the dorsal raphe nucleus. The authors felt that the patient developed serotonin syndrome from the combination of rasagiline, ziprasidone, and tramadol, despite the likely protective effects of the patient's DBS and levodopa therapy (El-Okdi, Lumbrezer, Karanovic, Ghose, & Assaly, 2014).

Given its rarity and high morbidity and mortality, it is not surprising that there have been no prospective trials examining treatment for serotonin syndrome. However, case reports suggest that rapid discontinuation of the causal agents is necessary to stop the symptoms of serotonin syndrome within 24 h, as well as supportive treatment for the secondary manifestations of serotonin syndrome that lead to morbidity and mortality (Feighner, Boyer, Tyler, & Neborsky, 1990; Sternbach, 1991; White & Simpson, 1981). Cyproheptadine, an antihistamine and serotonin antagonist, has been used in severe cases (Robottom, Weiner, et al., 2011).

4. DOPAMINE AGONIST WITHDRAWAL SYNDROME

Patients on dopamine agonists frequently experience side effects such as hallucinations, cognitive dysfunction, and impulse control disorders. If these side effects become debilitating, it may be necessary to taper patients off their dopamine agonists. However, it has been observed that a subset of patients who undergo tapering of dopamine agonists develop anxiety, diaphoresis, agitation, and a craving for the dopamine agonists. This putative withdrawal syndrome has been termed dopamine agonist withdrawal syndrome (DAWS) (Yu & Fernandez, 2017). Other symptoms include psychiatric symptoms, insomnia, daytime fatigue, autonomic symptoms, and generalized pain. DAWS was first described in 2010 in the course of a prospective study looking at the motor and nonmotor symptoms of patients with Parkinson's disease (Rabinak & Nirenberg, 2010). Out of the 93 patients studied, 26 underwent dopamine agonist taper and 5 of these patients developed a constellation of symptoms felt to comprise DAWS. The onset of the symptoms in all patients correlated with the withdrawal of dopamine agonists, which is why it was felt less likely to be simply a nonmotor symptom of PD itself. The patients who developed DAWS could not be treated with benzodiazepines, antidepressants, or cognitive behavioral therapy.

Based upon the initial study of 93 patients and another retrospective chart review study of 71 patients, the frequency of DAWS is estimated to be between 15% and 19% (Pondal et al., 2013; Rabinak & Nirenberg, 2010). Nearly all of the patients who experience DAWS were being treated for PD, although there was one patient who had been on a dopamine agonist for restless legs syndrome. The course of DAWS can be self-limited, but can also last for months to years. Risk factors for the development of DAWS include the presence of an impulse control disorder, being on higher doses of dopamine agonists, and a history of DBS. All of the patients who developed DAWS in the 2010 study had undergone dopamine agonist taper because of an impulse control disorder. Three of the patients who developed DAWS in that study ended up remaining on dopamine agonists and dealing with impulse control disorders instead. A more recent study of a larger number of patients with DAWS found that fewer than half of the patients had an impulse control disorder (Yu & Fernandez, 2017). DAWS does not appear to be associated with any one particular type of dopamine agonist, the speed of taper, or whether there was total vs partial withdrawal of the dopamine agonist. Thus far, no standard treatment has been established for DAWS, given the small number of people it has been described in. It is perhaps more important at this time for clinicians to be aware of the possibility of DAWS in patients for whom a dopamine agonist is being considered, especially if they seem to have a predisposition for developing an impulse control disorder.

It has been postulated that DAWS may occur because dopamine plays a key role in both reward and motor circuits. While dopamine agonists help with the motor symptoms of PD by stimulating the damaged nigrostriatal pathways that are involved in motor circuitry, they are concomitantly stimulating reward circuits of the mesocorticolimbic pathways that are intact. Thus this overstimulation of reward circuits may lead to the symptoms of addiction and craving similar to how other addictive substances activate the dopaminergic reward system.

5. DYSTONIC STORM

Dystonic storm occurs when a patient experiences sustained muscle contractions that are unremitting, continuous, and generalized. Dystonic storm occurs as a complication of primary dystonia (e.g., DYT-1 dystonia) or secondary dystonia, which can occur in patients with cerebral palsy or with pantothenate kinase-associated neurodegeneration (PKAN) (Frucht, 2014; Robottom, Factor, & Weiner, 2011). It is precipitated by infection, trauma, or changes in medication. Dystonic storm is not only terrifying and painful for patients but can also result in severe rhabdomyolysis causing myoglobinuria (Jankovic & Penn, 1982).

Treatment for dystonic storm must begin promptly, given that the side effects from sustained, severe muscle contractions can lead to lasting end-organ damage. The first line of treatment involves supportive management of the metabolic derangements that can occur from rhabdomyolysis, such as aggressive fluid administration and correction of electrolyte imbalances, as well as bulbar weakness and impairment of respiratory function, all of which may require ICU admission. Many treatments have been tried for dystonic storm in an attempt to stop muscle contractions, including baclofen, anticholinergics, benzodiazepines, and antidopaminergic medications (Frucht, 2014; Manji et al., 1998). The effort to stop the sustained dystonia may require sedation, leading to intubation, which may have already been required secondary to problems with respiration. Interestingly, there is a report in the literature of dystonic storm treated with intrathecal baclofen that allowed the patient to be taken off sedation (Dalvi, Fahn, & Ford, 1998).

Most medical treatments for dystonic storm have yielded mixed results, with no particular treatment consistently effective at stopping the sustained muscle contractions. Pallidotomy was initially observed to be effective in treating levodopa-induced dyskinesias, and pallidal stimulation mimicked lesioning of the globus pallidus. Studies of Parkinson's patients showed that stimulation of the subthalamic nucleus was superior to GPi stimulation for levodopa-induced dyskinesias (Krack et al., 1998). A subsequent study treating seven dystonia patients with bilateral GPi DBS showed good effect in eliminating dyskinesia. The six children in the study subsequently managed to walk without assistance and were reported to be "functionally normal" (Coubes, Roubertie, Vayssiere, Hemm, & Echenne, 2000).

Refractory dystonic storm is now routinely treated with bilateral DBS of the globus pallidus interna (GPi). A patient with PKAN developed generalized dystonia that left him wheelchair dependent. The patient was treated with benzodiazepines, baclofen, anticholinergics, botulinum toxin, and tetrabenazine with no effect and in fact made his dystonia worse leading to dystonic storm. The patient required sedation and intubation to control the dystonic storm, and 3 days after intubation, the patient underwent implantation of bilateral DBS electrodes in the GPi. Following initiation of DBS, the patient's dystonia improved. Interestingly, when the DBS system was inadvertently disconnected, the patient went back into dystonic storm that again did not remit until DBS could be resumed (Grandas, Fernandez-Carballal, Guzman-de-Villoria, & Ampuero, 2011). A paper examining 12 patients with dystonia treated with GPi DBS showed good effect of DBS on the dystonias, and in the three patients who developed dystonic storm, GPi stimulation was an effective treatment as well (Zorzi et al., 2005). Another study of five patients with DYT-1 dystonia who developed dystonic storm was effectively treated with DBS (Ben-Haim et al., 2016). The study also looked at 18 cases in the literature of patients with dystonic storm treated with DBS. Seventeen of the 18 patients had remission of dystonic storm with DBS while of the 20 cases of dystonic storm identified in the literature treated solely with medications, only 13 had remission.

As dystonic storm can be debilitating to patients with dystonia, patients with PD frequently experience disabling nonmotor symptoms such as anxiety disorders and diaphoresis that in their persistence cause a great deal of long-term dysfunction (Witjas et al., 2002). A recent study suggests the prevalence of anxiety disorders in PD is thought to be 31%, while the prevalence of patients with PD and panic disorders are 6.8% (Broen, Narayen, Kuijf, Dissanayaka, & Leentjens, 2016). Parkinson's patients with more severe symptoms appear more likely to have anxiety disorders, and younger Parkinson's patients (<62 years old) are more likely to have coexistent anxiety disorders (Dissanayaka et al., 2010). The onset of anxiety in PD patients is older than would be normally expected for anxiety disorders de novo (Richard, 2005). A case report in 2008 described a patient who presented with panic attacks and was subsequently diagnosed with Parkinson's disease. The patient's PD and panic attacks were subsequently both successfully treated with L-dopa (Matsui et al., 2006). Panic attacks have also been described as a complication of long-term L-dopa therapy (Vázquez, Jiménez-Jiménez, Garcia-Ruiz, & Garcia-Urra, 1993), more frequently occurring during "off" periods and should be managed by trying to decrease off periods (Lieberman, 1998). Viewing and treating anxiety disorders in Parkinson's patients as an extension of their Parkinson's disease rather than as a separate disease state may be important in better treating nonmotor symptoms of PD. This is important both for improving the quality of life in general in Parkinson's patients and also to better understand the whole pathophysiology of Parkinson's disease.

6. CONCLUSION

Prompt recognition of the movement disorder emergencies reviewed in this chapter is critical to management and to preventing morbidity and mortality.

REFERENCES

Addonizio, G., & Susman, V. L. (1987). ECT as a treatment alternative for patients with symptoms of neuroleptic malignant syndrome. *The Journal of Clinical Psychiatry*, *48*(3), 102–105.

Ben-Haim, S., Flatow, V., Cheung, T., Cho, C., Tagliati, M., & Alterman, R. L. (2016). Deep brain stimulation for status dystonicus: A case series and review of the literature. *Stereotactic and Functional Neurosurgery*, *94*(4), 207–215. http://dx.doi.org/10.1159/000446191.

Broen, M. P., Narayen, N. E., Kuijf, M. L., Dissanayaka, N. N., & Leentjens, A. F. (2016). Prevalence of anxiety in Parkinson's disease: A systematic review and meta-analysis. *Movement Disorders*, *31*(8), 1125–1133. http://dx.doi.org/10.1002/mds.26643.

Caroff, S. N., & Mann, S. C. (1993). Neuroleptic malignant syndrome. *The Medical Clinics of North America*, 77(1), 185–202.

Castillo, E., Rubin, R. T., & Holsboer-Trachsler, E. (1989). Clinical differentiation between lethal catatonia and neuroleptic malignant syndrome. *The American Journal of Psychiatry*, *146*(3), 324–328. http://dx.doi.org/10.1176/ajp.146.3.324.

Chaudhuri, K. R., & Schapira, A. H. (2009). Non-motor symptoms of Parkinson's disease: Dopaminergic pathophysiology and treatment. *Lancet Neurology*, *8*(5), 464–474. http://dx.doi.org/10.1016/s1474-4422(09)70068-7.

Coubes, P., Roubertie, A., Vayssiere, N., Hemm, S., & Echenne, B. (2000). Treatment of DYT1-generalised dystonia by stimulation of the internal globus pallidus. *Lancet*, *355*(9222), 2220–2221. http://dx.doi.org/10.1016/s0140-6736(00)02410-7.

Dalvi, A., Fahn, S., & Ford, B. (1998). Intrathecal baclofen in the treatment of dystonic storm. *Movement Disorders*, *13*(3), 611–612. http://dx.doi.org/10.1002/mds.870130344.

Dissanayaka, N. N., Sellbach, A., Matheson, S., O'Sullivan, J. D., Silburn, P. A., Byrne, G. J., et al. (2010). Anxiety disorders in Parkinson's disease: Prevalence and risk factors. *Movement Disorders*, *25*(7), 838–845. http://dx.doi.org/10.1002/mds.22833.

El-Okdi, N. S., Lumbrezer, D., Karanovic, D., Ghose, A., & Assaly, R. (2014). Serotonin syndrome after the use of tramadol and ziprasidone in a patient with a deep brain stimulator for Parkinson disease. *American Journal of Therapeutics*, *21*(4), e97–e99. http://dx.doi.org/10.1097/MJT.0b013e3182456d88.

Factor, S. A. (2007). Fatal Parkinsonism-hyperpyrexia syndrome in a Parkinson's disease patient while actively treated with deep brain stimulation. *Movement Disorders*, *22*(1), 148–149. http://dx.doi.org/10.1002/mds.21172.

Feighner, J. P., Boyer, W. F., Tyler, D. L., & Neborsky, R. J. (1990). Adverse consequences of fluoxetine-MAOI combination therapy. *The Journal of Clinical Psychiatry*, *51*(6), 222–225.

Friedman, J. H., Feinberg, S. S., & Feldman, R. G. (1985). A neuroleptic malignantlike syndrome due to levodopa therapy withdrawal. *JAMA*, *254*(19), 2792–2795.

Frucht, S. J. (2014). Treatment of movement disorder emergencies. *Neurotherapeutics*, *11*(1), 208–212. http://dx.doi.org/10.1007/s13311-013-0240-3.

Graham, P. M., & Ilett, K. F. (1988). Danger of MAOI therapy after fluoxetine withdrawal. *Lancet*, *2*(8622), 1255–1256.

Grandas, F., Fernandez-Carballal, C., Guzman-de-Villoria, J., & Ampuero, I. (2011). Treatment of a dystonic storm with pallidal stimulation in a patient with PANK2 mutation. *Movement Disorders*, *26*(5), 921–922. http://dx.doi.org/10.1002/mds.23586.

Gurrera, R. J., Caroff, S. N., Cohen, A., Carroll, B. T., DeRoos, F., Francis, A., et al. (2011). An international consensus study of neuroleptic malignant syndrome diagnostic criteria using the Delphi method. *The Journal of Clinical Psychiatry*, *72*(9), 1222–1228. http://dx.doi.org/10.4088/JCP.10m06438.

Hasan, S., & Buckley, P. (1998). Novel antipsychotics and the neuroleptic malignant syndrome: A review and critique. *The American Journal of Psychiatry*, *155*(8), 1113–1116. http://dx.doi.org/10.1176/ajp.155.8.1113.

Hébant, B., Guillaume, M., Desbordes, M., Gaillon, G., Maltête, D., & Lefaucheur, R. (2016). Combination of paroxetine and rasagiline induces serotonin syndrome in a parkinsonian patient. *Revue Neurologique (Paris)*, *172*(12), 788–789. http://dx.doi.org/10.1016/j.neurol.2016.10.002.

Jankovic, J., & Penn, A. S. (1982). Severe dystonia and myoglobinuria. *Neurology*, *32*(10), 1195–1197.

Kadowaki, T., Hashimoto, K., Suzuki, K., Watanabe, Y., & Hirata, K. (2011). Case report: Recurrent parkinsonism-hyperpyrexia syndrome following discontinuation of subthalamic deep brain stimulation. *Movement Disorders*, *26*(8), 1561–1562. http://dx.doi.org/10.1002/mds.23596.

Kipps, C. M., Fung, V. S., Grattan-Smith, P., de Moore, G. M., & Morris, J. G. (2005). Movement disorder emergencies. *Movement Disorders*, *20*(3), 322–334. http://dx.doi.org/10.1002/mds.20325.

Krack, P., Pollak, P., Limousin, P., Hoffmann, D., Xie, J., Benazzouz, A., et al. (1998). Subthalamic nucleus or internal pallidal stimulation in young onset Parkinson's disease. *Brain*, *121*(Pt. 3), 451–457.

Kuno, S., Mizuta, E., & Yamasaki, S. (1997). Neuroleptic malignant syndrome in parkinsonian patients: Risk factors. *European Neurology*, *38*(Suppl. 2), 56–59.

Kurlan, R., Hamill, R., & Shoulson, I. (1984). Neuroleptic malignant syndrome. *Clinical Neuropharmacology*, 7(2), 109–120.

Levenson, J. L. (1985). Neuroleptic malignant syndrome. *The American Journal of Psychiatry*, *142*(10), 1137–1145. http://dx.doi.org/10.1176/ajp.142.10.1137.

Lieberman, A. (1998). Managing the neuropsychiatric symptoms of Parkinson's disease. *Neurology*, *50*(Suppl. 6), S33–S38. discussion S44–S48.

Luchini, F., Lattanzi, L., Bartolommei, N., Cosentino, L., Litta, A., Kansky, C., et al. (2013). Catatonia and neuroleptic malignant syndrome: Two disorders on a same spectrum? Four case reports. *The Journal of Nervous and Mental Disease*, *201*(1), 36–42. http://dx.doi.org/10.1097/NMD.0b013e31827ab24b.

Manji, H., Howard, R. S., Miller, D. H., Hirsch, N. P., Carr, L., Bhatia, K., et al. (1998). Status dystonicus: The syndrome and its management. *Brain*, *121*(Pt. 2), 243–252.

Matsui, H., Udaka, F., Oda, M., Tamura, A., Kubori, T., Nishinaka, K., et al. (2006). Parkinson's disease following panic disorder. *The Journal of Neuropsychiatry and Clinical Neurosciences*, *18*(1), 130. http://dx.doi.org/10.1176/jnp.18.1.130.

Oates, J. A., & Sjoerdsma, A. (1960). Neurologic effects of tryptophan in patients receiving a monoamine oxidase inhibitor. *Neurology*, *10*, 1076–1078.

Panisset, M., Chen, J. J., Rhyee, S. H., Conner, J., & Mathena, J. (2014). Serotonin toxicity association with concomitant antidepressants and rasagiline treatment: Retrospective study (STACCATO). *Pharmacotherapy*, *34*(12), 1250–1258. http://dx.doi.org/10.1002/phar.1500.

Parkinson Study Group. (1993). Effects of tocopherol and deprenyl on the progression of disability in early Parkinson's disease. *The New England Journal of Medicine*, *328*(3), 176–183. http://dx.doi.org/10.1056/nejm199301213280305.

Pfeiffer, R. F., & Sucha, E. L. (1989). "On-off"-induced lethal hyperthermia. *Movement Disorders*, *4*(4), 338–341. http://dx.doi.org/10.1002/mds.870040408.

Pondal, M., Marras, C., Miyasaki, J., Moro, E., Armstrong, M. J., Strafella, A. P., et al. (2013). Clinical features of dopamine agonist withdrawal syndrome in a movement disorders clinic. *Journal of Neurology, Neurosurgery, and Psychiatry*, *84*(2), 130–135. http://dx.doi.org/10.1136/jnnp-2012-302684.

Rabinak, C. A., & Nirenberg, M. J. (2010). Dopamine agonist withdrawal syndrome in Parkinson disease. *Archives of Neurology*, *67*(1), 58–63. http://dx.doi.org/10.1001/archneurol.2009.294.

Radomski, J. W., Dursun, S. M., Reveley, M. A., & Kutcher, S. P. (2000). An exploratory approach to the serotonin syndrome: An update of clinical phenomenology and revised diagnostic criteria. *Medical Hypotheses*, *55*(3), 218–224. http://dx.doi.org/10.1054/mehy.2000.1047.

Richard, I. H. (2005). Anxiety disorders in Parkinson's disease. *Advances in Neurology*, *96*, 42–55.

Robottom, B. J., Factor, S. A., & Weiner, W. J. (2011). Movement disorders emergencies Part 2: Hyperkinetic disorders. *Archives of Neurology*, *68*(6), 719–724. http://dx.doi.org/10.1001/archneurol.2011.117.

Robottom, B. J., Weiner, W. J., & Factor, S. A. (2011). Movement disorders emergencies. Part 1: Hypokinetic disorders. *Archives of Neurology*, *68*(5), 567–572. http://dx.doi.org/10.1001/archneurol.2011.84.

Rodriguez, M. E., Luquin, M. R., Lera, G., Delgado, G., Salazar, J. M., & Obeso, J. A. (1990). Neuroleptic malignant syndrome treated with subcutaneous lisuride infusion. *Movement Disorders*, *5*(2), 170–172. http://dx.doi.org/10.1002/mds.870050215.

Sandyk, R. (1986). L-dopa induced "serotonin syndrome" in a parkinsonian patient on bromocriptine. *Journal of Clinical Psychopharmacology*, *6*(3), 194–195.

Sato, Y., Asoh, T., Metoki, N., & Satoh, K. (2003). Efficacy of methylprednisolone pulse therapy on neuroleptic malignant syndrome in Parkinson's disease. *Journal of Neurology, Neurosurgery, and Psychiatry*, *74*(5), 574–576.

Sechi, G. P., Tanda, F., & Mutani, R. (1984). Fatal hyperpyrexia after withdrawal of levodopa. *Neurology*, *34*(2), 249–251.

Shalev, A., & Munitz, H. (1986). The neuroleptic malignant syndrome: Agent and host interaction. *Acta Psychiatrica Scandinavica*, *73*(4), 337–347.

Smith, K. M., Eyal, E., & Weintraub, D. (2015). Combined rasagiline and antidepressant use in Parkinson disease in the ADAGIO study: Effects on nonmotor symptoms and tolerability. *JAMA Neurology*, *72*(1), 88–95. http://dx.doi.org/10.1001/jamaneurol.2014.2472.

Sternbach, H. (1991). The serotonin syndrome. *The American Journal of Psychiatry*, *148*(6), 705–713. http://dx.doi.org/10.1176/ajp.148.6.705.

Vázquez, A., Jiménez-Jiménez, F. J., Garcia-Ruiz, P., & Garcia-Urra, D. (1993). "Panic attacks" in Parkinson's disease. A long-term complication of levodopa therapy. *Acta Neurologica Scandinavica*, *87*(1), 14–18.

Wang, H. C., & Hsieh, Y. (2001). Treatment of neuroleptic malignant syndrome with subcutaneous apomorphine monotherapy. *Movement Disorders*, *16*(4), 765–767.

Watson, W. A., Litovitz, T. L., Rodgers, G. C., Jr., Klein-Schwartz, W., Reid, N., Youniss, J., et al. (2005). 2004 Annual report of the American Association of Poison Control Centers Toxic Exposure Surveillance System. *The American Journal of Emergency Medicine*, *23*(5), 589–666. http://dx.doi.org/10.1016/j.ajem.2005.05.001.

Weiss, D. M. (1995). Serotonin syndrome in Parkinson disease. *The Journal of the American Board of Family Practice*, *8*(5), 400–402.

White, K., & Simpson, G. (1981). Combined MAOI-tricyclic antidepressant treatment: A reevaluation. *Journal of Clinical Psychopharmacology*, *1*(5), 264–282.

Witjas, T., Kaphan, E., Azulay, J. P., Blin, O., Ceccaldi, M., Pouget, J., et al. (2002). Non-motor fluctuations in Parkinson's disease: Frequent and disabling. *Neurology*, *59*(3), 408–413.

Yu, X. X., & Fernandez, H. H. (2017). Dopamine agonist withdrawal syndrome: A comprehensive review. *Journal of the Neurological Sciences*, *374*, 53–55. http://dx.doi.org/10.1016/j.jns.2016.12.070.

Zorzi, G., Marras, C., Nardocci, N., Franzini, A., Chiapparini, L., Maccagnano, E., et al. (2005). Stimulation of the globus pallidus internus for childhood-onset dystonia. *Movement Disorders*, *20*(9), 1194–1200. http://dx.doi.org/10.1002/mds.20510.

SECTION 3

Therapeutics and Nonmotor Symptoms in Parkinson's Disease

CHAPTER THIRTY-FIVE

Nonmotor Effects of Conventional and Transdermal Dopaminergic Therapies in Parkinson's Disease

Ryul Kim, Beomseok Jeon[1]
Seoul National University, College of Medicine, Seoul, South Korea
[1]Corresponding author: e-mail address: brain@snu.ac.kr

Contents

Abstract

Nonmotor symptoms (NMS) are an integral component of Parkinson's disease (PD). Because the burden and range of NMS are key determinants of quality of life for patients and caregivers, their management is a crucial issue in clinical practice. Although a range of NMS have a dopaminergic pathophysiological basis, this fact is underrecognized, and thus, they are often regarded as dopamine unresponsive symptoms. However, substantial evidence indicates that many NMS respond to oral and transdermal dopaminergic therapies. In contrast, certain NMS are exacerbated or even precipitated by dopaminergic drugs and these unwanted effects may be seriously dangerous. Therefore, a dopaminergic strategy for NMS should be based on a consideration of the benefits vs the risks in individual patients with PD.

International Review of Neurobiology, Volume 134
ISSN 0074-7742
http://dx.doi.org/10.1016/bs.irn.2017.05.008

1. INTRODUCTION

Nonmotor symptoms (NMS) are now recognized to be an integral part of PD (Chaudhuri, Healy, & Schapira, 2006). Patients with PD experience a variety of NMS including neuropsychiatric, sleep-related, autonomic, gastrointestinal, and sensory symptoms (Barone et al., 2009; Martinez-Martin et al., 2007). While there tend to be more NMS in more advanced PD, some symptoms such as constipation, olfactory deficit, rapid eye movement (REM) sleep behavior disorder (RBD), and depression might precede the onset of motor symptoms by many years (Chaudhuri, Odin, Antonini, & Martinez-Martin, 2011). Even though the burden and range of NMS are key determinants of quality of life for patients and caregivers, they are often ignored and thus remain undertreated (Chaudhuri et al., 2010; Martinez-Martin, Rodriguez-Blazquez, Kurtis, & Chaudhuri, 2011; Schrag, Jahanshahi, & Quinn, 2000; Sjodahl, Hagell, & Nilsson, 2012). A large retrospective study reported that 21% of patients with PD initially presented with NMS; however, they were more likely to experience a delay in diagnosis and a high proportion of misdiagnosis (O'Sullivan et al., 2008).

Although the pathophysiology of NMS is complex because of their diversity, nondopaminergic dysfunction is thought to be a key mechanism in the development of NMS (Chaudhuri et al., 2006). Indeed, many NMS have a poor response to dopaminergic therapies (Chaudhuri & Schapira, 2009; Kim et al., 2009), and nondopaminergic therapies have an integral role in the treatment of NMS including neuropsychiatric, sleep-related, and autonomic symptoms (Fox, Brotchie, & Lang, 2008; Seppi et al., 2011; Zesiewicz et al., 2010). However, it should be noted that NMS are not necessarily synonymous with nondopaminergic symptoms. A recent systematic review investigated a possible underlying dopaminergic basis of NMS in PD based on imaging techniques and found in vivo evidence of a dopaminergic pathophysiological basis in 12 NMS (Qamar et al., 2017). Additionally, numerous clinical trials in patients with PD have shown the beneficial effect of dopaminergic therapies on certain aspects of NMS. These findings strongly suggest at least a dopaminergic contribution to NMS in PD, which is well exemplified by a previous review by Chaudhuri and Schapira (2009).

In contrast, a range of NMS can be aggravated or even precipitated by dopaminergic treatment in patients with PD. These involve psychiatric and behavioral, sleep-related, and autonomic side effects related to dopaminergic therapies, a withdrawal syndrome such as Parkinson hyperpyrexia syndrome and dopamine agonist withdrawal syndrome, and drug interaction such as serotonin syndrome. Certain adverse reactions, including peripheral edema and fibrotic reactions, are specifically related to the use of dopamine agonists (Antonini, Tolosa, Mizuno, Yamamoto, & Poewe, 2009). These unwanted effects of dopaminergic therapies are important to recognize because they can have a considerable impact on the quality of life of patients (Park & Stacy, 2011).

In this chapter, we first investigate the potential effect of oral and transdermal dopaminergic therapies on NMS in patients with PD. Next, we review several domains of NMS that might be aggravated or even precipitated by dopaminergic drugs. These are summarized in Table 1.

2. THE EFFECT OF DOPAMINERGIC THERAPIES ON NMS

2.1 Neuropsychiatric Symptoms

2.1.1 Depression

Depression is one of the most common neuropsychiatric symptoms in patients with PD, and clinically significant depressive symptoms occur in 40%–50% of those patients (Reijnders, Ehrt, Weber, Aarsland, & Leentjens, 2008). Dysfunction of a combination of dopaminergic and nondopaminergic systems such as serotonin and noradrenaline has been implicated in the etiology of depression in PD. Several imaging studies have shown that depression inversely correlates with dopaminergic availability (Di Giuda et al., 2012; Hesse et al., 2009; Rektorova, Srovnalova, Kubikova, & Prasek, 2008; Remy, Doder, Lees, Turjanski, & Brooks, 2005; Weintraub et al., 2005).

Many studies have examined the antidepressant effect of dopaminergic therapies involving levodopa, dopamine agonists, and monoamine oxidase-B (MAO-B) inhibitors for treatment of depression in PD. In particular, pramipexole (a nonergot D_2/D_3 agonist) has been well established as a therapeutic option for depression in PD based on the following three randomized clinical trials. Rektorova et al. (2003) conducted an 8-month, randomized, open-label, parallel-group study comparing pramipexole (1.5–4.5 mg/day) with pergolide (an ergot $D_1/D_2/D_3$ agonist, 1.5–4.5 mg/day) as an add-on therapy to the levodopa treatment in 41 PD patients with depression but no dementia. The Zung self-rating depression scale scores were decreased significantly in both groups, whereas the Montgomery–Asberg depression rating scale scores were significantly

Table 1 The Responsiveness of Nonmotor Symptoms to Dopaminergic Treatment in Parkinson's Disease

	Responsive to DRT	Unresponsive to DRT	Exacerbated by DRT
Neuropsychiatric symptoms	Depression Off period-related anxiety Apathy Cognitive impairment[a]	Dementia Confusion	Psychosis Dopamine dysregulation syndrome Punding Impulse control disorders
Autonomic dysfunction	Urgency (detrusor overactivity) Nocturia Erectile dysfunction	Frequency Sweating	Orthostatic hypotension Parkinson hyperpyrexia syndrome[b] Serotonin syndrome
Gastrointestinal dysfunction	Constipation[a] Unsatisfactory voiding of bowel	Ageusia Dysphagia Reflux, vomiting Fecal incontinence	Diarrhea Nausea
Sleep disorders	Restless legs syndrome Periodic limb movement disorder REM sleep behavior disorder[a] Insomnia	NREM sleep movement disorders Vivid dreaming Sleep-disordered breathing	Excessive daytime sleepiness
Sensory symptoms	Musculoskeletal pain Akathitic pain Central pain Off period-related dystonic pain	Radiculo-neuropathic pain Paresthesia Olfactory dysfunction Visual dysfunction	
Other symptoms	Fatigue		Peripheral edema Fibrotic reactions[c] Blurred vision Weight gain

[a]There is conflicting evidence.
[b]On withdrawal of dopaminergic medication.
[c]Association with ergot dopamine agonists.
DRT, dopamine replacement therapy; *REM*, rapid eyeball movement; *NREM*, non-REM.
Modified from Chaudhuri, K. R., & Schapira, A. H. (2009). Non-motor symptoms of Parkinson's disease: Dopaminergic pathophysiology and treatment. *The Lancet Neurology, 8*(5), 464–474.

decreased only in the pramipexole group. Another randomized study compared pramipexole (1.5–4.5 mg/day) with sertraline (an antidepressant, 50 mg/day) in 67 PD patients with major depression who did not have motor complications (Barone et al., 2006). In both groups, the Hamilton depression rating scale score was decreased at 12 weeks; however, a higher proportion of patients were improved in the pramipexole group (61%) than in the sertraline group (27%, $P = 0.006$). The authors suggested that the antidepressant effect of pramipexole is linked with its dopamine receptor agonism at the limbic dopamine D_3 receptors. Lastly, Barone et al. (2010) did a large double-blind, placebo-controlled, randomized study of pramipexole (0.375–3.0 mg/day) to treat depressive symptoms in patients with PD. Of the 323 included patients, 296 were randomly assigned to pramipexole ($n = 144$) or placebo ($n = 152$). At 12 weeks, the Beck Depression Inventory (BDI) scores decreased by an adjusted mean of 5.9 (SE 0.5) points in the pramipexole group and 4.0 (SE 0.5) points in the placebo group with a between group difference of 1.9 (95% CI: 0.5–3.4; $P = 0.01$). Path analysis suggested that a direct effect of pramipexole on depression accounted for 80% of the total treatment effect ($P = 0.04$). Based on these studies, the Movement Disorder Society Task Force Guidelines of 2011 concluded that pramipexole is efficacious for the treatment of depressive symptoms in PD (Seppi et al., 2011). Similarly, a potential antidepressant effect of prolonged-release ropinirole (a nonergot D_2/D_3 agonist) was reported in the EASE-PD (efficacy and safety evaluation in PD) adjunct study (Pahwa et al., 2007). In this 24-week, double-blind, placebo-controlled study of 393 patients with PD, the BDI scores as a secondary outcome measure were significantly improved with ropinirole compared with the placebo ($P = 0.01$).

There is conflicting evidence for use of transdermal rotigotine (a nonergot $D_1/D_2/D_3$ agonist) to treat depression in patients with PD. The RECOVER (randomized evaluation of the 24-h coverage: efficacy of rotigotine) study was a double-blind, placebo-controlled, randomized trial that compared the effect of rotigotine ($n = 190$) ranging from 2 to 16 mg/day and a placebo ($n = 97$) on early morning function and nocturnal sleep disturbances in 287 PD patients with unsatisfactory early morning impairment (Trenkwalder et al., 2011). The treatment with rotigotine was titrated to an optimal dose over 1–8 weeks with subsequent dose maintenance for 4 weeks. Depression was assessed by the BDI score as a secondary outcome measure. At the end of the treatment, rotigotine was significantly superior to the placebo in reducing the BDI score with a between group difference of 2.01 (95% CI: 0.47–3.55; $P = 0.011$). This was further shown in a post hoc analysis of individual items of the NMS scale from the RECOVER study (Chaudhuri et al., 2013). On the other hand, another randomized controlled trial using 380 patients with PD evaluated the antidepressant effect of transdermal rotigotine in which there was no difference in the change of the Hamilton depression rating scale as a primary outcome measure at the end of the treatment (Chung, Asgharnejad, Bauer, Ramirez, & Jeon, 2016).

The effect of rasagiline (a MAO-B inhibitor) on depression in PD remains unclear. The ADAGIO (attenuation of disease progression with Azilect given once-daily) study was an 18-month, double-blind, placebo-controlled trial evaluating whether rasagiline may have a disease modifying effect in 1176 early PD patients. The patients were randomly assigned to rasagiline 1 or 2 mg/day for 72 weeks or corresponding placebo for 36 weeks, followed by rasagiline 1 or 2 mg/day for 36 weeks (Olanow et al., 2009). A post hoc analysis of the ADAGIO study assessed the effect of rasagiline on NMS in patients with PD and found that the depression score of the nonmotor experiences of daily living was less deteriorated with rasagiline 1 mg/day compared to the placebo ($P < 0.05$), but not with rasagiline 2 mg/day (Poewe, Hauser, & Lang, 2015). Alternatively, in a recent 12-week, double-blind, placebo-controlled study of 123 PD patients with moderate depressive symptoms, there was no significant difference in the change of the BDI score as a primary outcome measure between rasagiline 1 mg/day and the placebo (Barone et al., 2015).

2.1.2 Anxiety

Anxiety is another frequent NMS, usually coexisting with depression. The prevalence of anxiety disorders has been estimated at between 20% and 49%, with generalized anxiety disorder, panic disorder, and social phobia being the most common (Gallagher & Schrag, 2012). There is evidence that dopaminergic dysfunction in the striatum is attributed to anxiety in PD (Qamar et al., 2017).

Anxiety can arise as either a dopamine-dependent event or a secondary reaction to other symptoms mainly motor disturbances. A previous review described that 88% of PD patients with anxiety disorder experienced its symptoms during the wearing-off period (Poewe, 2008), which suggests that dopaminergic treatment can improve anxiety by preventing wearing off. In fact, a double-blind, randomized, crossover study showed that treatment with levodopa significantly decreased anxiety in patients with wearing off compared with patients without wearing off ($P = 0.02$) (Kulisevsky et al., 2007). On the other hand, fatigue can be aggravated by the administration of high doses of levodopa particularly in patients with motor fluctuations (Coakeley, Martens, & Almeida, 2014). Dissanayaka et al. (2010) described that dyskinesia increased the risk of anxiety, although the dosage of levodopa had no relationship to anxiety.

2.1.3 Apathy

Apathy refers to a lack of motivation characterized by decreased goal-directed behavior and cognition with a prevalence of about 40% in patients with PD (Den Brok et al., 2015; Starkstein & Leentjens, 2008). Apathy usually occurs with cognitive dysfunction and depression; however, some patients develop apathy regardless of their presence. In a study on MPTP-lesioned monkeys, dopaminergic dysfunction within the ventral tegmental–nucleus accumbens pathway was shown to predict apathetic behaviors (Brown et al., 2012). Additionally, there is evidence from imaging studies that apathy in PD is associated with mesocorticolimbic dopaminergic denervation (Reijnders et al., 2010; Thobois et al., 2010).

A double-blind, placebo-controlled, randomized trial was conducted in 37 patients with PD presenting with apathy following subthalamic nucleus stimulation (Thobois et al., 2013). Patients were given either a piribedil (a nonergot D_2/D_3 agonist) up to 300 mg/day ($n = 19$) or a placebo ($n = 18$). At 12 weeks, the Starkstein Apathy Scale score was reduced by 34.6% in patients on piribedil compared with 3.2% in patients on the placebo ($P = 0.015$). This result is in agreement with a previous noncontrolled study showing the benefits of ropinirole in the treatment of postoperative apathy in patients with PD (Czernecki et al., 2008). There are several reports of transdermal rotigotine improving PD-associated apathy (Antonini et al., 2015; Chaudhuri et al., 2013; Chung et al., 2016; Trenkwalder et al., 2011); however, this effect was not observed in a recent randomized blinded trial by Hauser et al. (2016) to assess the effect of rotigotine on apathy as a primary outcome measure.

2.1.4 Cognitive Impairment and Dementia

Cognitive impairment occurs in 20%–57% of patients with PD within the first 3–5 years after diagnosis (Kehagia, Barker, & Robbins, 2010), and about 80% of those develop dementia in the advanced stage (Aarsland, Andersen, Larsen, & Lolk, 2003; Hely, Reid, Adena, Halliday, & Morris, 2008). Cognitive deficit in PD is mainly mediated by frontostriatal dopaminergic dysfunction in the early stage when a component of the cognitive dysfunction can be improved with dopaminergic treatment (Kehagia et al., 2010). Indeed, several studies in patients with PD receiving dopaminergic drugs suggest a positive effect on some cognitions involving frontal lobe-related tasks such as working memory and executive functions (Fera et al., 2007; Hanagasi et al., 2011; Kulisevsky et al., 2000; Lange et al., 1992; Miah, Olde Dubbelink, Stoffers, Deijen, & Berendse, 2012; Molloy et al., 2006). However, both cholinergic and noradrenergic dysfunctions can also affect cognition in this stage, and cholinergic deficit has an essential role in developing into dementia from cognitive impairment (Kehagia et al., 2010). Accordingly, dopaminergic treatment may not affect cognitive function in early PD, and its benefit on cognition is limited in patients with advanced PD who have dementia.

Several studies reported that dopaminergic therapies have a detrimental effect on cognitive performance (Barone et al., 2015; Brusa et al., 2003, 2013, 2005; Morrison, Borod, Brin, Halbig, & Olanow, 2004). Kehagia et al. (2010) described that this unwanted effect on cognition is explained by the dopamine overdose hypothesis. The ventral striatum is relatively spared and thus has an adequate dopamine concentration particularly in the early stage of PD. In that case, dopaminergic therapy can lead to side effects on cognitive function that rely on ventral striatal dopamine signaling, whereas it restores the dopamine concentration in the severely depleted dorsal striatum which can improve cognition associated with the dorsal frontostriatal circuitry. This hypothesis is supported by a recent functional imaging study (MacDonald et al., 2011).

Dopamine agonists may differently affect cognition in PD patients according to their pharmacological characteristics. A previous study in a population of nondemented PD patients showed that rotigotine, cabergoline (an ergot D_2/D_3 agonist), and pergolide do not affect cognition (Brusa et al., 2013, 2005), whereas another study in a similar population showed that pramipexole may slightly impair short-term verbal memory, attention, and executive function (Brusa et al., 2003).

2.2 Autonomic Dysfunction

2.2.1 Gastrointestinal Dysfunction

Constipation is one of the main gastrointestinal dysfunction in patients with PD resulting from decreased transport or disturbed anorectal evacuation (Pfeiffer, 2003). Evidence suggests that severe dopaminergic loss in the enteric nervous system is attributed to constipation in PD (Singaram et al., 1995), whereas its dopaminergic pathogenesis in the central nervous system remains unclear.

There is still uncertainty about the effect of dopaminergic therapies on constipation in PD. A small open-label study examined the effect of levodopa on anorectal constipation in 19 early PD patients who did not take any antiparkinsonian medication by using the quantitative lower-gastrointestinal autonomic test (Tateno et al., 2011). The authors reported that levodopa significantly lessened the first sensation during rectal filling ($P < 0.05$), the amplitude in the paradoxical sphincter contraction upon defecation ($P < 0.01$), and the postdefecation residuals ($P < 0.05$) after a 3-month administration period. However, other studies have shown that dopaminergic therapies aggravated constipation associated with PD (Müller et al., 2013; Pagano, Tan, Haider, Bautista, & Tagliati, 2015).

2.2.2 Sexual Dysfunction

Erectile dysfunction is one of the most common sexual dysfunction in male PD patients (Meco, Rubino, Caravona, & Valente, 2008). Although the cause of erectile dysfunction in PD is largely unknown, the dopaminergic system is associated with arousal-related vasodilatation of penile erectile structures (Palma & Kaufmann, 2014).

There is little evidence to support the effect of dopaminergic treatment on erectile function in PD. A 6-month, open-label, prospective study in 14 male PD patients showed that pergolide significantly improved erectile function (Pohanka, Kanovsky, Bares, Pulkrabek, & Rektor, 2004); however, it is not recommended for use in PD because of its adverse effects such as restrictive valvular heart disease and pulmonary fibrosis. Potential benefits of ropinirole and levodopa have been reported by a case report (Fine & Lang, 1999; Jimenez-Jimenez et al., 1999).

2.2.3 Urinary Dysfunction

Detrusor overactivity is considered the major cause of urinary symptoms associated with PD (Blackett, Walker, & Wood, 2009). It is clinically translated to urinary urgency or frequency in patients with PD. Both nigrostriatal and ventral tegmental area–mesolimbic dopaminergic fibers have an important role in normal bladder control and their dysfunction may result in symptoms of an overactive bladder in PD (Sakakibara, Uchiyama, Yamanishi, Shirai, & Hattori, 2008; Winge & Fowler, 2006). Growing evidence indicates that dopamine receptors have different functions on bladder control; D_1 receptors inhibit the micturition reflex, whereas D_2 receptors activate it (Seki et al., 2001; Yoshimura, Yoshida, Sasa, & Takaori, 1992). Thus, detrusor overactivity can be caused by a combination of underactive D_1 activity presumably with a hyperactive micturition reflex by D_2

stimulation. This is supported by previous studies reporting that detrusor overactivity was improved with pergolide, which has an affinity for the D_1 receptor subtype, but not with bromocriptine (an ergot D_2 agonist) which has no affinity for the D_1 receptor subtype (Kuno, Mizuta, Yamasaki, & Araki, 2004; Yoshimura, Mizuta, Yoshida, & Kuno, 1998).

Levodopa can worsen or improve detrusor overactivity. There are anecdotal reports of levodopa improving detrusor overactivity in patients with PD (Mostile & Jankovic, 2009). Inversely, two previous studies have shown that levodopa exacerbates detrusor overactivity in the filling phase, although it improves bladder emptying through increased detrusor contractility (Brusa et al., 2006; Uchiyama, Sakakibara, Hattori, & Yamanishi, 2003).

2.3 Sleep Disorders

2.3.1 RBD

RBD is characterized by a loss of muscle inhibition during REM sleep, which leads to dream-enacting behaviors and unpleasant dreams (Gagnon et al., 2002). Recent findings strongly suggest that RBD is the most robust biomarker of prodromal PD (Berg et al., 2015). It is related to degeneration of brain stem nuclei including the pedunculopontine and periceruleal nucleus that are linked to the dopaminergic ventral tegmental area in the midbrain (Chaudhuri & Schapira, 2009). Several imaging studies on RBD have shown a decrease in striatal dopaminergic bindings, which has been implicated in its pathogenesis (Albin et al., 2000; Eisensehr et al., 2000, 2003; Iranzo et al., 2010, 2011). These findings suggest that dopaminergic therapy can improve RBD in patients with PD; however, there is little evidence to support the use of dopaminergic drugs to treat RBD in PD, and the results are controversial. In a prospective study of 11 PD patients with RBD, pramipexole did not improve both subjective RBD symptoms and objective video-polysomnography parameters (Kumru et al., 2008). Previous reports on patients with PD have shown that levodopa can improve RBD symptoms (Tan, Salgado, & Fahn, 1996; Yamauchi et al., 2003), whereas a small pilot study on quantitative electromyography analysis during REM period reported an increase in both phasic twitching and tonic activity after treatment with levodopa (Garcia-Borreguero et al., 2002). These conflicting results indicate that dopaminergic mechanism does not have a major role in RBD pathogenesis in PD. A potential effect of the rotigotine patch on RBD has been reported but not confirmed by additional clinical trial evidence (Wang et al., 2016).

2.3.2 Restless Legs Syndrome and Periodic Limb Movement Disorder

Restless legs syndrome (RLS) refers to an urge to move the legs accompanied by uncomfortable and unpleasant sensations that worsen during night rest and disappears or improves with movement (Poewe & Högl, 2004). The descending supraspinal diencephalospinal dopaminergic neurons originating from the hypothalamus are believed to have an integral role in the pathogenesis of RLS. In patients with PD, these neurons degenerate along with nigrostriatal neurons, which lead to the development of RLS (Rijsman, Schoolderman, Rundervoort, & Louter, 2014). On the other hand, periodic limb movement disorder (PLMS) is a sleep-related phenomenon consisting of rhythmic movements of the legs most often occurring during non-REM sleep. It frequently occurs in association with RLS (Manconi et al., 2011). Nigrostriatal dopaminergic dysfunction has been implicated in the pathogenesis of PLMS (Poewe & Högl, 2004).

Nonergot derivative dopamine agonists such as pramipexole, ropinirole, and rotigotine have become a first-line treatment for RLS and PLMS in the general population (Vignatelli et al., 2006). Dopamine agonists targeting the dopamine D_3 receptor subtype have a higher efficacy for RLS and PLMS than a drug that preferentially targets the D_2 receptor subtype (Manconi et al., 2011). Unfortunately, there is little evidence that dopaminergic therapies can improve RLS and/or PLMS in patients with PD. An open-label study investigated the effect of cabergoline on PLMS in 15 patients with PD. The patients underwent two nights of polysomnography before and after 6–8 weeks of treatment with cabergoline (3–6 mg/day). The PLM index was significantly decreased with cabergoline from 34.9 to 6.7 per hour ($P = 0.02$), although there was an increased number of awakenings and stage shifts (Hogl, Rothdach, Wetter, & Trenkwalder, 2003). The potential benefits of the rotigotine patch were noted in a case report (Liguori et al., 2015). A double-blind, placebo-controlled trial evaluating nighttime movements in 11 elderly PD patients showed that levodopa with carbidopa given at bedtime decreased the frequency of spontaneous movements in bed from 43 to 28–33 per night (Leeman et al., 1987).

2.3.3 Insomnia

Both sleep onset and sleep maintenance insomnia are common in PD (Gomez-Esteban et al., 2006). Sleep maintenance insomnia is related to a variety of problems such as nocturnal akinesia and off period-associated motor and NMS (Chaudhuri & Schapira, 2009). Several studies have shown that dopamine agonists can be of benefit for sleep disturbance.

The CLEOPATRA-PD (clinical efficacy of pramipexole and transdermal rotigotine in advanced PD) study was a randomized, double-blind, double dummy placebo-controlled trial that evaluated the efficacy of pramipexole (up to 4.5 mg/day) and transdermal rotigotine (up to 16 mg/day) compared with a placebo in 506 patients with advanced PD (Poewe et al., 2007). The results showed that both dopamine agonists had a similar and significant improvement in the Parkinson's disease sleep scale (PDSS) as a secondary outcome measure at 6 months. The RECOVER study focusing on nocturnal disabilities and early morning akinesia in patients with PD showed that there was a greater improvement with transdermal rotigotine compared with a placebo on the UPDRS Part III scores in the early morning ($P = 0.0002$) and the PDSS-2 total score ($P < 0.0001$) (Trenkwalder et al., 2011). The authors concluded that transdermal rotigotine leads to improvement of sleep maintenance insomnia by reducing nocturnal PD-specific symptoms. This is further demonstrated by the results of Giladi, Fichtner, Poewe, and Boroojerdi (2010). The beneficial effect of prolonged-release ropinirole on sleep quality was reported in the EASE-PD study (Pahwa et al., 2007).

A few studies investigated the relationship between treatment with levodopa and sleep quality in patients PD. A double-blind, crossover study in 40 patients with PD reported that sustained-release levodopa with carbidopa improved nocturnal akinesia, but had no effect on the number of arousals, the total sleep duration, and the time to fall asleep (Stocchi, Barbato, Nordera, Berardelli, & Ruggieri, 1998). Another study on 15 PD patients with disruptive nocturnal symptoms showed that sustained-release levodopa with benserazide improved nocturnal akinesia and total time awake at 12 months (Van den Kerchove, Jacquy, Gonce, & De Deyn, 1993). Leeman et al. (1987) also reported that levodopa with carbidopa administered at bedtime resulted in improved subjective sleep quality. In contrast, a study using polysomnography showed that levodopa had no impact on the structure of sleep including total sleep time and the number of awakenings in patients with PD (Wailke, Herzog, Witt, Deuschl, & Volkmann, 2011).

2.4 Sensory Symptoms

2.4.1 Pain

Patients with PD frequently experience various types of pain ranging from 30% to 83% (Wasner & Deuschl, 2012). Pain in PD can be categorized into musculoskeletal, dystonic, radiculo–neuropathic, central, and akathitic pain based on their clinical presentation (Ha & Jankovic, 2012).

Musculoskeletal pain is the most common which is associated with the presence of rigidity, akinesia, and pathologic postures such as camptocormia (Beiske, Loge, Ronningen, & Svensson, 2009; Ha & Jankovic, 2012; Sophie & Ford, 2012). Dopaminergic treatment can improve both musculoskeletal pain induced by rigidity and akinesia and akathitic pain during the off period (Del Sorbo & Albanese, 2012; Ha & Jankovic, 2012). Striatal hand and foot deformities tend to show a less predictable response to dopaminergic therapy (Ashour, Tintner, & Jankovic, 2005).

Central pain of PD is related to a decrease of the pain threshold in which dopamine is thought to have an integral role (Sophie & Ford, 2012; Wasner & Deuschl, 2012). This is supported by a positron emission tomography study showing that pain in PD patients activated the right insula and prefrontal and left anterior cingulate cortices during the off period, and levodopa reduced pain-induced activation in these areas (Brefel-Courbon et al., 2005). Two previous studies reported that patients with PD had lower pain thresholds compared with healthy controls during the off period; however, administration of levodopa normalized the thresholds (Brefel-Courbon et al., 2005; Gerdelat-Mas et al., 2007). Another study showed that PD patients with central pain had hyperalgesia and less habituation of sympathetic sudomotor response to repetitive pain stimuli than PD patients without pain and healthy controls (Schestatsky et al., 2007). These abnormalities were improved with levodopa.

Dystonic pain has been reported by up to 40% of patients with PD (Beiske et al., 2009). It can occur in association with either a complication from the dopaminergic treatment such as peak-dose dystonia and diphasic dystonia or a wearing-off period, particularly in the early morning (Ha & Jankovic, 2012; Wasner & Deuschl, 2012). Accordingly, dopaminergic treatment can improve or aggravate dystonic pain.

Several studies investigated the effect of transdermal rotigotine on pain in patients with PD. The RECOVER study reported that transdermal rotigotine markedly improved the Likert pain scale measuring any type of pain (Trenkwalder et al., 2011), and its post hoc analysis suggested that the effect of rotigotine on pain is partly attributable to the benefits in motor function and sleep disturbances, such as dystonic symptoms at night or in the early morning (Kassubek et al., 2014). A randomized controlled pilot study on PD patients with chronic pain showed that transdermal rotigotine significantly decreased the "fluctuation-related pain" domain of the King's PD pain scale (KPPS) at 12 weeks, whereas the Likert pain scale as a primary measure and other domains of the KPPS were not improved (Rascol et al., 2016).

2.4.2 Olfactory Dysfunction

Olfactory impairment typically occurs several years before the onset of motor symptoms and might be a useful marker predicting progression to PD (Haehner, Hummel, & Reichmann, 2010; Ross et al., 2008). Patients with PD can have deficits in all three main modalities of the smell tests: odor detection threshold, discrimination, and identification (Saifee, Lees, & Silveira-Moriyama, 2010). Abnormalities in dopamine uptake in the striatum have been reported (Berendse, Roos, Raijmakers, & Doty, 2011; Bohnen, Gedela, Herath, Constantine, & Moore, 2008).

Only few studies investigated the relationship between levodopa and olfactory function in patients with PD. Quinn, Rossor, and Marsden (1987) conducted olfactory threshold tests on 78 patients with PD and reported no change between the on and off periods. Similarly, a further study failed to show the effect of levodopa on olfactory function (Saifee et al., 2010).

The results of rasagiline on olfactory dysfunction in PD are conflicting. A previous study on 17 PD patients reported that rasagiline was associated with a significant improvement in olfaction (Alvarez & Grogan, 2012), which is further supported by a study on an alpha-synuclein mouse model of PD (Petit et al., 2013). By contrast, Haehner et al. (2013) conducted a double-blind, placebo-controlled, randomized study comparing rasagiline (1 mg/day) with a placebo in 34 patients with PD. The results showed that there was no significant difference in olfactory function between the groups at 120 days. A similar result was reported in another randomized controlled trial of rasagiline (1 mg/day) as an add-on to dopamine agonist therapy in early PD patients (Hauser, Silver, Choudhry, Eyal, & Isaacson, 2014). A recent cross-sectional observational study did not show better olfactory function with rasagiline in patients with PD, although secondary analyses suggested a beneficial effect of rasagiline in those with a shorter disease duration (Haehner et al., 2015).

2.5 Other Symptoms

2.5.1 Fatigue

Fatigue usually refers to difficulty initiating or sustaining voluntary activity. It is believed to result from a complex interaction between the underlying disease process, peripheral control system, central control systems, and environmental factors (Chaudhuri & Behan, 2004). Fatigue can be either mental, physical, or both; however, the severity of the mental fatigue does not correlate with physical fatigue, suggesting that the two conditions are caused by independent mechanisms (Lou et al., 2003).

Although several neuroimaging studies found no correlation between fatigue and dopaminergic dysfunction (Chou, Kotagal, & Bohnen, 2016; Pavese, Metta, Bose, Chaudhuri, & Brooks, 2010; Schifitto et al., 2008), there have been numerous attempts to use dopaminergic therapies for the treatment of fatigue in PD. Schifitto et al. (2008) investigated fatigue in a cohort of previously untreated patients with early PD enrolled in the ELLDOPA (earlier vs later levodopa) trial. This double-blind, placebo-controlled, randomized study enrolled a total of 361 PD patients who were assigned to receive placebo or levodopa 150, 300, or 600 mg/day for 40 weeks, followed by a 2-week medication washout period. At 42 weeks, there was a greater increase in the fatigue severity scale in the placebo group (0.75 points) than in the groups receiving 150 (0.30 points), 300 (0.36 points), and 600 mg/day (0.33 points, $P = 0.03$) of levodopa. This finding is in agreement with the results of other studies (Lou et al., 2003; Ziv et al., 1998).

There are anecdotal reports of dopamine agonists improving fatigue in patients with PD (Hauser et al., 2010; Kieburtz, 2011; Morita et al., 2011), whereas a small prospective study showed that dopamine agonists did not improve both physical and mental fatigue (Oved et al., 2006).

Rasagiline is considered as a therapeutic option for fatigue in PD. In a substudy of the ADAGIO, rasagiline at 1 or 2 mg/day significantly improved the Parkinson fatigue scale at 36 weeks compared with the placebo (1 mg/day of rasagiline mean difference: 0.14, $P < 0.01$; 2 mg of rasagiline: 0.19, $P < 0.0001$) (Rascol et al., 2011). Lim et al. (2015) reported that rasagiline at 1 mg/day provided a significant improvement in the modified fatigue impact scale ($P = 0.003$), fatigue severity scale ($P = 0.018$), and physical and mental fatigue testing score ($P = 0.002$) at 12 weeks compared with the placebo.

3. THE UNWANTED EFFECT OF DOPAMINERGIC THERAPIES ON NMS

3.1 Psychiatric and Behavioral NMS

3.1.1 Impulse Control Disorder

Impulse control disorder (ICD) is defined as a failure to resist an impulse, or as the drive or temptation to carry out a harmful act (Voon, Potenza, & Thomsen, 2007). In PD, ICDs typically involve hypersexuality, pathological gambling, compulsive shopping, or compulsive eating (Antonini et al., 2009). Reward-seeking behavior is mediated by the amygdala and nucleus accumbens that receive dopaminergic projection from the ventral tegmental area, which may be attributed to the pathological mechanism of ICD (Zhang et al., 2014).

A cross-sectional study in 3090 patients with PD reported that 14% of the patients had an ICD (gambling in 5%, compulsive sexual behavior in 3.5%, compulsive buying in 5.7%, and binge-eating disorder in 4.3%), and 4% had comorbid ICDs (Weintraub et al., 2010). Additionally, the patients treated with dopamine agonists had a higher frequency of ICDs compared with those not treated with dopamine agonists (17.1% vs 6.9%, odds ratio: 2.72, $P < 0.001$); however, there was no significant difference between pramipexole (17.7%) and ropinirole (15.5%) in association with ICDs. Levodopa was associated with ICDs, although to a lesser extent than dopamine agonists. Another prospective study of 396 PD patients reported that the lifetime prevalence of pathologic gambling was 3.4%, which was increased to 7.2% for dopamine agonists (Voon et al., 2006). Similarly, pathologic gambling was not associated with the type of dopamine agonist. On the other hand, a recent observational study of 425 patients with PD reported that the frequency of ICDs was significantly lower with transdermal rotigotine (4.9%) compared with pramipexole (13.8%) and ropinirole (13.9%) (Rizos et al., 2016). In addition, the rate of ICDs for prolonged-release pramipexole (6.6%) was significantly lower than for immediate-release pramipexole (19.0%). The authors suggest a relatively low risk of ICDs with long-acting or transdermal dopamine agonists. These results are in line with the results of Garcia-Ruiz et al. (2014) who reported that rotigotine had a significantly lower rate of ICD compared to other short-acting dopamine agonists (19% vs 42%, $P < 0.01$).

3.1.2 Punding

Punding is characterized by a complex, prolonged, purposeless, and stereotyped behavior (Moller et al., 2008). Evidence suggests that abnormal dopaminergic transmission plays a causative role in the pathogenesis of punding (Spencer, Rickards, Fasano, & Cavanna, 2011). Evans et al. (2004) reported that punding was present in 34% of 50 PD patients who received a high levodopa equivalent dose (LED) ($\geq$800 mg/day), whereas none was suspected of punding in 53 patients receiving LED lower than 800 mg/day. The authors suggested a link between high doses of dopaminergic treatment and punding. This is further supported by a study by Miwa (2007). A cross-sectional study found that the severity of punding correlates with the dose of dopamine agonists ($P < 0.002$) but not with the dose of levodopa ($P > 0.25$) (Lawrence et al., 2007). However, discordant results have been reported by Miyasaki, Al Hassan, Lang, and Voon (2007) who described that punding was observed in only 1.4% of 291 PD patients and none had punding in 40 patients receiving LED at more than 800 mg/day.

3.1.3 Psychosis

Although visual hallucinations are the most common psychotic symptom in PD, auditory, tactile, and olfactory hallucinations are also relatively common (Fenelon, Soulas, Zenasni, & Cleret de Langavant, 2010). A large-scale study of 250 patients with PD described that 26% had psychotic symptoms with 52.3% of those having hallucinations and/or delusions (Mack et al., 2012). Possible mechanisms of PD psychosis include sleep, cognition, changes in visual function, and medication effects (Ffytche et al., 2017).

Although clinical experience suggests a strong link between dopaminergic treatment and PD psychosis, evidence is lacking to support their direct causality. A prospective cohort study of 230 patients with PD who were followed up for 12 years identified higher LED at baseline as a risk factor for PD psychosis (odds ratio: 1.26 per 100 mg, $P = 0.01$) (Forsaa et al., 2010). Morgante et al. (2012) reported that the prescription of dopamine agonist at baseline was associated with the development of PD psychosis over a 24-month period. One proposed mechanism to explain their relationship is that chronic dopaminergic therapy leads to overstimulation or hypersensitivity of the mesocorticolimbic D_2/D_3 receptors (Wolters, 1999). In contrast, there was no difference in LED between PD patients with and without psychosis in several cross-sectional studies (Aarsland, Larsen, Cummins, & Laake, 1999; Papapetropoulos, Argyriou, & Ellul, 2005; Sanchez-Ramos, Ortoll, & Paulson, 1996). These findings are in agreement with other prospective studies reporting that the use of levodopa or dopamine agonists, and LED did not increase the risk of PD psychosis (De la Riva, Smith, Xie, & Weintraub, 2014; Gibson et al., 2013; Goetz, Stebbins, & Ouyang, 2011; Zhu, Van Hilten, Putter, & Marinus, 2013).

3.2 Other NMS

3.2.1 Orthostatic Hypotension

Neurogenic orthostatic hypotension is defined as a sustained decrease in systolic blood pressure by at least 20 mm Hg or in diastolic blood pressure by 10 mm Hg within 3 min after rising from a supine to a standing position, which is not associated with cardiogenic causes or hypovolemia (Freeman et al., 2011). The prevalence of neurogenic orthostatic hypotension in

PD was estimated to be approximately 30% (Espay et al., 2016). Previous reports have shown that dopaminergic therapies can lead to hypotensive effects and thus aggravate or even cause orthostatic hypotension in patients with PD (Goldstein et al., 2005; Linden, Diehl, & Berlit, 1997; Reid, Calne, George, & Vakil, 1972). A recent prospective study of 17 patients with PD reported that oral administration of levodopa led to a significant decrease in mean arterial pressure (−15%, $P < 0.001$), cardiac stroke volume (−13%, $P < 0.01$), and measures of cardiac contractility (−18%, $P < 0.001$), whereas systemic vascular resistance, heart rate, and ventilator parameters remained changed (Noack, Schroeder, Heusser, & Lipp, 2014). The authors suggested that the hypoactive effect of levodopa is caused by a negative cardiac inotropism rather than peripheral vasodilatation. Other dopaminergic drugs including dopamine agonists and MAO-B inhibitors can increase the risk of hypotension (Churchyard, Mathias, & Lees, 1999; Etminan, Gill, & Samii, 2003).

3.2.2 Excessive Daytime Sleepiness

Excessive daytime sleepiness (EDS) is defined as a disabling tendency to fall asleep during normal activities affecting around 33% of patients with PD (Arnulf & Leu-Semenescu, 2009; Chaudhuri & Logishetty, 2009). The mechanism of EDS is multifactorial and includes disease-related factors such as dysfunction of the sleep–wake cycle because of pathological changes in the brain stem and hypothalamus, sleep disorders, and sedative side effects of medications (Comella, 2007). Previous studies have shown that both a higher LED and a use of dopamine agonist were significant predictors of EDS (Breen, Williams-Gray, Mason, Foltynie, & Barker, 2013; Gjerstad, Alves, Wentzel-Larsen, Aarsland, & Larsen, 2006; O'Suilleabhain & Dewey, 2002; Tholfsen, Larsen, Schulz, Tysnes, & Gjerstad, 2015; Valko et al., 2010; Zhu, Van Hilten, & Marinus, 2016). Specifically, Razmy, Lang, and Shapiro (2004) reported that LED rather than the type of dopamine agonist is the most important predisposing factor of EDS in PD patients treated with dopamine agonists.

3.2.3 Fibrotic Reactions With Ergot Dopamine Agonists

The development of pleuropulmonary, pericardial, retroperitoneal, or heart valve fibrosis is specifically related to ergot-derived dopamine agonists (Moller et al., 2008). A meta-analysis from seven cross-sectional studies investigated a relationship between dopamine agonists and the incidence of heart valve disease in patients with PD. The study included 477 patients with ergot dopamine agonists, 127 patients with nonergot dopamine agonists such as pramipexole and ropinirole, and 364 control patients. The results showed that moderate to severe valvular changes were observed in 26% with ergot dopamine agonists, in 10% with nonergot dopamine agonists, and in 10% of the control patients. Additionally, previous studies reported that a higher cumulative dose of ergot dopamine agonists was associated with the severity of valvular change (Van Camp et al., 2004; Zanettini et al., 2007).

3.2.4 Peripheral Edema

Peripheral edema occurs as a complication of dopamine agonists, particularly pramipexole, or even levodopa (Antonini et al., 2009; Blackard, 1993; Tan & Ondo, 2000). A post hoc analysis of data from the CALM-PD trial of 301 PD patients showed that the 4-year incidence of development or worsening of edema was estimated to be 45%, and it is associated with the initial treatment with pramipexole (hazard ratio: 3.18, 95% CI: 1.95–5.18, $P < 0.0001$). The occurrence of edema tended to emerge more frequently after a 2-year follow-up, which suggests that the duration of treatment with pramipexole is associated with the development of edema. Another study by Kleiner-Fisman and Fisman (2007) reported a 16% prevalence of peripheral edema presumably induced by pramipexole in patients with PD. Dopamine is regarded as an import regulator of the sympathetic nervous system, aldosterone secretion, and ATP-medicated sodium and potassium channels and thus can result in peripheral side effects.

4. CONCLUSIONS

The treatment of NMS in patients with PD is a crucial issue in clinical practice given their impact on the quality of life for patients and caregivers. Although various NMS in PD have a dopaminergic pathophysiological basis, this fact is often unrecognized and they are regarded as dopamine unresponsive symptoms. However, substantial evidence indicates that many NMS respond to oral and transdermal dopaminergic therapies including levodopa, dopamine agonists, and MAO-B inhibitors. Conversely, certain NMS are exacerbated or caused by dopaminergic drugs and these unwanted effects may be seriously dangerous. Therefore, a dopaminergic strategy for NMS should be based on a consideration of the benefits vs the risks in individual patients with PD.

ACKNOWLEDGMENTS

Author's contributions. Dr. R. Kim did the literature search and wrote the first draft of the manuscript. Dr. B. Jeon made critical revisions to the manuscript.

Conflict of Interest: No conflicting relationship exists for the authors.

REFERENCES

Aarsland, D., Andersen, K., Larsen, J. P., & Lolk, A. (2003). Prevalence and characteristics of dementia in Parkinson disease: An 8-year prospective study. *Archives of Neurology*, *60*(3), 387–392.

Aarsland, D., Larsen, J. P., Cummins, J. L., & Laake, K. (1999). Prevalence and clinical correlates of psychotic symptoms in Parkinson disease: A community-based study. *Archives of Neurology*, *56*(5), 595–601.

Albin, R. L., Koeppe, R. A., Chervin, R. D., Consens, F. B., Wernette, K., Frey, K. A., et al. (2000). Decreased striatal dopaminergic innervation in REM sleep behavior disorder. *Neurology*, *55*(9), 1410–1412.

Alvarez, M. V., & Grogan, P. M. (2012). Hyposmia in Parkinson's disease. *Psychiatry and Clinical Neurosciences*, *66*(4), 370.

Antonini, A., Bauer, L., Dohin, E., Oertel, W. H., Rascol, O., Reichmann, H., et al. (2015). Effects of rotigotine transdermal patch in patients with Parkinson's disease presenting with non-motor symptoms—Results of a double-blind, randomized, placebo-controlled trial. *European Journal of Neurology*, *22*(10), 1400–1407.

Antonini, A., Tolosa, E., Mizuno, Y., Yamamoto, M., & Poewe, W. H. (2009). A reassessment of risks and benefits of dopamine agonists in Parkinson's disease. *The Lancet Neurology*, *8*(10), 929–937.

Arnulf, I., & Leu-Semenescu, S. (2009). Sleepiness in Parkinson's disease. *Parkinsonism & Related Disorders*, *15*(Suppl. 3), S101–S104.

Ashour, R., Tintner, R., & Jankovic, J. (2005). Striatal deformities of the hand and foot in Parkinson's disease. *The Lancet Neurology*, *4*(7), 423–431.

Barone, P., Antonini, A., Colosimo, C., Marconi, R., Morgante, L., Avarello, T. P., et al. (2009). The PRIAMO study: A multicenter assessment of nonmotor symptoms and their impact on quality of life in Parkinson's disease. *Movement Disorders*, *24*(11), 1641–1649.

Barone, P., Poewe, W., Albrecht, S., Debieuvre, C., Massey, D., Rascol, O., et al. (2010). Pramipexole for the treatment of depressive symptoms in patients with Parkinson's disease: A randomised, double-blind, placebo-controlled trial. *The Lancet Neurology*, *9*(6), 573–580.

Barone, P., Santangelo, G., Morgante, L., Onofrj, M., Meco, G., Abbruzzese, G., et al. (2015). A randomized clinical trial to evaluate the effects of rasagiline on depressive symptoms in non-demented Parkinson's disease patients. *European Journal of Neurology*, *22*(8), 1184–1191.

Barone, P., Scarzella, L., Marconi, R., Antonini, A., Morgante, L., Bracco, F., et al. (2006). Pramipexole versus sertraline in the treatment of depression in Parkinson's disease: A national multicenter parallel-group randomized study. *Journal of Neurology*, *253*(5), 601–607.

Beiske, A. G., Loge, J. H., Ronningen, A., & Svensson, E. (2009). Pain in Parkinson's disease: Prevalence and characteristics. *Pain*, *141*(1–2), 173–177.

Berendse, H. W., Roos, D. S., Raijmakers, P., & Doty, R. L. (2011). Motor and non-motor correlates of olfactory dysfunction in Parkinson's disease. *Journal of the Neurological Sciences*, *310*(1–2), 21–24.

Berg, D., Postuma, R. B., Adler, C. H., Bloem, B. R., Chan, P., Dubois, B., et al. (2015). MDS research criteria for prodromal Parkinson's disease. *Movement Disorders*, *30*(12), 1600–1611.

Blackard, W. G. (1993). Edema—An infrequently recognized complication of bromocriptine and other ergot dopaminergic drugs. *American Journal of Medicine*, *94*(4), 445.

Blackett, H., Walker, R., & Wood, B. (2009). Urinary dysfunction in Parkinson's disease: A review. *Parkinsonism & Related Disorders*, *15*(2), 81–87.

Bohnen, N. I., Gedela, S., Herath, P., Constantine, G. M., & Moore, R. Y. (2008). Selective hyposmia in Parkinson disease: Association with hippocampal dopamine activity. *Neuroscience Letters*, *447*(1), 12–16.

Breen, D. P., Williams-Gray, C. H., Mason, S. L., Foltynie, T., & Barker, R. A. (2013). Excessive daytime sleepiness and its risk factors in incident Parkinson's disease. *Journal of Neurology, Neurosurgery, and Psychiatry*, *84*(2), 233–234.

Brefel-Courbon, C., Payoux, P., Thalamas, C., Ory, F., Quelven, I., Chollet, F., et al. (2005). Effect of levodopa on pain threshold in Parkinson's disease: A clinical and positron emission tomography study. *Movement Disorders*, *20*(12), 1557–1563.

Brown, C. A., Campbell, M. C., Karimi, M., Tabbal, S. D., Loftin, S. K., Tian, L. L., et al. (2012). Dopamine pathway loss in nucleus accumbens and ventral tegmental area predicts apathetic behavior in MPTP-lesioned monkeys. *Experimental Neurology*, *236*(1), 190–197.

Brusa, L., Bassi, A., Stefani, A., Pierantozzi, M., Peppe, A., Caramia, M. D., et al. (2003). Pramipexole in comparison to l-dopa: A neuropsychological study. *Journal of Neural Transmission*, *110*(4), 373–380.

Brusa, L., Pavino, V., Massimetti, M. C., Bove, R., Iani, C., & Stanzione, P. (2013). The effect of dopamine agonists on cognitive functions in non-demented early-mild Parkinson's disease patients. *Functional Neurology*, *28*(1), 13–17.

Brusa, L., Petta, F., Pisani, A., Miano, R., Stanzione, P., Moschella, V., et al. (2006). Central acute D2 stimulation worsens bladder function in patients with mild Parkinson's disease. *Journal of Urology*, *175*(1), 202–206.

Brusa, L., Tiraboschi, P., Koch, G., Peppe, A., Pierantozzi, M., Ruggieri, S., et al. (2005). Pergolide effect on cognitive functions in early-mild Parkinson's disease. *Journal of Neural Transmission*, *112*(2), 231–237.

Chaudhuri, A., & Behan, P. O. (2004). Fatigue in neurological disorders. *Lancet*, *363*(9413), 978–988.

Chaudhuri, K. R., Healy, D. G., & Schapira, A. H. (2006). Non-motor symptoms of Parkinson's disease: Diagnosis and management. *The Lancet Neurology*, *5*(3), 235–245.

Chaudhuri, K. R., & Logishetty, K. (2009). Dopamine receptor agonists and sleep disturbances in Parkinson's disease. *Parkinsonism & Related Disorders*, *15*(Suppl. 4), S101–S104.

Chaudhuri, K. R., Martinez-Martin, P., Antonini, A., Brown, R. G., Friedman, J. H., Onofrj, M., et al. (2013). Rotigotine and specific non-motor symptoms of Parkinson's disease: Post hoc analysis of RECOVER. *Parkinsonism & Related Disorders*, *19*(7), 660–665.

Chaudhuri, K. R., Odin, P., Antonini, A., & Martinez-Martin, P. (2011). Parkinson's disease: The non-motor issues. *Parkinsonism & Related Disorders*, *17*(10), 717–723.

Chaudhuri, K. R., Prieto-Jurcynska, C., Naidu, Y., Mitra, T., Frades-Payo, B., Tluk, S., et al. (2010). The nondeclaration of nonmotor symptoms of Parkinson's disease to health care professionals: An international study using the nonmotor symptoms questionnaire. *Movement Disorders*, *25*(6), 704–709.

Chaudhuri, K. R., & Schapira, A. H. (2009). Non-motor symptoms of Parkinson's disease: Dopaminergic pathophysiology and treatment. *The Lancet Neurology*, *8*(5), 464–474.

Chou, K. L., Kotagal, V., & Bohnen, N. I. (2016). Neuroimaging and clinical predictors of fatigue in Parkinson disease. *Parkinsonism & Related Disorders*, *23*, 45–49.

Chung, S. J., Asgharnejad, M., Bauer, L., Ramirez, F., & Jeon, B. (2016). Evaluation of rotigotine transdermal patch for the treatment of depressive symptoms in patients with Parkinson's disease. *Expert Opinion on Pharmacotherapy*, *17*(11), 1453–1461.

Churchyard, A., Mathias, C. J., & Lees, A. J. (1999). Selegiline-induced postural hypotension in Parkinson's disease: A longitudinal study on the effects of drug withdrawal. *Movement Disorders*, *14*(2), 246–251.

Coakeley, S., Martens, K. E., & Almeida, Q. J. (2014). Management of anxiety and motor symptoms in Parkinson's disease. *Expert Review of Neurotherapeutics*, *14*(8), 937–946.

Comella, C. L. (2007). Sleep disorders in Parkinson's disease: An overview. *Movement Disorders*, *22*(Suppl. 17), S367–S373.

Czernecki, V., Schupbach, M., Yaici, S., Levy, R., Bardinet, E., Yelnik, J., et al. (2008). Apathy following subthalamic stimulation in Parkinson disease: A dopamine responsive symptom. *Movement Disorders*, *23*(7), 964–969.

De la Riva, P., Smith, K., Xie, S. X., & Weintraub, D. (2014). Course of psychiatric symptoms and global cognition in early Parkinson disease. *Neurology*, *83*(12), 1096–1103.

Del Sorbo, F., & Albanese, A. (2012). Clinical management of pain and fatigue in Parkinson's disease. *Parkinsonism & Related Disorders*, *18*(Suppl. 1), S233–S236.

Den Brok, M. G., Van Dalen, J. W., Van Gool, W. A., Moll van Charante, E. P., De Bie, R. M., & Richard, E. (2015). Apathy in Parkinson's disease: A systematic review and meta-analysis. *Movement Disorders*, *30*(6), 759–769.

Di Giuda, D., Camardese, G., Bentivoglio, A. R., Cocciolillo, F., Guidubaldi, A., Pucci, L., et al. (2012). Dopaminergic dysfunction and psychiatric symptoms in movement disorders: A 123I-FP-CIT SPECT study. *European Journal of Nuclear Medicine and Molecular Imaging*, *39*(12), 1937–1948.

Dissanayaka, N. N., Sellbach, A., Matheson, S., O'Sullivan, J. D., Silburn, P. A., Byrne, G. J., et al. (2010). Anxiety disorders in Parkinson's disease: Prevalence and risk factors. *Movement Disorders*, *25*(7), 838–845.

Eisensehr, I., Linke, R., Noachtar, S., Schwarz, J., Gildehaus, F. J., & Tatsch, K. (2000). Reduced striatal dopamine transporters in idiopathic rapid eye movement sleep behaviour disorder. Comparison with Parkinson's disease and controls. *Brain*, *123*(6), 1155–1160.

Eisensehr, I., Linke, R., Tatsch, K., Kharraz, B., Gildehaus, J. F., Wetter, C. T., et al. (2003). Increased muscle activity during rapid eye movement sleep correlates with decrease of striatal presynaptic dopamine transporters. IPT and IBZM SPECT imaging in subclinical and clinically manifest idiopathic REM sleep behavior disorder, Parkinson's disease, and controls. *Sleep*, *26*(5), 507–512.

Espay, A. J., LeWitt, P. A., Hauser, R. A., Merola, A., Masellis, M., & Lang, A. E. (2016). Neurogenic orthostatic hypotension and supine hypertension in Parkinson's disease and related synucleinopathies: Prioritisation of treatment targets. *The Lancet Neurology*, *15*(9), 954–966.

Etminan, M., Gill, S., & Samii, A. (2003). Comparison of the risk of adverse events with pramipexole and ropinirole in patients with Parkinson's disease: A meta-analysis. *Drug Safety*, *26*(6), 439–444.

Evans, A. H., Katzenschlager, R., Paviour, D., O'Sullivan, J. D., Appel, S., Lawrence, A. D., et al. (2004). Punding in Parkinson's disease: Its relation to the dopamine dysregulation syndrome. *Movement Disorders*, *19*(4), 397–405.

Fenelon, G., Soulas, T., Zenasni, F., & Cleret de Langavant, L. (2010). The changing face of Parkinson's disease-associated psychosis: A cross-sectional study based on the new NINDS-NIMH criteria. *Movement Disorders*, *25*(6), 763–766.

Fera, F., Nicoletti, G., Cerasa, A., Romeo, N., Gallo, O., Gioia, M. C., et al. (2007). Dopaminergic modulation of cognitive interference after pharmacological washout in Parkinson's disease. *Brain Research Bulletin*, *74*(1–3), 75–83.

Ffytche, D. H., Creese, B., Politis, M., Chaudhuri, K. R., Weintraub, D., Ballard, C., et al. (2017). The psychosis spectrum in Parkinson disease. *Nature Reviews Neurology*, *13*(2), 81–95.

Fine, J., & Lang, A. E. (1999). Dose-induced penile erections in response to ropinirole therapy for Parkinson's disease. *Movement Disorders*, *14*(4), 701–702.
Forsaa, E. B., Larsen, J. P., Wentzel-Larsen, T., Goetz, C. G., Stebbins, G. T., Aarsland, D., et al. (2010). A 12-year population-based study of psychosis in Parkinson disease. *Archives of Neurology*, *67*(8), 996–1001.
Fox, S. H., Brotchie, J. M., & Lang, A. E. (2008). Non-dopaminergic treatments in development for Parkinson's disease. *The Lancet Neurology*, 7(10), 927–938.
Freeman, R., Wieling, W., Axelrod, F. B., Benditt, D. G., Benarroch, E., Biaggioni, I., et al. (2011). Consensus statement on the definition of orthostatic hypotension, neurally mediated syncope and the postural tachycardia syndrome. *Clinical Autonomic Research*, *21*(2), 69–72.
Gagnon, J. F., Bedard, M. A., Fantini, M. L., Petit, D., Panisset, M., Rompre, S., et al. (2002). REM sleep behavior disorder and REM sleep without atonia in Parkinson's disease. *Neurology*, *59*(4), 585–589.
Gallagher, D. A., & Schrag, A. (2012). Psychosis, apathy, depression and anxiety in Parkinson's disease. *Neurobiology of Disease*, *46*(3), 581–589.
Garcia-Borreguero, D., Caminero, A. B., De la Llave, Y., Larrosa, O., Barrio, S., Granizo, J. J., et al. (2002). Decreased phasic EMG activity during rapid eye movement sleep in treatment-naïve Parkinson's disease: Effects of treatment with levodopa and progression of illness. *Movement Disorders*, *17*(5), 934–941.
Garcia-Ruiz, P. J., Martinez-Castrillo, J. C., Alonso-Canovas, A., Herranz-Barcenas, A., Vela, L., Sanchez-Alonso, P., et al. (2014). Impulse control disorder in patients with Parkinson's disease under dopamine agonist therapy: A multicentre study. *Journal of Neurology, Neurosurgery, and Psychiatry*, *85*(8), 840–844.
Gerdelat-Mas, A., Simonetta-Moreau, M., Thalamas, C., Ory-Magne, F., Slaoui, T., Rascol, O., et al. (2007). Levodopa raises objective pain threshold in Parkinson's disease: A RIII reflex study. *Journal of Neurology, Neurosurgery, and Psychiatry*, *78*(10), 1140–1142.
Gibson, G., Mottram, P. G., Burn, D. J., Hindle, J. V., Landau, S., Samuel, M., et al. (2013). Frequency, prevalence, incidence and risk factors associated with visual hallucinations in a sample of patients with Parkinson's disease: A longitudinal 4-year study. *International Journal of Geriatric Psychiatry*, *28*(6), 626–631.
Giladi, N., Fichtner, A., Poewe, W., & Boroojerdi, B. (2010). Rotigotine transdermal system for control of early morning motor impairment and sleep disturbances in patients with Parkinson's disease. *Journal of Neural Transmission*, *117*(12), 1395–1399.
Gjerstad, M. D., Alves, G., Wentzel-Larsen, T., Aarsland, D., & Larsen, J. P. (2006). Excessive daytime sleepiness in Parkinson disease: Is it the drugs or the disease? *Neurology*, *67*(5), 853–858.
Goetz, C. G., Stebbins, G. T., & Ouyang, B. (2011). Visual plus nonvisual hallucinations in Parkinson's disease: Development and evolution over 10 years. *Movement Disorders*, *26*(12), 2196–2200.
Goldstein, D. S., Eldadah, B. A., Holmes, C., Pechnik, S., Moak, J., Saleem, A., et al. (2005). Neurocirculatory abnormalities in Parkinson disease with orthostatic hypotension: Independence from levodopa treatment. *Hypertension*, *46*(6), 1333–1339.
Gomez-Esteban, J. C., Zarranz, J. J., Lezcano, E., Velasco, F., Ciordia, R., Rouco, I., et al. (2006). Sleep complaints and their relation with drug treatment in patients suffering from Parkinson's disease. *Movement Disorders*, *21*(7), 983–988.
Ha, A. D., & Jankovic, J. (2012). Pain in Parkinson's disease. *Movement Disorders*, *27*(4), 485–491.
Haehner, A., Habersack, A., Wienecke, M., Storch, A., Reichmann, H., & Hummel, T. (2015). Early Parkinson's disease patients on rasagiline present with better odor discrimination. *Journal of Neural Transmission*, *122*(11), 1541–1546.

Haehner, A., Hummel, T., & Reichmann, H. (2010). Olfactory function in Parkinson's disease. *European Neurological Review*, *5*(1), 26–29.

Haehner, A., Hummel, T., Wolz, M., Klingelhofer, L., Fauser, M., Storch, A., et al. (2013). Effects of rasagiline on olfactory function in patients with Parkinson's disease. *Movement Disorders*, *28*(14), 2023–2027.

Hanagasi, H. A., Gurvit, H., Unsalan, P., Horozoglu, H., Tuncer, N., Feyzioglu, A., et al. (2011). The effects of rasagiline on cognitive deficits in Parkinson's disease patients without dementia: A randomized, double-blind, placebo-controlled, multicenter study. *Movement Disorders*, *26*(10), 1851–1858.

Hauser, R. A., Schapira, A. H., Rascol, O., Barone, P., Mizuno, Y., Salin, L., et al. (2010). Randomized, double-blind, multicenter evaluation of pramipexole extended release once daily in early Parkinson's disease. *Movement Disorders*, *25*(15), 2542–2549.

Hauser, R. A., Silver, D., Choudhry, A., Eyal, E., & Isaacson, S. (2014). Randomized, controlled trial of rasagiline as an add-on to dopamine agonists in Parkinson's disease. *Movement Disorders*, *29*(8), 1028–1034.

Hauser, R. A., Slawek, J., Barone, P., Dohin, E., Surmann, E., Asgharnejad, M., et al. (2016). Evaluation of rotigotine transdermal patch for the treatment of apathy and motor symptoms in Parkinson's disease. *BMC Neurology*, *16*, 90.

Hely, M. A., Reid, W. G., Adena, M. A., Halliday, G. M., & Morris, J. G. (2008). The Sydney multicenter study of Parkinson's disease: The inevitability of dementia at 20 years. *Movement Disorders*, *23*(6), 837–844.

Hesse, S., Meyer, P. M., Strecker, K., Barthel, H., Wegner, F., Oehlwein, C., et al. (2009). Monoamine transporter availability in Parkinson's disease patients with or without depression. *European Journal of Nuclear Medicine and Molecular Imaging*, *36*(3), 428–435.

Hogl, B., Rothdach, A., Wetter, T. C., & Trenkwalder, C. (2003). The effect of cabergoline on sleep, periodic leg movements in sleep, and early morning motor function in patients with Parkinson's disease. *Neuropsychopharmacology*, *28*(10), 1866–1870.

Iranzo, A., Lomena, F., Stockner, H., Valldeoriola, F., Vilaseca, I., Salamero, M., et al. (2010). Decreased striatal dopamine transporter uptake and substantia nigra hyperechogenicity as risk markers of synucleinopathy in patients with idiopathic rapid-eye-movement sleep behaviour disorder: A prospective study. *The Lancet Neurology*, *9*(11), 1070–1077.

Iranzo, A., Valldeoriola, F., Lomena, F., Molinuevo, J. L., Serradell, M., Salamero, M., et al. (2011). Serial dopamine transporter imaging of nigrostriatal function in patients with idiopathic rapid-eye-movement sleep behaviour disorder: A prospective study. *The Lancet Neurology*, *10*(9), 797–805.

Jimenez-Jimenez, F. J., Tallon-Barranco, A., Cabrera-Valdivia, F., Gasalla, T., Orti-Pareja, M., & Zurdo, M. (1999). Fluctuating penile erection related with levodopa therapy. *Neurology*, *52*(1), 210.

Kassubek, J., Chaudhuri, K. R., Zesiewicz, T., Surmann, E., Boroojerdi, B., Moran, K., et al. (2014). Rotigotine transdermal system and evaluation of pain in patients with Parkinson's disease: A post hoc analysis of the RECOVER study. *BMC Neurology*, *14*, 42.

Kehagia, A. A., Barker, R. A., & Robbins, T. W. (2010). Neuropsychological and clinical heterogeneity of cognitive impairment and dementia in patients with Parkinson's disease. *The Lancet Neurology*, *9*(12), 1200–1213.

Kieburtz, K. (2011). Twice-daily, low-dose pramipexole in early Parkinson's disease: A randomized, placebo-controlled trial. *Movement Disorders*, *26*(1), 37–44.

Kim, H. J., Park, S., Cho, Y., Hong, K., Cho, J., Seo, S., et al. (2009). Nonmotor symptoms in de novo Parkinson disease before and after dopaminergic treatment. *Journal of the Neurological Sciences*, *287*(1), 200–204.

Kleiner-Fisman, G., & Fisman, D. N. (2007). Risk factors for the development of pedal edema in patients using pramipexole. *Archives of Neurology, 64*(6), 820–824.

Kulisevsky, J., Garcia-Sanchez, C., Berthier, M. L., Barbanoj, M., Pascual-Sedano, B., Gironell, A., et al. (2000). Chronic effects of dopaminergic replacement on cognitive function in Parkinson's disease: A two-year follow-up study of previously untreated patients. *Movement Disorders, 15*(4), 613–626.

Kulisevsky, J., Pascual-Sedano, B., Barbanoj, M., Gironell, A., Pagonabarraga, J., & Garcia-Sanchez, C. (2007). Acute effects of immediate and controlled-release levodopa on mood in Parkinson's disease: A double-blind study. *Movement Disorders, 22*(1), 62–67.

Kumru, H., Iranzo, A., Carrasco, E., Valldeoriola, F., Marti, M. J., Santamaria, J., et al. (2008). Lack of effects of pramipexole on REM sleep behavior disorder in Parkinson disease. *Sleep, 31*(10), 1418–1421.

Kuno, S., Mizuta, E., Yamasaki, S., & Araki, I. (2004). Effects of pergolide on nocturia in Parkinson's disease: Three female cases selected from over 400 patients. *Parkinsonism & Related Disorders, 10*(3), 181–187.

Lange, K. W., Robbins, T. W., Marsden, C. D., James, M., Owen, A. M., & Paul, G. M. (1992). L-Dopa withdrawal in Parkinson's disease selectively impairs cognitive performance in tests sensitive to frontal lobe dysfunction. *Psychopharmacology, 107*(2), 394–404.

Lawrence, A. J., Blackwell, A. D., Barker, R. A., Spagnolo, F., Clark, L., Aitken, M. R., et al. (2007). Predictors of punding in Parkinson's disease: Results from a questionnaire survey. *Movement Disorders, 22*(16), 2339–2345.

Leeman, A. L., O'Neill, C. J., Nicholson, P. W., Deshmukh, A. A., Denham, M. J., Royston, J. P., et al. (1987). Parkinson's disease in the elderly: Response to and optimal spacing of night time dosing with levodopa. *British Journal of Clinical Pharmacology, 24*(5), 637–643.

Liguori, C., Placidi, F., Stefani, A., Mercuri, N. B., Marciani, M. G., Stanzione, P., et al. (2015). Rotigotine effect on sleep in a de novo Parkinson's disease patient affected by periodic limb movement disorder. *Parkinsonism & Related Disorders, 21*(12), 1476–1478.

Lim, T. T., Kluger, B. M., Rodriguez, R. L., Malaty, I. A., Palacio, R., Jr., Ojo, O. O., et al. (2015). Rasagiline for the symptomatic treatment of fatigue in Parkinson's disease. *Movement Disorders, 30*(13), 1825–1830.

Linden, D., Diehl, R. R., & Berlit, P. (1997). Sympathetic cardiovascular dysfunction in long-standing idiopathic Parkinson's disease. *Clinical Autonomic Research, 7*(6), 311–314.

Lou, J. S., Kearns, G., Benice, T., Oken, B., Sexton, G., & Nutt, J. (2003). Levodopa improves physical fatigue in Parkinson's disease: A double-blind, placebo-controlled, crossover study. *Movement Disorders, 18*(10), 1108–1114.

MacDonald, P. A., MacDonald, A. A., Seergobin, K. N., Tamjeedi, R., Ganjavi, H., Provost, J. S., et al. (2011). The effect of dopamine therapy on ventral and dorsal striatum-mediated cognition in Parkinson's disease: Support from functional MRI. *Brain, 134*(5), 1447–1463.

Mack, J., Rabins, P., Anderson, K., Goldstein, S., Grill, S., Hirsch, E. S., et al. (2012). Prevalence of psychotic symptoms in a community-based Parkinson disease sample. *The American Journal of Geriatric Psychiatry, 20*(2), 123–132.

Manconi, M., Ferri, R., Zucconi, M., Clemens, S., Giarolli, L., Bottasini, V., et al. (2011). Preferential D2 or preferential D3 dopamine agonists in restless legs syndrome. *Neurology, 77*(2), 110–117.

Martinez-Martin, P., Rodriguez-Blazquez, C., Kurtis, M. M., & Chaudhuri, K. R. (2011). The impact of non-motor symptoms on health-related quality of life of patients with Parkinson's disease. *Movement Disorders, 26*(3), 399–406.

Martinez-Martin, P., Schapira, A. H., Stocchi, F., Sethi, K., Odin, P., MacPhee, G., et al. (2007). Prevalence of nonmotor symptoms in Parkinson's disease in an international setting; study using nonmotor symptoms questionnaire in 545 patients. *Movement Disorders*, *22*(11), 1623–1629.

Meco, G., Rubino, A., Caravona, N., & Valente, M. (2008). Sexual dysfunction in Parkinson's disease. *Parkinsonism & Related Disorders*, *14*(6), 451–456.

Miah, I. P., Olde Dubbelink, K. T., Stoffers, D., Deijen, J. B., & Berendse, H. W. (2012). Early-stage cognitive impairment in Parkinson's disease and the influence of dopamine replacement therapy. *European Journal of Neurology*, *19*(3), 510–516.

Miwa, H. (2007). Stereotyped behavior or punding in Parkinson's disease. *Journal of Neurology*, *254*(5), 61–67.

Miyasaki, J. M., Al Hassan, K., Lang, A. E., & Voon, V. (2007). Punding prevalence in Parkinson's disease. *Movement Disorders*, *22*(8), 1179–1181.

Moller, J. C., Eggert, K. M., Unger, M., Odin, P., Chaudhuri, K. R., & Oertel, W. H. (2008). Clinical risk-benefit assessment of dopamine agonists. *European Journal of Neurology*, *15*(Suppl. 2), 15–23.

Molloy, S. A., Rowan, E. N., O'Brien, J. T., McKeith, I. G., Wesnes, K., & Burn, D. J. (2006). Effect of levodopa on cognitive function in Parkinson's disease with and without dementia and dementia with Lewy bodies. *Journal of Neurology, Neurosurgery, and Psychiatry*, 77(12), 1323–1328.

Morgante, L., Colosimo, C., Antonini, A., Marconi, R., Meco, G., Pederzoli, M., et al. (2012). Psychosis associated to Parkinson's disease in the early stages: Relevance of cognitive decline and depression. *Journal of Neurology, Neurosurgery, and Psychiatry*, *83*(1), 76–82.

Morita, A., Okuma, Y., Kamei, S., Yoshii, F., Yamamoto, T., Hashimoto, S., et al. (2011). Pramipexole reduces the prevalence of fatigue in patients with Parkinson's disease. *Internal Medicine*, *50*(19), 2163–2168.

Morrison, C. E., Borod, J. C., Brin, M. F., Halbig, T. D., & Olanow, C. W. (2004). Effects of levodopa on cognitive functioning in moderate-to-severe Parkinson's disease (MSPD). *Journal of Neural Transmission*, *111*(10 – 11), 1333–1341.

Mostile, G., & Jankovic, J. (2009). Treatment of dysautonomia associated with Parkinson's disease. *Parkinsonism & Related Disorders*, *15*(Suppl. 3), S224–S232.

Müller, B., Assmus, J., Larsen, J. P., Haugarvoll, K., Skeie, G. O., & Tysnes, O. B. (2013). Autonomic symptoms and dopaminergic treatment in de novo Parkinson's disease. *Acta Neurologica Scandinavica*, *127*(4), 290–294.

Noack, C., Schroeder, C., Heusser, K., & Lipp, A. (2014). Cardiovascular effects of levodopa in Parkinson's disease. *Parkinsonism & Related Disorders*, *20*(8), 815–818.

Olanow, C. W., Rascol, O., Hauser, R., Feigin, P. D., Jankovic, J., Lang, A., et al. (2009). A double-blind, delayed-start trial of rasagiline in Parkinson's disease. *New England Journal of Medicine*, *361*(13), 1268–1278.

O'Suilleabhain, P. E., & Dewey, R. B., Jr. (2002). Contributions of dopaminergic drugs and disease severity to daytime sleepiness in Parkinson disease. *Archives of Neurology*, *59*(6), 986–989.

O'Sullivan, S. S., Williams, D. R., Gallagher, D. A., Massey, L. A., Silveira-Moriyama, L., & Lees, A. J. (2008). Nonmotor symptoms as presenting complaints in Parkinson's disease: A clinicopathological study. *Movement Disorders*, *23*(1), 101–106.

Oved, D., Ziv, I., Treves, T. A., Paleacu, D., Melamed, E., & Djaldetti, R. (2006). Effect of dopamine agonists on fatigue and somnolence in Parkinson's disease. *Movement Disorders*, *21*(8), 1257–1261.

Pagano, G., Tan, E. E., Haider, J. M., Bautista, A., & Tagliati, M. (2015). Constipation is reduced by beta-blockers and increased by dopaminergic medications in Parkinson's disease. *Parkinsonism & Related Disorders*, *21*(2), 120–125.

Pahwa, R., Stacy, M. A., Factor, S. A., Lyons, K. E., Stocchi, F., Hersh, B. P., et al. (2007). Ropinirole 24-hour prolonged release: Randomized, controlled study in advanced Parkinson disease. *Neurology*, *68*(14), 1108–1115.

Palma, J., & Kaufmann, H. (2014). Autonomic disorders predicting Parkinson's disease. *Parkinsonism & Related Disorders*, *20*(Suppl. 1), S94–S98.

Papapetropoulos, S., Argyriou, A. A., & Ellul, J. (2005). Factors associated with drug-induced visual hallucinations in Parkinson's disease. *Journal of Neurology*, *252*(10), 1223–1228.

Park, A., & Stacy, M. (2011). Dopamine-induced nonmotor symptoms of Parkinson's disease. *Parkinson's Disease, 2011* (2011), 6, Article ID 485063.

Pavese, N., Metta, V., Bose, S. K., Chaudhuri, K. R., & Brooks, D. J. (2010). Fatigue in Parkinson's disease is linked to striatal and limbic serotonergic dysfunction. *Brain*, *133*(11), 3434–3443.

Petit, G. H., Berkovich, E., Hickery, M., Kallunki, P., Fog, K., Fitzer-Attas, C., et al. (2013). Rasagiline ameliorates olfactory deficits in an alpha-synuclein mouse model of Parkinson's disease. *PLoS One*, *8*(4). e60691.

Pfeiffer, R. F. (2003). Gastrointestinal dysfunction in Parkinson's disease. *The Lancet Neurology*, *2*(2), 107–116.

Poewe, W. (2008). Non-motor symptoms in Parkinson's disease. *European Journal of Neurology*, *15*(Suppl. 1), 14–20.

Poewe, W., Hauser, R. A., & Lang, A. (2015). Effects of rasagiline on the progression of nonmotor scores of the MDS-UPDRS. *Movement Disorders*, *30*(4), 589–592.

Poewe, W., & Högl, B. (2004). Akathisia, restless legs and periodic limb movements in sleep in Parkinson's disease. *Neurology*, *63*(Suppl. 3), S12–S16.

Poewe, W. H., Rascol, O., Quinn, N., Tolosa, E., Oertel, W. H., Martignoni, E., et al. (2007). Efficacy of pramipexole and transdermal rotigotine in advanced Parkinson's disease: A double-blind, double-dummy, randomised controlled trial. *The Lancet Neurology*, *6*(6), 513–520.

Pohanka, M., Kanovsky, P., Bares, M., Pulkrabek, J., & Rektor, I. (2004). Pergolide mesylate can improve sexual dysfunction in patients with Parkinson's disease: The results of an open, prospective, 6-month follow-up. *European Journal of Neurology*, *11*(7), 483–488.

Qamar, M. A., Sauerbier, A., Politis, M., Carr, H., Loehrer, P., & Chaudhuri, K. R. (2017). Presynaptic dopaminergic terminal imaging and non-motor symptoms assessment of Parkinson's disease: Evidence for dopaminergic basis? *npj Parkinson's Disease*, *3*(1), 5.

Quinn, N. P., Rossor, M. N., & Marsden, C. D. (1987). Olfactory threshold in Parkinson's disease. *Journal of Neurology, Neurosurgery, and Psychiatry*, *50*(1), 88–89.

Rascol, O., Fitzer-Attas, C. J., Hauser, R., Jankovic, J., Lang, A., Langston, J. W., et al. (2011). A double-blind, delayed-start trial of rasagiline in Parkinson's disease (the ADAGIO study): Prespecified and post-hoc analyses of the need for additional therapies, changes in UPDRS scores, and non-motor outcomes. *The Lancet Neurology*, *10*(5), 415–423.

Rascol, O., Zesiewicz, T., Chaudhuri, K. R., Asgharnejad, M., Surmann, E., Dohin, E., et al. (2016). A randomized controlled exploratory pilot study to evaluate the effect of rotigotine transdermal patch on Parkinson's disease-associated chronic pain. *The Journal of Clinical Pharmacology*, *56*(7), 852–861.

Razmy, A., Lang, A. E., & Shapiro, C. M. (2004). Predictors of impaired daytime sleep and wakefulness in patients with Parkinson disease treated with older (ergot) vs newer (nonergot) dopamine agonists. *Archives of Neurology*, *61*(1), 97–102.

Reid, J. L., Calne, D. B., George, C. F., & Vakil, S. D. (1972). The action of L(-)-dopa on baroreflexes in Parkinsonism. *Clinical Science*, *43*(6), 851–859.

Reijnders, J. S., Ehrt, U., Weber, W. E., Aarsland, D., & Leentjens, A. F. (2008). A systematic review of prevalence studies of depression in Parkinson's disease. *Movement Disorders*, *23*(2), 183–189.

Reijnders, J. S., Scholtissen, B., Weber, W. E., Aalten, P., Verhey, F. R., & Leentjens, A. F. (2010). Neuroanatomical correlates of apathy in Parkinson's disease: A magnetic resonance imaging study using voxel-based morphometry. *Movement Disorders*, *25*(14), 2318–2325.

Rektorova, I., Rektor, I., Bares, M., Dostal, V., Ehler, E., Fanfrdlova, Z., et al. (2003). Pramipexole and pergolide in the treatment of depression in Parkinson's disease: A national multicentre prospective randomized study. *European Journal of Neurology*, *10*(4), 399–406.

Rektorova, I., Srovnalova, H., Kubikova, R., & Prasek, J. (2008). Striatal dopamine transporter imaging correlates with depressive symptoms and tower of London task performance in Parkinson's disease. *Movement Disorders*, *23*(11), 1580–1587.

Remy, P., Doder, M., Lees, A., Turjanski, N., & Brooks, D. (2005). Depression in Parkinson's disease: Loss of dopamine and noradrenaline innervation in the limbic system. *Brain*, *128*(6), 1314–1322.

Rijsman, R. M., Schoolderman, L. F., Rundervoort, R. S., & Louter, M. (2014). Restless legs syndrome in Parkinson's disease. *Parkinsonism & Related Disorders*, *20*(Suppl. 1), S5–S9.

Rizos, A., Sauerbier, A., Antonini, A., Weintraub, D., Martinez-Martin, P., Kessel, B., et al. (2016). A European multicentre survey of impulse control behaviours in Parkinson's disease patients treated with short- and long-acting dopamine agonists. *European Journal of Neurology*, *23*(8), 1255–1261.

Ross, G. W., Petrovitch, H., Abbott, R. D., Tanner, C. M., Popper, J., Masaki, K., et al. (2008). Association of olfactory dysfunction with risk for future Parkinson's disease. *Annals of Neurology*, *63*(2), 167–173.

Saifee, T., Lees, A. J., & Silveira-Moriyama, L. (2010). Olfactory function in Parkinson's disease in ON versus OFF states. *Journal of Neurology, Neurosurgery, and Psychiatry*, *81*(11), 1293–1295.

Sakakibara, R., Uchiyama, T., Yamanishi, T., Shirai, K., & Hattori, T. (2008). Bladder and bowel dysfunction in Parkinson's disease. *Journal of Neural Transmission*, *115*(3), 443–460.

Sanchez-Ramos, J. R., Ortoll, R., & Paulson, G. W. (1996). Visual hallucinations associated with Parkinson disease. *Archives of Neurology*, *53*(12), 1265–1268.

Schestatsky, P., Kumru, H., Valls-Sole, J., Valldeoriola, F., Marti, M. J., Tolosa, E., et al. (2007). Neurophysiologic study of central pain in patients with Parkinson disease. *Neurology*, *69*(23), 2162–2169.

Schifitto, G., Friedman, J. H., Oakes, D., Shulman, L., Comella, C. L., Marek, K., et al. (2008). Fatigue in levodopa-naive subjects with Parkinson disease. *Neurology*, *71*(7), 481–485.

Schrag, A., Jahanshahi, M., & Quinn, N. (2000). What contributes to quality of life in patients with Parkinson's disease? *Journal of Neurology, Neurosurgery, and Psychiatry*, *69*(3), 308–312.

Seki, S., Igawa, Y., Kaidoh, K., Ishizuka, O., Nishizawa, O., & Andersson, K. E. (2001). Role of dopamine D1 and D2 receptors in the micturition reflex in conscious rats. *Neurourology and Urodynamics*, *20*(1), 105–113.

Seppi, K., Weintraub, D., Coelho, M., Perez-Lloret, S., Fox, S. H., Katzenschlager, R., et al. (2011). The movement disorder society evidence-based medicine review update: Treatments for the non-motor symptoms of Parkinson's disease. *Movement Disorders*, *26*(Suppl. 3), S42–S80.

Singaram, C., Ashraf, W., Gaumnitz, E. A., Torbey, C., Sengupta, A., Pfeiffer, R., et al. (1995). Dopaminergic defect of enteric nervous system in Parkinson's disease patients with chronic constipation. *Lancet*, *346*(8979), 861–864.

Sjodahl, H. C., Hagell, P., & Nilsson, M. H. (2012). Motor and non-motor predictors of illness-related distress in Parkinson's disease. *Parkinsonism & Related Disorders*, *18*(3), 299–302.

Sophie, M., & Ford, B. (2012). Management of pain in Parkinson's disease. *CNS Drugs*, *26*(11), 937–948.

Spencer, A. H., Rickards, H., Fasano, A., & Cavanna, A. E. (2011). The prevalence and clinical characteristics of punding in Parkinson's disease. *Movement Disorders*, *26*(4), 578–586.

Starkstein, S. E., & Leentjens, A. F. (2008). The nosological position of apathy in clinical practice. *Journal of Neurology, Neurosurgery, and Psychiatry*, *79*(10), 1088–1092.

Stocchi, F., Barbato, L., Nordera, G., Berardelli, A., & Ruggieri, S. (1998). Sleep disorders in Parkinson's disease. *Journal of Neurology*, *245*(Suppl. 1), S15–S18.

Tan, E. K., & Ondo, W. (2000). Clinical characteristics of pramipexole-induced peripheral edema. *Archives of Neurology*, *57*(5), 729–732.

Tan, A., Salgado, M., & Fahn, S. (1996). Rapid eye movement sleep behavior disorder preceding Parkinson's disease with therapeutic response to levodopa. *Movement Disorders*, *11*(2), 214–216.

Tateno, F., Sakakibara, R., Yokoi, Y., Kishi, M., Ogawa, E., Uchiyama, T., et al. (2011). Levodopa ameliorated anorectal constipation in de novo Parkinson's disease: The QL-GAT study. *Parkinsonism & Related Disorders*, *17*(9), 662–666.

Thobois, S., Ardouin, C., Lhommee, E., Klinger, H., Lagrange, C., Xie, J., et al. (2010). Non-motor dopamine withdrawal syndrome after surgery for Parkinson's disease: Predictors and underlying mesolimbic denervation. *Brain*, *133*(4), 1111–1127.

Thobois, S., Lhommee, E., Klinger, H., Ardouin, C., Schmitt, E., Bichon, A., et al. (2013). Parkinsonian apathy responds to dopaminergic stimulation of D2/D3 receptors with piribedil. *Brain*, *136*(5), 1568–1577.

Tholfsen, L. K., Larsen, J. P., Schulz, J., Tysnes, O. B., & Gjerstad, M. D. (2015). Development of excessive daytime sleepiness in early Parkinson disease. *Neurology*, *85*(2), 162–168.

Trenkwalder, C., Kies, B., Rudzinska, M., Fine, J., Nikl, J., Honczarenko, K., et al. (2011). Rotigotine effects on early morning motor function and sleep in Parkinson's disease: A double-blind, randomized, placebo-controlled study (RECOVER). *Movement Disorders*, *26*(1), 90–99.

Uchiyama, T., Sakakibara, R., Hattori, T., & Yamanishi, T. (2003). Short-term effect of a single levodopa dose on micturition disturbance in Parkinson's disease patients with the wearing-off phenomenon. *Movement Disorders*, *18*(5), 573–578.

Valko, P. O., Waldvogel, D., Weller, M., Bassetti, C. L., Held, U., & Baumann, C. R. (2010). Fatigue and excessive daytime sleepiness in idiopathic Parkinson's disease differently correlate with motor symptoms, depression and dopaminergic treatment. *European Journal of Neurology*, *17*(12), 1428–1436.

Van Camp, G., Flamez, A., Cosyns, B., Weytjens, C., Muyldermans, L., Van Zandijcke, M., et al. (2004). Treatment of Parkinson's disease with pergolide and relation to restrictive valvular heart disease. *Lancet*, *363*(9416), 1179–1183.

Van den Kerchove, M., Jacquy, J., Gonce, M., & De Deyn, P. P. (1993). Sustained-release levodopa in parkinsonian patients with nocturnal disabilities. *Acta Neurologica Belgica*, *93*(1), 32–39.

Vignatelli, L., Billiard, M., Clarenbach, P., Garcia-Borreguero, D., Kaynak, D., Liesiene, V., et al. (2006). EFNS guidelines on management of restless legs syndrome and periodic limb movement disorder in sleep. *European Journal of Neurology*, *13*(10), 1049–1065.

Voon, V., Hassan, K., Zurowski, M., Duff-Canning, S., de Souza, M., Fox, S., et al. (2006). Prospective prevalence of pathologic gambling and medication association in Parkinson disease. *Neurology*, *66*(11), 1750–1752.

Voon, V., Potenza, M. N., & Thomsen, T. (2007). Medication-related impulse control and repetitive behaviors in Parkinson's disease. *Current Opinion in Neurology*, *20*(4), 484–492.
Wailke, S., Herzog, J., Witt, K., Deuschl, G., & Volkmann, J. (2011). Effect of controlled-release levodopa on the microstructure of sleep in Parkinson's disease. *European Journal of Neurology*, *18*(4), 590–596.
Wang, Y., Yang, Y., Wu, H., Lan, D., Chen, Y., & Zhao, Z. (2016). Effects of rotigotine on REM sleep behavior disorder in Parkinson disease. *Journal of Clinical Sleep Medicine*, *12*(10), 1403–1409.
Wasner, G., & Deuschl, G. (2012). Pains in Parkinson disease—Many syndromes under one umbrella. *Nature Reviews Neurology*, *8*(5), 284–294.
Weintraub, D., Koester, J., Potenza, M. N., Siderowf, A. D., Stacy, M., Voon, V., et al. (2010). Impulse control disorders in Parkinson disease: A cross-sectional study of 3090 patients. *Archives of Neurology*, *67*(5), 589–595.
Weintraub, D., Newberg, A. B., Cary, M. S., Siderowf, A. D., Moberg, P. J., Kleiner-Fisman, G., et al. (2005). Striatal dopamine transporter imaging correlates with anxiety and depression symptoms in Parkinson's disease. *Journal of Nuclear Medicine*, *46*(2), 227–232.
Winge, K., & Fowler, C. J. (2006). Bladder dysfunction in Parkinsonism: Mechanisms, prevalence, symptoms, and management. *Movement Disorders*, *21*(6), 737–745.
Wolters, E. C. (1999). Dopaminomimetic psychosis in Parkinson's disease patients: Diagnosis and treatment. *Neurology*, *52*(Suppl. 3), S10–S13.
Yamauchi, K., Takehisa, M., Tsuno, M., Kaneda, Y., Taniguchi, T., Ohno, H., et al. (2003). Levodopa improved rapid eye movement sleep behavior disorder with diffuse Lewy body disease. *General Hospital Psychiatry*, *25*(2), 140–142.
Yoshimura, N., Mizuta, E., Yoshida, O., & Kuno, S. (1998). Therapeutic effects of dopamine D1/D2 receptor agonists on detrusor hyperreflexia in 1-methyl-4-phenyl-1,2,3,6-tetrahydropyridine-lesioned parkinsonian cynomolgus monkeys. *Journal of Pharmacology and Experimental Therapeutics*, *286*(1), 228–233.
Yoshimura, N., Yoshida, O., Sasa, M., & Takaori, S. (1992). Dopamine D-1 receptor-mediated inhibition of micturition reflex by central dopamine from the substantia nigra. *Neurourology and Urodynamics*, *11*(5), 535–545.
Zanettini, R., Antonini, A., Gatto, G., Gentile, R., Tesei, S., & Pezzoli, G. (2007). Valvular heart disease and the use of dopamine agonists for Parkinson's disease. *New England Journal of Medicine*, *356*(1), 39–46.
Zesiewicz, T. A., Sullivan, K. L., Arnulf, I., Chaudhuri, K. R., Morgan, J. C., Gronseth, G. S., et al. (2010). Practice parameter: Treatment of nonmotor symptoms of Parkinson disease: Report of the quality standards subcommittee of the American Academy of Neurology. *Neurology*, *74*(11), 924–931.
Zhang, G., Zhang, Z., Liu, L., Yang, J., Huang, J., Xiong, N., et al. (2014). Impulsive and compulsive behaviors in Parkinson's disease. *Frontiers in Aging Neuroscience*, *6*, 318.
Zhu, K., Van Hilten, J. J., & Marinus, J. (2016). Course and risk factors for excessive daytime sleepiness in Parkinson's disease. *Parkinsonism & Related Disorders*, *24*, 34–40.
Zhu, K., Van Hilten, J. J., Putter, H., & Marinus, J. (2013). Risk factors for hallucinations in Parkinson's disease: Results from a large prospective cohort study. *Movement Disorders*, *28*(6), 755–762.
Ziv, I., Avraham, M., Michaelov, Y., Djaldetti, R., Dressler, R., Zoldan, J., et al. (1998). Enhanced fatigue during motor performance in patients with Parkinson's disease. *Neurology*, *51*(6), 1583–1586.

CHAPTER THIRTY-SIX

Infusional Therapies, Continuous Dopaminergic Stimulation, and Nonmotor Symptoms

Ulrika Mundt-Petersen[1], Per Odin
Lund University, Faculty of Medicine, Lund, Sweden
[1]Corresponding author: e-mail address: ulrika.mundt-petersen@med.lu.se

Contents

International Review of Neurobiology, Volume 134
ISSN 0074-7742
http://dx.doi.org/10.1016/bs.irn.2017.05.036

Abstract

Pump-based Parkinson (PD) therapies, including subcutaneous apomorphine infusion (CSA) and levodopa-carbidopa intestinal gel (LCIG), presently constitute the most effective pharmacological treatments available for advanced PD. These therapies are based on a more constant delivery of the dopaminergic drug resulting in a more continuous dopaminergic stimulation and a more stable treatment effect. This can be detected as reduction of time in off, reduction of dyskinesia frequency and severity, as well as increase of time in on without troublesome dyskinesias. A number of open-label studies now suggest that also the nonmotor PD symptomatology can improve under CSA and LCIG therapy. The most consistent improvements are seen concerning sleep, mood, and apathy, gastrointestinal symptoms, and urological symptoms. But also cardiovascular symptoms, perception, attention, and sexual function might show beneficial effects when moving from conventional therapies to pump treatment. Further there might be negative influences on some parts of the nonmotor symptomatology through side effects of CSA and LCIG therapy. In this chapter, we review the present knowledge about these aspects of the pump-based therapies. This information might be valuable when deciding on advanced therapy for individual patients.

1. INTRODUCTION

The most effective treatment for early Parkinson's disease (PD) is considered to be peroral medication with levodopa (Olanow & Koller, 1998). Long-term levodopa intake, however, results in clinical fluctuations and dyskinesias. Studies show that within 2–5 years of levodopa treatment more than 50% of patients develop motor fluctuations, and the risk of dyskinesia increases by 10% per year (LeWitt, 2014; Schapira & Obeso, 2006; Schrag, 2000). Three device-aided therapies constitute the most effective treatment strategies in advanced PD: deep brain stimulation (DBS), levodopa-carbidopa intestinal gel (LCIG), and continuous subcutaneous apomorphine infusion (CSA). In the following, we will focus on the pump-based therapies, LCIG and CSA, and especially the effects of these treatments on the nonmotor symptomatology.

2. BACKGROUND: SUBCUTANEOUS APOMORPHINE INFUSION

2.1 Methodology

The nonergot dopamine agonist apomorphine is a nonselective agonist for dopamine D_2 and D_1 receptors. Apomorphine has an anti-PD effect which is similar to that of L-dopa. Apomorphine is infused with the help of a small infusion pump, and a needle that delivers the apomorphine subcutaneously, mostly on the stomach or thighs. CSA is normally used as a 16-h daytime treatment, but can be given over 24 h in the case the patient needs night-time effect. The dose needed varies interindividually and has to be titrated for each patient. It is quite stable over time intraindividually. The majority of patients with CSA use some oral L-dopa in parallel, but most often at daily doses that have been reduced with around 50% compared to before the start of CSA treatment (Trenkwalder et al., 2015). The possibility to reduce L-dopa seems to be essential for achieving a less pulsative and thereby more effective treatment; in an overview of 19 clinical studies, Bhidayasiri et al. (2015) described everything from a mean increase of L-dopa dose of 16% to a mean decrease of the L-dopa dose with 81%. The initiation and adjustment of CSA is often handled during a 1–2 weeks inpatient stay, but some centers initiate the treatment in an outpatient setting. The total daily dose of CSA can be divided into three parts: the morning dose, the maintenance dose, and the extra bolus dose at demand. Most patients have maintenance doses between 4 and 7 mg/h. During the first part of CSA treatment, patients may get the peripherally acting dopamine D_2 antagonist domperidone 10 mg three times daily to reduce otherwise common side effects including nausea and orthostatic hypotension.

2.2 The Effect on Motor Symptoms

As previously suggested, CSA seems to be the most effective in patients who tolerate larger dose increases of apomorphine and either considerable reductions or total removal of daytime L-dopa medication (Bhidayasiri et al., 2015). The documented effects of CSA have first and foremost been a reduction of the time spent in off. In a review of 24 studies with in total 552 patients the mean reduction of off time was 59.3% (Todorova & Chaudhuri, 2016). Bhidayasiri et al. (2015) demonstrate mean improvements of "off" time between 40% and 85% in 19 different studies involving patients with disease duration from 10 to 19 years and total daily apomorphine doses between 31 and 162 mg. In the most successful cases off time can be totally eliminated. Apart from reduction of off time, several other effects on motor symptoms have also been reported. Katzenschlager, Hughes, and Evans (2005) demonstrated an improvement also concerning dyskinesias; the amount of dyskinesias produced by a certain dose of L-dopa or apomorphine was reduced with around 40% after 6 months of apomorphine infusion compared to before start of infusion. Improvements have also been seen on UPDRS parts II, III, and/or IV, indicating an improvement of ADL function and motor symptoms, while decreasing motor complications (Drapier, Eusebio, Degos, Vérin, & Durif, 2016; Martinez-Martin et al., 2015; Pålhagen et al., 2016; Rambour, Moreau, Salleron, & Devos, 2014; Todorova et al., 2013).

2.3 The Effect on Quality of Life

Several studies show significant improvements of health-related quality of life (HrQoL) after 6 months of CSA treatment. This has been demonstrated with the PDQ-8 (Martinez-Martin et al., 2011, 2015) as well as the PDQ-39 instruments (Drapier et al., 2016). In direct comparison with LCIG, CSA seems to give a similar effect on HrQoL (Martinez-Martin et al., 2015).

2.4 Safety

Subcutaneous nodules occur in most patients treated with CSA and are the most common adverse event. The clinical experience is that skin nodules can be counteracted/treated through a number of measures (Trenkwalder et al., 2015); however, evidence for these methods are limited. The measures recommended include good skin hygiene at the infusion site, switching between infusion sites, using lower concentrations of apomorphine, applying ultrasound treatment, or using silicone dressings. It is reported that subcutaneous nodules are more common in men than in women and occur more frequently with higher infusion rate (Borgemeester, Drent, & van Laar, 2015). The risk for noduli formation may be different with different apomorphine preparations and studies are ongoing comparing available apomorphines in this respect.

Visual hallucinations occur in about 20% of patients (Borgemeester et al., 2015). Mostly the hallucinations are mild, but more severe visual hallucinations also occur in about 10% of patients (Trenkwalder et al., 2015). Orthostatic hypotension can be induced or amplified by CSA treatment and can often be treated with an adequate fluid/salt intake, compression socks, or careful positional changes, but sometimes pharmacological measures must be considered. Impulse control disorders do occur with CSA treatment; such symptoms were reported in 9.8% of the patients in a 6-month observational study (Martinez-Martin et al., 2015). Hemolytic anemia does occur in 1%–2% of patients with CSA and can be a reason for treatment discontinuation (Borgemeester et al., 2015; Trenkwalder et al., 2015). Other severe adverse events that do lead to discontinuation are, e.g., severe cases of subcutaneous nodules, visual hallucinations, or orthostatic hypotension—each being the cause for discontinuation in about 3% of patients (Borgemeester et al., 2015).

There is still little information concerning the typical duration of apomorphine treatment. The clinical experience seems to be that this is typically a treatment for a limited time period of some months to years, but in single patients it can also be a life-long therapy.

2.5 The Indications, the Contraindications, and the "Ideal" Patient

Like other advanced PD therapies, CSA should be considered in patients who have problems with motor fluctuations and/or dyskinesias in spite of optimized peroral and/or transdermal therapy. An international expert group (Odin et al., 2015) recommends advanced therapy in patients who experience more than 1–2 h of off symptomatology or troublesome dyskinesias in spite of at least five doses of peroral L-dopa per day. The patients should also have tried dopamine agonists, MAO-B inhibitors and COMT inhibitors. In contrast to DBS, high age, slight-moderate dementia, and depression are not considered as contraindications for CSA. Lack of L-dopa effect and insufficient capacity of patient/caregivers to handle a pump-based therapy are the main contraindications for the therapy. Other symptoms like insomnia, depression, nonmotor fluctuations (e.g., anxiety, pain, clouded thinking, apathy), dysartria, and restless legs can rather be seen as an indication for the therapy (Odin et al., 2015). CSA is the most reversible of the advanced PD therapies and this might be an advantage in patients where the decision for advanced therapy is not so clear. The low degree of invasivity is sometimes seen as an advantage, especially among younger patients.

3. EFFECTS ON NONMOTOR PD SYMPTOMS

3.1 Effect of CSA on Total Nonmotor Symptom Burden

The nonmotor symptom scale (NMSS) is the only validated tool to give an overall estimation of nonmotor symptoms (NMS) in PD (Martinez-Martin et al., 2009). It estimates the frequency and severity of 30 different NMS in nine different domains. NMSS has been used for evaluation of CSA effects in two studies (Martinez-Martin et al., 2011, 2015) (Table 1).

In the first study, Martinez-Martin et al. (2011) performed a nonrandomized, open-label investigation in 17 patients where CSA was compared to standard PD therapy. NMSS scores, Unified Parkinson's disease Rating Scale (UPDRS) III (motor) and IV (motor complications), as well as HrQoL measures were assessed before initiation of the therapy and after 6 months on pump treatment. CSA treatment resulted in highly significant

Table 1 Effects Registered of Apomorphine on Nonmotor Symptoms, as Measured With the Nonmotor Symptom Scale (NMSS) in Two Different Studies, Both Open-Label

	Martinez-Martin et al. (2011)	Martinez-Martin et al. (2015)
Length of follow-up	12 mo	6 mo
Number of patients at baseline	17	43
Number of patients at follow-up	17	43
NMSST, before	105.94	82.37
NMSST, after	56.94	56.21
NMSST, % improvement	46%*	32%*
Cardiovascular, *P*	0.03	NS
Sleep/fatigue, *P*	0.0003	0.024
Mood/cognition, *P*	0.0005	0.0003
Perception/hallucinations, *P*	0.04	0.003
Attention/memory, *P*	0.006	0.003
Gastrointestinal, *P*	0.002	0.003
Urinary, *P*	0.001	0.002
Sexual, *P*	NS	NS
Miscellaneous, *P*	0.0003	NS

P=P-value for change at follow-up vs baseline; NS=not significant; mo=months; *Significant (P<0.05). The cardiovascular section of NMSS includes questions on orthostatic hypotension and syncope; the sleep/fatigue section includes questions on daytime sleepiness, fatigue, insomnia, restless legs, and REM-sleep behavior disorder; the mood/cognition section includes questions on apathia, anhedonia, and depression; the perception/hallucinations section includes questions on hallucinations and delusions; the attention/memory section includes questions on attention and memory; the gastrointestinal section includes questions on salivation, dysphagia, and constipation; the urinary section includes questions on urgency, frequency, and nocturia; the sexual section includes questions on lack of interest in sex and difficulties performing sex; the miscellaneous section includes questions on pain, difficulties to taste/smell, weight change, and excessive sweating.

improvements of UPDRS III (P=0.0003), UPDRS IV (P=0.0003), PDQ-8 (Parkinson's Disease Questionnaire, P=0.001), and NMSS total score (P=0.0003). Apomorphine was well tolerated in patients with visual hallucinations, illusions, and paranoid ideations. Significant improvements were seen in the following NMSS subscores: sleep, mood/apathy, perception, attention, gastrointestinal, urinary, and miscellany. The only subscore not showing significant improvement was the sexual one. The Effect Size was large (>0.8) concerning the sleep, moderate (0.5–0.79) concerning the mood/apathy, urinary, sexual, and miscellany, and small (0.20–0.49) concerning the perceptual, attention, and gastrointestinal subscores. The lowest numbers needed to treat (NNT) were seen for the sleep (1.71), mood/apathy (2.43), and urinary (2.83) subscores. Among the single items of NMSS significant improvements ($P<0.025$) were seen concerning light-headedness, fatigue, difficult falling asleep, restless legs, lost interest in surrounding, lack motivation, nervous, sad, difficult experiencing pleasure, problems with concentration, dribbling saliva, urgency, frequency, nocturia, and excessive sweating. The following items did not show significant improvement: light-headedness, fainting, daytime sleeping, flat mood, hallucinations, delusions, double vision, forget recent events, forget doing things, swallowing, constipation, altered interest in sex, problems having sex, unexplained pains, lost taste/smell, and change in weight.

In the second EuroInf study, Martinez-Martin et al. (2015) performed and open-label prospective, observational, 6-month comparative study between CSA and LCIG. This was done in 43 patients on CSA (mean age 62.3 years, disease duration 14 years) and 44 patients on LCIG (mean age 62.7 years, disease duration 16.1 years). NMSS, UPDRS III, UPDRS IV, and PDQ-8 scores were evaluated before start of therapy and after 6 months on pump treatment. CSA treatment resulted in highly significant improvements for UPDRS III ($P<0.0001$), UPDRS IV ($P<0.0001$), PDQ-8 (Parkinson's Disease Questionnaire, $P<0.0001$), and NMSS total score ($P=0.0007$). Significant improvements were seen in the following NMSS subscores: sleep/fatigue, mood/apathy, perceptual, attention/memory, gastrointestinal, and urinary. No significant improvement was seen in the cardiovascular, sexual, and miscellaneous subscores. The Effect Size was small (0.20–0.49) concerning the cardiovascular, sleep/fatigue, mood/apathy, perceptual, attention/memory, gastrointestinal, and miscellaneous subscores. The lowest NNT values were seen for the sleep/fatigue (2.05), mood/apathy (2.53), gastrointestinal (2.87), and urinary (2.87) subscores.

3.2 Apomorphine and Neuropsychiatric Symptoms

CSA may have a potential positive effect in the treatment of some neuropsychiatric symptoms of PD patients, including psychosis (Antonini et al., 2011; Chaudhuri, Abbott, & Millac, 1991; Ellis et al., 1997; Garcia Ruiz et al., 2008; van Laar, Postma, & Drent, 2010) and in some cases also cognitive functions (Yarnall et al., 2016). Early studies rather suggested that CSA could increase the risk for psychosis in PD patients (Muhiddin, Roche, & Pearce, 1994; Pollak, Champay, Hommel, Perret, & Benabid, 1989; Ruggieri, Stocchi, Carta, & Agnoli, 1989), but in more recent investigations it seems that CSA has been safe and even might improve psychotic symptomatology (Antonini et al., 2011; Chaudhuri et al., 1991; Ellis et al., 1997; Garcia Ruiz et al., 2008; van Laar et al., 2010). The possible mechanisms behind this might include the possibility to reduce or stop other PD drugs with high risk for psychotic side effect, but also the fact that CSA itself has a relatively low tendency to induce hallucinations and other psychotic symptoms. Some authors have claimed that the serotonin antagonistic effect of the piperidine moiety (Borgemeester, Lees, & van Laar, 2016; Chaudhuri et al., 1991) might play a role. Also that the potential ability of CSA to increase lower order visual perception could be an explanation (Borgemeester et al., 2016).

The cognitive status of the patient is a key factor when deciding on the use of advanced PD therapies. Cognitive decline and dementia are regarded as contraindications for the use of DBS. CSA on the other hand is regarded as safe, at least in cases with mild-to-moderate cognitive dysfunction/dementia. In an investigation with 23 advanced PD patients, Drapier et al. (2012) showed that there were no significant negative effects of CSA on cognitive tests assessing executive functions, visuospatial construction, or memory (after 12 months of treatment). Lately it has been suggested that CSA can have a role for amyloid deposition in the brain. This was first shown in an investigation in an Alzheimer's disease mouse model, where the authors demonstrated that treatment of the mice with apomorphine resulted in improvement of memory function and decrease in intraneural A-beta and p-tau levels (Himeno et al., 2011). Then Yarnall et al. demonstrated in cognitive normal human PD cases, that there was significantly reduced brain A-beta deposition in those treated with apomorphine before death compared to controls. Further investigations are now ongoing.

The NMSS-based investigations indicate a positive effect of LCIG therapy on the mood/cognition subscores (Martinez-Martin et al., 2011, 2015). There have been indications of a positive effect of CSA treatment on mood also in earlier studies (Morgante et al., 2004).

3.3 Apomorphine and Gastrointestinal Symptoms

Two studies have analyzed the role of apomorphine treatment for dysphagia. Tison et al. (1996) investigated the effect of apomorphine on dysphagia using videofluoroscopy in eight PD patients. They concluded that apomorphine improved swallowing abnormalities and swallowing time in a subgroup of PD patients. Hunter, Crameri, Austin, Woodward, and Hughes (1997) compared the effect of levodopa/carbidopa and apomorphine in 15 patients using barium swallowing methodology. They demonstrated a significant effect of apomorphine on the pharyngeal phase for semisolids only. Further there are results speaking for improvement of drooling of saliva with CSA (Martinez-Martin et al., 2011; Suttrup & Warnecke, 2016).

A few authors have investigated the effect of apomorphine on defecatory dysfunction. Mathers et al. (1989) found a functional improvement in 4 out of 6 patients with chronic constipation. Edwards, Quigley, Harned, Hofman, and Pfeiffer (1993) found improvements of defecographic abnormalities following administration of apomorphine in a group of eight PD patients.

3.4 Apomorphine and Sleep

Apart from the results achieved with NMSS referred earlier, a few further studies address the possible positive effects of CSA on sleep. Reuter, Ellis, and Chaudhuri (1999) suggested positive effects of CSA on nocturnal symptoms or restless legs syndrome. Priano et al. (2003) showed improvement concerning periodic limb movements, motor symptoms, and sleep architecture with polysomnography. Garcia Ruiz et al. (2008) applied actigraphy to show improvement of severe insomnia and sleep fragmentation in one patient achieving CSA. Borgemeester et al. (2016) reported a retrospective study showing improvement of sleep problems in 17 out of 20 patients.

3.5 Apomorphine and Sensory Symptoms

There is so far very limited information available concerning the effect of apomorphine on sensory symptoms of PD, including pain, olfactory, and visual disturbances. There are studies suggesting a positive effect of apomorphine on off period pain (Coelho, Ferreira, Rosa, & Sampaio, 2008; Factor, Brown, & Molho, 2000). But a randomized controlled trial with apomorphine failed to show any effect on pain processing in PD compared to placebo (Dellapina et al., 2011).

3.6 Apomorphine and Urinary and Sexual Dysfunction

The published information so far indicate that apomorphine may have positive effects on urinary dysfunction, but the evidence for this is still very limited. Christmas et al. (1988) demonstrated improvement of voiding efficacy following apomorphine in PD patients. Aranda and Cramer (1993) found that apomorphine could improve overall detrusor activity. In a further study CSA pumps were reported to improve lower urinary tract symptoms, but the same effect was seen with other types of dopaminergic treatment and DBS.

Considering that apomorphine has been used for the treatment of erectile dysfunction in males and hypoactive sexual desire in females (review: Bronner & Vodušek, 2011; Ribarič, 2012), it does not seem unlikely that CSA might have effects on the sexual functioning of PD patients. The data from the NMSS-based studies (Martinez-Martin et al., 2011, 2015) do not show a clear result in this respect, and further studies focusing these effects would be of value.

3.7 Apomorphine and Impulse Control Disorders

There is so far only limited data concerning CSA and ICD. In a large retrospective study on hypersexuality, Garcia Ruiz et al. (2008) reported only one case in a group of 82 patients followed for 19.93 months. Smaller open-label studies suggest a low rate of ICD in patients on CSA (Magennis, Cashell, O'Brien, & Lynch, 2012; Todorova et al., 2013). A recent 3-year prospective study seems to confirm this (Todorova, Bitsara, Teleg, & Chaudhuri, 2015); 41 patients on CSA were prospectively screened for ICD every 3 months; seven new ICD cases were diagnosed, but discontinuation of CSA only became necessary in one case. The other six cases as well as previous cases all improved or resolved without stopping the CSA therapy. In the open-label prospective trial by Martinez-Martin et al. (2015), only four new cases of ICD developed under CSA, and none required discontinuation of the therapy. According to a recent review of adverse event reports about ICD received by the US Food and Drug Administration from 2003 to 2012 apomorphine appears to have less risk for ICD compared to oral dopamine agonists like pramipexole and ropinirole (83).

4. BACKGROUND: LEVODOPA-CARBIDOPA INTESTINAL GEL

4.1 Methodology

LCIG is a cellulose gel containing L-dopa and the decarboxylase inhibitor carbidopa at concentrations of 20 and 5 mg/mL, respectively. It is delivered with a portable infusion pump. The treatment response is often evaluated during a nasoduodenal phase before initiating the final equipment. During LCIG treatment, the gel is infused continuously through the abdominal wall with the help of a percutaneous endoscopic gastrostomy (PEG) that has been inserted in local anesthesia. The gel then passes through the ventricle and duodenum within a tube that has its distal end in proximal part of jejunum.

LCIG can be used as a monotherapy, but it is not uncommon to use additional dopamine agonists, COMT inhibitors, and/or amantadine during the treatment. Most patients are on a 16-h regimen and only use the infusion during the waking day. Some patients also use the infusion during the night and are therefore on a continuous treatment. It is recommendable to reduce the maintenance dose by around 30% during sleep if patients are on a 24-h treatment regime. The total daily dose consists of three parts: a morning bolus dose, a continuous dose, and an extra bolus dose at demand. The morning dose makes it possible to quickly reach an effective plasma concentration, typically consists of 100–200 mg L-dopa and has an effect within 10–30 min. The continuous dose consists of 20–200 mg/h L-dopa that is infused throughout the day. The continuous dose is titrated after response to the clinical symptomatology; a process that is most often handled during a 1- to 2-week inpatient stay. An outpatient-unit-based start of the therapy is routine in some centers. The extra bolus doses consist of 10–40 mg L-dopa that the patient can use when he or she feels the need for additional effect and normally has effect within 10–20 min.

4.2 The Effect on Motor Symptoms

The effects of LCIG treatment have been investigated in three randomized controlled trials (RCTs) and in a number of observational studies.

The first RCT of LCIG included 12 patients and reported a significantly increased number of observations in the near-normal motor state (including mild off state and mild dyskinesia) during LCIG infusion compared to controlled-release levodopa monotherapy (Nyholm et al., 2003). The estimated mean difference was 19%. UPDRS parts I and II were unchanged but part IV was significantly reduced with LCIG.

The second RCT demonstrated significant reduction of observations in moderate to severe off state with LCIG compared to individually selected conventional therapies in 21 patients (Nyholm et al., 2005). UPDRS parts II and IV were significantly improved with LCIG whereas part III in on state was unchanged.

The third RCT, a double-blind double-dummy 12-week study allocated 37 patients to LCIG and 34 to immediate-release oral levodopa/carbidopa. The number of hours in patient-rated off state was significantly lower with LCIG compared to immediate-release levodopa (Olanow et al., 2014). Patients spent a mean of 4.0 h less in off (controls: −2.1 h) and 4.1 h more in on without troublesome dyskinesia (controls: +1.9 h). UPDRS part II was significantly improved with LCIG whereas part III in on state was unchanged.

In open-label or retrospective studies, the mean reduction of time spent in off ranges from 40% to 80% (Antonini et al., 2013; Antonini, Yegin, Preda, Bergmann, & Poewe, 2015; Chang et al., 2016; Fernandez et al., 2015; Lopiano et al., 2016; Martinez-Martin et al., 2015; Pickut, van der Linden, Dethy, Van De Maele, & de Beyl, 2014; Reddy et al., 2012; Valldeoriola et al., 2016; Volkmann et al., 2013; Zibetti, Merola, Artusi, & Rizzi, 2014), which corresponds to the 64% reduction of time in off reported in the randomized controlled trial (Olanow et al., 2014).

Timpka et al. (2016) demonstrated an improvement also concerning dyskinesias; the amount of dyskinesias produced by a certain dose of L-dopa was reduced with around 70% after 6 months of LCIG infusion compared to before start of infusion. An antidyskinetic effect has been further strengthened in recent publications from Antonini et al. (2016) and Juhász et al. (2017).

Open-label trials also present moderate evidence for beneficial effects on UPDRS parts II, III, and/or IV indicating effects on ADL function, motor symptoms, and improvement of motor complications (Wirdefeldt, Odin, & Nyholm, 2016).

The long-term effects of LCIG are still being investigated; however, recently there have been an increasing number of long-term open-label or registry-based studies presenting satisfactory and durable effects of long-term treatment (Buongiorno et al., 2015; Valldeoriola et al., 2016; Zibetti et al., 2014).

4.3 The Effect on Quality of Life

Health-related quality of life is commonly used as a secondary outcome in studies on LCIG. The treatment has repeatedly led to significant improvements in mean health-related quality of life in patients with advanced PD. The double-blind double-dummy RCT reported improved health-related quality of life, with a difference between LCIG and immediate-release levodopa of −7.0 for the PDQ39 summary score at 12 weeks of follow-up (Olanow et al., 2014). Another RCT also reported improved quality of life (median PDQ39 summary score 25 after 3 weeks LCIG treatment, vs 35 after 3 weeks of conventional treatment) (Nyholm et al., 2005). In total 12 of the 15 open-label studies that assessed health-related quality of life reported statistically significant improvement with LCIG, mostly of similar magnitude as the RCTs (Antonini et al., 2008, 2013, 2015; Bohlega et al., 2015; Caceres-Redondo et al., 2014; Chang et al., 2016; Fasano, Ricciardi, Lena, Bentivoglio, & Modugno 2012; Fernandez et al., 2015; Foltynie et al., 2013; Honig et al., 2009; Martinez-Martin et al., 2015; Pålhagen et al., 2012; Reddy et al., 2012; Sensi et al., 2014; Zibetti et al., 2013). The cause for the positive effect on health-related quality of life is probably multifactorial, but is likely to be a consequence of the reduced time in off and reduction of both motor and nonmotor complications.

4.4 Safety

LCIG is efficacious and the discontinuation rate is low, but adverse events and complications are common. Safety data from four studies on LCIG were recently merged and analyzed (Lang et al., 2015). The patients included had been treated with LCIG for a median of 911 days and the total exposure was 963 patient-years. The study made a distinction between procedure/device and nonprocedure/device adverse events. Adverse events that were associated to the procedure and/or device occurred in 76% of patients and complications of device insertion and abdominal pain were the most common—occurring in 41% and 36% of patients, respectively. Most patients, i.e., 92%, had some form of nonprocedure/device adverse event. Insomnia and falls were the most common, both reported by 23% of patients. Complications to device insertion and abdominal pain were also the most common serious procedure/device adverse events with a prevalence of 8% and 4%, respectively. Among non/procedure/device adverse events, pneumonia and PD symptoms were the most commonly reported in 5% and 2% of the patients. 0.5% of the participants died from a cause "possibly related" to the drug or device.

Polyneuropathy was reported in 5.8% of patients. The occurrence of peripheral neuropathy during LCIG treatment has been debated recently and there is growing evidence for the need of an active monitoring of symptoms of neuropathy (Devigili, Rinaldo, Lettieri, & Eleopra, 2016; Merola et al., 2016). The need for supplementation with vitamin B12 and folic acid should be evaluated regularly, but patients who do not improve from supplementation should be considered for discontinuation of LCIG.

4.5 The Indications, the Contraindications, and the "Ideal" Patient

Patients with advanced PD and severe motor fluctuations or dyskinesia despite optimized oral or transdermal treatment may come in question for LCIG. Like for CSA LCIG can, according to Odin et al. (2015), be considered in patients who experience more than 1–2 h of off symptomatology or troublesome dyskinesias in spite of at least 5 doses of peroral L-dopa per day. High age, dementia, and depression are not considered as contraindications for LCIG. Lack of L-dopa effect, insufficient capacity of patient/caregivers to handle a pump-based therapy and contraindications for abdominal surgery are the main contraindications for LCIG therapy. Other symptoms like hallucinations/delusions, impulse control disorders, insomnia, depression, nonmotor fluctuations (e.g., anxiety, pain, clouded thinking, apathy), dysarthria and restless legs can rather be seen as indication for the therapy (Odin et al., 2015). The fact that LCIG can be given as monotherapy might for many patients with complex medication schedules be seen as an advantage of the therapy.

5. LCIG: EFFECTS ON NONMOTOR PD SYMPTOMS

5.1 Effect of LCIG on Total Nonmotor Symptom Burden

Several open-label noncontrolled studies have evaluated the effects of LCIG on NMS using validated instruments, mostly the NMSS. A RCT with NMS as primary outcome is presently performed (Table 2).

The first report (Honig et al., 2009) included 22 PD patients followed for 6 months. The NMSS total score improved with 56% from 89.9 (baseline) to 39.4 points (6 months) ($P=0.0001$). Statistically significant improvement was seen in six of nine domains, including the cardiovascular, sleep/fatigue, attention/memory, gastrointestinal, urinary, and miscellaneous domains. The improvement of the NMSS total score correlated highly significantly with the change in HrQoL (PDQ-8).

In a study involving 17 patients treated over 6 months with LCIG and a comparator group of nine patients not given LCIG because of funding restrictions, Reddy et al. (2012) registered an improvement of NMSS total score in the LCIG group ($P=0.004$), but not in the comparator group. Improvements were seen in the subscores for sleep/fatigue, gastrointestinal,

Table 2 Effects Registered of LCIG on Nonmotor Symptoms, as Measured With the Nonmotor Symptom Scale (NMSS) in 10 Different Studies, All Open-Label

	Honig et al. (2009)	Reddy et al. (2012)	Fasano et al. (2012)	Sensi et al. (2014)		Caceres-Redondo et al. (2014)	Antonini et al. (2015)		Martinez-Martin et al. (2015)	Bohlega et al. (2015)	Ricciardi et al. (2016)[a]	Juhász et al. (2017)
Length of follow-up	6 mo	6 mo	25 mo	6 mo	24 mo	24 mo	6 mo	12 mo	6 mo	6 mo	26 mo	24 mo
Number of patients at baseline	22	17	14	28		29	375		44	20	8	34
Number of patients at follow-up	22	17	14	28	17	16	71[b]	59[b]	44	18	8	34
NMSST, before	89.9	113.9		51.8	51.8		75.3	75.3	90.95	237.1	134.0	88.9
NMSST, after	39.4			44.6	38.0		54.8	53.1	53.66	81.6	75	72.0
NMSST, % improvement	56%*	40%*	14%[c]	14%[c]	27%[c]	17%*	27%*	29%*	41%*	65%*	44%	19%*
Cardiovascular, *P*	0.0004	NS	NS	NS	NS	NS		NS	0.0076		NS	0.018
Sleep/fatigue, *P*	0.0001	0.02	NS	NS	NS	<0.05		0.0001	<0.0001		0.02	NS
Mood/cognition, *P*	0.02	NS	NS	NS	NS	NS	0.0426	NS	0.021		0.04	0.029
Perception/hallucinations, *P*	NS	NS	NS	NS	NS	NS		NS	0.010		0.04	NS

Continued

Table 2 Effects Registered of LCIG on Nonmotor Symptoms, as Measured With the Nonmotor Symptom Scale (NMSS) in 10 Different Studies, All Open-Label—cont'd

	Honig et al. (2009)	Reddy et al. (2012)	Fasano et al. (2012)	Sensi et al. (2014)	Caceres-Redondo et al. (2014)	Antonini et al. (2015)	Martinez-Martin et al. (2015)	Bohlega et al. (2015)	Ricciardi et al. (2016)	Juhász et al. (2017)
Attention/memory, *P*	0.002	NS	NS	NS NS	NS	NS	0.011		NS	NS
Gastrointestinal, *P*	0.0003	0.0012	NS	NS NS	<0.05	0.0096	<0.0001		NS	NS
Urinary, *P*	0.002	0.0015	NS	NS NS	NS	0.0199	0.0001		0.04	NS
Sexual, *P*	0.03	0.026	NS	NS NS	NS	NS	0.014		NS	NS
Miscellaneous, *P*	0.0004	NS	NS	NS NS	NS	NS	0.0008		NS	NS

[a]In the Ricciardi study, patients received LCIG infusion around the clock (24 h/day) throughout the study period.
[b]Interim analysis.
[c]Not significant.
P = P-value for change at follow-up vs baseline; NS = not significant; mo = months; *Significant ($P < 0.05$). The cardiovascular section of NMSS includes questions on orthostatic hypotension and syncope; the sleep/fatigue section includes questions on daytime sleepiness, fatigue, insomnia, restless legs, and REM-sleep behavior disorder; the mood/cognition section includes questions on apathia, anhedonia, and depression; the perception/hallucinations section includes questions on hallucinations and delusions; the attention/memory section includes questions on attention and memory; the gastrointestinal section includes questions on salivation, dysphagia, and constipation; the urinary section includes questions on urgency, frequency, and nocturia; the sexual section includes questions on lack of interest in sex and difficulties performing sex; the miscellaneous section includes questions on pain, difficulties to taste/smell, weight change, and excessive sweating.

urinary, and sexual function. The difference in change from baseline to follow-up between the LCIG group and the control group was significant for NMSS total score, as well as, the cardiovascular, sleep, gastrointestinal, urinary, sexual, and miscellaneous subscores ($P < 0.027$ for all).

Fasano et al. (2012) monitored 14 patients over a mean of 24.9 months. The authors found a statistical trend to improvement of the NMSS total score and the domain perception/hallucinations.

Sensi et al. (2014) followed 28 patients with LCIG treatment for 6 months, and 17 of these for 2 years. There was a tendency to improvement of NMSS total score from 51.8 at baseline to 44.6 after 6 months and 38.0 after 2 years, but these changes were not statistically significant. There were also no significant improvements among the NMSS subscores.

A long-term follow-up was performed by Caceres-Redondo et al. (2014). Twenty-nine patients were evaluated at baseline and 16 of these reached the follow-up evaluation at 24 months. A statistically significant improvement was registered for NMSS total score, as well as for the subscores for sleep/fatigue and gastrointestinal. Six further subscores showed a nonsignificant trend for improvement (mood/cognition, cardiovascular, perception/hallucinations, attention/memory, urinary, and miscellaneous). The sexual subscore showed a nonsignificant trend for worsening.

In a prospective long-term study, 375 patients were included and the 12 months interim results of the first 172 patients have been evaluated (Antonini et al., 2015). Compared to the baseline value (75.3), the NMSS total score was significantly reduced both at month 6 (54.8; $P = 0.0001$) and at month 12 (53.1; $P = 0.0014$). At 12 months, the subscores for the sleep/fatigue, gastrointestinal, and urinary domains were significantly reduced. The mood/cognition subscore was significantly improved after 6 months only.

The EuroInf study (Martinez-Martin et al., 2015; see earlier) included 44 patients on LCIG (mean age 62.7 years, disease duration 16.1 years). LCIG treatment resulted in highly significant improvements for UPDRS III ($P < 0.0001$), UPDRS IV ($P < 0.0001$), PDQ-8 (Parkinson's Disease Questionnaire, $P < 0.0001$), and NMSS total score ($P < 0.0001$). Significant improvements were seen in all NMSS subscores including the cardiovascular, sleep/fatigue, mood/apathy, perceptual, attention/memory, gastrointestinal, urinary, sexual functioning, and miscellaneous subscores. The Effect Size was moderate (0.5–0.79) concerning the sleep/fatigue, gastrointestinal, urinary, and miscellany subscores, and small (0.20–0.49) concerning the cardiovascular, mood/apathy, perceptual, attention/memory, and sexual functioning subscores. The lowest NNT values (<3) were seen for the cardiovascular (2.59), sleep/fatigue (1.63), mood/apathy (2.59), gastrointestinal (1.83), urinary (2.11), and miscellaneous (1.83) subscores.

Juhász et al. (2017) reported on 12-month outcomes of 34 patients starting LCIG therapy. The total score of the NMSS improved significantly ($P=0.027$). The cardiovascular and mood sections showed significant improvements.

Ricciardi et al. (2016) reported on eight patients who received 24h LCIG therapy over a mean of 26 months because of severe nocturnal akinesia. They registered significant improvements of the sleep/fatigue, mood/cognition, perception, and urinary subscores of NMSS, but not of the total NMSS score.

In comparison to the effects on motor symptoms, the evidence strength for the overall effects on nonmotor is, however, still weak and in need of further investigation (Wirdefeldt et al., 2016).

5.2 LCIG and Neuropsychiatric Symptoms

Concerning cognitive function authors report no worsening after starting LCIG (Fasano et al., 2012; Pickut et al., 2014). In single patients' improvement of cognitive function has been monitored (Sánchez-Castañeda et al., 2010). Long-term LCIG does not seem to change the clinical course of cognitive deterioration and worsening of dementia has been reported (Caceres-Redondo et al., 2014; Sensi et al., 2014; Zibetti et al., 2014).

Several NMSS-based investigations indicate a positive effect of LCIG therapy on the mood/cognition subscores (Honig et al., 2009; Juhász et al., 2017; Martinez-Martin et al., 2015; Ricciardi et al., 2016). There are, however, so far no investigations closer examining the LCIG effects on mood and depression.

Hallucinations and other psychotic symptoms seem to be less common under LCIG therapy relative conventional PD therapy. This is reflected by a significant improvement of the perception/hallucination subscore seen in the Martinez-Martin et al. (2015) and Ricciardi et al. (2016) studies.

5.3 LCIG and Gastrointestinal Symptoms

According to the NMSS results, gastrointestinal problems might belong to those commonly improving in patients receiving LCIG therapy. This can involve improvement of dribbling of saliva, swallowing, and constipation, monitored by NMSS, but probably also further gastrointestinal symptoms.

There is, however, so far a lack of dedicated studies examining these symptomatic effects of LCIG therapy.

At the same time, many of the side effects and complications of LCIG therapy concern the gastrointestinal tract. According to the evidence-based review of Wirdefeldt et al. (2016) (1244 patients from 25 studies), the most common complications were: dislocation of tube (160 reports); complication of device insertion (147 reports); abdominal pain (133 reports); irritation, granulation, or erythema at stoma (118 reports); infection at stoma (117 reports); occlusion, kinking, or obstruction of tube (117 reports); procedural pain (78 reports); constipation (58 reports); PEG internal retention failure (24 reports); peritonitis (20 reports); pneumoperitoneum (24 reports); problems leading to replacement of PEG (13 reports); accidental removal of tube (12 reports); and pump malfunction (27 reports). The most common gastrointestinal adverse events (with at least 10 AE reports) related to the LCIG infusion were nausea (65 reports) and weight loss (31 reports).

5.4 LCIG and Sleep

Improvement of sleep has, apart from NMSS, also been documented with the Parkinson's disease sleep scale, PDSS, version 1 and 2 (Fasano et al., 2012; Honig et al., 2009; Juhász et al., 2017; Ricciardi et al., 2016; Zibetti et al., 2014). According to Zibetti et al. (2014), the PDSS-2 subscores for disturbed sleep, motor symptoms at night, and Parkinson symptoms at night improved. According to the Epworth Sleepiness Scale, ESS, also measures of daytime sleepiness improved (Zibetti et al., 2014). Sleep seems to improve even if patients only get LCIG infusion daytime (Zibetti et al., 2014), but might further improve with 24h LCIG infusion (Nyholm et al., 2005).

5.5 LCIG and Sensory Symptoms

Both NMSS results and clinical experience indicate that LCIG treatment can result in improvement on some type of PD-related pain, but there is so far a fundamental lack of studies addressing this aspect of the therapy.

5.6 LCIG and Urinary and Sexual Dysfunction

Several of the results from studies using NMSS indicate a consistent effect on urological symptoms when applying LCIG therapy (Antonini et al., 2015; Honig et al., 2009; Martinez-Martin et al., 2015; Reddy et al., 2012; Ricciardi et al., 2016). The symptoms monitored by NMSS are urgency, frequency, and nocturia. There are so far no studies further exploring these effects of the therapy. Concerning sexual dysfunction one NMSS-based study (Martinez-Martin et al., 2015) indicate positive effects of LCIG treatment, but so far there are no other results confirming this.

5.7 LCIG and Impulse Control Disorders

Concerning impulse control disorders, ICD, LCIG has been reported to improve such symptoms in some cases; Catalán et al. (2013) reported on eight patients with severe ICD and Dopamine Dysregulation Syndrome, DDS, who all improved under LCIG treatment (mean treatment time: 25 weeks). Punding improved in all five and disappeared in 1 of 5 patients who had this. Todorova et al. (2015) reported on eight patients with active ICD, who all improved under LCIG over 3 years; the ICD symptoms disappeared in 6 out of 8 and there were no new cases of ICD under LCIG therapy. Fasano et al. (2012) reported of improvements of ICD symptoms in seven patients under 24 months ICD therapy; the symptoms disappeared in four of the patients. Ricciardi et al. (2016) reported on reduction of frequency and severity of ICD symptoms in eight patients treated with 24h LCIG infusion; the symptoms disappearing in 3 out of 6 patients presenting with such symptoms at start of therapy. However, one patient in this series had to stop LCIG treatment because of dopamine dysregulation problems and psychotic symptoms.

6. CONCLUSIONS

Both CSA and LCIG treatments may result in significant effects concerning nonmotor PD symptoms. This includes improvements of NM symptoms, but also, in some cases, side effects leading to appearance or worsening of NMS. There is, however, a lack of randomized controlled studies monitoring these effects and the evidence base for the effects is still not strong. There is also a need for studies focusing the effects on individual NMS, also examining the mechanisms of effect. Such mechanisms might include reduction of off time, a stronger dopaminergic effect, CSA/LCIG-specific effects, side effects of the treatments, as well as, a possibility to reduce other parts of the patients PD treatment, thereby potentially modifying effects and side effects of those treatments (Storch et al., 2015). An improved understanding of NMS fluctuations might contribute to the understanding of these mechanisms. Finally, it would be valuable with randomized trials comparing the advanced therapies, also concerning nonmotor effects, thereby forming a more solid base for choice of treatment for the individual patient.

REFERENCES

Antonini, A., Fung, V. S., Boyd, J. T., Slevin, J. T., Hall, C., Chatamra, K., et al. (2016). Effect of levodopa-carbidopa intestinal gel on dyskinesia in advanced Parkinson's disease patients. *Movement Disorders*, *31*(4), 530–537. http://dx.doi.org/10.1002/mds.26528. Epub 2016 Jan 28.

Antonini, A., Isaias, I. U., Rodolfi, G., Landi, A., Natuzzi, F., Siri, C., et al. (2011). A 5-year prospective assessment of advanced Parkinson disease patients treated with subcutaneous apomorphine infusion or deep brain stimulation. *Journal of Neurology*, *258*(4), 579–585.

Antonini, A., Mancini, F., Canesi, M., Zangaglia, R., Isaias, I. U., Manfredi, L., et al. (2008). Duodenal levodopa infusion improves quality of life in advanced Parkinson's disease. *Neurodegenerative Diseases*, *5*(3–4), 244–246.

Antonini, A., Odin, P., Opiano, L., Tomantschger, V., Pacchetti, C., Pickut, B., et al. (2013). Effect and safety of duodenal levodopa infusion in advanced Parkinson's disease: A retrospective multicenter outcome assessment in patient routine care. *Journal of Neural Transmission*, *120*(11), 1553–1558. http://dx.doi.org/10.1007/s00702-013-1026-9.

Antonini, A., Yegin, A., Preda, C., Bergmann, L., & Poewe, W. (2015). Global long-term study on motor and non-motor symptoms and safety of levodopa-carbidopa intestinal gel in routine care of advanced Parkinson's disease patients; 12-month interim outcomes. *Parkinsonism & Related Disorders*, *21*(3), 231. http://dx.doi.org/10.1016/j.parkreldis.2014.12.012.

Aranda, B., & Cramer, P. (1993). Effects of apomorphine and L-dopa on the parkinsonian bladder. *Neurourology and Urodynamics*, *12*(3), 203–209.

Bhidayasiri, R., Chaudhuri, K. R., LeWitt, P., Martin, A., Boonpang, K., & van Laar, T. (2015). Effective delivery of apomorphine in the management of Parkinson disease: Practical considerations for clinicians and Parkinson nurses. *Clinical Neuropharmacology*, *38*(3), 89–103.

Bohlega, S., Abou Al-Shaar, H., Alkhairallah, T., Al-Ajlan, F., Hasan, N., & Alkahtani, K. (2015). Levodopa-carbidopa intestinal gel infusion therapy in advanced Parkinson's disease: Single middle eastern center experience. *European Neurology*, *74*(5–6), 227–236. http://dx.doi.org/10.1159/000442151.

Borgemeester, R. W. K., Drent, M., & van Laar, T. (2015). Motor and non-motor outcomes of continuous apomorphine infusion in 125 Parkinson's disease patients. *Parkinsonism & Related Disorders*, *23*, 17–22. http://dx.doi.org/10.1016/j.parkreldis.2015.11.013.

Borgemeester, R. W. K., Lees, A. J., & van Laar, T. (2016). Parkinson's disease, visual hallucinations and apomorphine: A review of the available evidence. *Parkinsonism & Related Disorders*, *27*, 35. http://dx.doi.org/10.1016/j.parkreldis.2016.04.023.

Bronner, G., & Vodušek, D. B. (2011). Management of sexual dysfunction in Parkinson's disease. *Therapeutic Advances in Neurological Disorders*, *4*(6), 375–383. http://dx.doi.org/10.1177/1756285611411504.

Buongiorno, M., Antonelli, F., Cámara, A., Puente, V., de Fabregues-Nebot, O., Hernandez-Vara, J., et al. (2015). Long-term response to continuous duodenal infusion of levodopa/carbidopa gel in patients with advanced Parkinson disease: The Barcelona registry. *Parkinsonism and Related Disorders*, *21*(8), 871–876. http://dx.doi.org/10.1016/j.parkreldis.2015.05.014.

Caceres-Redondo, M. T., Carrillo, F., Lama, M. J., Huertas-Fernandez, I., Vargas-Gonzalez, L., Carballo, M., et al. (2014). Long-term levodopa/carbidopa intestinal gel in advanced Parkinson's disease. *Journal of Neurology*, *261*(3), 561–569.

Catalán, M. J., de Pablo-Fernández, E., Villanueva, C., Fernández-Diez, S., Lapeña-Montero, T., García-Ramos, R., et al. (2013). Levodopa infusion improves impulsivity and dopamine dysregulation syndrome in Parkinson's disease. *Movement Disorders*, *28*(14), 2007–2010. http://dx.doi.org/10.1002/mds.25636. Epub 2013 Oct 10.

Chang, F. C. F., Kwan, V., Van Der Poorten, D., Mahant, N., Wolfe, N., Ha, A. D., et al. (2016). Intraduodenal levodopa-carbidopa intestinal gel infusion improves both motor performance and quality of life in advanced Parkinson's disease. *Journal of Clinical Neuroscience, 25*, 41–45.

Chaudhuri, K. R., Abbott, R. J., & Millac, P. A. (1991). Subcutaneous apomorphine for parkinsonian patients with psychiatric side effects on oral treatment. *Journal of Neurology, Neurosurgery, and Psychiatry, 54*(4), 372–373.

Christmas, T. J., Kempster, P. A., Chapple, C. R., Frankel, J. P., Lees, A. J., Stern, G. M., et al. (1988). Role of subcutaneous apomorphine in parkinsonian voiding dysfunction. *The Lancet, 2*(8626–8627), 1451–1453.

Coelho, M., Ferreira, J., Rosa, M., & Sampaio, C. (2008). Treatment options for non-motor symptoms in late-stage Parkinson's disease. *Expert Opinion on Pharmacotherapy, 9*(4), 523–535.

Dellapina, E., Gerdelat-Mas, A., Ory-Magne, F., Pourcel, L., Galitzky, M., Calvas, F., et al. (2011). Apomorphine effect on pain threshold in Parkinson's disease: A clinical and positron emission tomography study. *Movement Disorders, 26*(1), 153–157.

Devigili, G., Rinaldo, S., Lettieri, C., & Eleopra, R. (2016). Levodopa/carbidopa intestinal gel therapy for advanced Parkinson disease: An early toxic effect for small nerve fibers? *Muscle and Nerve, 54*(5), 970–972. http://dx.doi.org/10.1002/mus.25377.

Drapier, S., Eusebio, A., Degos, B., Vérin, M., & Durif, F. (2016). Quality of life in Parkinson's disease improved by apomorphine pump: The OPTIPUMP cohort study. *Journal of Neurology, 263*(6), 1111. http://dx.doi.org/10.1007/s00415-016-8106-3.

Drapier, S., Gillioz, A. S., Leray, E., Peron, J., Rouaud, T., Marchand, A., et al. (2012). Apomorphine infusion in advanced Parkinson's patients with subthalamic stimulation contraindications. *Parkinsonism & Related Disorders, 18*(1), 40–44.

Edwards, L. L., Quigley, E. M., Harned, R. K., Hofman, R., & Pfeiffer, R. F. (1993). Defecatory function in Parkinson's disease: Response to apomorphine. *Annals of Neurology, 33*(5), 490–493.

Ellis, C., Lemmens, G., Parkes, J. D., Abbott, R. J., Pye, I. F., Leigh, P. N., et al. (1997). Use of apomorphine in parkinsonian patients with neuropsychiatric complications to oral treatment. *Parkinsonism & Related Disorders, 3*(2), 103–107.

Factor, S. A., Brown, D. L., & Molho, E. S. (2000). Subcutaneous apomorphine injections as a treatment for intractable pain in Parkinson's disease. *Movement Disorders, 15*(1), 167–169.

Fasano, A., Ricciardi, L., Lena, F., Bentivoglio, A. R., & Modugno, N. (2012). Intrajejunal levodopa infusion in advanced Parkinson's disease: Long-term effects on motor and non-motor symptoms and impact on patient's and caregiver's quality of life. *European Review for Medical and Pharmacological Sciences, 16*(1), 79–89.

Fernandez, H. H., Standaert, D. G., Hauser, R. A., Lang, A. E., Fung, V. S., Klostermann, F., et al. (2015). Levodopa-carbidopa intestinal gel in advanced Parkinson's disease: Final 12-month, open-label results. *Movement Disorders, 30*(4), 500–509.

Foltynie, T., Magee, C., James, C., Webster, G. J., Lees, A. J., & Limousin, P. (2013). Impact of duodopa on quality of life in advanced Parkinson's disease: A UK case series. *Parkinson's Disease, 2013*, 362908.

Garcia Ruiz, P. J., Ignacio, A. S., Pensado, B. A., Garcia, A. C., Frech, F. A., Lopez, M. A., et al. (2008). Efficacy of long-term continuous subcutaneous apomorphine infusion in advanced Parkinson's disease with motor fluctuations: A multicenter study. *Movement Disorders, 23*(8), 1130–1136.

Himeno, E., Ohyagi, Y., Ma, L., Nakamura, N., Miyoshi, K., Sakae, N., et al. (2011). Apomorphine treatment in Alzheimer mice promoting amyloid-beta degradation. *Annals of Neurology, 69*(2), 248–256.

Honig, H., Antonini, A., Martinez-Martin, P., Forgacs, I., Faye, G. C., Fox, T., et al. (2009). Intrajejunal levodopa infusion in Parkinson's disease: A pilot multicenter study of effects on nonmotor symptoms and quality of life. *Movement Disorders, 24*(10), 1468–1474.

Hunter, P. C., Crameri, J., Austin, S., Woodward, V., & Hughes, A. J. (1997). Response of parkinsonian swallowing dysfunction to dopaminergic stimulation. *Journal of Neurology, Neurosurgery, and Psychiatry*, *63*(5), 579–583.

Juhász, A., Aschermann, Z., Acs, P., Janszky, J., Kovács, M., Makkos, A., et al. (2017). Levodopa/carbidopa intestinal gel can improve both motor and nonmotor experiences of daily living in Parkinson's disease: An open-label study. *Parkinsonism & Related Disorders*, *37*, 79–86. http://dx.doi.org/10.1016/j.parkreldis.2017.02.001.

Katzenschlager, R., Hughes, A., & Evans, A. (2005). Continuous subcutaneous apomorphine therapy improves dyskinesias in Parkinson's disease: A prospective study using single-dose challenges. *Movement Disorders*, *20*(2), 151. http://dx.doi.org/10.1002/mds.20276.

Lang, A. E., Rodriguez, R. L., Boyd, J. T., Chouinard, S., Zadikoff, C., Espay, A. J., et al. (2015). Integrated safety of levodopa-carbidopa intestinal gel from prospective clinical trials. *Movement Disorders*, *31*(4), 538–546.

LeWitt, P. A. (2014). Levodopa therapy for Parkinson's disease: Pharmacokinetics and pharmacodynamics. *Movement Disorders*, *30*(1), 64. http://dx.doi.org/10.1002/mds.26082.

Lopiano, L., Modugno, N., Sensi, M., Meco, G., Cannas, A., Gusmaroli, G., et al. (2016). Motor outcomes in patients with advanced Parkinson's disease treated with levodopa/carbidopa intestinal gel in Italy: An interim analysis from the GREENFIELD observational study. *Neurological Sciences*, *37*(11), 1785–1792. http://dx.doi.org/10.1007/s10072-016-2664-0.

Magennis, B., Cashell, A., O'Brien, D., & Lynch, T. (2012). An audit of apomorphine in the management of complex idiopathic Parkinson's disease in Ireland. *Movement Disorders*, *27*(Suppl. 1), 144.

Martinez-Martin, P., Reddy, P., Antonini, A., Henriksen, T., Katzenschlager, R., Odin, P., et al. (2011). Chronic subcutaneous infusion therapy with apomorphine in advanced Parkinson's disease compared to conventional therapy: A real life study of non motor effect. *Journal of Parasitic Diseases*, *1*(2), 197–203. http://dx.doi.org/10.3233/JPD-2011-11037.

Martinez-Martin, P., Reddy, P., Katzenschlager, R., Antonini, A., Todorova, A., Odin, P., et al. (2015). EuroInf: A multicenter comparative observational study of apomorphine and levodopa infusion in Parkinson's disease. *Movement Disorders*, *30*(4), 510–516. http://dx.doi.org/10.1002/mds.26067.

Martinez-Martin, P., Rodriguez-Blazquez, C., Abe, K., Bhattacharyya, K. B., Bloem, B. R., Carod-Artal, F. J., et al. (2009). International study on the psychometric attributes of the non-motor symptoms scale in Parkinson disease. *Neurology*, *73*(19), 1584–1591.

Mathers, S. E., Kempster, P. A., Law, P. J., Frankel, J. P., Bartram, C. I., Lees, A. J., et al. (1989). Anal sphincter dysfunction in Parkinson's disease. *Archives of Neurology*, *46*(10), 1061–1064.

Merola, A., Romagnolo, A., Zibetti, M., Bernardini, A., Cocito, D., & Lopiano, L. (2016). Peripheral neuropathy associated with levodopa-carbidopa intestinal infusion: A long-term prospective assessment. *European Journal of Neurology*, *23*(3), 501–509. http://dx.doi.org/10.1111/ene.12846.

Morgante, L., Basile, G., Epifanio, A., Spina, E., Antonini, A., Stocchi, F., et al. (2004). Continuous apomorphine infusion (CAI) and neuropsychiatric disorders in patients with advanced Parkinson's disease: A follow-up of two years. *Archives of Gerontology and Geriatrics. Supplement*, (9), 291–296.

Muhiddin, K. A., Roche, M. T., & Pearce, V. R. (1994). Subcutaneous apomorphine infusion in Parkinson's disease: Does it have a role? *Postgraduate Medical Journal*, *70*(823), 344–346.

Nyholm, D., Askmark, H., Gomes-Trolin, C., Knutson, T., Lennernas, H., Nystrom, C., et al. (2003). Optimizing levodopa pharmacokinetics: Intestinal infusion versus oral sustained-release tablets. *Clinical Neuropharmacology*, *26*(3), 156–163.

Nyholm, D., Nilsson Remahl, A. I., Dizdar, N., Constantinescu, R., Holmberg, B., & Askmark, H. (2005). Duodenal levodopa infusion monotherapy vs oral polypharmacy in advanced Parkinson disease. *Neurology*, *64*(2), 216–223.

Odin, P., Chaudhuri, K. R., Slevin, J. T., Volkmann, J., Dietrichs, E., Martinez-Martin, P., et al. (2015). Collective physician perspectives on non-oral medication approaches for the management of clinically relevant unresolved issues in Parkinson's disease: Consensus from an international survey and discussion program. *Parkinsonism & Related Disorders*, *21*(10), 1133. http://dx.doi.org/10.1016/j.parkreldis.2015.07.020.

Olanow, C. W., Kieburtz, K., Odin, P., Espay, A. J., Standaert, D. G., Fernandez, H. H., et al. (2014). Continuous intrajejunal infusion of levodopa-carbidopa intestinal gel for patients with advanced Parkinson's disease: A randomised, controlled, double-blind, double-dummy study. *The Lancet. Neurology*, *13*(2), 141–149. http://dx.doi.org/10.1016/s1474-4422(13)70293-x.

Olanow, C. W., & Koller, W. C. (1998). An algorithm (decision tree) for the management of Parkinson's disease: Treatment guidelines. *Neurology*, *50*(3 Suppl. 3), S1. http://dx.doi.org/10.1212/wnl.50.3_suppl_3.s1.

Pålhagen, S. E., Dizdar, N., Hauge, T., Holmberg, B., Jansson, R., Linder, J., et al. (2012). Interim analysis of long-term intraduodenal levodopa infusion in advanced Parkinson disease. *Acta Neurologica Scandinavica*, *126*(6), e29–e33.

Pålhagen, S. E., Sydow, O., Johansson, A., Nyholm, D., Holmberg, B., Widner, H., et al. (2016). Levodopa-carbidopa intestinal gel (LCIG) treatment in routine care of patients with advanced Parkinson's disease: An open-label prospective observational study of effectiveness, tolerability and healthcare costs. *Parkinsonism & Related Disorders*, *29*, 17. http://dx.doi.org/10.1016/j.parkreldis.2016.06.002.

Pickut, B. A., van der Linden, C., Dethy, S., Van De Maele, H., & de Beyl, D. Z. (2014). Intestinal levodopa infusion: The Belgian experience. *Neurological Sciences*. http://dx.doi.org/10.1007/s10072-013-1612-5.

Pollak, P., Champay, A. S., Hommel, M., Perret, J. E., & Benabid, A. L. (1989). Subcutaneous apomorphine in Parkinson's disease. *Journal of Neurology, Neurosurgery, and Psychiatry*, *52*(4), 544.

Priano, L., Albani, G., Brioschi, A., Guastamacchia, G., Calderoni, S., Lopiano, L., et al. (2003). Nocturnal anomalous movement reduction and sleep microstructure analysis in parkinsonian patients during 1-night transdermal apomorphine treatment. *Neurological Sciences*, *24*(3), 207–208.

Rambour, M., Moreau, C., Salleron, J., & Devos, D. (2014). Continuous subcutaneous infusion of apomorphine in Parkinson's disease: Retrospective analysis of a series of 81 patients. *Revue Neurologique*, *170*(3), 205. http://dx.doi.org/10.1016/j.neurol.2013.10.012.

Reddy, P., Martinez-Martin, P., Rizos, A., Martin, A., Faye, G. C., Forgacs, I., et al. (2012). Intrajejunal levodopa versus conventional therapy in Parkinson disease: Motor and non-motor effects. *Clinical Neuropharmacology*, *35*(5), 205–207. http://dx.doi.org/10.1097/WNF.0b013e3182613dea.

Reuter, I., Ellis, C. M., & Chaudhuri, K. R. (1999). Nocturnal subcutaneous apomorphine infusion in Parkinson's disease and restless legs syndrome. *Acta Neurologica Scandinavica*, *100*(3), 163–167.

Ribarič, S. (2012). The pharmacological properties and therapeutic use of apomorphine. *Molecules*, *17*(5), 5289–5309. http://dx.doi.org/10.3390/molecules17055289.

Ricciardi, L., Bove, F., Espay, K. J., Lena, F., Modugno, N., Poon, Y.-Y., et al. (2016). In N. Gálvez-Jiménez, et al. (Eds.), *Parkinson's disease: Current and future therapeutics and clinical trials* (pp. 597–598). Cambridge, UK: Published by Cambridge University Press, © Cambridge University Press.

Ruggieri, S., Stocchi, F., Carta, A., & Agnoli, A. (1989). Side-effects of subcutaneous apomorphine in Parkinson's disease. *The Lancet*, *1*(8637), 566.

Sánchez-Castañeda, C., Campdelacreu, J., Miró, J., Juncadella, M., Jaumà, S., & Calopa, M. (2010). Cognitive improvement after duodenal levodopa infusion in cognitively impaired Parkinson's disease patients. *Progress in Neuropsychopharmacology and Biological Psychiatry*, *34*(1), 250–251. http://dx.doi.org/10.1016/j.pnpbp.2009.10.021.

Schapira, A. H. V., & Obeso, J. (2006). Timing of treatment initiation in Parkinson's disease: A need for reappraisal? *Annals of Neurology*, *59*(3), 559. http://dx.doi.org/10.1002/ana.20789.

Schrag, A. (2000). Dyskinesias and motor fluctuations in Parkinson's disease: A community-based study. *Brain*, *123*(11), 2297. http://dx.doi.org/10.1093/brain/123.11.2297.

Sensi, M., Preda, F., Trevisani, L., Contini, E., Gragnaniello, D., Capone, J. G., et al. (2014). Emerging issues on selection criteria of levodopa carbidopa infusion therapy: Considerations on outcome of 28 consecutive patients. *Journal of Neural Transmission (Vienna)*, *121*(6), 633–642.

Storch, A., Schneider, C. B., Klingelhöfer, L., Odin, P., Fuchs, G., Jost, W. H., et al. (2015). Quantitative assessment of non-motor fluctuations in Parkinson's disease using the Non-Motor Symptoms Scale (NMSS). *Journal of Neural Transmission (Vienna)*, *122*(12), 1673–1684. http://dx.doi.org/10.1007/s00702-015-1437-x. Epub 2015 Aug 12.

Suttrup, I., & Warnecke, T. (2016). Dysphagia in Parkinson's disease. *Dysphagia*, *31*(1), 24–32.

Timpka, J., Fox, T., Fox, K., Honig, H., Odin, P., Martinez-Martin, P., et al. (2016). Improvement of dyskinesias with l-dopa infusion in advanced Parkinson's disease. *Acta Neurologica Scandinavica*, *133*(6), 451–458. http://dx.doi.org/10.1111/ane.12483.

Tison, F., Wiart, L., Guatterie, M., Fouillet, N., Lozano, V., Henry, P., et al. (1996). Effects of central dopaminergic stimulation by apomorphine on swallowing disorders in Parkinson's disease. *Movement Disorders*, *11*(6), 729–732.

Todorova, A., Bitsara, C., Teleg, E. S. A., & Chaudhuri, K. R. (2015). Impulse control dysfunction and dopamine agonist withdrawal syndrome in genetic Parkinson's: A case report. *Basal Ganglia*, *5*(2–3), 57–58.

Todorova, A., & Chaudhuri, K. R. (2016). In N. Gálvez-Jiménez, et al. (Eds.), *Parkinson's disease: Current and future therapeutics and clinical trials* (pp. 48–62). Cambridge, UK: Published by Cambridge University Press, Cambridge University Press.

Todorova, A., Martin, A., Okai, D., Samuel, M., Brown, R., David, A., et al. (2013). Assessment of impulse control disorders in Parkinson's patients with infusion therapies: A single center experience. *Movement Disorders*, *28*(Suppl. 1), 366.

Trenkwalder, C., Chaudhuri, K. R., Ruiz, P. J. G., LeWitt, P., Katzenschlager, R., Sixel-Döring, F., et al. (2015). Expert consensus group report on the use of apomorphine in the treatment of Parkinson's disease—Clinical practice recommendations. *Parkinsonism & Related Disorders*, *21*(9), 1023–1030. http://dx.doi.org/10.1016/j.parkreldis.2015.06.012.

Valldeoriola, F., Grandas, F., Santos-Garcia, D., Regidor, I., Catalan, M. J., Arbelo, J. M., et al. (2016). Long-term effectiveness of levodopa-carbidopa intestinal gel in 177 Spanish patients with advanced Parkinson's disease. *Neurodegenerative Disease Management*, *6*(4), 289–298. http://dx.doi.org/10.2217/nmt-2016-0021.

van Laar, T., Postma, A. G., & Drent, M. (2010). Continuous subcutaneous infusion of apomorphine can be used safely in patients with Parkinson's disease and preexisting visual hallucinations. *Parkinsonism & Related Disorders*, *16*(1), 71–72.

Volkmann, J., Albanese, A., Antonini, A., Chaudhuri, K. R., Clarke, C. E., de Bie, R. M., et al. (2013). Selecting deep brain stimulation or infusion therapies in advanced Parkinson's disease: An evidence-based review. *Journal of Neurology*, *260*(11), 2701–2714. http://dx.doi.org/10.1007/s00415-012-6798-6.

Wirdefeldt, K., Odin, P., & Nyholm, D. (2016). Levodopa–carbidopa intestinal gel in patients with Parkinson's disease: A systematic review. *CNS Drugs*, *30*(5), 381. http://dx.doi.org/10.1007/s40263-016-0336-5.

Yarnall, A. J., Lashley, T., Ling, H., Lees, A. J., Coleman, S. Y., O'Sullivan, S. S., et al. (2016). Apomorphine: A potential modifier of amyloid deposition in Parkinson's disease? *Movement Disorders*, *31*(5), 668–675.

Zibetti, M., Merola, A., Artusi, C. A., & Rizzi, L. (2014). Levodopa/carbidopa intestinal gel infusion in advanced Parkinson's disease: A 7-year experience. *European Journal of Neurology*, *21*(2), 312. http://dx.doi.org/10.1111/ene.12309.

Zibetti, M., Merola, A., Ricchi, V., Marchisio, A., Artusi, C. A., Rizzi, L., et al. (2013). Long-term duodenal levodopa infusion in Parkinson's disease: A 3-year motor and cognitive follow-up study. *Journal of Neurology*, *260*(1), 105–114.

CHAPTER THIRTY-SEVEN

Deep Brain Stimulation and Nonmotor Symptoms

Elliot Hogg, Jeffrey Wertheimer, Sarah Graner, Michele Tagliati[1]
Cedars-Sinai Medical Center, Los Angeles, CA, United States
[1]Corresponding author: e-mail address: michele.tagliati@cshs.org

Contents

Abstract

Deep brain stimulation (DBS) is currently the treatment of choice for advanced Parkinson's disease (PD). Several brain targets, including the subthalamic nucleus and the globus pallidus internus, have been successfully employed, with excellent motor outcomes. Despite less established knowledge, DBS may be a powerful tool for managing a wide variety of nonmotor symptoms (NMS) in PD patients, either directly or indirectly due to motor benefit or reduction of dopaminergic drug load. After an assessment of global nonmotor outcomes of DBS, as measured by currently available clinical scales and questionnaires, this chapter will address DBS effects on four main NMS categories: *neurobehavioral*, including cognitive and neuropsychiatric symptoms, *autonomic dysfunction*, including orthostatic hypotension, constipation, and urinary dysfunction, *sleep disturbances*, including insomnia, REM sleep behavior disorder, and restless leg syndrome, to conclude with *sensory symptoms*, mainly focusing on pain. An overall positive impact of DBS on most NMS emerges from the reviewed studies.

International Review of Neurobiology, Volume 134
ISSN 0074-7742
http://dx.doi.org/10.1016/bs.irn.2017.05.022

However, current opinion on the effect of DBS on NMS in PD needs to be tempered by the relatively low number of cases and the lack of large, controlled, specifically designed studies for most NMS categories.

1. INTRODUCTION

Despite advances in Parkinson's disease (PD) treatment, nonmotor symptoms (NMS) continue to represent a major barrier in advanced PD. NMS are a strikingly diverse set of symptoms, which are extremely common and appear across all phases of the disease. They may predate the development of motor symptoms (Lim & Lang, 2010) and evolve inexorably over the course of the disease, with upward of 50% of patients experiencing NMS by 20 years (Chaudhuri, Healy, & Schapira, 2006; Hely, Reid, Adena, Halliday, & Morris, 2008).

As the neuropathology of PD evolves across widespread regions of the central and peripheral nervous systems, NMS grow increasingly disabling (Lim & Lang, 2010), although they remain largely underrecognized by health care providers (Muzerengi et al., 2006; Parsons, Rogers, Braaten, Woods, & Troster, 2006). Understanding the role of deep brain stimulation (DBS) in managing these symptoms is important not only to preserve the quality of life in advanced PD, but even to improve survival, as NMS can be a significant predictor of mortality (Chaudhuri et al., 2006; Schrag, Sauerbier, & Chaudhuri, 2015).

DBS is currently considered the most effective treatment for advanced PD, particularly when frequent motor fluctuations and dyskinesias start affecting the quality of life (Bronstein et al., 2011). Current practice of DBS for PD encompasses two extremely well-tested targets, the subthalamic nucleus (STN) and globus pallidus internus (GPi), both with strong evidence for significant motor outcomes (Deuschl et al., 2006; Follett et al., 2010; Odekerken et al., 2013; Okun et al., 2012). However, our knowledge of nonmotor outcomes after DBS is less established, in great part because most of the large trials that established the success of DBS therapy for PD were designed to predominantly assess motor outcomes, and did not include comprehensive analysis of potential nonmotor effects (Anderson, Burchiel, Hogarth, Favre, & Hammerstad, 2005; Follett et al., 2010; Kahn et al., 2012; Nakamura, Christine, Starr, & Marks, 2007; Tir et al., 2007; Weaver, Follett, Hur, Ippolito, & Stern, 2005).

Nevertheless, DBS may be a powerful tool for managing NMS as striking improvements in sensory, cognitive, and dysautonomic symptoms after bilateral STN DBS have been reported (Witjas, Kaphan, Regis, et al., 2007). The presence of nonmotor fluctuations (NMF) or nonmotor OFF periods has been suggested as a reason for recommending STN DBS for patients (Witjas, Kaphan, et al., 2007). However, treatment response remains debated and even levodopa-responsive NMF may not improve after STN DBS, even in patients with excellent motor response (Borgohain, Kandadai, Jabeen, & Kannikannan, 2012; Cersosimo, Piedimonte, Raina, & Micheli, 2007). As DBS is considered particularly effective for levodopa-responsive symptoms, it is worth remembering that levodopa-unresponsive NMS are perhaps the greatest challenge in current PD care (Lim & Lang, 2010).

While direct effects of DBS on NMS are debated, some of the benefits may be indirect. For example, some NMS appear to be caused or exacerbated by dopaminergic therapy and may therefore benefit from the reduced medication load associated with STN DBS. This can be especially helpful for psychosis, orthostatic hypotension (OH), daytime somnolence, or impulse control disorders (ICDs) (Ardouin et al., 2006; Bandini, Primavera, Pizzorno, & Cocito, 2007; Lim & Lang, 2010; Molina et al., 2000; Witjas et al., 2005; Witjas, Regis, Delphini, Peragut, & Azulay, 2007). In other cases reducing medications may lead to new-onset NMS, such as apathy, depression, or restless legs syndrome (RLS) (Kedia, Moro, Tagliati, Lang, & Kumar, 2004; Lhommée et al., 2012; Okun et al., 2005; Thobois et al., 2010). Given the increasing role of neurostimulation in the care of advanced PD, an understanding of the effects (both for good and ill) will be vital in the care of PD NMS moving forward.

In this chapter, we will review important studies addressing NMS in relation to DBS therapy. After an assessment of the general nonmotor outcomes of DBS, as measured by currently available clinical scales and questionnaires, we will address DBS effects in four main NMS categories: neurobehavioral, autonomic, sleep, and sensory (Chaudhuri et al., 2006; Kurtis, Rajah, Delgado, & Dafsari, 2017; Lim & Lang, 2010).

2. NONMOTOR OUTCOMES OF DBS

Before delving into specific NMS, it is worthwhile to take an overall view of the general nonmotor outcomes of DBS. Several nonmotor symptoms scale (NMSS) are available to encompass a broad cross section of PD patient experience (Chaudhuri & Martinez-Martin, 2008). These have provided a good opportunity to look at global DBS impact on NMS across several studies and across different centers and over time.

In a study of 24 PD patients followed for 1 year after bilateral STN DBS and evaluated using the nonmotor symptom questionnaire (NMSQ), subjects reported an average decrease in the number of NMS from 12 before surgery to 7 after chronic stimulation, with autonomic symptoms showing the greatest reduction. Overall 27/30 items on the NMSQ were less frequently reported postsurgery, suggesting that DBS was able to significantly decrease NMS with improved quality of life (Nazzaro, Pahwa, & Lyons, 2011). Similar results were reported by a smaller study following 10 PD patients, 9 of whom with STN implants and one with GPi DBS, for an average of 12 months. Outcomes revealed that the NMSQ improved an average of 3.7 points, with the largest improvement in mood, miscellaneous symptoms (pain, anosmia, weight change, and sweating), and cardiovascular/falls subscores (Hwynn et al., 2011). A significant average improvement of 1.7 symptoms in the NMSQ was reported in a group of 56 PD patients with bilateral STN DBS followed for at least 1 year and compared to a control group on best medical therapy (BMT) (Mridula et al., 2015). Symptoms in the cardiovascular, gastrointestinal, and sleep domains were significantly less frequent, while sexual disturbances were significantly more frequent in the DBS group (Mridula et al., 2015).

Using the NMSS, 10 PD patients were followed over 3–6 months following bilateral STN DBS, resulting in a significant 36% reduction of mean NMSS scores following surgery, with particular benefits in the domains of sleep, mood/cognition, urinary, sexual, and miscellaneous (Reich et al., 2011). In a group of 25 consecutive STN DBS patients, followed 12 months postoperatively, most domains of NMSS improved in parallel with motor symptoms, with sleep benefiting the most (Deli et al., 2015). Using a structured questionnaire designed to assess NMF, a significant reduction of the total number of reported NMF was observed in 40 PD patients examined 12 months following STN DBS, with pain/sensory (84.2%) and dysautonomic and cognitive fluctuations (>60%) showing the best response (Witjas, Kaphan, et al., 2007). Using a similar questionnaire, a significant reduction in the total number of NMF was also reported in 17 PD patients followed over 2 years after STN DBS. Interestingly, turning OFF stimulation in the OFF-medication state worsened motor function but not NMF (Ortega-Cubero et al., 2013).

Finally, the multicenter, prospective, international EuroInf study used NMSS and NMSQ in 60 STN DBS patients who were followed for 6 months postoperatively. Post hoc analyses showed a significant reduction of sleep/fatigue and miscellaneous domains, perceptual problems/hallucinations, and urinary scores, with an overall moderate effect size on the NMSS, suggesting that STN DBS improves the NMS burden in PD (Dafsari et al., 2016). This global impact on NMS is illustrated in radar graphic demonstrating the effect of DBS in this large cohort (see Fig. 1). A summary of nonmotor outcomes of DBS is presented in Table 1. While the overall positive impact on most NMS is clear, it is important to remember that the number of patients evaluated is still relatively low.

3. NEUROBEHAVIORAL PROFILES AND DBS

Although in contemporary times DBS therapy is thought to be a relatively safe intervention from cognitive and psychiatric perspective for the appropriate DBS candidate, potential adverse neurobehavioral and neuropsychiatric outcomes have been observed since ablative surgical intervention was introduced for a treatment of PD. Ablative operations dating back to the 1940s and 1950s, including pallidotomy, thalamotomy, and subthalamotomy, were associated with significant cognitive morbidity (de Bie et al., 2002; Laitinen, 2000). However, more recent unilateral ablative procedures, regardless of the target (thalamus, STN, or GPi), were thought to be significantly safer from a cognitive standpoint (Baron, Vitek, Bakay, et al., 1996; Troster & Woods, 2003). Some cognitive decrements have been identified with unilateral subthalamotomy (i.e., left-sided stimulation ensued in decrements in selective attention, response inhibition, and memory for verbal information; Patel et al., 2003). Unilateral pallidotomy commonly resulted in verbal fluency deficits (i.e., Cahn, Sullivan, Shear, et al., 1998; Lacritz, Cullum, Frol, Dewey, & Giller, 2000; Troster, Woods, Fields, et al., 2003). Although major cognitive changes following ablative procedures have been rarely reported in current times, mild and typically transient changes in attention, memory, and executive function decrements may occur occasionally (Troster & Fields, 2008).

Since the inception of DBS therapy as an alternative to ablative procedures, attention toward neurobehavioral sequelae following surgical intervention has burgeoned. Although a few studies failed to report adverse effects of DBS on behavior (Ardouin, Pillon, Peiffer, et al., 1999;

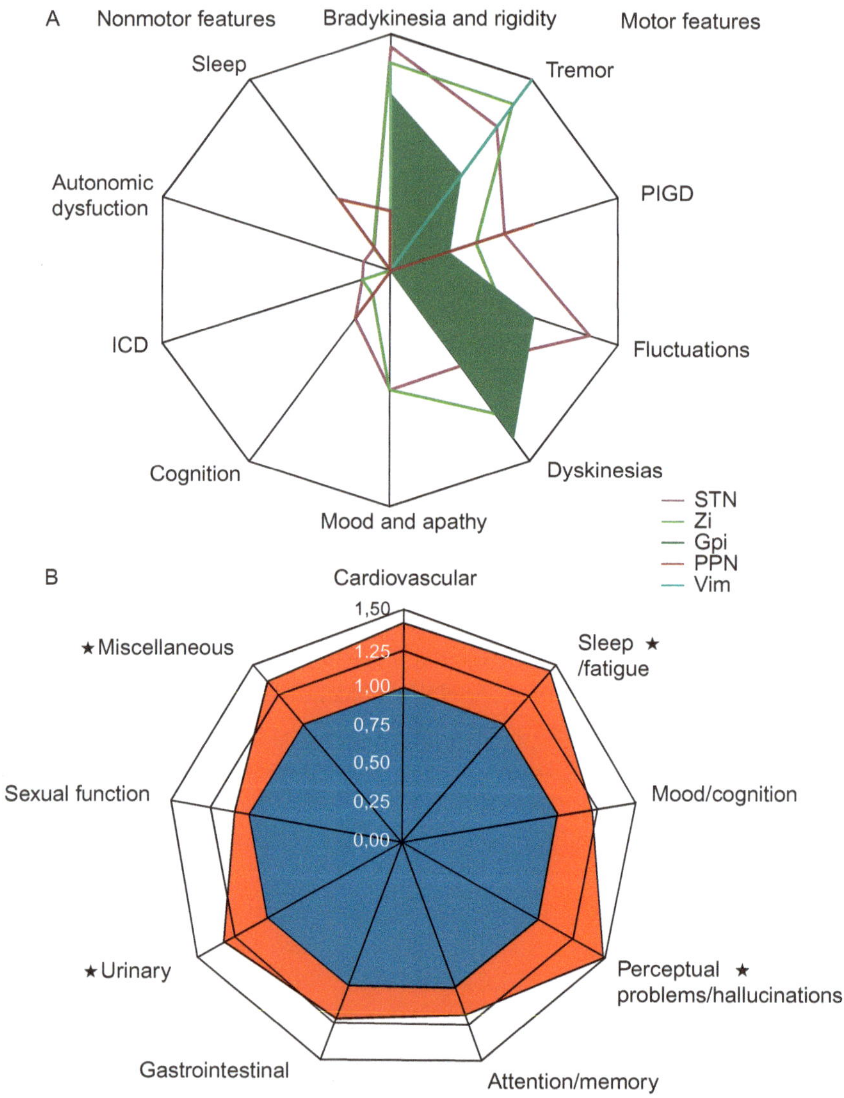

Fig. 1 Synoptic diagrams of (A) different motor and nonmotor effects of deep brain stimulation at various targets and (B) DBS effects on different subdomains of the non-motor symptoms scale (NMSS), where the *blue area* represents baseline values and the *red* reflects NMSS domain improvement at 2–6 months of follow-up. *Panel (A): From Fasano, A., Daniele, A., & Albanese, A. (2012). Treatment of motor and non-motor features of Parkinson's disease with deep brain stimulation.* Lancet Neurology, 11, *429–442, with permission. Panel (B): From Dafsari, H. S., Reddy, P., Herchenbach, C., Wawro, S., Petry-Schmelzer, J. N., Visser-Vandewalle, V., et al. (2016). Beneficial effects of bilateral subthalamic stimulation on non-motor symptoms in Parkinson's disease.* Brain Stimulation, *9(1), 78–85, with permission.*

Table 1 Selected Studies Evaluating the Impact of DBS on Nonmotor Symptoms

Source	Nazzaro et al. (2011)[a]	Reich et al. (2011)	Hwynn et al. (2011)	Deli et al. (2015)	Dafsari et al. (2016)	Total
Number of subjects (target)	24 (STN)	10 (STN)	9 (STN) 1 (GPi)	25 (STN)	60 (STN)	129
Postoperative follow-up interval	12 months	3–6 months	6 months	12 months	6 months	8-month weighted average follow-up interval
NMSS subsections:	% Change	% Change	% Change	% Change	% Change	Weighted average % change
CV and falls	−28.6	−7.1	−50.0	−41.9	−36.5	−34.8
Sleep/fatigue	−35.6	−40.4	−0.0	−36.8	−43.7	−37.2
Mood/cognition	−68.7	−38.2	−10.0	−27.6	−15.0	−28.9
Hallucination/perception	−100	−31.4	0.0	33.3	−48.1	−36.9
Attention/memory	−30.3	−20.7	−10.0	−30.2	−17.1	−21.8
GI	−32.4	−1.3	−30.0	−39.0	−23.3	−26.8
Urinary	−41.2	−36.2	10.0	−44.9	−27.0	−31.0
Sexual	−36.7	−44.2	4.4	11.6	−8.9	−11.8
Miscellaneous	−60	−63.2	−30.0	−48.1	−34.9	−43.95

[a]Based on a conversion of NMSQ to NMSS: Rukmini et al. (2015) and Chaudhuri et al. (2006).

Perozzo, Rizzone, Bergamasco, Castelli, et al., 2001; Schneider, Habel, Volkmann, et al., 2003), neurobehavioral complications of DBS surgery have included transient confusion, circumscribed cognitive decrements, depression, suicidal tendencies, anxiety, apathy, hypomania, ICDs, and psychosis (Burn & Troster, 2004; Rothlind, York, Carlson, et al., 2015; Voon, Kubu, Krack, Houeto, & Troster, 2006). The range of adverse neurobehavioral outcomes following surgical intervention is summarized in Table 2. In the following sections, we will review neurobehavioral symptoms associated with DBS surgery in PD, including outcomes described for both GPi and STN surgical targets.

Table 2 Possible Neurobehavioral Effect of Deep Brain Stimulation for Parkinson's Disease

DBS Target	Possible Beneficial Effect	Possible Side Effect
Subthalamic nucleus	Improved depression	Depression
	Improved anxiety	Anxiety
	Improved cognition (i.e., visuospatial functioning)	Apathy
		Hypomania
		Psychosis
		Hypersexuality
		Dopamine dysregulation syndrome
		Suicidality
		Reduced cognition (word fluency; executive control; learning and memory)
Globus pallidus internus	Improved depression	Depression
	Improved anxiety	Anxiety
	Improved cognition (i.e., mild memory improvement; visuospatial functioning)	Hypomania
		Reduced cognition (word fluency; executive control; learning and memory)

3.1 Cognition

Cognitive impairment is common in PD, with report of cognitive deficits manifesting for some people with Parkinson's prior to or at the time of diagnosis (Elgh et al., 2009; Pont-Sunyer, Hotter, Gaig, Seppi, et al., 2015). Cognitive deficits in the early stages of PD tend to correlate with frontostriatal circuit dysfunction with particular manifestation commensurate with dysfunction in the prefrontal cortex (Bondi, Kaszniak, Bayles, & Vance, 1993; Taylor, Saint-Cyr, & Lang, 1986). Although cognitive impairment is often perceived as heterogeneous for individuals with PD, a predominant pattern of cognitive dysfunction in PD involves slowed processing speed, difficulties with concentrated attention and executive control, memory, and visuospatial functions (Litvan et al., 2012; Marras, Troster, Kulisevsky, & Stebbins, 2014). Cognitive correlates with prefrontal dysfunction may account for many of the cognitive deficits manifested in PD (Dubois, Boller, Pillon, & Agid, 1991). Mild cognitive impairment (PD-MCI) is common in nondemented patients, ranging from 19% to 38% (mean prevalence, 27%; Litvan et al., 2012). Long-term longitudinal studies report that most PD patients develop dementia (PDD), with dementia affecting up to 80% of patients with PD over the course of their illness (Aarsland, Anderson, Larsen, Lolk, & Kragh-Sorensen, 2003). Dementia is the most frequent exclusion criterion for DBS surgery (Bronstein et al., 2011), and conceptualization of cognitive profiles for individuals who are deemed appropriate candidates for DBS is based on individuals without cognitive difficulties or with mild to moderate cognitive difficulties not reaching classification of dementia.

Since the early years of DBS therapy for individuals with PD, there have been mixed findings regarding the impact of DBS on cognition. Improvements in cognitive processes, such as enhanced cognitive flexibility and visuomotor sequencing, have been reported in early studies (Ardouin et al., 1999; Pillon, Ardouin, Damier, et al., 2000). However, adverse cognitive changes have also been reported and are commonly cited in the literature. In a randomized controlled study comparing bilateral DBS vs BMT for individuals with advanced PD, there was a statistically significant treatment difference between baseline and follow-up, where the DBS group demonstrated mild decrements in performance as compared to the BMT on composite measures of working memory, processing speed, phonemic fluency, and delayed recall on a memory for visual information task (Weaver et al., 2009). The likelihood of significant decline in neuropsychological functioning increases with DBS (STN and GPi combined) as compared to BMT, though affecting a small minority of patients who also appear to respond less optimally to DBS (Rothlind et al., 2015). Compared to the group receiving BMT, the combined DBS group had significantly greater mean reductions at 6-month follow-up in performance on multiple measures of processing speed and working memory. Eleven percent of the DBS group displayed reliable decline by multiple indicators in two or more cognitive domains, a significantly higher rate than in the BMT group (3%).

Research has reported that unilateral GPi DBS is relatively safe from a neurobehavioral standpoint (for review, see Fields & Troster, 2000), although postoperative declines in verbal fluency have been noted as a possibility (Merello, Nouzeilles, Kuzis, et al., 1999). Bilateral GPi DBS has also been reported to have minimal, if any adverse outcomes on cognition (Ardouin et al., 1999; Follett et al., 2010). In contrast, other research has demonstrated increased probability of cognitive adverse side effects for STN DBS when comparing unilateral pallidotomy and bilateral STN DBS (Gironell, Kulisevsky, Rami, et al., 2003; Smeding, Esselink, Schmand, et al., 2005) and bilateral pallidal and STN DBS (Anderson et al., 2005; Volkmann, Allert, Voges, et al., 2001).

Findings regarding possible declines and/or improvements in global cognitive abilities, memory, attention, and executive functions after STN DBS are somewhat inconsistent, and the majority of studies show only circumscribed, if any, cognitive changes (Deuschl et al., 2006; Kleiner-Fisman et al., 2006). Cognitive decrements following DBS implantation mainly manifest in decrements involving attention and phonemic fluency tasks (Foki, Hitzl, Pirker, et al., 2017). As in pallidotomy and GPi DBS, verbal fluency decrements appear to be the most commonly reported change for STN DBS (Limousin et al., 1997). Mild-to-moderate declines in verbal fluency that persist over time have been found in 30%–50% of patients after having DBS implantation (Funkiewiez, Ardouin, Caputo, et al., 2004). In a meta-analysis of cognitive sequelae following STN DBS (Parsons et al., 2006), after adjusting for heterogeneity of variance in study effect sizes, the analyses revealed significant, albeit small, declines in executive functions and verbal learning and memory. Moderate declines were reported in semantic and phonemic verbal fluency, after controlling for age, disease duration, stimulation parameters, and change in dopaminergic dose. Such findings may not be surprising given that the DBS electrodes traverse through the prefrontal cortex and subcortical white matter when implanted (Weintraub & Burn, 2011). Furthermore, the variability regarding cognitive sequelae following surgery appears to be related to the angle of the surgical trajectory and the proximity of the electrodes to the target (York, Wilde, Simpson, & Jankovic, 2009). In a controlled study using constant-current bilateral STN stimulation, 101 PD patients received stimulation immediately after device implantation, while 35 started stimulation 3 months after surgery (Okun et al., 2012). The study showed a good cognitive safety profile, with the exception of decrements in verbal fluency and Stroop task scores, which appeared to be related to device implantation, although stimulation likely had an additive effect (Tröster, Jankovic, Tagliati, Peichel, & Okun, 2017). Modifying stimulation parameters has been found to reverse the cognitive decrement occurring following DBS therapy (Frankemolle, Wu, Noecker, et al., 2010), indicating that some DBS-induced cognitive decrements may be modifiable.

Comparison of the two DBS targets, STN and GPi, has demonstrated mixed results. The COMPARE trial showed no significant differences on cognitive outcome measures between STN DBS and GPi DBS in the optimal DBS state (Okun et al., 2009), similar to a larger randomized controlled trial (Odekerken et al., 2013). Other efforts revealed few significant group differences, with the STN DBS group showing greater reductions in processing speed when compared to the GPi, and the GPi DBS group displaying lower mean performance on one measure of learning and memory that placed greater demands on executive control (Rothlind et al., 2015). Another randomized controlled study found that the GPi and STN DBS groups had similar slight decrements in neurocognitive function at a 24-month follow-up, with the exception of a greater decrement in processing speed for the STN DBS group (Follett et al., 2010).

In conclusion, the neuropsychological literature suggests that cognitive declines associated with DBS surgery are minimal when compared to the robust motor improvements (Follett et al., 2010; Kumar, Lozano, Sime, et al., 1999; Weaver et al., 2009). Despite some cognitive risks, both STN and GPi targets seem to be relatively safe from a cognitive vantage point for those PD patients who are determined to be good candidates (Bronstein et al., 2011; Follett et al., 2010; Kurtis et al., 2017; Parsons et al., 2006; Rothlind et al., 2015). As noted in a critical review of the evidence investigating the effect of DBS on cognition, "The main risk factors associated with cognitive decline and surgery have been subdivided into (1) preoperative risk factors: impaired attention, higher levodopa equivalent dosages and motor impairment, older age, higher motor, and severe axial symptoms; (2) intraoperative risk factors such as surgical electrode trajectories (through the frontal lobe and head of the caudate nucleus); and (3) postoperative factors including stimulation, parameters, electrode location, and medication changes" (Kurtis et al., 2017).

3.2 Neuropsychiatric Complications

Similar to the relative cognitive safety of DBS, post-DBS psychiatric status may also improve (Voon et al., 2006). However, psychiatric complications following DBS therapy have been documented and may include emotional reactivity, depression, apathy, hypomania, hypersexuality, disinhibition, and psychosis.

3.2.1 Emotional Reactivity, Depression, and Anxiety

Individuals with PD who have a history of psychiatric disturbance may be vulnerable to emotional hyperreactivity, psychiatric decompensation, and social maladjustment following DBS (Houeto, Mesnage, Mallet, et al., 2002). Base rates for depression are relatively high in PD, with major depression occurring in 5%–20%, with an additional 10%–30% of individuals with PD experiencing minor or subsyndromal depression (Allain, Schuck, & Manduit, 2000; Tandberg, Larsen, Aarsland, Laake, & Cummings, 1996). Depression in PD likely results from a complex interaction of psychological and neurobiological factors (Weintraub & Burn, 2011). In some studies, there were no between-group (DBS vs BMT) differences in mood post-DBS surgery (Deuschl et al., 2006; Weaver et al., 2009). In other studies, however, depression has been reported following DBS and may represent a consequence of stimulation or new found awareness of symptoms that may have been present preoperatively (Bejjani, Damier, Arnulf, Thivard, et al., 1999; Frank, Samanta, Moustafa, & Sherman, 2007; Hälbig et al., 2009; Houeto et al., 2002; Saint-Cyr, Trepanier, Kumar, Lozano, & Lang, 2000; Schüpbach, Gargiulo, Welter, et al., 2006; Smeding, Speelman, Koning-Haanstra, et al., 2006). No significant mood changes were reported between STN and GPi stimulation in the optimal DBS state (Okun et al., 2009). In another randomized control study, those patients who received STN DBS were more likely to experience worsening in depressive symptoms as compared to GPi stimulation, suggesting that either STN stimulation or reduction of PD medications can contribute to depression (Follett et al., 2010).

In a systematic review, depression was observed in 8% of the individuals who underwent STN DBS, and there has been report of suicide in several cases (Temel et al., 2006). Prevalence of suicide attempts after DBS has been found to range from 0.5% to 2.9% (Krack et al., 2003). The reported increased rate of suicide in patients with PD who have undergone STN DBS underscores the need for accurate preoperative psychiatric assessment and treatment of depression as well as the need for careful and detailed postoperative follow-up (Voon, Krack, Lang, et al., 2008). Depression may be directly related to DBS therapy—notably stimulation and lead placement, but depression symptoms may also be related to reduced dopaminergic medications after surgery, reactionary symptoms, or psychosocial factors (Berozzo, Rizzzone, Bergamasco, et al., 2001; Thobois et al., 2010; Voon et al., 2006).

A high prevalence of anxiety has been reported in individuals with PD, with ranges reported to be between 20% and 75% of individuals with PD (Aarsland et al., 1999; Menza & Dobkin, 2006; Menza, Robertson-Hoffman, & Bonapace, 1993). Some studies have found improvement in anxiety symptoms (Daniele, Albanese, Contarino, et al., 2003; Witt, Daniels, Reiff, et al., 2008), while others have reported worsening of anxiety following DBS therapy (Castelli et al., 2006; Houeto et al., 2002; Voon et al., 2006). Some individuals presented with reactionary anxiety following DBS implantation, such as having a fear that the stimulator would suddenly fail (Houeto et al., 2002).

3.2.2 Apathy

Apathy, defined as general feelings of low motivation, lack of interest and effort, and indifference (Reijnders et al., 2010; Starkstein & Leentjens, 2008), is also common in PD, independent of depression, with prevalence estimates ranging from 16% to 70% (Aarsland, Marsh, & Schrag, 2009; Leentjens et al., 2008). Apathy is linked to various areas in the brain that are also affected by PD (Aarsland et al., 2009; Borek, Amick, & Friedman, 2006; Reijnders et al., 2010). A meta-analysis of lesions affecting the basal ganglia conveyed that apathy was the most common behavioral disturbance in these individuals (Bhatia & Marsden, 1994). Studies looking at apathy in individuals with PD who have undergone DBS reported mixed results, with some individuals experiencing more and others experiencing less apathy than patients on BMT (Contarino et al., 2007). Incidence of apathy following STN DBS therapy has been reported to be 12%–25% (Krack et al., 2003). Apathy increased linearly from baseline to 6 months post-DBS in one study, independently of the surgical target, while apathy in a control group did not change over the same interval (Kirsch-Darrow et al., 2011). As with depression and anxiety, apathy can occur after DBS surgery as a delayed dopamine withdrawal syndrome (Thobois et al., 2010).

3.2.3 Hypomania

Hypomania, defined as persistent disinhibition and pervasive euphoric mood, can occur in 4%–15% of individuals who undergo DBS implantation, usually occurring within the first 3 months postoperatively (Herzog, Volkmann, Krack, et al., 2003; Krack et al., 2003; Volkmann et al., 2001). Hypomania has been observed prior to initiation of STN stimulation, suggesting a specific role of surgery or microlesion effect (Romito, Raja, Daiel, et al., 2002). In a selected number of patients, reversible and reproducible hypomanic symptoms were reported after stimulation of specific anatomical contacts, involving preferentially the right hemisphere (Ulla, Thobois, Llorca, et al., 2011). Finally, the synergistic interaction between stimulation and dopaminergic medications affecting similar circuitries likely plays an important role in postoperative hypomania (Voon et al., 2006).

3.2.4 Psychosis/Hallucinations

Psychosis is not uncommon in individuals with PD, with one study reporting hallucinations in 27% (Aarsland et al., 1999). Hallucinations are generally visual and appear to be more likely associated with underlying PD and dopaminergic medications (Parent, Levesque, & Parent, 2001). Some research, however, has documented psychosis and transient hallucinations as being associated with DBS (Diederich, Alesch, & Goetz, 2000; Schüpbach, Chastan, Welter, et al., 2005).

3.2.5 Impulsive Behavior

ICDs (i.e., gambling, buying, sexual behavior, eating, repetitive purposeless motor acts) are known to be relatively common in PD (Voon & Fox, 2007), particularly in association with D2-selective dopamine agonist treatment (Weintraub & Burn, 2011). In a large multisite observational study, an ICD was identified in 14% of PD patients, and 4% had more than one ICD (Weintraub, Koester, Potena, et al., 2010). In a cross-sectional research design, there were no differences between a DBS group and non-DBS group on compulsivity; however, impulsivity scores were significantly higher in the DBS patients (Hälbig et al., 2009). STN DBS has been associated with improvement in ICD symptoms (Ardouin et al., 2006), but there is also evidence that ICDs may begin or worsen transiently post-DBS (Smeding et al., 2007). A review of the literature did not identify a significant difference in ICDs between patients receiving dopaminergic medication and patients on DBS in four large studies, although nine case reports detailed up to 56% of patients that developed ICDs after DBS (Demetriades,

Rickards, & Cavanna, 2011). Another review of the evidence investigating the effect of DBS on NMS of PD noted that varying DBS outcomes on ICDs may be attributable to the differing pathophysiological mechanisms where, for example, some ICDs that are related to hyperdopaminergic states may improve with medication adjustments, while, in a subset of patients with greater mesolimbic degeneration, DBS-induced ICDs may ensue from alterations in the limbic-basal ganglia loop and lead to hyperactivation of the ventral striatum (Kurtis et al., 2017).

3.2.6 Managing Expectations

Managing patient expectations for DBS outcomes is also relevant to neurobehavioral outcomes. The magnitude of the STN DBS effect appears to be significantly higher when the patients are aware of the stimulation condition than when they are blinded to it (De la Fuente-Fernández, 2004). Other research found that clinical benefit was heightened when the patients were advised that the stimulation was ON, whereas clinical worsening was further potentiated when the patients were advised that the stimulation was OFF, a response modulated by a nocebo effect (Mercado et al., 2006). PD patients who undergo STN DBS have specific hopes and wishes regarding a successful outcome of surgery that does not simply depend on motor improvement (Maier et al., 2013). Individuals with preoperative apathy and depression may result in postoperative negative subjective outcomes, which may ensue in unimproved quality of life. Moreover, preoperative apathy and depression scores might be helpful tools to identify "risk candidates" for STN DBS, providing the opportunity of an accompanying intervention (i.e., psychosocial support and counseling) prior to and after DBS intervention.

Recognition of cognitive and psychiatric comorbidities, especially cognitive impairment, depression, anxiety, apathy, ICD, and psychosis prior to surgery, is important since such symptoms confound surgical outcomes. A neuropsychological and neuropsychiatric examination should be performed in DBS candidates to identify problems and predisposing factors, in addition to highlighting strengths that may help identify optimal therapeutic outcomes, such as cognitive reserve (Stern, 2002), expectation management (Maier et al., 2013), factors associated with resilience (Frazier & Marsh, 2006; Freedman, 2008; Rasmussen, Scheier, & Greenhouse, 2009), and perceived adequate social support systems (MacCarthy & Brown, 1989; Simpson, Haines, Lekwuwa, Wardle, & Crawford, 2006). Recommendations for treatment of active cognitive disturbance and psychiatric problems prior to surgical intervention have been proposed (Susatia, Foote, Ward, &

Okun, 2011), because such symptoms may worsen following DBS surgery and obviate best outcomes. Additionally, interpreting etiological causes for cognitive disturbance following DBS implantation requires understanding of the broader constellation of symptoms for individuals with PD. It is important to prevent misattribution of cognitive decrement directly to DBS-induced cognitive changes when contributions from other factors should be considered, such as the natural progression of PD, medication effects, sleep disturbance, pain, and psychiatric symptoms. The neuropsychological and neuropsychiatric profile helps elucidate the risk factors for cognitive and/or psychiatric changes following DBS therapy. In the final analysis, it is incumbent upon the treatment team to delineate the risk factors, create expectations about possible outcomes, and help the patient and family make an educated decision about whether or not to pursue DBS. The implications related to the risks for changes in cognition are significant, but the cognitive and psychiatric profile has to be considered in the larger context of the patient's degree of motor disability (and possibility for functional improvement) and the type of care and resources available for a particular patient. A multidisciplinary team approach for assessing DBS candidacy, educating about DBS therapy outcomes vs PD-induced symptom progression, and managing motor and NMS longitudinally can help facilitate optimal treatment recommendations with the goal of improving quality of life for the person with Parkinson's and care providers.

4. AUTONOMIC DYSFUNCTION

4.1 Orthostatic Hypotension

OH is commonly defined as a drop in blood pressure (BP) when standing and is often associated with a sensation of lightheadedness and other well-described symptoms (Senard et al., 1997). OH can be identified clinically as a BP fall of at least 20 mm Hg systolic or 10 mm Hg diastolic when standing as compared to lying positing or during head-up tilt (HUT) testing (Low, 2008). OH is very common in PD (Ludwig et al., 2007; Senard et al., 1997), and its prevalence increases with disease duration and severity (Senard et al., 1997), with over half of patients likely to be affected at 20 years, although many remain asymptomatic (Hely et al., 2008; Senard et al., 1997). OH hampers activities of daily life in patients with PD, independently of motor impairment (Magerkurth, Schnitzer, & Braune, 2005). Like other NMS, OH may predate the motor features of PD by many years (Goldstein, 2006; Kaufmann, Nahm, Purohit, & Wolfe, 2004), although early OH presentation in the course of PD may suggest an alternative diagnosis, like multiple system atrophy (Low, 2008) or progressive supranuclear palsy (Wenning et al., 1999). OH can be induced or exacerbated by dopaminergic medications (Senard et al., 1997) and is frequently associated with clinostatic hypertension, a combination described as the "Hyp–Hyp phenomenon" (Lagi & Spini, 2010). Despite these challenges, identification of OH is important because it is associated with physical frailty and increased mortality (Luukinen, Koski, Laippala, & Kivelä, 1999; Masaki et al., 1998). OH may also be predictive of other NMS including with the greatest concern being cognitive decline (Elmståhl & Rosén, 1997).

The incidence of OH before and after DBS surgery is usually not reported in major controlled trials. Therefore, our knowledge is derived from smaller studies designed to address the influence of DBS on autonomic function, which have shown conflicting outcomes. Three advanced PD patients with STN DBS (one unilateral, two bilateral) were found to have a significant increase in heart rate during a brief round of stimulation 6 months following surgery, suggesting a cardiac regulatory effect (Kaufmann, Bhattacharya, Voustianiouk, & Gracies, 2002). Improvement of postural hypotension was reported in 14 PD patients treated with bilateral STN DBS and studied with HUT testing. STN DBS resulted in a 5-bpm increase in heart rate and a 5-mm Hg increase in arterial BP after a 3-min HUT test, suggesting a beneficial effect in peripheral vasoconstriction and baroreflex sensitivity, with resulting BP stabilization (Stemper et al., 2006). In another study, eight PD subjects studied 6 months after STN DBS reported a subjective reduction in postural dizziness and syncope, which correlated with a correspondent, statistically significant reduction of orthostatic systolic BP drop from 51.6 mm Hg before to 17.3 mm Hg after DBS surgery (Tateno et al., 2015). Of note, two other studies failed to show any STN DBS benefit on HUT, suggesting that STN stimulation had only minor effects on autonomic functions (Holmberg, Corneliusson, & Elam, 2005; Ludwig et al., 2007). The finding that heart rate variability (as measured from 24-h ECG recordings) did not change 12 months after bilateral STN surgery seems to further support this latter view (Erola, Haapaniemi, Heikkinen, Huikuri, & Myllyä, 2006).

Long-term effects of DBS in OH may vary, also in relationship with the reduced DRT allowed by stimulation. A substantial but statistically nonsignificant reduction in the percentage of patients with OH (from 45.8% preoperatively to 12.5% 6 months postoperatively) was found in a study assessing cardiovascular autonomic function in 24 STN DBS patients (Trachani et al., 2012). Reduced pharmacological treatment did not prevent the development of cardiovascular autonomic failure in a group of 11 PD patients with STN DBS, which showed similar heart rate variability but increased incidence of pathological autonomic test results when compared with a group of PD subjects left on best medical treatment for an equivalent follow-up of 1 year (Holmberg et al., 2005). Similar outcomes were reported by a long-term follow-up study of PD patients after DBS, which showed no substantial NMS changes despite a significant improvement in motor symptoms, with 36% of patients progressively developing severe postural hypotension during the course of follow-up (Merola et al., 2011). Nonmotor features of advanced PD, including OH, were mostly unchanged 24 months after STN DBS in a group of 36 patients, without evidence that DBS may protect against the later development of OH (Zibetti et al., 2007). Importantly, patients with DBS who had OH were at higher risk of death than non-OH PD patients (Ryu et al., 2016).

The mechanism by which DBS may impact autonomic features appears to be dependent on the anatomic regions stimulated. The basal ganglia are connected to areas involved in regulation of the autonomic nervous system (ANS), and therefore may affect ANS functions (Braak & Braak, 2000; Ludwig et al., 2007). Stimulation of the globus pallidum pars externa (GPe) and the STN increases heart and respiratory rates in cats, which links the basal ganglia to central command of autonomic function (Ángyán, 1994).

Stimulating anteromedial or lateral to the STN may produce several autonomic effects (Pollak et al., 2002). Stimulation of the anterior limb of the internal capsule may lead to sweating, elevated respiratory, or heart rate, as well as cold and hot sensations (Okun et al., 2007; Shapira et al., 2006). STN stimulation has been shown to produce objective findings of sweating, flushing, in addition to heart rate and BP increases (Sauleau et al., 2005). Interestingly, these findings are not lateralizing and appear to occur even at locations providing the most improvement in motor control, suggesting that they may be inherent to the STN rather than adjacent structures (Sauleau et al., 2005). Hypertensive crisis can be produced by stimulation with a medially placed electrode in the subthalamus (Sánchez-Ferro, Benito-León, & Gómez-Esteban, 2013), suggesting unwanted activation of nearby structures such as the posterior hypothalamus, which may lead to stimulation-induced autonomic effects (Lipp et al., 2005).

The underlying mechanism beyond these autonomic effects remains unclear. Two postulated theories include either indirect limbic system activation or indirect stimulation of the central autonomic network, elements of which can occur near the STN (Hyam, Brittain, et al., 2012; Hyam, Kringelbach, Silburn, Aziz, & Green, 2012).

In conclusion, the evidence relating an effect of DBS on autonomic function in PD is mixed and virtually derived only from STN DBS studies, while GPi stimulation does not appear to have an effect on OH (Fasano, Daniele, & Albanese, 2012). The available data support a neutral or beneficial effect of STN DBS on OH (see Table 3), possibly due to its action on central pathways controlling autonomic responses (Benedetti et al., 2004). Clearly, more research is needed to clarify the effects of DBS treatment on OH.

4.2 Constipation

Constipation, defined as decreased stool frequency, with less than three bowel movements per week (Pfeiffer, 2003), is common in PD and often precedes the development of motor symptoms (Abbott et al., 2001, 2007;

Table 3 Selected Studies Evaluating the Impact of DBS on Cardiovascular Autonomic Functions and Orthostatic Hypotension

Source	*N*	Target	Follow-up	Outcome
Kaufmann et al. (2002)	3	STN	6 months	Increased HR
Holmberg et al. (2005)	11	STN	12 months	No change HRV, BP Increased percentage abnormal HUT
Sauleau et al. (2005)	17	STN	Intraoperative	Increased HR and BP
Erola et al. (2006)	14	STN	12 months	No change HRV
Stemper et al. (2006)	14	STN	7 months	Increased HR and BP
Ludwig et al. (2007)	14	STN	Unknown	No change HRV, HR, BP
Zibetti et al. (2007)	36	STN	24 months	No OH variations
Trachani et al. (2012)	24	STN		Decreased percentage OH
Tateno et al. (2015)	8	STN	6 months	Less severe OH drop

BP, blood pressure; *HR*, heart rate; *HRV*, heart rate variability; *HUT*, head-up tilt; *OH*, orthostatic hypotension; *STN*, subthalamic nucleus.

Chaudhuri, Odin, Antonini, & Martinez-Martin, 2011). The risk of constipation increases with disease duration and decreases with beta-blocker use (Krogh, Ostergaard, Sabroe, & Laurberg, 2008; Pagano, Tan, Haider, Bautista, & Tagliati, 2015). Constipation appears linked to the underlying neuropathology of PD by α-synuclein deposition in enteric nervous system (Petrovitch et al., 2009). It has been proposed that PD pathology may start in the enteric nervous system and spread to the brain along the vagus nerve (Braak & Del Tredici, 2008).

Few studies have examined the effect of DBS on constipation (Kurtis et al., 2017). Bilateral STN DBS appeared to significantly improve constipation in a group of 36 PD patients evaluated at 12 and 24 months after surgery (Zibetti et al., 2007). Similar results were reported in 20 PD patients using MDS-UPDRS part I scores showing significant improvement in constipation 6 months post-STN DBS (Chou, Taylor, & Patil, 2013). STN DBS was also shown to improve gastric emptying in 16 PD patients evaluated with a 13C-acetate breath test 3 months after surgery, while dopaminergic medication had no effect, suggesting that a centrally mediated effect on gastrointestinal function may be modulated by DBS to improve gastrointestinal symptoms in PD (Arai et al., 2012).

Long-term studies reported conflicting outcomes. A group of 56 PD patients with bilateral STN DBS for at least 1 year showed a significantly lower NMS quest scores, particularly in relation to constipation, when compared to controls (Mridula et al., 2015). Another cohort of 30 PD patients with bilateral STN DBS for 1 year demonstrated a trend toward improvement in constipation as measured by the MDS-UPDRS (Jafari, Pahwa, Nazzaro, Arnold, & Lyons, 2016). In other studies, however, STN stimulation did not appear to prevent the development of constipation in PD patients with long progression of disease. The prevalence of patients requiring pharmacological treatment for constipation rose significantly at 5 years and at >7 years after STN DBS in a cohort of 19 PD patient with 30 years progression of disease (Merola et al., 2011). Finally, a study following 26 PD patients for an average of 11 years, with assessments at 1, 5, and 11 years after implantation, reported that five patients developed new constipation by their last follow-up, while two of the four subjects with severe constipation at baseline improved during the same period (Rizzone et al., 2014). In conclusion, while there is substantial evidence supporting a positive effect for STN DBS on constipation in PD (Table 4), such benefit may not be long lasting.

Table 4 Selected Studies Evaluating the Impact of DBS on Constipation

Source	Subjects (Target)	Follow-up	Scale/Tool	Effect
Zibetti et al. (2007)	36 (STN)	24 months	UPDRS and clinical charts	Improvement in constipation
Chou et al. (2013)	20 (STN)	6 months	MDS-UPDRS	Improvement in constipation
Arai et al. (2012)	16 (STN)	3 months	^{13}C-Acetate breath test	Improved gastric emptying
Mridula et al. (2015)	56 (STN)	12 months	NMS quest	Reduced prevalence of constipation
Jafari et al. (2016)	30 (STN)	12 months	MDS-UPDRS	Trend toward improvement in constipation

MDS-UPDRS, Movement Disorders Society-Unified Parkinson Disease Rating Scale; *NMS*, nonmotor symptoms; *STN*, subthalamic nucleus.

4.3 Urinary Dysfunction

Urinary dysfunction is common in PD and may, like many other NMS, predate motor symptoms (Hely et al., 2008). Urinary incontinence is described in 71% of PD patients after 20 years of disease progression (Postuma, Lang, Massicotte-Marquez, & Montplaisir, 2006). Even in early, untreated PD 64%–87% of patients suffer from lower urinary tract symptoms (LUTS), which can be severe in a third of them (Kim, Jeon, & Paek, 2015). Urinary dysfunction has been linked to reduced bladder capacity due to detrusor hyperreflexia with involuntary contractions of the bladder (Araki, Kitahara, Oida, & Kuno, 2000). DBS may impact urinary symptoms via its action on the D1–GABAergic direct pathway in the basal ganglia, which is an inhibitor of micturition. There may also be downstream effects of stimulation on cortical control of bladder tone (Herzog et al., 2006; Sakakibara et al., 2012).

An early study on the effect of STN DBS on voiding evaluated urodynamic changes in five PD patients ON and OFF stimulation, finding an increase in bladder capacity and reflex volume during the ON state, which was interpreted as a confirmation of the basal ganglia's role in voiding control (Finazzi-Agro et al., 2003). In another study, 16 PD patients with STN DBS (irrespective of urinary symptoms) were assessed

6–29 months postsurgery. STN stimulation led to a normalization of the micturition reflex and improvement in bladder capacity, which was reversed in the OFF state, suggesting that STN DBS can reverse "overactive bladder" (Seif et al., 2004). Another class IV study used the International Prostate Symptom Score (IPSS) as well as pressure flow studies (PFS) on six patients following chronic STN DBS. Total IPSS value, involuntary detrusor contraction, and bladder capacity all significantly improved during STN stimulation, suggesting once more that STN DBS improves urinary function in PD (Shimizu et al., 2007). A prospective investigation of 16 PD patients evaluated before and after STN DBS implantation with the IPSS and the Danish Prostate Symptom Score (DanPSS), as well as urodynamics studies, showed an improvement in the overactive bladder symptoms along with the "troublesome" symptoms of overactive bladder. Urodynamic parameters, however, did not change significantly (Winge et al., 2007).

Nine PD patients with bilateral STN DBS were studied during dynamic bladder filling and empty bladder conditions in the ON and OFF stimulation state, while changes in regional cerebral blood flow (rCBF) were measured by PET. Results showed that ON-stimulation state was associated with significant changes in rCBF in the posterior thalamus and the insular cortex, suggesting that STN DBS can facilitate the support sensory perception as it relates to urinary control (Herzog et al., 2008). A small study, using a phone questionnaire to target a group of early-onset PD patients with STN DBS, reported an improvement of bladder function following implantation, suggesting that LUTS may particularly benefit from STN DBS in early-onset PD patients (Halim, Baumgartner, & Binder, 2011). However, it is important to keep in mind that despite the positive findings of the above studies in chronically stimulated patients (Table 5), NMS, including incontinence, continue to progress. Following STN DBS 20% of PD patients with a median duration of disease of 30 years required a catheter or a diaper at 3 years, 37% at 5 years and more than 50% at more than 7 years after surgery (Merola et al., 2011).

Thalamic DBS, typically reserved for patients with essential tremor and, less often, severe PD tremor, may worsen urinary symptoms although evidence is limited to a single study evaluating seven ET patients. In this small group, thalamic stimulation was associated with a significant decrease in bladder volume at first desire to void, and patients reported a strong desire to void. Other parameters were nonsignificant (Kessler et al., 2008).

Table 5 Selected Studies Evaluating the Impact of DBS on Urinary Dysfunction

Source	Subjects (Target)	Follow-up	Scale/Tool	Outcome
Finazzi-Agro et al. (2003)	5 (STN)	4–9 months	Urodynamic evaluation	Increased bladder capacity and reflex volume
Seif et al. (2004)	16 (STN)	6–29 months	Urodynamic evaluation	Normalized micturition reflex and improved bladder capacity
Herzog et al. (2006)	11 (STN)	Unknown	PET during urodynamics	Significant rCBF increase in anterior cingulate cortex with full bladder, further enhanced during STN DBS OFF
Shimizu et al. (2007)	6 (STN)	Unknown	IPSS, PFS	Improved total IPSS, involuntary detrusor contraction, and bladder capacity
Winge et al. (2007)	16 (STN)	6–10 months	IPSS, DanPSS, urodynamics	Improved overactive bladder. Urodynamics nonsignificant

DanPSS, Danish Prostate Symptom Score; *IPSS*, International Prostate Symptom Score; *PET*, positron emission tomography; *PFS*, pressure flow studies.

5. SLEEP DISORDERS

Sleep disorders comprise a diverse array of symptoms and are extremely common in PD (Kurtis, Rodriguez-Blazquez, Martinez-Martin, & ELEP Group, 2013). These include sleep-disordered breathing/sleep apnea, parasomnias, REM sleep behavior disorder (RBD), sleepwalking, sleep attacks, insomnia, excessive daytime sleepiness (EDS), RLS, periodic limb movements of sleep, and vivid dreaming (Zesiewicz et al., 2010). Sleep disorders may be predictive of later development of PD (Abbott et al., 2005) and increase in prevalence with disease duration; up to 60%–90% of PD patients experience sleep disturbances (Kurtis et al., 2013). Sleep problems are exacerbated or even caused by several factors including age-related brain changes, nighttime motor symptoms, cognitive impairment, mood disorders, obstructive sleep apnea, and side effects of DRT, especially, levodopa and dopamine agonists (Kurtis et al., 2013). Dopaminergic drugs used commonly in PD may be partially to blame for insomnia and EDS.

There is increasing evidence that DBS, and especially STN DBS, may improve sleep quality, although RBD or EDS may not be equally affected (Kurtis et al., 2017). The mechanism by which this occurs is not clear, but it may be due to a number of factors including medication reduction, reduced sleep fragmentation, and improved nighttime mobility (Antonini, Landi, Mariani, DeNotaris, & Pezzoli, 2004).

5.1 Insomnia, EDS, and Sleep Quality

EDS is very common in PD and is experienced by 70% of PD patients after 20 years of progression (Hely et al., 2008; Zibetti et al., 2007). EDS is a complex disorder with many contributing factors, including pathologic brain changes, concomitant sleep disorders, and DRT, especially dopamine agonists (Zesiewicz et al., 2010). In addition, depression, fatigue, cognitive impairment, and autonomic dysfunction, including cardiovascular, urinary, and thermoregulatory abnormalities, can contribute to insomnia and hypersomnia (Kurtis et al., 2013). Several sleep scales are available for the determination of sleep quality such as the Pittsburgh sleep quality index (PSQI) questionnaire (Buysse, Reynolds, Monk, Berman, & Kupfer, 1989), the Epworth Sleepiness Scale (Johns, 1991), or the Parkinson's Disease Sleep Scale (PDSS, Chaudhuri et al., 2002).

Several studies documented sleep quality improvements in patients with advanced PD undergoing STN DBS (Table 6), although measures of insomnia were not usually a primary outcome (Zesiewicz et al., 2010). One group studied the sleep/wake cycle in 10 PD patients 1 month before and 3 months after initiation of STN DBS. With STN stimulation, subjective and objective sleep qualities measures were improved. Total sleep time increased significantly, as well as sleep efficiency and deep slow-wave sleep. Convincingly, these improvements were reversed when stimulation was stopped. Sleep improvement was explained by the decrease in motor disturbances and reduction in DRT (Monaca et al., 2004). Another small study addressed subjective sleep quality and polysomnographic architecture in 11 patients with advanced PD following chronic bilateral STN DBS. Seven of the eight patients who rated their sleep quality as unsatisfactory before DBS reported significant improvement following surgery, with polysomnography documented increase in uninterrupted sleep and a 38%

Table 6 Selected Studies Evaluating the Impact of DBS on Sleep Quality

Source	Subjects (Target)	Follow-up	Scale/Tool	Percent Change
Monaca et al. (2004)	10 (STN)	3 months	Polysomnography	Increased total sleep time, sleep efficiency, and deep slow-wave sleep
Lyons and Pahwa (2006)	43 (STN)	24 months	ESS	Increased sleep time, reduced subjective sleep problems, no change in EDS
Volkmann et al. (2009)	20 (GPi) 45 (STN)	6 months 3 years	SIP	Significant improvement in their sleep and rest (SR) SIP subscore
Iranzo, Valldeoriola, Santamaria, Tolosa, and Rumia (2002)	11 (STN)	6 months	Polysomnography	Increase in uninterrupted sleep, 38% decrease in the arousal index, improved nocturnal mobility, no change in RBD
Arnulf, Bejjani, Garma, Bonnet, Houeto, et al. (2000)	10 (STN)	3–6 months	Polysomnography	Decrease in wakefulness after sleep onset. REM sleep behavior disorders did not improve
Hjort, Østergaard, and Dupont (2004)	6 (STN)	3 months	PDSS	Significant improvement in PDSS, sleep quality and motor symptoms nocturia, sleep fragmentation, and EDS did not change
Nishida et al. (2011)	16 (STN)	6–10 months	PDSS and polysomnography	Total sleep scale scores and daytime sleepiness subscale scores significantly improved, decreased wakefulness after sleep onset, decreased RBD
Amara et al. (2012)	53 (STN) unilateral	6 months	PSQI questionnaire	Unilateral STN DBS improves subjective sleep quality

EDS, excessive daytime sleepiness; *ESS*, Epworth Sleepiness Scale; *PDSS*, Parkinson's Disease Sleep Scale; *RBD*, REM sleep behavior disorder; *SIP*, sickness impact profile.

decrease in the arousal index. Nocturnal mobility also improved, but there was no change in RBD. These findings were again interpreted as a combination of increased nocturnal mobility and reduction of sleep fragmentation associated with DBS (Iranzo et al., 2002). An improvement in total sleep efficiency and a total sleep time increase, in addition to a decrease in wakefulness after sleep onset, was described in 10 PD patients with insomnia treated with STN DBS. Yet again, RBDs did not improve nor did periodic leg movements, suggesting that insomnia in PD may arise primarily from nighttime motor disability (Arnulf, Bejjani, Garma, Bonnet, Houeto, et al., 2000).

Using the PDSS, researchers compared 10 patients with STN DBS to a control group discovering a significant improvement in PDSS after 3 months of STN stimulation as compared to no change in the controls. Overall sleep quality and motor symptoms improved, whereas nocturia, sleep fragmentation, and EDS did not change, despite medication reduction (Hjort et al., 2004). Other groups have reported similar results. Ten PD patients were monitored with the PDSS and polysomnography before and after STN DBS, when total sleep scale scores and daytime sleepiness subscale scores significantly improved (Nishida et al., 2011). The improved total sleep scale score correlated with decreased wakefulness after sleep onset. Different than other observations, this study showed a significantly increased amount of normal rapid eye movement sleep and decreased abnormal rapid eye movement sleep without atonia, which correlated with improved daytime sleepiness. Sleep improvement did not significantly correlate with resolution of motor complication or reduced dopaminergic dosages, suggesting that STN DBS may have beneficial effects on sleep disturbance in advanced PD by restoring sleep architecture and normal rapid eye movement (Nishida et al., 2011).

In a longer follow-up study, 43 PD patients were followed for 24 months after STN DBS using the Epworth Sleepiness Scale. Increased sleep time was significantly correlated with improvements in bradykinesia but not rigidity. Subjective sleep problems were overall reduced after STN DBS, but there was no change in EDS at 6, 12, or 24 months after surgery, despite a decrease in antiparkinsonian medications (Lyons & Pahwa, 2006).

More research is needed to determine the value of GPi DBS for sleep quality. In one study, all 20 patients with GPi DBS demonstrated a statistically significant improvement in their sleep and rest (SR) subscore, suggesting that GPi DBS may, like STN stimulation, improve sleep quality (Volkmann et al., 2009).

5.2 REM Sleep Behavior Disorder

RBD is defined by the presence of complex motor behaviors occurring during REM sleep (Zibetti et al., 2010). Anatomically, REM sleep and wakefulness in particular appear to be mediated by the pedunculopontine tegmental nucleus (PPTg) and surrounding areas (Stefani et al., 2010). Like other NMS, RBD often presents before the development of motor symptoms and is predictive of future development of PD (Olson, Boeve, & Silber, 2000; Postuma et al., 2006). RBD is common in PD and is an important prognostic indicator associated with higher incidence of cognitive decline (Vendette et al., 2007), autonomic impairment (Postuma, Gagnon, Vendette, Charland, & Montplaisir, 2008a), and falls (Zibetti et al., 2010). RBD appears to be less likely in patients with tremor-predominant PD and in those with a strong response to levodopa (Postuma, Gagnon, Vendette, Charland, & Montplaisir, 2008b).

There is little evidence supporting beneficial effects of DBS on RBD, and RBD appears to be associated with an increased risk of developing disability over time following STN DBS for PD (Rizzone et al., 2014). Twelve out of 41 (29%) PD patients with RBD treated with STN DBS had significantly poorer outcome 1 and 3 years after surgery as compared to patients without RBD, in particular when also experiencing axial symptoms (Zibetti et al., 2010). A larger study retrospectively evaluating 90 PD patients with PD, of which 52.2% had RBD preoperatively, found that 17.7% more patients had developed RBD de novo 1 year after STN DBS, raising the possibility that STN DBS is a provoking factor in the development of RBD (Kim, Paek, et al., 2015). In this case, the Unified Parkinson Disease Rating Scale (UPDRS) axial scores for the "ON" state were lower in the RBD group than in the non-RBD group following DBS.

5.3 Restless Legs Syndrome/Periodic Leg Movements Disorder

RLS is a common disorder in the general population affecting up to 25% of adults and is even more common in PD (Ondo, Vuong, & Jankovic, 2002). The physiologic changes underpinning RLS are complex and poorly understood, but they have been attributed to an impairment of central dopaminergic transmission. There is conflicting SPECT imaging evidence pointing to a central striatal dopaminergic dysfunction (Eisensehr et al., 2001; Michaud, Soucy, Chabli, Lavigne, & Montplaisir, 2002) with normal results in other studies (Linke et al., 2004). One postulated mechanism for the effects of DBS in RLS may be alteration in basal ganglia neuronal firing leading to secondary effects on the thalamus (Rye & DeLong, 1999) or via an impaired diencephalospinal dopamine pathway (Driver-Dunckley et al., 2006; Rye, 2004). Thalamic activation with RLS sensory leg symptoms has been revealed by functional magnetic resonance imaging (Bucher, Seelos, Oertel, Reiser, & Trenkwalder, 1997; Driver-Dunckley et al., 2006).

The role of DBS in RLS is confounded by the decrease in dopaminergic medication common associated with STN DBS therapy. Eleven out of 195 patients with STN DBS reported the new occurrence of problematic RLS postoperatively in one study (Kedia et al., 2004). In the subgroup that developed RLS, antiparkinsonian medications were reduced by 74%, as opposed to only 40.4% in the group that did not develop RLS. In all patients, RLS symptoms improved with customized pharmacologic treatment.

Given the potential for the development of RLS, it is recommended to assess for RLS in PD patients undergoing medication reduction following STN DBS, especially if those reductions are >50% of their preoperative levodopa equivalents (Chahine, Ahmed, & Sun, 2011). A few studies have assessed the risk of "unmasking" RLS by medication reduction after STN DBS. A small study of six advanced PD patients, who were able to decrease levodopa equivalent dosage by an average of 56%, showed marked improvement of RLS symptoms and objective scores after STN DBS, suggesting that STN DBS may actually treat RLS in advanced PD (Driver-Dunckley et al., 2006). In another cohort of 17 PD patients, 6 of whom were diagnosed with RLS preoperatively and were followed for 1 and 6 months postoperatively, no patients developed RLS following DBS surgery. In addition, IRLSSG rating scale scores improved, as compared to baseline, in those that had preoperative RLS (Chahine et al., 2011).

Similar to RLS, periodic leg movements disorder (PLMD) is a common condition that is responsive to dopaminergic treatment. The prevalence of PLMD has been reported as 3.9% in the general population, with increasing prevalence with age, and in women (Ohayon & Roth, 2002). PLMD is often associated with PD (Covassin, Neikrug, et al., 2012). The use of levodopa/carbidopa probably decreases the frequency of spontaneous nighttime leg movements, and there is level B evidence for its use as a treatment of PLM in patients with PD (Zesiewicz et al., 2010). There are no large studies of PLMD in DBS, but there are case reports of improvement with GPi stimulation (Klassen, Lyons, & Evidente, 2013).

In conclusion, while there is evidence supporting the improvement of sleep quality after STN DBS, RBD appears to be negatively affected by DBS. There is similar conflict regarding the effect of DBS on RLS and PLMD (Table 7). Only a few studies have addressed the effects of GPi

Table 7 Selected Studies Evaluating the Impact of DBS on Restless Syndrome

Source	Subjects (Target)	Follow-up	Scale/Tool	Mean Medication Reduction	Outcome
Kedia et al. (2004)	195 (STN) No RLS preop	3–12 months	Self-reported prevalence	79% for RLS 40.4% non-RLS	11 New cases of RLS postoperatively
Driver-Dunckley et al. (2006)	25 (STN) 6 RLS preop	3–24 months	IRLS[a]	56%	84% IRLS improvement
Chahine et al. (2011)	17 (STN) 6 RLS preop	6 months	IRLS[a]	50%	40% IRLS improvement
Nazzaro et al. (2011)	24 (STN) 13 RLS preop	12 months	NMSQ	54%	33.4% RLS prevalence reduction

[a]International Restless Legs Syndrome Study Group (2003).
IRLS, International RLS Scale; *NMSQ*, nonmotor symptoms questionnaire.

DBS on sleep quality in PD patients and reported subjective improvement of daytime sleepiness even though these patients did not reduce DRT (Volkmann et al., 2009). Ventrointermediate nucleus (Vim) DBS does not influence sleep architecture or sleep spindles (Arnulf, Bejjani, Garma, Bonnet, Damier, et al., 2000). The observation that REM behavior disorder is improved after stimulation of the pedunculopontine nucleus has not been confirmed (Lim et al., 2009).

6. SENSORY SYMPTOMS

Despite the classical definition of "movement disorder," sensory symptoms including pain, dysesthesia, and/or numbness are frequent in PD (Ford, 1997; Kim et al., 2008). Pain in particular can have a significantly negative role in the quality of life and is reported in over >40% of PD patients (Ford, 1997; Goetz, Tanner, Levy, Wilson, & Garron, 1986). Pain is often located in the limbs most affected by typical motor symptoms (Cury et al., 2014; Goetz et al., 1986).

Multiple studies have suggested that DBS treatment can relieve pain (Table 8), particularly in those subjects who experienced pain upon wearing OFF (Juri, Rodríguez-Oroz, Burguera, Guridi, & Obeso,

Table 8 Selected Studies Evaluating the Impact of DBS on Sensory Functions and Pain

Source	Subjects (Target)	Follow-up	Scale/Tool	Outcomes
Erola et al. (2005)	29 (STN)	12 months	PDQ-39 and the Finnish version of the NHP	Bodily discomfort subscale of PDQ-39 significantly improved
Kim et al. (2008)	29 (STN)	3–6 months	Survey of pain in the "ON" and "OFF" state	Preoperative pain improved in 20/23 OFF pain decreased in 12/18 and resolved in 5/18
Hwynn et al. (2011)	9 (STN) 1 (GPi)	One time evaluation	NMS questionnaire	Large mean decrease of NMS scores in miscellaneous symptoms including pain
Maruo et al. (2011)	17 (STN)	One time test	Quantitative sensory testing	Temperature sensations disturbed in PD patients STN DBS improves temperature sensation
Spielberger, Wolf, Kress, Seppi, and Poewe (2011)	15 (STN)	One time occurrence	Thermal detection Thermal pain thresholds	No influence of STN DBS or L-dopa on thermal detection thresholds or thermal pain sensitivity
Dellapina et al. (2012)	24 (STN)	12 months	Experimental nociceptive stimulations using $H_2^{15}O$ PET in both DBS OFF and ON conditions	Increased subjective heat pain threshold Reduced pain-induced cerebral activity in the somatosensory cortex (BA 40) in patients with pain No effect in pain-free patients

de Andrade et al. (2012)	25 (STN)	12 months	Short McGill Pain Quest; Brief Pain Inventory; Neuropathic Pain Symptom Inventory	Decreased chronic pain prevalence
Kim, Jeon, Lee, Paek, and Kim (2012)	21 (STN)	24 months	Survey	Beneficial effect on pain in PD persists up to 24 months
Sürücü, Baumann-Vogel, Uhl, Imbach, and Baumann (2013)	14 (STN)	41 months	Survey	Higher reduction of pain severity than achieved with dopaminergic treatment in 8/14 patients
Marques et al. (2013)	19 (STN)	One time evaluation	Evaluated pain thresholds (thermal and mechanical) by survey	Clinical pain alleviation after STN DBS could be attributable to a direct central modulation of pain perception, via increased mechanical pain and tolerance thresholds
Pellaprat et al. (2014)	58 (STN)	12 months	Short McGill Pain Quest; PDQ-39 bodily discomfort subscore; UPDRSII item 17	Substantial beneficial effect on pain, independent of motor effects and mood
Cury et al. (2014)	44 (STN)	12 months	Specific scales and questionnaires	Reduced prevalence of pain Significant improvements in pain intensity, NMS, and quality of life
Jung et al. (2015)	24 (STN)	8 years	Survey	Beneficial effect on pain in PD persists up to 8 years

2010; Kim et al., 2008). A recent review found overall class III evidence for a benefit of DBS on pain (Kurtis et al., 2017). A significant improvement of the mean bodily discomfort score of the PDQ-39 was reported by several authors in parallel with the improvement of time spent ON (Deuschl et al., 2006; Erola et al., 2005). In a study of 36 PD patients with bilateral STN DBS, sensory complaints at 24-month follow-up improved for larger part of the day due to the increased time spent ON (Zibetti et al., 2007). However, significant pain improvement was not correlated with motor outcomes, depression scores, or levodopa reduction in other studies (Marques et al., 2013; Pellaprat et al., 2014). Some observations have suggested that response of pain symptoms to DBS can be predicted by the response of pain to levodopa (Sürücü et al., 2013), although others have failed to confirm such correlation (Gierthmuhlen et al., 2010; Spielberger et al., 2011).

In a study specifically focused on the effect of DBS on pain, 23 out of 29 (79%) PD patients reported pain prior to surgery and 18 (62%) had pain as an integral part of their motor fluctuations. After DBS, 20 out of 23 study subjects experienced improvement in pain, including 12 out of 18 in the fluctuating pain group, of which 5 reported complete resolution of OFF time pain (Kim et al., 2008). Similar results were reported in a group of 41 PD patients subjects followed for 1 year after STN DBS, who reported an impressive reduction of pain prevalence from 70% to 21% after surgery, paralleled by quality of life improvement (Cury et al., 2014).

In addition to reducing pain in a significant number of PD patients, DBS has been reported to increase sensitivity to thermal stimuli and reduce sensitivity to mechanical and thermal pain (de Andrade et al., 2012; Maruo et al., 2011). STN DBS significantly increased subjective heat pain threshold and reduced pain-induced cerebral activity in the somatosensory cortex in eight PD patients with pain before surgery, whereas it had no effect in pain-free patients (Dellapina et al., 2012). However, thermal pain thresholds were unchanged by STN DBS in two other studies following 17 and 15 PD patients for over 36 months after surgery (Gierthmuhlen et al., 2010; Spielberger et al., 2011). Dystonic and musculoskeletal pain appear to respond to DBS better than central and neuropathic pain (Cury et al., 2014).

Long-term follow-up studies have confirmed that STN DBS can improve pain associated with PD, with a persistent beneficial effect up to 8 years (Jung et al., 2015). However, new painful symptoms, particularly musculoskeletal aching, may develop and negatively affect quality of life (Jung et al., 2015; Kim et al., 2012).

REFERENCES

Aarsland, D., Anderson, K., Larsen, J. P., Lolk, A., & Kragh-Sorensen, P. (2003). Prevalence and characteristics of dementia in Parkinson's disease—An 8-year prospective study. *Archives of Neurology, 60*, 387–392.

Aarsland, D., Larsen, J., Lim, N., Janvin, C., Karlsen, K., Tandberg, E., et al. (1999). Range of neuropsychiatric disturbances in patients with Parkinson's disease. *Journal of Neurology, Neurosurgery, and Psychiatry, 67*, 492–496.

Aarsland, D., Marsh, L., & Schrag, A. (2009). Neuropsychiatric symptoms in Parkinson's disease. *Movement Disorders, 24*(15), 2175–2186.

Abbott, R. D., Petrovitch, H., White, L. R., Masaki, K. H., Tanner, C. M., Curb, J. D., et al. (2001). Frequency of bowel movements and the future risk of Parkinson's disease. *Neurology, 57*, 456–462.

Abbott, R. D., Ross, G. W., Petrovitch, H., Tanner, C. M., Davis, D. G., Masaki, K. H., et al. (2007). Bowel movement frequency in late-life and incidental Lewy bodies. *Movement Disorders, 22*, 1581–1586.

Abbott, R. D., Ross, G. W., White, L. R., Tanner, C. M., Masaki, K. H., Nelson, J. S., et al. (2005). Excessive daytime sleepiness and subsequent development of Parkinson disease. *Neurology, 65*, 1442–1446.

Allain, J., Schuck, S., & Manduit, N. (2000). Depression in Parkinson's disease. *British Medical Journal, 320*, 1287–1288.

Amara, A. W., Standaert, D. G., Guthrie, S., Cutter, G., Watts, R. L., & Walker, H. C. (2012). Unilateral subthalamic nucleus deep brain stimulation improves sleep quality in Parkinson's disease. *Parkinsonism & Related Disorders, 18*(1), 63–68.

Anderson, V. C., Burchiel, K. J., Hogarth, P., Favre, J., & Hammerstad, J. P. (2005). Pallidal vs. subthalamic nucleus deep brain stimulation in Parkinson disease. *Archives of Neurology, 62*, 554–560.

Ángyán, L. (1994). Somatomotor and cardiorespiratory responses to basal ganglia stimulation in cats. *Physiology & Behavior, 56*, 167–173.

Antonini, A., Landi, A., Mariani, C., DeNotaris, R., & Pezzoli, G. (2004). Deep brain stimulation and its effect on sleep in Parkinson's disease. *Sleep Medicine, 5*, 211–214.

Arai, E., Arai, M., Uchiyama, T., Higuchi, Y., Aoyagi, K., Yamanaka, Y., et al. (2012). Subthalamic deep brain stimulation can improve gastric emptying in Parkinson's disease. *Brain, 135*, 1478–1485.

Araki, I., Kitahara, M., Oida, T., & Kuno, S. (2000). Voiding dysfunction and Parkinson's disease: Urodynamic abnormalities and urinary symptoms. *Journal of Urology, 164*, 1640–1643.

Ardouin, C., Pillon, B., Peiffer, E., et al. (1999). Bilateral subthalamic or pallidal stimulation or Parkinson's disease affects neither memory nor executive functions: A consecutive series of 62 patients. *Annals of Neurology, 46*, 217–223.

Ardouin, C., Voon, V., Worbe, Y., Abouazar, N., Czernecki, V., Hosseini, H., et al. (2006). Pathological gambling in Parkinson's disease improves on chronic subthalamic nucleus stimulation. *Movement Disorders, 21*, 1941–1946.

Arnulf, I., Bejjani, B. P., Garma, L., Bonnet, A. M., Damier, P., Pidoux, B., et al. (2000a). Effect of low and high frequency thalamic stimulation on sleep in patients with Parkinson's disease and essential tremor. *Journal of Sleep Research, 9*, 55–62.

Arnulf, I., Bejjani, B. P., Garma, L., Bonnet, A. M., Houeto, J. L., Damier, P., et al. (2000b). Improvement of sleep architecture in PD with subthalamic nucleus stimulation. *Neurology, 55*, 1732–1734.

Bandini, F., Primavera, A., Pizzorno, M., & Cocito, L. (2007). Using STN DBS and medication reduction as a strategy to treat pathological gambling in Parkinson's disease. *Parkinsonism & Related Disorders, 13*, 369–371.

Baron, M. S., Vitek, J. L., Bakay, R., et al. (1996). Treatment of advanced Parkinson's disease by posterior GPi pallidotomy: 1-Year results of a pilot study annals of neurology. *Annals of Neurology, 40*, 355–366.

Bejjani, B., Damier, P., Arnulf, I., Thivard, L., et al. (1999). Transient acute depression induced by high-frequency deep brain stimulation. *The New England Journal of Medicine, 340*, 1476–1480.

Benedetti, F., Colloca, L., Lanotte, M., Bergamasco, B., Torre, E., & Lopiano, M. (2004). Autonomic and emotional responses to open and hidden stimulations of the human subthalamic region. *Brain Research Bulletin, 63*, 203–211.

Berozzo, P., Rizzzone, M., Bergamasco, B., et al. (2001). Deep brain stimulation of subthalamic nucleus: Behavioral modifications and familiar relations. *Neurological Sciences, 22*, 81–82.

Bhatia, K. P., & Marsden, C. D. (1994). The behavioral and motor consequences of focal lesions of the basal ganglia in man. *Brain, 117*, 859–876.

Bondi, M. W., Kaszniak, A. W., Bayles, K. A., & Vance, K. T. (1993). Contributions of frontal system dysfunction to memory and perceptual abilities in Parkinson's disease. *Neuropsychology, 7*, 89–102.

Borek, L. L., Amick, M. M., & Friedman, J. H. (2006). Non-motor aspects of Parkinson's disease. *CNS Spectrums, 11*(7), 541–554.

Borgohain, R., Kandadai, R. M., Jabeen, A., & Kannikannan, M. A. (2012). Nonmotor outcomes in Parkinson's disease: Is deep brain stimulation better than dopamine replacement therapy? *Therapeutic Advances in Neurological Disorders, 5*(1), 23–41.

Braak, H., & Braak, E. (2000). Pathoanatomy of Parkinson's disease. *Journal of Neurology, 247*(Suppl. 2), II3–II10.

Braak, H., & Del Tredici, K. (2008). Invited article: Nervous system pathology in sporadic Parkinson disease. *Neurology, 70*, 1916–1925.

Bronstein, J. M., Tagliati, M., Alterman, R. L., Lozano, A. M., Volkmann, J., Stefani, A., et al. (2011). Deep brain stimulation for Parkinson's disease: An expert consensus and review of key issues. *Archives of Neurology, 68*, 165–171.

Bucher, S. F., Seelos, K. C., Oertel, W. H., Reiser, M., & Trenkwalder, C. (1997). Cerebral generators involved in the pathogenesis of the restless legs syndrome. *Annals of Neurology, 41*, 639–645.

Burn, D. J., & Troster, A. I. (2004). Neuropsychiatric complications of medical and surgical therapies for Parkinson's disease. *Journal of Geriatric Psychiatry and Neurology, 17*(3), 172–180.

Buysse, D. J., Reynolds, C. F., Monk, T. H., Berman, S. R., & Kupfer, D. J. (1989). The Pittsburgh sleep quality index: A new instrument for psychiatric practice and research. *Psychiatry Research, 28*(2), 193–213.

Cahn, D., Sullivan, E. V., Shear, P. K., et al. (1998). Neuropsychological and motor functioning after unilateral anatomically guided posterior ventral pallidotomy. Preoperative performance and three-month follow-up. *Neuropsychiatry, Neuropsychology, and Behavioral Neurology, 1*, 136–145.

Castelli, L., Perozzo, P., Zibetti, M., Crivelli, B., Morabito, U., Lanotte, M., et al. (2006). Chronic deep brain stimulation of the subthalamic nucleus for Parkinson's disease: Effects on cognition, mood, anxiety and personality traits. *European Neurology, 55*, 136–144.

Cersosimo, M. G., Piedimonte, F., Raina, G. B., & Micheli, F. E. (2007). Bilateral STN-DBS fails to improve non-motor fluctuations in a PD patient. *Parkinsonism & Related Disorders, 13*(8), 537–538.

Chahine, L. M., Ahmed, A., & Sun, Z. (2011). Effects of STN DBS for Parkinson's disease on restless legs syndrome and other sleep-related measures. *Parkinsonism & Related Disorders, 17*(3), 208–211.

Chaudhuri, K. R., Healy, D. G., & Schapira, A. H. (2006). Non-motor symptoms of Parkinson's disease: Diagnosis and management. *Lancet Neurology, 5*, 235–245.

Chaudhuri, K., & Martinez-Martin, P. (2008). Quantitation of non-motor symptoms in Parkinson's disease. *European Journal of Neurology, 15*(Suppl. 2), 2–8.
Chaudhuri, K. R., Odin, P., Antonini, A., & Martinez-Martin, P. (2011). Parkinson's disease: The non-motor issues. *Parkinsonism & Related Disorders, 17*, 717–723.
Chaudhuri, K. R., Pal, S., DiMarco, A., Whately-Smith, C., Bridgman, K., Mathew, R., et al. (2002). The Parkinson's disease sleep scale: A new instrument for assessing sleep and nocturnal disability in Parkinson's disease. *Journal of Neurology, Neurosurgery, and Psychiatry, 73*, 629–635.
Chou, K. L., Taylor, J. L., & Patil, P. G. (2013). The MDS-UPDRS tracks motor and nonmotor improvement due to subthalamic nucleus deep brain stimulation in Parkinson disease. *Parkinsonism & Related Disorders, 19*, 966–969.
Contarino, M. F., Daniele, A., Sibilia, A. H., Romito, L. M., Bentivoglio, A. R., Gainotti, G., et al. (2007). Cognitive outcome 5 years after bilateral chronic stimulation of subthalamic nucleus in patients with Parkinson's disease. *Journal of Neurology, Neurosurgery, and Psychiatry, 78*(3), 248–252.
Covassin, N., Neikrug, A. B., et al. (2012). Clinical correlates of periodic limb movements in sleep in Parkinson's disease. *Journal of the Neurological Sciences, 316*, 131–136.
Cury, R. G., Galhardoni, R., Fonoff, E. T., dos Santos Ghilardi, M. G., Fonoff, F., Arnaut, D., et al. (2014). Effects of deep brain stimulation on pain and other nonmotor symptoms in Parkinson disease. *Neurology, 83*(16), 1403–1409.
Dafsari, H. S., Reddy, P., Herchenbach, C., Wawro, S., Petry-Schmelzer, J. N., Visser-Vandewalle, V., et al. (2016). Beneficial effects of bilateral subthalamic stimulation on non-motor symptoms in Parkinson's disease. *Brain Stimulation, 9*(1), 78–85.
Daniele, A., Albanese, A., Contarino, M. F., et al. (2003). Cognitive and behavioural effects of chronic stimulation of the subthalamic nucleus in patients with Parkinson's disease. *Journal of Neurology, Neurosurgery, and Psychiatry, 74*, 175–182.
de Andrade, D. C., Lefaucheur, J. P., Galhardoni, R., Ferreira, K. S., Paiva, A. R. B., Bor-Seng-Shu, E., et al. (2012). Subthalamic deep brain stimulation modulates small fiber-dependent sensory thresholds in Parkinson's disease. *Pain, 153*(5), 1107–1113.
de Bie, R. M., de Haan, R. J., Schuurman, P. R., Esselink, R. A., Bosch, D. A., & Speelman, J. D. (2002). Morbidity and mortality following pallidotomy in Parkinson's disease: A systematic review. *Neurology, 58*, 1008–1012.
De la Fuente-Fernández, R. (2004). Uncovering the hidden placebo effect in deep-brain stimulation for Parkinson's disease. *Parkinsonism & Related Disorders, 10*(3), 125–127.
Deli, G., Aschermann, Z., Ács, P., Bosnyák, E., Janszky, J., Faludi, B., et al. (2015). Bilateral subthalamic stimulation can improve sleep quality in Parkinson's disease. *Journal of Parkinson's Disease, 5*(2), 361–368.
Dellapina, E., Ory-Magne, F., Regragui, W., Thalamas, C., Lazorthes, Y., Rascol, O., et al. (2012). Effect of subthalamic deep brain stimulation on pain in Parkinson's disease. *Pain, 153*(11), 2267–2273.
Demetriades, P., Rickards, H., & Cavanna, A. E. (2011). Impulse control disorders following deep brain stimulation of the subthalamic nucleus in Parkinson's disease: Clinical aspects. *Parkinson's Disease, 2011*, 1–9.
Deuschl, G., Schade-Brittinger, C., Krack, P., Volkmann, J., Schäfer, H., Bötzel, K., et al. (2006). A randomized trial of deep brain stimulation for Parkinson's disease. *New England Journal of Medicine, 362*, 2077–2091.
Diederich, N. J., Alesch, F., & Goetz, C. G. (2000). Visual hallucinations induced by deep brain stimulation in Parkinson's disease. *Clinical Neuropharmacology, 23*(5), 287–289.
Driver-Dunckley, E., Evidente, V. G., Adler, C. H., Hillman, R., Hernandez, J., Fletcher, G., et al. (2006). Restless legs syndrome in Parkinson's disease patients may improve with subthalamic stimulation. *Movement Disorders, 21*, 1287–1289.
Dubois, B., Boller, F., Pillon, B., & Agid, Y. (1991). Cognitive deficits in Parkinson's disease. In F. Boller & J. Grafman (Eds.), *Handbook of neuropsychology: Vol. 5*. Amsterdam: Elsevier.

Eisensehr, I., Wetter, T. C., Linke, R., Noachtar, S., Lindeiner, H. V., Gildehaus, F. J., et al. (2001). Normal IPT and IBZM SPECT in drug-naive and levodopa-treated idiopathic restless legs syndrome. *Neurology*, *57*(7), 1307–1309.

Elgh, E., Domellöf, M., Linder, J., Edström, M., Stenlund, H., & Forsgren, L. (2009). Cognitive function in early Parkinson's disease: A population-based study. *European Journal of Neurology*, *16*, 1278–1284.

Elmståhl, S., & Rosén, I. (1997). Postural hypotension and EEG variables predict cognitive decline: Results from a 5-year follow-up of healthy elderly women. *Dementia and Geriatric Cognitive Disorders*, *8*(3), 180–187.

Erola, T., Haapaniemi, T., Heikkinen, E., Huikuri, H., & Myllyä, V. (2006). Subthalamic nucleus deep brain stimulation does not alter long-term heart rate variability in Parkinson's disease. *Clinical Autonomic Research*, *16*(4), 286–288.

Erola, T., Karinen, P., Heikkinen, E., Tuominen, J., Haapaniemi, T., Koivukangas, J., et al. (2005). Bilateral subthalamic nucleus stimulation improves health-related quality of life in Parkinsonian patients. *Parkinsonism & Related Disorders*, *11*(2), 89–94.

Fasano, A., Daniele, A., & Albanese, A. (2012). Treatment of motor and non-motor features of Parkinson's disease with deep brain stimulation. *Lancet Neurology*, *11*, 429–442.

Fields, J., & Troster, A. (2000). Cognitive outcomes after deep brain stimulation for Parkinson's disease: A review of initial studies and recommendations for future research. *Brain and Cognition*, *42*(2), 268–293.

Finazzi-Agro, E., Peppe, A., D'amico, A., Petta, F., Mazzone, P., Stanzione, P., et al. (2003). Effects of subthalamic nucleus stimulation on urodynamic findings in patients with Parkinson's disease. *The Journal of Urology*, *169*(4), 1388–1391.

Foki, T., Hitzl, D., Pirker, W., et al. (2017). Assessment of individual cognitive changes after deep brain stimulation surgery in Parkinson's disease sing the Neuropsychological Test Battery Vienna short version. *Wiener Klinische Wochenschrift*, 1–8.

Follett, K. A., Weaver, F. M., Stern, M., Hur, K., Harris, C. L., Luo, P., et al.CSP 468 Study Group, (2010). Pallidal versus subthalamic deep-brain stimulation for Parkinson's disease. *New England Journal of Medicine*, *362*, 2077–2091.

Ford, B. (1997). Pain in Parkinson's disease. *Clinical Neuroscience (New York, NY)*, *5*(2), 63–72.

Frank, M. J., Samanta, J., Moustafa, A. A., & Sherman, S. J. (2007). Hold your horses: Impulsivity, deep brain stimulation, and medication in parkinsonism. *Science*, *318*(5854), 1309–1312.

Frankemolle, A. M. M., Wu, J., Noecker, A. M., et al. (2010). Reversing cognitive-motor impairments in Parkinson's disease patients using a computational modelling approach to deep brain stimulation programming. *Brain*, *133*, 746–761.

Frazier, L., & Marsh, L. (2006). Coping. In M. Menza & L. Marsh (Eds.), *Psychiatric issues in Parkinson's disease: A practical guide* (pp. 235–247). United Kingdom: Taylor & Francis.

Freedman, R. (2008). Editorial. Coping, resilience, and outcome. *American Journal of Psychiatry*, *165*, 1505–1506.

Funkiewiez, A., Ardouin, C., Caputo, E., et al. (2004). Long term effects of bilateral subthalamic nucleus stimulation on cognitive function, mood, and behaviour in Parkinson's disease. *Journal of Neurology, Neurosurgery, and Psychiatry*, *75*, 834–839.

Gierthmuhlen, J., Arning, P., Binder, A., Herzog, J., Deuschl, G., Wasner, G., et al. (2010). Influence of deep brain stimulation and levodopa on sensory signs in Parkinson's disease. *Movement Disorders*, *25*, 1195–1202.

Gironell, A., Kulisevsky, J., Rami, L., et al. (2003). Effects of pallidotomy and bilateral subthalamic stimulation on cognitive function in Parkinson disease. *Journal of Neurology*, *250*, 917–923.

Goetz, C. G., Tanner, C. M., Levy, M., Wilson, R. S., & Garron, D. C. (1986). Pain in Parkinson's disease. *Movement Disorders*, *1*(1), 45–49.

Goldstein, D. S. (2006). Orthostatic hypotension as an early finding in Parkinson's disease. *Clinical Autonomic Research*, *16*, 46–54.

Hälbig, T. D., Tse, W., Frisina, P. G., Baker, B. R., Hollander, E., Shapiro, H., et al. (2009). Subthalamic deep brain stimulation and impulse control in Parkinson's disease. *European Journal of Neurology*, *16*(4), 493–497.

Halim, A., Baumgartner, L., & Binder, D. K. (2011). Effect of deep brain stimulation on autonomic dysfunction in patients with Parkinson's disease. *Journal of Clinical Neuroscience*, *18*(6), 804–806.

Hely, M. A., Reid, W. G., Adena, M. A., Halliday, G. M., & Morris, J. G. (2008). The Sydney multicenter study of Parkinson's disease: The inevitability of dementia at 20 years. *Movement Disorders*, *23*(6), 837–844.

Herzog, J., Volkmann, J., Krack, P., et al. (2003). Two-year follow-up of subthalamic deep brain stimulation in Parkinson's disease. *Movement Disorders*, *18*, 1332–1337.

Herzog, J., Weiss, P. H., Assmus, A., Wefer, B., Seif, C., Braun, P. M., et al. (2006). Subthalamic stimulation modulates cortical control of urinary bladder in Parkinson's disease. *Brain*, *129*, 3366–3375.

Herzog, J., Weiss, P. H., Assmus, A., Wefer, B., Seif, C., Braun, P. M., et al. (2008). Improved sensory gating of urinary bladder afferents in Parkinson's disease following subthalamic stimulation. *Brain*, *131*(1), 132–145.

Hjort, N., Østergaard, K., & Dupont, E. (2004). Improvement of sleep quality in patients with advanced Parkinson's disease treated with deep brain stimulation of the subthalamic nucleus. *Movement Disorders*, *19*(2), 196–199.

Holmberg, B., Corneliusson, O., & Elam, M. (2005). Bilateral stimulation of nucleus subthalamicus in advanced Parkinson's disease: No effects on, and of, autonomic dysfunction. *Movement Disorders*, *20*(8), 976–981.

Houeto, J. L., Mesnage, V., Mallet, L., et al. (2002). Behavioural disorders, Parkinson's disease and subthalamic stimulation. *Journal of Neurology, Neurosurgery, and Psychiatry*, *72*, 701–707.

Hwynn, N., Ul Haq, I., Malaty, I. A., Resnick, A. S., Dai, Y., Foote, K. D., et al. (2011). Effect of deep brain stimulation on Parkinson's nonmotor symptoms following unilateral DBS: A pilot study. *Parkinson's Disease, 2011*.

Hyam, J. A., Brittain, J. S., Paterson, D. J., Davies, R. J., Aziz, T. Z., & Green, A. L. (2012). Controlling the lungs via the brain: A novel neurosurgical method to improve lung function in humans. *Neurosurgery*, *70*(2), 469–478.

Hyam, J. A., Kringelbach, M. L., Silburn, P. A., Aziz, T. Z., & Green, A. L. (2012). The autonomic effects of deep brain stimulation—A therapeutic opportunity. *Nature Reviews Neurology*, *8*(7), 391–400.

International Restless Legs Syndrome Study Group. (2003). Validation of the International Restless Legs Syndrome Study Group rating scale for restless legs syndrome. *Sleep Medicine*, *4*(2), 121–132.

Iranzo, A., Valldeoriola, F., Santamaria, J., Tolosa, E., & Rumia, J. (2002). Sleep symptoms and polysomnographic architecture in advanced Parkinson's disease after chronic bilateral subthalamic stimulation. *Journal of Neurology, Neurosurgery, and Psychiatry*, *72*(5), 661–664.

Jafari, N., Pahwa, R., Nazzaro, J. M., Arnold, P. M., & Lyons, K. E. (2016). MDS-UPDRS to assess non-motor symptoms after STN DBS for Parkinson's disease. *International Journal of Neuroscience*, *126*(1), 25–29.

Johns, M. W. (1991). A new method for measuring daytime sleepiness: The Epworth sleepiness scale. *Sleep*, *14*(6), 540–545.

Jung, Y. J., Kim, H. J., Jeon, B. S., Park, H., Lee, W. W., & Paek, S. H. (2015). An 8-year follow-up on the effect of subthalamic nucleus deep brain stimulation on pain in Parkinson disease. *JAMA Neurology*, *72*(5), 504–510.

Juri, C., Rodríguez-Oroz, M. C., Burguera, J. A., Guridi, J., & Obeso, J. A. (2010). Pain and dyskinesia in Parkinson's disease. *Movement Disorders, 25*(1), 130–132.
Kahn, E., D'Haese, P. F., Dawant, B., Allen, L., Kao, C., Charles, P. D., et al. (2012). Deep brain stimulation in early stage Parkinson's disease: Operative experience from a prospective randomised clinical trial. *Journal of Neurology, Neurosurgery, and Psychiatry, 83*, 164–170. Published Online First: 02 September 2011.
Kaufmann, H., Bhattacharya, K. F., Voustianiouk, A., & Gracies, J. M. (2002). Stimulation of the subthalamic nucleus increases heart rate in patients with Parkinson disease. *Neurology, 59*(10), 1657.
Kaufmann, H., Nahm, K., Purohit, D., & Wolfe, D. (2004). Autonomic failure as the initial presentation of Parkinson disease and dementia with Lewy bodies. *Neurology, 63*(6), 1093–1095.
Kedia, S., Moro, E., Tagliati, M., Lang, A. E., & Kumar, R. (2004). Emergence of Restless Legs Syndrome (RLS) during subthalamic stimulation for Parkinson's disease. *Neurology, 63*, 2410–2412.
Kessler, T. M., Burkhard, F. C., Z'Brun, S., Stibal, A., Studer, U. E., Hess, C. W., et al. (2008). Effect of thalamic deep brain stimulation on lower urinary tract function. *European Urology, 53*(3), 607–612.
Kim, H. J., Jeon, B. S., Lee, J. Y., Paek, S. H., & Kim, D. G. (2012). The benefit of subthalamic deep brain stimulation for pain in Parkinson disease: A 2-year follow-up study. *Neurosurgery, 70*(1), 18–24.
Kim, H. J., Jeon, B. S., & Paek, S. H. (2015). Nonmotor symptoms and subthalamic deep brain stimulation in Parkinson's disease. *Journal of Movement Disorders, 8*(2), 83–91.
Kim, H. J., Paek, S. H., Kim, J. Y., Lee, J. Y., Lim, Y. H., & Kim, M. R. (2015). Rapid eye movement sleep behavior disorder after bilateral subthalamic stimulation in Parkinson's disease. *Journal of Clinical Neuroscience, 22*(2), 315–319.
Kim, H. J., Paek, S. H., Kim, J. Y., Lee, J. Y., Lim, Y. H., Kim, M. R., et al. (2008). Chronic subthalamic deep brain stimulation improves pain in Parkinson disease. *Journal of Neurology, 255*(12), 1889–1894.
Kirsch-Darrow, L., Zahodne, L. B., Marsiske, M., Okun, M., Foote, K., & Bowers, D. (2011). The trajectory of apathy after deep brain stimulation: From pre-surgery to 6 months post-surgery in Parkinson's disease. *Parkinsonism & Related Disorders, 17*(3), 182–188.
Klassen, B. T., Lyons, M. K., & Evidente, V. G. (2013). Bilateral globus pallidus internus deep brain stimulation in Parkinson's disease with periodic limb movement disorder. *Turkish Neurosurgery, 23*(1), 104–106.
Kleiner-Fisman, G., Herzog, J., Fisman, D. N., Tamma, F., Lyons, K. E., Pahwa, R., et al. (2006). Subthalamic nucleus deep brain stimulation: Summary and meta-analysis of outcomes. *Movement Disorders, 21*(S14), S290–S304.
Krack, P., Batir, A., Van Blercom, N., Chabardes, S., Fraix, V., Ardouin, C., et al. (2003). Five-year follow-up of bilateral stimulation of the subthalamic nucleus in advanced Parkinson's disease. *New England Journal of Medicine, 349*, 1925–1934.
Krogh, K., Ostergaard, K., Sabroe, S., & Laurberg, S. (2008). Clinical aspects of bowel symptoms in Parkinson's disease. *Acta Neurologica Scandinavica, 117*(1), 60–64.
Kumar, R., Lozano, A. M., Sime, E., et al. (1999). Comparative effects of unilateral and bilateral subthalamic nucleus deep brain stimulation. *Neurology, 53*(3), 561–566.
Kurtis, M. M., Rajah, T., Delgado, L. F., & Dafsari, H. S. (2017). The effect of deep brain stimulation on the non-motor symptoms of Parkinson's disease: A critical review of the current evidence. *npj Parkinson's Disease, 3*, 16024.
Kurtis, M. M., Rodriguez-Blazquez, C., Martinez-Martin, P., & ELEP Group. (2013). Relationship between sleep disorders and other non-motor symptoms in Parkinson's disease. *Parkinsonism & Related Disorders, 19*(12), 1152–1155.

Lacritz, L. H., Cullum, C. M., Frol, A. B., Dewey, R. B., Jr., & Giller, C. A. (2000). Neuropsychological outcome following unilateral stereotactic pallidotomy in intractable Parkinson's disease. *Brain and Cognition*, *42*(3), 364–378.

Lagi, A., & Spini, S. (2010). Clinostatic hypertension and orthostatic hypotension. *Clinical Cardiology*, *33*(6), E10–E15.

Laitinen, L. (2000). Behavioral complications of early pallidotomy. *Brain and Cognition*, *42*, 313–323.

Leentjens, A. F., Dujardin, K., Marsh, L., Martinez-Martin, P., Richard, I. H., Starkstein, S. E., et al. (2008). Apathy and anhedonia rating scales in Parkinson's disease: Critique and recommendations. *Movement Disorders*, *23*(14), 2004–2014.

Lhommée, E., Klinger, H., Thobois, S., Schmitt, E., Ardouin, C., Bichon, A., et al. (2012). Subthalamic stimulation in Parkinson's disease: Restoring the balance of motivated behaviours. *Brain*, *135*(5), 1463–1477.

Lim, S. Y., & Lang, A. E. (2010). The nonmotor symptoms of Parkinson's disease—An overview. *Movement Disorders*, *25*(S1), S123–S130.

Lim, A. S., Moro, E., Lozano, A. M., Hamani, C., Dostrovsky, J. O., Hutchison, W. D., et al. (2009). Selective enhancement of rapid eye movement sleep by deep brain stimulation of the human pons. *Annals of Neurology*, *66*(1), 110–114.

Limousin, P., Greene, J., Pollak, P., Rothwell, J., Benabid, A. L., & Frackowiak, R. (1997). Changes in cerebral activity pattern due to subthalamic nucleus or internal pallidum stimulation in Parkinson's disease. *Annals of Neurology*, *42*(3), 283–291.

Linke, R., Eisensehr, I., Wetter, T. C., Gildehaus, F. J., Pöpperl, G., Trenkwalder, C., et al. (2004). Presynaptic dopaminergic function in patients with restless legs syndrome: Are there common features with early Parkinson's disease? *Movement Disorders*, *19*(10), 1158–1162.

Lipp, A., Tank, J., Trottenberg, T., Kupsch, A., Arnold, G., & Jordan, J. (2005). Sympathetic activation due to deep brain stimulation in the region of the STN. *Neurology*, *65*(5), 774–775.

Litvan, I., Goldman, J., Troster, A. I., Schmand, B. A., Weintraub, D., Petersen, R. C., et al. (2012). Diagnostic criteria for mild cognitive impairment in Parkinson's disease: Movement Disorder Society Task Force guidelines. *Movement Disorders*, *27*(3), 349–356.

Low, P. A. (2008). Prevalence of orthostatic hypotension. *Clinical Autonomic Research*, *18*(1), 8–13.

Ludwig, J., Remien, P., Guballa, C., Binder, A., Binder, S., Schattschneider, J., et al. (2007). Effects of subthalamic nucleus stimulation and levodopa on the autonomic nervous system in Parkinson's disease. *Journal of Neurology, Neurosurgery, and Psychiatry*, *78*(7), 742–745.

Luukinen, H., Koski, K., Laippala, P., & Kivelä, S. L. (1999). Prognosis of diastolic and systolic orthostatic hypotension in older persons. *Archives of Internal Medicine*, *159*(3), 273–280.

Lyons, K. E., & Pahwa, R. (2006). Effects of bilateral subthalamic nucleus stimulation on sleep, daytime sleepiness, and early morning dystonia in patients with Parkinson disease. *Journal of Neurosurgery*, *104*(4), 502–505.

MacCarthy, B., & Brown, R. (1989). Psychosocial factors in Parkinson's disease. *British Journal of Clinical Psychology*, *28*, 41–52.

Magerkurth, C., Schnitzer, R., & Braune, S. (2005). Symptoms of autonomic failure in Parkinson's disease: Prevalence and impact on daily life. *Clinical Autonomic Research*, *15*(2), 76–82.

Maier, F., Lewis, C. J., Horstkoetter, N., Eggers, C., Kalbe, E., Maarouf, M., et al. (2013). Patients' expectations of deep brain stimulation, and subjective perceived outcome related to clinical measures in Parkinson's disease: A mixed-method approach. *Journal of Neurology, Neurosurgery, and Psychiatry*, *84*, 1273–1281.

Marques, A., Chassin, O., Morand, D., Pereira, B., Debilly, B., Derost, P., et al. (2013). Central pain modulation after subthalamic nucleus stimulation: A crossover randomized trial. *Neurology, 81*(7), 633–640.

Marras, C., Troster, A., Kulisevsky, J., & Stebbins, G. T. (2014). The tools of the trade: A state of the art "how to Assess Cognition" in the patient with Parkinson's disease. *Movement Disorders, 29*(5), 584–596.

Maruo, T., Saitoh, Y., Hosomi, K., Kishima, H., Shimokawa, T., Hirata, M., et al. (2011). Deep brain stimulation of the subthalamic nucleus improves temperature sensation in patients with Parkinson's disease. *Pain, 152*(4), 860–865.

Masaki, K. H., Schatz, I. J., Burchfiel, C. M., Sharp, D. S., Chiu, D., Foley, D., et al. (1998). Orthostatic hypotension predicts mortality in elderly men. *Circulation, 98*(21), 2290–2295.

Menza, M., & Dobkin, R. (2006). Anxiety. In M. Menza & L. Marsh (Eds.), *Psychiatric issues in Parkinson's disease: A practical guide* (pp. 139–153). United Kingdom: Taylor & Francis.

Menza, M., Robertson-Hoffman, D., & Bonapace, A. (1993). Parkinson's disease and anxiety: Comorbidity with depression. *Biological Psychiatry, 34*, 465–470.

Mercado, R., Constantoyannis, C., Mandat, T., Kumar, A., Schulzer, M., Stoessl, A. J., et al. (2006). Expectation and the placebo effect in Parkinson's disease patients with subthalamic nucleus deep brain stimulation. *Movement Disorders, 21*, 1457–1461.

Merello, M., Nouzeilles, M. I., Kuzis, G., et al. (1999). Unilateral radiofrequency lesion versus electrostimulation of posteroventral pallidum: A prospective randomized comparison. *Movement Disorders, 14*(1), 50–56.

Merola, A., Zibetti, M., Angrisano, S., Rizzi, L., Ricchi, V., Artusi, C. A., et al. (2011). Parkinson's disease progression at 30 years: A study of subthalamic deep brain-stimulated patients. *Brain, 134*, 2074–2084.

Michaud, M., Soucy, J. P., Chabli, A., Lavigne, G., & Montplaisir, J. (2002). SPECT imaging of striatal pre- and postsynaptic dopaminergic status in restless legs syndrome with periodic leg movements of sleep. *Journal of Neurology, 249*, 164–170.

Molina, J. A., Sáinz-Artiga, M. J., Fraile, A., Jiménez-Jiménez, F. J., Villanueva, C., Ortí-Pareja, M., et al. (2000). Pathologic gambling in Parkinson's disease: A behavioral manifestation of pharmacologic treatment? *Movement Disorders, 15*(5), 869–872.

Monaca, C., Ozsancak, C., Jacquesson, J. M., Poirot, I., Blond, S., Destee, A., et al. (2004). Effects of bilateral subthalamic stimulation on sleep in Parkinson's disease. *Journal of Neurology, 251*(2), 214–218.

Mridula, K. R., Borgohain, R., Jabeen, S. A., Padmaja, G., Bandaru, V. S., Ankathi, P., et al. (2015). Comparison of frequencies of non motor symptoms in Indian Parkinson's disease patients on medical management versus deep brain stimulation: A case-control study. *Iranian Journal of Neurology, 14*(2), 86.

Muzerengi, S., Lewis, H., Edwards, M., Kipps, E., Bahl, A., Martinez-Martin, P., et al. (2006). Non-motor symptoms in Parkinson's disease: An underdiagnosed problem. *Future Medicine, 2*, 967–982.

Nakamura, K., Christine, C. W., Starr, P. A., & Marks, W. J. (2007). Effects of unilateral subthalamic and pallidal deep brain stimulation on fine motor functions in Parkinson's disease. *Movement Disorders, 22*(5), 619–626.

Nazzaro, J. M., Pahwa, R., & Lyons, K. E. (2011). The impact of bilateral subthalamic stimulation on non-motor symptoms of Parkinson's disease. *Parkinsonism & Related Disorders, 17*(8), 606–609.

Nishida, N., Murakami, T., Kadoh, K., Tohge, R., Yamanegi, M., Saiki, H., et al. (2011). Subthalamic nucleus deep brain stimulation restores normal rapid eye movement sleep in Parkinson's disease. *Movement Disorders, 26*(13), 2418–2422.

Odekerken, V., Van Laar, T., Staal, M., Mosch, A., Hoffmann, C. F., Nijssen, P. C., et al. (2013). Subthalamic nucleus versus globus pallidus bilateral deep brain stimulation for advanced Parkinson's disease (NSTAPS study): A randomised controlled trial. *The Lancet, 12*(1), 37–44.

Ohayon, M. M., & Roth, T. (2002). Prevalence of restless legs syndrome and periodic limb movement disorder in the general population. *Journal of Psychosomatic Research, 53*(1), 547–554.

Okun, M. S., Fernandez, H. H., Wu, S. S., Kirsch-Darrow, L., Bowers, D., Bova, F., et al. (2009). Cognition and mood in Parkinson's disease in subthalamic nucleus versus globus pallidus interna deep brain stimulation: The COMPARE trial. *Annals of Neurology, 65*, 586–595.

Okun, M. S., Gallo, B. V., Mandybur, G., Jagid, J., Foote, K., Revilla, F., et al. (2012). A randomized, controlled trial of constant current subthalamic DBS in Parkinson's disease. *Lancet Neurology, 11*, 140–149.

Okun, M. S., Mann, G., Foote, K. D., Shapira, N. A., Bowers, D., Springer, U., et al. (2007). Deep brain stimulation in the internal capsule and nucleus accumbens region: Responses observed during active and sham programming. *Journal of Neurology, Neurosurgery, and Psychiatry, 78*(3), 310–314.

Okun, M. S., Tagliati, M., Pourfar, M., Fernandez, H. H., Rodriguez, R. L., Alterman, R. L., et al. (2005). Management of referred deep brain stimulation failures: A retrospective analysis from 2 movement disorders centers. *Archives of Neurology, 62*(8), 1250–1255.

Olson, E. J., Boeve, B. F., & Silber, M. H. (2000). Rapid eye movement sleep behavior disorder: Demographic, clinical and laboratory findings in 93 cases. *Brain, 123*, 331–339.

Ondo, W. G., Vuong, K. D., & Jankovic, J. (2002). Exploring the relationship between Parkinson disease and restless legs syndrome. *Archives of Neurology, 59*, 421–424.

Ortega-Cubero, S., Clavero, P., Irurzun, C., Gonzalez-Redondo, R., Guridi, J., Obeso, J. A., et al. (2013). Effect of deep brain stimulation of the subthalamic nucleus on non-motor fluctuations in Parkinson's disease: Two-year' follow-up. *Parkinsonism & Related Disorders, 19*(5), 543–547.

Pagano, G., Tan, E. E., Haider, J. M., Bautista, A., & Tagliati, M. (2015). Constipation is reduced by beta-blockers and increased by dopaminergic medications in Parkinson's disease. *Parkinsonism & Related Disorders, 21*(2), 120–125.

Parent, A., Levesque, M., & Parent, M. (2001). A re-evaluation of the current model of the basal ganglia. *Parkinsonism & Related Disorders*, 7, 193–198.

Parsons, T. D., Rogers, S. A., Braaten, A. J., Woods, S. P., & Troster, A. I. (2006). Cognitive sequelae of subthalamic nucleus deep brain stimulation in Parkinson's disease: A meta-analysis. *Lancet Neurology, 5*, 578–588.

Patel, N. K., Heywood, P., O'Sullivan, K., McCarter, R., Love, S., & Gill, S. S. (2003). Unilateral subthalamotomy in the treatment of Parkinson's disease. *Brain, 126*, 1136–1145.

Pellaprat, J., Ory-Magne, F., Canivet, C., Simonetta-Moreau, M., Lotterie, J. A., Radji, F., et al. (2014). Deep brain stimulation of the subthalamic nucleus improves pain in Parkinson's disease. *Parkinsonism & Related Disorders, 20*(6), 662–664.

Perozzo, P., Rizzone, M., Bergamasco, B., Castelli, L., et al. (2001). Deep brain stimulation of the subthalamic nucleus in Parkinson's disease: Comparison of pre -and postoperative neuropsychological evaluation. *Journal of Neurological Sciences, 192*, 9–15.

Petrovitch, H., Abbott, R. D., Ross, G. W., Nelson, J., Masaki, K. H., Tanner, C. M., et al. (2009). Bowel movement frequency in late-life and substantia nigra neuron density at death. *Movement Disorders, 24*(3), 371–376.

Pfeiffer, R. F. (2003). Gastrointestinal dysfunction in Parkinson's disease. *The Lancet Neurology, 2*(2), 107–116.

Pillon, B., Ardouin, C., Damier, P., et al. (2000). Neuropsychological changes between "off" and "on" STN or GPi stimulation in Parkinson's disease. *Neurology, 55*, 411–418.

Pollak, P., Krack, P., Fraix, V., Mendes, A., Moro, E., Chabardes, S., et al. (2002). Intraoperative micro-and macrostimulation of the subthalamic nucleus in Parkinson's disease. *Movement Disorders, 17*(S3), S155–S161.

Pont-Sunyer, C., Hotter, A., Gaig, C., Seppi, K., et al. (2015). The onset of nonmotor symptoms in Parkinson's disease (the ONSET PD study). *Movement Disorders, 30*(2), 229–237.

Postuma, R. B., Gagnon, J. F., Vendette, M., Charland, K., & Montplaisir, J. (2008a). Manifestations of Parkinson disease differ in association with REM sleep behavior disorder. *Movement Disorders*, *23*(12), 1665–1672.

Postuma, R. B., Gagnon, J. F., Vendette, M., Charland, K., & Montplaisir, J. (2008b). REM sleep behaviour disorder in Parkinson's disease is associated with specific motor features. *Journal of Neurology, Neurosurgery, and Psychiatry*, *79*(10), 1117–1121.

Postuma, R. B., Lang, A. E., Massicotte-Marquez, J., & Montplaisir, J. (2006). Potential early markers of Parkinson disease in idiopathic REM sleep behavior disorder. *Neurology*, *66*(6), 845–851.

Rasmussen, H., Scheier, M. F., & Greenhouse, J. B. (2009). Optimism and physical health: A meta-analytic review. *Annals of Behavioral Medicine*, *37*, 239–256.

Reich, M. M., Chaudhuri, K. R., Ashkan, K., Hulse, N., Costello, A., Moriarty, J., et al. (2011). Changes in the non-motor symptom scale in Parkinson's disease after deep brain stimulation. *Basal Ganglia*, *1*(3), 131–133.

Reijnders, R. L., Scholtissen, B., Weber, W. E., Aalten, P., Verhey, F. R., & Leentjens, A. F. (2010). Neuroanatomical correlates of apathy in Parkinson's disease: A magnetic resonance imaging study using voxel-based morphometry. *Movement Disorders*, *30*(25), 2318–2325.

Rizzone, M. G., Fasano, A. L., Daniele, A., Zibetti, M., Merola, A., Rizzi, L., et al. (2014). Long-term outcome of subthalamic nucleus DBS in Parkinson's disease: From the advanced phase towards the late stage of the disease? *Parkinsonism & Related Disorders*, *20*(4), 376–381.

Romito, L. M., Raja, M., Daiel, A., et al. (2002). Transient mania with hypersexuality after surgery for high frequency stimulation of the subthalamic nucleus in Parkinson's disease. *Movement Disorders*, *17*, 1371–1374.

Rothlind, J. C., York, M. K., Carlson, K., et al. (2015). Neuropsychological changes following deep brain stimulation surgery for Parkinson's disease: Comparisons of treatment at pallidal and subthalamic targets versus best medical therapy. *Journal of Neurology, Neurosurgery, and Psychiatry*, *86*(6), 622–629.

Rukmini, M. K., Borgohain, R., Jabeen, S. A., Padmaja, G., Bandaru, V. S., Ankathi, P., et al. (2015). Comparison of frequencies of non motor symptoms in Indian Parkinson's disease patients on medical management versus deep brain stimulation: A case-control study. *Iranian Journal of Neurology*, *14*(2), 86–93.

Rye, D. B. (2004). Parkinson's disease and RLS: The dopaminergic bridge. *Sleep Medicine*, *5*(3), 317–328.

Rye, D. B., & DeLong, M. R. (1999). Amelioration of sensory limb discomfort of restless legs syndrome by pallidotomy. *Annals of Neurology*, *46*(5), 800–801.

Ryu, H. S., Kim, M. S., You, S., Kim, M. J., Kim, Y. J., Kim, J., et al. (2016). Mortality of advanced Parkinson's disease patients treated with deep brain stimulation surgery. *Journal of the Neurological Sciences*, *369*, 230–235.

Saint-Cyr, J. A., Trepanier, L. L., Kumar, R., Lozano, A. M., & Lang, A. E. (2000). Neuropsychological consequences of chronic bilateral stimulation of the subthalamic nucleus in Parkinson's disease. *Brain*, *123*, 2091–2108.

Sakakibara, R., Tateno, F., Kishi, M., Tsuyuzaki, Y., Uchiyama, T., & Yamamoto, T. (2012). Pathophysiology of bladder dysfunction in Parkinson's disease. *Neurobiology of Disease*, *46*(3), 565–571.

Sánchez-Ferro, Á., Benito-León, J., & Gómez-Esteban, J. C. (2013). The management of orthostatic hypotension in Parkinson's disease. *Frontiers in Neurology*, *4*, 64.

Sauleau, P., Raoul, S., Lallement, F., Rivier, I., Drapier, S., Lajat, Y., et al. (2005). Motor and non motor effects during intraoperative subthalamic stimulation for Parkinson's disease. *Journal of Neurology*, *252*(4), 457–464.

Schneider, F., Habel, U., Volkmann, J., et al. (2003). Deep brain stimulation of the subthalamic nucleus enhances emotional processing in Parkinson disease. *Archives of General Psychiatry*, *60*, 296–302.

Schrag, A., Sauerbier, A., & Chaudhuri, K. R. (2015). New clinical trials for nonmotor manifestations of Parkinson's disease. *Movement Disorders*, *30*, 1490–1504.

Schüpbach, W. M. M., Chastan, N., Welter, M. L., et al. (2005). Stimulation of the subthalamic nucleus in Parkinson's disease: A 5 year follow up. *Journal of Neurology, Neurosurgery, and Psychiatry*, *76*, 1640–1644.

Schüpbach, M., Gargiulo, M., Welter, M. L., et al. (2006). Neurosurgery in Parkinson disease: A distressed mind in a repaired body? *Neurology*, *66*(12), 1811–1816.

Seif, C., Herzog, J., van der Horst, C., Schrader, B., Volkmann, J., Deuschl, G., et al. (2004). Effect of subthalamic deep brain stimulation on the function of the urinary bladder. *Annals of Neurology*, *55*(1), 118–120.

Senard, J. M., Rai, S., Lapeyre-Mestre, M., Brefel, C., Rascol, O., Rascol, A., et al. (1997). Prevalence of orthostatic hypotension in Parkinson's disease. *Journal of Neurology, Neurosurgery, and Psychiatry*, *63*(5), 584–589.

Shapira, N. A., Okun, M. S., Wint, D., Foote, K. D., Byars, J. A., Bowers, D., et al. (2006). Panic and fear induced by deep brain stimulation. *Journal of Neurology, Neurosurgery, and Psychiatry*, 77(3), 410–412.

Shimizu, N., Matsumoto, S., Mori, Y., Yoshioka, N., Uemura, H., Nakano, N., et al. (2007). Effects of deep brain stimulation on urodynamic findings in patients with Parkinson's disease. *Hinyokika kiyo. Acta urologica Japonica*, *53*(9), 609–612.

Simpson, J., Haines, K., Lekwuwa, G., Wardle, J., & Crawford, T. (2006). Social support and psychological outcome in people with Parkinson's disease: Evidence for a specific pattern of associations. *British Journal of Clinical Psychology*, *45*, 585–590.

Smeding, J. M., Esselink, R. A., Schmand, B., et al. (2005). Unilateral pallidotomy versus bilateral subthalamic nucleus stimulation in PD: A comparison of neuropsychological effects. *Journal of Neurology*, *252*, 176–182.

Smeding, H. M. M., Goudriaan, A. E., Foncke, E. M. J., Schuurman, P. R., Speelman, J. D., & Schmand, B. (2007). Pathological gambling after bilateral subthalamic nucleus stimulation in Parkinson disease. *Journal of Neurology, Neurosurgery, and Psychiatry*, *78*, 517–519.

Smeding, H. M., Speelman, J. D., Koning-Haanstra, M., et al. (2006). Neuropsychological effects of bilateral STN stimulation in Parkinson disease: A controlled study. *Neurology*, *66*(12), 1830–1836.

Spielberger, S., Wolf, E., Kress, M., Seppi, K., & Poewe, W. (2011). The influence of deep brain stimulation on pain perception in Parkinson's disease. *Movement Disorders*, *26*(7), 1367–1368.

Starkstein, S. E., & Leentjens, A. F. (2008). The nosological position of apathy in clinical practice. *Journal of Neurology, Neurosurgery, and Psychiatry*, *79*(10), 1088–1092.

Stefani, A., Pierantozzi, M., Ceravolo, R., Brusa, L., Galati, S., & Stanzione, P. (2010). Deep brain stimulation of pedunculopontine tegmental nucleus (PPTg) promotes cognitive and metabolic changes: A target-specific effect or response to a low-frequency pattern of stimulation? *Clinical EEG and Neuroscience*, *41*(2), 82–86.

Stemper, B., Berić, A., Welsch, G., Haendl, T., Sterio, D., & Hilz, M. J. (2006). Deep brain stimulation improves orthostatic regulation of patients with Parkinson disease. *Neurology*, *67*, 1781–1785.

Stern, Y. (2002). What is cognitive reserve? Theory and research application of the reserve concept. *Journal of International Neuropsychological Society*, *8*, 448–460.

Sürücü, O., Baumann-Vogel, H., Uhl, M., Imbach, L. L., & Baumann, C. R. (2013). Subthalamic deep brain stimulation versus best medical therapy for L-dopa responsive pain in Parkinson's disease. *Pain*, *154*(8), 1477–1479.

Susatia, F., Foote, K., Ward, H., & Okun, M. (2011). Assessing patient outcome and troubleshooting deep brain stimulation. In W. Marks Jr., (Ed.), *Deep brain stimulation management* (pp. 91–130). New York: Cambridge University Press.

Tandberg, E., Larsen, J. P., Aarsland, D., Laake, K., & Cummings, J. L. (1996). The occurrence of depression in Parkinson's disease. A community-based study. *Archives of Neurology, 53*, 175–179.

Tateno, F., Sakakibara, R., Nagao, T., Tuyusaki, Y., Aiba, Y., Tateno, H., et al. (2015). Deep brain stimulation ameliorates postural hypotension in Parkinson's disease. *Journal of the American Geriatrics Society, 63*(10), 2186–2189.

Taylor, A. E., Saint-Cyr, J. A., & Lang, A. (1986). Frontal lobe dysfunction in Parkinson's disease. *Brain, 109*, 845–883.

Temel, Y., Kessels, A., Tan, S., Topdag, A., Boon, P., & Visser-Vandewalle, V. (2006). Behavioural changes after bilateral subthalamic stimulation in advanced Parkinson disease: A systematic review. *Parkinsonism & Related Disorders, 12*, 265–272.

Thobois, S., Ardouin, C., Lhommée, E., Klinger, H., Lagrange, C., Xie, J., et al. (2010). Non-motor dopamine withdrawal syndrome after surgery for Parkinson's disease: Predictors and underlying mesolimbic denervation. *Brain, 133*, 1111–1127.

Tir, M., Devos, D., Blond, S., Touzet, G., Reyns, N., Duhamel, A., et al. (2007). Exhaustive, one-year follow-up of subthalamic nucleus deep brain stimulation in a large, single-center cohort of Parkinsonian patients. *Neurosurgery, 61*(2), 297–305.

Trachani, E., Constantoyannis, C., Sakellaropoulos, G. C., Stavrinou, M. L., Nikiforidis, G., & Chroni, E. (2012). Heart rate variability in Parkinson's disease unaffected by deep brain stimulation. *Acta Neurologica Scandinavica, 126*(1), 56–61.

Troster, A., & Fields, J. (2008). Parkinson's disease, progressive supranuclear palsy, corticobasal degeneration, and related disorders of the frontostriatal system. In J. Morgan & J. Ricker (Eds.), *Textbook of clinical neuropsychology* (pp. 536–577). United Kingdom: Taylor & Francis.

Tröster, A. I., Jankovic, J., Tagliati, M., Peichel, D. L., & Okun, M. S. (2017). Neuropsychological outcomes in a controlled trial of constant current deep brain stimulation for Parkinson's disease. *Movement Disorders, 32*, 433–440.

Troster, A., & Woods, S. (2003). Neuropsychological aspects of Parkinson's disease and parkinsonian syndromes. In R. Pahwa, K. E. Lyons, & W. C. Koller (Eds.), *Handbook of Parkinson's disease.* (3rd ed.). New York: Marcel Dekker.

Troster, A. I., Woods, S. P., Fields, J. A., et al. (2003). Verbal fluency declines after pallidotomy: An interaction between task and lesion laterality. *Applied Neuropsychology, 10*, 69–75.

Ulla, M., Thobois, S., Llorca, P. M., et al. (2011). Contact dependent reproducible hypomania induced by deep brain stimulation in Parkinson's disease: Clinical, anatomical and functional imaging study. *Journal of Neurology, Neurosurgery, and Psychiatry, 82*(6), 607–614.

Vendette, M., Gagnon, J. F., Decary, A., Massicotte-Marquez, J., Postuma, R. B., Doyon, J., et al. (2007). REM sleep behavior disorder predicts cognitive impairment in Parkinson disease without dementia. *Neurology, 69*(19), 1843–1849.

Volkmann, J., Albanese, A., Kulisevsky, J., Tornqvist, A. L., Houeto, J. L., Pidoux, B., et al. (2009). Long-term effects of pallidal or subthalamic deep brain stimulation on quality of life in Parkinson's disease. *Movement Disorders, 24*(8), 1154–1161.

Volkmann, J., Allert, N., Voges, J., et al. (2001). Safety and efficacy of pallidal or subthalamic nucleus stimulation in advanced PD. *Neurology, 56*, 548–551.

Voon, V., & Fox, S. H. (2007). Medication-elated impulse control and repetitive behaviors in Parkinson disease. *Archives of Neurology, 64*, 1089–1096.

Voon, V., Krack, P., Lang, A. E., et al. (2008). A multicentre study on suicide outcomes following subthalamic stimulation for Parkinson's disease. *Brain, 131*, 2720–2728.

Voon, V., Kubu, C., Krack, P., Houeto, J., & Troster, A. (2006). Deep brain stimulation: Neuropsychological and neuropsychiatric issues. *Movement Disorders, 21*(14), S305–S327.

Weaver, F., Follett, K., Hur, K., Ippolito, D., & Stern, M. (2005). Deep brain stimulation in Parkinson disease: A metaanalysis of patient outcomes. *Journal of Neurosurgery, 103*(6), 956–967.

Weaver, F. M., Follett, K., Stern, M., Hur, K., Harris, C., Marks, W. J., et al. (2009). Bilateral deep brain stimulation vs. best medical therapy for patients with advanced Parkinson disease: A randomized clinical trial. *Journal of the American Medical Association, 301*, 63–73.
Weintraub, D., & Burn, D. (2011). Parkinson's disease: The quintessential neuropsychiatric disorder. *Movement Disorders, 26*(6), 1022–1031.
Weintraub, D., Koester, J., Potena, M. N., et al. (2010). Impulse control disorder in Parkinson disease: A cross-sectional study of 3090 patients. *Archives of Neurology, 67*, 589–595.
Wenning, G. K., Scherfler, C., Granata, R., Bösch, S., Verny, M., Chaudhuri, K. R., et al. (1999). Time course of symptomatic orthostatic hypotension and urinary incontinence in patients with postmortem confirmed parkinsonian syndromes: A clinicopathological study. *Journal of Neurology, Neurosurgery, and Psychiatry, 67*(5), 620–623.
Winge, K., Nielsen, K. K., Stimpel, H., Lokkegaard, A., Jensen, S. R., & Werdelin, L. (2007). Lower urinary tract symptoms and bladder control in advanced Parkinson's disease: Effects of deep brain stimulation in the subthalamic nucleus. *Movement Disorders, 22*(2), 220–225.
Witjas, T., Baunez, C., Henry, J. M., Delfini, M., Regis, J., Cherif, A. A., et al. (2005). Addiction in Parkinson's disease: Impact of subthalamic nucleus deep brain stimulation. *Movement Disorders, 20*(8), 1052–1055.
Witjas, T., Kaphan, E., Regis, J., et al. (2007). Effects of chronic subthalamic stimulation on nonmotor fluctuations in Parkinson's disease. *Movement Disorders, 22*, 1729–1734.
Witjas, T., Regis, J., Delphini, M., Peragut, J. C., & Azulay, J. P. (2007). Dopamine dysregulation syndrome: Impact of subthalamic chronic stimulation. *Movement Disorders, 22*(Suppl. 16), S182.
Witt, K., Daniels, C., Reiff, J., et al. (2008). Neuropsychological and psychiatric changes after deep brain stimulation for Parkinson's disease: A randomized, multicenter study. *Lancet Neurology, 7*, 605–614.
York, M., Wilde, E., Simpson, R., & Jankovic, J. (2009). Relationship between neuropsychological outcome and DBS surgical trajectory and electrode location. *Journal of the Neurological Sciences, 287*, 159–171.
Zesiewicz, T. A., Sullivan, K. L., Arnulf, I., Chaudhuri, K. R., Morgan, J. C., Gronseth, G. S., et al. (2010). Practice parameter: Treatment of nonmotor symptoms of Parkinson disease report of the quality standards subcommittee of the American Academy of Neurology. *Neurology, 74*(11), 924–931.
Zibetti, M., Rizzi, L., Colloca, L., Cinquepalmi, A., Angrisano, S., Castelli, L., et al. (2010). Probable REM sleep behaviour disorder and STN-DBS outcome in Parkinson's disease. *Parkinsonism & Related Disorders, 16*(4), 265–269.
Zibetti, M., Torre, E., Cinquepalmi, A., Rosso, M., Ducati, A., Bergamasco, B., et al. (2007). Motor and nonmotor symptom follow-up in parkinsonian patients after deep brain stimulation of the subthalamic nucleus. *European Neurology, 58*, 218–223.

FURTHER READING

American Academy of Neurology (AAN). (2009). *Parkinson's disease physician performance measurement set* (p. 45). St. Paul, MN: American Academy of Neurology (AAN).
Mikos, A., Zahodne, L., Okun, M. S., Foote, K., & Bowers, D. (2010). Cognitive declines after unilateral deep brain stimulation surgery in Parkinson's disease: A controlled study using reliable change, part II. *The Clinical Neuropsychologist, 24*(2), 235–245.
Strauss, E., Sherman, E., & Spreen, O. (2006). *A compendium of neuropsychological tests: Administration, norms, and commentary* (pp. 477–499). Oxford, UK: Oxford University Press.

CHAPTER THIRTY-EIGHT

Noninvasive Brain Stimulation and Implications for Nonmotor Symptoms in Parkinson's Disease

Irena Rektorová*,†, Ľubomíra Anderková*,†,1

*School of Medicine and Movement Disorders Center, St. Anne's University Hospital, Masaryk University, Brno, Czech Republic

†Applied Neuroscience Research Group, Central European Institute of Technology, CEITEC MU, Masaryk University, Brno, Czech Republic

[1]Corresponding author: e-mail address: lubomira.anderkova@gmail.com

Contents

Abstract

Transcranial noninvasive brain stimulation includes both repetitive transcranial magnetic stimulation (rTMS) and transcranial direct current stimulation (tDCS). TMS uses a rapidly changing magnetic field to induce currents and action potentials in underlying brain tissue, whereas tDCS involves the application of weak electrical currents to modulate neuronal membrane potential. In this chapter, we provide a literature review with a focus on the therapeutic potential of both techniques in the treatment of nonmotor symptoms of Parkinson's disease (PD). On the whole, the results of studies are rather preliminary but promising as they show some positive effects of rTMS and tDCS particularly on depressive symptoms and cognitive dysfunctions in PD. More carefully controlled trials with standardized methodology, adequately sized and well-characterized samples, and the inclusion of multimodal approaches are warranted in the future.

Transcranial noninvasive brain stimulation (NIBS) includes repetitive transcranial magnetic stimulation (rTMS) as well as transcranial direct current stimulation (tDCS) and transcranial alternating current stimulation (tACS); the latter two techniques are referred to as transcranial current stimulation (TCS). TMS uses a rapidly changing magnetic field to induce currents

International Review of Neurobiology, Volume 134
ISSN 0074-7742
http://dx.doi.org/10.1016/bs.irn.2017.05.009

and action potentials in underlying brain tissue, whereas tDCS involves the application of weak (1–2 mA) electrical currents to modulate neuronal membrane potential. Depending on the stimulation parameters, both TMS and tDCS can be used to excite (>5 Hz TMS, anodal tDCS) or inhibit (<1 Hz TMS, cathodal tDCS) the underlying cortical tissue when applied over the primary motor cortex (M1), but the aftereffects vary across subjects (Fox et al., 2014). Of note, such results may not be valid when other cortical areas are targeted, e.g., both 1 and 25 Hz rTMS-induced excitatory aftereffects when applied over the inferior frontal gyrus (Balaz, Srovnalova, Rektorova, & Rektor, 2010; Srovnalova, Marecek, & Rektorova, 2011). Online stimulation refers to the condition in which a person is executing a task (motor, cognitive, etc.) while receiving rTMS/TCS. The off-line approach is when stimulation occurs before a task, but some NIBS-induced aftereffects may interrelate with the final results. It has been demonstrated that distinct behavioral aftereffects may outlast the duration of multiple sessions of stimulation by weeks or months and thus may have therapeutic potential (Rabey et al., 2013). So far, rTMS has received FDA approval only for the treatment of pharmacoresistant depression (George et al., 1995; Pascual-Leone, Rubio, Pallardo, & Catala, 1996; Schonfeldt-Lecuona, Cardenas-Morales, Freudenmann, Kammer, & Herwig, 2010). In the current review, we focus on the therapeutic potential of rTMS and tDCS in the treatment of nonmotor symptoms of Parkinson's disease (PD).

Before reviewing the literature, several additional remarks have to be made in order to better understand the potentials and limitations of both NIBS techniques and the current state of the art. The effect of rTMS decreases with the distance from the stimulating coil, with an immediate effect on underlying brain tissue to a depth of only approximately 2 cm beneath the scalp. However, changes in distant interconnected regions can also be observed. There is evidence that the behavioral aftereffects of rTMS depend on the number of pulses and sessions applied, the rate of application, and the intensity of each stimulus, as well as on the current state of the brain and the cortical region(s) targeted by the coil. There is no universal agreement on the rTMS parameters to be used to enhance or inhibit specific brain function, mainly due to a lack of understanding of the mechanisms responsible for the lasting modifications of cortical excitability and due to the within-subject and between-subject variability of rTMS-induced effects. Several neurotransmitter systems can be modulated by rTMS, e.g., the application of rTMS to the motor cortex or dorsolateral prefrontal cortex (DLPFC) may increase or decrease the release of monoamines (particularly dopamine) in different cortical and subcortical areas of the brain interconnected with the stimulated area (Cho & Strafella, 2009; Strafella, Paus, Barrett, & Dagher, 2001; Strafella, Paus, Fraraccio, & Dagher, 2003). In addition, high-frequency rTMS applied over the prefrontal cortex may act via the stimulation of the glutamatergic prefrontal neurons (Michael et al., 2003) and may increase neurotrophic factors in the brain (Brunoni, Boggio, & Fregni, 2008). Other mechanisms of action have also been proposed, including changes in the effectiveness of synapses between cortical neurons (long-term potentiation (LTP) and long-term depression (LTD)).

The tDCS modulates cortical excitability through the application of weak electrical currents in the form of direct current (DC) brain polarization. Depending on DC polarity, neuronal firing rates increase or decrease, presumably due to DC-induced changes in resting membrane potentials (for review, see Floel, 2014). The aftereffects of tDCS on cortical excitability are modulated by *N*-methyl-D-aspartate receptor-dependent processes (Nitsche et al., 2004), and by LTP- and LTD-like mechanisms (Floel, 2014; Nitsche et al., 2003). Similar to rTMS, the aftereffects of tDCS depend on the stimulation protocols and on the precise electrode placement.

In general, both techniques may modulate the abnormal brain reorganization caused by distinct brain pathology or they may interact with the normal processes of brain plasticity and induce or enhance compensatory mechanisms leading to increased brain reserves and thus enhance brain resilience (Poletti, Emre, & Bonuccelli, 2011; Priori, Hallett, & Rothwell, 2009). The results of the studies employing functional MRI (fMRI), fluoro-deoxy-glucose PET, EEG, and other neuroimaging and electrophysiology methods in PD patients may help in formulating hypotheses to be tested by using NIBS. However, it has to be stressed that it is more difficult to induce clear behavioral and clinically relevant benefits of NIBS than to induce changes in cortical plasticity or brain activation/resting state network connectivity changes (Hallett & Chokroverty, 2005).

1. METHODS

A systematic search for articles written in English before 2017 was conducted in the Web of Science and PubMed databases. A wide range of keywords was used: *rTMS*, *TMS*, *tDCS*, *magnetic stimulation*, *electric stimulation*, *nonmotor*, *depression*, *cognition*, and *Parkinson*. We combined these keywords and carefully reviewed all of the titles and abstracts of the resulting articles. We excluded the meeting abstracts, editorial material, and articles that were not written in English. We then focused on our predefined selection criteria: application of rTMS/tDCS, involvement of human subjects, inclusion of patients with PD, and evaluation of nonmotor symptoms. Articles that did not meet these criteria were excluded. Our aim was to include original controlled studies dealing with the therapeutic application of rTMS/tDCS and assessing nonmotor aftereffects as one of the studied outcomes. For elimination of the placebo effects we specifically selected only studies where the effects were controlled by either sham stimulation or active stimulation over control region. We identified 31 articles meeting our focus and field of interest.

The articles were divided into two groups according to the stimulation method used: rTMS or tDCS. Articles were listed chronologically and then alphabetically.

2. RESULTS

We identified 31 controlled studies published before 2017 dealing with the application of NIBS (20 for rTMS and 11 for tDCS) for nonmotor symptoms in PD. The mean clinical population was about 27 subjects per study (range of 9–106 participants), the mean age was 69 years, and the mean disease duration was 8 years. This review includes 837 subjects altogether, of which 800 had PD, 24 had PD with mild cognitive impairment (PD-MCI), and 13 had Lewy body dementia (LBD). Information about all of the studies is shown in Tables 1 and 2.

Most of the rTMS studies used figure-of-eight coils; six studies used round coils (Arias et al., 2010; Benninger et al., 2011, 2012; Kimura et al., 2011; Makkos et al., 2016; Okabe et al., 2003). There was a certain uniformity in the selection of the targeted areas. The DLPFC was stimulated in 10 rTMS and 8 tDCS studies. Researchers stimulated this area either bilaterally (Benninger et al., 2011; Doruk et al., 2014; Forogh et al., 2017; Manenti et al., 2016; Srovnalova et al., 2012; Swank et al., 2016) or unilaterally with a left-sided preponderance (Biundo et al., 2015; Boggio et al., 2006, 2005; Cardoso et al., 2008; Elder et al., 2016; Fregni et al., 2006; Pal et al., 2010; Pereira et al., 2013; Sedlackova et al., 2009; Shin et al., 2016); in one study, only the right DLPFC was stimulated (Koch et al., 2004). Motor regions such as the primary motor cortex, supplementary motor area, dorsal premotor cortex (PMd), or stimulation with a round coil with the center over the vertex were also frequently used (Arias et al., 2010; Benninger et al., 2011, 2012, 2010; Boggio et al., 2006; Ferrucci et al., 2016;

Table 1 Studies Using rTMS for Nonmotor Symptoms in Parkinson's Disease

Study	Subjects (*n*, Means of Age, Disease Duration, and HYS)	Stimulation Parameters: Coil/Position	Stimulation Parameters: Number of Pulses, Sessions	Stimulation Parameters: Frequency, Intensity	Outcome Measures	Results
Flamez et al. (2016)	15 PD; age 69 years; disease duration 14 years; HYS 3	Figure 8/L + R M1	1000, single or multiple (10 sess.)	1 Hz, 90% RMT	UPDRS, dyskinesias, GDS, D-KEFS	No significant effect
Makkos et al. (2016)	44 PD; age 67 years; disease duration 7 years; HYS 2.5	Round/L + R M1	600, 10 sess. in 10 days	5 Hz, 90% RMT	MADRS, BDI, UPDRS, HYS, TUG, NMSQ, MMSE, TMT, Stroop test, ADL, ESS, PDSS, HRQoL	↑ Depression rating scales (MADRS, BDI) ↑ UPDRS, NMSQ, HRQoL
Shin, Youn, Chung, and Sohn (2016)	18 PD; age 68 years; disease duration 6 years; HYS 2.5	Figure 8/L DLPFC	600, 10 sess. in 10 days	5 Hz, 90% RMT	HAMD, MADRS, BDI, UPDRS	↑ Depression rating scales (MADRS, HAMD, BDI)
Maruo et al. (2013)	21 PD; age 63 years; disease duration 12 years; HYS 3	Figure 8/L + R M1	1000, 3 sess. in 3 days	10 Hz, 110% RMT	UPDRS, SEMS, VAS, 10 mWT, FT, MADRS, apathy, QST	↑ Motor symptoms only
Shirota, Ohtsu, Hamada, Enomoto, and Ugawa (2013)	106 PD; age 67 years; disease duration 8 years; HYS 3	Figure 8/SMA	1000, 8 sess. in 8 weeks	1 or 10 Hz, 110% AMT	UPDRS, HAMD, apathy, NMSQ	↑ Motor symptoms only
Benninger et al. (2012)	26 PD; age 64 years; disease duration 9 years; HYS 2.5	Round/L + R M1	300, 8 sess. in 2 weeks	50 Hz, 80% AMT	Gait, bradykinesia, UPDRS, FAB, BDI, Health survey, SRTT	No significant effect

Continued

Table 1 Studies Using rTMS for Nonmotor Symptoms in Parkinson's Disease—cont'd

		Stimulation Parameters				
Study	**Subjects (*n*, Means of Age, Disease Duration, and HYS)**	**Coil/ Position**	**Number of Pulses, Sessions**	**Frequency, Intensity**	**Outcome Measures**	**Results**
Srovnalova, Marecek, Kubikova, and Rektorova (2012)	10 PD; age 66 years; disease duration 5 years; HYS na	Figure 8/ L/R DLPFC	600, single	25 Hz, 80% RMT	TOL	↑ Total time after R DLPFC
Benninger et al. (2011)	26 PD; age 64 years; disease duration 8.5 years; HYS 2.5	Round/L + R M1, L +R DLPFC	600, 8 sess. in 2 weeks	iTBS, 80% AMT	Gait, bradykinesia, UPDRS, FAB, BDI, Health survey, SRTT	↑ Depression rating scale (BDI)
Kimura et al. (2011)	12 PD; age 69 years; disease duration 8.5 years; HYS 3	Round/ M1 + SMA	60, 4 sess. in 4 weeks	0.2 Hz, 700 V	MMSE, HAMD, WAIS-R, HYS, UPDRS, ADL, Actigraph	↑ Motor symptoms only
Srovnalova et al. (2011)	10 PD; age 66 years; disease duration 5 years; HYS na	Figure 8/L + R IFG	600, single	25 Hz, 80% RMT	Stroop test, FAB	↑ All Stroop subtests
Arias, Vivas, Grieve, and Cudeiro (2010)	18 PD; age na; disease duration na; HYS na	Round/ vertex	100, 10 sess. in 10 days	1 Hz, 90% RMT	HAMD, UPDRS, PDSS, Actigraphy	Only placebo effects
Pal, Nagy, Aschermann, Balazs, and Kovacs (2010)	22 PD; age 68 years; disease duration 6 years; HYS 2	Figure 8/L DLPFC	600, 10 sess.	5 Hz, 90% RMT	MMSE, BDI, MADRS, Stroop test, TMT, UPDRS, HYS, TUG, ADL, VAS, ESS	↑ Depression rating scales (MADRS, BDI) ↑ Accuracy of Stroop test

Sedlackova, Rektorova, Srovnalova, and Rektor (2009)	10 PD; age 64 years; disease duration 8 years; HYS na	Figure 8/L PMd, L DLPFC, L OCC	1350, single	10 Hz, 100% RMT	VFT, TMT, Digit Span	No significant effect
Cardoso et al. (2008)	21 PD; age 65 years; disease duration 11 years; HYS 2.5	Figure 8/L DLPFC	3750, 12 sess. in 4 weeks	5 Hz, 120% RMT	MMSE, HAMD, BDI, HYS, UPDRS, ADL, fMRI Ekman's faces task	↑ Depression rating scales (HAMD, BDI) ↑ MMSE, ADL fMRI changes in depression-related neuronal network
Hamada, Ugawa, Tsuji, and Effectiveness of rTMS on Parkinson's Disease Study Group, Japan (2008)	98 PD; age 66 years; disease duration 8 years; HYS 3	Figure 8/ SMA	1000, 8 sess. in 8 weeks	5 Hz, 110% AMT	UPDRS, HAMD, VAS	↑ Motor symptoms only
Fregni et al. (2006)	26 PD; age 66 years; disease duration 7 years; HYS 2	Figure 8/L DLPFC	3000, 10 sess. in 2 weeks	15 Hz, 110% RMT	MMSE, HAMD, BDI, HYS, UPDRS, ADL	↑ Depression rating scales (HAMD, BDI) and changes in regional cerebral blood flow using SPECT
Boggio et al. (2005)	25 PD; age 65 years; disease duration 7 years; HYS na	Figure 8/L DLPFC	3000, 10 sess. in 2 weeks	15 Hz, 110% RMT	TMT-B, WCST, COWA-FAS, Stroop test, HVOT, Digit	↑ WCST, Stroop test, HVOT ↑ Depression rating

Continued

Table 1 Studies Using rTMS for Nonmotor Symptoms in Parkinson's Disease—cont'd

Study	Subjects (*n*, Means of Age, Disease Duration, and HYS)	Stimulation Parameters: Coil/Position	Stimulation Parameters: Number of Pulses, Sessions	Stimulation Parameters: Frequency, Intensity	Outcome Measures	Results
					Span, RCPM, BDI, HAMD, UPDRS	scales (HAMD, BDI)
Fregni et al. (2004)	42 PD; age 66 years; disease duration 8 years; HYS 2	Figure 8/L DLPFC	3000, 10 sess. in 2 weeks	15 Hz, 110% RMT	MMSE, HAMD, BDI, HYS, UPDRS, ADL	↑ Depression rating scales (HAMD, BDI) ↑ MMSE, ADL
Koch et al. (2004)	10 PD; age 62 years; disease duration 5 years; HYS <2	Figure 8/R DLPFC, SMA	250, single	5 Hz, 100% RMT	Time reproduction task	↑ Time reproduction after R DLPFC
Okabe, Ugawa, Kanazawa, and Effectiveness of rTMS on Parkinson's Disease Study Group (2003)	85 PD; age 67 years; disease duration 9 years; HYS 3	Round/vertex, inion	100, 8 sess. in 8 weeks	0.2 Hz, 110% AMT	UPDRS, HAMD, VAS	Only placebo effects

10mWT, 10 min walking test; *ADL*, Activities of Daily Living; *BDI*, Beck Depression Inventory; *COWA-FAS*, Controlled Oral Word Association Test; *D-KEFS*, Delis-Kaplan Executive Function System; *DLPFC*, dorsolateral prefrontal cortex; *ESS*, Epworth Sleepiness Scale; *FAB*, frontal assessment battery; *FT*, finger tapping; *GDS*, Geriatric Depression Scale; *HAMD*, Hamilton Depression Scale; *HRQoL*, health-related quality of life; *HVOT*, Hooper Visual Organization Test; *HYS*, Hoehn–Yahr Scale; *IFG*, inferior frontal gyri; *iTBS*, intermittent theta burst stimulation; *M1*, primary motor cortex; *MADRS*, Montgomery–Asberg Depression Rating Scale; *MMSE*, Mini-Mental State Examination; *n*, number of subjects; *na*, not available; *NMSQ*, nonmotor symptoms questionnaire; *OCC*, occipital cortex; *PD*, patients with Parkinson's disease; *PDSS*, Parkinson's Disease Sleep Scale; *PMd*, dorsal premotor cortex; *QST*, quantitative sensory testing; *R/AMT*, resting/active motor threshold; *RCPM*, Raven's colored progressive matrices; *SEMS*, self-assessment motor scale; *sess.*, sessions; *single*, single session for each targeted area; *SMA*, supplementary motor area; *SRTT*, serial reaction time task; *TMT(-B)*, Trail-Making Test(-part B); *TOL*, Tower of London; *TUG*, timed up and go test; *UPDRS*, Unified Parkinson's Disease Rating Scale; *VAS*, Visual Analog Scale; *VFT*, Verbal Fluency Test; *WAIS-R*, Wechsler Adult Intelligence Scale Revised; *WCST*, Wisconsin Card Sorting Test; ↑ improvement of.

Table 2 Studies Using tDCS for Nonmotor Symptoms in Parkinson's Disease

Study	Subjects (*n*, Means of Age, Disease Duration, and HYS)	Stimulation Parameters			Outcome Measures	Results
		Position (Anode/Cathode)	Duration (min), Number of Sessions	Intensity/Current Density (mA/cm^2)		
Forogh et al. (2017)	23 PD; age 63 years; disease duration na; HYS na	L DLPFC/R DLPFC	20, 8 sess. in 2 weeks + occupational therapy	2/0.06	FSI, ESS	↑ Fatigue
Elder et al. (2016)	13 LBD; age 65 years; disease duration 8 years; HYS na	L DLPFC (F3–FP1)/R deltoid	20, single	2.8/0.08	Attentional and visuoperceptual tasks	↑ Attention (choice reaction time and digit vigilance)
Ferrucci et al. (2016)	9 PD; age 74 years; disease duration 11 years; HYS 2.5	L + R M1 (C3 +C4) or L +R cerebellum/R deltoid	20, 5 sess. in 5 days	2/0.06	UPDRS, PDQ, BDI, Word Recall, VAT, SRTT	↑ Dyskinesias only
Manenti et al. (2016)	20 PD; age 69 years; disease duration 7 years; HYS 2	L or R DLPFC (F3–F7 or F4–F8)/contralateral SOA	25, 10 sess. in 2 weeks + physical therapy	2/0.06	MMSE, BDI, HYS, UPDRS, PDQ, RBDSQ, PD-CRS, Digit Span, CANTAB PAL/RTI, TMT, FAB, semantic fluency, motor tasks	↑ PD-CRS, semantic fluency

Continued

Table 2 Studies Using tDCS for Nonmotor Symptoms in Parkinson's Disease—cont'd

Study	Subjects (*n*, Means of Age, Disease Duration, and HYS)	Stimulation Parameters			Outcome Measures	Results
		Position (Anode/Cathode)	Duration (min), Number of Sessions	Intensity/Current Density (mA/cm^2)		
Schabrun, Lamont, and Brauer (2016)	16 PD; age 68 years; disease duration 6 years; HYS 2	L M1/R SOA	20, 9 sess. in 3 weeks + dual task gait training	2/0.06	Dual task gait training, TUG, bradykinesia, SRTT	↑ Correct responses while performing the dual task TUG only
Swank, Mehta, and Criminger (2016)	10 PD; age 69 years; disease duration 8 years; HYS 2	L DLPFC/R DLPFC	20, single + single/dual task gait activity	2	TUG, dual task TUG, PDQ	Reduce of dual task cost during gait
Biundo et al. (2015)	24 PD-MCI; age 71 years; disease duration na, HYS na	L DLPFC (F3)/R SOA	20, 16 in 4 weeks + cognitive training	2	UPDRS, PDQ, BDI, STAI, MoCA, RBANS	Transient decrease of attention subtest and trend for memory increasing
Doruk, Gray, Bravo, Pascual-Leone, and Fregni (2014)	18 PD; age 61 years; disease duration na; HYS na	L or R DLPFC (F3 or F4)/contralateral SOA	20, 10 sess. in 2 weeks	2/0.06	TMT, WCST, PCL, WM, Stroop test, HVOT, Digit Span, RCPM, BDI, HAMD, HAMA, UPDRS, and other motor tasks	↑ TMT-B after L&R DLPFC ↑ BDI after L DLPFC

Pereira et al. (2013)	16 PD; age 62 years; disease duration na; HYS 1.5	L DLPFC (F3) or L TPC (P3-T5)/R SOA	20, single	2/0.06	fMRI semantic and phonemic fluency tasks	↑ Phonemic fluency after DLPFC stimulation, changes in functional connectivity
Benninger et al. (2010)	25 PD; age 64 years; disease duration 10 years; HYS 2.5	M1 + PMC or PFC/mastoids	20, 8 sess. in 2.5 weeks	2/0.02	Gait, bradykinesia, UPDRS, BDI, Health survey, SRTT, self-assessment	↑ Motor symptoms only
Boggio et al. (2006)	18 PD; age 61 years; disease duration 13 years; HYS 2.5	L DLPFC (F3) or M1/R SOA	20, single	1 or 2/0.03 or 0.06	3-Back letter working memory paradigm	↑ Correct responses and less errors after L DLPFC 2 mA stimulation

BDI, Beck Depression Inventory; *CANTAB PAL/RTI*, CANTAB Paired Associative Learning/Reaction Time Index; *DLPFC*, dorsolateral prefrontal cortex; *ESS*, Epworth Sleepiness Scale; *FAB*, frontal assessment battery; *FSI*, fatigue severity index; *HAMA*, Hamilton Anxiety Scale; *HAMD*, Hamilton Depression Scale; *HVOT*, Hooper Visual Organization Test; *HYS*, Hoehn–Yahr Scale; *LBD*, patients with Lewy body dementia; *M1*, primary motor cortex; *MoCA*, montreal cognitive assessment; *n*, number of subjects; *na*, not available; *PD*, patients with Parkinson's disease; *PD-CRS*, Parkinson's Disease Cognitive Rating Scale; *PD-MCI*, Parkinson's disease patients with mild cognitive impairment; *PDQ*, Parkinson's disease questionnaire; *PFC*, prefrontal cortex; *PMC*, premotor cortex; *RBANS*, repeatable battery for the assessment of neuropsychological status; *RBDSQ*, REM sleep behavior disorder screening questionnaire; *RCPM*, Raven colored progressive matrices; *sess.*, sessions; *single*, single session for each targeted area; *SOA*, supraorbital area; *SRTT*, serial reaction time task; *STAI*, State-Trait Anxiety Inventory; *TMT(-B)*, Trail-Making Test(-part B); *TPC*, temporoparietal cortex; *TUG*, timed up and go test; *UPDRS*, Unified Parkinson's Disease Rating Scale; *VAT*, visual attention task; *WCST*, Wisconsin Card Sorting Test; ↑ improvement of.

Flamez et al., 2016; Hamada et al., 2008; Kimura et al., 2011; Koch et al., 2004; Makkos et al., 2016; Maruo et al., 2013; Okabe et al., 2003; Schabrun et al., 2016; Sedlackova et al., 2009; Shirota et al., 2013). Other cortical regions were stimulated less frequently, including the inferior frontal gyri (IFG) (Srovnalova et al., 2011), cerebellum (Ferrucci et al., 2016), and temporoparietal cortex (Pereira et al., 2013). The high occurrence of DLPFC stimulation might be due to fact that this coil localization has been identified as having antidepressant effects (George et al., 1995) and has therefore been intensively studied in recent years. The cathode in tDCS stimulation was frequently positioned at the contralateral supraorbital area (Biundo et al., 2015; Boggio et al., 2006; Doruk et al., 2014; Manenti et al., 2016; Pereira et al., 2013; Schabrun et al., 2016) or the deltoid (Elder et al., 2016; Ferrucci et al., 2016). The cathodal electrode on the mastoids (Benninger et al., 2010) or bilateral DLPFC electrode placement were also used (Forogh et al., 2017; Swank et al., 2016).

The stimulation parameters vary from study to study. For rTMS, the number of pulses ranged from 100 to 3750 per day. In some studies, the effect of a single rTMS session was measured (Anderkova & Rektorova, 2014; Flamez et al., 2016; Koch et al., 2004; Sedlackova et al., 2009; Srovnalova et al., 2012, 2011); other researchers assessed rTMS effects after repeated sessions over up to 12 days of stimulation (Cardoso et al., 2008). In most of the publications, high-frequency rTMS (ranging from 5 to 25 Hz) was used. Low-frequency rTMS ($\leq$1 Hz) was applied in five studies (Arias et al., 2010; Flamez et al., 2016; Kimura et al., 2011; Okabe et al., 2003; Shirota et al., 2013). The intensity of rTMS pulses was between 120% and 80% of the resting motor threshold, which corresponds to the lowest intensity capable of inducing motor-evoked potentials of the relaxed hand muscle in at least 5 out of 10 trials. tDCS stimulation was usually applied for 20 min with 2 mA intensity (ranging from 1 to 2.8 mA) with effects assessed after single or multiple sessions (up to 16).

Several articles deal with the various cognitive effects of rTMS/tDCS on patients with PD. Some focus on motor deficits and depression as a primary outcome, assessing the cognitive functions of rTMS only as a secondary outcome.

Srovnalova et al. (2012, 2011) published two randomized, sham-controlled studies with a crossover design that focused solely on cognitive functions in PD. In one study, the authors showed that high-frequency rTMS of the right DLPFC induced significant improvement in the Tower of London task performance, as expressed by total problem-solving time.

No significant change was detected after left DLPFC rTMS or sham stimulation. In the second study, the researchers demonstrated improvement in all Stroop subtests after active high-frequency rTMS applied sequentially over both the left and right IFG. The results indicate a nonspecific enhancement of the cognitive processing speed. No significant change was induced by active or sham rTMS on Frontal Assessment Battery scores. Previously, researchers from the same lab (Sedlackova et al., 2009) found no significant effect of left high-frequency stimulation of the PMd and DLPFC on cognitive functions as measured by the Verbal Fluency Test—category, the Trail-Making Test (TMT), or the Digit Span Test in cognitively normal PD patients.

Koch et al. (2004) demonstrated that high-frequency rTMS improves time perception as measured by a time reproduction task in patients with PD. Pal et al. (2010) studied the effects of left high-frequency DLPFC rTMS in PD patients with depression. Significant improvements in depression rating scales—Beck Depression Inventory (BDI) and Montgomery–Asberg Depression Rating Scale (MADRS)—and in Stroop test accuracy were identified. Other cognitive tests, such as the Mini-Mental State Examination (MMSE) and the TMT, revealed no change after the stimulation.

Similar sample groups (depressed PD patients) were used in two other studies that compared the effects of rTMS and fluoxetine (Boggio et al., 2005; Fregni et al., 2004). High-frequency rTMS of the left DLPFC significantly improved Stroop test performance after 2 weeks of treatment. There was also a significant improvement in the Hooper Visual Organization Test (HVOT) after 2 weeks of stimulation; the improvement persisted for up to 8 weeks after the initial stimulation. In the 8 weeks after stimulation, there was a significant decrease in perseverative errors in the Wisconsin Card Sorting Test (WCST) compared to the initial values. In the other study, significant change was observed in Activities of Daily Living (ADL) scores 8 weeks after rTMS in comparison with the baseline scores. There was also a significant difference in MMSE scores after 2 weeks of treatment, with more improvement in the group that received rTMS than in the group that had taken fluoxetine (Fregni et al., 2004). The authors concluded that rTMS is effective in treating depression (BDI, Hamilton Depression Scale—HAMD) and in improving some aspects of cognition in PD patients with depression, and the effect seems to be similar to or even better than the effects of fluoxetine.

Improved depression rating scale scores after rTMS were seen in several other studies (Benninger et al., 2011; Cardoso et al., 2008; Fregni et al., 2006; Makkos et al., 2016; Shin et al., 2016). In a study by Makkos et al.

(2016) depression as well as nonmotor symptoms, UPDRS, and health-related quality of life improved. Cardoso et al. (2008) observed fMRI BOLD changes in depression-related neuronal networks and improved MMSE and ADL scores. In a SPECT study by Fregni et al. (2006), antidepressant treatment with either rTMS or fluoxetine was associated with similar changes in regional cerebral blood flow in the frontal–limbic network. Motor symptom improvement only (with no effect on nonmotor symptoms) was found in four studies (Hamada et al., 2008; Kimura et al., 2011; Maruo et al., 2013; Shirota et al., 2013). These results are not very surprising as either primary or supplementary motor areas were stimulated in these studies. No significant effects or only placebo effects of the stimulation were also found (Arias et al., 2010; Benninger et al., 2012; Flamez et al., 2016; Okabe et al., 2003).

tDCS stimulation showed several beneficial effects in cognitive functions, such as improved attention (but not visuoperceptual function) in LBD (Elder et al., 2016); semantic fluency and global cognitive functioning (Manenti et al., 2016) using a combination of stimulation and physical therapy; executive functions (TMT-B) together with depressive symptoms (Doruk et al., 2014); phonemic fluency and changes in functional connectivity (Pereira et al., 2013); working memory (Boggio et al., 2006); and a dual task characterized by motor-cognitive interplay (Schabrun et al., 2016; Swank et al., 2016). In the PD-MCI group, only a transient decrease in the attention subtest scores and strong trend for memory improvement after tDCS were found (Biundo et al., 2015). Forogh et al. (2017) found significant effects of tDCS on fatigue but not on daytime sleepiness. Motor symptom improvement only, with no effect on nonmotor scales, after tDCS was found in two studies (Benninger et al., 2010; Ferrucci et al., 2016).

3. DISCUSSION

This review summarizes the results of articles that studied the effects of rTMS and tDCS on nonmotor symptoms of patients with PD. On the whole, the results of the published works show some positive effects of rTMS and tDCS, particularly on depressive symptoms and cognitive dysfunctions in PD.

The published literature shows a great variability in simulation parameters, study protocol designs, and outcome measures. In many cases, the sample sizes were underpowered. In general, high-frequency rTMS protocols seem to produce more significant effects on various studied outcomes than low-frequency stimulation protocols. Applying accelerated (intensified) protocols with higher numbers of rTMS sessions (including even more sessions per day) in individual patients might be a way forward which should be tested in future studies.

Despite the aforementioned limitations, the overall results of the reviewed studies suggest that rTMS has beneficial and long-lasting antidepressant effects in PD populations, particularly when high-frequency stimulation is targeted over the left DLPFC and when repeated session protocols are used. The rTMS-induced effect was similar to that of antidepressant medication, with considerably fewer side effects. Future work should examine the possible additive effects of both rTMS and antidepressant medication.

The effects of rTMS on cognition in PD patients are less prominent mainly due to the heterogeneity in study outcomes. In general, there is a tendency for the nonspecific enhancement of cognitive speed or time to finish cognitive tasks. More studies with carefully designed protocols and reasonably powered populations are needed to shed more light on the cognitive aftereffects of rTMS. While improvements in the motor symptoms of PD have been reported (for a recent meta-analysis, see Chou, Hickey, Sundman, Song, & Chen, 2015), no persistent and meaningful effect on nonmotor symptoms was demonstrated after stimulation of the motor cortices.

The effects of tDCS have been even more varied. The study results revealed some benefit of this method on various nonmotor symptoms of PD such as impaired attention, executive functions, memory, and fatigue. Only one study (Doruk et al., 2014) showed an effect on depressive symptoms in the PD population. Future studies will have to include PD patients with a current depressive episode and compare the possible antidepressive effects of tDCS with sham stimulation and/or with antidepressant medication. A potential online combination with either cognitive training or another behavioral activity is possible (mainly due to the ease of application of tDCS) and strongly encouraged with the aim of enhancing specific functions of interest more effectively.

Both TMS and TCS can now be combined with a variety of neuroimaging and electrophysiological methods which can inform subsequent NIBS, providing information about where, when, and how to stimulate the brain. Moreover, neuroimaging and electrophysiology can provide readouts (i.e., indices or measurements) of neuronal activity, which make it possible to assess the changes caused by NIBS and the neural underpinnings of its behavioral aftereffects. Combining these methods and even online monitoring of the state of the brain has been addressed and encouraged by several authors (Bergmann, Karabanov, Hartwigsen, Thielscher, & Siebner, 2016; Fox et al., 2014) with an aim to optimally tailor the NIBS treatment to meet the distinct needs of individual patients. Stimulating multiple targets (Lomarev et al., 2006; Rabey et al., 2013) or combining invasive and noninvasive stimulation techniques (Fox et al., 2014; Udupa et al., 2016) may represent a meaningful way forward if based on firm hypotheses supported by neuroimaging or neurophysiology results.

In summary, the results of studies including NIBS for modulating the nonmotor symptoms of PD are still preliminary. However, they encourage further study in order to more precisely define the role of NIBS in the treatment of highly specific and homogenous PD patient subgroups or even individual PD subjects. To establish these methods as relevant treatment tools in PD, more well-designed studies demonstrating positive, reproducible, and long-lasting effects with meaningful behavioral outcomes and impacts on quality of life or activities of daily living are needed. More carefully controlled trials with standardized methodology, adequately sized and characterized samples, and the inclusion of multimodal approaches are warranted in the future.

ACKNOWLEDGMENT

This work was supported by the project Central European Institute of Technology 2020 (LQ1601) by the MEYS CR and by the 16-31868A grant of the Czech Ministry of Health.

REFERENCES

Anderkova, L., & Rektorova, I. (2014). Cognitive effects of repetitive transcranial magnetic stimulation in patients with neurodegenerative diseases—Clinician's perspective. *Journal of the Neurological Sciences*, *339*(1–2), 15–25. http://dx.doi.org/10.1016/j.jns.2014.01.037.

Arias, P., Vivas, J., Grieve, K. L., & Cudeiro, J. (2010). Double-blind, randomized, placebo controlled trial on the effect of 10 days low-frequency rTMS over the vertex on sleep in Parkinson's disease. *Sleep Medicine*, *11*(8), 759–765. http://dx.doi.org/10.1016/j.sleep.2010.05.003.

Balaz, M., Srovnalova, H., Rektorova, I., & Rektor, I. (2010). The effect of cortical repetitive transcranial magnetic stimulation on cognitive event-related potentials recorded in the subthalamic nucleus. *Experimental Brain Research*, *203*(2), 317–327. http://dx.doi.org/10.1007/s00221-010-2232-4.

Benninger, D. H., Berman, B. D., Houdayer, E., Pal, N., Luckenbaugh, D. A., Schneider, L., et al. (2011). Intermittent theta-burst transcranial magnetic stimulation for treatment of Parkinson disease. *Neurology*, *76*(7), 601–609. http://dx.doi.org/10.1212/WNL.0b013e31820ce6bb.

Benninger, D. H., Iseki, K., Kranick, S., Luckenbaugh, D. A., Houdayer, E., & Hallett, M. (2012). Controlled study of 50-Hz repetitive transcranial magnetic stimulation for the treatment of Parkinson disease. *Neurorehabilitation and Neural Repair*, *26*(9), 1096–1105. http://dx.doi.org/10.1177/1545968312445636.

Benninger, D. H., Lomarev, M., Lopez, G., Wassermann, E. M., Li, X. B., Considine, E., et al. (2010). Transcranial direct current stimulation for the treatment of Parkinson's disease. *Journal of Neurology, Neurosurgery, and Psychiatry*, *81*(10), 1105–1111. http://dx.doi.org/10.1136/jnnp.2009.202556.

Bergmann, T. O., Karabanov, A., Hartwigsen, G., Thielscher, A., & Siebner, H. R. (2016). Combining non-invasive transcranial brain stimulation with neuroimaging and electrophysiology: Current approaches and future perspectives. *NeuroImage*, *140*, 4–19. http://dx.doi.org/10.1016/j.neuroimage.2016.02.012.

Biundo, R., Weis, L., Fiorenzato, E., Gentile, G., Giglio, M., Schifano, R., et al. (2015). Double-blind randomized trial of t-DCS versus sham in Parkinson patients with mild cognitive impairment receiving cognitive training. *Brain Stimulation*, *8*(6), 1223–1225. http://dx.doi.org/10.1016/j.brs.2015.07.043.

Boggio, P. S., Ferrucci, R., Rigonatti, S. P., Covre, P., Nitsche, M., Pascual-Leone, A., et al. (2006). Effects of transcranial direct current stimulation on working memory in patients with Parkinson's disease. *Journal of the Neurological Sciences*, *249*(1), 31–38. http://dx.doi.org/10.1016/j.jns.2006.05.062.

Boggio, P., Fregni, F., Bermpohl, F., Mansur, C., Rosa, M., Rumi, D., et al. (2005). Effect of repetitive TMS and fluoxetine on cognitive function in patients with Parkinson's disease and concurrent depression. *Movement Disorders*, *20*(9), 1178–1184. http://dx.doi.org/10.1002/mds.20508.

Brunoni, A. R., Boggio, P. S., & Fregni, F. (2008). Can the 'yin and yang' BDNF hypothesis be used to predict the effects of rTMS treatment in neuropsychiatry? *Medical Hypotheses*, *71*(2), 279–282. http://dx.doi.org/10.1016/j.mehy.2008.02.011.

Cardoso, E. F., Fregni, F., Maia, F. M., Boggio, P. S., Myczkowski, M. L., Coracini, K., et al. (2008). rTMS treatment for depression in Parkinson's disease increases BOLD responses in the left prefrontal cortex. *International Journal of Neuropsychopharmacology*, *11*(2), 173–183. http://dx.doi.org/10.1017/s1461145707007961.

Cho, S. S., & Strafella, A. P. (2009). rTMS of the left dorsolateral prefrontal cortex modulates dopamine release in the ipsilateral anterior cingulate cortex and orbitofrontal cortex. *PLoS One*, *4*(8), 8. http://dx.doi.org/10.1371/journal.pone.0006725.

Chou, Y. H., Hickey, P. T., Sundman, M., Song, A. W., & Chen, N. K. (2015). Effects of repetitive transcranial magnetic stimulation on motor symptoms in Parkinson disease: A systematic review and meta-analysis. *JAMA Neurology*, *72*(4), 432–440. http://dx.doi.org/10.1001/jamaneurol.2014.4380.

Doruk, D., Gray, Z., Bravo, G. L., Pascual-Leone, A., & Fregni, F. (2014). Effects of tDCS on executive function in Parkinson's disease. *Neuroscience Letters*, *582*, 27–31. http://dx.doi.org/10.1016/j.neulet.2014.08.043.

Elder, G. J., Firbank, M. J., Kumar, H., Chatterjee, P., Chakraborty, T., Dutt, A., et al. (2016). Effects of transcranial direct current stimulation upon attention and visuoperceptual function in Lewy body dementia: A preliminary study. *International Psychogeriatrics*, *28*(2), 341–347. http://dx.doi.org/10.1017/s1041610215001180.

Ferrucci, R., Cortese, F., Bianchi, M., Pittera, D., Turrone, R., Bocci, T., et al. (2016). Cerebellar and motor cortical transcranial stimulation decrease levodopa-induced dyskinesias in Parkinson's disease. *Cerebellum*, *15*(1), 43–47. http://dx.doi.org/10.1007/s12311-015-0737-x.

Flamez, A., Cordenier, A., De Raedt, S., Michiels, V., Smetcoren, S., Van Merhaegen-Wieleman, A., et al. (2016). Bilateral low frequency rTMS of the primary motor cortex may not be a suitable treatment for levodopa-induced dyskinesias in late stage Parkinson's disease. *Parkinsonism & Related Disorders*, *22*, 54–61. http://dx.doi.org/10.1016/j.parkreldis.2015.11.009.

Floel, A. (2014). tDCS-enhanced motor and cognitive function in neurological diseases. *NeuroImage*, *85*, 934–947. http://dx.doi.org/10.1016/j.neuroimage.2013.05.098.

Forogh, B., Rafiei, M., Arbabi, A., Motamed, M. R., Madani, S. P., & Sajadi, S. (2017). Repeated sessions of transcranial direct current stimulation evaluation on fatigue and daytime sleepiness in Parkinson's disease. *Neurological Sciences*, *38*(2), 249–254. http://dx.doi.org/10.1007/s10072-016-2748-x.

Fox, M. D., Buckner, R. L., Liu, H. S., Chakravarty, M. M., Lozano, A. M., & Pascual-Leone, A. (2014). Resting-state networks link invasive and noninvasive brain stimulation across diverse psychiatric and neurological diseases. *Proceedings of the National Academy of Sciences of the United States of America*, *111*(41), E4367–E4375. http://dx.doi.org/10.1073/pnas.1405003111.

Fregni, F., Ono, C. R., Santos, C. M., Bermpohl, F., Buchpiguel, C., Barbosa, E. R., et al. (2006). Effects of antidepressant treatment with rTMS and fluoxetine on brain perfusion in PD. *Neurology*, *66*(11), 1629–1637. http://dx.doi.org/10.1212/01.wnl.0000218194.120504.60.

Fregni, F., Santos, C., Myczkowski, M., Rigolino, R., Gallucci-Neto, J., Barbosa, E., et al. (2004). Repetitive transcranial magnetic stimulation is as effective as fluoxetine in the treatment of depression in patients with Parkinson's disease. *Journal of Neurology, Neurosurgery, and Psychiatry*, *75*(8), 1171–1174. http://dx.doi.org/10.1136/jnnp.2003.027060.

George, M. S., Wassermann, E. M., Williams, W. A., Callahan, A., Ketter, T. A., Basser, P., et al. (1995). Daily repetitive transcranial magnetic stimulation (rTMS) improves mood in depression. *Neuroreport*, *6*(14), 1853–1856. http://dx.doi.org/10.1097/00001756-199510020-00008.

Hallett, M., & Chokroverty, S. (2005). *Magnetic stimulation in clinical neurophysiology*. Oxford, United Kingdom: Elsevier Butterworth-Heinemann.

Hamada, M., Ugawa, Y., Tsuji, S., & Effectiveness of rTMS on Parkinson's Disease Study Group, Japan. (2008). High-frequency rTMS over the supplementary motor area for treatment of Parkinson's disease. *Movement Disorders*, *23*(11), 1524–1531. http://dx.doi.org/10.1002/mds.22168.

Kimura, H., Kurimura, M., Kurokawa, K., Nagaoka, U., Arawaka, S., Wada, M., et al. (2011a). A comprehensive study of repetitive transcranial magnetic stimulation in Parkinson's disease. *ISRN Neurology*, *2011*, 845453. http://doi.org/10.5402/2011/845453.

Koch, G., Oliveri, M., Brusa, L., Stanzione, P., Torriero, S., & Caltagirone, C. (2004). High-frequency rTMS improves time perception in Parkinson disease. *Neurology*, *63*(12), 2405–2406. doi:63/12/2405 [pii].

Lomarev, M. P., Kanchana, S., Bara-Jimenez, W., Iyer, M., Wassermann, E. M., & Hallett, M. (2006). Placebo-controlled study of rTMS for the treatment of Parkinson's disease. *Movement Disorders*, *21*(3), 325–331. http://dx.doi.org/10.1002/mds.20713.

Makkos, A., Pal, E., Aschermann, Z., Janszky, J., Balazs, E., Takacs, K., et al. (2016). High-frequency repetitive transcranial magnetic stimulation can improve depression in Parkinson's disease: A randomized, double-blind, placebo-controlled study. *Neuropsychobiology*, *73*(3), 169–177. http://dx.doi.org/10.1159/000445296.

Manenti, R., Brambilla, M., Benussi, A., Rosini, S., Cobelli, C., Ferrari, C., et al. (2016). Mild cognitive impairment in Parkinson's disease is improved by transcranial direct current stimulation combined with physical therapy. *Movement Disorders*, *31*(5), 715–724. http://dx.doi.org/10.1002/mds.26561.

Maruo, T., Hosomi, K., Shimokawa, T., Kishima, H., Oshino, S., Morris, S., et al. (2013). High-frequency repetitive transcranial magnetic stimulation over the primary foot motor

area in Parkinson's disease. *Brain Stimulation, 6*(6), 884–891. http://dx.doi.org/10.1016/j.brs.2013.05.002.
Michael, N., Gosling, M., Reutemann, M., Kersting, A., Heindel, W., Arolt, V., et al. (2003). Metabolic changes after repetitive transcranial magnetic stimulation (rTMS) of the left prefrontal cortex: A sham-controlled proton magnetic resonance spectroscopy (H-1 MRS) study of healthy brain. *European Journal of Neuroscience, 17*(11), 2462–2468. http://dx.doi.org/10.1046/j.1460-9568.2003.02683.x.
Nitsche, M. A., Fricke, K., Henschke, U., Schlitterlau, A., Liebetanz, D., Lang, N., et al. (2003). Pharmacological modulation of cortical excitability shifts induced by transcranial direct current stimulation in humans. *The Journal of Physiology, 553*(1), 293–301. http://dx.doi.org/10.1113/jphysiol.2003.049916.
Nitsche, M. A., Grundey, J., Liebetanz, D., Lang, N., Tergau, F., & Paulus, W. (2004). Catecholaminergic consolidation of motor cortical neuroplasticity in humans. *Cerebral Cortex, 14*(11), 1240–1245. http://dx.doi.org/10.1093/cercor/bhh085.
Okabe, S., Ugawa, Y., Kanazawa, I., & Effectiveness of rTMS on Parkinson's Disease Study Group (2003). 0.2-Hz Repetitive transcranial magnetic stimulation has no add-on effects as compared to a realistic sham stimulation in Parkinson's disease. *Movement Disorders, 18*(4), 382–388. http://dx.doi.org/10.1002/mds.10370.
Pal, E., Nagy, F., Aschermann, Z., Balazs, E., & Kovacs, N. (2010). The impact of left prefrontal repetitive transcranial magnetic stimulation on depression in Parkinson's disease: A randomized, double-blind, placebo-controlled study. *Movement Disorders, 25*(14), 2311–2317. http://dx.doi.org/10.1002/mds.23270.
Pascual-Leone, A., Rubio, B., Pallardo, F., & Catala, M. D. (1996). Rapid-rate transcranial magnetic stimulation of left dorsolateral prefrontal cortex in drug-resistant depression. *Lancet, 348*(9022), 233–237. http://dx.doi.org/10.1016/s0140-6736(96)01219-6.
Pereira, J. B., Junque, C., Bartres-Faz, D., Marti, M. J., Sala-Llonch, R., Compta, Y., et al. (2013). Modulation of verbal fluency networks by transcranial direct current stimulation (tDCS) in Parkinson's disease. *Brain Stimulation, 6*(1), 16–24. http://dx.doi.org/10.1016/j.brs.2012.01.006.
Poletti, M., Emre, M., & Bonuccelli, U. (2011). Mild cognitive impairment and cognitive reserve in Parkinson's disease. *Parkinsonism & Related Disorders, 17*(8), 579–586. http://dx.doi.org/10.1016/j.parkreldis.2011.03.013.
Priori, A., Hallett, M., & Rothwell, J. C. (2009). Repetitive transcranial magnetic stimulation or transcranial direct current stimulation? *Brain Stimulation, 2*(4), 241–245. http://dx.doi.org/10.1016/j.brs.2009.02.004.
Rabey, J. M., Dobronevsky, E., Aichenbaum, S., Gonen, O., Marton, R. G., & Khaigrekht, M. (2013). Repetitive transcranial magnetic stimulation combined with cognitive training is a safe and effective modality for the treatment of Alzheimer's disease: A randomized, double-blind study. *Journal of Neural Transmission, 120*, 813–819. http://dx.doi.org/10.1007/s00702-012-0902-z.
Schabrun, S. M., Lamont, R. M., & Brauer, S. G. (2016). Transcranial direct current stimulation to enhance dual-task gait training in Parkinson's disease: A pilot RCT. *PLoS One, 11*(6), 14. http://dx.doi.org/10.1371/journal.pone.0158497.
Schonfeldt-Lecuona, C., Cardenas-Morales, L., Freudenmann, R. W., Kammer, T., & Herwig, U. (2010). Transcranial magnetic stimulation in depression—Lessons from the multicentre trials. *Restorative Neurology and Neuroscience, 28*(4), 569–576. http://dx.doi.org/10.3233/rnn-2010-0561.
Sedlackova, S., Rektorova, I., Srovnalova, H., & Rektor, I. (2009). Effect of high frequency repetitive transcranial magnetic stimulation on reaction time, clinical features and cognitive functions in patients with Parkinson's disease. *Journal of Neural Transmission, 116*(9), 1093–1101. http://dx.doi.org/10.1007/s00702-009-0259-0.

Shin, H. W., Youn, Y. C., Chung, S. J., & Sohn, Y. H. (2016). Effect of high-frequency repetitive transcranial magnetic stimulation on major depressive disorder in patients with Parkinson's disease. *Journal of Neurology*, *263*(7), 1442–1448. http://dx.doi.org/10.1007/s00415-016-8160-x.

Shirota, Y., Ohtsu, H., Hamada, M., Enomoto, H., & Ugawa, Y. (2013). Supplementary motor area stimulation for Parkinson disease: A randomized controlled study. *Neurology*, *80*(15), 1400–1405. http://dx.doi.org/10.1212/WNL.0b013e31828c2f66.

Srovnalova, H., Marecek, R., Kubikova, R., & Rektorova, I. (2012). The role of the right dorsolateral prefrontal cortex in the Tower of London task performance: Repetitive transcranial magnetic stimulation study in patients with Parkinson's disease. *Experimental Brain Research*, *223*(2), 251–257. http://dx.doi.org/10.1007/s00221-012-3255-9.

Srovnalova, H., Marecek, R., & Rektorova, I. (2011). The role of the inferior frontal gyri in cognitive processing of patients with Parkinson's disease: A pilot rTMS study. *Movement Disorders*, *26*(8), 1545–1548. http://dx.doi.org/10.1002/mds.23663.

Strafella, A. P., Paus, T., Barrett, J., & Dagher, A. (2001). Repetitive transcranial magnetic stimulation of the human prefrontal cortex induces dopamine release in the caudate nucleus. *Journal of Neuroscience*, *21*(15), 4.

Strafella, A. P., Paus, T., Fraraccio, M., & Dagher, A. (2003). Striatal dopamine release induced by repetitive transcranial magnetic stimulation of the human motor cortex. *Brain*, *126*, 2609–2615. http://dx.doi.org/10.1093/brain/awg268.

Swank, C., Mehta, J., & Criminger, C. (2016). Transcranial direct current stimulation lessens dual task cost in people with Parkinson's disease. *Neuroscience Letters*, *626*, 1–5. http://dx.doi.org/10.1016/j.neulet.2016.05.010.

Udupa, K., Bahl, N., Ni, Z., Gunraj, C., Mazzella, F., Moro, E., et al. (2016). Cortical plasticity induction by pairing subthalamic nucleus deep-brain stimulation and primary motor cortical transcranial magnetic stimulation in Parkinson's disease. *Journal of Neuroscience*, *36*(2), 396–404. http://dx.doi.org/10.1523/jneurosci.2499-15.2016.

CHAPTER THIRTY-NINE

Botulinum Toxin Therapy for Nonmotor Aspects of Parkinson's Disease

Jarosław Sławek[*,†,1], **Mariusz Madaliński**[‡]
[*]Medical University of Gdańsk, Gdańsk, Poland
[†]St. Adalbert Hospital, "Copernicus" Ltd., Gdańsk, Poland
[‡]The Pennine Acute Hospitals NHS Trust, Manchester, United Kingdom
[1]Corresponding author: e-mail address: jaroslawek@gumed.edu.pl

Contents

Abstract

The medical treatment of Parkinson's disease (PD) is symptomatic both in terms of motor and nonmotor aspects. The nonmotor symptoms therapy should be taken into account as many of them negatively influence the quality of life and are treatable.

International Review of Neurobiology, Volume 134
ISSN 0074-7742
http://dx.doi.org/10.1016/bs.irn.2017.04.012

Botulinum neurotoxin (BoNT) offers effective treatment for drooling and bladder dysfunctions and alternative treatment for constipation and pain related to parkinsonism. BoNT for drooling is probably the best-documented therapy for nonmotor symptoms by clinical trials. Ultrasound-guided injections into parotid and submandibular glands can be recommended as the first line treatment, especially when anticholinergic oral medications are not indicated due to the risk of confusion, cognitive decline, or psychosis. BoNT therapy of constipation and neurogenic bladder is not so well evidenced and the latter indication is based on positive effects in clinical trials in multiple sclerosis and posttraumatic spine injuries. BoNT injections into the salivary glands or bladder are effective and safe with relatively low risk of complications. Authors reviewed the literature on the use of BoNT for all nonmotor aspects of PD, showed their position in guidelines and systematic reviews and gave the practical remarks on injection techniques and dosing.

1. INTRODUCTION

Clinical presentation of Parkinson's disease (PD) consists of a number of nonmotor symptoms for which effective symptomatic treatments are available. These symptoms include neuropsychiatric ones, such as psychosis and depression, or autonomic, such as orthostatic hypotension. For some of them, including drooling, urinary urgency and incontinence, constipation and pain, injections of Botulinum neurotoxin (BoNT) can be offered as a therapeutic modality. BoNT, blocking the release of acetylcholine, is effective at the level of neuromuscular junction not only in striated (dystonia, spasticity), but also in smooth muscles (overactive bladder). It may also inhibit the cholinergic parasympathetic secretomotor fibers of salivary glands (drooling, sialorrhea) or sympathetic nerve fibers that stimulate eccrine sweat glands, when injected intradermally (hyperhidrosis). There are seven serotypes (A–G) of BoNT and each acts at the specific part of the so-called SNARE (soluble *N*-ethylmaleimide-sensitive factor attachment receptor) protein complex, including synaptobrevin, syntaxin, and SNAP-25 (synaptosomal-associated protein 25), which are responsible for neurotransmitter release into the synaptic cleft. Light chain of BoNT cleaves these proteins, which results in muscle relaxation or glandular secretion inhibition (Truong & Hallett, 2013). After BoNT injection, the clinical improvement lasts 3–4 months when injected into striate muscles and up to 6–9 months in case of smooth muscles or eccrine sweat glands. The lack of permanent effect is related to neuronal sprouting with formation of temporal new synapses, and after a longer period of time even the recovery of neurotransmitter release in originally blocked terminals

Table 1 Commercially Available BoNT Preparations (According to FDA Recommendation)

Trade Names	Unique, Nonproprietary Names	Serotype
Botox	OnabotulinumtoxinA	A
Dysport	AbobotulinumtoxinA	A
Xeomin	IncobotulinumtoxinA	A
NeuroBloc/MyoBloc	RimabotulinumtoxinA	B

FDA, food and drug administration.

(de Paiva, Meunier, Molgó, Aoki, & Dolly, 1999). There are two commercially available serotypes of BoNT: A and B. Four commercially available products have their unique names and international units which are not interchangeable among products (Table 1).

2. BoNT FOR DROOLING

The commonly used term sialorrhea should be referred to clinical conditions with excessive production of saliva. Drooling, defined as excessive pooling and poor control of saliva in the oral cavity, seems to be a more proper term for the majority of neurological entities with impaired salivary clearance, mostly due to swallowing problems (Dand & Sakel, 2010). PD is the most common etiology in adults, and the conditions with drooling as the clinical problem are shown in Table 2.

2.1 Prevalence

The most recent prevalence study on drooling in PD performed in China showed that it is a common problem with the rate of 52.7% (273 out of 518 examined patients). The most common correlations were the late-onset of the disease, higher levodopa equivalent daily doses, higher incidences of dysarthria, dysphagia and fluctuations, higher scores for the UPDRS part III, NMSS (Nonmotor Symptoms Scale), HAMD and HAMA (Hamilton Depression and Anxiety scales, respectively), and higher scores for the mobility, activities of daily life, stigma and communication subdomains of the PDQ-39. The logistic regression analysis indicated that the strongest correlation was with dysarthria and dysphagia (Ou et al., 2015). The review paper by Srivanitchapoom et al. summarized 15 prevalence studies performed between 1991 and 2012. Different groups used various tools and

Table 2 Neurological Conditions Accompanied by Drooling

- Parkinson's disease
- Atypical parkinsonism (MSA, PSP, CBD, DLB)
- Amyotrophic lateral sclerosis
- Cerebrovascular diseases (pseudobulbar, bulbarpalsy)
- Head injuries
- Muscle diseases (oculopharyngeal muscular dystrophy)
- Facial nerve paralysis
- Cerebral palsy
- Mental retardation

CBD, corticobasal degeneration; DLB, dementia with Lewy bodies; MSA, multisystem atrophy; PSP, progressive supranuclear palsy.

Table 3 Complications Related to Drooling

- Poor oral and perioral hygiene (dermatitis)
- Oral odour
- Eating difficulty
- Speaking problems
- Respiratory tract infections
- Aspiration pneumonia
- Social embarrassment
- Depression and anxiety
- Poor quality of life

the populations ranged from 44 up to 943 participants. The prevalence of drooling ranged widely from 10% to 84% of patients.

Hypersalivation may significantly impact the quality of life of PD patients. Moreover, it may result even in life-threatening consequences, like aspiration pneumonia (Table 3) (Srivanitchapoom, Pandey, & Hallett, 2014). Bloom et al. described the patient with severe drooling presenting with debilitating perioral dermatitis with erythematous, demarcated plaques, who did not tolerate oral anticholinergics and was successfully treated with BoNT (Bloem, Kalf, van der Kerkhof, & Zwarts, 2009).

2.2 Mechanisms

Saliva production and release is controlled by three pairs of salivary glands: parotid, submandibular, and sublingual. Under normal conditions, 70% of saliva is produced by sublingual and submandibular glands. Stimulation (e.g., eating) increases mainly the parotid gland activity (Stuchell & Mandel, 1988). In PD, poor control of saliva may be multifactorial

Table 4 Different Mechanisms Responsible for Drooling in PD Patients (Srivanitchapoom et al., 2014)

- Dysphagia
- Lingual bradykinesia
- Upper esophageal dysphagia
- Hypomimia
- Abnormal flexed posture
- Increasing salivary flow rate

and mainly related not only to swallowing dysfunction (impairment of oropharyngeal phase), but also to impaired motor control of the tongue, upper esophageal dysmotility, hypomimia with involuntary mouth opening, and stooped posture with dropped head, which promotes saliva release from the mouth space (Table 4) (Srivanitchapoom et al., 2014).

Drooling in PD is probably not related to overproduction of saliva. Study with the use of Tc-99 m scintigraphy to measure the parotid gland activity showed the same salivary production in PD patients and in healthy controls (Nicaretta, de Rosso, Maliska, & Costa, 2008). Another study, using video-laryngoscopy and barium swallowing test, showed correlation between dysphagia and drooling severity (Nobrega et al., 2008).

2.3 Assessment

The assessment of hypersalivation is not easy and based on subjective assessment scales like VAS (Visual Analog Scale) rated by patient and/or family. Movement Disorders Society (MDS) recommends the use of Drooling Severity and Frequency Scale (DSFS), Drooling Rating Scale, and Sialorrhea Clinical Scale for PD (SCS-PD) (Evatt et al., 2009). Objective measures used in several studies also include cotton rolls weighs or video-laryngoscopy to assess dysphagia and, indirectly, the risk of drooling.

2.4 Therapy

Before BoNT era only a limited number of interventions were available, including oral or sublingual anticholinergics and radiotherapy. Anticholinergic medications work by blocking the parasympathetic innervation of the salivary flow. The commonly used agents include atropine, glycopyrrolate, ipratropium bromide (Rickert & Blitzer, 2012). The problem with oral anticholinergics is that they are contraindicated in patients with glaucoma, may provoke behavioral changes with hallucinations and cognitive impairment, confusion, drowsiness, blurred vision, urinary retention, and may influence the gastrointestinal (GI) motility resulting in constipation. Many of these symptoms are present at advanced stages of PD; therefore, those medications are not useful. Approximately 30% of patients are unable to tolerate them (Rickert & Blitzer, 2012). Other oral medications that are possibly effective in sialorrhea include orally administered adrenergic receptor agonists: clonidine (α-2) and modafinil (α-1) (Chou, Evatt, Hinson, & Kompoliti, 2007). On the other hand, adrenergic receptors antagonists, such as clozapine used in PD for the treatment of psychosis, may be associated with increased drooling. Other medications that may promote drooling are cholinesterase inhibitors (rivastigmine, donepezil, galantamine) and quetiapine. Neuroleptics (clozapine, quetiapine) should be used for as short a period of time as possible, so discontinuation may improve drooling. On the other hand, cholinesterase inhibitors used for the treatment of PD-related dementia should usually be maintained for many years. Therefore, other treatment strategies, with better tolerability and safety profiles, like locally injected BoNT, are preferred in PD patients. Glycopyrrolate according to MDS recommendation is the only medication considered to be effective and probably clinically useful (Seppi, Weintraub, & Coelho, 2011). Glycopyrrolate can be rated "efficacious" for the very short-term treatment of sialorrhea in PD. Although efficacious in 1-week treatment, the study provides insufficient evidence for the treatment of sialorrhea in PD exceeding 1 week.

According to the group of MDS experts, BoNT-A and BoNT-B are considered "efficacious" for the treatment of sialorrhea in PD. As far as safety is concerned, they have an acceptable risk with specialized monitoring. The practice implications are that they are "clinically useful" for the treatment of sialorrhea in PD (Seppi et al., 2011). This chapter was published in 2011 and is the last version of MDS evidence-based review on this topic. Nevertheless, since that time only one additional study (randomized, placebo-controlled, using rimabotulinumtoxinB) was identified. Similar to the previous studies on BoNT-Å and -B, this study was short lasting (20 weeks) and confirmed statistically significant therapeutic effect of BoNT-B at week 4 (Chinnapongse, Gullo, Nemeth, Zhang, & Griggs, 2012).

Unfortunately, EFNS (European Federation of Neurological Societies) guidelines on therapy of PD did not mention drooling at all (Ferreira et al., 2013).

In 2015, Petracca et al. published literature overview on all studies performed with regards to sialorrhea in all studied indications. They identified 16 (out of 65) studies with PD patients included, but only 7 solely dedicated to sialorrhea accompanying PD. Only four studies, three using rimabotulinumtoxinB and one using onabotulinumtoxinA, were double-blind, placebo-controlled (all of them are summarized in Table 5) (Petracca et al., 2015).

Unfortunately, by the time this chapter was in preparation the results of a new, large, multicenter, placebo-controlled, double-blind study of Merz and using incobotulinumtoxinA had still not been published.

The majority of the studies were only short lasting. Petracca et al. published data on long-term assessment, but the study was not placebo-controlled and considered both, PD ($n = 33$) and ALS (amyotrophy lateral sclerosis, $n = 32$) patients. Patients were given intrasalivary, ultrasound-guided injections of 250 units of abobotulinumtoxinA or 2500 units of rimabotulinumtoxinB for at least two occasions (mean number of injections was 5 per patient, from the minimum of 2 up to the maximum of 16). The overall benefit was obtained in 282 out of 317 injections (89%), with the mean duration of 87 days (SD $\pm$ 36.8, range 30–240 days) and was similar for both serotypes. The repeated injections did not benefit the duration of the effect. Side effects were mild to moderate, more frequent in ALS, and included viscous saliva (5.4% for Abo and 9% for Rima); dry mouth (2% for Abo and 3.5% for Rima); and other less frequent than 1% were chewing difficulties, jaw dislocation, transitory paresis of facial nerve, pain, hematoma, mouth bleeding. Interestingly, dysphagia—the most expected complication in both groups of patients—was not noted, but reported in other 14 studies regarding various indications and doses (Petracca et al., 2015).

The same side effects were reported in other studies that were mainly uncontrolled (see the review by Srivanitchapoom et al. (2014). Transient swallowing difficulties were reported only in two double-blind, placebo-controlled studies performed by Lagalla et al. (2006, 2009). Comparing the studies, Lagalla et al. (2009) obtained significantly longer duration of benefit. When present, benefits lasted on average 19.2 $\pm$ 6.3 weeks in the BoNT-B group compared to 6.7 $\pm$ 1.4 weeks in controls. The review study summarizing several published papers confirmed that the clinical effect of both, BoNT-A and -B, starts approximately 1 week after the injection and lasts 3–5 months (Srivanitchapoom et al., 2014).

2.5 BoNT Injections: Technical Aspects

There are various techniques of BoNT injections published in the literature, as well as the wide range of doses. Most authors agree that intrasalivary gland

Table 5 Summary of Double-Blind Placebo-Controlled Studies on BoNT-A and BoNT-B in the Treatment of Drooling in PD

Authors	*n*	BoNT Type	Total Dose and Injected Glands	USG-guided Injections	Improvement	Duration of the Effect (weeks)	Complications
Ondo, Hunter, & Moore, 2004	16	B (Rima)	2500 units, parotid and submandibular	No	Yes	12–20	Dry mouth, diarrhea, neck pain, worsened gait
Lagalla, Millevolte, Capecci, Provinciali, & Ceravolo, 2006	32	A (Ona)	100 units, parotid	No	Yes: 88% of pts and 5% of placebo	No information	Mild transitory swallowing difficulties in one patient
Lagalla, Millevolte, Capecci, Provinciali, & Ceravolo, 2009	36	B (Rima)	4000 units, parotid	No	Yes: 78% responders: moderate (44.4%), very good (33.3%)	19.2 ± 6.3	Mild transient dysphagia (three patients) and mild weakness of chewing (one patient)
Chinnapongse et al., 2012	54	B (Rima)	1500, 2500, 3500 units, parotid and submandibular	No	Yes: (93%, 92%, and 100% for all three doses, respectively)	15	Dry mouth (15%)

n, number of patients; PD, Parkinson's disease.

ultrasound-guided injections are more effective and safe (however, there are no evidence-based data) than blind injections based on anatomical landmarks. BoNT is injected into both, parotid and submandibular glands. The largest is the parotid gland located in pre- and subauricular region, partially on the masseter muscle. The submandibular gland lies within the so-called submandibular triangle between two bellies of the digastric muscle and inferior margin of the mandible. Mylohyoid muscle divides the superficial and deep lobes (Truong, Olthoff, & Laskawi, 2013). BoNT can be injected solely to parotid, but in order to obtain a better result additional injection to submandibular gland is recommended. According to some authors, ultrasound-guided injection reduces the risk of complications. Using EMG guidance, one can avoid unintended intramuscular injection. In case of injecting submandibular gland, EMG guidance enables identification of mylohyoid muscle and the injection of the deep lobe. The structures to avoid within the region of the parotid gland are facial nerve, maxillary and superficial temporal artery branches, as well as retromandibular vein (Rickert & Blitzer, 2012).

The anatomical considerations and location of salivary glands with typical injection points are presented in Figs. 1 and 2A, B.

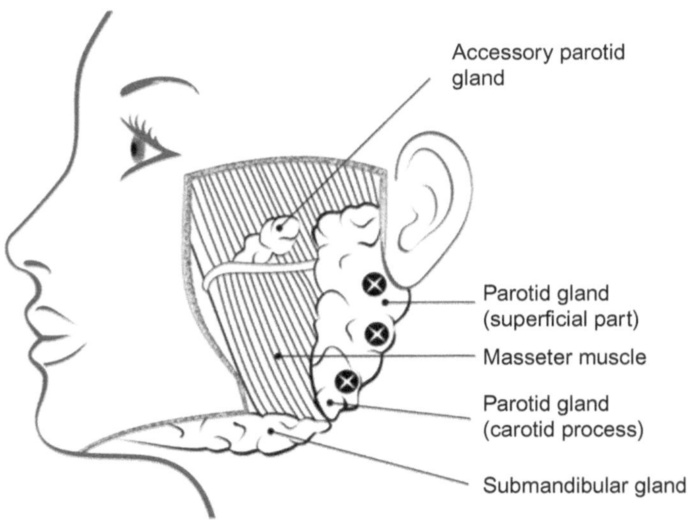

Fig. 1 The anatomical position of salivary glands and injection points. *With a kind permission of Sławek, J., & Rudzińska, M. (Eds.) (2016). Botulinum Toxin in neurological practice. Gdańsk, Poland: Via Medica Publishing Co.*

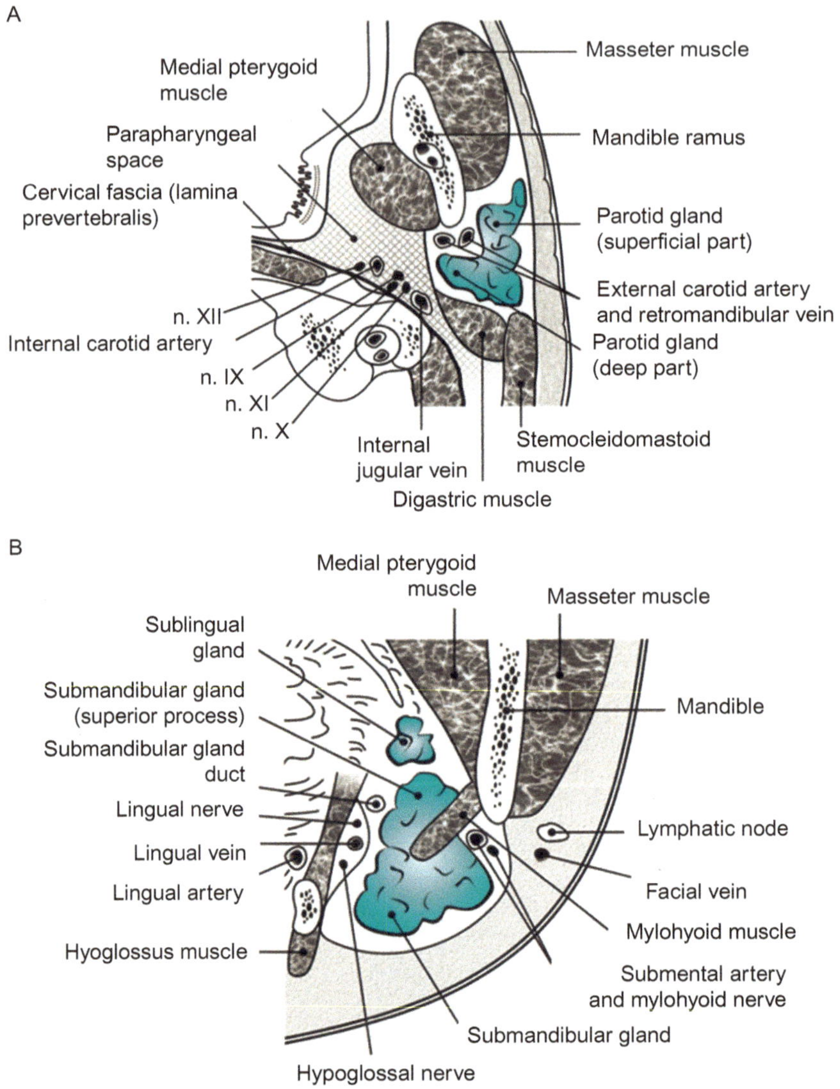

Fig. 2 (A and B) The anatomical surrounding structures that have to be considered when injecting parotid and submandibular glands. *With kind permission of Sławek, J., & Rudzińska, M. (Eds.) (2016). Botulinum Toxin in neurological practice. Gdańsk, Poland: Via Medica Publishing Co.*

According to Truong et al., the recommended dose is 22.5 units of onabotulinumtoxinA/incobotulinumtoxinA into each parotid gland at three locations, and 15 units of the same preparation for each submandibular gland at one or two sites (Lagalla et al., 2009). However, referring to the published studies, onabotulinumtoxinA was effective in doses from 5 to 50 units for parotid and 5 units for submandibular gland, and rimabotulinumtoxinB was effectively injected in doses 500–2500 units and 250 units per parotid and submandibular gland, respectively (Friedman & Potulska, 2001; Lipp, Trottenberg, Schink, Kupsch, & Arnold, 2003; Porta, Gamba, Bertacchi, & Vaj, 2001; Srivanitchapoom et al., 2014; Truong et al., 2013). The most common solution used was 2 mL of normal saline per one vial of BoNT.

However, it is important to note that the ideal dose of BoNT has not yet been identified (Rickert & Blitzer, 2012).

2.6 Summary

Both BoNT-A and -B seem to be safe and effective therapeutic options for disabling drooling in PD patients. As BoNT acts peripherally within salivary glands, it is free of unacceptable side effects in the population of advanced PD, such as cognitive, psychotic, and GI (constipation) complications. Injections are relatively easy, even based on anatomical landmarks, but ultrasound-guided injections are recommended.

3. BOTULINUM TOXIN FOR NEUROGENIC BLADDER

Bladder dysfunction of the lower urinary tract, called LUTS (lower urinary tract symptoms) consists of irritative (nocturia, urgency, frequency), obstructive or voiding (incomplete emptying, hesitancy, double voiding, and poor stream), storage (incontinence), or mixed symptoms (Sakakibara et al., 2012). In patients with PD, all variants may be present due to different reasons, not always related to neurodegenerative disease itself (like benign prostatic hyperplasia, prostatic cancer, stress, postdeliveries, or idiopathic incontinence). However, other disease-related factors, such as cognitive decline and dementia or the use of PD-related and -unrelated medications (e.g., diuretics used for comorbidities) may influence bladder function. LUTS in PD is associated with poor quality of life (Liu, Uchiyama, Sakakibara, & Yamamoto, 2015). Nocturia, the most common form of LUTS in PD, is associated with sleep disturbances, falls, hip fractures, and increased mortality (Sakakibara, Panicker, Finazzi-Agro, Iacovelli, & Bruschini, 2016). If left untreated, it may result in recurrent lower urinary tract infections and kidney damage.

3.1 Prevalence

In unselected PD cohorts, LUTS is significantly more frequent than in healthy controls, ranging from 27% to 85% of patients (McDonald, Winge, & Burn, 2017). Urinary problems are present even at the early stages of the disease (58% of patients have bladder overactivity revealed in urodynamic testing), but at that time they may be misinterpreted as unrelated to motor symptoms (Uchiyama et al., 2011). PRIAMO study dedicated to nonmotor symptoms among 1072 PD patients showed LUTS in 57.3% of them (Barone et al., 2009). The most common LUTS symptom is nocturia (57%–86%) followed by frequency (32%–71%), urgency (32%–68%), and urge incontinence (21%–40%) (McDonald et al., 2017). Comparing with MSA (multisystem atrophy) the urinary incontinence occurs later in PD (12 vs 144 months) (Wenning et al., 1999).

3.2 Mechanism

Urodynamic studies confirmed the crucial role of detrusor hyperactivity as the major dysfunction in PD. Sakakibara et al. have found that urinary symptoms in 72% of PD patients were attributed to detrusor overactivity (81%) and external sphincter relaxation problems (33%) (Sakakibara et al., 2016). Postganglionic parasympathetic M2/M3 cholinergic activity transmitted by pelvic nerves is the major excitatory mechanism causing detrusor contractions (McDonald et al., 2017; Sakakibara et al., 2012). Activation of D1 receptors within the basal ganglia tonically inhibits and activation of D2 receptors facilitates the micturition reflex (McDonald et al., 2017).

3.3 Assessment

There is a lot of questionnaires that may be useful in everyday practice, but it seems that before the decision on effective treatment other causes should be excluded as a rule, and urodynamic study as an objective type of measurement should be performed. It seems to be especially important in patients with suspected detrusor overactivity in whom BoNT treatment is planned.

3.4 Therapy

The role of dopaminergic therapy is uncertain. Acute administration of levodopa may worsen bladder functions, but chronic use increased bladder capacity and volume at first sensation of its filling (Brusa et al., 2007). The same promising results were obtained in a clinical trial with rasagiline

(Brusa et al., 2014). There are no data on the role of dopamine agonists, except for the improvement noted after apomorphine (Aranda & Cramer, 1993). The mainstream of oral therapy for overactive bladder consists of antimuscarinic agents and mirabegron (beta-3 adrenergic receptor agonist), see Table 6.

Antimuscarinic medications use was assessed in large cohorts of multiple sclerosis (MS) and spinal cord injury populations. In the study published by Manack et al., the majority of patients were treated with oxybutynine and tolterodine (86.7%), but the therapy was stopped after the mean of 209 days due to lack of efficacy or intolerability. Lack of treatment was responsible for increased risk (21.4%) of hospitalization, infection, nephrolithiasis, and renal insufficiency (Manack et al., 2011). In PD patients, especially those with advanced disease, the use of nonselective anticholinergic/antimuscarinic oral medications is controversial as they may exacerbate cognitive decline, trigger psychosis, confusion, dry mouth, fractures, and constipation, all symptoms that may occur in this cohort.

Mirabegron seems to be superior to oral antimuscarinic agents, but so far no clinical trial in PD has been performed (McDonald et al., 2017).

Desmopressin is effective especially for nocturia; however, it may cause hyponatremia and headache and should rather be avoided in the elderly. This is a possible option after lack of effect or intolerance of other treatments (McDonald et al., 2017). Alpha-adrenergic receptor antagonists, such as prazosin, doxazosin, tamsulosin, when used for improvement of voiding

Table 6 The Potential Treatments of Overactive Bladder in PD Patients (Sakakibara et al., 2016)

Antimuscarinic agents
– Oxybutynine
– Tolterodine
– Propiverine
– Trospium
– Solifenacin
– Darifenacin
– Fesoterodine
Adrenergic agonists
– Mirabegron
Desmopressin
Botulinum neurotoxin type A
Deep brain stimulation

(with concomitant PD prostate hyperplasia) have to be administered with caution, as they decrease arterial blood pressure and may exacerbate orthostatic hypotension and trigger syncope, falls, and fractures.

For patients resistant to oral medications, or due to unacceptable side effects, local injections of BoNT seem to be a treatment of choice. It is now FDA (Food and Drug Administration agency) approved method of treatment with no specific indication (for urinary incontinence in adult neurological patients with inadequate response or intolerance of antimuscarinic medications). Although both, ona- and abobotulinumtoxinA, were used in clinical trials, greater body of evidence favored onabotulinumtoxinA (Botox) and only this preparation is officially approved and licensed for this indication. The majority of clinical trials included mainly patients who underwent spine injuries and those with MS. These double-blind and placebo-controlled clinical trials, performed in large cohorts of patients proved the efficacy and good tolerability of BoNT (Cruz et al., 2011; Ehren et al., 2007; Ginsberg et al., 2012; Schurch et al., 2005). They used 200–300 units (Cruz et al., 2011; Ginsberg et al., 2012; Schurch et al., 2005) of onabotulinumtoxinA and 500 units of abobotulinumtoxinA (Ehren et al., 2007). The continence improvement rate was very high and similar for both preparations (about 86%), mean increase of bladder volume was 40%–178% and the reduction of muscle detrusor pressure was 22%–51%. What is interesting and valuable (as the technique requires cystoscopy) is the long improvement from 6 up to even 9 months. The method seems to be useful in all clinical entities with neurogenic bladder as an accompanying symptom (Table 7).

Recently published study by Nitti et al. proved the efficacy of lower doses (100 units of onabotulinumtoxinA). BoNT significantly decreased daily frequency of urinary incontinence episodes vs placebo and 22.9% vs

Table 7 Potential Indications for Local Injections of BoNT Into Bladder Detrusor Muscle

Multiple sclerosis
Brain and spine injuries
Cerebral palsy
Diabetic neuropathy
Cerebrovascular disease
Myelomeningocele
Parkinsonism
– Parkinson's disease
– Multisystem atrophy

6.5% of patients became completely continent. A larger proportion of onabotulinumtoxinA than placebo-treated patients reported a positive response on the treatment benefit scale (60.8% vs 29.2%). All the other overactive bladder symptoms improved compared to placebo. OnabotulinumtoxinA improved patient health-related quality of life across multiple measures. Uncomplicated urinary tract infection was the most common adverse event. A 5.4% rate of urinary retention was observed. The most common adverse events were infections, but they were also common in the placebo group, which means that the procedure and not medication itself is a responsible factor (Nitti et al., 2017).

Lower doses may result in lower rates of potential adverse events and lower cost of the procedure.

The most troublesome adverse event of this treatment is the need of self-catheterization in some patients who develop postvoid residual volumes. In a group studied by Ginsberg et al., among 60% of patients who did not use catheterization before BoNT injection 35%–42% had to do it after the procedure, for doses of 200 and 300 units, respectively (Ginsberg et al., 2012). All patients should be informed of such risk prior to the therapy, although this side effect is reversible.

Unfortunately, up to date no controlled studies were published for overactive bladder in population of PD patients. In an open-label study, Anderson et al. used onabotulinumtoxinA in 20 PD patients who did not respond to oral anticholinergics. The outcome measures were King's Health Questionnaire, voiding symptom score, and 3-day voiding diary. Free uroflowmetry with postvoid ultrasounds and cystometrogram pressure/flow studies were performed as well. Participants were followed up at 1, 3, and 6 months after cytoscopic injection of a low dose of BoNT (100 units in 10 mL of normal saline). The obtained results were meaningful: moderate to marked symptom relief at 3 months and a 50% incontinence decrease over 6 months relative to pretreatment period was reported in 59% patients; five patients failed to complete the 6-month endpoint. No urinary retention required catheterization. However, troublesome problems like number of voids per day and number of episodes of nocturia did not change significantly (Anderson, Orenberg, & Glowe, 2014). Another study conducted by Giannantoni et al. (2009) included patients with both, PD ($n = 4$) and MSA ($n = 2$). To avoid the urinary dysfunction due to prostatic hypertrophy, only females refractory to oral anticholinergics were included in the study. All patients received 200 units of onabotulinumtoxinA under cytoscopic guidance and all achieved improvement after 3 months in terms of both, daytime and night-time frequency. No patients had further episodes of urgency and urge urinary incontinence during the 5-month follow-up. Moreover, urodynamic examination revealed improvement in all tested variables. There were no systemic side effects and intermittent catheterization was required only in those with MSA (Giannantoni et al., 2009). Two years later the same group published results of another small open study performed in eight PD patients who were injected only 100 units of onabotulinumtoxinA. At months 1, 3, and 6, the authors observed decreased daytime and night-time urinary frequency with decreased number of urinary incontinence episodes. The quality of life scores also improved. In two patients, postvoid residual urine volume developed (>300 mL) and they required intermittent self-catheterization twice a day for 2 months (Giannantoni et al., 2011).

One study assessed the efficacy of abobotulinumtoxinA in a dose of 500 units (diluted in 30 mL) in a cohort of 16 PD patients that were followed for 12 months. Voiding frequency improved in all patients, except for six who were not assessed due to almost complete initial incontinence. None of the patients needed intermittent catheterization after the procedure. Bladder capacity increased by the 3rd month, and the quality of life scores of both patient and caregiver improved. The efficacy started to wear off between month 9 and 12. Twelve patients were offered the repeated injection, but only four decided to continue. There were no systemic or central adverse events (Kulaksizoglu & Parman, 2014).

The latest version of MDS evidence-based recommendations does not include the overactive bladder recommendations (Seppi et al., 2011). According to EFNS therapeutic guidelines (Table 8), BoNT is at the end

Table 8 EFNS Guidelines for the Treatment of Urinary Disturbance in PD (Ferreira et al., 2013)

- When symptoms appear suddenly: exclude urinary tract infection
- Nocturia: reduce intake of fluid after 6 pm. Sleep with head-up tilt of bed to reduce urine production
- Night-time dopaminergic therapy should be optimized (GPP)
- Use anticholinergic drugs (GPP): trospium chloride (10–20 mg 2–3 times daily) and tolterodine (2 mg twice daily). Trospium is less apt to penetrate the blood–brain barrier. Cognition may worsen
- Botulinum toxin type A injected in the detrusor muscle

EFNS, European Federation of Neurological Societies; PD, Parkinson's disease of the list of suggested interventions since the only Class IV studies were performed for PD as a particular indication (Ferreira et al., 2013).

3.5 Technical Aspects

Before injection urinalysis is recommended to exclude the infection. The so-called prophylactic antibiotic therapy 1–3 days before and after the procedure is recommended in products characteristic of onabotulinumtoxinA (Botox). The procedure is performed via cystoscopy under sedation and local anesthesia (intravesical 2% lidocaine, 30–50 mL), and injections are multiple—approximately 30 sites (each 1 mL) 1 cm apart, to bladder detrusor muscle avoiding the trigone region (Fig. 3).

Doses range from 100 to 300 units (dilution from 10 to 30 mL of normal saline), although doses used in double-blind, placebo-controlled studies were 200 and 300 units of onabotulinumtoxinA. So far no such trials have been conducted in PD patients. Therefore, the data from open-label studies and for other indications should be taken into consideration when therapy is offered (Schurch & Carda, 2013).

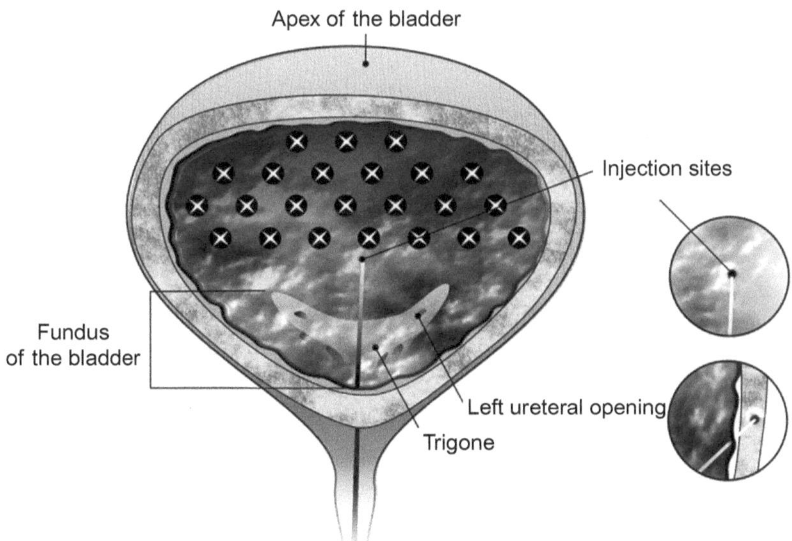

Fig. 3 Multipoint injection of BoNT into the bladder detrusor. *With kind permission of Allergan.*

3.6 Summary

LUTS are common in PD and may influence the quality of life. Night-time frequency disrupts sleep and may result in night falls and confusion. However, the etiology of urinary problems seems to be multifactorial and requires a thorough examination. It is likely that detrusor muscle overactivity is related to PD itself. Therefore, in PD patients antimuscarinic agents are the first-line treatment, but potentially serious adverse effects typically limit their use. Mirabegron seems to be a promising new medication, however, up till now there have been no studies regarding PD. Only few open-label studies were performed for overactive bladder with the use of BoNT in PD, but as the treatment is local and without central effects it seems to be a reasonable option for patients refractory or intolerable of oral medications. Extrapolating the data from studies on other indications, which received the FDA approval, it seems that intravesical BoNT could be a safe and effective therapeutic option, however, still requiring new studies on dosing and efficacy in PD patients.

4. BoNT FOR GI DYSFUNCTIONS

The pathophysiology of GI dysfunction in PD is multifactorial. Lewy bodies containing alpha-synuclein have been identified in autonomic neurons supplying the GI tract. Constipation is one of the nonmotor PD symptoms, which may precede the symptomatic phase of the disease and what may suggest the early involvement of GI tract. Therefore, GI dysfunction is probably related to the PD itself, but other factors need to be considered like diet, medications (e.g., with anticholinergic properties which may decrease GI motility), or infections (*Helicobacter pylori*). GI dysfunctions like gastroparesis are responsible for delayed gastric emptying and motor fluctuations due to erratic levodopa absorption. Moreover, this process may be also related to diet protein and probably to other factors like *H. pylori* infection as well (Narożańska et al., 2014). Although, BoNT is not a gold standard for therapy of those dysfunctions, it may constitute the therapeutic option in selected cases and more studies are required to establish clear indications. BoNT injection is safe, and gastroenterologists and surgeons use the toxin to treat a variety of GI disorders, as it may improve symptoms in many patients who have failed conventional therapies (Table 9) (Brisinda, Sivestrini, Bianco, & Maria, 2015; Madalinski et al., 2002, 2001). The toxin not only can improve symptoms, but can also minimize side effects associated with the use of systemic medications.

Table 9 Indications for BoNT Injection in the Alimentary Tract

A. The Upper Gastrointestinal Tract
– The esophagus
– Cricopharyngeal achalasia
– Esophageal achalasia
– Diffuse esophageal spasm
– Noncardiac chest pain suspected to be on the basis of a dysfunctional esophagus
The stomach
– Gastroparesis
The biliary tract
– Sphincter Oddi dysfunction (*diagnostically and therapeutically*)
B. The Lower Gastrointestinal Tract
– Pelvis and anorectal disorders
– Pelvic floor dyssynergis
– Chronic or acute anal fissure
– Chronic idiopathic anal pain
– After pull-through surgery for Hirschsprung's disease (*diagnostically and therapeutically*)

4.1 Upper GI Tract

4.1.1 Gastroparesis

Impaired gastric motility leads to early satiety, sensation of postprandial fullness, belching, nausea, vomiting, and upper abdominal pain. Regardless of the symptoms generated by the delayed gastric emptying, which also may be caused by other multiple reasons, the prevalence of delayed gastric emptying in PD is very high (70%–100% patients), but the prevalence of symptomatic gastroparesis in PD is unknown (Heetun & Quigley, 2012). There is no expert agreement on the use of the term gastroparesis—for grossly impaired emptying of the stomach or any delayed gastric emptying. According to American College of Gastroenterology, the diagnosis should be based on the combination of symptoms of gastroparesis, absence of gastric outlet obstruction or ulceration, and delay in gastric emptying. Scintigraphic gastric emptying of solids is the standard for gastric emptying evaluation and the diagnosis. Alternative method is gastric emptying assessment with wireless capsule and ^{13}C breath test using octanoate or spirulina incorporated into a solid metal. The most reliable parameter for diagnosis of gastroparesis is stomach retention of solids at 4 h measured by scintigraphy (medications which can affect gastric emptying should be stopped minimum 48 h before testing) (Camilleri, Parkman, Shafi, Abell, & Gerson, 2013).

Endoscopic management of gastroparesis most commonly involves intrapyloric BoNT injection (other options are mentioned in the Table 10). BoNT directly affects muscle tone and contractility by irreversible binding to cholinergic endings, decreases contractile secondary to decreased acetylcholine release, and reduces substance P immune-reactivity, and also disrupts pyloric myoelectric activity. The pylorus is injected in a radial pattern at or within 2 cm of the pylorus with total dose of 100–200 units of ona- or incobotulinumtoxinA. Although, BoNT injection reduced symptoms and accelerated gastric emptying in diabetic and idiopathic gastroparesis, it has not been confirmed in double-blinded randomized controlled studies (McCarty & Rustagi, 2015). There is also a very

Table 10 Types of Gastrointestinal Dysfunction in Parkinson's Disease and Possible Alternative to BoNT Treatments

Type of Dysfunction	Treatment Alternatives
Oro-pharyngeal dysfunction (AO dysphagia)	Chewing gum, sucking on hard candy (for mild symptoms) For more advanced: Speech and position therapies Anticholinergic drugs, agonist of selective alpha 1 and alpha 2-adrenergic receptor or muscarinic receptor, radiotherapy, surgical sectioning of the chorda tympani nerves, salivary gland/duct surgeries, percutaneous endoscopic gastrostomy, or surgical gastrojejunostomy
Esophageal motility	Nonpharmacological techniques (e.g., swallowing training), optimizing dopaminergic medication, percutaneous endoscopic gastrostomy, or surgical gastrojejunostomy
Gastroparesis	Domperidone, erythromycin, gastric stimulator implantation, transpyloric stenting, Endoscopic pyloromyotomy, percutaneous endoscopic jejunostomy Electric stimulation, surgery
Slow transit constipation	Increase fluid intake, dietary fiber, increase exercise, bulk-forming laxatives, stool softeners, stimulant laxatives, saline laxatives, hyperosmotic laxatives Abdominal massage, functional magnetic stimulation
Outlet type constipation	High fiber diet, laxatives, biofeedback training, balloon retraining, partial myectomy of the puborectalis muscle

limited data of only few patients with PD treated; the only two cases have been published so far. A female aged 78 and a male aged 65—both with persistent nausea and weight loss, and PD diagnosis for 10 and 6 years, respectively. The patients were treated with 100 units of onabotulinumtoxinA in the pyloric sphincter. The female reported complete resolution of her nausea and persistent discomfort in the upper abdomen (5 weeks after the BoTN injection) and the male presented with significant improvement in the nausea and the abdominal pain as well (2 months after the injection). Both patients gained some weight and gastric emptying scans showed a great improvement 3 months after injection (the male gastric emptying was even normal) (Gil, Hwynn, Fabian, Joseph, & Fernandez, 2011). This type of treatment may be potentially useful for improving the stomach reflex and better levodopa availability. Nevertheless, the way of intervention seems to be a disadvantage.

4.2 Lower GI Tract

Constipation may start up to 10 or even 20 years before motor symptoms of PD and is recognized as the most common manifestation of autonomic dysfunction in PD (70%–80% affected patients) (Essa & Hamdy, 2016; Kim & Sung, 2015). The future risk of PD increases fourfold in men who had less than one bowel movement a day, compared to individuals who had more than one. Chronic constipation in patients with PD may be due to prolonged colonic transit time and/or to outlet obstruction (48% of patients have mixed type of constipation). Chronic slow-transit constipation is caused by involvement of the myenteric plexus neurons and alteration in propulsive motor activity of the large bowel and seems to reflect degeneration of the enteric nervous system (ENS) and in the central nervous system (Su, Gandhy, Barlow, & Triadafilopoulos, 2016).

4.2.1 Functional Outlet Constipation

Dyssynergic defecation (also known as pelvic floor dyssynergia) contributes to a failure of appropriate relaxation of the puborectal muscle during straining or even its pathological contraction, which intensifies the flap-valve reaction of the anorectal angle and resulting in an obstruction to the onward stool passage. In healthy individuals, the puborectalis muscle and the external anal sphincter relax to straighten the anorectal angle and open the anal canal, which is controlled by the brain, sacral, and the ENS playing the crucial role in modulating the peristaltic reflex (Fig. 4).

Analysis of patients with PD and constipation investigated using high-resolution manometry, balloon expulsion testing, and wireless motility capsule (which assessed colonic transit) showed not only impaired relaxation or paradoxical anal contraction. In patients with PD, dyssynergic defecation consists also of inadequate propulsive force during attempted defecation. Sue et al.'s study showed that 89% patients with PD and functional constipation suffered from type II and IV of pelvic floor dyssynergia (Table 11), and 57% had overlap constipation (Albanese et al., 1997). Paradoxical contraction of the striated anal sphincter and the puborectalis muscle during defecation with inadequate rectal contraction seems to be the most prevalent abnormality in patients with PD and constipation; therefore, a therapy with BoTN might be a therapeutic approach to the outlet constipation in PD.

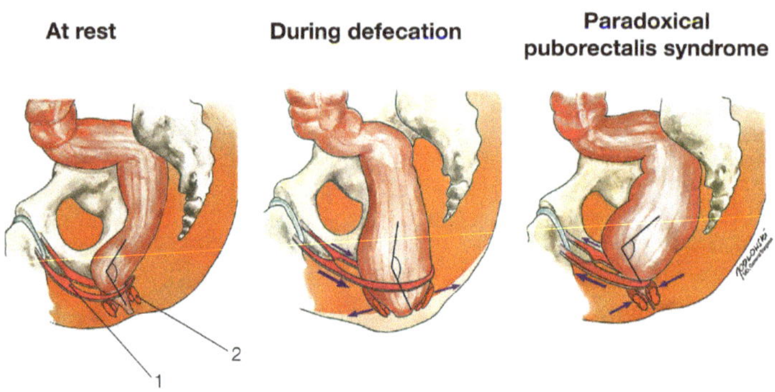

Fig. 4 Features of pelvic floor dyssynergia. The arrows indicate contraction or relaxation of the puborectalis muscle (1) and the external anal sphincter (2). At rest, the puborectalis is contracted and pulls the rectum forward. The act of defecation requires coordinated relaxation of the puborectalis muscle (1), the internal and external anal sphincter (2), with descent of the pelvic floor when an intraabdominal pressure increases, and then straightening of the anorectal angle to be successful. Paradoxical puborectalis syndrome occurs when the muscle does not relax during defecation; it actually contracts harder, creating an even sharper angle in the rectum, resulting in difficulty emptying the rectum.

Table 11 Types of Pelvic Floor Dyssynergia

Type of Dyssynergia	Rectal Propulsion	Anal Muscles
I	Adequate	Paradoxical contraction
II	Impaired	Paradoxical contraction
III	Adequate	Incomplete relaxation
IV	Impaired	Incomplete relaxation

The study by Sue et al. also revealed poor rectal sensation and lack rectoanal inhibitory reflex (suggestive of autonomic and ENS neuropathy) in 12%–30% cases (Albanese et al., 1997).

In 1997, Albanese et al. published for the first time an injection of BoNT for pelvic floor dyssynergia in a patient with PD (Albanese, Brisinda, Bentivoglio, & Maria, 2003). The muscle was palpated and verified by recording with unipolar electromyographic injection needle attached to the syringe containing the BoTN. In the next study, Albanese and colleagues reported injection of a total 30 units onabotulinumtoxinA in two sites in 10 patients and good result in 9/10 of them; clinical benefit lasted for approximately 12 weeks (Albanese et al., 2003).

In 2005, Cadeddu et al. published 18 patients with outlet type constipation in PD, where symptomatic improvement was noted at 2 months evaluation in 10 patients after initial injection of 100 units (onabotulinumtoxinA) in two sites on either side of puborectalis muscle under ultrasonographic guidance (Cadeddu et al., 2005). At 2 months, eight patients were retreated with 200 units and four of them had a further symptomatic improvement. At 4 months, six patients suffered symptomatic recurrence and these patients were retreated successfully with higher doses up to 200–300 units.

Contrary to Albanese and Cadeddu, we applied 125 units of abobotulinumtoxinA for spastic pelvic floor syndrome without transrectal ultrasonography or electromyographic control, which was a novel approach for BoTN injection over 10 years ago. During a rectal examination, patients were asked to pinpoint the most painful area during anal sphinters distention and BoTN was injected by anoscope in so identified the area near the posterior midline in the proximal part of the anus. This method of BoTN injection affects part of the external anal sphincter at the anorectal ring, subsequently leading to relaxation of the sphinters complex. All 13 patients (without PD diagnosis) reported subjective improvement and the results were monitored using anal manometry and a balloon expulsion test for 5 months (Madalinski, Chelstowska, & Dlugosinska, 2004).

A systemic review of the literature for the assessment of BoTN as a therapeutic method for outlet obstruction, including comparative and noncomparative trials was published. The paper concerns only studies with a sample size of at least 15 patients and excluding studies with follow-up shorter than 6 months. Authors analyzed 11 articles, but only 7 studies met the eligibility criteria (Emile et al., 2016; Farid et al., 2009a, 2009b;

Hompes et al., 2012; Maria, Cadeddu, Brandara, Marniga, & Brisinda, 2006; Shafik & El-Sibai, 1998; Zhang et al., 2014), comprising 189 patients (108 females and 81 males); female to male ratio was 1.3:1 with the median age 41.4. The median dose of applied BoTN-A per procedure was 100 units (range from 20 to 100 units); the median percentage of subjects who reported initial improvement of symptoms was 77.4% (range 37.5%–86.7%). After 4 months, this percentage diminished to a median of 46% (range 25%–100%). Anal pressure decreased in two studies (in 28.5% and 70.8% of patients); the rest of included studies showed no significant changes in anorectal manometry. Rates of improvement were evaluated also by balloon expulsion, EMG, and defecography, and ranged between 37.5%–80%, 54%–86.7%, and 25%–86.6%, respectively. Three studies used anorectal ultrasonography-guided technique, one study EMG-guided technique, and the rest used manual palpation with the index finger, but neither ultrasonography nor EMG-guided technique did not add significant value regarding initial or a long-term improvement. The site of injection varied, but the BoTN applying in a combination of lateral and posterior injection did not offer better results than lateral injection alone (either at 5 and 7 o'clock, or at 3 and 9 o'clock—viewed in the lithotomy position).

Complication rate across the studies concerned 7.4% of patients: a fecal incontinence in a minor grade for a few weeks (11 patients—reported by two studies only), an anal fissure (2 patients), and a rectal prolapse (1 patient) (Mills, Bahroo, & Pagan, 2015).

4.3 Summary

MDS evidence-based recommendations and EFNS guidelines do not mention BoNT as the potential therapeutic option (Ferreira et al., 2013). The only recommended as possibly efficacious are macrogol (for constipation) and domperidone for levodopa or dopamine agonists related gastric motility problems as nausea, vomiting, and anorexia (Seppi et al., 2011).

The only two studies included in this analysis assessing efficacy and safety of BoTN for pelvic floor dyssynergia were randomized comparative trials, and the optimal treatment of outlet constipation is still debatable. It is hard to extrapolate those results to PD patients, but the method can be taken into consideration in case of other treatments failure. Evidence of BoNT effectiveness for the treatment of constipation in PD has been unproven and so far, BoTN efficacy has been demonstrated in only two open-label studies (Mills et al., 2015). We think that a simple method of BoTN administration by a finger guidance will be beneficial for further studies and development.

5. BoNT FOR PAIN RELATED TO PD

Pain is one of the most common nonmotor symptoms of PD. According to the systematic review by Broen et al., it may affect 40%–85% of PD patients (Broen, Braaksma, & Patijn, 2012). In Japanese population, pain was the third most prevalent symptom after constipation (85.4%) and sleep problems (73.7%), with other sensory disturbances occurring in 72.7% of patients (Maeda et al., 2017). In a group of PD patients treated with BoNT between 1995 and 2014 retrospectively reviewed by Bruno et al., pain was the main indication for the treatment (50.6% of patients). Other indications included: functional impairment resulting from dystonia (26.25%), sialorrhea (18.75%), freezing of gait, and camptocormia. Eighty one percent of patients with pain reported benefits after the first injection and the effect was maintained till the last recorded visit (Bruno, Fox, Mancini, & Miyasaki, 2016).

The origin of the pain may be related directly to PD, as it is in case of the shoulder pain preceding PD diagnosis or foot spasms and other off-period sensory sensations related to off-dystonia. However, it may also affect the elderly PD patients in an indirect way, as it is with the low back pain related to limited mobility and postural deformities like camptocormia, or fulfill the criteria of a central pain (Chaudhuri & Schapira, 2009). Other pain syndromes, such as peripheral neuropathic pain, oral and genital pain, may occur more frequently in PD. In a recently published study, Szadejko et al. reported that when compared to healthy subjects PD patients had a higher frequency of paresthesia (29.4% vs 11.8%); impaired pain perception (40.0% vs 19.6%) and impaired light touch response (38.0% vs 16.3%). Neuropathy in PD may be related to levodopa treatment and higher homocysteine and/or elevated methylmalonic acid levels (Szadejko et al., 2016). Therefore, pain in PD may be of musculoskeletal, dystonic, neuropathic, central, and restless legs-associated or akathetic origin (Szadejko et al., 2016). The most common are musculoskeletal (70%) and dystonic (40%) pains (Beiske, Loge, Ronningen, & Svensson, 2009; Ford, 2009). Due to variability of underlying etiologies, the treatment of pain in PD is difficult. If the symptoms are "dopaminergic" they can be ameliorated with proper adjustment of dopaminergic therapy (e.g., shoulder pain at early stage, end-of-dose foot dystonia, off-period pain). The musculoskeletal pain may additionally benefit from exercises, nonsteroid antiinflammatory agents, or opioids. In refractory cases, subthalamic DBS may be beneficial (Maruo, Saitoh, & Hosomi, 2011). BoNT offers an additional therapeutic option in patients with pain related to foot or hand dystonia that is refractory to dopaminergic therapy or even DBS (Cordivari, Misra, Catania, & Lees, 2001; Jankovic, 2004, 2009). In a review by Perez et al. among emerging analgesic therapies only foot dystonia was mentioned as a potentially effective indication for BoNT therapy (Perez-Lloret et al., 2012). However, the evidence for this treatment remains inconsistent. Authors describe the upcoming double-blind, placebo-controlled clinical trial with the use of incobotulinumtoxinA for this indication. The paper was released in 2012 and so far the results of this study remain unpublished.

The very first trial on foot dystonia in 22 patients with PD was reported by Pacchetti et al. in 1995. BoNT (onabotulinumtoxinA) was injected into tibialis posterior, tibialis anterior, gastrocnemius, flexor digitorum longus, and extensor hallucis longus with a median dose of 40 units for each muscle, distributed in two sites. In all patients, the pain improved within 10 days, whereas in 21 patients the pain disappeared completely for 4 months (range, 3–7 months); a concomitant improvement in intensity of the dystonic spasm was also observed (Pacchetti, Albani, Martignoni, Godi, & Alfonsi, 1995).

In 2016, Gupta and Visvanathan published a study on foot injections of BoNT in six patients with painful foot dystonia who did not improve after DBS surgery. Dystonic foot and ankle muscles were identified and injected with 250–400 units of onabotulinumtoxinA. Three weeks afterward significant improvements were noted in both, dystonia and pain (Gupta & Visvanathan, 2016).

Recent MDS evidence-based review did not mention pain at all and EFNS guidelines mentioned BoNT treatment as an option for off-period and early morning dystonias at the level of GPP (good practice point) (Ferreira et al., 2013; Seppi et al., 2011).

6. BOTULINUM TOXIN FOR HYPERHIDROSIS

Hyperhidrosis or excessive sweating is relatively common in PD, but rarely it is focal, limited to palms, axilla, or feet. Usually, it is generalized and related to off states (e.g., during the nights) or is a side effect of the severe choreatic dyskinesia with high energy output. BoNT-A, due to the block of acetylocholine release from sympathetic nerve fibers, stimulating eccrine sweat glands is a highly effective therapy. Injections are intradermal, multiple and require a local anesthesia (e.g., by lidocaine, but ice-pack use is more safe and practical, making injections less painful and decrease the local bleeding). The effect is longer than intramuscular—up to 6–9 months. BoNT-A was studied in primary hyperhidrosis and is licensed for this indication.

Table 12 Potential Indications for Motor Problems Associated With Parkinson's Disease and Atypical Parkinsonisms (Bonanni, Thomas, Varanese, Scorrano, & Onofrj, 2007; Dupeyron et al., 2015; Mills et al., 2015)

Jaw and hand tremor
Blepharospasm and lid opening apraxia (PD and atypical parkinsonism as PSP)
Retrocollis (PSP)
Pisa syndrome (MSA)
Camptocormia
Lateral axial dystonia

MSA, multisystem atrophy; PD, Parkinson's disease; PSP, progressive supranuclear palsy.

According to the evidence-based system, as highly effective procedure is classified at the level A. There are no studies in PD; therefore, specific recommendation in this group of patients is not possible.

7. BOTULINUM TOXIN FOR OTHER INDICATIONS IN PD

Although the topic of this chapter includes mainly nonmotor aspects of PD and the use of BoNT, there are many more other potential indications for BoNT injections related to motor problems. The majority of studies are open-label case series and require further investigations. The potential new indications are presented in Table 12.

REFERENCES

Albanese, A., Brisinda, G., Bentivoglio, A. R., & Maria, G. (2003). Treatment of outlet obstruction constipation in Parkinson's disease with botulinum neurotoxin A. *The American Journal of Gastroenterology*, *98*, 1439–1440.

Albanese, A., Maria, G., Bentivoglio, A. R., Brisinda, G., Cassetta, E., & Tonali, P. (1997). Severe constipation in Parkinson's disease relieved by botulinum toxin. *Movement Disorders*, *12*, 764–766.

Anderson, R. U., Orenberg, E. K., & Glowe, P. (2014). Onabotulinumtoxin A office treatment for neurogenic bladder incontinence in Parkinson's disease. *Urology*, *83*(1), 22–27.

Aranda, B., & Cramer, P. (1993). Effects of apomorphine and L-dopa on the parkinsonian bladder. *Neurourology and Urodynamics*, *12*(3), 203–209.

Barone, P., Antonini, A., Colosimo, C., Marconi, R., Morgante, L., Avarello, T. P., et al. (2009). The PRIAMO study: A multicenter assessment of nonmotor symptoms and their impact on quality of life in Parkinson's disease. *Movement Disorders*, *24*(11), 1641–1649.

Beiske, A. G., Loge, J. H., Ronningen, A., & Svensson, E. (2009). Pain in Parkinson's disease: Prevalence and characteristics. *Pain*, *141*, 173–177.

Bloem, B. R., Kalf, J. G., van der Kerkhof, P. C. M., & Zwarts, M. J. (2009). Debilitating consequences of drooling. *Journal of Neurology*, *256*, 1382–1383.

Bonanni, L., Thomas, A., Varanese, S., Scorrano, V., & Onofrj, M. (2007). Botulinum toxin treatment of lateral axial dystonia in parkinsonism. *Movement Disorders, 22*(14), 2097–2103.

Brisinda, G., Sivestrini, N., Bianco, G., & Maria, G. (2015). Treatment of gastrointestinal sphincters spasms with botulinum toxin A. *Toxins (Basel), 7*, 1882–1916.

Broen, M. P., Braaksma, M. M., & Patijn, J. (2012). Prevalence of pain in Parkinson's disease: A systematic review using the modified QUADAS tool. *Movement Disorders, 27*(4), 480–484.

Bruno, V. A., Fox, S. H., Mancini, D., & Miyasaki, J. M. (2016). Botulinum toxin use in refractory pain and other symptoms in parkinsonism. *The Canadian Journal of Neurological Sciences, 43*(5), 697–702.

Brusa, L., Musco, S., Bernardi, G., Iani, C., Pierantozzi, M., Stanzione, P., et al. (2014). Rasagiline effect on bladder disturbances in early mild Parkinson's disease patients. *Parkinsonism & Related Disorders, 20*(8), 931–932.

Brusa, L., Petta, F., Pisani, A., Moschella, V., Iani, C., Stanzione, P., et al. (2007). Acute vs chronic effects of L-dopa on bladder function in patients with mild Parkinson disease. *Neurology, 68*(18), 1455–1459.

Cadeddu, F., Bentivoglio, A. R., Brandara, F., Marniga, G., Brisinda, G., & Maria, G. (2005). Outlet type constipation in Parkinson's disease: Results of botulinum toxin treatment. *Alimentary Pharmacology & Therapeutics, 22*, 997–1003.

Camilleri, M., Parkman, H. P., Shafi, M. A., Abell, T. L., & Gerson, L. (2013). Clinical guideline: Management of gastroparesis. *The American Journal of Gastroenterology, 108*, 18–37.

Chaudhuri, K. R., & Schapira, A. H. (2009). Non-motor symptoms of Parkinson's disease: Dopaminergic pathophysiology and treatment. *Lancet Neurology, 8*, 464–474.

Chinnapongse, R., Gullo, K., Nemeth, P., Zhang, Y., & Griggs, L. (2012). Safety and efficacy of botulinum toxin type B for treatment of sialorrhea in Parkinson's disease: A prospective double-blind trial. *Movement Disorders, 27*(2), 219–226.

Chou, K. L., Evatt, M., Hinson, V., & Kompoliti, K. (2007). Sialorrhea in Parkinson's disease: A review. *Movement Disorders, 26*, 2306–2313.

Cordivari, C., Misra, V. P., Catania, S., & Lees, A. J. (2001). Treatment of dystonic clenched fist with botulinum toxin. *Movement Disorders, 16*, 907–913.

Cruz, F., Herschorn, S., Aliotta, P., Brin, M., Thompson, C., Lam, W., et al. (2011). Efficacy and safety of onabotulinumtoxinA in patients with urinary incontinence due to neurogenic detrusor overactivity: A randomised, double-blind, placebo-controlled trial. *European Urology, 60*(4), 742–750.

Dand, P., & Sakel, M. (2010). The management of drooling in motor neuron disease. *International Journal of Palliative Nursing, 16*, 560–564.

de Paiva, A., Meunier, F. A., Molgó, J., Aoki, K. R., & Dolly, J. O. (1999). Functional repair of motor endplates after botulinum neurotoxin type a poisoning: Biphasic switch of synaptic activity between nerve sprouts and their parent terminals. *Proceedings of the National Academy of Sciences of the United States of America, 96*(6), 3200–3205.

Dupeyron, A., Viollet, E., Coroian, F., Gagnard, C., Renard, D., & Castelnovo, G. (2015). Botulinum toxin-a for treatment of Pisa syndrome: A new target muscle. *Parkinsonism & Related Disorders, 21*(6), 669–670.

Ehren, I., Volz, D., Farrelly, E., Berglund, L., Brundin, L., Hultling, C., et al. (2007). Efficacy and impact of botulinum toxin a on quality of life in patients with neurogenic detrusor overactivity: A randomised, placebo-controlled, double-blind study. *Scandinavian Journal of Urology and Nephrology, 41*(4), 335–340.

Emile, S. H., Elfeki, H. A., Elbanna, H. G., Youssef, M., Thabet, W., El-Hamed, T. M. A., et al. (2016). Efficacy and safety of botulinum toxin in treatment of anismus: A systematic review. *World Journal of Gastrointestinal Pharmacology and Therapeutics, 7*, 453–462.

Essa, H., & Hamdy, S. (2016). Evaluating the scope of gastrointestinal symptoms of Parkinson's disease: A review of the evidence. *Neuroenterology*, *4*, 1–12.

Evatt, M. L., Chaudhuri, K. R., Chou, K. L., Cubo, E., Hinson, V., Kompoliti, K., et al. (2009). Dysautonomia rating scales in Parkinson's disease: Sialorrhea, dysphagia, and constipation—Critique and recommendations by movement disorders task force on rating scales for Parkinson's disease. *Movement Disorders*, *24*(5), 635–646.

Farid, M., El Monem, H. A., Omar, W., El Nakeeb, A., Fikry, A., Youssef, T., et al. (2009a). Comparative study between biofeedback retraining and botulinum neurotoxin in the treatment of anismus patients. *International Journal of Colorectal Disease*, *24*, 115–120.

Farid, M., Youssef, T., Mahdy, T., Omar, W., Moneim, H. A., El Nakeeb, A., et al. (2009b). Comparative study between botulinum toxin injection and partial division of puborectalis for treating anismus. *International Journal of Colorectal Disease*, *24*, 327–334.

Ferreira, J. J., Katzenschlager, R., Bloem, B. R., Bonuccelli, U., Burn, D., Deuschl, G., et al. (2013). Summary of the recommendations of the EFNS/MDS-ES review on therapeutic management of Parkinson's disease. *European Journal of Neurology*, *20*(1), 5–15.

Ford, B. (2009). Parkinson disease: Pain in Parkinson disease: The hidden epidemic. *Nature Reviews. Neurology*, *5*, 242–243.

Friedman, A., & Potulska, A. (2001). Quantitative assessment of parkinsonian sialorrhea and results of treatment with botulinum toxin. *Parkinsonism & Related Disorders*, 7(4), 329–332.

Giannantoni, A., Conte, A., Proietti, S., Giovannozzi, S., Rossi, A., Fabbrini, G., et al. (2011). Botulinum toxin type A in patients with Parkinson's disease and refractory overactive bladder. *The Journal of Urology*, *186*(3), 960–964.

Giannantoni, A., Rossi, A., Mearini, E., Del Zingaro, M., Porena, M., & Berardelli, A. (2009). Botulinum toxin A for overactive bladder and detrusor muscle overactivity in patients with Parkinson's disease and multiple system atrophy. *The Journal of Urology*, *182*(4), 1453–1457.

Gil, R. A., Hwynn, N., Fabian, T., Joseph, S., & Fernandez, H. H. (2011). Botulinum toxin type A for the treatment of gastroparesis in Parkinson's disease patients. *Parkinsonism & Related Disorders*, *17*, 285–287.

Ginsberg, D., Gousse, A., Keppenne, V., Sievert, K. D., Thompson, C., Lam, W., et al. (2012). Phase 3 efficacy and tolerability study of onabotulinumtoxinA for urinary incontinence from neurogenic detrusor overactivity. *The Journal of Urology*, *187*(6), 2131–2139.

Gupta, A. D., & Visvanathan, R. (2016). Botulinum toxin for foot dystonia in patients with Parkinson's disease having deep brain stimulation: A case series and a pilot study. *Journal of Rehabilitation Medicine*, *48*(6), 559–562.

Heetun, Z. S., & Quigley, E. M. M. (2012). Gastroparesis and Parkinson's disease: A systemic review. *Parkinsonism & Related Diseases*, *18*, 433–440.

Hompes, R., Harmston, C., Wijffels, N., Jones, O. M., Cunningham, C., & Lindsey, I. (2012). Excellent response rate of anismus to botulinum toxin if rectal prolapse misdiagnosed as anismus ('pseudoanismus') is excluded. *Colorectal Disease*, *14*, 224–230.

Jankovic, J. (2004). Botulinum toxin in clinical practice. *Journal of Neurology, Neurosurgery, and Psychiatry*, *75*, 951–957.

Jankovic, J. (2009). Disease-oriented approach to botulinum toxin use. *Toxicon*, *54*, 614–623.

Kim, J. S., & Sung, H. Y. (2015). Gastrointestinal autonomic dysfunction in patients with Parkinson's disease. *Journal of Movement Disorders*, *8*, 76–82.

Kulaksizoglu, H., & Parman, Y. (2014). Use of botulinum toxin-A for the treatment of overactive bladder symptoms in patients with Parkinsons's disease. *Parkinsonism & Related Disorders*, *16*(8), 531–534.

Lagalla, G., Millevolte, M., Capecci, M., Provinciali, L., & Ceravolo, M. G. (2006). Botulinum toxin type A for drooling in Parkinson's disease: A double-blind, randomized, placebo-controlled study. *Movement Disorders*, *21*(5), 704–707.

Lagalla, G., Millevolte, M., Capecci, M., Provinciali, L., & Ceravolo, M. G. (2009). Long-lasting benefits of botulinum toxin type B in Parkinson's disease-related drooling. *Journal of Neurology, 256*(4), 563–567.

Lipp, A., Trottenberg, T., Schink, T., Kupsch, A., & Arnold, G. (2003). A randomized trial of botulinum toxin A for treatment of drooling. *Neurology, 61*(9), 1279–1281.

Liu, Z., Uchiyama, T., Sakakibara, R., & Yamamoto, T. (2015). Underactive and overactive bladders are related to motor function and quality of life in Parkinson's disease. *International Urology and Nephrology, 47*(5), 751–757.

Madalinski, M., Chelstowska, M., & Dlugosinska, J. (2004). In *New method of botulinum toxin injection for the treatment of spastic pelvic floor syndrome XI Polish Congress of Gastroenterology. Warszawa, Poland* (p. 280). 2004, Poster V17.

Madalinski, M., Sławek, J., Duzynski, W., Zbytek, B., Jagiello, K., Adrich, Z., et al. (2002). Side effects of botulinum toxin injection for benign anal disorders. *European Journal of Gastroenterology & Hepatology, 14*, 853–856.

Madalinski, M., Sławek, J., Zbytek, B., Duzynski, W., Adrich, Z., Jagiello, K., et al. (2001). Topical nitrates and the higher doses of botulinum toxin for chronic anal fissure. *Hepato-Gastroenterology, 48*, 977–979.

Maeda, T., Shimo, Y., Chiu, S. W., Yamaguchi, T., Kashihara, K., Tsuboi, Y., et al. (2017). Clinical manifestations of nonmotor symptoms in 1021 Japanese Parkinson's disease patients from 35 medical centers. *Parkinsonism & Related Disorders, 38*, 54–60.

Manack, A., Motsko, S. P., Haag-Molkenteller, C., Dmochowski, R. R., Goehring, E. L., Jr., Nguyen-Khoa, B. A., et al. (2011). Epidemiology and healthcare utilization of neurogenic bladder patients in a US claims database. *Neurourology and Urodynamics, 30*(3), 395–401.

Maria, G., Cadeddu, F., Brandara, F., Marniga, G., & Brisinda, G. (2006). Experience with type A botulinum toxin for treatment of outlet-type constipation. *The American Journal of Gastroenterology, 101*, 2570–2575.

Maruo, T., Saitoh, Y., & Hosomi, K. (2011). Deep brain stimulation of the subthalamic nucleus improves temperature sensation in patients with Parkinson's disease. *Pain, 152*, 860–865.

McCarty, T. R., & Rustagi, T. (2015). Endoscopic treatment of gastroparesis. *World Journal of Gastroenterology, 21*, 6842–6849.

McDonald, C., Winge, K., & Burn, D. J. (2017). Lower urinary tract symptoms in Parkinson's disease: Prevalence, etiology and management. *Parkinsonism & Related Diseases, 35*, 8–16.

Mills, R., Bahroo, L., & Pagan, F. (2015). An update on the use of botulinum toxin therapy in Parkinson's disease. *Current Neurology and Neuroscience Reports, 15*, 511.

Narożańska, E., Białecka, M., Adamiak-Giera, U., Gawrońska-Szklarz, B., Sołtan, W., Schinwelski, M., et al. (2014). Pharmacokinetics of levodopa in patients with Parkinson disease and motor fluctuations depending on the presence of helicobacter pylori infection. *Clinical Neuropharmacology, 37*(4), 96–99.

Nicaretta, D. H., de Rosso, A. L., Maliska, C., & Costa, M. M. (2008). Scitigraphic analysis of the parotid glands in patients with sialorrhea and Parkinson's disease. *Parkinsonism & Related Diseases, 14*, 338–341.

Nitti, V. W., Dmochowski, R., Herschorn, S., Sand, P., Thompson, C., Nardo, C., et al. (2017). EMBARK study group. OnabotulinumtoxinA for the treatment of patients with overactive bladder and urinary incontinence: Results of a phase 3, randomized, placebo controlled trial. *The Journal of Urology, 197*(2S), S216–S223.

Nobrega, A. C., Rodrigues, B., Torres, A. C., Scarpel, R. D., Neves, C. A., & Melo, A. (2008). Is drooling secondary to a swallowing disorder in patients with Parkinson's disease? *Parkinsonism & Related Diseases, 14*, 243–245.

Ondo, W. G., Hunter, C., & Moore, W. (2004). A double-blind placebo-controlled trial of botulinum toxin B for sialorrhea in Parkinson's disease. *Neurology, 62*(1), 37–40.

Ou, R., Guo, X., Wei, Q., Cao, B., Yang, J., Song, W., et al. (2015). Prevalence and clinical correlates of drooling in Parkinson disease: A study on 518 Chinese patients. *Parkinsonism & Related Diseases*, *3*, 211–215.

Pacchetti, C., Albani, G., Martignoni, E., Godi, L., & Alfonsi, E. (1995). "Off" painful dystonia in Parkinson's disease treated with botulinum toxin. *Movement Disorders*, *10*(3), 333–336.

Perez-Lloret, S., Rey, M. V., Dellapina, E., Pellaprat, J., Brefel-Courbon, C., & Rascol, O. (2012). Emerging analgesic drugs for Parkinson's disease. *Expert Opinion on Emerging Drugs*, *17*(2), 157–171.

Petracca, M., Guidubaldi, A., Ricciardi, L., Ialongo, T., Del Grande, A., Mulas, D., et al. (2015). Botulinum toxin A and B in sialorrhea: Long term data and literature overview. *Toxicon*, *107*, 129–140.

Porta, M., Gamba, M., Bertacchi, G., & Vaj, P. (2001). Treatment of sialorrhoea with ultrasound guided botulinum toxin type a injection in patients with neurological disorders. *Journal of Neurology, Neurosurgery, and Psychiatry*, *70*(4), 538–540.

Rickert, S. M., & Blitzer, A. (2012). Sialorrhea. In A. Blitzer, B. E. Benson, & J. Guss (Eds.), *Botulinum neurotoxin for head and neck disorders* (pp. 213–221). New York, Stuttgart: Thieme.

Sakakibara, R., Panicker, J., Finazzi-Agro, E., Iacovelli, V., & Bruschini, H. (2016). Parkinson's disease subcomittee, the neurourology promotion committee in the international continence society. A guideline for the management of bladder dysfunction in Parkinson's disease and other gait disorders. *Neurourology and Urodynamics*, *35*(5), 551–563.

Sakakibara, R., Tateno, F., Kishi, M., Tsuyuzaki, Y., Uchiyama, T., & Yamamoto, T. (2012). Pathophysiology of bladder dysfunction in Parkinson's disease. *Neurobiology of Disease*, *46*(3), 565–571.

Schurch, B., & Carda, S. (2013). Botulinum neurotoxin applications in urological disorders. In D. Truong, M. Hallett, C. Zachary, & D. Dressler (Eds.), *Manual of botulinum toxin therapy* (2nd ed., pp. 288–294). New York: Cambridge University Press.

Schurch, B., de Sèze, M., Denys, P., Chartier-Kastler, E., Haab, F., Everaert, K., et al. (2005). Botulinum toxin type a is a safe and effective treatment for neurogenic urinary incontinence: Results of a single treatment, randomized, placebo controlled 6-month study. *The Journal of Urology*, *174*(1), 196–200.

Seppi, K., Weintraub, D., & Coelho, M. (2011). The Movement Disorder Society evidence-based medicine review update: Treatments for non-motor symptoms of Parkinson's disease. *Movement Disorders*, *26*(Suppl. 3), S42–S80.

Shafik, A., & El-Sibai, O. (1998). Botulin toxin in the treatment of nonrelaxing puborectalis syndrome. *Digestive Surgery*, *15*, 347–351.

Srivanitchapoom, P., Pandey, S., & Hallett, M. (2014). Drooling in Parkinson's disease: A review. *Parkinsonism & Related Diseases*, *20*, 1109–1118.

Stuchell, R. N., & Mandel, I. D. (1988). Salivary gland dysfunction and swallowing disorders. *Otolaryngologic Clinics of North America*, *21*, 649–661.

Su, A., Gandhy, R., Barlow, C., & Triadafilopoulos, G. (2016). Utility of high-resolution anorectal manometry and wireless motility capsule in the evaluation of patients with Parkinson's disease and chronic constipation. *BMJ Open Gastroenterology*, *3*, 1–8.

Szadejko, K., Dziewiatowski, K., Szabat, K., Robowski, P., Schinwelski, M., Sitek, E., et al. (2016). Polyneuropathy in levodopa-treated Parkinson's patients. *Journal of the Neurological Sciences*, *371*, 36–41.

Truong, D., & Hallett, M. (2013). Pharmacology of botulinum neurotoxins. In D. Truong, M. Hallett, C. Zachary, & D. Dressler (Eds.), *Manual of botulinum toxin therapy* (2nd ed., pp. 12–15). New York: Cambridge University Press.

Truong, D., Olthoff, A., & Laskawi, R. (2013). Botulinum neurotoxin therapy for laryngeal muscle hyperactivity syndromes. In D. Truong, M. Hallett, C. Zachary, & D. Dressler (Eds.), *Manual of botulinum toxin therapy* (2nd ed., pp. 85–100). New York: Cambridge University.

Uchiyama, T., Sakakibara, R., Yamamoto, T., Ito, T., Yamaguchi, C., Awa, Y., et al. (2011). Urinary dysfunction in early and untreated Parkinson's disease. *Journal of Neurology, Neurosurgery, and Psychiatry*, *82*(12), 1382–1386.

Wenning, G. K., Scherfler, C., Granata, R., Bösch, S., Verny, M., Chaudhuri, K. R., et al. (1999). Time course of symptomatic orthostatic hypotension and urinary incontinence in patients with postmortem confirmed parkinsonian syndromes: A clinicopathological study. *Journal of Neurology, Neurosurgery, and Psychiatry*, *67*(5), 620–623.

Zhang, Y., Wang, Z. N., He, L., Gao, G., Zhai, Q., Yin, Z. T., et al. (2014). Botulinum toxin type-a injection to treat patients with intractable anismus unresponsive to simple biofeedback training. *World Journal of Gastroenterology*, *20*, 12602–12607.

CHAPTER FORTY

Nutrition and Nonmotor Symptoms of Parkinson's Disease

Laurie K. Mischley[1]
Bastyr University Research Institute, Kenmore, WA, United States
[1]Corresponding author: e-mail address: lmischley@bastyr.edu

Contents

Abstract

To date, no guidelines exist for the screening, evaluation, and management of nutritional status in PD. Dozens of studies demonstrate an association between diet in adulthood with subsequent risk of developing PD. Individuals with PD are at increased risk of malnutrition due to the increased metabolic demands and disease pathophysiology. Risk of malnutrition is further complicated by anosmia, swallowing difficulties, constipation, and drug–nutrient interactions. An emerging body of evidence suggests that the

International Review of Neurobiology, Volume 134
ISSN 0074-7742
http://dx.doi.org/10.1016/bs.irn.2017.04.013

intestinal tract is affected early in the disease, creating therapeutic opportunities for early intervention. Dietary modification and nutritional supplementation may improve symptoms of constipation, depression, insomnia, dystonia, and help prevent cognitive dysfunction. This review summarizes the state of the science related to nutrition and nonmotor symptoms of PD.

1. INTRODUCTION

Nutrition is the study of that which must be exogenously obtained for human health and survival. A nutrient is considered *essential* if it "serves an indispensable physiologic function, but cannot be synthesized endogenously at an adequate rate by healthy subjects" (Chipponi, Bleier, Santi, & Rudman, 1982). The determination of dietary essentiality had traditionally been established through classic feeding studies using purified diets with, or without, the nutrient being studied. Over time, if a nutrient is essential, a deficiency syndrome will emerge as signs and symptoms of impaired growth, function, biochemical alterations, or symptoms of illness become apparent. *Conditionally essential* nutrients are those which are ordinarily synthesized in the body in sufficient amounts to meet physiological requirements, but must be exogenously supplied during periods when physiologic demands are greater than biosynthetic capacity. Unique nutritional requirements may additionally occur as a result of genetic polymorphisms, impaired absorption, or the presence of metabolic disease.

Thus far, a set of Parkinson's specific nutritional recommendations has not been developed. The US Department of Agriculture (USDA) Food and Nutrition Information Center (FNIC) maintains an information database of Dietary Reference Intakes (DRIs) for vitamins, minerals, and macronutrients developed by the Institute of Medicine (IOM) of the National Academy of Science (NAS). The DRI levels have largely replaced the Recommended Daily Intake or Reference Daily Intake (RDI) system still used for product labeling. The RDI is the intake considered to be sufficient to meet the needs of 97.5% (2 standard deviations below the mean) of the healthy population.

The quantification of the minimum required dose to prevent deficiency symptoms is determined by incremental refeeding until the dose resulting in syndrome resolution is reached (Chipponi et al., 1982).

In classic deprivation studies, the stages of deficiency are as follows:
Stage 1 Deficiency: Physiologic function continues normally while stores are being depleted. Adipose tissue, bone, muscle, storage forms, etc. act to maintain serum concentrations, e.g., depletion of calcium from bones.
Stage 2 Deficiency: The depletion of body stores results in biochemical alterations, although clinical symptoms are not yet apparent, e.g., elevated homocysteine in response to vitamin B12 or folic acid insufficiency.
Stage 3 Deficiency: In addition to biological perturbations, clinical symptoms become apparent, e.g., dementia and dermatitis as a result of niacin deficiency.

2. EVIDENCE FOR NUTRITIONAL DEFICIENCIES IN PD

2.1 Coenzyme Q10

Coenzyme Q10, or ubiquinone, is a fat-soluble coenzyme found primarily in the mitochondria of eukaryotic cells. It is an essential component of the electron transport chain, and thus the generation of ATP. In 2012, a case-controlled study of 22 patients with PD and 88 age- and gender-matched controls compared the frequency of nutritional deficiencies using a Functional Intracellular Assay (FIA). Thirty-four percent of individuals with PD were deficient in coenzyme Q10, compared to only 8.5% of controls ($P = 0.006$) (Mischley, Allen, & Bradley, 2012).

2.2 Glutathione

Glutathione is a tripeptide comprised of glycine, cysteine, and glutamic acid. Essential for life (Winkler et al., 2011), glutathione is both endogenously synthesized and fortified by diet. As the most abundant low molecular weight thiol, GSH plays critical roles as a reducing agent and in cellular detoxification (Forman, Zhang, & Rinna, 2009). "Parkinson's Disease: A disorder due to nigral glutathione deficiency?" was published in 1982 following the discovery that postmortem glutathione concentrations were depleted in the substantia nigra of PD samples compared to age-matched controls (Perry, Godin, & Hansen, 1982). These findings were replicated in a follow-up study in which the authors concluded the most reasonable explanation for the glutathione deficiency is that a neurotoxic compound that can be conjugated with glutathione is accumulating in the substantia nigra (Perry & Yong, 1986). More recently, among 58 individuals with idiopathic PD, the lower the whole blood glutathione concentration, the more severe their PD symptoms, as measured by the Unified PD Rating Scale (UPDRS) score ($P = 0.02$, 95% CI: −13.96, −1.14) (Mischley et al., 2016).

2.3 Lithium

Lithium is an essential mineral found in all cells of the human body. Lithium cannot be synthesized endogenously, and thus must be obtained through diet, fortification, or supplementation. It is found in trace amounts in soil, where it is unevenly distributed in the Earth's crust. The majority of lithium comes from the consumption of water, vegetables, and grains, with the average consumption in the United States estimated to range between 650 and 3100 μg/day (Schrauzer, 2002). Low lithium concentrations in the drinking water have been associated with depression in several studies (Giotakos, Nisianakis, Tsouvelas, & Giakalou, 2013; Schopfer & Schrauzer, 2011). In the CNS, lithium is instrumental in maintaining homeostasis of the second messenger cAMP (Montezinho et al., 2004) and plays a role in the regulation of autophagy (Yao, Zhao, Khan, & Yang, 2013), which has been shown to be dysregulated in neurodegenerative disease (Rosello, Warnes, & Meier, 2012). In a rodent model, lithium enhances the clearance of intracellular α-synuclein (Motoi, Shimada, Ishiguro, & Hattori, 2014). A clinical case series of lithium status in 80 patients with PD found 63.8% of samples were deficient in lithium, defined as greater than 2 standard deviations below the laboratory reference range (Mischley, 2013).

2.4 Thiamine

Thiamine, or vitamin B1, is an essential cofactor for several key enzymes required for brain oxidative metabolism. Thiamine is naturally found in high concentrations in the human substantia nigra (Baker et al., 1984). In thiamine deficiency, there are reduced levels of striatal dopamine, and intrastriatal administration of thiamine has been shown to enhance the release of dopamine (Yamashita, Zhang, & Nakamura, 1993). In 24 PD patients and 40 matched controls, PD patients had lower concentrations of free thiamine in the cerebrospinal fluid, and this depletion was not seen in individuals taking levodopa (Jimenez-Jimenez et al., 1999). Recently, an open-label pilot study of 100 mg intramuscular thiamine twice weekly in 50 patients with PD reported significant improvement in motor and nonmotor symptoms, which was sustained during the 3–27-month follow-up period (Costantini & Fancellu, 2016; Costantini et al., 2015).

3. PD INCREASES RISK OF MALNUTRITION

A systematic review of the literature based on available longitudinal data revealed that having PD is a significant risk factor for malnutrition (OR: 2.450; $P=0.047$), as were increased age, constipation, poor-to-moderate self-reported health status, cognitive decline, dementia, apathy, dysphagia, and institutionalization, all of which are commonly seen in PD (Moreira et al., 2016). Using the Mini Nutritional Assessment (MNA), a validated screening tool for nutritional status, the prevalence of malnutrition in patients with PD ranges from 15% to 22% (Wang et al., 2010).

4. DIET AND RISK OF PD INCIDENCE

Several prospective longitudinal studies have been conducted in which diet and nutrition have been periodically assessed throughout adulthood and incidence of PD documented within the cohort. Collectively, these data are consistent in suggesting vegetables, fruits, nuts, seeds, and fish contain phytochemicals and other constituents with neuroprotective potential (Gao, Cassidy, Schwarzschild, Rimm, & Ascherio, 2012; Gao et al., 2007; Jiang, Ju, Jiang, & Zhang, 2014; Qi & Li, 2014). On a molecular level, a decreased risk of PD has been associated with increased flavonoid, carotenoid, and nicotine content of the diet (Alcalay et al., 2012; Gao et al., 2012; Miyake et al., 2011; Nielsen, Franklin, Longstreth, Swanson, & Checkoway, 2013).

Dairy consumption has been associated with an increased risk of PD in four of five prospective observational studies to explore the role of diet and risk of PD incidence (Chen et al., 2007; Chen, Zhang, Hernan, Willett, & Ascherio, 2002; Kyrozis et al., 2013; Park et al., 2005). The association was not observed in the Japanese cohort (Miyake et al., 2011). Some studies suggest the association is stronger in men and greatest with cream and ice cream. Hypotheses for the association include diary's ability to lower uric acid levels (Zgaga et al., 2012) or the presence of a neurotoxicant in dairy products, such as organochlorine pesticide residues (Fitzmaurice, Rhodes, Cockburn, Ritz, & Bronstein, 2014). Lactose deficiency, the inability to digest the milk sugar lactase, is increasingly common with age. Not only does continued consumption of lactase lead to intestinal inflammation and symptoms of gas, bloating, constipation, and/or diarrhea, in one study up to 90% of lactose-deficient individuals had evidence of small intestinal bacterial overgrowth (SIBO) (Almeida et al., 2008), a condition found in over 50% individuals with PD and fewer than 10% of controls (Gabrielli et al., 2011).

5. AN INTESTINAL DISEASE THAT SPREADS TO THE BRAIN?

The gastrointestinal tract is affected early in the course of PD. Lack of thirst, dehydration, constipation, impaired olfaction, and irritable bowel syndrome have been documented decades prior to the onset of motor symptoms (Lai, Liao, Lin, & Sung, 2014). On a cellular level, the earliest signs of pathology occur in the intestinal track, with α-synuclein aggregations in the salivary glands and intestinal mucosa at least a decade prior to the onset of PD motor symptoms (Adams-Carr et al., 2016; Mukherjee, Biswas, & Das, 2016). There is an emerging body of literature suggesting the pathology of PD may spread from the gut to the brain via the vagus nerve; a large Danish epidemiological study demonstrated that individuals who have had their entire vagus nerve severed (as a treatment for peptic ulcer) had an almost 50% reduction in PD (Svensson et al., 2015).

5.1 Digestive Enzyme Insufficiency

While attention has turned to the vagus nerve as a potential channel for propagation of disease, little attention has been given to the role of the vagus nerve in maintaining gastrointestinal health. In response to smell, taste, chewing, and anticipation of food, the vagus nerve stimulates the release of hydrochloric acid (HCl) from gastric parietal cells. Hypochlorhydria/achlorhydria is the reduction/absence, respectively, of HCl in the gastric mucosa. Achlorhydria can result in impairment of absorption of vitamin B12 (Holt, 2007), folic acid (MacKenzie & Russell, 1976), and is associated with SIBO (Husebye, Skar, Hoverstad, & Melby, 1992; Pereira, Gainsborough, & Dowling, 1998). Among patients who had become refractory to levodopa therapy, 22/38 patients (58%) were found to have reduced gastric acid. Following 2 weeks of coadministration of 30 mL of lemon juice, an attempt at acidifying the stomach to increase medication availability, with each dose of levodopa, there was a measurable increase in circulating L-dopa concentration after 60 and 180 min, with corresponding improvements in rigidity, akinesia, and gait. Although rarely discussed, this finding supports the hypothesis that hypochlorhydria is prevalent and relevant in PD and should be considered in those with erratic medication responsiveness.

HCl production can be measured via a variety of techniques including gastric acid aspiration tests, Heidelberg testing (pH sensor on a swallowed capsule), and via endoscopy (Ghosh, Lewis, Axon, & Everett, 2011). Pancreatic insufficiency can be evaluated using fecal elastase testing. In a PD patient with suspected hypochlorhydria, enzymatic supplementation (HCl, betaine, pepsin) resolved lower abdominal pain in a published case report. Upon discontinuation of the enzyme supplement, the abdominal pain returned, and was again relieved when the enzyme supplement was restarted (Kines & Krupczak, 2016).

5.2 Celiac Disease

Celiac disease is an autoimmune disease in which the ingestion of the dietary protein gluten leads to damage of the small intestine. Celiac disease may present at any age. Interestingly, it is not uncommon for patients with celiac disease to have no gastrointestinal symptoms, while neurological symptoms are common. In celiac patients with neurological disorders, there is a high prevalence of antineuronal antibodies (Hadjivassiliou et al., 2010). A 2014 case report of a 75-year-old male without gastrointestinal symptoms presented with a 1-year history of gait disorder, instability, and fatigue, with hypomimia, bradykinesia, rigidity, and postural instability on physical examination. Single-photon emission computed tomography (SPECT) DaT-SCAN showed a pronounced decrease of left putamen radiotracer uptake and moderate decrease in the contralateral putamen. Clinical and SPECT data led to a diagnosis of PD. Evidence of malabsorption led to the evaluation for celiac disease, for which the patient was positive (anti-gliadin ab IgA, IgG, anti-TTG, + antiendomysial antibodies). Following the gastroenterologist-prescribed gluten-free diet, the patient had clinically verified "almost complete remission of symptoms" through the 18 months of follow-up, while SPECT evidence showed continued evidence of degeneration (Di Lazzaro, Capone, Cammarota, Di Giuda, & Ranieri, 2014). Celiac disease should be suspected in any individual with unexplained evidence of malabsorption, low serum iron or folic acid, anemia, or osteoporosis (Holt, 2007). The diagnosis is made if elevated titers of antitissue transglutaminase or antiendomysial antibodies are identified.

5.3 *Helicobacter pylori*

Almost one-third of individuals with PD are infected with *Helicobacter pylori*, a common infection in the human upper digestive tract. Infected individuals do not report gastrointestinal symptoms more than uninfected individuals, although they tend to be older and have worse motor scores (Tan et al., 2015). When the serum of 30 *H. pylori* positive and 30 negative PD patients was compared, researchers identified the presence of 13 autoantibodies against proteins essential for neurological function in the group with *H. pylori* (Suwarnalata et al., 2016). Because *H. pylori* affects levodopa absorption and infection may aggravate the degenerative course of PD, eradication is essential. A study of 27 individuals with PD and *H. pylori* infection demonstrated statistically significant improvements in levodopa onset time, overall ON duration, UPDRS scores, and PDQ-39, a quality-of-life measure, 12 weeks following eradication (Hashim et al., 2014). *H. pylori* screening should be performed on patients with gas, bloating, nausea, loss of appetite, belching, gnawing or burning abdominal pain, or bad breath.

6. SYMPTOM-SPECIFIC NUTRITIONAL CONSIDERATIONS

6.1 Constipation

Constipation has been shown to affect approximately 85% of individuals with PD and increases the risk of malnutrition (Maeda et al., 2017; Wang et al., 2010). In some cases, lack of thirst and constipation have been shown to precede motor symptoms by more than 18 years (Ueki & Otsuka, 2004). In addition to physical discomfort, constipation decreases appetite, interferes with the uptake of medications, leads to the reabsorption of toxicants, and is associated with growth of abnormal intestinal microflora, a condition known as dysbiosis, which can further perpetuate delayed gastric emptying. In a study of 51 drug-naïve individuals with PD and 20 health age-mated controls, serum 25-hydroxyvitamin D3 concentrations were decreased in the group with delayed gastric emptying time. The authors concluded vitamin D insufficiency may be implicated in the pathogenesis of delayed gastric emptying in PD (Kwon et al., 2016).

Probiotic supplements have been shown to improve symptoms of bloat, pain, and improved stool consistency in patients with PD. Use of a *Lactobacillus casei* Shirota supplement in 40 constipated PD patients for 5 weeks resulted in an increase in number of days per week with normal consistency stools ($P < 0.01$), reduction in number of days participants felt bloated

($P < 0.01$), reduction in abdominal pain ($P < 0.01$), and reduction in sensation of incomplete emptying ($P < 0.01$) (Cassani et al., 2011). Another study looked at the effectiveness of a fermented milk containing probiotics and prebiotics in PD patients with Rome III-confirmed constipation. The results of the randomized, double-blind, placebo-controlled trial demonstrated the probiotic/prebiotic supplement increased the number of complete bowel movements per week ($P = 0.002$), the primary endpoint (Barichella et al., 2016).

6.2 Cognitive Impairment

Abdominal obesity and impaired olfaction at diagnosis were associated with reduced Mini-Mental Status Examination 3 years following PD diagnosis (Vikdahl, Domellof, Forsgren, & Haglin, 2015). Not specific to PD, a recent meta-analysis of 21 cohort studies investigated associations between fish and polyunsaturated fatty acid intake and risk of cognitive impairment. Among the 181,580 participants with 4438 follow-up periods ranging from 2.1 to 21 years, dietary fish was associated with lower risk of cognitive impairment and dementia (Zhang, Chen, Qiu, et al., 2016). Individuals should be encouraged to consume several servings of nonfried fish per week and DHA supplementation should be considered.

Homocysteine, a normal product of human metabolism, is neurotoxic at high concentrations, accelerating dopaminergic degeneration, and is an established risk factor for developing cognitive impairment in PD (Xie, Feng, Peng, Xiao, & Zhang, 2017). Increase in homocysteine can be due to deficiency of vitamins B12, B6, folic acid, or betaine (trimethylglycine). It is well established that levodopa interferes with folic acid uptake and utilization, and there is a dose–response relationship between levodopa use and homocysteine concentrations (Paul & Borah, 2016). Individuals on levodopa or with symptoms of cognitive decline should have homocysteine concentrations measured annually; individuals with concentrations greater than 10 μmol/L should be prescribed a homocysteine-lowering supplement.

6.3 Depression and Anxiety

Individuals with PD and compromised nutritional status are more likely to be affected by mood disorders, such as anxiety and depression (Fereshtehnejad et al., 2014). A clinical case series of 117 consecutive Chinese patients found that Geriatric Depression Scale score was the second most important predictor of nutritional impairment, following constipation (OR, 1.166; 95% CI, 1.042–1.304; $P = 0.007$) (Wang et al., 2010). A 2008 pilot study of 31 individuals with PD and depression was randomized to receive omega-3 fish oil or mineral oil capsules for 12 weeks; whether or not they were taking antidepressants, individuals randomized to receive the omega-3 fish oil capsules showed a significant decrease in symptoms of depression, as measured by the Montgomery-Asberg Rating Scale (MADRS) and the Clinical Global Impression Scale (CGI) (da Silva et al., 2008).

Anxiety disorders are common in PD, occurring in approximately a quarter of patients (Dissanayaka et al., 2010). There is even evidence that having anxiety increases an individuals' risk of developing PD. The extent to which anxiety may be an early PD symptom or share an underlying pathophysiology with PD is yet to be determined (Weisskopf, Chen, Schwarzschild, Kawachi, & Ascherio, 2003). Poor nutritional status, as measured by the MNA, is inversely correlated with serum cortisol (Haglin & Backman, 2016). There is an emerging body of literature elucidating the relationship between diet and mood, likely mediated by the intestinal microbiota. There is evidence that vegans report less stress and anxiety than omnivores (Beezhold, Radnitz, Rinne, & DiMatteo, 2015). Taken together, these data suggest a plant-based diet, including fish, and possibly an omega-3 fatty acid supplement, should be prescribed to PD patients with anxiety or depression.

6.4 Sleep Disorders

While most humans can synthesize sufficient amounts of melatonin to meet physiological needs, there is evidence that individuals with PD inadequately produce this hormone (De Pablo-Fernandez et al., 2017). Thus, like dopamine, it may be exogenously required, or a conditionally essential nutrient. As already discussed, there is a substantial body of literature suggesting glutathione production is inadequate in PD. Because of the interdependent circadian rhythms of glutathione and melatonin, insufficient glutathione synthesis during the day will interfere with melatonin production at night, and vice versa (Pablos et al., 1998). In an uncontrolled trial of 30 individuals with PD, 3 mg of melatonin was associated with a 21% decrease on Parkinson Fatigue Scale ($P < 0.05$), improvements in sleep (PD Sleep Scale) ($P < 0.05$), decrease in state anxiety (Spielberger's scale) ($P < 0.05$), and improvement in quality of life (PDQ-39) ($P < 0.05$)

(Datieva, Rosinskaia, & Levin, 2013). While there was no change in polysomnography, individuals with PD randomized to receive 3 mg melatonin reported improved quality of sleep ($P = 0.03$) over placebo after 4 weeks of supplementation (Medeiros et al., 2007). A meta-analysis of nine randomized controlled trials suggests exogenous melatonin is effective for improving overall sleep quality and the clinical and neurophysiological aspects of rapid eye movement sleep behavior disorder in PD (Zhang, Chen, Su, et al., 2016).

6.5 Dystonia

A study seeking to evaluate whether lithium might be an effective treatment for dyskinesia found that, while not effective for dyskinesia, there was an unexpected improvement in dystonia, a state of sustained muscle contraction common in PD, in four of the study participants. The authors stated, "three patients on levodopa reported relief of muscular cramps and painful toe curling only when taking lithium carbonate … another reported a welcome reduction in oculogyric crises" (McCaul & Stern, 1974). Following the report that lithium carbonate was capable of reducing dystonia, seven patients with disabling dystonia were recruited for a double-blind, placebo-controlled study of 600 mg slow-release lithium carbonate, administered at bedtime. The study was designed to include three phases: the first phase was an open-label trial lasting at least 3 months. During phases 2 and 3, individuals were randomized to receive either lithium or placebo in a crossover fashion, with both phases 2 and 3 designed to last for 2 months, followed by phase 4, an open-label portion of the study. The authors reported, "Painful dystonic cramps were much reduced or abolished in 19 of 21 treatment periods with lithium (13 open, 6 blind). They returned to (6) or toward (1) pretreatment severity in all 7 placebo periods. The beneficial effect appeared 1–16 days after starting and disappeared 1–14 days after stopping lithium" (Quinn & Marsden, 1986).

7. DRUG–NUTRIENT INTERACTIONS

7.1 Levodopa

Dietary protein interferes with levodopa uptake and patients should be advised to take levodopa on an empty stomach, or with a small meal low in protein (Robertson et al., 1991). Constipation also interferes with uptake of levodopa, a major determinant of medication effectiveness. If constipated, individuals with erratic response to levodopa focus improving bowel transit time; suggestions include a plant-based diet, increased fluid intake, and vitamin D (Kwon et al., 2016) and probiotic supplementation (Barichella et al., 2016; Cassani et al., 2011) should be considered. As previously discussed, a highly acidic gastric environment is required for activation of levodopa, and individuals with PD are at increased risk of HCl insufficiency; other symptoms of malabsorption, e.g., weight loss and osteopenia, should warrant an evaluation of HCl adequacy.

Not all drug–nutrient interactions are negative, as several studies suggest CDP-choline has a levodopa-sparing effect (Cubells & Hernando, 1988). 500–1200 mg of CDP-choline has been shown to permit a 30%–50% reduction in levodopa dose, respectively (Cubells & Hernando, 1988; Eberhardt, Birbamer, Gerstenbrand, Rainer, & Traegner, 1990), which has been attributed to an increase in pre- and postsynaptic dopamine receptors and an increase in the synthesis of tyrosine hydroxylase (Marti Masso & Urtasun, 1991). For individuals suffering from levodopa-induced dyskinesia (LID), high-dose omega-3 DHA (approximately 5 g) was associated with a 30%–50% reduction in LID in a primate model of LID and a clinical case series (Mischley & McNames, 2013; Samadi et al., 2006).

Perhaps most importantly, levodopa interferes with the uptake and metabolism of folic acid, as discussed in Section 6.2. Given the high risk of dementia in PD and the lack of effective treatments, prevention strategies should include utilization of levodopa-sparing therapies, e.g., CDP-choline, and annual homocysteine screening.

8. ASSESSMENT OF NUTRITIONAL STATUS

Evaluation of nutritional status should always begin with a dietary history. Providers should be trained to ask about diet, specifically looking for high-calorie foods low in nutritional content (e.g., fried foods, sugar-sweetened beverages) as a risk factor for malnutrition. The majority of the patient's diet should be comprised of fresh fruits, vegetables, nuts and seeds, whole grains, beans, and nonfried fish.

Physical exam is an essential component of nutritional evaluation. While reduced body mass index is well known to be associated with malnutrition, obesity, too, is associated with malnutrition and adipose tissue is a major source of inflammatory cytokines (Gulland, 2016). As there is no comprehensive way to screen for nutritional adequacy, clinicians must look for signs and symptoms associated with the nutrients in questions. For example, when mineral deficiencies are suspected, such a calcium or magnesium insufficiency, tapping on the facial nerve can result in a positive Chvostek sign, in which the facial muscles on the ipsilateral side of the face will twitch.

The same is true for laboratory assessments of nutritional status; there is no single test, or even panel of tests, one can order to determine nutritional adequacy. Providers must understand the nutrient's physiological role(s) to order the correct tests; for example, neurological and/or psychological abnormalities associated with vitamin B12 deficiency can occur in spite of normal serum B12 concentrations (Andres et al., 2005; Lindenbaum, Rosenberg, Wilson, Stabler, & Allen, 1994). Thus, a comprehensive screening for vitamin B12 status includes serum cobalamin, methylmalonic acid, homocysteine, and mean cell volume (Wong, 2015). Blood is not the only biospecimen used for nutritional assessment, either. For example, the gold standard for assessing lithium status is hair (Anke, Arnhold, Groppel, & Krause, 1991), urine for iodine (Markou et al., 2002), and breath tests for the detection of SIBO (Rezaie et al., 2017). While reference ranges do not yet exist, stool tests are commercially available that use polymerase chain reaction methodology to provide insight into the patient's intestinal microbiome. As more research becomes available on the association between dysbiosis and PD, stool tests are likely to grow in popularity.

9. CONCLUSIONS

Two-hundred years after the original description of PD, attention is finally turning to the potential roles of diet and nutrition in disease etiology and progression. At first glance, it is difficult to imagine such a complex, debilitating syndrome could be notably influences by something as inexpensive and simple as a food or supplement. Those individuals hesitant to entertain the possibility are reminded that 3 million people died of scurvy before it was discovered that the disease could be cured with limes. Especially poignant is the fact that scurvy was successfully treated with limes for 200 years *before* the ascorbic acid molecule was discovered. Nutritional interventions have been so successful at curing disease that clinicians take the need for nutritional evaluation for granted, and many would have a hard time recognizing pellagra if a patient presented with it. This 200th year anniversary of PD is likely to herald enhancements in diagnostic and treatment modalities, as well as foster emerging paradigms, including the development of nutritional guidelines for the screening, evaluation, and management of PD.

REFERENCES

Adams-Carr, K. L., Bestwick, J. P., Shribman, S., Lees, A., Schrag, A., & Noyce, A. J. (2016). Constipation preceding Parkinson's disease: A systematic review and meta-analysis. *Journal of Neurology, Neurosurgery, and Psychiatry*, *87*(7), 710–716. http://dx.doi.org/10.1136/jnnp-2015-311680.

Alcalay, R. N., Gu, Y., Mejia-Santana, H., Cote, L., Marder, K. S., & Scarmeas, N. (2012). The association between Mediterranean diet adherence and Parkinson's disease. *Movement Disorders: Official Journal of the Movement Disorder Society*, *27*(6), 771–774. http://dx.doi.org/10.1002/mds.24918.

Almeida, J. A., Kim, R., Stoita, A., McIver, C. J., Kurtovic, J., & Riordan, S. M. (2008). Lactose malabsorption in the elderly: Role of small intestinal bacterial overgrowth. *Scandinavian Journal of Gastroenterology*, *43*(2), 146–154. http://dx.doi.org/10.1080/00365520701676617.

Andres, E., Affenberger, S., Vinzio, S., Kurtz, J. E., Noel, E., Kaltenbach, G., et al. (2005). Food-cobalamin malabsorption in elderly patients: Clinical manifestations and treatment. *The American Journal of Medicine*, *118*(10), 1154–1159. http://dx.doi.org/10.1016/j.amjmed.2005.02.026.

Anke, M., Arnhold, W., Groppel, B., & Krause, U. (1991). The biological importance of lithium. In G. N. Schrauzer (Ed.), *Lithium in biology and medicine* (pp. 149–167). Weinheim: VCH Verlagsgesellschaft.

Baker, H., Frank, O., Chen, T., Feingold, S., DeAngelis, B., & Baker, E. (1984). Vitamin content of some normal human brain segments. *Journal of Neuroscience Research*, *11*(4), 419–435. http://dx.doi.org/10.1002/jnr.490110409.

Barichella, M., Pacchetti, C., Bolliri, C., Cassani, E., Iorio, L., Pusani, C., et al. (2016). Probiotics and prebiotic fiber for constipation associated with Parkinson disease: An RCT. *Neurology*, *87*(12), 1274–1280. http://dx.doi.org/10.1212/WNL.0000000000003127.

Beezhold, B., Radnitz, C., Rinne, A., & DiMatteo, J. (2015). Vegans report less stress and anxiety than omnivores. *Nutritional Neuroscience*, *18*(7), 289–296. http://dx.doi.org/10.1179/1476830514Y.0000000164.

Cassani, E., Privitera, G., Pezzoli, G., Pusani, C., Madio, C., Iorio, L., et al. (2011). Use of probiotics for the treatment of constipation in Parkinson's disease patients. *Minerva Gastroenterologica e Dietologica*, *57*(2), 117–121.

Chen, H., O'Reilly, E., McCullough, M. L., Rodriguez, C., Schwarzschild, M. A., Calle, E. E., et al. (2007). Consumption of dairy products and risk of Parkinson's disease. *American Journal of Epidemiology*, *165*(9), 998–1006. http://dx.doi.org/10.1093/aje/kwk089.

Chen, H., Zhang, S. M., Hernan, M. A., Willett, W. C., & Ascherio, A. (2002). Diet and Parkinson's disease: A potential role of dairy products in men. *Annals of Neurology*, *52*(6), 793–801. http://dx.doi.org/10.1002/ana.10381.

Chipponi, J. X., Bleier, J. C., Santi, M. T., & Rudman, D. (1982). Deficiencies of essential and conditionally essential nutrients. *The American Journal of Clinical Nutrition*, *35*(5 Suppl), 1112–1116.

Costantini, A., & Fancellu, R. (2016). An open-label pilot study with high-dose thiamine in Parkinson's disease. *Neural Regeneration Research*, *11*(3), 406–407. http://dx.doi.org/10.4103/1673-5374.179047.

Costantini, A., Pala, M. I., Grossi, E., Mondonico, S., Cardelli, L. E., Jenner, C., et al. (2015). Long-term treatment with high-dose thiamine in Parkinson disease: An open-label pilot study. *Journal of Alternative and Complementary Medicine*, *21*(12), 740–747. http://dx.doi.org/10.1089/acm.2014.0353.

Cubells, J. M., & Hernando, C. (1988). Clinical trial on the use of cytidine diphosphate choline in Parkinson's disease. *Clinical Therapeutics*, *10*(6), 664–671.

da Silva, T. M., Munhoz, R. P., Alvarez, C., Naliwaiko, K., Kiss, A., Andreatini, R., et al. (2008). Depression in Parkinson's disease: A double-blind, randomized, placebo-controlled pilot study of omega-3 fatty-acid supplementation. *Journal of Affective Disorders*, *111*(2–3), 351–359. http://dx.doi.org/10.1016/j.jad.2008.03.008.

Datieva, V. K., Rosinskaia, A. V., & Levin, O. S. (2013). The use of melatonin in the treatment of chronic fatigue syndrome and circadian rhythm disorders in Parkinson's disease. *Zhurnal Nevrologii I Psikhiatrii Imeni S.S. Korsakova*, *113*(7 Pt. 2), 77–81.

De Pablo-Fernandez, E., Breen, D. P., Bouloux, P. M., Barker, R. A., Foltynie, T., & Warner, T. T. (2017). Neuroendocrine abnormalities in Parkinson's disease. *Journal of Neurology, Neurosurgery, and Psychiatry*, *88*(2), 176–185. http://dx.doi.org/10.1136/jnnp-2016-314601.

Di Lazzaro, V., Capone, F., Cammarota, G., Di Giuda, D., & Ranieri, F. (2014). Dramatic improvement of parkinsonian symptoms after gluten-free diet introduction in a patient with silent celiac disease. *Journal of Neurology*, *261*(2), 443–445. http://dx.doi.org/10.1007/s00415-014-7245-7.

Dissanayaka, N. N., Sellbach, A., Matheson, S., O'Sullivan, J. D., Silburn, P. A., Byrne, G. J., et al. (2010). Anxiety disorders in Parkinson's disease: Prevalence and risk factors. *Movement Disorders: Official Journal of the Movement Disorder Society*, *25*(7), 838–845. http://dx.doi.org/10.1002/mds.22833.

Eberhardt, R., Birbamer, G., Gerstenbrand, F., Rainer, E., & Traegner, H. (1990). Citicoline in the treatment of Parkinson's disease. *Clinical Therapeutics*, *12*(6), 489–495.

Fereshtehnejad, S. M., Ghazi, L., Shafieesabet, M., Shahidi, G. A., Delbari, A., & Lokk, J. (2014). Motor, psychiatric and fatigue features associated with nutritional status and its effects on quality of life in Parkinson's disease patients. *PLoS One*, *9*(3), e91153. http://dx.doi.org/10.1371/journal.pone.0091153.

Fitzmaurice, A. G., Rhodes, S. L., Cockburn, M., Ritz, B., & Bronstein, J. M. (2014). Aldehyde dehydrogenase variation enhances effect of pesticides associated with Parkinson disease. *Neurology*, *82*(5), 419–426. http://dx.doi.org/10.1212/WNL.0000000000000083.

Forman, H. J., Zhang, H., & Rinna, A. (2009). Glutathione: Overview of its protective roles, measurement, and biosynthesis. *Molecular Aspects of Medicine*, *30*(1–2), 1–12. http://dx.doi.org/10.1016/j.mam.2008.08.006.

Gabrielli, M., Bonazzi, P., Scarpellini, E., Bendia, E., Lauritano, E. C., Fasano, A., et al. (2011). Prevalence of small intestinal bacterial overgrowth in Parkinson's disease. *Movement Disorders: Official Journal of the Movement Disorder Society*, *26*(5), 889–892. http://dx.doi.org/10.1002/mds.23566.

Gao, X., Cassidy, A., Schwarzschild, M. A., Rimm, E. B., & Ascherio, A. (2012). Habitual intake of dietary flavonoids and risk of Parkinson disease. *Neurology*, *78*(15), 1138–1145. http://dx.doi.org/10.1212/WNL.0b013e31824f7fc4.

Gao, X., Chen, H., Fung, T. T., Logroscino, G., Schwarzschild, M. A., Hu, F. B., et al. (2007). Prospective study of dietary pattern and risk of Parkinson disease. *The American Journal of Clinical Nutrition*, *86*(5), 1486–1494.

Ghosh, T., Lewis, D. I., Axon, A. T., & Everett, S. M. (2011). Review article: Methods of measuring gastric acid secretion. *Alimentary Pharmacology & Therapeutics*, *33*(7), 768–781. http://dx.doi.org/10.1111/j.1365-2036.2010.04573.x.

Giotakos, O., Nisianakis, P., Tsouvelas, G., & Giakalou, V. V. (2013). Lithium in the public water supply and suicide mortality in Greece. *Biological Trace Element Research*, *156*(1–3), 376–379. http://dx.doi.org/10.1007/s12011-013-9815-4.

Gulland, A. (2016). Malnutrition and obesity coexist in many countries, report finds. *BMJ*, *353*, i3351. http://dx.doi.org/10.1136/bmj.i3351.

Hadjivassiliou, M., Sanders, D. S., Grunewald, R. A., Woodroofe, N., Boscolo, S., & Aeschlimann, D. (2010). Gluten sensitivity: From gut to brain. *Lancet Neurology*, *9*(3), 318–330. http://dx.doi.org/10.1016/S1474-4422(09)70290-X.

Haglin, L., & Backman, L. (2016). Covariation between plasma phosphate and daytime cortisol in early Parkinson's disease. *Brain and Behavior*, *6*(12), e00556. http://dx.doi.org/10.1002/brb3.556.

Hashim, H., Azmin, S., Razlan, H., Yahya, N. W., Tan, H. J., Manaf, M. R., et al. (2014). Eradication of Helicobacter pylori infection improves levodopa action, clinical symptoms and quality of life in patients with Parkinson's disease. *PLoS One*, *9*(11), e112330. http://dx.doi.org/10.1371/journal.pone.0112330.

Holt, P. R. (2007). Intestinal malabsorption in the elderly. *Digestive Diseases*, *25*(2), 144–150. http://dx.doi.org/10.1159/000099479.

Husebye, E., Skar, V., Hoverstad, T., & Melby, K. (1992). Fasting hypochlorhydria with gram positive gastric flora is highly prevalent in healthy old people. *Gut*, *33*(10), 1331–1337.

Jiang, W., Ju, C., Jiang, H., & Zhang, D. (2014). Dairy foods intake and risk of Parkinson's disease: A dose-response meta-analysis of prospective cohort studies. *European Journal of Epidemiology*, *29*(9), 613–619. http://dx.doi.org/10.1007/s10654-014-9921-4.

Jimenez-Jimenez, F. J., Molina, J. A., Hernanz, A., Fernandez-Vivancos, E., de Bustos, F., Barcenilla, B., et al. (1999). Cerebrospinal fluid levels of thiamine in patients with Parkinson's disease. *Neuroscience Letters*, *271*(1), 33–36.

Kines, K., & Krupczak, T. (2016). Nutritional interventions for gastroesophageal reflux, irritable bowel syndrome, and hypochlorhydria: A case report. *Integrative Medicine (Encinitas, Calif.)*, *15*(4), 49–53.

Kwon, K. Y., Jo, K. D., Lee, M. K., Oh, M., Kim, E. N., Park, J., et al. (2016). Low serum vitamin D levels may contribute to gastric dysmotility in de novo Parkinson's disease. *Neurodegenerative Diseases*, *16*(3–4), 199–205. http://dx.doi.org/10.1159/000441917.

Kyrozis, A., Ghika, A., Stathopoulos, P., Vassilopoulos, D., Trichopoulos, D., & Trichopoulou, A. (2013). Dietary and lifestyle variables in relation to incidence of Parkinson's disease in Greece. *European Journal of Epidemiology*, *28*(1), 67–77. http://dx.doi.org/10.1007/s10654-012-9760-0.

Lai, S. W., Liao, K. F., Lin, C. L., & Sung, F. C. (2014). Irritable bowel syndrome correlates with increased risk of Parkinson's disease in Taiwan. *European Journal of Epidemiology*. http://dx.doi.org/10.1007/s10654-014-9878-3.

Lindenbaum, J., Rosenberg, I. H., Wilson, P. W., Stabler, S. P., & Allen, R. H. (1994). Prevalence of cobalamin deficiency in the Framingham elderly population. *The American Journal of Clinical Nutrition*, *60*(1), 2–11.

MacKenzie, J. F., & Russell, R. I. (1976). The effect of pH on folic acid absorption in man. *Clinical Science and Molecular Medicine*, *51*(4), 363–368.

Maeda, T., Shimo, Y., Chiu, S. W., Yamaguchi, T., Kashihara, K., Tsuboi, Y., et al. (2017). Clinical manifestations of nonmotor symptoms in 1021 Japanese Parkinson's disease patients from 35 medical centers. *Parkinsonism & Related Disorders*, *38*, 54–60. http://dx.doi.org/10.1016/j.parkreldis.2017.02.024.

Markou, K. B., Georgopoulos, N. A., Anastasiou, E., Vlasopoulou, B., Lazarou, N., Vagenakis, G. A., et al. (2002). Identification of iodine deficiency in the field by the rapid urinary iodide test: Comparison with the classic Sandell-Kolthoff reaction method. *Thyroid*, *12*(5), 407–410. http://dx.doi.org/10.1089/105072502760043495.

Marti Masso, J. F., & Urtasun, M. (1991). Citicoline in the treatment of Parkinson's disease. *Clinical Therapeutics*, *13*(2), 239–242.

McCaul, J. A., & Stern, G. M. (1974). Letter: Lithium in Parkinson's disease. *Lancet*, *1*(7866), 1117.

Medeiros, C. A., Carvalhedo de Bruin, P. F., Lopes, L. A., Magalhaes, M. C., de Lourdes Seabra, M., & de Bruin, V. M. (2007). Effect of exogenous melatonin on sleep and motor dysfunction in Parkinson's disease. A randomized, double blind, placebo-controlled

study. *Journal of Neurology, 254*(4), 459–464. http://dx.doi.org/10.1007/s00415-006-0390-x.
Mischley, L. K. (2013). *Lithium deficiency in Parkinson's disease.* Master thesis. ProQuest, School of Public Health, Department of Epidemiology, University of Washington.
Mischley, L. K., Allen, J., & Bradley, R. (2012). Coenzyme Q10 deficiency in patients with Parkinson's disease. *Journal of the Neurological Sciences, 318*(1–2), 72–75. http://dx.doi.org/10.1016/j.jns.2012.03.023.
Mischley, L. K., & McNames, J. (2013). A case report of fish oil for the treatment of levodopa-induced dyskinesia [abstract]. *Movement Disorders, 28*(Suppl. 1), 994.
Mischley, L. K., Standish, L. J., Weiss, N. S., Padowski, J. M., Kavanagh, T. J., White, C. C., et al. (2016). Glutathione as a biomarker in Parkinson's disease: Associations with aging and disease severity. *Oxidative Medicine and Cellular Longevity, 2016.* article ID 9409363. http://dx.doi.org/10.1155/2016/9409363.
Miyake, Y., Tanaka, K., Fukushima, W., Sasaki, S., Kiyohara, C., Tsuboi, Y., et al. (2011). Lack of association of dairy food, calcium, and vitamin D intake with the risk of Parkinson's disease: A case-control study in Japan. *Parkinsonism & Related Disorders, 17*(2), 112–116. http://dx.doi.org/10.1016/j.parkreldis.2010.11.018.
Montezinho, L. P., B Duarte, C., Fonseca, C. P., Glinka, Y., Layden, B., Mota de Freitas, D., et al. (2004). Intracellular lithium and cyclic AMP levels are mutually regulated in neuronal cells. *Journal of Neurochemistry, 90*(4), 920–930. http://dx.doi.org/10.1111/j.1471–4159.2004.02551.x.
Moreira, N. C., Krausch-Hofmann, S., Matthys, C., Vereecken, C., Vanhauwaert, E., Declercq, A., et al. (2016). Risk factors for malnutrition in older adults: A systematic review of the literature based on longitudinal data. *Advances in Nutrition,* 7(3), 507–522. http://dx.doi.org/10.3945/an.115.011254.
Motoi, Y., Shimada, K., Ishiguro, K., & Hattori, N. (2014). Lithium and autophagy. *ACS Chemical Neuroscience, 5*(6), 434–442. http://dx.doi.org/10.1021/cn500056q.
Mukherjee, A., Biswas, A., & Das, S. K. (2016). Gut dysfunction in Parkinson's disease. *World Journal of Gastroenterology, 22*(25), 5742–5752. http://dx.doi.org/10.3748/wjg.v22.i25.5742.
Nielsen, S. S., Franklin, G. M., Longstreth, W. T., Swanson, P. D., & Checkoway, H. (2013). Nicotine from edible Solanaceae and risk of Parkinson disease. *Annals of Neurology, 74*(3), 472–477. http://dx.doi.org/10.1002/ana.23884.
Pablos, M. I., Reiter, R. J., Ortiz, G. G., Guerrero, J. M., Agapito, M. T., Chuang, J. I., et al. (1998). Rhythms of glutathione peroxidase and glutathione reductase in brain of chick and their inhibition by light. *Neurochemistry International, 32*(1), 69–75.
Park, M., Ross, G. W., Petrovitch, H., White, L. R., Masaki, K. H., Nelson, J. S., et al. (2005). Consumption of milk and calcium in midlife and the future risk of Parkinson disease. *Neurology, 64*(6), 1047–1051. http://dx.doi.org/10.1212/01.WNL.0000154532.98495.BF.
Paul, R., & Borah, A. (2016). L-DOPA-induced hyperhomocysteinemia in Parkinson's disease: Elephant in the room. *Biochimica et Biophysica Acta, 1860*(9), 1989–1997. http://dx.doi.org/10.1016/j.bbagen.2016.06.018.
Pereira, S. P., Gainsborough, N., & Dowling, R. H. (1998). Drug-induced hypochlorhydria causes high duodenal bacterial counts in the elderly. *Alimentary Pharmacology & Therapeutics, 12*(1), 99–104.
Perry, T. L., Godin, D. V., & Hansen, S. (1982). Parkinson's disease: A disorder due to nigral glutathione deficiency? *Neuroscience Letters, 33*(3), 305–310.
Perry, T. L., & Yong, V. W. (1986). Idiopathic Parkinson's disease, progressive supranuclear palsy and glutathione metabolism in the substantia nigra of patients. *Neuroscience Letters, 67*(3), 269–274.

Qi, H., & Li, S. (2014). Dose-response meta-analysis on coffee, tea and caffeine consumption with risk of Parkinson's disease. *Geriatrics & Gerontology International, 14*(2), 430–439. http://dx.doi.org/10.1111/ggi.12123.

Quinn, N., & Marsden, C. D. (1986). Lithium for painful dystonia in Parkinson's disease. *Lancet, 1*(8494), 1377.

Rezaie, A., Buresi, M., Lembo, A., Lin, H., McCallum, R., Rao, S., et al. (2017). Hydrogen and methane-based breath testing in gastrointestinal disorders: The North American Consensus. *American Journal of Gastroenterology, 112*, 775–784. http://dx.doi.org/10.1038/ajg.2017.46.

Robertson, D. R., Higginson, I., Macklin, B. S., Renwick, A. G., Waller, D. G., & George, C. F. (1991). The influence of protein containing meals on the pharmacokinetics of levodopa in healthy volunteers. *British Journal of Clinical Pharmacology, 31*(4), 413–417.

Rosello, A., Warnes, G., & Meier, U. C. (2012). Cell death pathways and autophagy in the central nervous system and its involvement in neurodegeneration, immunity and central nervous system infection: To die or not to die—That is the question. *Clinical and Experimental Immunology, 168*(1), 52–57. http://dx.doi.org/10.1111/j.1365-2249.2011.04544.x.

Samadi, P., Gregoire, L., Rouillard, C., Bedard, P. J., Di Paolo, T., & Levesque, D. (2006). Docosahexaenoic acid reduces levodopa-induced dyskinesias in 1-methyl-4-phenyl-1,2,3,6-tetrahydropyridine monkeys. *Annals of Neurology, 59*(2), 282–288. http://dx.doi.org/10.1002/ana.20738.

Schopfer, J., & Schrauzer, G. N. (2011). Lithium and other elements in scalp hair of residents of Tokyo Prefecture as investigational predictors of suicide risk. *Biological Trace Element Research, 144*(1–3), 418–425. http://dx.doi.org/10.1007/s12011-011-9114-x.

Schrauzer, G. N. (2002). Lithium: Occurrence, dietary intakes, nutritional essentiality. *Journal of the American College of Nutrition, 21*(1), 14–21.

Suwarnalata, G., Tan, A. H., Isa, H., Gudimella, R., Anwar, A., Loke, M. F., et al. (2016). Augmentation of autoantibodies by Helicobacter pylori in Parkinson's disease patients may be linked to greater severity. *PLoS One, 11*(4), e0153725. http://dx.doi.org/10.1371/journal.pone.0153725

Svensson, E., Horvath-Puho, E., Thomsen, R. W., Djurhuus, J. C., Pedersen, L., Borghammer, P., et al. (2015). Vagotomy and subsequent risk of Parkinson's disease. *Annals of Neurology, 78*(4), 522–529. http://dx.doi.org/10.1002/ana.24448.

Tan, A. H., Mahadeva, S., Marras, C., Thalha, A. M., Kiew, C. K., Yeat, C. M., et al. (2015). Helicobacter pylori infection is associated with worse severity of Parkinson's disease. *Parkinsonism & Related Disorders, 21*(3), 221–225. http://dx.doi.org/10.1016/j.parkreldis.2014.12.009.

Ueki, A., & Otsuka, M. (2004). Life style risks of Parkinson's disease: Association between decreased water intake and constipation. *Journal of Neurology, 251*(Suppl. 7), vII18–23. http://dx.doi.org/10.1007/s00415-004-1706-3.

Vikdahl, M., Domellof, M. E., Forsgren, L., & Haglin, L. (2015). Olfactory function, eating ability, and visceral obesity associated with MMSE three years after Parkinson's disease diagnosis. *The Journal of Nutrition, Health & Aging, 19*(9), 894–900. http://dx.doi.org/10.1007/s12603-015-0573-1.

Wang, G., Wan, Y., Cheng, Q., Xiao, Q., Wang, Y., Zhang, J., et al. (2010). Malnutrition and associated factors in Chinese patients with Parkinson's disease: Results from a pilot investigation. *Parkinsonism & Related Disorders, 16*(2), 119–123. http://dx.doi.org/10.1016/j.parkreldis.2009.08.009.

Weisskopf, M. G., Chen, H., Schwarzschild, M. A., Kawachi, I., & Ascherio, A. (2003). Prospective study of phobic anxiety and risk of Parkinson's disease. *Movement Disorders:*

Official Journal of the Movement Disorder Society, *18*(6), 646–651. http://dx.doi.org/10.1002/mds.10425.

Winkler, A., Njalsson, R., Carlsson, K., Elgadi, A., Rozell, B., Abraham, L., et al. (2011). Glutathione is essential for early embryogenesis—Analysis of a glutathione synthetase knockout mouse. *Biochemical and Biophysical Research Communications*, *412*(1), 121–126. http://dx.doi.org/10.1016/j.bbrc.2011.07.056.

Wong, C. W. (2015). Vitamin B12 deficiency in the elderly: Is it worth screening? *Hong Kong Medical Journal*, *21*(2), 155–164. http://dx.doi.org/10.12809/hkmj144383.

Xie, Y., Feng, H., Peng, S., Xiao, J., & Zhang, J. (2017). Association of plasma homocysteine, vitamin B12 and folate levels with cognitive function in Parkinson's disease: A meta-analysis. *Neuroscience Letters*, *636*, 190–195. http://dx.doi.org/10.1016/j.neulet.2016.11.007.

Yamashita, H., Zhang, Y.-X., & Nakamura, S. (1993). The effects of thiamine and its phosphate esters on dopamine release in the rat striatum. *Neuroscience Letters*, *158*, 229–231.

Yao, H., Zhao, D., Khan, S. H., & Yang, L. (2013). Role of autophagy in prion protein-induced neurodegenerative diseases. *Acta Biochimica et Biophysica Sinica (Shanghai)*, *45*(6), 494–502. http://dx.doi.org/10.1093/abbs/gmt022.

Zgaga, L., Theodoratou, E., Kyle, J., Farrington, S. M., Agakov, F., Tenesa, A., et al. (2012). The association of dietary intake of purine-rich vegetables, sugar-sweetened beverages and dairy with plasma urate, in a cross-sectional study. *PLoS One*, 7(6), e38123. http://dx.doi.org/10.1371/journal.pone.0038123.

Zhang, Y., Chen, J., Qiu, J., Li, Y., Wang, J., & Jiao, J. (2016). Intakes of fish and polyunsaturated fatty acids and mild-to-severe cognitive impairment risks: A dose-response meta-analysis of 21 cohort studies. *The American Journal of Clinical Nutrition*, *103*(2), 330–340. http://dx.doi.org/10.3945/ajcn.115.124081.

Zhang, W., Chen, X. Y., Su, S. W., Jia, Q. Z., Ding, T., Zhu, Z. N., et al. (2016). Exogenous melatonin for sleep disorders in neurodegenerative diseases: A meta-analysis of randomized clinical trials. *Neurological Sciences*, *37*(1), 57–65. http://dx.doi.org/10.1007/s10072-015-2357-0.

CHAPTER FORTY-ONE

Complementary and Alternative Medicine and Exercise in Nonmotor Symptoms of Parkinson's Disease

Indu Subramanian[1]
UCLA/West LA VA, Los Angeles, CA, United States
[1]Corresponding author: e-mail address: ISubramanian@mednet.ucla.edu

Contents

Abstract

The use of complementary and alternative medicine (CAM) therapy in nonmotor symptoms (NMS) for Parkinson disease (PD) is growing worldwide. Well-performed, systematic evidence-based research is largely lacking in this area and many studies include various forms of CAM with small patient numbers and a lack of standardization of the approaches studied. Taichi, Qigong, dance, yoga, mindfulness, acupuncture, and other CAM therapies are reviewed and there is some evidence for the following: Taichi in sleep and PDQ39; dance in cognition, apathy, and a mild trend to improved fatigue; yoga in PDQ39; and acupuncture in depression, PDQ39, and sleep.

International Review of Neurobiology, Volume 134
ISSN 0074-7742
http://dx.doi.org/10.1016/bs.irn.2017.05.037

Exercise including occupational therapy (OT) and physical therapy (PT) has been studied in motor symptoms of PD and balance but only with small studies with a mounting evidence base for use of exercise in NMS of PD including PDQ39, sleep, fatigue, depression, and some subsets of cognition. Studies of OT and PT largely show some benefit to depression, apathy, and anxiety. Sustainability of an improvement has not been shown given short duration of follow up. Finding optimal control groups and blind for these interventions is also an issue. This is a very important area of study since patients want to be self-empowered and they want guidance on which form of exercise is the best. Additionally, evidence for PT and OT in NMS would give added weight to get these interventions covered through medical insurance.

ABBREVIATIONS

AS Apathy Scale
BAI Beck Anxiety Index
BDI Beck Depression Inventory
Brooks spatial task
ESS Epworth Sleep Scale
EWB emotional well-being score
FAB frontal assessment battery
Flanker task
FSS Fatigue Severity Scale
GDS Geriatric Depression Scale
HADS Hospital Anxiety and Depression Scale
HAM-D Hamilton Rating Scale for Depression
Krups Fatigue Scale
LARS Lille Apathy Rating Scale
LPDQ Levine–Pilowsky Depression Questionnaire
MADRS Montgomery–Asberg Depression Rating Scale
MAQ Multidimensional Anxiety Questionnaire
MoCA Montreal Cognitive Assessment
MRT mental rotation task
NMSS Nonmotor Symptoms Scale
PDQ8 Parkinson Disease Questionnaire Short Form (PDQ8)
PDQ39 Parkinson Disease Questionnaire (PDQ39) Bodily Discomfort, Mobility, Activities of Daily Living, Emotional Well Being, Communication, Cognitive Impairment, Stigma, and Social Support
PDSS PD Sleep Scale
PFS-16 Parkinson Disease Fatigue Scale
PRM pattern recognition memory
rev. Corsi block
SDS Zung Self-Rating Depression Scale
SES Self-Efficacy Scale
SF-12 12-item short form health survey
SF-36 QOL Short Form Health Survey (SF-36)
SRM spatial recognition memory

SWM spatial working memory
Stroop
TMT-A and -B trails making task A and B
verbal fluency
Visual Analog Scale for pain
Vitality Plus Scale
WHO QOL

1. PART 1: COMPLEMENTARY AND ALTERNATIVE MEDICINE (CAM) IN NONMOTOR SYMPTOMS (NMS) OF PARKINSON DISEASE (PD)

There is an increasing interest in the use of CAM in patients with PD worldwide. A 2001 survey of outpatient PD clinics in the United States found that 40% of patients were using CAM therapies (Rajendran, Thompson, & Reich, 2001) compared to 60% in Dublin, Ireland in 2008 (Murphy, Rogers, Hutchinson, & Tubridy, 2008), and 61% in Singapore in 2006 (Tan, Lau, Jamora, & Chan, 2006). CAM therapy is even more commonly sought out in Korea where up to 76% sought out CAM treatments in 2012 (Kim & Jeon, 2012), in contrast to 38% in Argentina in 2010 (Pecci et al., 2010) and 46% in India in 2016 (Pandit et al., 2016).

CAM is a broad term that is made up of a number of unrelated treatments and practices but globally may be connected with a holistic approach to wellness. The National Center for Complementary and Alternative Medicine (NCCAM) classifies CAM interventions into (1) natural products such as herbals, vitamins, minerals, and probiotics; (2) mind body practices such as acupuncture, massage, meditation, movement therapies, relaxation techniques, Tai Chi, and Yoga; and (3) alternative systems such as Chinese medicine, homeopathy, or Ayurveda. Over $14 billion dollars per year are spent on CAM therapy in the United States alone. There has been a good review by Bega in 2014 (Bega & Zadikoff, 2014) on motor issues in PD and CAM. In this review, we will focus on CAM therapies and their effects on NMS in PD.

1.1 Qigong and Taichi in NMS of PD

Tai Chi is an ancient Chinese martial arts tradition combining deep breaths and relaxation with slow movements that are fluid with some holding of postures. Tai Chi has been found to be both safe and feasible in PD patients.

Referring to Table 1, three trials compared forms of Tai Chi to controls looking largely at PDQ39 and some cognitive measures. Kurt et al. (2017) had 40 patients randomized to Aichi vs land-based exercise for 5 weeks and found improved PDQ39. Nocera et al. (2013) studied 16 patients doing Taichi vs 5 control patients for 16 weeks and found improved PDQ39 total and in the emotional well-being score (EWB) but not in digit span, verbal fluency, Stroop, or trail A and B tests. Li et al. (2014) randomized patients to Taichi vs resistance training vs stretching and showed improved vitality plus scale and PDQ8.

Qigong is an older practice and a predecessor to Taichi. The focus is on meditation and slow movement. Referring to Table 1, there are four studies that looked at Qigong with nonmotor measures ranging from sleep, fatigue, depression, cognition, and PDQ39. Three out of four studies were controlled studies. Schmitz-Hubsch et al. (2006) studied 56 patients, 32 in qigong and 24 in a control group. Qigong was performed for 90 min sessions over 6 months. There were some issues with the study design but there was no clear improvement in Montgomery–Asberg Depression Rating Scale (MADRS) and PDQ39 in this study. There was a trend to improved daytime sleepiness and constipation in the Qigong group. Xiao and Zhuang (2016) studied Qigong in 100 patients randomized to Qigong plus walking vs walking alone. There was an improvement in the PD sleep scale (PDSS) but not in the PD fatigue scale (PFS-16). Burini et al. (2006) studied Qigong in a randomized control trial with crossover design in 26 patients. There were no differences seen in BDI, fatigue, or PDQ39 with this trial. Wassom et al. (2015) did a small trial with pre- and posttest comparison with a 6-week Qigong intervention. There was a mild change in sleep scale but not in other PDQ39, EWB, digit span, verbal fluency, Stroop, or trails A and B.

As a summary, Taichi has some increasing evidence in helping sleep and PDQ39 and 8.

1.2 Dance in NMS of PD

The broad classification of dance includes many forms of movement that may involve individual, partnered, or group dancing and is enriched by aspects of social engagement and music.

Tango dance includes complex movement, social interaction, and incorporates challenging movements from a balance, cognitive, and motor learning standpoint. Referring to Table 2, there were six trials of Tango looking

Table 1 Qigong and Taichi in NMS of PD

Author/Year Number of Patients/Arms	Intervention Description	Nonmotor Measures	Outcome of Study
Schmitz-Hubsch et al. (2006) 56 patients—32 patients vs 24 controls	Qigong 90 min, weekly for 2 month Then 2 month pause, then 2 month Qigong; control no treatment	MADRS PDQ39	No improvement in these measures ? improved constipation and daytime sleeping in QG group
Xiao and Zhuang (2016) 100 patients	Qigong baduanjin 45 min 4 × /week + daily walking 30 min vs walking only for 6 months	PDSS-2 PFS-16	PDSS-2 improved
Burini et al. (2006) 26 patients—active treatment (AT) 1 + Qigong (QG) 2 or QG1 + AT2 (20 sessions each with 2 month interval)	Qigong 3 × /week, 7 week RCT crossover	BDI Fatigue PDQ39	No difference in these measures
Wassom, Lyons, Pahwa, and Liu (2015) 7 patients	Qigong-6 week Pre- and posttest comparison	PDSS PDQ39 EWB Digit span Verbal fluency Stroop Trails A and B	Mild change in PDSS, not in others
Kurt, Buyukturan, Buyukturan, Erdem, and Tuncay (2017) 40 patients	Aichi vs land-based exercise; 5 weeks	PDQ39	Better in Ai-chi group
Nocera, Amano, Vallabhajosula, and Hass (2013) 21 patients	15 Taichi, 6 controls 60 min Taichi 3 × / week; 16 week Taichi	PDQ39 EWB Digit span Verbal fluency Stroop Trails A and B	PDQ39 total and emotional well-being Improved in Taichi group Not the other measures
Li et al. (2014) 195 patients	60 min 2 × /week, 24 weeks, RCT Taichi vs resistance vs stretching	PDQ8 Vitality plus scale	PDQ8 and VPS improved in Taichi group

Table 2 Dance on NMS in PD

Author/Year Number of Patients/Arms	Intervention Description	Nonmotor Measures	Outcome of Study
Rios Romenets, Anang, Fereshtehnejad, Pelletier, and Postuma (2015) 33 patients	Tango vs self-directed exercise 2×/week 12 week, 1 h, RCT	MOCA BDI Krups fatigue SS AS PDQ39	MOCA and Fatigue Severity Scale (FSS) nonsignificant trend to improvement in tango group
Hackney and Earhart (2009b)	Tango/ ballroom vs Taichi 13 weeks, 2×/week, 60 min	PDQ39	Improved social support PDQ39 SI with tango and not others including control
Duncan and Earhart (2014) Tango vs none 10 patients	Tango-12 week, 1 h, 2×/week, 2 years RCT	UPDRS2	UPDRS 2 improved
McKee and Hackney (2013) 33 patients—24 vs 9	Tango vs group lecture	PDQ39 BDI MOCA SF-12 FAB Rev. Corsi block Brooks spatial task	Improved executive function/spatial cognition-Brooks and FAB
Hashimoto, Takabatake, Miyaguchi, Nakanishi, and Naitou (2015) 46 patients	Dance 60 min, 1 time per week 12 week vs control	Mood/AS FAB MRT (mental rotation task) Self-rated depression scale	Improved AS and SDS
Volpe, Signorini, Marchetto, Lynch, and Morris (2013) 24 patients	Irish set vs PT 6 months, 1 time per week, 90 min	PDQ39	No difference in PDQ39

at a range of NMS including depression, apathy, PDQ39, and cognitive measures. Rios Romenets et al. (2015) randomized 33 patients to tango vs self-directed exercise for 12 weeks and found that there was a trend to improved Montreal Cognitive Assessment (MOCA) and Krups fatigue scale. There was no change in BDI, Apathy Scale (AS), or PDQ39. Hackney and Earhart (2009a) studied tango/ballroom dance compared to Taichi for 13 weeks and found that PDQ39 social isolation was improved with tango but no other parts of the PDQ39. Duncan and Earhart (2014) studied 10 patients randomized to tango vs usual activity for 12 weeks. The UPDRS 2 scale improved with the tango group. McKee and Hackney (2013) studied 33 patients who were randomized to 24 weeks of tango vs 9 patients who attended lectures. The tango group had improved executive function on frontal assessment battery (FAB) and spatial cognition on Brooks spatial task. Hashimoto et al. (2015) studied 46 patients with dance vs a control and showed improved apathy on the AS and self-reported depression on the Zung Self-Rating Depression Scale (SDS). No changes were seen on mood scale, mental rotation task, or FAB. Volpe et al. (2013) studied 24 patients with Irish set dancing vs physical therapy (PT) for 6 months. There was no difference in the groups with PDQ39. Dance appears to have some evidence in cognition, apathy, and mild trend to improved fatigue.

1.3 Yoga/Mindfulness on NMS in PD

Yoga is a unique exercise/CAM therapy which includes movement, breath work, and meditation. It is part of the ancient Ayurvedic tradition and has been around for over 2000 years. There have been four trials in yoga only three of which have controls looking at mood and PDQ39. Sharma, Robbins, Wagner, and Colgrove (2015) did a RCT of yoga vs control with measures of SF-36 depression, QOL, and Geriatric Depression Scale (GDS). The only positive result was in SF-36 depression. Ni studied 26 patients randomized to power yoga vs control. There was an improvement in PDQ39. Bega, Gonzalez-Latapi, Zadikoff, and Simuni (2014) did a study on yoga vs resistance training in a small group of PD patients. There was no change in PDQ39 in this study that had a large dropout rate. Boulgarides, Barakatt, and Coleman-Salgado (2014) did an 8-week yoga intervention with pre- and postintervention scales. There was an improvement in the Hospital Anxiety and Depression Scale (HADS).

Table 3 Yoga/Mindfulness on NMS in PD

Author/Year Number of Patients/ Arms	Intervention Description	Nonmotor Measures	Outcome of Study
Sharma et al. (2015) 8 patients and 5 controls	Yoga–RCT 2×/week, 12 weeks	QOL SF36-depression GDS	SF-36 depression improved
Ni et al. (2016) 26 patient	Power Yoga–RCT Yoga 3 months 2×/week	PDQ39	PDQ39 overall improved score
Bega et al. (2016) 17 patients	Yoga vs resistance exercise—12 weeks, 2×/week, 1 h	PDQ39	No difference in groups
Boulgarides et al. (2014) 10 patients	Yoga—8 week yoga pre- and postcomparison	HADS	Improved HADS depression subscale
Advocat et al. (2016)	Mindfulness—6 weeks vs wait list control—24/35 vs 33/37	PDQ39	Significant improvement in PDQ39-ADL

Mindfulness is a form of meditation and is an approach that is reproducible and easily taught. It's feasibility in the PD population is only recently being explored. Mindfulness was studied in a small study by Advocat et al. (2016) with 24 patients vs 33 patients on a wait list control. There was a significant improvement in PDQ39 ADL and a trend to improvement in PDQ39SI. Yoga may help PDQ39 and mood (Table 3).

1.4 Miscellaneous CAM Interventions on NMS in PD

Active theater was studied by Modugno et al. (2010) vs PT in 20 patients. Epworth sleep scale (ESS), PDQ39, HADS, and UPDRS2 were measured. There was an improvement in all of these measures.

Music therapy has been implicated in imaging improvements in fMRI gray matter volume in motor and auditory cortex in animal and human studies, music is shown to release neurochemicals and hormones. Increased limbic activity in the orbitofrontal cortex and increased mesolimbic dopamine release has been triggered by music. Pleasant music in particular has been shown to increase PET and fMRI in the nucleus accumbens and midbrain nuclei that can inhibit pain. Thus, music therapy has been of interest in treating NMS in PD patients.

Music therapy was studied in three studies, two of which had controls and looked at mood, QOL, and PDQ39. Pacchetti et al. (2000) studied 32 patients with 3 month of musical therapy vs controls. They studied QOL and a happiness measure which both improved. Elefant, Baker, Lotan, Lagesen, and Skeie (2012) studied a 20-week active music therapy intervention with a single 1 h session per week in 10 patients on MADRS depression and anxiety scores and did not show a change. A drum circle randomized control trial study for 6 weeks in 10 patients vs 10 control patients showed improved PDQ39 but not depression in the GDS by Pantelyat, Syres, Reichwein, and Willis (2016).

Donoyama and Ohkoshi (2012) and Donoyama, Suoh, and Ohkoshi (2014) studied Japanese Massage for 30 min in a single patient and showed improved pain on a visual analog scale. This study did not have a control. Brefel studied spa therapy in 31 PD patients randomized to early vs late spa therapy for 3 weeks including massage and PT. They found that PDQ39, SF-36, GHQ-28 improved right after intervention but this was not sustained at 20 weeks. Johns, Blake, and Sinclair (2010) looked at PDQ39 (Brefel-Courbon et al., 2003) pre and post 8 reflexology sessions in 16 PD patients and showed improvement. The Alexander technique uses hand contact to assess and manipulate changes in muscle activity. Stallibrass, Sissons, and Chalmers (2002) studied Alexander technique vs massage vs control. They used BDI at a 6-month follow up and the attitudes to self-scale. Both of these scales improved with the Alexander technique.

Cash and Lageman (2015) studied expressive writing vs nonexpressive writing in 28 patient and 15 caregivers of PD patients. He found that there was improved anxiety in both groups regardless of type of writing. Combs et al. (2011) studied boxing drills (with a bit of stretching, strength, and endurance) in a case series of six patients. There was no control group. There is improved PDQOL with boxing. Herz et al. (2013) studied Wi playing over 4 weeks for three sessions per week including two games of tennis and bowling and one game of boxing per session. There was improved PDQ39 and a trend toward improved HAM.

Nordic walking was studied in two studies looking at nonmotor issues. Cugusi et al. (2015) studied 20 patients with Nordic walking for 12 weeks and measured PFS-16, Beck Depression Inventory-II (BDI-II), Starkstein Apathy Scale, and Nonmotor Symptoms Scale (NMSS), and PDQ8. There was an improvement in PFS-16, BDI-II, SAS, PDQ8, and NMSS. Reuter et al. (2011) studied Nordic walking vs walking/flexibility exercises in 90 patients in a RTC for 6 months, 3 times per week for 70 min sessions. There was no difference between groups in the PDQ39 (Table 4).

1.5 Acupuncture for NMS in PD

There are 11 studies looking at acupuncture, only 1 of which had a sham control, 7 had a control group that did not get an intervention, and 3 had pre- and postintervention measures. Referring to Table 5, these studies looked at sleep, fatigue, PDQ39, anxiety depression, and some looked at bladder function and constipation. Eng et al. (2006) studied Tuina massage/acupuncture/instrument-delivered Qigong in 25 patients for 6 months with improved BDI and PDQ39 pre- vs postintervention. Shulman et al. (2002) did an open label trial on 20 patients and showed improved sleep and rest. Looking at the RCTs, Christian did a RCT with acupuncture in 14 patients looking at PDQ8, PDQ39, and GDS with no significant change in these measures. Chen, Zhang, Li, and Holscher (2015) studied 40 patients with acupuncture 2 times per week along with Western medicine vs Western medicine alone for 18 weeks and 36 weeks in 13 patients. They studied BDI-II, Beck Anxiety Index (BAI), WHOqol, and UPDRS and showed improvements in the BDI-II, WHOqol, and UPDRS-mind, behave, and mood. Cho et al. (2012) studied 22 patients in a randomized single blind control trial looking at BDI and PDQOL and showed improved BDI. Zhang et al. (2015) did a RCT for 8 weeks with acupuncture + madopar/tolteridine (M/T) vs M/T alone and showed improved bladder function in the acupuncture group. Kluger et al. (2016) did a study on 94 patients with acupuncture vs sham acupuncture 2 times per week for 6 months. He found no difference in the modified fatigue impact scale between the groups but both groups improved on this measure. Acupuncture may help depression, PDQ39, and sleep.

Table 4 Miscellaneous CAM Interventions on NMS in PD

Author/Year Number of Patients/ Arms	Intervention Description	Nonmotor Measures	Outcome of Study
Modugno et al. (2010) 40 patients	20 patients in active theater vs 10 patients in PT vs 10 patients in control group 3 years	PDQ39 ESS HADS UPDRS1	PDQ39 scale, Epworth sleepiness scale, and Hamilton depression, UPDRS 1 improvement; improved daytime sleepiness, depression
Donoyama and Ohkoshi (2012) 1 patient	Massage Japanese 30 min	Visual analog scale for pain	Pain improved by this scale
Brefel-Courbon et al. (2003) 31 patients	Spa therapy—massage daily/physical therapy × 3 weeks, delayed start in other group RCT	PDQ39 SF-36 GHQ-28	PDQ39 improved SF-36 improved GHQ-28 improved at week 4 but not sustained improvement at 20 weeks
Johns et al. (2010) 16 patients	Reflexology—pre- and posttreatment 8 treatment sessions	PDQ39	PDQ39 improved
Pacchetti et al. (2000) 32 patients	Music 3 months RCT	QOL Happiness measure	Happiness measure; QOL improved
Elefant et al. (2012) 10 patients	Music—60 min, once per week, 20 weeks	MADRS	No change in MADRS at 10 weeks or 20 weeks
Pantelyat et al. (2016) 20 patients	Drum circle—6 weeks; 10 patients vs 10 controls	PDQ39 GDS	PDQ39 improved No mood change
Cugusi et al. (2015) 20 patients	Nordic walking 2×/week, 1 h, 12 weeks	PFS-16, Beck Depression Inventory-II, Starkstein AS, and NMSS, PDQ8	PFS-16, BDI-II, SAS, PDQ8, NMSS improved No other scales

Continued

Table 4 Miscellaneous CAM Interventions on NMS in PD—cont'd

Author/Year Number of Patients/ Arms	Intervention Description	Nonmotor Measures	Outcome of Study
Reuter et al. (2011) 90 patients RTC	Nordic walking, walking, flexibility 6 months, 3×/week, 70 min	PDQ39	No difference in PDQ39
Stallibrass et al. (2002)	Alexander technique vs massage vs control, 6 month	Beck Depression Inventory, attitudes to self-scale	BDI and ASS improved
Cash and Lageman (2015) 28 patients and 15 caregivers expressive vs other writing	Expressive writing vs other writing	Anxiety-Multidimensional Anxiety Questionnaire (MAQ) QOL	Improved anxiety in MAQ in both groups
Combs et al. (2011) Case series 6 patients	Boxing—24–36 sessions, 90 min boxing drills, stretching, strength, endurance	PDQOL	PDQOL improved
Herz et al. (2013)	Wii-4 weeks (3×/week) each session included 2 games of tennis, 2 games of bowling, and 1 game of boxing	HAM PDQ39	Improved PDQ39 Trend toward improvement in HAM

2. PART 2: EXERCISE ON NMS IN PD

There have been various areas of investigation in the nonmotor realm in PD using exercise. Here we describe exercise not largely using PT or occupational therapy (OT) guidance. Many of these interventions involve group classes with a personal trainer guiding the class. Studies have looked

Table 5 Acupuncture for NMS in PD

Author/Year Number of Patients/Arms	**Intervention Description**	**Nonmotor Measures**	**Outcome of Study**
Eng, Lyons, Greene, and Pahwa (2006) 25 patients	6 months Tuina massage Acupuncture QIGONG instrument pre- and postintervention	BDI PDQ39	BDI and PDQ39 improvement
Shulman et al. (2002) 20 patients	Open label	Anecdotal sleep and rest reports	Improved sleep and rest
Cristian, Katz, Cutrone, and Walker (2005) 14 patients	RCT of acupuncture vs no treatment	PDQ8 PD39 GDS	No significant change in these measures
Chen, Change, et al. (2015) 40 patients	40 patients, 2 × /week vs Western medicine only for 18 weeks, 13 patients had acupuncture for 36 week RCT	BDI-II BAI WHOqol UPDRS-mind, behave, mood	Improved BDI-II Improved WHOqol UPDRS-MBM
Cho et al. (2012) 22 patients	Randomized single blind control trial	BDI PDQOL	Improved BDI
Chen, Feng, and Zhang (2012) 60 patients	RCT 8 weeks; madopar/ tolteridine (M/T) vs M/T + acupuncture	Bladder function	Improved bladder Function in acupuncture group
Kluger et al. (2016) 94 patients	RCT of acupuncture vs sham points 2 × /week	Fatigue impact scale	No change in fatigue; modified fatigue impact scale vs sham—both groups were less fatigued

at PDQ39, cognition, mood, sleep, fatigue, and apathy. The description of the studies is under the heading of the NMS measures. In this review, we again focus on the NMS described in Tables 6– 9. In the next session, the focus is on specific OT and PT intervention and NMS outcomes.

Table 6 Exercise Effects on Quality of Life as Measured by PDQ39

Author/Year Number of Patients/ Arms	Intervention Description	Nonmotor Measures	Outcome of Study
Morberg, Jensen, Bode, and Wermuth (2014) 12 patients and 12 controls	High-intensity personalized physical training—32 weeks, 2–3 ×/week	PDQ39 total Body discomfort EWB	Body discomfort and EWB subscore improved; not total PDQ39
Schenkman et al. (2012) 121 patients	Flexibility individualized followed by group with physical therapy vs aerobic exercise by trainer, vs home-based exercise; 30 min; 16 month RCT	PDQ39	No difference in groups
Park et al. (2014) 31 patients	Early vs late start 1 h 3 ×/week, 48 weeks vs 24 weeks	PDQ39 BDI	Early start improved on BDI
Rodrigues de Paula, Teixeira-Salmela, Coelho de Morais Faria, Rocha de Brito, and Cardoso (2006) 20 patients	36 group sessions, 75 min, 3 ×/week for 3 months, aerobic conditioning/muscle strengthening (pre- and postintervention comparison)	QOL-Nottingham health proof	Improved social interaction, emotional reaction, physical ability, NHP-S sleep improved
Corcos et al. (2013) 51 patients	20 in weightlifting (PRE) vs 18 modified fitness counts (MFC) (stretch, balance, strengthening) 2 ×/week for 24 months	PDQ39	PDQ39 improved in PRE vs MFC
Dibble, Hale, Marcus, Gerber, and LaStayo (2009) 20 patients with PD vs aged match control	12 week High-intensity eccentric resistance training	PDQ39	PDQ39 improved in exercise group
van Nimwegen et al. (2013) 540 patients	ParkFit (inc. regular physical therapy) vs control	PDQ39	No change in PDQ39

Table 7 Exercise Effects on Sleep/Fatigue in PD

Author/Year Number of Patients/ Arms	Intervention Description	Nonmotor Measures	Outcome of Study
Frazzitta et al. (2015) 89 patients +49 controls	RCT-89 patients vs 49 controls; 28 days—three 1-h sessions of cardio, stretching, balance, gait with OT for ADLs	PD sleep scale	Improved PDSS for sleep quality, motor symptoms, daytime somnolence
Nascimento et al. (2014 34 patients	RCT vs usual are 3×/week, 6 month, muscular resistance, balance/motor coordination, aerobic fitness	MSQ sleep	Improved sleep disturbance (insomnia, daytime somnolence)
Winward et al. (2012) 39 patients	Gym vs nothing RCT 30–45 min, 12 weeks, varied frequency	Fatigue SS	Nonsignificant improvement

Table 8 Exercise Effects on Depression in PD

Author/Year Number of Patients/ Arms	Intervention Description	Nonmotor Measures	Outcome of Study
Uc et al. (2014) 49 patients	45 min aerobic walking no control 3×/week for 6 months at home	GDS Flanker task, MOCA	Decreased depression Improved flanker task
Bridgewater and Sharpe (1998) 26 patients	20–30 min aerobic pilot RCT vs usual care	LPDQ depression	Improved mood in exercise group
Nadeau, Pourcher, and Corbeil (2014) 45 patients	Speed vs mixed treadmill exercise, RCT 24 weeks, 1 h, 3×/week	BDI MMSE UPDRS1	No significant improvement

2.1 Exercise Effects on Quality of Life as Measured by PDQ39 in PD

Referring to Table 6, Morberg et al. (2014) did a study on high-intensity personalized physical training on 12 patients vs 12 controls. The sessions were 2–3 times per week for 32 weeks. PDQ39 was measured and

Table 9 Exercise Effects on Cognition in PD

Author/Year Number of Patients/Arms	Intervention Description	Nonmotor Measures	Outcome of Study
Picelli et al. (2016) 17 patients	9 in intervention group, 8 controls RCT, 12 sessions, 45 min, treadmill 3×/week, 4 weeks	FAB	FAB improved
David et al. (2015) 51 patients	2×/week; 24 month 60–90 min progressive resistance training (PRET) vs stretching, balance, nonprogressive control	BTA Digit span Stroop	Improved in all three for PRET Only digit span, Stroop improved in stretch group No diff. at 12 vs 24 month
Pohl, Dizdar, and Hallert (2013) 18 patients	Ronnie Gardner rhythm and music method RCT 2×/week, 1 h, 6 weeks	Cognition	No change
Ridgel, Kim, Fickes, Muller, and Alberts (2011)	Passive leg cycling Pre- and posttest comparison	TMT-A and B	Improved TMT-B scores and times for both tests improved
Cruise et al. (2011) 28 patients	RCT, 60 min aerobic exercise 2×/week, 3 months vs usual care	MMSE PRM, SRM, SOC, SWM	Exercise group had improved executive function
Shulman et al. (2013) 67 patients	RCT of 30 min high intensity treadmill vs 50 min low intensity treadmill vs stretching/resistance 3×/week for 3 months	BDI PFS-16	No change in any group

showed improved body discomfort and well-being scores but not total PDQ39. Park et al. (2014) looked at 31 patients randomized to early vs late start exercise 1 h 3 times per week for 48 weeks vs 24 weeks. They measured PDQ39 and BDI and found that there was an improvement in the Beck score in the early start group. Schenkman et al. (2012) looked at 121 patients randomized to one of three arms: flexibility (individualized followed by group class supervised by PT) vs aerobic exercise supervised by an exercise trainer vs home-based exercise looking at PDQ39. There was no difference between the groups with this quality of life measure. Rodrigues de Paula et al. (2006) studied 20 patients with 36 group sessions lasting 75 min each, 3 times per week for 3 months including pre- and postintervention measures. The intervention included aerobic conditioning and muscle strengthening. The Nottingham health profile was used to measure quality of life. There was improved social interacting, emotional reaction, physical ability, and sleep with a trend toward improved pain after the intervention. Corcos et al. (2013) studied 51 patients—20 in progressive resistance exercise, 18 in modified fitness class (stretch, balance, and strengthening) 2 times per week for 24 months. They measured PDQ39 and found an improvement in this measure in the progressive resistance exercise group. Dibble et al. (2009) studied 20 patients with high-intensity eccentric resistance training for 12 weeks and found that PDQ39 improved in the exercise group. van Nimwegen et al. (2011, 2013) studied 540 patients with ParkFit intervention (including regular PT) vs control. There was no change in PDQ39 in this study.

In summary, exercise improves QOL measured by PDQ39 in a few studies.

2.2 Exercise Effects on Sleep/Fatigue in PD

Frazzitta et al. (2015) reported on rehab. intervention of 28 days of a combination of three 1-h session of cardio, stretching, balance, and gait along with OT for ADLs in 89 patients vs 49 controls and measure the PDSS. They showed that PDSS for sleep quality, motor symptoms, and daytime somnolence improved. Nascimento et al. (2014) studied 34 patients in a RCT for 3 times per week for 6 months using muscular resistance, balance/motor coordination, and aerobic fitness vs usual care and looked at MSQ sleep. They showed that there was improved sleep disturbance in the intervention including insomnia and daytime somnolence. Winward et al. (2012) studied 39 patients in a RCT with gym vs no exercise and found that there was no significant improvement in the fatigue SS. In summary, exercise improves sleep in a couple of studies.

2.3 Exercise Effects on Depression in PD

Uc et al. (2014) studied 49 patients with a 45 min aerobic walking intervention with no control. The patients walked 3 times per week for 6 months at home. They looked at the GDS and MOCA and the flanker task and showed improved flanker task and decreased depression with walking. Nadeau et al. (2014) studied 45 patients with speed vs mixed treadmill in a RCT trial for 24 weeks. The patients did the exercise for 24 weeks for 1 h per session for 3 sessions per week. BDI, MMSE, and UPDRS1 were studied but were not changed. Bridgewater and Sharpe (1998) studied 26 patients doing 20–30 min aerobic exercise vs usual care in a RCT. They showed improved mood in the exercise group on Levine–Pilowsky Depression Questionnaire (LPDQ) and depression. Hence, exercise improves depression in a couple of small studies.

2.4 Exercise Effects on Cognition in PD

Picelli et al. (2016) looked at 17 patients, 9 in the intervention group looking at 12 sessions of treadmill for 45 min, 3 times per week for 4 weeks. They measured a FAB which improved with exercise. David et al. (2015) studied 51 patients for 2 sessions per week, 60–90 min progressive resistance training (PRET) vs stretching, balance, nonprogressive training for 24 month. They studied the BTA, digit span, and Stroop and showed improvements in all three for PRET but only the digit span and Stroop in the stretch group. There was no difference between 12 and 24 months. Pohl et al. (2013) studied the Ronnie Gardner Rhythm and music method and showed no change in cognition. Ridgel et al. (2011) studied passive leg cycling as the intervention with pre- and postintervention measures of the trails making test A and B with improved TMT-B scores and improved times for performing both types of trail tests.

Cruise et al. (2011) performed a RCT on 28 patients for 60 min 2 times per week for 3 months doing aerobic exercise vs usual care. They looked at MMSE, pattern recognition memory (PRM), spatial recognition memory (SRM), SOC, and spatial working memory (SWM). They found that the exercise group improved with executive function. Shulman et al. (2013) studied 67 patients in a RCT of 30 min of high-intensity treadmill vs 50 min of low-intensity treadmill vs stretching and resistance 3 times per week for 3 months. Outcome measures included BDI, PFS-16, and fatigue and showed no change in any group. Almost all of the trials reviewed, showed some modest improvement in a subsection of cognition in many patients with PD.

3. PART 3: THERAPY: PT AND OT EFFECTS ON NMS

PT is defined as a therapy for the preservation, enhancement, or restoration of movement and physical function impaired or threatened by disease, injury, or disability that utilizes therapeutic exercise, physical modalities (such as massage and electrotherapy), assistive devices, and patient education and training—called also physiotherapy. This therapy is administered to the patient by a physical therapist usually on an individual basis. OT is a therapy based on engagement in meaningful activities of daily life (as self-care skills, education, work, or social interaction) especially to enable or encourage participation in such activities despite impairments or limitations in physical or mental functioning. This therapy is administered by an occupational therapist usually in a one-on-one basis. This is in contrast to the studies described in the exercise section above which were usually group classes with a personal trainer as the instructor.

Teixeira-Machado et al. (2015) studied Feldenkrais PT 2 times per week for 1 h for 25 weeks in an RCT in 30 patients. They measured BDI and found an improvement in this scale in the PT group. Dashtipour studied 60 min of aerobic/resistance training vs LSVT big therapy in 11 patients. They looked at BAI and BDI as outcome measures and found reduced depression in the combined exercise group.

King et al. (2015) performed a RCT study on 58 patients with exercise that was individualized by a PT vs a group class intervention based on an agility boot camp for PD including Taichi, boxing, lunges, kayaking, agility course, and Pilates vs an unsupervised home exercise program for three sessions per week for 4 weeks. They looked at PDQ39, LARS (Lille Apathy Rating Scale), depression, Self-Efficacy Scale (SES) and found that LARS, depression, and SES were best in patients who had the individual PT. Cholewa, Boczarska-Jedynak, and Opala (2013) studied 70 patients in a RTC of PT for 1 h per week for 12 weeks and found an improvement in UPDRS 1. Miyai et al. (2000) did a similar study looking at 24 patients using body weight supported treadmill vs PT 3 times per week for 45 min for 1 month and found no change in UPDRS 1 between groups.

Sturkenboom et al. (2014) studied OT vs control in a RCT looking at fatigue scale and BDI with no change in these outcomes. Tomlinson et al. (2014) studied 39 patients in a RCT of OT vs control. The OT group got 45 min 3 times per month for 2 months. HADS and anxiety scale did not improved.

Table 10 Physical Therapy/Occupational Therapy Effects on NMS in PD

Author/Year Number of Patients/Arms	Intervention Description	Nonmotor Measures	Outcome of Study
Teixeira-Machado et al. (2015) 30 patients	Feldenkrais PT 2×/week 1 h 25 weeks RCT	BDI	BDI improved in PT group
Dashtipour et al. (2015) 11 patients	60 min aerobic/resistance vs LSVT BIG PT intervention	BAI BDI	Reduced depression in combined exercise group
King et al. (2015) 58 patients	RCT exercise individual PT vs group class on agility boot camp 3 week for 4 weeks	PDQ39, LARS, depression, SES	LARS, depression, and SES improved in PT group
Cholewa et al. (2013) 70 patients	PT RCT 1 h for 12 weeks	UPDRS 1	UPDRS1 improved in PT group
Miyai et al. (2000) 24 patients	Body weight supported treadmill vs PT 3×/week, 45 min, for 1 month	UPDRS 1	No change in UPDRS1
Sturkenboom et al. (2014) 191 patients	OT vs control—10 week, home-based OT vs usual care 124 patients vs 67 controls (RCT 2:1)	Euroqol 5 dimension Fatigue SS BDI	Euroqol 5 improved in OT group
Tomlinson et al. (2014) 39 patients	OT RCT 45 min 3×/month for 2 months	HADS—anxiety/depression	No change

In summary, studies of OT and PT largely show some benefit to depression, apathy, and anxiety (Table 10).

4. CONCLUSIONS

The areas of complementary/alternative medicine and exercise show promise in the treatment of NMS in PD. Well-performed, systematic evidence-based research studies are largely lacking in this area and many studies include various forms of CAM with small patient numbers and a lack of standardization of approaches studied. Taichi, Qigong, dance, yoga, mindfulness, acupuncture, and other CAM therapies were reviewed and there is some evidence for the following: Taichi in sleep and PDQ39; dance in cognition, apathy, and a mild trend to improved fatigue; yoga in PDQ39; and acupuncture in depression, PDQ39, and sleep. Further studies need to be performed in order to properly counsel patients on which type of CAM therapy could be utilized to help the various NMS.

Exercise including OT and PT has been studied in motor symptoms of PD and balance but there are some small studies with a mounting evidence base for use of exercise in NMS of PD including PDQ39, sleep, fatigue, depression, and some subsets of cognition. Studies of OT and PT largely show some benefit to depression, apathy, and anxiety. Sustainability of an improvement has not been shown given a short duration of follow up. The optimal way to control and blind for these interventions is also an issue. Hence, further studies in exercise including OT and PT are warranted and needed in order to counsel patients on which exercise would be beneficial for their individual motor and NMS. Additionally, evidence for PT and OT in NMS would give added weight to getting these interventions covered through medical insurance.

The future is bright for these fast-paced, ever-changing areas of CAM therapy, and exercise in the treatment of NMS for PD. A basic understanding of these areas is vital for clinicians in order to counsel patients who are hungry to try these novel interventions.

REFERENCES

Advocat, J., Enticott, J., Vandenberg, B., Hassed, C., Hester, J., & Russell, G. (2016). The effects of a mindfulness-based lifestyle program for adults with Parkinson's disease: A mixed methods, wait list controlled randomised control study. *BMC Neurology*, *16*, 166. http://dx.doi.org/10.1186/s12883-016-0685-1.

Bega, D., Gonzalez-Latapi, P., Zadikoff, C., & Simuni, T. (2014). A review of the clinical evidence for complementary and alternative therapies in Parkinson's disease. *Current Treatment Options in Neurology*, *16*(10), 314. http://dx.doi.org/10.1007/s11940-014-0314-5.

Bega, D., Stein, J., Zadikoff, C., Simuni, T., Corcos, D., et al. (2016). Yoga versus resistance training in mild to moderate severity Parkinson's disease: A 12-week pilot study. *Journal of Yoga & Physical Therapy*, *6*, 222. http://dx.doi.org/10.4172/2157-7595.1000222.

Bega, D., & Zadikoff, C. (2014). Complementary & alternative management of Parkinson's disease: An evidence-based review of eastern influenced practices. *Journal of Movement Disorders*, 7(2), 57–66. http://dx.doi.org/10.14802/jmd.14009.

Boulgarides, L. K., Barakatt, E., & Coleman-Salgado, B. (2014). Measuring the effect of an eight-week adaptive yoga program on the physical and psychological status of individuals with Parkinson's disease. A pilot study. *International Journal of Yoga Therapy*, *24*, 31–41.

Brefel-Courbon, C., Desboeuf, K., Thalamas, C., Galitzky, M., Senard, J. M., Rascol, O., et al. (2003). Clinical and economic analysis of spa therapy in Parkinson's disease. *Movement Disorders*, *18*(5), 578–584. http://dx.doi.org/10.1002/mds.10404.

Bridgewater, K. J., & Sharpe, M. H. (1998). Trunk muscle performance in early Parkinson's disease. *Physical Therapy*, *78*(6), 566–576.

Burini, D., Farabollini, B., Iacucci, S., Rimatori, C., Riccardi, G., Capecci, M., et al. (2006). A randomised controlled cross-over trial of aerobic training versus qigong in advanced Parkinson's disease. *Europa Medicophysica*, *42*(3), 231–238.

Cash, T. V., & Lageman, S. K. (2015). Randomized controlled expressive writing pilot in individuals with Parkinson's disease and their caregivers. *BMC Psychology*, *3*, 44. http://dx.doi.org/10.1186/s40359-015-0101-4.

Chen, F. P., Chang, C. M., Shiu, J. H., Chiu, J. H., Wu, T. P., Yang, J. L., et al. (2015a). A clinical study of integrating acupuncture and Western medicine in treating patients with Parkinson's disease. *The American Journal of Chinese Medicine*, *43*(3), 407–423. http://dx.doi.org/10.1142/S0192415X15500263.

Chen, Y. L., Feng, W. J., & Zhang, X. L. (2012). Parkinson's disease combined with overactive bladder syndrome treated with acupuncture and medication. *Zhongguo Zhen Jiu*, *32*(3), 215–218.

Chen, Y., Zhang, Y., Li, L., & Holscher, C. (2015b). Neuroprotective effects of geniposide in the MPTP mouse model of Parkinson's disease. *European Journal of Pharmacology*, *768*, 21–27. http://dx.doi.org/10.1016/j.ejphar.2015.09.029.

Cho, S. Y., Shim, S. R., Rhee, H. Y., Park, H. J., Jung, W. S., Moon, S. K., et al. (2012). Effectiveness of acupuncture and bee venom acupuncture in idiopathic Parkinson's disease. *Parkinsonism & Related Disorders*, *18*(8), 948–952. http://dx.doi.org/10.1016/j.parkreldis.2012.04.030.

Cholewa, J., Boczarska-Jedynak, M., & Opala, G. (2013). Influence of physiotherapy on severity of motor symptoms and quality of life in patients with Parkinson disease. *Neurologia i Neurochirurgia Polska*, *47*(3), 256–262.

Combs, S. A., Diehl, M. D., Staples, W. H., Conn, L., Davis, K., Lewis, N., et al. (2011). Boxing training for patients with Parkinson disease: A case series. *Physical Therapy*, *91*(1), 132–142. http://dx.doi.org/10.2522/ptj.20100142.

Corcos, D. M., Robichaud, J. A., David, F. J., Leurgans, S. E., Vaillancourt, D. E., Poon, C., et al. (2013). A two-year randomized controlled trial of progressive resistance exercise for Parkinson's disease. *Movement Disorders*, *28*(9), 1230–1240. http://dx.doi.org/10.1002/mds.25380.

Cristian, A., Katz, M., Cutrone, E., & Walker, R. H. (2005). Evaluation of acupuncture in the treatment of Parkinson's disease: A double-blind pilot study. *Movement Disorders*, *20*(9), 1185–1188. http://dx.doi.org/10.1002/mds.20503.

Cruise, K. E., Bucks, R. S., Loftus, A. M., Newton, R. U., Pegoraro, R., & Thomas, M. G. (2011). Exercise and Parkinson's: Benefits for cognition and quality of life. *Acta Neurologica Scandinavica*, *123*(1), 13–19. http://dx.doi.org/10.1111/j.1600-0404.2010.01338.x.

Cugusi, L., Solla, P., Serpe, R., Carzedda, T., Piras, L., Oggianu, M., et al. (2015). Effects of a Nordic walking program on motor and non-motor symptoms, functional performance and body composition in patients with Parkinson's disease. *NeuroRehabilitation*, *37*(2), 245–254. http://dx.doi.org/10.3233/NRE-151257.

Dashtipour, K., Johnson, E., Kani, C., Kani, K., Hadi, E., Ghamsary, M., et al. (2015). Effect of exercise on motor and nonmotor symptoms of Parkinson's disease. *Parkinsons Disease*, *2015*, 586378. http://dx.doi.org/10.1155/2015/586378.

David, F. J., Robichaud, J. A., Leurgans, S. E., Poon, C., Kohrt, W. M., Goldman, J. G., et al. (2015). Exercise improves cognition in Parkinson's disease: The PRET-PD randomized, clinical trial. *Movement Disorders*, *30*(12), 1657–1663. http://dx.doi.org/10.1002/mds.26291.

Dibble, L. E., Hale, T. F., Marcus, R. L., Gerber, J. P., & LaStayo, P. C. (2009). High intensity eccentric resistance training decreases bradykinesia and improves quality of life in persons with Parkinson's disease: A preliminary study. *Parkinsonism & Related Disorders*, *15*(10), 752–757. http://dx.doi.org/10.1016/j.parkreldis.2009.04.009.

Donoyama, N., & Ohkoshi, N. (2012). Effects of traditional Japanese massage therapy on various symptoms in patients with Parkinson's disease: A case-series study. *Journal of Alternative and Complementary Medicine*, *18*(3), 294–299. http://dx.doi.org/10.1089/acm.2011.0148.

Donoyama, N., Suoh, S., & Ohkoshi, N. (2014). Effectiveness of Anma massage therapy in alleviating physical symptoms in outpatients with Parkinson's disease: A before-after study. *Complementary Therapies in Clinical Practice*, *20*(4), 251–261. http://dx.doi.org/10.1016/j.ctcp.2014.07.010.

Duncan, R. P., & Earhart, G. M. (2014). Are the effects of community-based dance on Parkinson disease severity, balance, and functional mobility reduced with time? A 2-year prospective pilot study. *Journal of Alternative and Complementary Medicine*, *20*(10), 757–763. http://dx.doi.org/10.1089/acm.2012.0774.

Elefant, C., Baker, F. A., Lotan, M., Lagesen, S. K., & Skeie, G. O. (2012). The effect of group music therapy on mood, speech, and singing in individuals with Parkinson's disease—A feasibility study. *Journal of Music Therapy*, *49*(3), 278–302.

Eng, M. L., Lyons, K. E., Greene, M. S., & Pahwa, R. (2006). Open-label trial regarding the use of acupuncture and yin tui na in Parkinson's disease outpatients: A pilot study on efficacy, tolerability, and quality of life. *Journal of Alternative and Complementary Medicine*, *12*(4), 395–399. http://dx.doi.org/10.1089/acm.2006.12.395.

Frazzitta, G., Maestri, R., Ferrazzoli, D., Riboldazzi, G., Bera, R., Fontanesi, C., et al. (2015). Multidisciplinary intensive rehabilitation treatment improves sleep quality in Parkinson's disease. *Journal of Clinical Movement Disorders*, *2*, 11. http://dx.doi.org/10.1186/s40734-015-0020-9.

Hackney, M. E., & Earhart, G. M. (2009a). Health-related quality of life and alternative forms of exercise in Parkinson disease. *Parkinsonism & Related Disorders*, *15*(9), 644–648. http://dx.doi.org/10.1016/j.parkreldis.2009.03.003.

Hackney, M. E., & Earhart, G. M. (2009b). Short duration, intensive tango dancing for Parkinson disease: An uncontrolled pilot study. *Complementary Therapies in Medicine*, *17*(4), 203–207. http://dx.doi.org/10.1016/j.ctim.2008.10.005.

Hashimoto, H., Takabatake, S., Miyaguchi, H., Nakanishi, H., & Naitou, Y. (2015). Effects of dance on motor functions, cognitive functions, and mental symptoms of Parkinson's disease: A quasi-randomized pilot trial. *Complementary Therapies in Medicine*, *23*(2), 210–219. http://dx.doi.org/10.1016/j.ctim.2015.01.010.

Herz, N. B., Mehta, S. H., Sethi, K. D., Jackson, P., Hall, P., & Morgan, J. C. (2013). Nintendo Wii rehabilitation ("Wii-hab") provides benefits in Parkinson's disease. *Parkinsonism & Related Disorders*, *19*(11), 1039–1042. http://dx.doi.org/10.1016/j.parkreldis.2013.07.014.

Johns, C., Blake, D., & Sinclair, A. (2010). Can reflexology maintain or improve the well-being of people with Parkinson's disease? *Complementary Therapies in Clinical Practice*, *16*(2), 96–100. http://dx.doi.org/10.1016/j.ctcp.2009.10.003.

Kim, J. Y., & Jeon, B. S. (2012). Complementary and alternative medicine in Parkinson's disease patients in Korea. *Current Neurology and Neuroscience Reports*, *12*(6), 631–632. http://dx.doi.org/10.1007/s11910-012-0309-0.

King, L. A., Wilhelm, J., Chen, Y., Blehm, R., Nutt, J., Chen, Z., et al. (2015). Effects of group, individual, and home exercise in persons with Parkinson disease: A randomized clinical trial. *Journal of Neurologic Physical Therapy*, *39*(4), 204–212. http://dx.doi.org/10.1097/NPT.0000000000000101.

Kluger, B. M., Rakowski, D., Christian, M., Cedar, D., Wong, B., Crawford, J., et al. (2016). Randomized, controlled trial of acupuncture for fatigue in Parkinson's disease. *Movement Disorders*, *31*(7), 1027–1032. http://dx.doi.org/10.1002/mds.26597.

Kurt, E. E., Buyukturan, B., Buyukturan, O., Erdem, H. R., & Tuncay, F. (2017). Effects of Ai Chi on balance, quality of life, functional mobility, and motor impairment in patients with Parkinson's disease. *Disability and Rehabilitation*, 1–10. http://dx.doi.org/10.1080/09638288.2016.1276972.

Li, F., Harmer, P., Liu, Y., Eckstrom, E., Fitzgerald, K., Stock, R., et al. (2014). A randomized controlled trial of patient-reported outcomes with Tai Chi exercise in Parkinson's disease. *Movement Disorders*, *29*(4), 539–545. http://dx.doi.org/10.1002/mds.25787.

McKee, K. E., & Hackney, M. E. (2013). The effects of adapted tango on spatial cognition and disease severity in Parkinson's disease. *Journal of Motor Behavior*, *45*(6), 519–529. http://dx.doi.org/10.1080/00222895.2013.834288.

Miyai, I., Fujimoto, Y., Ueda, Y., Yamamoto, H., Nozaki, S., Saito, T., et al. (2000). Treadmill training with body weight support: Its effect on Parkinson's disease. *Archives of Physical Medicine and Rehabilitation*, *81*(7), 849–852. http://dx.doi.org/10.1053/apmr.2000.4439.

Modugno, N., Iaconelli, S., Fiorlli, M., Lena, F., Kusch, I., & Mirabella, G. (2010). Active theater as a complementary therapy for Parkinson's disease rehabilitation: A pilot study. *ScientificWorldJournal*, *10*, 2301–2313. http://dx.doi.org/10.1100/tsw.2010.221.

Morberg, B. M., Jensen, J., Bode, M., & Wermuth, L. (2014). The impact of high intensity physical training on motor and non-motor symptoms in patients with Parkinson's disease (PIP): A preliminary study. *NeuroRehabilitation*, *35*(2), 291–298. http://dx.doi.org/10.3233/NRE-141119.

Murphy, S. M., Rogers, A., Hutchinson, M., & Tubridy, N. (2008). Counting the cost of complementary and alternative therapies in an Irish neurological clinic. *European Journal of Neurology*, *15*(12), 1380–1383. http://dx.doi.org/10.1111/j.1468-1331.2008.02320.x.

Nadeau, A., Pourcher, E., & Corbeil, P. (2014). Effects of 24 week of treadmill training on gait performance in Parkinson's disease. *Medicine and Science in Sports and Exercise*, *46*(4), 645–655. http://dx.doi.org/10.1249/MSS.0000000000000144.

Nascimento, C. M., Ayan, C., Cancela, J. M., Gobbi, L. T., Gobbi, S., & Stella, F. (2014). Effect of a multimodal exercise program on sleep disturbances and instrumental activities of daily living performance on Parkinson's and Alzheimer's disease patients. *Geriatrics & Gerontology International*, *14*(2), 259–266. http://dx.doi.org/10.1111/ggi.12082.

Ni, M., Signorile, J. F., Mooney, K., Balachandran, A., Potiaumpai, M., Luca, C., et al. (2016). Comparative effect of power training and high-speed yoga on motor function in older patients with Parkinson disease. *Archives of Physical Medicine and Rehabilitation*, *97*(3), 345–354, e315. http://dx.doi.org/10.1016/j.apmr.2015.10.095.

Nocera, J. R., Amano, S., Vallabhajosula, S., & Hass, C. J. (2013). Tai Chi exercise to improve non-motor symptoms of Parkinson's disease. *Journal of Yoga & Physical Therapy*, *3*. http://dx.doi.org/10.4172/2157-7595.1000137.

Pacchetti, C., Mancini, F., Aglieri, R., Fundaro, C., Martignoni, E., & Nappi, G. (2000). Active music therapy in Parkinson's disease: An integrative method for motor and emotional rehabilitation. *Psychosomatic Medicine*, *62*(3), 386–393.

Pandit, A. K., Vibha, D., Srivastava, A. K., Shukla, G., Goyal, V., & Behari, M. (2016). Complementary and alternative medicine in Indian Parkinson's disease patients. *Journal of Traditional and Complementary Medicine*, *6*(4), 377–382. http://dx.doi.org/10.1016/j.jtcme.2015.03.009.

Pantelyat, A., Syres, C., Reichwein, S., & Willis, A. (2016). DRUM-PD: The use of a drum circle to improve the symptoms and signs of Parkinson's disease (PD). *Movement Disorders—Clinical Practice*, *3*(3), 243–249. http://dx.doi.org/10.1002/mdc3.12269.

Park, A., Zid, D., Russell, J., Malone, A., Rendon, A., Wehr, A., et al. (2014). Effects of a formal exercise program on Parkinson's disease: A pilot study using a delayed start design. *Parkinsonism & Related Disorders*, *20*(1), 106–111. http://dx.doi.org/10.1016/j.parkreldis.2013.10.003.

Pecci, C., Rivas, M. J., Moretti, C. M., Raina, G., Ramirez, C. Z., Diaz, S., et al. (2010). Use of complementary and alternative therapies in outpatients with Parkinson's disease in Argentina. *Movement Disorders*, *25*(13), 2094–2098. http://dx.doi.org/10.1002/mds.23235.

Picelli, A., Varalta, V., Melotti, C., Zatezalo, V., Fonte, C., Amato, S., et al. (2016). Effects of treadmill training on cognitive and motor features of patients with mild to moderate Parkinson's disease: A pilot, single-blind, randomized controlled trial. *Functional Neurology*, *31*(1), 25–31.

Pohl, P., Dizdar, N., & Hallert, E. (2013). The Ronnie Gardiner rhythm and music method—A feasibility study in Parkinson's disease. *Disability and Rehabilitation*, *35*(26), 2197–2204. http://dx.doi.org/10.3109/09638288.2013.774060.

Rajendran, P. R., Thompson, R. E., & Reich, S. G. (2001). The use of alternative therapies by patients with Parkinson's disease. *Neurology*, *57*(5), 790–794.

Reuter, I., Mehnert, S., Leone, P., Kaps, M., Oechsner, M., & Engelhardt, M. (2011). Effects of a flexibility and relaxation programme, walking, and nordic walking on Parkinson's disease. *Journal of Aging Research*, *2011*, 232473. http://dx.doi.org/10.4061/2011/232473.

Ridgel, A. L., Kim, C. H., Fickes, E. J., Muller, M. D., & Alberts, J. L. (2011). Changes in executive function after acute bouts of passive cycling in Parkinson's disease. *Journal of Aging and Physical Activity*, *19*(2), 87–98.

Rios Romenets, S., Anang, J., Fereshtehnejad, S. M., Pelletier, A., & Postuma, R. (2015). Tango for treatment of motor and non-motor manifestations in Parkinson's disease: A randomized control study. *Complementary Therapies in Medicine*, *23*(2), 175–184. http://dx.doi.org/10.1016/j.ctim.2015.01.015.

Rodrigues de Paula, F., Teixeira-Salmela, L. F., Coelho de Morais Faria, C. D., Rocha de Brito, P., & Cardoso, F. (2006). Impact of an exercise program on physical, emotional, and social aspects of quality of life of individuals with Parkinson's disease. *Movement Disorders*, *21*(8), 1073–1077. http://dx.doi.org/10.1002/mds.20763.

Schenkman, M., Hall, D. A., Baron, A. E., Schwartz, R. S., Mettler, P., & Kohrt, W. M. (2012). Exercise for people in early- or mid-stage Parkinson disease: A 16-month randomized controlled trial. *Physical Therapy*, *92*(11), 1395–1410. http://dx.doi.org/10.2522/ptj.20110472.

Schmitz-Hubsch, T., Pyfer, D., Kielwein, K., Fimmers, R., Klockgether, T., & Wullner, U. (2006). Qigong exercise for the symptoms of Parkinson's disease: A randomized, controlled pilot study. *Movement Disorders*, *21*(4), 543–548. http://dx.doi.org/10.1002/mds.20705.

Sharma, N. K., Robbins, K., Wagner, K., & Colgrove, Y. M. (2015). A randomized controlled pilot study of the therapeutic effects of yoga in people with Parkinson's disease. *International Journal of Yoga*, *8*(1), 74–79. http://dx.doi.org/10.4103/0973-6131.146070.

Shulman, L. M., Katzel, L. I., Ivey, F. M., Sorkin, J. D., Favors, K., Anderson, K. E., et al. (2013). Randomized clinical trial of 3 types of physical exercise for patients with Parkinson disease. *JAMA Neurology*, *70*(2), 183–190. http://dx.doi.org/10.1001/jamaneurol.2013.646.

Shulman, L. M., Wen, X., Weiner, W. J., Bateman, D., Minagar, A., Duncan, R., et al. (2002). Acupuncture therapy for the symptoms of Parkinson's disease. *Movement Disorders*, *17*(4), 799–802. http://dx.doi.org/10.1002/mds.10134.

Stallibrass, C., Sissons, P., & Chalmers, C. (2002). Randomized controlled trial of the Alexander technique for idiopathic Parkinson's disease. *Clinical Rehabilitation*, *16*(7), 695–708.

Sturkenboom, I. H., Graff, M. J., Hendriks, J. C., Veenhuizen, Y., Munneke, M., Bloem, B. R., et al. (2014). Efficacy of occupational therapy for patients with Parkinson's disease: A randomised controlled trial. *Lancet Neurology*, *13*(6), 557–566. http://dx.doi.org/10.1016/S1474-4422(14)70055-9.

Tan, L. C., Lau, P. N., Jamora, R. D., & Chan, E. S. (2006). Use of complementary therapies in patients with Parkinson's disease in Singapore. *Movement Disorders*, *21*(1), 86–89. http://dx.doi.org/10.1002/mds.20662.

Teixeira-Machado, L., Araujo, F. M., Cunha, F. A., Menezes, M., Menezes, T., & Melo DeSantana, J. (2015). Feldenkrais method-based exercise improves quality of life in individuals with Parkinson's disease: A controlled, randomized clinical trial. *Alternative Therapies in Health and Medicine*, *21*(1), 8–14.

Tomlinson, C. L., Herd, C. P., Clarke, C. E., Meek, C., Patel, S., Stowe, R., et al. (2014). Physiotherapy for Parkinson's disease: A comparison of techniques. *Cochrane Database of Systematic Reviews*, *6*, CD002815. http://dx.doi.org/10.1002/14651858.CD002815.pub2.

Uc, E. Y., Doerschug, K. C., Magnotta, V., Dawson, J. D., Thomsen, T. R., Kline, J. N., et al. (2014). Phase I/II randomized trial of aerobic exercise in Parkinson disease in a community setting. *Neurology*, *83*(5), 413–425. http://dx.doi.org/10.1212/WNL.0000000000000644.

van Nimwegen, M., Speelman, A. D., Hofman-van Rossum, E. J., Overeem, S., Deeg, D. J., Borm, G. F., et al. (2011). Physical inactivity in Parkinson's disease. *Journal of Neurology*, *258*(12), 2214–2221. http://dx.doi.org/10.1007/s00415-011-6097-7.

van Nimwegen, M., Speelman, A. D., Overeem, S., van de Warrenburg, B. P., Smulders, K., Dontje, M. L., et al. (2013). Promotion of physical activity and fitness in sedentary patients with Parkinson's disease: Randomised controlled trial. *BMJ*, *346*, f576. http://dx.doi.org/10.1136/bmj.f576.

Volpe, D., Signorini, M., Marchetto, A., Lynch, T., & Morris, M. E. (2013). A comparison of Irish set dancing and exercises for people with Parkinson's disease: A phase II feasibility study. *BMC Geriatrics*, *13*, 54. http://dx.doi.org/10.1186/1471-2318-13-54.

Wassom, D. J., Lyons, K. E., Pahwa, R., & Liu, W. (2015). Qigong exercise may improve sleep quality and gait performance in Parkinson's disease: A pilot study. *The International Journal of Neuroscience*, *125*(8), 578–584. http://dx.doi.org/10.3109/00207454.2014.966820.

Winward, C., Sackley, C., Meek, C., Izadi, H., Barker, K., Wade, D., et al. (2012). Weekly exercise does not improve fatigue levels in Parkinson's disease. *Movement Disorders*, *27*(1), 143–146. http://dx.doi.org/10.1002/mds.23966.

Xiao, C. M., & Zhuang, Y. C. (2016). Effect of health Baduanjin Qigong for mild to moderate Parkinson's disease. *Geriatrics & Gerontology International*, *16*(8), 911–919. http://dx.doi.org/10.1111/ggi.12571.

Zhang, G., Xiong, N., Zhang, Z., Liu, L., Huang, J., Yang, J., et al. (2015). Effectiveness of traditional Chinese medicine as an adjunct therapy for Parkinson's disease: A systematic review and meta-analysis. *PLoS One*, *10*(3), e0118498. http://dx.doi.org/10.1371/journal.pone.0118498.

CHAPTER FORTY-TWO

Speech, Voice, and Communication

Julia A. Johnson[1]
Kings College Hospital NHS Foundation Trust, London, United Kingdom
[1]Corresponding author: e-mail address: juliajohnson@nhs.net

Contents

Abstract

Communication changes are an important feature of Parkinson's and include both motor and nonmotor features. This chapter will cover briefly the motor features affecting speech production and voice function before focusing on the nonmotor aspects. A description of the difficulties experienced by people with Parkinson's when trying to communicate effectively is presented along with some of the assessment tools and therapists' treatment options. The idea of clinical heterogeneity of PD and subtyping patients with different communication problems is explored and suggestions are made on how this may influence clinicians' treatment methods and choices so as to provide personalized therapy programmes. The importance of encouraging and supporting people to maintain social networks, employment, and leisure activities is stated as the key to achieving sustainability. Finally looking into the future, the emergence of new technologies is seen as providing further possibilities to support therapists in the goal of helping people with Parkinson's to maintain good communication skills throughout the course of the disease.

International Review of Neurobiology, Volume 134
ISSN 0074-7742
http://dx.doi.org/10.1016/bs.irn.2017.04.014

1. INTRODUCTION

James Parkinson described speech and voice characteristics in his 1817 essay on the shaking palsy (Parkinson, 1817) stating that for one patient: "It was with difficulty he uttered a few words" and then showing progression of symptoms by explaining that after three and a half years and a fall the same patient had deteriorated to a level where: "What words he still could utter were monosyllables, and these came out, after much struggle, in a violent expiration, and with such a low voice and indistinct articulation, as hardly to be understood but by those who were constantly with him. He fetched his breath rather hard;" (Parkinson, 2002, p. 231). These descriptions from 200 years ago of what Parkinson called a "malady" still have resonance for our observations today. This chapter will explore the voice and speech difficulties, experienced by people with Parkinson's, focusing in particular on communication skills and how they relate to other nonmotor symptoms.

2. CLASSICAL DESCRIPTION

Between 70% and 90% of people with Parkinson's have some form of communication difficulty either self-reported or laboratory tested (Ho, Iansek, Marigliani, et al., 1998; Logeman, Fisher, Boshes, & Blonsky, 1978). The perceptual difficulties usually start at the level of laryngeal function with voice changes.

The classical Parkinson's speech deficit is referred to as a "hypokinetic dysarthria." Darley, Aronson, and Brown (1969) defined dysarthria as "any combination of disorders of respiration, phonation, resonance, articulation or prosody that may cause muscle slowness, weakness, inco-ordination or alteration in muscle tone." The most commonly reported perceptual features of hypokinetic dysarthria described by Darley et al. (1969) in their clusters of deviant speech dimensions are

1. Reduced loudness
2. Monoloudness and monopitch
3. Breathiness and harshness
4. Reduced prosody
5. Imprecise articulation

These features are put into context alongside all the dysarthrias in their seminal text on Motor Speech Disorders first published in 1975 (Darley, Aronson, & Brown, 1975). Their tenet was that the dysarthrias can be distinguished largely by how the speech sounds. The hypokinetic dysarthria is related to the core motor features of rigidity, bradykinesia, and tremor. Hypo = less and kinetic = movement. Muscular rigidity can explain manifestation such as the blank facial expression, monotony of pitch and loudness, reduced volume, and imprecise articulation. Bradykinesia explains the slowness in responses due to a difficulty initiating movements. Tremor can be present in the voice, tongue, jaw, or lips. Gillivan-Murphy, Carding, and Miller (2016) found that although tremor may be heard in the voice it was likely to be related to oscillatory movement in structures across the vocal tract rather than just the vocal folds. Direct Examination of the larynx using techniques such as fibreoptic nasendoscopy and stroboscopy show that the vocal cords lose their fine flexibility. This impacts on pitch range and control. Vocal cord adduction also becomes weaker, sometimes visualized as bowing of the vocal cords or a glottal chink, which results in greater escape of air during phonation and breathy or hoarse voice quality. At the articulatory level, speech sounds may become imprecise and this, along with voice changes, contributes to lower speech intelligibility.

The fact that these classical features of dysarthria may be caused by the motor aspects of Parkinson's is supported by descriptions of deep brain stimulation (DBS). Changes in the location of the electrodes along with bipolar directional steering and amplitude of settings can influence the severity of the dysarthria postsurgery. Careful placement of the electrodes and settings for the stimulation is therefore required to optimize speech clarity and intelligibility (Little et al., 2016; Reker, Dembek, Becker, Visser-Vandewalle, & Timmermann, 2016).

3. COMMUNICATION

The nonmotor aspects of speech and voice are better explored under the broader heading of "communication." Communication difficulties may emerge during the premotor phase manifesting years before the classic motor symptoms. Voice changes have been explored by some investigators as a possible diagnostic symptom during the prodromal phase. The changes in communication are insidious in onset and not confined to the mid or later stages of the disease as once thought.

Communication effectiveness can be impaired on several levels. A reduction in facial muscle activity results in the "mask like" facial expression, which gives an impression of flat affect, despite an apparently normal experience of emotion. This is likely to contribute to the perception of depression, anxiety, and detachment. Communication partners sensing this may withdraw, leading to communication breakdown and isolation for the person with Parkinson's. The mask like face has also been shown to impact on the impressions of healthcare professionals who may misjudge a person's mood and willingness to engage (Pentland, Pitkairn, & Gray, 1987). This could therefore influence the success of relationships. Some people with Parkinson's, in addition to having reduced facial expression may fail to recognize the nuances of their listener's expressions, which is reported as impaired emotional recognition (Robert et al., 2014). This contributes to a reluctance to engage in social intercourse and increased isolation. In some cases, eye contact may be disrupted with a reduced eye blink reflex or eyes fixed closed for several seconds. Gestural movements, such as head nodding and shaking may also be reduced.

Even when patients are able to produce a normal vocal loudness and speech, they may not use it in conversation. Many patients fear they are shouting if they use what a listener would perceive as normal loudness levels. Patients may be able to shout out clearly and loudly in a sudden emergency situation, but then return to mumbled inaudible speech when in general conversation. Patients often accuse their partners, or even their pets, of being deaf rather than acknowledging that it is their own vocal loudness, which is reduced. Ho, Bradshaw, and Iansek (2000) showed how patients overestimated the volume of their speech during both reading and conversation when compared to normal controls. Their study contrasting the volume level of speech production with perceived volume provided support for the view that people with Parkinson's have "sensory anomalies in the perception of speech volume and in the sensorimotor integration involved in appropriately scaling speech volume." Ho, Bradshaw, and Inset (1999) found that patients were able to regulate their speech volume to some extent in response to the distance of their listener, but not so efficiently in response to background noise such as when speaking at a cocktail party. Control subjects were shown to easily demonstrate the Lombard effect, automatically speaking louder when competing against increasing levels of background noise, but subjects with Parkinson's were less able to respond appropriately to these implicit cues. In clinical practice when working on vocal loudness, patients usually manage to recruit adequate volumes in automatic sequences such as counting and reading familiar phrases and sentences. However, they complain that it is harder to achieve good volume in conversational tasks when they need to think about "what" they are saying rather than "how" they are saying it. This is supported by Mobes et al.'s study

(Mobes, Joppich, Stiebritz, Dengler, & Schroder, 2008) that examined emotional speech in Parkinson's. This study, suggested that alterations of emotional processing contribute to the speech changes (such as monotony), and not solely the underlying motor impairment. Parkinson's speech has been described as a mumbling speech with trailing off near the end of a sentence.

Communication changes impact upon an individual and their social life even when there is no discernible neurological dysarthria or change in a person's intelligibility of speech. In Miller et al.'s qualitative study (Miller, Noble, Jones, & Burn, 2006) some key themes emerged, which support a broader impact on communication than can be explained by core motor features. Themes that emerged from their in-depth interviews highlight the "demoralising effect of losing track mid-sentence, the impact on self-perception and family dynamics as well as the sheer physical and mental effort involved in staying in the conversation." 82% of patients in Miller's study on intelligibility changes (Miller et al., 2007) reported being unhappy with how they spoke.

There has been a recent upsurge in the importance of social engagement on health outcomes. Those with communication difficulties are at more risk of developing social isolation. Studies have shown that cognitive function declines rapidly when communication with others is not regularly maintained and Andrew, Mitnitski, Kirkland, and Rockwood (2012) found that absolute mortality risk rose with increasing social vulnerability.

It is also important to take into account the physical and sensory declines that accompany normal aging and comorbidities as the population living with Parkinson's increases with age (Hindle, 2010; Huber, Darling, Francis, & Zhang, 2012). There is a risk that communication difficulties may lead to poorer healthcare access and utilization, isolation, depression, reduced independence, and increased caregiver burden.

4. LANGUAGE

As part of communication, language may be simply described as the act of finding the appropriate words in a timely manner and combining these into coherent sentences (semantics and syntax). Language represents cortical and subcortical network activity. Language production and comprehension deficits have been noted in Parkinson's patients and this is an area being actively investigated and is the subject of an ever increasing number of publications. Parkinson's patients may use shorter, less complex sentences and produce more grammatically wrong or incomplete utterances compared with matched controls (Walsh & Smith, 2011). Difficulties with word retrieval have been reported to be worse for verbs than for nouns (Salmazo-Silva et al., 2017). There is discussion in the literature as to whether the language changes in Parkinson's represent true language impairments, or impairments in other cognitive functions such as attention and memory. They may even be an element of adaptation for the motor speech difficulties.

5. SUBTYPES

Parkinson's essay illustrated a series of cases under the heading of "shaking palsy." There were individual variations, but the similarities in the presentations allowed all patients to be placed under one common diagnostic label. The idea is now being proposed that Parkinson's is a heterogeneous syndrome rather than a specific disease (Titova, Padmakumar, Lewis, & Chaudhuri, 2016). This is supported by the wide range of communication difficulties with which individual patients present. Looking at a typical clinical cohort, it may be helpful to place patients into subtypes with a similar collection of signs and symptoms. Some suggested speech subtypes may be

1. Festination of speech

   This group would cover the patients presenting with oral festination known as palilalia. They exhibit short rushes of speech leading to a dysfluency. In some patients, this festination or urgency of speech may be heard as a real stammer/stutter. Although this stammer can emerge as the disease progresses on closer questioning many patients describe previous dysfluency in childhood with either a continuity with Parkinson's or a reemergence and exacerbation of symptoms (Shahed & Jankovic, 2001). This may be more common following deep brain stimulation of the subthalmic nucleus (DBS-STN). A possible link between festination and tremor has been suggested, but no correlation was found in a study by Moreau, Ozsancak, Blatt, Derambure, et al. (2007) who instead found a correlation with freezing of gait.

2. Language

   This group have reduced grammatical complexity in speech and greater problems with lexical retrieval.

3. Anxiety

   Elevated levels of anxiety would be a specific symptom for this group. While anxiety is a normal emotion, anxiety problems can cause people serious psychological distress and affect their physical health. This group of patients may be continually seeking out reassurance and new therapies. These patients typically become very dependent on their medical and therapy team. They may find it difficult to take on self-management advice and recommendations. While this is not a specific speech or language presentation, it does impact on the patient's management and their therapeutic expectations.

4. Depression

   Patients with significant depression may find it difficult to engage with the therapeutic process due to feelings of hopelessness and despair. This group often have even greater reductions in vocal loudness.

5. Cognitive Impairment

   This group would be distinguished from those with specific language disturbances by more generalized cognitive impairment and likely memory disturbances. Decreased cognitive functioning is explored in depth in the chapter "Cognition in PD" written by Lewis (Ch ID 00043). For some this phenomenon may be part of a global process of dementia. Other patients may experience a particular pattern of deficits, which are specific to Parkinson's disease. These deficits may be very subtle and sometimes undetectable in a straightforward conversation. They include slow thinking and slow learning, problems shifting cognitive sets, problems of internal cueing, and procedural memory.

   Subtyping gives some clarity to understanding individual presentation however of course no subtype is necessarily exclusive. Adjusting therapy programmes to suit the different presentations will be discussed in Section 7.

6. ASSESSMENTS

Baseline assessment is a core requirement to monitor patient performance and provide targeted treatment programmes. In patients whose symptoms change depending on where they are in their medication cycle, the assessments should be carried out in both a "drug on" and "drug off" state. The best tool for this is arguably video recordings taken on diagnosis and at regular 6–12 monthly intervals. Patients themselves are in the best position to store these recordings in this modern era of mobile technology. Many healthcare providers struggle with the requirements for storage of this data and also with its collection over the projected lifespan of an ever expanding group of people with communication impairments, particularly since those with Parkinson's will hopefully live for many years. Video recordings of storytelling or picture descriptions can be used to capture aspects of language function such as sentence length and grammatical structure as well as motor and nonmotor speech performance. Videos of conversation with a partner will contribute even richer data for analysis.

Assessments such as the Robertson dysarthria Profile (Defazio, Guerrieri, Liuzzi, Gigante, & di Nicola, 2016) and Frenchay dysarthria assessment (Enderby & Palmer, 2008) provide a useful analysis of the total impairments in speech production including both subjective and objective measures. These can be supplemented by specific acoustic measures for vocal loudness and pitch range. Simple hand held devices and apps are available for easy use in a clinical setting such as the "Speech Tool" iTunes app (Nolan et al., 2012), which provides average volume ratings through a tab marked "How Loud Am I."

Assessment of Intelligibility needs to be included because maintenance of speech intelligibility is vital for successful spoken communication. Tests such as the Sentence Intelligibility Test section of the Assessment of Intelligibility of Dysarthric Speech (Yorkston, Beukelman, & Hakel, 1996) are recommended. However, these single word and sentence intelligibility tests do not capture the variability of Parkinson's speech. When asked to read words and sentences aloud in this type of testing situation, patients tend to recruit their best performance, which is often not indicative of their intelligibility in conversational speech. Less formal rating scales based on recordings taken during activities such as reading aloud, storytelling, conversation, or picture description are often used in preference in clinical practice.

The speech evaluation parameter contained in the Unified Parkinson's Disease Rating Scale part III (UPDRS-III) (Goetz, Tilley, et al., 2008) can be used as a brief measure during medical examination. This gives a rating from 0, indicating normal speech, to 4, indicating severe unintelligible speech. The nonmotor questionnaire (NMSQuest) (Chaudhuri, Martinez Martin, Schapira, et al., 2006) also contains questions such as "Difficulty concentrating or staying focussed," which may indicate communication changes requiring further investigation.

Nonmotor aspects of speech and voice are further examined using tools such as the Voice Handicap Index (Jacobson, Johnson, Grywalski, et al., 1997) or the shortened 10 item version, which can be quickly administered in clinic (Rosen, Lee, Osborne, Zullo, & Murry, 2004) and includes questions such as "People have difficulty understanding me in a noisy room" followed by a scale from 0 to 4 (0, never; 1, almost never; 2, sometimes; 3, almost always; 4, always) to capture the patients' views about their voice.

Another tool, which was specifically developed to assess the psychological and social consequences of dysarthria is The Dysarthria Impact Profile (DIP) (Walshe, Peach, & Miller, 2009). It has a total of 48 statements covering four specific areas of inquiry selected for use in the profile following in depth interviews with patients:

1. Section A: The effect of dysarthria on the speaker as a person, which includes statements such as "Because of my speech I am more dependent on people now than I was before"
2. Section B: Acceptance of dysarthria with statements such as "I would rather miss out on conversation than admit to having a speech problem"
3. Section C: Perception of others' reaction to speech, i.e., "People treat me as if I am stupid because they can't understand me"
4. Section D: How dysarthria affects communication with others, i.e., "Although I have difficulty speaking I do not avoid communicating with people I know."

Some statements are negatively worded and some statements are positively worded. The DIP was designed to be completed by the person with dysarthria alone or with assistance. A summated 5-point rating scale is used ranging from "strongly agree" to "strongly disagree." A useful section at the end looks at the dysarthria relative to other worries and concerns the patient may have. In this section, the patient is asked to list four things, apart from speech, that upsets or worries them. Speech is then added to this list and the patient is asked to put the five things in order with what they are most concerned about at the top and what they are least concerned about at the bottom. This helps the multidisciplinary team to prioritize interventions for the patient.

It is important to capture the level of communication participation. The Social Network Inventory (Blackstone & Berg, 2012) provides a useful framework for this to allow documentation of the numbers of contacts patients have on diagnosis within the different categories of life partners, close relatives and good friends, acquaintances, paid workers, and unfamiliar partners. This inventory can be monitored to see how networks may be reducing or changing over time.

The Dysarthria-in-Interaction Profile (Bloch & Tuomainen, 2017) is a new tool, under development which can be used to identify different features of interaction between people with dysarthria and their individual conversation partners (dyads) during normal day to day dialogue. This profile provides a mechanism for describing interactions which take place, analyzing how an interaction may break down and then providing a framework to record and assist optimizing interactions between specific dyads.

A comprehensive assessment of communication includes health-related quality of life, psychological well-being, self-efficacy for the patients, and carers plus carer stress, which are discussed in other chapters.

Finally, getting an objective measure of laryngeal activity by an Ear Nose and Throat specialist to examine the vocal fold structure and movement using instrumentation such as fibreoptic laryngoscopy or stroboscopy is important so as to guide treatment and exclude any accompanying laryngeal pathology.

7. TREATMENT

Neuropharmacological and neurosurgical treatments are reported as having a variable and unpredictable impact on improving speech production (Pinto et al., 2004). The Cochrane Review (Herd et al., 2012) stated "There is no consensus amongst therapists on which SLT method to use or whether it should be a combination of methods." Speech and Language Therapy interventions are therefore selected by most clinicians on an individual patient basis including techniques such as vocal training, speech control training, breathing control training, auditory feedback alteration therapy, articulation therapy, or singing training. These programmes rely largely on the muscle and brain training that comes with repetition of activity. The general agreement is that the patient needs to be taught to put in more effort when speaking in order to ensure they are fully heard and understood by their listeners.

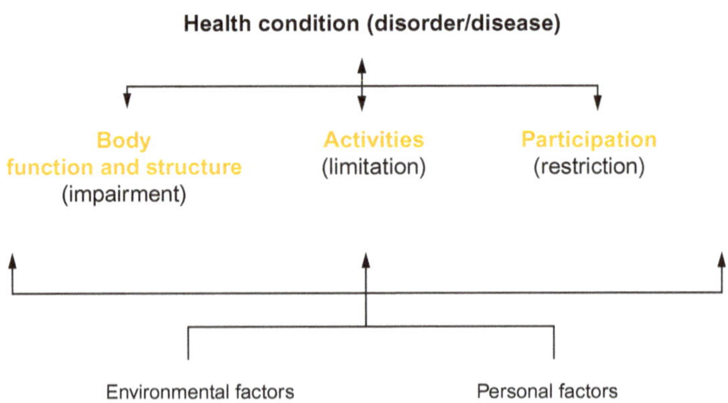

Fig. 1 The international classification of functioning disability and health (ICF) (W.H.O. 2001).

The International Classification of Functioning Disability (ICF) and Health (W.H.O. 2001) provides a framework for working with dysarthria. The ICF Model in Fig. 1 is taken from the World Health Organisation publications material: http://www.who.int/classifications/drafticfpracticalmanual2.pdf.

Using this framework, "impairment"-based treatment can focus on the core speech production parameters of respiration, phonation, velopharyngeal function, and articulation. "Activity" treatment works on any limitations with a focus on intelligibility, rate, and naturalness. "Participation" work encourages improved speech when taking part in general activities of daily living such as shopping, dealing with bank, negotiating contacts with telephone companies, employment, maintaining, and forging relationships.

The treatment programme with the largest evidence base is Lee Silverman Voice Treatment (LSVT®) (Ramig et al., 2001). This is an intensive treatment programme given in 16 hourly treatment sessions over 4 weeks. A comprehensive LSVT programme covers all stages of the ICF focusing specific exercises on the impairment of reduced vocal loudness and working with patients to achieve natural sounding, clearly understood speech. The modeled voice/speech is then incorporated into a hierarchy of individualized activities. During the phase referred to as "calibration," the patient is encouraged to feel comfortable using an appropriate clear voice with normal loudness in all activities of daily living, thus working on "participation."

The plan of developing more personalized therapies and individualized medicine to deal with the heterogeneity of Parkinson's suggested by Titova et al. (2016) is also true for speech therapy. The specific programmes such as LSVT may need to be adapted to meet the needs of some of the patient subgroups and symptoms. More extensive rate control strategies are advised for those with palilalia. For example, encouraging a patient to point to the first letter of each word on an alphabet or keyboard as they say it may help to restrict speed of utterances. Closer attention to articulation with repetition of speech sounds, sequences of sounds (i.e., "pataka"), and multisyllabic words may be required for patients who have undergone DBS treatment. Patients with a specific language impairment may benefit from conversation analytic approach with therapy undertaken with key partners to work on pragmatics and strategies to optimize repair in conversation (Griffiths, Barnes, Britten, & Wilkinson, 2011; Saldert, Ferm, & Bloch, 2014). Those with cognitive impairment may need additional focus on cognitive training programmes alongside the voice/speech-based activities.

Exercise programmes need to be adjusted and simplified to allow for cognitive difficulties. For example, using retrieval-based memory strategies, which have been shown to train efficient encoding of text information in healthy older adults (Rogalski, Altmann, & Rosenbek, 2014).

Many patients are happy to self-manage with the skills learnt following a programme such as LSVT®, but those with high anxiety may be continually seeking more. Improvements have been reported following specialist rehabilitation in the community with reduced anxiety and improved psychological well-being for carers (Gage, Grainger, Ting, et al., 2014). The improvements were sustained for longer when trained care assistants were employed to provide continuing support. Several studies report gains in speech and voice parameters following a set programme of therapy. The challenge is maintaining the gains over the life span for people living with Parkinson's. There is a paucity of long-term studies to guide us and continuity of care is a challenge. Therefore, looking at a person's lifestyle and personality to help them identify their social networks and continued opportunities to focus on optimizing their communication skills over the years should be part of the therapy remit. Multiple chances for practicing speaking are important. The success of SLT will need to be through ensuring and trying to facilitate ongoing opportunities for continued speech practice to enhance communication. This may be through religious activities, dramatic societies, providing loud vocal support while watching or participating in sporting activities, book groups, and poetry readings. To improve speech, people have to SPEAK. There is a natural tendency to talk less when speech is impaired. Disuse then leads to muscle atrophy/deconditioning. There is therefore an element of "USE IT OR LOSE IT." Encouraging story telling is part of sustaining gains made in therapy. Story telling is an important part of how we communicate with each other and helps us to forge relationships.

Some services provide group treatment such as "The LOUD Crowd" either as a follow-up to help maintain improved communication after an intensive one to one programme or as an alternative to individual treatment. Group treatment sessions are a cost effective way of providing maintenance programmes. They provide an opportunity to enhance interaction strategies.

If people enjoy singing this should be encouraged. There is evidence that group singing can be beneficial for people living with Parkinson's (Evans, Canavan, Foy, Langfordd, & Proctor, 2012; Kreutz, Bongard, Rohrmann, et al., 2004). One study showed significant improvements in emotional wellbeing throughout the duration of the programme. Choral singing as well as exercising laryngeal muscles may also help provide a sense of wellbeing thus impacting on other mood-related nonmotor factors (Pacchetti et al., 2000). As mooted by Plato: "Music flatters the senses and calms the soul."

8. INTO THE FUTURE

New technology promises high potential in this area both for gathering outcome data in assessments and supporting treatment programmes. Parkinson's patients have been shown to benefit from use of special smartphone applications. In one multicentre randomized controlled trial carried out in England and Scotland (Lakshminarayana et al., 2017), patients using a health monitoring app to promote self-management, showed a 10% improvement in treatment adherence and significantly improved patients' perception of the quality of their clinical consultation compared to a control group. There were no confounding effects of gender, number of comorbidities or age.

Treatment programmes have been shown to be effective when delivered through a videoconferencing application (Constantinescu et al., 2011) and via Skype. Individual or group speech therapy can be supplemented with technology-based intervention in order to enhance the frequency of training. The effectiveness of this type of training is based upon reliable and frequent usage of technology. The effective use of new technologies is based on acceptance of technology by both patients and therapists and requires an approach that combines attractiveness of the system, intuitive usability, and convincing effectiveness (Ritterfeld, Muehlhaus, Frieg, & Bilda, 2016). As when learning to play a musical instrument or learning a new language, mastery behavior is generally associated with perseverance, whereas helpless behavior is associated with task avoidance. Therapy time spent supporting patients to use the new technologies will therefore be beneficial in the long run. Patients have to believe that using more effort when speaking is useful and getting feedback on their loudness levels may help with this. For example, one of the patients using the "Speech Tool" app (Nolan et al., 2012), which provides feedback of average speech volume when talking and carrying out exercises, stated "It convinced me that I haven't been speaking at full volume, I am now trying to maintain a satisfactory level using feedback from the app." As well as using the technology to support patients in their efforts to use optimal speech in their daily lives, these technologies can also allow uploading of speech samples to a remote server for analysis by speech therapists and permit interactive monitoring and adjustment of home-based practice. In order for this to be successful therapists will need to be adequately trained with the skills and knowledge to operate new technologies as they emerge.

Speech recognition software such as Dragon Dictate or Siri, which provide written text from speech input, can be helpful in overcoming writing and typing difficulties. These programmes are also a useful therapy tool because they encourage people to speak clearly in order for the words to be recognized and transcribed correctly. However, the software does not deal well with pauses typical in the speech of people with Parkinson's and may start predicting words to fill the context or randomly guessing words from environmental noises, which is frustrating even for those without any communication difficulty.

Assistive technology devices and software can be used to supplement speech or as an alternative communication tool for those with a severe dysarthria whose speech is not intelligible despite use of effort. Text-based and picture or icon-based systems are available. In diseases where total speech loss is common, such as motor neuron disease, voice banking, and message banking are becoming increasingly popular using tools such as CereVoice Me www.cereproc.com/en/products/cerevoiceme, ModelTalker (www.modeltalker.org), and My-own-voice (www.acapela-group.com/voices/voice-replacement). The personally recorded synthetic voice can then be downloaded on to a device for use to support communication.

9. CONCLUSION

As stated by Douglas, Brown, and Barry (2004) "Communication is a dynamic and complex process; it is the medium through which we establish and maintain our social self at home, at work, in our community and in society generally." Communication difficulties can impact on confidence and lead to isolation, relationship strain and stress, which can then exacerbate other Parkinson's symptoms. It is therefore an important nonmotor symptom to fully investigate and provide support for, throughout a patient's journey.

REFERENCES

Andrew, M. K., Mitnitski, A., Kirkland, S. A., & Rockwood, K. (2012). The impact of social vulnerability on the survival of the fittest older adults. *Age and Aging*, *41*(2), 161–165.

Blackstone, S. W., & Berg, M. H. (2012). *Social networks: A communication inventory for individuals with complex communication needs and their communication partners*. Verona, WI: Attainment Company, Inc. (updated version).

Bloch, S., & Tuomainen, J. (2017). Progressive dysarthria and augmentative and alternative communication in conversation: Establishing the reliability of the dysarthria-in-interaction profile. *International Journal of Language and Communication Disorders, 52*(1), 3–9.
Chaudhuri, K. R., Martinez Martin, P., Schapira, A. H. V., et al. (2006). An international multicenter pilot study of the first comprehensive self completed non-motor symptoms questionnaire for Parkinson's disease: The NMSQuest study. *Movement Disorders, 21*, 916–923.
Constantinescu, G., Theodoros, D., Russell, T., Ward, E., Wilson, S., & Wootton, R. (2011). Treating disordered speech and voice in Parkinson's disease online: A randomized controlled non-inferiority trial. *International Journal of Language & Communication Disorders, 46*(1), 1–16.
Darley, F. L., Aronson, A. E., & Brown, J. R. (1969). Clusters of deviant speech dimensions in the dysarthrias. *Journal of Speech and Hearing Research, 12*, 462–469.
Darley, F. L., Aronson, A. E., & Brown, J. R. (1975). *Motor speech disorders*. Philadelphia: Saunders[chapter 8].
Defazio, G., Guerrieri, M., Liuzzi, D., Gigante, A. F., & di Nicola, V. (2016). Assessment of voice and speech symptoms in early Parkinson's disease by the Robertson dysarthria profile. *Neurological Sciences, 37*, 443–449. http://dx.doi.org/10.1007/s10072-015-2422-8.
Douglas, J., Brown, L., & Barry, S. (2004). The evidence base for the treatment of aphasia after stroke. In S. Reilly, J. Douglas, & J. Oates (Eds.), *Evidence based practice in speech pathology* (pp. 37–58). London: Whurr Publishers Ltd. chapter 3.
Enderby, P., & Palmer, R. (2008). *Frenchay dysarthria assessment* (2nd ed.). Austin, TX: Pro-Ed.
Evans, C., Canavan, M., Foy, C., Langfordd, R., & Proctor, R. (2012). Can group singing provide effective speech therapy for people with Parkinson's disease? *Arts and Health, 4*(1), 83–95. http://dx.doi.org/10.1080/17533015.2011.584883.
Gage, H., Grainger, L., Ting, S., et al. (2014). Specialist rehabilitation for people with Parkinson's disease in the community: A randomised controlled trial. In *Health Services and Delivery Research*. Southampton (UK): NIHR Journals Library. No. 2.15. http://dx.doi.org/10.3310/hsdr02510. Available from: https://www.ncbi.nlm.nih.gov/books/NBK263782/.
Gillivan-Murphy, P., Carding, P., & Miller, N. (2016). Vocal tract characteristics in Parkinson's disease. *Current Opinion in Otolaryngology & Head and Neck Surgery, 24*(3), 175–182.
Goetz, C. G., Tilley, B. C., et al. (2008). Movement Disorder Society-sponsored revision of the unified Parkinson's disease rating scale (MDS-UPDRS): Scale presentation and clinimetric testing results. *Movement Disorders, 23*(15), 2129–2170.
Griffiths, S., Barnes, R., Britten, N., & Wilkinson, R. (2011). Investigating interactional competencies in Parkinson's disease: The potential benefits of a conversation analytic approach. *International Journal of Language and Communication Disorders, 46*, 397–509.
Herd, C. P., Tomlinson, C. L., Deane, K. H. O., Brady, M. C., Smith, C. H., Sackley, C. M., et al. (2012). Comparison of speech and language therapy techniques for speech problems in Parkinson's disease. *Cochrane Database of Systematic Reviews*, (8). 8CD002814. http://dx.doi.org/10.1002/14651858.CD002814.pub2.
Hindle, J. V. (2010). Ageing, neurodegeneration and Parkinson's disease. *Age and Ageing, 39*(2), 156.
Ho, A. K., Bradshaw, J. L., & Iansek, R. (2000). Volume perception in parkinsonian speech. *Movement Disorders, 15*(6), 1125–1131.
Ho, A. K., Bradshaw, J. L., & Inset, R. (1999). Regulation of parkinsonian speech volume: The effect of interlocutor distance. *Journal of Neurology, Neurosurgery, and Psychiatry, 67*, 199–202.
Ho, A. K., Iansek, R., Marigliani, C., et al. (1998). Speech impairment in a large sample of patients with PD. *Behavioural Neurology, 11*, 131–137.

Huber, J. E., Darling, M., Francis, E. J., & Zhang, D. (2012). Impact of typical aging and Parkinson's disease on the relationship among breath pausing, syntax, and punctuation. *American Journal of Speech-Language Pathology, 21*(4), 368–379[Online].

Jacobson, B., Johnson, A., Grywalski, C., et al. (1997). The voice handicap index (VHI): Development and validation. *American Journal of Speech-Language Pathology, 6*, 66–69.

Kreutz, G., Bongard, S., Rohrmann, S., et al. (2004). Effects of choir singing or listening on secretory immunoglobulin a, cortisol and emotional state. *Journal of Behavioural Medicine, 27*(6), 623–635.

Lakshminarayana, R., Wang, D., Burn, D., Chaudhuri, K. R., Galtrey, C., Valle, G. N., et al. (2017). Using a smartphone-based self-management platform to support medication adherence and clinical consultation in Parkinson's disease. *Nature Parkinson's Journal Parkinson's Disease, 3*. http://dx.doi.org/10.1038/s41531-016-0003-z.

Little, S., Tripoliti, E., Beudel, M., Pogosyan, A., Cagnan, H., Herz, D., et al. (2016). Adaptive deep brain stimulation for Parkinson's disease demonstrates reduced speech side effects compared to conventional stimulation in the acute setting. *Journal of Neurology, Neurosurgery, and Psychiatry, 87*(12), 1388–1389. http://dx.doi.org/10.1136/jnnp-2016-313518.

Logeman, J., Fisher, H., Boshes, B., & Blonsky, E. (1978). Frequency and co-occurrence of vocal tract dysfunction in the speech of a large sample of Parkinson patients. *The Journal of Speech and Hearing Disorders, 43*, 47–57.

Miller, N., Allcock, L., Jones, D., Noble, E., Hildreth, A. J., & Burn, D. J. (2007). Prevalence and pattern of perceived intelligibility changes in Parkinson's disease. *Journal of Neurology, Neurosurgery, and Psychiatry, 78*(11), 1188–1190.

Miller, N., Noble, E., Jones, D., & Burn, D. (2006). Life with communication changes in Parkinson's disease. *Age and Ageing, 35*, 235–239.

Mobes, J., Joppich, G., Stiebritz, F., Dengler, R., & Schroder, C. (2008). Emotional Speech in Parkinson's Disease. *Movement Disorders, 23*(6), 824–829.

Moreau, C., Ozsancak, C., Blatt, J. L., Derambure, P., et al. (2007). Oral festination in Parkinson's disease: Biomechanical analysis and correlation with festination and freezing of gait. *Movement Disorders, 22*(10), 1503–1506.

Nolan, P., Hoskins, S., Johnson, J., Powell, V., Chaudhuri, K. R., & Eglin, R. (2012). Implicit theory manipulations affecting efficacy of a smartphone application aiding speech therapy for Parkinson's patients. *Studies in Health Technology and Informatics, 181*, 138–142.

Pacchetti, C., Mancini, F., Aglieri, R., Fundaro, C., Martignoni, E., & Nappi, G. (2000). Active music therapy in Parkinson's disease: An integrative method for motor and emotional rehabilitation. *Psychosomatic Medicine, 62*, 386–393.

Parkinson, J. (1817). *An essay on the shaking palsy*. London: Sherwood, Neely, and Jones.

Parkinson J. (2002) An essay on the shaking palsy. 1817 The Journal of Neuropsychiatry and Clinical Neurosciences. 14, 2, 223–36; PMID: 11983801, Discussion 222.

Pentland, B., Pitkairn, T. K., & Gray, T. K. (1987). The effects of reduced expression in Parkinson's disease in impression formation by health professionals. *Clinical Rehabilitation, 1*(4), 307–313.

Pinto, S., Ozsancak, C., Tripoliti, E., Thobois, S., Limousin-Dowsey, P., & Auzou, P. (2004). Treatments for dysarthria in Parkinson's disease. *Lancet Neurology, 3*(9), 547–556. Review.

Ramig, L. O., Sapir, S., Countryman, S., Pawlas, A. A., O'Brien, C., Hoehn, M., et al. (2001). Intensive voice treatment (LSVT) for individuals with Parkinson's disease: A 2-year follow-up. *Journal of Neurology, Neurosurgery and Psychiatry, 71*, 493–498.

Reker, P., Dembek, T. A., Becker, J., Visser-Vandewalle, V., & Timmermann, L. (2016). Directional deep brain stimulation: A case of avoiding dysarthria with bipolar directional

current steering. *Parkinsonism & Related Disorders*, *31*, 156–158. http://dx.doi.org/10.1016/j.parkreldis.2016.08.007.
Ritterfeld U., Muehlhaus J. Frieg H., Bilda K. (2016). Developing a technology-based speech intervention for acquired dysarthria: a psychological approach. International Conference on Computers Helping People with Special Needs. Lecture notes in computer science 06 July. 9758; pp. 93–100.
Robert, G., Le Jeune, F., Dondaine, T., Drapier, S., Péron, J., Lozachmeur, C., et al. (2014). Apathy and impaired emotional facial recognition networks overlap in Parkinson's disease: A PET study with conjunction analyses. *Journal of Neurology, Neurosurgery and Psychiatry*, *85*(10), 1153–1158. 0022-3050.
Rogalski, Y., Altmann, L. J. P., & Rosenbek, J. C. (2014). Retrieval practice and testing improve memory in older adults. *Aphasiology*, *28*(4), 381–400. http://dx.doi.org/10.1080/02687038.2013.870965.
Rosen, C., Lee, A., Osborne, J., Zullo, T., & Murry, T. (2004). Development and validation of the voice handicap index-10. *Laryngoscope*, *114*(9), 1549–1556.
Saldert, C., Ferm, U., & Bloch, S. J. (2014). Semantic trouble sources and their repair in conversations affected by Parkinson's disease disease. *International Journal of Language and Communication Disorders*, *49*, 710–721.
Salmazo-Silva, H., de Mattos Pimenta Parentea, M. A., Rocha, M. S., Baradel, R. R., Cravo, A. M., Sato, J. R., et al. (2017). Lexical-retrieval and semantic memory in Parkinson's disease: The question of noun and verb dissociation. *Brain and Language*, *165*, 10–20.
Shahed, J., & Jankovic, J. (2001). Re-emergence of childhood stuttering in Parkinson's disease: A hypothesis. *Movement Disorders*, *16*(1), 114–118.
Titova, N., Padmakumar, C., Lewis, S. J. G., & Chaudhuri, K. R. (2016). Parkinson's: A syndrome rather than a disease? *Journal of Neural Transmission*. http://dx.doi.org/10.1007/s00702-016-1667-6.
Walsh, B., & Smith, A. (2011). Linguistic complexity, speech production, and comprehension in Parkinson's disease: Behavioral and physiological indices. *Journal of Speech, Language, and Hearing Research*, *54*(3), 787–802.
Walshe, M., Peach, R. K., & Miller, N. (2009). Dysarthria impact profile: Development of a scale to measure psychosocial effects. *International Journal of Language & Communication Disorders*, *44*(5), 693–715.
Yorkston, K. M., Beukelman, D. R., & Hakel, M. (1996). *The speech intelligibility test*. Lincoln, NE: Tice Technology.

CHAPTER FORTY-THREE

Swallowing Dysfunctions in Parkinson's Disease

Janine A. Simons[1]
University of Lübeck, Lübeck, Germany
[1]Corresponding author: e-mail address: janine.simons@neuro.uni-luebeck.de

Contents

Abstract

Dysphagia is a very frequent and highly relevant symptom in Parkinson's disease (PD) for quality of life, morbidity, and remaining lifetime, which is unfortunately widely underdiagnosed and underestimated regarding patients' centered care. Especially in early stages, the causal association between disease and swallowing disabilities remains unnoticed, which may be accounted for by the inability of caregivers and physicians to detect subtle swallowing problems and by the low self-awareness among PD patients. In order to prevent patients from serious negative consequences for health issues (e.g., aspiration pneumonia or malnutrition) as well as for negative impact on their quality of life, it is on the highest importance of managing dysphagia timely and working closely together in a multidisciplinary team, who all are involved in the patients' care system.

This chapter includes background information on epidemiology, pathophysiology, and symptomatology of swallowing disorders in PD. This is followed by a summary of the clinical course and health treats, adequate diagnostic procedures for early identification of dysphagia as well as effective treatment strategies. The conclusion provides recommendations for clinical practice routine.

International Review of Neurobiology, Volume 134
ISSN 0074-7742
http://dx.doi.org/10.1016/bs.irn.2017.05.026

1. BACKGROUND OF SWALLOWING DISABILITIES

In 1817, James Parkinson already mentioned dysphagia as a vital non-motor symptom (NMS) (Parkinson, 2002), which is categorized as autonomic and gastrointestinal dysfunction (Chaudhuri, Healy, & Schapira, 2006; Pfeiffer, 2003), but only within the last years swallowing disabilities has gained increasing clinical significance and has also come into focus of research.

1.1 Epidemiology

Dysphagia frequently accompany Parkinson's disease (PD), occurring in up to 100% of patients in advanced disease stages, but frequency data vary widely according to underlying classification, measuring method, and disease stage. Based on a pooled prevalence, oropharyngeal dysphagia is evaluated with 35% subjectively (95% CI: 28–41) and even with 82% (77–87) when objective, instrumental measures were taken into account (Kalf, de Swart, Bloem, & Munneke, 2012). When PD patients' self-reports have been matched with the results of fiberoptic endoscopic evaluation of swallowing (FEES) or video fluoroscopic swallowing study (VFSS), for example, oropharyngeal disorders have become evident in over 50% of subjectively nondysphagic PD patients (Bird, Woodward, Gibson, Phyland, & Fonda, 1994; Fuh et al., 1997), and silent aspiration additionally occurred in up to 15% (Ali et al., 1996). Swallowing impairment does not appear only in the late stage of the disease, whereby mild oropharyngeal symptoms as well as esophageal dysfunctions are very common in early stage PD and can even be the first presenting sign of the disease, marking the prodromal stage of PD (Noyce et al., 2012; Potulska, Friedman, Krolicki, & Spychala, 2003; Sung et al., 2010; Thomas & Haigh, 1995). Although aspiration episodes might occur also in the early stages, severe dysphagia with chronic aspiration and relevant clinical complications is usually more connected with advanced stages of PD, and occurrence within the first year after disease onset is predictably related to atypical Parkinsonism (Müller et al., 2001).

1.2 Pathoanatomy and Pathophysiology

Whereas the symptomatology of neurogenic PD-associated dysphagia is well described, its multidimensional neural pathophysiology has still not been sufficiently and fully investigated, and needs to be further explored. In general, the control of swallowing is a complex semiautomatic, repetitive medullary program, modified by bolus volume and consistency, and by involvement of peripheral, afferent as well as central feedback. Dysfunctions in the brainstem, affecting the swallowing central pattern generator and degeneration of the substantia nigra seem to play an important pathophysiological role in PD, and disturbances of dopaminergic mechanism as well as nondopaminergic neural networks may be a major causal factor (Chaudhuri et al., 2006; Suntrup et al., 2013). Beside the dopaminergic basal ganglia system, which is predominately affected by the neuropathological process controlling the supramedullary swallowing system (Leopold & Daniels, 2010) also lewy bodies occur in nondopaminergic brainstem and cortical areas (according to Braak staging), primarily involved in the control of swallowing (Braak et al., 2006). In addition, it can be assumed that the disease likely develops in the gut and moves in a rostral projection to medulla areas. Gastrointestinal manifestation appears during disease progression. Postmortem investigations showed the presence of alpha-synuclein in peripheral sensory as well as motor nerves innervating the pharyngeal muscles that were stronger impaired in dysphagic than in nondysphagic PD patients (Mu, Sobotka, Chen, et al., 2013). Cholinergic dysfunction may also be associated with dysphagia in early PD and abnormal value of short-latency afferent inhibition could play a role in predicting the risk of dysphagia serving as good biomarker (Lee et al., 2015). Another pathological factor seems to be a reduced concentration of substance P in the sputum of dysphagic PD patients resulting in disturbed protective reflexes (i.e., swallowing or throat-clearing/coughing) and (silent) aspiration (Ebihara et al., 2003; Troche, Brandimore, Okun, Davenprot, & Hegland, 2014).

1.3 Pathomechanism and Symptomatology

All four swallowing phases are driven by a complex, sequential response along a continuum of mostly automaticity and only little arbitrary control. Disturbances in PD can affect any phase of the swallowing process (preoral, oral, pharyngeal, or esophageal) as well as adjacent respiratory, olfactory, and salivary systems. There is a consensus in literature about the fact that swallowing dysfunctions caused by slowed movement execution, restricted range of motion, reduced physical strength, and probably impaired perception.

Table 1 shows the wide range of pathomechanism of swallowing and consequential symptoms' characteristics in PD, allocated to the different swallowing phases.

Table 1 Pathomechanism and Symptoms of Dysphagia in PD

Phases of Swallowing	Swallowing Disturbance Pattern/ Pathomechanism	Symptoms/Signs	Main Findings
Preoral phase/oral preparation phase	– Reduced sense of smell and taste – Reduced sensory and tactile–kinesthetic orofacial perception – Lack of proprioception of the tongue – Reduced oral strength/strength endurance – Disturbed motion dynamics of jaw muscles – Disturbed tongue movements, difficulties in forming "tongue spoon," tongue rigidity, tongue tremor	– Insufficient saliva production – Incomplete chewing/ mastication and abnormal bolus preparation/formation with rigor or akinesia – Choking, hawking, coughing	– Anterior bolus leakage and drooling – Posterior leakage/ premature bolus spillage/ pharyngeal pooling – Predeglutitive penetration or aspiration/silent aspiration – Difficulty/inability bolus preparation and swallowing
Oral phase/oral propulsion/oral processing and transportation	– Reduced oral bolus control – Reduced oropharyngeal bolus transport with festination of the tongue onto the soft palate	– Repetitive rocking and rolling festination-type motion of the tongue – Piecemeal deglutition – Choking, hawking, coughing	– Posterior leakage/ premature bolus spillage/ pharyngeal pooling/ – Predeglutitive penetration or aspiration/silent aspiration – Oral residues – Delayed oral transit time

Pharyngeal phase	– Disturbed swallow triggering – Reduced velopharyngeal closure, reduced backward movement of the tongue, and reduced elevation of velum, hyoid bone, and larynx – Reduced contraction of pharyngeal structures – Disturbed pharyngeal bolus transport – Increased hypopharyngeal intrabolus pressure – Reduced/missing pharyngeal sensitivity – Reduced laryngeal sensitivity – Disturbed coordination of breathing and swallowing – Insufficient laryngeal vestibule closure, dysfunction of epiglottis, incomplete closure of arytenoids, false cords, and true vocal cords	– Liquid/dribbling out of the nose – Pharyngeal residues – Pharyngeal pooling/cricopharyngeal pooling – Choking, hawking, coughing – Delayed pharyngeal transit time	– Nasal penetration – Delayed pharyngeal swallow/pharyngeal pooling – Reduced rate of spontaneous swallows/saliva pooling – Delayed pharyngeal transit time – Somatosensory deficits, reduced airway protection – Postdeglutitive penetration or aspiration – Silent aspiration – Reduced pharyngeal clearance
Esophageal phase	– Insufficient cricopharyngeal muscle relaxation/hypertonic sphincter with high hypopharyngeal intrabolus pressure – Disturbed opening of the upper esophageal sphincter	– Laryngeal/pharyngeal redness/inflammation (arytenoids, vocal folds, and pharyngeal wall)	– Cricopharyngeal dysfunction/dysfunction of the upper esophageal sphincter – Gastroesophageal reflux

Continued

Table 1 Pathomechanism and Symptoms of Dysphagia in PD—cont'd

Phases of Swallowing	Swallowing Disturbance Pattern/ Pathomechanism	Symptoms/Signs	Main Findings
	– Impaired suprahyoid muscle performance – Diffuse/tertiary/multiple simultaneous esophageal contractions and spasms – Weak bolus propulsion, hypomobility of esophagus, loss of peristaltic waves – Exhibited esophagogastric junction outflow obstruction	– Acid reflux, acid/sour regurgitation, air burping, epigastric pain, gastric upset	– Esophageal motility dysfunctions, esophageal or cardia spasm, achalasia – Zenker's diverticula
Swallow-associated symptoms/secondary handicaps	– Reduced swallow frequency – Levodopa-induced xerostomia – Disturbed expiratory muscles, reduced force of glottal/supraglottal explosion – Dopamine-induced swallow difficulties – Disadvantageous head and body postures – Disturbed hand-mouth coordination – Motor disabilities (freezing phenomenon, tremor/rigor, akinesia, hyperkinesia, restless legs/restlessness) – Psychomental stresses (e.g., limited perception and l attention, anxiety, depression, dementia, fatigue, exhaustion, insomnia, medical induces psychosis)	- Variety dysphagia symptoms (see all phases) Additionally: - Pseudohypersalivation - Hyposalivation	– Drooling – Difficulty swallowing/ disturbed swallow triggering – Difficulty swallowing specific consistencies/ mixed consistencies or pills/tablets – Reduced laryngeal and pharyngeal clearance – On/off medication fluctuations – Secondary enhanced swallow problems and health threats

Other clinical complications and comorbidities possibly influencing swallow performance and nutritional status:

- Health problems (e.g., physical weakness/frailty, weight loss/malnutrition/sarcopenia, dehydration, lung infections/pneumonia)
- Other nonmotor problems (e.g., hypokinetic-rigid dysarthrophonia)
- Other gastrointestinal dysfunctions (e.g., obstipation, diarrhea, gastroparesis)
- Others

According to (Alfonsi, Versino, Merlo, Pacchetti, et al., 2007; Ali et al., 1996; Bird et al., 1994; Castell et al., 2001; Chou, Evatt, Hinson, & Kompoliti, 2007; Davydov & Botts, 2000; Ebihara et al., 2003; Edwards, Quigley, Hofman, & Pfeiffer, 1993; Edwards, Quigley, & Pfeiffer, 1992; Johnston, Li, Castell, & Castell, 1995; Kalf, Bloem, & Munneke, 2012; Leopold & Kagel, 1997; Leow, Beckert, Anderson, & Huckabee, 2012; Leslie, Drinnan, Ford, & Wilson, 2005; Logemann, 1998; Mari et al., 1997; Marks, Turner, O'Sullivan, Deighton, & Lees, 2001; Moreau, Ozsancak, Blatt, et al., 2007; Nóbrega et al., 2008; Pehlivan et al., 1996; Pfeiffer, 2003; Pinnington, Muhiddin, Ellis, & Playford, 2000; Proulx, de Courval, Wiseman, & Panisset, 2005; Rodrigues, Nóbrega, Sampaio, Argolo, & Melo, 2011; Su, Gandhy, Barlow, & Triadafilopoulos, 2017; Troche, Huebner, Rosenbek, et al., 2011; Tumilasci et al., 2006; Umemoto, Tsuboi, Kitashima, et al., 2011).

2. CLINICAL IMPLICATIONS OF DYSPHAGIA AND RELEVANCE FOR QUALITY OF LIFE

Dysphagia is an important risk factor for hospitalization and premature mortality and causes dehydration and malnutrition, aspiration and aspiration pneumonia as well as a reduction of health-related quality of life.

Half of PD patients with dysphagia are aspirating and oftentimes silent aspiration occurs. The proportion of silent aspiration without reflectory clearing reaction is reported to be higher in PD than in other neurological diseases (Bird et al., 1994; Leslie et al., 2005; Mari et al., 1997). Pulmonary long-term complications like pneumonia, especially aspiration pneumonia due to swallowing disorders, is one of the main causes of death in all parkinsonian syndromes (Fernandez & Lapane, 2002; Hely et al., 1999; Müller et al., 2001; Pennington, Snell, Lee, & Walker, 2010; Pfeiffer, 2003; Wang, You, Chen, & Cai, 2002). In the CamPalGN, 10-year follow-up study cohort 63 of 142 PD patients (44%) died during the follow-up period and pneumonia was the commonest cause of mortality with 33% (Williams-Gray et al., 2013). Additionally, in a 3-year follow-up study in 155.237 nursing home residents with a diagnosis of PD aspiration pneumonia was the strongest predictor for mortality (RR 1.58; 95% CI: 0.97–2.56) (Fernandez & Lapane, 2002). A two-times higher risk of mortality has been shown in PD patients in contrast to elderly persons without any neurological diseases (OR 2.1; CI: 1.4–3.1), particularly caused by pneumonia and cachexia (D'amelio et al., 2006). The average survival time after begin of chronic dysphagia as well as dysarthria was further determined to be 15–24 month, postmortem verified (Müller et al., 2001).

Furthermore, swallowing dysfunction has an important impact of life quality in PD, impair social interaction and participation, and lead to psychosocial burden (Leow, Huckabee, Anderson, & Beckert, 2010; Manor, Balas, Giladi, Mootanah, & Cohen, 2009; Plowman-Prine et al., 2009; van Hooren et al., 2015).

Dysphagia causes depression and exhaustion, and gives a feeling of shame, fear, fatigue, anger, and resignation as well as isolation or exclusion because of the variety of oropharyngeal symptoms (e.g., saliva drooling, choking, coughing, or breathing problems), thus requiring dietary regulations or nutritional restrictions with abandonment of certain food consistencies. Relatives and partners may also develop a feeling of frustration, helplessness, and being overwhelmed because of the restrictions in daily living. Activities like restaurant visits, dinner with family and friends, private events, invitations to birthday parties, or professional appointments are a major barrier. Therefore, swallowing problems significantly reduces peoples' sense of wellbeing, contentment, and pleasure of selecting and consuming food. Negative changes in self-confidence, self-image, or understanding of roles and social function as well as impairments of social interaction and communication behavior may appear apart from the general burden due to the underlying disease. Emotions and attitude regarding general food intake, the demand for food and liquids, and the duration for meals as well as the safety of eating and self-care quality may further strongly be influenced in a negative manner.

The multicentre Priamo study group ($n=1.072$) discovered prevalence of different NMS in PD patients with relevance to quality of life, that has been evaluated by the Unified Parkinson's Disease Rating Scale, motor part (UPDRS III) or the self-reporting Parkinson's Disease Questionnaire (PDQ-39; dimensions: mobility, activity of daily living, emotional wellbeing, stigmatism, social support, cognition, communication, and physical complaints). Gastrointestinal dysfunctions were one of the most frequent NMS (61%), including saliva drooling (31%) and swallowing difficulties (16%), and respiratory symptoms have also been reported (20%), including coughing (11%) (Barone et al., 2009). Assessments via PDQ-39, swallowing-specific quality of life questionnaire (SWAL-QOL; dimensions: burden, eating duration, eating desire, symptom frequency, food selection, communication, fear, mental health, social, fatigue) and Beck Depression Inventory revealed a high number of psychosocial symptoms as long-term effects of swallowing dysfunction as well as a reduction in the overall quality of life (Plowman-Prine et al., 2009). In addition, PD patients were shown to be more impaired in the swallow-specific life quality in comparison to the healthy control group. Significantly, differences have been presented nearly in all subcategories of the SWAL-QOL whereas strongest impairments have been identified in the categories appetite, food selection, and duration of meals with disease progression that also might have negative consequences for the nutritional state (Leow et al., 2010).

Swallowing dysfunctions are one of the main factors for unintentional weight loss and malnutrition in PD associated with reduced food intake, increased catabolism, and energy expenditure (Kalf, de Swart, et al., 2012). Food intake is reduced due to chewing and swallowing problems with prolonged mealtimes, tiredness during the meals, or avoidance of solid food, for example. Drinking only in little sips, choking during/after drinking, or compelling necessity for thickened drinks instead of thin fluids contribute to insufficient fluid intake. Neuropsychiatric symptoms like depression and PD-associated dementia additionally increase the risk for malnutrition and dehydration. With the progression of dysphagia, malnutrition (protein energy malnutrition, sarcopenia) and dehydration are frequent clinically relevant complications (Barichella et al., 2013; Barichella, Cereda, & Pezzoli, 2009) that usually require tube feeding (enteral nutrition via nasogastric intubation or percutaneous endoscopic gastrostomy) or parenteral fluid replacement therapy, and additional oral nutritional supplements, if possible. Dependence from tube feeding may be understandably a negative issue for quality of life and independency. On the other hand, it might also help to improve life quality again when used only for accompanying oral-nutrition providing assistance to ensure both, the quantity of liquid and the number of calories or other essential nutrients, particularly micronutrients as well as the quality of enjoyment and the taste of good food and drinks in little amounts.

Apart from the health and nutritional status swallowing dysfunction negatively affects the oral administration of medication leading to pharyngeal residues of pills and tablets, aspiration or the inability to swallow those. Changes in drug efficacy or absence of the desired effects are the consequence. Improper medication intake (i.e., levodopa intake mixed with yoghurt or other protein-containing food/drinks for better deglutition, or swallowing tablets grounded in mortars sorely or together with other drugs) can also cause unintended side-effects.

In the very late stage of PD with severe dysphagia supply with tracheal cannulas might be indicated safeguarding breathing and protect the patients from aspirating their own saliva. Blocked tracheal cannulas deactivate the sense of smell and taste, affect the pharyngeal/laryngeal sensibility, and make it impossible for the patient to produce voice or to communicate. This intervention procedure is therefore a strong negative issue for quality of life and should be excluded as long as possible. Nevertheless, it should be borne in mind that cannulas do not fully prevent from developing respiratory/lung infections.

Furthermore, health-threatening complications and comorbidities caused by dysphagia also affects the national health budgets as they lead to intensive care, higher drug intake, readmissions, and increased length of hospital stays. PD-related healthcare costs due to gastrointestinal dysfunctions including dysphagia increased about more than 10% according to a US administrative health claims database, for instance (Makaroff, Gunn, Gervasoni, & Richy, 2011).

3. DYSPHAGIA ASSESSMENT

3.1 Self-Awareness and Clinical Identification

In contrast to motor problems and a variety of other NMSs inclusive saliva drooling, that early creates mental suffering and high burden the self-awareness for dysphagia is insufficient in PD patients, especially at the onset of the disease. The physician's or nurse's question regarding the presence of dysphagia (Do you have swallowing problems?) will be denied very likely by the most of the patients (Miller et al., 2009). Subclinical dysphagia with beginning oropharyngeal problems can easily be compensated by the patients themselves. For example, drinking in little sips or with chin-down posture helps to control the bolus and prevent premature spillage and aspiration for a while. Drinking while eating solids makes it easier to flush down food residues, that got stuck in the mouth or pharynx. Other consciously compensation strategies are cutting solid or fibrous food into smaller pieces or avoiding specific food or consistencies that might have caused any swallow problems in the early past. Due to those little modifications of food preparation and adapted swallowing early dysphagia symptoms also remain unnoticed for caregivers as well as for health professionals (Ali et al., 1996; Miller et al., 2009). Additionally, "silent dysphagia symptoms" (silent laryngeal penetration/aspiration without any sensibility and reactions to clear the throat/airways like hawking or coughing) are present in PD patients, including in an early stage. Furthermore, patients likely do not recognize beginning symptoms because of the lack of oral/pharyngeal/laryngeal tactile–kinesthetic and sensibility senses, or impaired awareness of changes in voice quality after swallowing indicating bolus penetration on the level of vocal folds, and also due to the fact that presenting symptoms might fluctuate in the early stages of dysphagia (partially associated with good responsiveness of levodopa-equivalent therapy in the initial phase). Swallowing disorders are therefore mostly diagnosed and treated very late in advanced stages, after they have become manifested, clinically relevant, or severe health threats occurred (like aspiration pneumonia/other airway infections, relevant weight loss, or dehydration) (Monte, Da Silva-Junior, Braga-Neto, Nobre e Souza, & Sales de Bruin, 2005; Volonte, Porta, & Comi, 2002). In addition, dysphagia might negatively influence the pharmacological therapy of PD patients reducing accordingly the drug effect, which needs to be taken into account and which requires appropriate adaptive treatment interventions. Not rare to see that insufficient medication intake is another major reason for clinical presentation.

However, swallowing impairment is usually underestimated and identified very late. Procedures for initiating and pursuing readily applicable diagnostic investigations as well as counteractive treatment are not implemented on time. Therefore, early screening for dysphagia with valid tools is absolutely required (Miller et al., 2009; Suttrup & Warnecke, 2016).

3.2 Clinical Predictors

Being able to clearly identify patients under risk this will fundamentally influence the patients' early allocation to required screening assessments and further swallowing examination and therefore lead to a big advantage in medical practice. In medical routine, there is no certainty about the right time patients with Parkinsonism should undergo screening for dysphagia. Several factors have been identified to be highly correlated with different swallowing dysfunctions in PD that raised the likelihood for dysphagia development and aspiration. The following clinical predictors should be taken into account (Table 2).

3.3 Dysphagia Screenings, Clinical and Instrumental Assessment

To assess dysphagia severity generically a variety of clinical as well as instrumental protocols have been developed, but for evaluation of overall

Table 2 Clinical Predictors for Aspiration-Associated Dysphagia in PD

Clinical Risk Factors
H&Y > 3
UDPDRS III > 26
BMI < 20/relevant unintended weight loss
Dementia
Depression
Saliva drooling, sialorrhea, reduced oral bolus control[a]
Dysarthria
Disease duration >10 years

[a]Patients' report being unable to hold food in the mouth.
BMI, body mass index kg/m^2; H&Y, Hoehn & Yahr staging; UPDRSIII, Unified Parkinson's Disease Rating Scale, motor part. According to (Cereda et al., 2014; Coelho et al., 2010; Han et al., 2011; Johnston et al., 1997; Lam et al., 2007; Müller et al., 2001; Perez-Lloret et al., 2012; Simons, Eisemann, Fietzek, & Katalinic, 2015; Warnecke et al., 2010).

collectives, especially for Parkinson syndromes there are only limited sufficiently valid scales and screening tools available.

3.3.1 Patient Questionnaires

Two PD-specific dysphagia screening tools have been published, the Swallowing disturbance questionnaire (SDQ) (Manor, Giladi, Cohen, Fliss, & Cohen, 2007) as well as the Munich Dysphagia Test—Parkinson's disease (MDT-PD) (Simons et al., 2014). Depending on the questionnaire's screening result clinical logopedic swallowing examination or further instrumental diagnostic procedure (FEES, VFSS) is indicated in order to evaluate the patient's characteristic symptoms and underlying pathomechanism. While the SDQ was criticized by the "Movement disorders task force on PD rating scales" (Evatt et al., 2009) because of its very limited clinimetric properties and poor clinical significance, the development of the MDT-PD self-reported outcome questionnaire was facilitated ($n=187$) which was validated including standardized protocols for clinical swallowing evaluation and FEES, thus resulting in very good predictive quality and high discriminatory abilities detecting both beginning oropharyngeal symptoms (sensitivity 82.4%, specificity 71.4%) as well as risk of laryngeal penetration or aspiration (90.0% and 85.7%). The questionnaire consists of 26 items, divided into four categories, and concerning (1) swallowing symptoms occurring during eating and drinking, (2) difficulty swallowing independent from food intake, (3) swallowing-specific and accompanying burden, and (4) swallow-related health and nutrition state questions. As the MDT-PD is easily to use in clinical routine, well accepted by the patients, and quickly to evaluate by using a web application (Fig. 1), its implementation in clinical practice has started within the last years. Until now, the screening questionnaire is available in German and English language. Additional cross-national translation studies are already running (e.g., French, Italian, Chinese, and Brazilian Portuguese) as well as validation for atypical Parkinsonian syndromes.

The well-established Nonmotor Symptoms Questionnaire (NMS-Quest) (Chaudhuri, Martinez-Martin, et al., 2006) consists of 30 items that globally screen for the full range of NMSs in PD. It also contains one question addressing swallowing difficulties with choking while drinking/eating (quest 3), one about saliva dribbling during daytime (1), and one accompanying question regarding unexplained weight loss (11). When applying NMS in a European multicentre cohort among 242 PD patients only 65 of them (27%) have been identified as potentially dysphagic (quest 3),

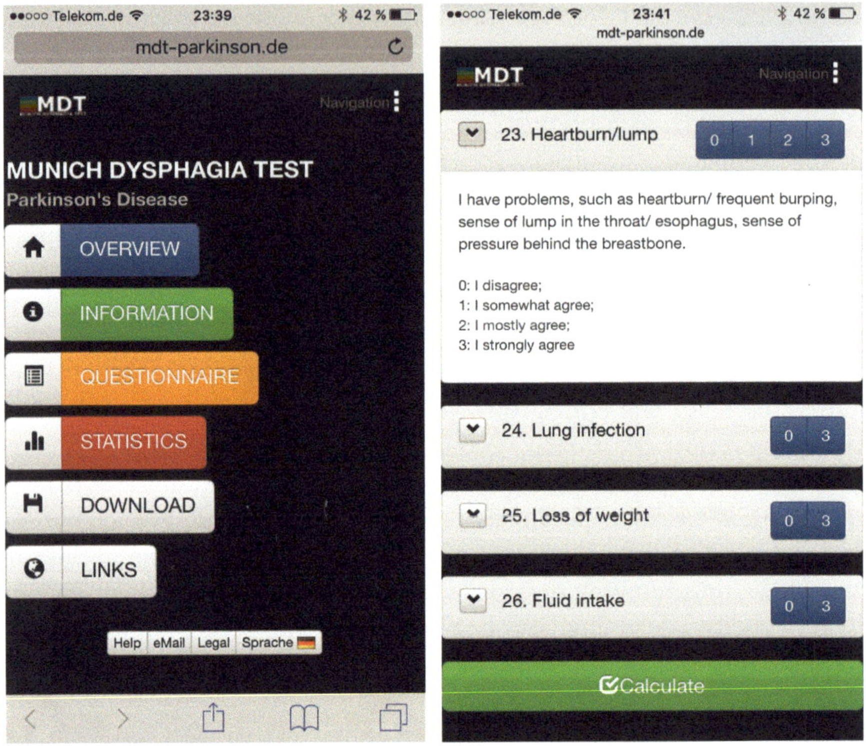

Fig. 1 MDT-PD web application. Screening result comes with a weighted sum score, its clinical interpretation and recommendation for further diagnostic and medical treatment procedures, if needed. Classifications: (0) no relevant dysphagia, (1) beginning oropharyngeal dysphagia, and (2) dysphagia with penetration/aspiration risk.

whereas the symptom drooling (1) was positively screened in 101 individuals (48%) and unexpected weight loss was assessed in 55 patients (23%) (Chaudhuri et al., 2010). This strengthens the belief of physicians that administering the NMS to patients at risk can help drawing first attention to swallowing associated problems, but it also displays that when using NMS alone dysphagia might be underestimated as the rate lies under the previously outlined pooled rate of subjectively evaluated dysphagia prevalence (Kalf, de Swart, et al., 2012).

Due to the fact that PD patients have a lack of self-awareness upon their dysphagia symptoms, especially in the beginning stage, swallowing dysfunction is further not sufficiently evaluated by the single item regarding chewing/swallowing problems in general (quest 10) and the question about drooling/saliva (9) by the 20-item patient questionnaire of the Movement

Disorder Society-sponsored revision of the UPDRS (MDS-UPDRS part II) (Goetz et al., 2006), and will therefore be likely remain undetected.

3.3.2 Clinical Screening and Evaluation

As the 50-mL water swallow test clinically screen for swallowing disorders has been shown to be insufficiently for reliable prediction of severe dysphagia in PD patients compared to instrumental swallowing evaluation (Lam et al., 2007), the 90-mL timed water swallow test (Suiter & Leder, 2008) shows good predictive value for aspiration by adequate stressing the swallowing process, but does not give any information on the swallowing performance for other consistencies, and is validated among subjects with various etiology including only a small group of PD patients, additionally. Moreover, assessing the maximum swallow volume and overall time to finish the amount of water should be taken into account, and pathological swallow symptoms and aspiration signs during/after drinking should be noticed. The 150-mL timed swallow test has been proved among PD patients even in off-drug state ($n=137$) (Miller et al., 2009), revealing a remarkable percentage of patients who were unable to complete the test (23%). In 84% swallow rate (mL/s) felt below the norm value in standard population about more than 1 SD (swallow-duration, s/swallow) and in 32% more than 2 SD. Bolus volume was significantly higher ($P<0.01$) in patients who succeed the test than in patients who failed and showed also a shorter duration per swallow ($P=0.01$). Additionally, a maximum swallow volume below 20 mL indicates swallowing dysfunction in PD (Potulska et al., 2003). Because of the insufficient association between patients' self-reports regarding their swallowing condition and the results of the timed water test, the application of valid screening questionnaires followed by further objectively/instrumentally swallow assessments are strongly recommend.

A comprehensive clinical assessment for evaluating the safety and efficacy of swallowing should usually comprise in a first step an anamnestic part including a nutritional status, then a functional exam of swallow-related muscles and structures (e.g., oral motor skills referring to muscle tone and strength, range of motion, speed, coordination, and dissociation of lips, tongue, mouth, and jowls as well as masticatory performance; further examination parameters include saliva management, palatal reflex, pharyngeal sensibility, choking reflex, voluntary cough, swallow triggering, or larynx elevation during swallowing), and thereafter swallow tests using different consistencies (e.g., thin and thickened liquids, puree textures, and solid or crumbly food).

Anyway, there is no Parkinson-specific, standardized clinical assessment recommended to be used nation-wide or even internationally. For validation of the MDT-PD questionnaire standard protocols for clinical swallow evaluation as well as for FEES were developed (Simons et al., 2014), adapted from established scales for neurogenic dysphagia originally designed for other etiology, and modified to use in PD patients. The clinical protocol comprises items regarding medication cycle, nutritional and swallow-related health state, followed by evaluation of orofacial dyskinesia/tremor, enoral saliva, and also drooling, using the Drooling Score Scale (Nóbrega et al., 2008) as well as oral and pharyngeal structures, functions, and sensory reflexes. Thereafter, swallow tests are included with liquids and food in sufficient quantity to display real-life condition, and finally the ingestion of placebo pills of different size. Evaluating the patient's swallowing performance at different times of the day as well as in medication on- and off-state may surely be quite helpful in order to reveal possible alterations due to motor ability and also vigilance or other individual conditions. The severity grades with symptoms' characterization has been conceptualized considering a modified 90-mL water swallow test or standard values of healthy people in the elderly, for instance.

Nevertheless, it should be kept in mind that clinical swallow assessments are an indispensable component in the evaluation of dysphagia, but they are also limited in their scope, as only the patient's reaction, specific symptoms, or indirect signs (e.g., coughing/hawking, changes in voice/wet voice, breathing problems, prolonged oral preparation time, delayed swallow reflex, reduced larynx movement, multiple swallows, or oral residues) can be used to detect and characterize also the invisible dysphagia findings like pharyngeal residues or aspiration. Especially, it is not possible to make a reliable estimate of silent aspiration. Therefore, instrumental examinations might often be a very important and useful adjunct that should be taken into account to complete dysphagia assessment and allowing to evaluate the specific nature of oropharyngeal and esophageal symptoms.

3.3.3 Instrumental Swallowing Evaluation

The instrumental diagnostic gold standard for neurogenic dysphagia assessment is considered to be the FEES (Langmore & Aviv, 2001) as well as the VFSS (Logemann, 1998). Depending on the underlying clinical question, FEES or VFSS might deliver certain advantages and diagnostically benefit. Therefore, both are valuable and should not be seen as alternative, but as complementary dynamic procedures providing a comprehensive analysis of oropharyngeal disturbance patterns.

For FEES a flexible rhynolaryngoscope (diameter 2.4–3.5 mm) is passed over the velum into the hypopharynx in order to examine the anatomical structures and physiological mechanism, and to test a series of swallows with different consistencies and food from everyday life that can be evaluated pre- and postdeglutitive. The effectiveness of trained swallow maneuvers or adaptations during the treatment period can be performed as well. FEES, as noninvasive diagnostic procedure, is often the method of choice for first-line dysphagia assessment. It allows to reliably examine the late oral and pharyngeal phase of swallowing, helps to understand the underlying pathophysiology, is well tolerated, easily repeatable, and can be performed at the bedside (Tohara et al., 2010). The endoscope is attached to a light source (LED/Xenon) and using an endoscope with a chip-camera instead of a fiberoptic endoscope enhances the resolution (acronym: flexible endoscopic evaluation of swallowing). Endoscopic images can be displayed on a monitor and digitally recorded (possibly added with recordings from the patient's head/body posture by a second camera—picture-in-picture), and also used as visual biofeedback method. The FEES with sensory testing (FEESST) is a standardized extension of the method, that quantitatively tests the sensory thresholds performed by supplied air puffs to the arytenoids and laryngeal mucosa in order to test the patient's reaction of laryngeal adductor reflex (Aviv, Kim, Sacco, et al., 1998; Aviv, Murry, Zschommler, Cohen, & Gartner, 2005). In accordance with Langmore-standard, FEES is performed in three steps: (1) anatomic–physiologic assessment, (2) direct examination of swallowing food and liquids, and (3) test of therapeutic maneuvers.

Many different protocols exist, but no standard protocol has been established so far, which can be used for a Parkinson-specific FEES assessment. Usually, the penetration–aspiration scale (PAS) (Rosenbek, Robbins, Roecker, Coyle, & Wood, 1996) is used to evaluate the severity on laryngeal bolus aspiration and the patient's ability on airway clearance, sometimes complemented by Murray's clearance guidelines (Murray, 1999). The first standardized PD-specific FEES protocol (PARK-FEES) was recently developed and tested for multiprofessional interrater reliability presenting good intercoder agreement (Simons et al., 2014; Simons, von Clarmann, & Warnecke, 2016). PARK-FEES consists of 10 parameters rating the patient's performance in the anatomic–physiological as well as in the swallow exams (e.g., pharyngeal secretion accumulation, premature bolus leakage, pharyngeal residues, laryngeal penetration/aspiration according to

PAS, spontaneously airway clearance ability). Severity of the dysphagia symptomatology is rated for each nutrition consistency individually and for both typical performance and potential outliner (maximum value). Standard swallow examples are the same like in the above-mentioned clinical PD-specific assessment (90-mL water, half slice of bread with crust, butter chocolate cookie, and two uncoated placebo pills/divisible tablets). Fluctuations of Levodopa on–off state have also to be considered. The FEES-Levodopa-Test (Warnecke et al., 2010, 2016) evaluates the responsiveness of pharmacological therapy with Levodopa in Parkinson-related dysphagia. First, patients are endoscopically examined in their medication off-state, whereby premature spillage, pharyngeal residues, and penetration/aspiration are assessed while swallowing liquids, semisolid, and solid food. After oral Levodopa application evaluation should be repeated in on-state. Improvement of swallowing performance resulting in a reduced score with a minimum of 30% indicates levodopa responsiveness. The test complements dysphagia diagnosis and might help to plan further therapy options.

Fig. 2 shows main findings of dysphagia among PD patients during FEES.

VFSS, which is based on radiological imaging at an X-ray department (sometimes also known as modified barium swallowing examination) was the first instrumental procedure for dysphagia assessments. Patients are asked to take barium mixed swallows of different nutrition consistencies during the examination process, whereby the bolus is visualized (25–30 frames/s) on its way from the oral cavity to the stomach. VFSS therefore allows effectively evaluating the oral and pharyngeal swallow phase including the upper esophageal sphincter (UES). According to Logemann-standard, the examination begins with lateral X-rays and ends with a short sequence of anterior–posterior X-rays. Images are recorded on video, or more commonly digitally

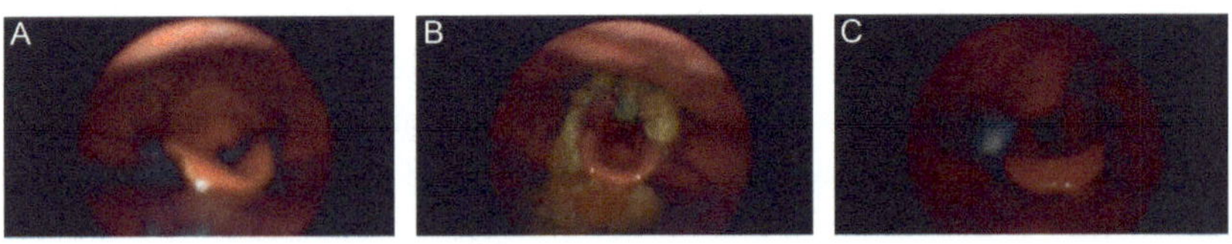

Fig. 2 Oropharyngeal findings under endoscopic view. (A) Premature spillage of thin liquid with predeglutitive laryngeal aspiration over the left aryepiglottic fold and cuneiform cartilage. (B) Bolus accumulation of bread in the epiglottic area, epiglottic vallecula, and piriform sinus with penetrating the posterior commissure/vocal cords over the cuneiform and comiculate cartilages, arytenoids, and postcricoid region. (C) Residue of a tablet in the right aryepiglottic fold.

(digital spot imaging, DSI). High frequency cinematography allowing 50–200 frames/s is reserved for scientific questions in the first place. Radiation dose for a usual diagnostic evaluation is about 3–5 mSv, and faster sequences even more. Evaluation should focus on, e.g., oral onset time, oral and pharyngeal transit time, triggering of swallow reflex, anterior–superior movement of hyoid bone, duration of velopharyngeal closure, amount of aspiration, amount of residues in epiglottic vallecula, posterior pharyngeal wall and piriform sinus, and duration and expanse of the UES opening. The previously mentioned PAS was originally developed for VFSS and is equally valid and reliable for both VFSS and FEES (Rosenbek et al., 1996).

High-resolution manometry (HRM) is another complementary instrumental diagnostic procedure, which is useful to assess disorders of esophageal peristalsis that could occur in combination with oropharyngeal symptoms or as isolated esophageal motor dysfunction. HRM offers a detailed evaluation of pressure and timing events during swallowing in the pharyngoesophageal segment. Transnasally used catheters with pressure sensors (spaced 1–2 cm apart) are positioned from the pharynx or upper esophagus to the stomach in order to generate pressure simultaneously. Pressure output is processed by a software that uses interpolation generating esophageal pressure topography plots representing esophageal motility and sphincter function on color-coded, pressure-space–time plots. In contrast to conventional manometry, the improved method of HRM across the length of the entire pharynx, starting from velopharynx (e.g., tongue base) to the cervical esophagus (e.g., UES) allows a more accurate and reliable diagnosis of pharyngeal dysfunctions and esophageal motility disorders as well as identification of former undetectable subtle findings (Carlson & Pandolfino, 2015; McCulloch, Hoffman, & Ciucci, 2010).

Future research should be focused on further reliable and PD-specific protocols for each of the mentioned diagnostic approaches. Those should serve as cross-national standard for dysphagia diagnosis and as the basis for comparable evaluation of functional changes or modification effects on swallowing abilities in medical practice as well as in clinical trials.

4. DYSPHAGIA TREATMENT

For rehabilitation of deglutition affected structures, sensory reflexes and functions several different therapy interventions have been developed and investigated over the past few years, which expand the repertoire of conventional logopedic therapy among patients with Parkinsonism.

Conservative functional logopedic training of the oropharyngeal complex consists of exercises to increase strength of motion of the tongue, hyoidale muscular, pharynx and larynx to ameliorate oral bolus control, to enhance coordination between swallowing and breathing, and to improve airway protection. In addition, swallowing maneuvers are trained to compensate oropharyngeal and laryngeal dysfunctions. Dietary adaptations including texture modification or preparation of food and liquids are aiming to avoid aspiration and make swallowing more efficient. Instrumental techniques can serve as visual/auditory biofeedback within the therapy process controlling performed maneuvers and applied dietary procedures. Whereas the possibilities of biofeedback training via VFSS is limited due to the ethical issues keeping radiation exposure to a minimum, FEES does not expose any radiation and additionally offers real-time biofeedback by placing the monitor clearly visible right next to the patient during endoscopy in order to facilitate monitoring and directly modifying the swallow process, or to train cleansing mechanisms. Further therapy approaches are based on pharmacological treatment or optimized drug therapy, respectively, and others comprise, for example, symptom-specific and apparatus-based interventions like breathing training with expiratory resistance. Novel technologies focus on invasive surgical interventions as well as on noninvasive experimental approaches that are intended to be used as alternative intervention or complementary to conventional dysphagia therapy. These have all been studied within several randomized controlled trials, case–control studies, or prospective cohort studies. Study design and therapy effectiveness were systematically reviewed regarding their underlying level of evidence (Baijens & Speyer, 2009; Michou, Baijens, Rofes, Sanz Cartagena, & Clavé, 2013; Michou & Hamdy, 2010; van Hooren, Baijens, Voskuilen, Oosterloo, & Kremer, 2014).

Table 3 summarizes the most relevant treatment approaches in PD actually discussed in the scientific community that will be subsequently outlined for debating its impact on cough and swallow outcomes.

The expiratory muscle straightening training (EMST) seems to be effective in reducing aspiration episodes by increasing cough volume (Pitts et al., 2009; Troche et al., 2010) and might be a promising adjunctive therapy optimizing pharmacological and logopedic treatment. Future studies should be focused on a broader spectrum of PD patients regarding disease stage and swallowing severity.

Table 3 Salient Dysphagia Treatment Approaches in PD

PD-Related Dysphagia Treatments	
Logopedic interventions and complemented advices	– Conservative rehabilitative treatment (e.g., functional trainings, compensatory swallowing maneuvers and head/body postures, and adaptive dietary strategies and texture modification) – Expiratory muscle straightening training (EMST) – Instrumental techniques used as biofeedback (FEES, VFSS); video-assisted swallowing therapy (VAST) – Gum-chewing – Thermal–tactile stimulation (TTS)
Pharmaceutical therapy	– Optimized dopaminergic therapy (levodopa, apomorphine, rotigotine) – Botulinum toxin (type A)
Operative procedures and novel technologies	– Surface neuromuscular electrical stimulation (NMES) – Deep brain stimulation (DBS) – Repetitive transcranial magnetic stimulation (rTMS)

Video-assisted swallowing therapy (VAST), based on a visual cuing mechanism from FEES recordings, might help reducing food residue compared to conventional swallowing exercises and improving sense of favor (Manor, Mootanah, Freud, Giladi, & Cohen, 2013), but needs to be investigated and validated for longer follow-up periods.

Although Lee Silverman voice treatment (LSVT), originally designed for dysarthria treatment in PD patients, has shown positive results on swallowing performance among a small study cohort in mild to moderate disease stage (El Sharkawi et al., 2002), evidence discovered in larger follow-up studies is also absolutely necessary.

Swallow frequency can be increased and swallow latency can be decreased by gum chewing (South, Somers, & Jog, 2010), but no evidence could be revealed influencing the swallowing physiology in prandial context until now.

Thermal–tactile stimulation (TTS) of the anterior faucial pillars to improve pharyngeal swallowing reflex (Regan, Walshe, & Tobin, 2010) seems to provide short-term sensory optimization of impaired glossopharyngeal and vagus nerves with improvement of motor function, but transferability into clinically beneficial effects is limited so far.

After (single) sessions of surface neuromuscular electrical stimulation (NMES) of the neck using different electrode positions (healthy controlled or against traditional logopedic therapy) only slight significant improvements on swallowing could be observed (Baijens et al., 2012, 2013; Heijnen, Speyer, Baijens, & Bogaardt, 2012), and although no side-effects have been noticed carefully consideration should be done in medical practice whether or not to use at that time.

Botulinum toxin (type A) injections in the parotid glands might provide positive effects on sialorrhea in PD, lasting for about 30 days (Nobrega, Rodrigues, & Melo, 2009), but no significant reduction of penetration/aspiration episodes could be revealed.

Central dopaminergic treatment using L-dopa (L-3,4-dihydroxyphenylalanine, levodopa) revealed inhomogeneous results on therapy effectiveness regarding swallowing functions, ranging from some improvement of beginning dysphagia symptoms during the first year after disease onset, over significant improvement of swallowing in the on-state comparing to off-state conditions, to no clear responsiveness at all, whereby for the bigger proportion of patients study results tend to no meaningful clinical improvement of swallowing, and moreover, different methodological quality can be found in the literature (Menezes & Melo, 2009; Muller et al., 2013; Warnecke, Hamacher, Oelenberg, & Dziewas, 2014). Recently, assessed neurophysiologic mechanism and changes in brain activation as response on swallow tests in on/off-levodopa state in PD patients by using repetitive transcranial magnetic stimulation (rTMS) reveals that levodopa therapy might even negatively affect swallowing (Michou et al., 2014). However, answering the question, if dopaminergic therapy using levodopa might improve the swallowing process in combination with conventional logopedic or other interventions large randomized controlled trials are necessary. The same applies for other dopaminergic therapy pilot trials using subcunaneous apomorphine stimulation (in combination with domperidone) or rotigotine transdermal patch that were shown to be an alternative medical option probably improving dysphagia in PD, based on VFSS evaluation among very small cohorts (Hirano et al., 2015; Tison et al., 1996).

There are further studies investigating chin-tuck maneuvers vs a dietary adaptation of thicken up liquids, which on the one hand do not indicate any clear superiority of one strategy over the other regarding a cumulative 3-month pneumonia incidence (Robbins et al., 2008), or on the other hand discovering, that by performing chin-tuck maneuver aspiration episodes can be reduced, respectively, and patients themselves reported to have aversion on thickened drinks (Logemann et al., 2008). This suggests preferring chin-tuck maneuver in clinical practice, whenever the patient is able to perform adequately and safely.

Whereas most study results investigating deep brain stimulation (DBS) in dysphagic PD patients revealed no clinically significant improvement on swallowing functions, some showed slight positive effects and others no effect, or even deterioration. Additionally, stimulating the subthalamic nucleus (STN) has been considering causing more impairment to swallowing function than globus pallidus inernus stimulation, but no experimental study examined direct comparison of both stimulation loci, and also no comparison exists between unilateral vs bilateral DBS surgery, or coincident stimulation effects on swallowing functions (Sapir, Ramig, & Fox, 2008; Troche, Brandimore, Foote, & Okun, 2013). A recently performed small trial investigating 60-Hz DBS of bilateral STN compared with routine 130 Hz revealed first results showing significant improvement of aspiration frequency assessed by VFSS (as well as freezing of gait) using low-frequency stimulation that persist over the 6-weeks study period (Xie et al., 2015). Increased esophageal body contractions and enhanced opening of the lower esophageal sphincter after using bilateral STN stimulation for at least 6 month was further noted in another newer study confirmed by HRM, which suggests the involvement of the nigrostriatal–striatonigral loop in the esophageal motility control (Derrey et al., 2015).

Finally, first evidence is grown that the noninvasive approach of rTMS of the mouth motor cortex improves swallowing, e.g., used in acute stroke patients (Shin et al., 2012), and preliminary data also showed positive treating outcomes on phonation in PD (Sapir et al., 2008). Therapy effects on swallowing in PD patients need yet to be further investigated. Additionally, stimulation area, frequency, intensity, and duration need to be studied yielding the best treatment efficacy.

Conclusively, several treatment options have been examined in the past years, but there is only little valid evidence for certain approaches. It seems that combined treatment approaches would be beneficial, but much more work is necessary to improve the dysphagia management, and further well-designed future studies with larger patient cohorts of mildly to severely affected patients regarding dysphagia and PD stage, and focus on longer-terms are strictly needed to reveal more evidence on certain therapy outcome effects for PD-related oropharyngeal and also esophageal dysphagia. In addition, treatment studies, especially those of operative procedures and novel technologies often concentrate exclusively on aspiration as outcome criteria to assess swallow improvement, but should also take milder oropharyngeal symptoms into account. Defining valid treatment strategies are still major need and focus of future research in order to enhance patient care and quality of life.

5. SUMMARY AND REFERRALS

Swallowing dysfunction in PD patients was shown to be very common, wide-spread, and clinically relevant in the context of NMSs. Oropharyngeal and esophageal symptoms affect early and late-stage PD causing high psychosocial burden and leading to various of life-threatening consequences that need intensive multiprofessional care and therefore also has an impact of national health budgets.

Understanding the nature of PD-related dysphagia and identifying reliable physiological as well as clinical marker causing dysphagia in PD enables to better identify patients at risk. In addition, finding biomarker for disturbed swallowing processes describing syndrome differences allowing stratification on Parkinson patients and can be used in settings of differential-diagnosis. Gained pathophysiological information and classified syndrome patterns of dysphagia in Parkinsonism might also help estimating disease prognosis.

Furthermore, diagnosis and treatment of swallowing dysfunctions in PD are very complex and have become a demonstrator for a challenging interdisciplinary collaboration. Standardized dysphagia screening programs as well as early intervention might be an important cornerstone in PD to maintain or even improve quality of life, prevent patients from severe health threats as long as possible, and help to significantly reduce malnutrition, aspiration pneumonia, and mortality rates.

In general, systematic screening for swallowing disorders among patients at risk in the elderly of different etiology is recommended, including assessment of the nutritional status, which might help to reduce pneumonia rates (up to 55%) and duration of hospitalization, if they are implemented in patient care programs (Baijens et al., 2016; Kaspar & Ekberg, 2012). For the same reason, research based and professionally consented position statements were defined providing best clinical practice among European healthcare centers for patients at risk of neurodegenerative associated dysphagia (Baijens et al., 2016; European Society for Swallowing Disorders, 2013).

Anyway, although early screening among PD patients is recently also highly suggested in several national guidelines for Parkinson syndromes and neurogenic dysphagia comprising administration of disease-specific and patient-based questionnaires as well as clinical screenings/"bedside tests" in the first step, and followed by further swallowing evaluations using standardized protocols, if necessary, this is not established as routinely applied examination procedure in daily medical practice until now. This shall emphasize that more clearly nation-wide recommendations, binding standards, and training courses for adequate patient care is indispensable. Therefore, a comprehensive set of clinical and instrumental diagnostics for oropharyngeal as well as esophageal pattern characterization should be performed when screened dysphagia-positive. Different consistencies with liquids and food from everyday life including medication intake should be tested in sufficient quantity, and also in comparison of on- and off-state condition as well as at different times of the day.

With regard to Parkinson-related dysphagia treatment (e.g., dietary adaptation, functional therapy, swallow maneuvers, and complement novel technologies like neural stimulation procedures), future research is strongly needed in order to increase evidence on certain therapy approaches. Being able to define the optimal strategy for PD patients or for specific subgroups of PD, respectively, there is a need of large randomized controlled trials assessing the effectiveness of different therapy options as well as the right time of treatment (early vs late), and therapeutic stability of clinical effects (short-term and long-term effects). In actual clinical practice treatment should be selected symptom-orientated on individual needs. Therefore, testing effectiveness of different options should be performed before starting permanent treatment, and considering individual possibilities as well as the patient's perspective. Shared decision making is also important with regard to nutritional management and dietary restrictions, weighting carefully potential clinical benefits against individual constraints.

Concluding statement: The primary goal of dysphagia research and management has always been, and still is to decrease negative consequences to health and to support high standard of quality of life. *With knowing comes caring, with caring comes change*!

REFERENCES

Alfonsi, E., Versino, M., Merlo, I. M., Pacchetti, C., et al. (2007). Electrophysiologic patterns of oral-pharyngeal swallowing in parkinsonian syndromes. *Neurology*, *68*(8), 583–590.

Ali, G. N., Wallace, K. L., Schwarz, R., DeCarle, D. J., Zagami, A. S., & Cook, I. J. (1996). Mechanism of oral-pharyngeal dysphagia in patients with Parkinson's disease. *Gastroenterology*, *110*(2), 383–392.

Aviv, J. E., Kim, T., Sacco, R. L., et al. (1998). FEESST: A new bedside endoscopic test of the motor and sensory components of swallowing. *The Annals of Otology, Rhinology, and Laryngology*, *107*(5 Pt. 1), 378–387.

Aviv, J., Murry, T., Zschommler, A., Cohen, M., & Gartner, C. (2005). Flexible endoscopic evaluation of swallowing with sensory testing: Patient characteristics and analysis of safety in 1,340 consecutive examinations. *The Annals of Otology, Rhinology, and Laryngology*, *114*(3), 173–176.

Baijens, L. W., Clave, P., Cras, P., Ekberg, O., Forster, A., Kolb, G. F., et al. (2016). European society for swallowing disorders—European union geriatric medicine society white paper: Oropharyngeal dysphagia as a geriatric syndrome. *Clinical Interventions in Aging*, *11*, 1403–1428. http://dx.doi.org/10.2147/cia.s107750.

Baijens, L. W., & Speyer, R. (2009). Effects of therapy for dysphagia in Parkinson's disease: Systematic review. *Dysphagia*, *24*(1), 91–102. http://dx.doi.org/10.1007/s00455-008-9180-1.

Baijens, L. W., Speyer, R., Passos, V. L., Pilz, W., Roodenburg, N., & Clave, P. (2012). The effect of surface electrical stimulation on swallowing in dysphagic Parkinson patients. *Dysphagia*, *27*(4), 528–537. http://dx.doi.org/10.1007/s00455-011-9387-4.

Baijens, L. W., Speyer, R., Passos, V. L., Pilz, W., van der Kruis, J., Haarmans, S., et al. (2013). Surface electrical stimulation in dysphagic Parkinson patients: A randomized clinical trial. *Laryngoscope*, *123*(11), E38–E44. http://dx.doi.org/10.1002/lary.24119.

Barichella, M., Cereda, E., Madio, C., Iorio, L., Pusani, C., Cancello, R., et al. (2013). Nutritional risk and gastrointestinal dysautonomia symptoms in Parkinson's disease outpatients hospitalised on a scheduled basis. *British Journal of Nutrition*, *110*(2), 347–353. http://dx.doi.org/10.1017/S0007114512004941.

Barichella, M., Cereda, E., & Pezzoli, G. (2009). Major nutritional issues in the management of Parkinson's disease. *Movement Disorders*, *24*(13), 1881–1892.

Barone, P., Antonini, A., Colosimo, C., Marconi, R., Morgante, L., Avarello, T. P., et al. (2009). The Priamo study: A multicenter assessment of nonmotor symptoms and their impact on quality of life in Parkinson's disease. *Movement Disorders*, *24*(11), 1641–1649.

Bird, M. R., Woodward, M. C., Gibson, E. M., Phyland, D. J., & Fonda, D. (1994). Asymptomatic swallowing disorders in eldery patients with Parkinson's disease: A description of findings on clinical examination and videoflouroscopy in sixteen patients. *Age and Aging*, *23*, 251–254.

Braak, H., Müller, C. M., Rüb, U., Ackermann, H., Bratzke, H., De Vos, R. A., et al. (2006). Pathology associated with sporadic Parkinson's disease—Where does it end? *Journal of Neural Transmission. Supplementum*, *70*, 89–97.

Carlson, D. A., & Pandolfino, J. E. (2015). High-resolution manometry in clinical practice. *Gastroenterology & Hepatology*, *11*(6), 374–384.

Castell, J. A., Johnston, B. T., Colcher, A., Li, Q., Gideon, R. M., & Castell, D. O. (2001). Manometric abnormalities of the oesophagus in patients with Parkinson's disease. *Neurogastroenterology and Motility: The Official Journal of the European Gastrointestinal Motility Society*, *13*(4), 361–364.

Cereda, E., Cilia, R., Klersy, C., Canesi, M., Zecchinelli, A. L., Mariani, C. B., et al. (2014). Swallowing disturbances in Parkinson's disease: A multivariate analysis of contributing factors. *Parkinsonism & Related Disorders*, *20*(12), 1382–1387. http://dx.doi.org/10.1016/j.parkreldis.2014.09.031.

Chaudhuri, K. R., Healy, D. G., & Schapira, H. V. (2006a). Non-motor symptoms of Parkinson's disease—Diagnosis and management. *The Lancet. Neurology*, *5*(3), 235–245.

Chaudhuri, K. R., Martinez-Martin, P., Schapira, A. H., Stocchi, F., Sethi, K., Odin, P., et al. (2006b). International multicenter pilot study of the first comprehensive self-completed nonmotor symptoms questionnaire for Parkinson's disease: The NMSQuest study. *Movement Disorders*, *21*(7), 916–923. http://dx.doi.org/10.1002/mds.20844.

Chaudhuri, K. R., Prieto-Jurcynska, C., Naidu, Y., Mitra, T., Frades-Payo, B., Tluk, S., et al. (2010). The nondeclaration of nonmotor symptoms of Parkinson's disease to health care professionals: An international study using the nonmotor symptoms questionnaire. *Movement Disorders*, *25*(6), 704–709. http://dx.doi.org/10.1002/mds.22868.

Chou, K. L., Evatt, M., Hinson, V., & Kompoliti, K. (2007). Sialorrhea in Parkinson's disease: A review. *Movement Disorders*, *22*(16), 2306–2313.

Coelho, M., Marti, M. J., Tolosa, E., Ferreira, J. J., Valldeoriola, F., Rosa, M., et al. (2010). Late-stage Parkinson's disease: The Barcelona and Lisbon cohort. *Journal of Neurology*, *257*(9), 1524–1532. http://dx.doi.org/10.1007/s00415-010-5566-8.

D'amelio, M., Ragonese, P., Morgante, L., Reggio, A., Callari, G., Salemi, G., et al. (2006). Long-term survival of Parkinson's disease: A population-based study. *Journal of Neurology*, *253*(1), 33–37.

Davydov, L., & Botts, S. R. (2000). Clozapine-induced hypersalviation. *The Annals of Pharmacotherapy*, *34*(5), 662–665.

Derrey, S., Chastan, N., Maltete, D., Verin, E., Dechelotte, P., Lefaucheur, R., et al. (2015). Impact of deep brain stimulation on pharyngo-esophageal motility: A randomized crossover study. *Neurogastroenterology and Motility*, *27*(9), 1214–1222. http://dx.doi.org/10.1111/nmo.12607.

Ebihara, S., Saito, H., Kanda, A., Nakajoh, M., Takahashi, H., Arai, H., et al. (2003). Impaired efficacy of cough in patients with Parkinson's disease. *Chest*, *124*(3), 1009–1015.

Edwards, L. L., Quigley, E. M., Hofman, R., & Pfeiffer, R. F. (1993). Gastrointestinal symptoms in Parkinson's disease: 18-month follow-up study. *Movement Disorders*, *8*(1), 83–86.

Edwards, L. L., Quigley, E. M., & Pfeiffer, R. F. (1992). Gastrointestinal dysfunction in Parkinson's disease: Frequency and pathophysiology. *Neurology*, *42*(4), 726–732.

El Sharkawi, A., Ramig, L., Logemann, J. A., Pauloski, B. R., Rademaker, A. W., Smith, C. H., et al. (2002). Swallowing and voice effects of Lee Silverman voice treatment (LSVT): A pilot study. *Journal of Neurology, Neurosurgery, and Psychiatry*, *72*(1), 31–36.

European Society for Swallowing Disorders. (2013). Position statements and meeting abstracts uniting Europe against dysphagia Barcelona, Spain Oct 25–27, 2012. *Dysphagia*, *28*, 288–330.

Evatt, M. L., Chaudhuri, K. R., Chou, K. L., Cubo, E., Hinson, V., Kompoliti, K., et al. (2009). Dysautonomia rating scales in Parkinson's disease: Sialorrhea, dysphagia, and constipation—Critique and recommendations by movement disorders task force on rating scales for Parkinson's disease. *Movement Disorders: Official Journal of the Movement Disorder Society*, *24*(5), 635–646.

Fernandez, H., & Lapane, K. (2002). Predictors of mortality among nursing home residents with a diagnosis of Parkinso's disease. *Medical Science Monitor*, *8*(4), 241–246.

Fuh, J., Lee, R., Wang, S., Lin, C., Wang, P., Chiang, J., et al. (1997). Swallowing difficulty in Parkinson's disease. *Clinical Neurology and Neurosurgery*, *99*(2), 106–112.

Goetz, C. G., Fahn, S., Martinez-Martin, P., Poewe, W., Sampaio, C., Stebbins, G. T., et al. (2006). Movement disorder society-sponsored revision of the unified Parkinson's disease rating scale (MDS-UPDRS): Process, format, and clinimetric testing plan. *Movement Disorders*, *22*(1), 41–47.

Han, M., Ohnishi, H., Nonaka, M., Yamauchi, R., Hozuki, T., Hayashi, T., et al. (2011). Relationship between dysphagia and depressive states in patients with Parkinson's disease. *Parkinsonism & Related Disorders*, *17*(6), 437–439.

Heijnen, B. J., Speyer, R., Baijens, L. W., & Bogaardt, H. C. (2012). Neuromuscular electrical stimulation versus traditional therapy in patients with Parkinson's disease and oropharyngeal dysphagia: Effects on quality of life. *Dysphagia*, *27*(3), 336–345. http://dx.doi.org/10.1007/s00455-011-9371-z.

Hely, M., Morris, J., Traficante, R., Reid, W., O'Sullivan, D., & Williamson, P. (1999). The Sydney multicentre study of Parkinson's disease: Progression and mortality at 10 years. *Journal of Neurology, Neurosurgery, and Psychiatry*, *67*(3), 300–307.

Hirano, M., Isono, C., Sakamoto, H., Ueno, S., Kusunoki, S., & Nakamura, Y. (2015). Rotigotine transdermal patch improves swallowing in Dysphagic patients with Parkinson's disease. *Dysphagia*, *30*(4), 452–456. http://dx.doi.org/10.1007/s00455-015-9622-5.

Johnston, B. T., Castell, J. A., Stumacher, S., Colcher, A., Gideon, R. M., Li, Q., et al. (1997). Comparison of swallowing function in Parkinson's disease and progressive supranuclear palsy. *Movement Disorders*, *12*(3), 322–327.

Johnston, B. T., Li, Q., Castell, J. A., & Castell, D. O. (1995). Swallowing and esophageal function in Parkinson's disease. *American Journal of Gastroenterology*, *90*(10), 1741–1746.

Kalf, J. G., Bloem, B. R., & Munneke, M. (2012b). Diurnal and nocturnal drooling in Parkinson's disease. *Journal of Neurology*, *259*(1), 119–123.

Kalf, J., de Swart, B., Bloem, B., & Munneke, M. (2012a). Prevalence of oropharyngeal dysphagia in Parkinson's disease: A meta-analysis. *Parkinsonism & Related Disorders*, *18*(4), 311–315. http://dx.doi.org/10.1016/j.parkreldis.2011.11.006.

Kaspar, K., & Ekberg, O. (2012). Identifying vulnerable patients: Role of the EAT-10 and the multidisciplinary team for early intervention and comprehensive dysphagia care. *Nestlé Nutrition Institute Workshop Series*, *72*, 19–31.

Lam, K., Kwai, Y., Lam, F., Kwong Lau, K., Kay Chan, Y., Yee Ling, K. E., et al. (2007). Simple clinical tests may predict severe oropharyngeal dysphagia in Parkinson's disease. *Movement Disorders*, *22*(5), 640–644.

Langmore, S. E., & Aviv, J. E. (2001). Endoscopic procedures to evaluate oropharyngeal swallowing. In S. E. Langmore (Ed.), *Endoscopic evaluation and treatment of swallowing disorders* (pp. 73–100). Stuttgart, New York: Thieme.

Lee, K. D., Koo, J. H., Song, S. H., Jo, K. D., Lee, M. K., & Jang, W. (2015). Central cholinergic dysfunction could be associated with oropharyngeal dysphagia in early Parkinson's disease. *Journal of Neural Transmission (Vienna, Austria: 1996)*, *122*(11), 1553–1561. http://dx.doi.org/10.1007/s00702-015-1427-z.

Leopold, N., & Daniels, S. (2010). Supranuclear control of swallowing. *Dysphagia*, *25*(3), 250–257.

Leopold, N. A., & Kagel, M. C. (1997). Pharyngo-esophageal dysphagia in Parkinson's disease. *Dysphagia*, *12*, 11–18.

Leow, L., Beckert, L., Anderson, T., & Huckabee, M. (2012). Changes in chemosensitivity and mechanosensitivity in aging and Parkinson's disease. *Dysphagia*, *27*(1), 106–114. http://dx.doi.org/10.1007/s00455-011-9347-z.

Leow, L., Huckabee, M. L., Anderson, T., & Beckert, L. (2010). The impact of dysphagia and quality of life and Parkinson's disease as measured by quality of life (SWAL-QOL) questionnaire. *Dysphagia*, *25*(3), 216–220.

Leslie, P., Drinnan, M. J., Ford, G. A., & Wilson, J. A. (2005). Swallow respiratory patterns and aging: Presbyphagia or dysphagia? *The Journals of Gerontology. Series A, Biological Sciences and Medical Sciences*, *60*(3), 391–395.

Logemann, J. A. (1998). *Evaluation and treatment of swallowing disorders*. Austin (Texas): Pro-Ed.

Logemann, J. A., Gensler, G., Robbins, J., Lindblad, A. S., Brandt, D., Hind, J. A., et al. (2008). A randomized study of three interventions for aspiration of thin liquids in patients with dementia or Parkinson's disease. *Journal of Speech, Language, and Hearing Research*, *51*(1), 173–183. http://dx.doi.org/10.1044/1092-4388(2008/013).

Makaroff, L., Gunn, A., Gervasoni, C., & Richy, F. (2011). Gastrointestinal disorders in Parkinson's disease: Prevalence and health outcomes in a US claims database. *Journal of Parkinson's Disease*, *1*(1), 65–74. http://dx.doi.org/10.3233/jpd-2011-001.

Manor, Y., Balas, M., Giladi, N., Mootanah, R., & Cohen, J. T. (2009). Anxiety, depression and swallowing disorders in patients with Parkinson's disease. *Parkinsonism and Related Disorders, 15*(6), 453–456.

Manor, Y., Giladi, N., Cohen, A., Fliss, D., & Cohen, J. (2007). Validation of a swallowing disturbance questionnaire for detecting dysphagia in patients with Parkinson's disease. *Movement Disorders, 22*(13), 1917–1921.

Manor, Y., Mootanah, R., Freud, D., Giladi, N., & Cohen, J. T. (2013). Video-assisted swallowing therapy for patients with Parkinson's disease. *Parkinsonism & Related Disorders, 19*(2), 207–211. http://dx.doi.org/10.1016/j.parkreldis.2012.10.004.

Mari, F., Matei, M., Ceravolo, M. G., Pisani, A., Montesini, A., & Provinciali, L. (1997). Predictive value of clinical indices in detecting aspiration in patients with neurological disorders. *Journal of Neurology, Neurosurgery, and Psychiatry, 63*(4), 456–460.

Marks, L., Turner, K., O'Sullivan, J., Deighton, B., & Lees, A. (2001). Drooling in Parkinson's disease: A novel speech and language therapie intervention. *International Journal of Language & Communication Disorders, 36*, 282–287.

McCulloch, T. M., Hoffman, M. R., & Ciucci, M. R. (2010). High-resolution manometry of pharyngeal swallow pressure events associated with head turn and chin tuck. *The Annals of Otology, Rhinology, and Laryngology, 119*(6), 369–376.

Menezes, C., & Melo, A. (2009). Does levodopa improve swallowing dysfunction in Parkinson's disease patients? *Journal of Clinical Pharmacy and Therapeutics, 34*(6), 673–676. http://dx.doi.org/10.1111/j.1365-2710.2009.01031.x.

Michou, E., Baijens, L., Rofes, L., Sanz Cartagena, P., & Clavé, P. (2013). Oropharyngeal swallowing disorders in Parkinson's disease: Revisited. *International Journal of Speech & Language Pathology and Audiology, 1*, 76–88. http://dx.doi.org/10.12970/2311-1917.2013.01.02.5.

Michou, E., & Hamdy, S. (2010). Dysphagia in Parkinson's disease: A therapeutic challenge? *Expert Review of Neurotherapeutics, 10*(6), 875–878. http://dx.doi.org/10.1586/ern.10.60.

Michou, E., Hamdy, S., Harris, M., Vania, A., Dick, J., Kellett, M., et al. (2014). Characterization of corticobulbar pharyngeal neurophysiology in dysphagic patients with Parkinson's disease. *Clinical Gastroenterology and Hepatology: The Official Clinical Practice Journal of the American Gastroenterological Association, 12*(12), 2037–2045, e2031–2034. http://dx.doi.org/10.1016/j.cgh.2014.03.020.

Miller, N., Allcock, L., Hildreth, A. J., Jones, D., Noble, E., & Burn, D. J. (2009). Swallowing problems in Parkinson disease: Frequency and clinical correlates. *Journal of Neurology, Neurosurgery, and Psychiatry, 80*(9), 1047–1049.

Monte, F. S., Da Silva-Junior, F. P., Braga-Neto, P., Nobre e Souza, M. A., & Sales de Bruin, V. M. (2005). Swallowing abnormalities and dyskinesia in Parkinson's disease. *Movement Disorders, 20*(4), 457–462.

Moreau, C., Ozsancak, C., Blatt, J. L., et al. (2007). Oral festination in Parkinson's disease: Biomechanical analysis and correlation with festination and freezing of gait. *Movement Disorders: Official Journal of the Movement Disorder Society, 22*(10), 1503–1506.

Mu, L., Sobotka, S., Chen, J., et al. (2013). Parkinson's disease affects peripheral sensory nerves in the pharynx. *Journal of Neuropathology and Experimental Neurology, 72*(7), 614–623.

Muller, B., Assmus, J., Larsen, J. P., Haugarvoll, K., Skeie, G. O., & Tysnes, O. B. (2013). Autonomic symptoms and dopaminergic treatment in de novo Parkinson's disease. *Acta Neurologica Scandinavica, 127*(4), 290–294. http://dx.doi.org/10.1111/ane.12010.

Müller, J., Wenning, G., Verny, M., McKee, A., Chaudhuri, K. R., Jellinger, K., et al. (2001). Progression of dysarthria and dysphagia in postmortem-confirmed parkinsonian disorders. *Archives of Neurology, 58*(2), 259–264.

Murray, J. (1999). *Manual of dysphagia assessment in adults. The laryngoscopic evaluation of swallowing or FEES.* San Diego, London: Singular Publishing Group.

Nobrega, A. C., Rodrigues, B., & Melo, A. (2009). Does botulinum toxin injection in parotid glands interfere with the swallowing dynamics of Parkinson's disease patients? *Clinical Neurology and Neurosurgery*, *111*(5), 430–432. http://dx.doi.org/10.1016/j.clineuro.2008.12.007.

Nóbrega, A. C., Rodrigues, B., Torres, A. C., Scarpel, R., Neves, C. A., & Melo, A. (2008). Is drooling secondary to a swallowing disorder in patients with Parkinson's disease? *Parkinsonism and Related Disorders*, *14*(3), 243–245.

Noyce, A., Silveira-Moriyama, L., Gilpin, P., Ling, H., Howard, R., & Lees, A. (2012). Severe dysphagia as a presentation of Parkinson's disease. *Movement Disorders*, *27*(3), 457–458.

Parkinson, J. (2002). An essay on the shaking palsy, 1817. *The Journal of Neuropsychiatry and Clinical Neurosciences*, *14*(2), 222–236.

Pehlivan, M., Yüceyar, N., Ertekin, C., Çelebi, G., Ertaş, M., Tulga, K., et al. (1996). An electronic device measuring the frequency of spontaneous swallowing: Digital phagometer. *Dysphagia*, *11*(4), 259–264. http://dx.doi.org/10.1007/BF00265212.

Pennington, S., Snell, K., Lee, M., & Walker, R. (2010). The cause of death in idiopathic Parkinson's disease. *Parkinsonism & Related Disorders*, *16*(7), 434–437.

Perez-Lloret, S., Negre-Pages, L., Ojero-Senard, A., Damier, P., Destee, A., Tison, F., et al. (2012). Oro-buccal symptoms (dysphagia, dysarthria, and sialorrhea) in patients with Parkinson's disease: Preliminary analysis from the French COPARK cohort. *European Journal of Neurology*, *19*(1), 28–37. http://dx.doi.org/10.1111/j.1468-1331.2011.03402.x.

Pfeiffer, R. F. (2003). Gastrointestinal dysfunction in Parkinson's disease. *Lancet Neurology*, *2*(2), 107–116.

Pinnington, L. L., Muhiddin, K. A., Ellis, R. E., & Playford, E. D. (2000). Noninvasive assessment of swallowing and respiration in Parkinson's disease. *Journal of Neurology*, *247*, 773–777.

Pitts, T., Bolser, D., Rosenbek, J., Troche, M., Okun, M. S., & Sapienza, C. (2009). Impact of expiratory muscle strength training on voluntary cough and swallow function in Parkinson disease. *Chest*, *135*(5), 1301–1308. http://dx.doi.org/10.1378/chest.08-1389.

Plowman-Prine, E., Sapienza, C., Okun, M., Pollock, S., Jacobson, C., Wu, S., et al. (2009). The relationship between quality of life and swallowing in Parkinson's disease. *Movement Disorders*, *24*(9), 1352–1358.

Potulska, A., Friedman, A., Krolicki, L., & Spychala, A. (2003). Swallowing disorders in Parkinson's disease. *Parkinsonism & Related Disorders*, *9*(6), 349–353.

Proulx, M., de Courval, F. P., Wiseman, M. A., & Panisset, M. (2005). Salivary production in Parkinson's disease. *Movement Disorders*, *20*(2), 204–207. http://dx.doi.org/10.1002/mds.20189.

Regan, J., Walshe, M., & Tobin, W. O. (2010). Immediate effects of thermal-tactile stimulation on timing of swallow in idiopathic Parkinson's disease. *Dysphagia*, *25*(3), 207–215. http://dx.doi.org/10.1007/s00455-009-9244-x.

Robbins, J., Gensler, G., Hind, J., Logeman, J. A., Lindblad, A. S., Brand, D., et al. (2008). Comparison of 2 interventions for liquid aspiration on pneumonia incidence. A randomized trial. *Annals of Internal Medicine*, *148*(7), 509–518.

Rodrigues, B., Nóbrega, C. A., Sampaio, M., Argolo, N., & Melo, A. (2011). Silent saliva aspiration in Parkinson's disease. *Movement Disorders*, *26*(1), 138–141.

Rosenbek, J. C., Robbins, J. A., Roecker, E. B., Coyle, J. L., & Wood, J. L. (1996). A penetration-aspiration scale. *Dysphagia*, *11*(2), 93–98.

Sapir, S., Ramig, L., & Fox, C. (2008). Speech and swallowing disorders in Parkinson disease. *Current Opinion in Otolaryngology & Head and Neck Surgery*, *16*(3), 205–210. http://dx.doi.org/10.1097/MOO.0b013e3282febd3a.

Shin, J., Yang, E., Cho, K., Barcenas, C. L., Kim, W. J., Min, Y., et al. (2012). Clinical application of repetitive transcranial magnetic stimulation in stroke rehabilitation. *Neural Regeneration Research*, 7(8), 627–634. http://dx.doi.org/10.3969/j.issn.1673-5374.2012.08.011.

Simons, J. A., Eisemann, N., Fietzek, U. M., & Katalinic, A. (2015). High-risk defining clinical parameters for dysphagia in Parkinson's disease. In *Movement Disorders: 30 (Suppl. 1). Abstracts of the Nineteenth International Congress of Parkinson's Disease and Movement Disorders, June 13–18, 2015; San Diego, California, USA* (pp. S422–S423). http://dx.doi.org/10.1002/mds.26295.

Simons, J. A., Fietzek, U. M., Waldmann, A., Warnecke, T., Schuster, T., & Ceballos-Baumann, A. O. (2014). Development and validation of a new screening questionnaire for dysphagia in early stages of Parkinson's disease. *Parkinsonism & Related Disorders*, *20*(9), 992–998. http://dx.doi.org/10.1016/j.parkreldis.2014.06.008.

Simons, J. A., von Clarmann, S., & Warnecke, T. (2016). Reliability of a newly developed protocol for fiberoptic endoscopic evaluation of swallowing in Parkinson's patients (PARK-FEES). In *Movement Disorders: 31 (Suppl. 2). Abstracts of the Twentieth International Congress of Parkinson's Disease and Movement Disorders* (pp. S517–S518). http://dx.doi.org/10.1002/mds.26688.

South, A. R., Somers, S. M., & Jog, M. S. (2010). Gum chewing improves swallow frequency and latency in Parkinson patients: A preliminary study. *Neurology*, *74*(15), 1198–1202.

Su, A., Gandhy, R., Barlow, C., & Triadafilopoulos, G. (2017). Clinical and manometric characteristics of patients with Parkinson's disease and esophageal symptoms. *Diseases of the Esophagus*, *30*(4), 1–6. http://dx.doi.org/10.1093/dote/dow038.

Suiter, D. M., & Leder, S. B. (2008). Clinical utility of the 3-ounce water swallow test. *Dysphagia*, *23*(3), 244–250.

Sung, H. Y., Kim, J. S., Lee, K. S., Kim, Y. I., Song, I. U., Chung, S. W., et al. (2010). The prevalence and patterns of pharyngoesophageal dysmotility in patients with early stage Parkinson's disease. *Movement Disorders*, *25*(14), 2361–2368. http://dx.doi.org/10.1002/mds.23290.

Suntrup, S., Teismann, I., Bejer, J., Suttrup, I., Winkels, M., Mehler, D., et al. (2013). Evidence for adaptive cortical changes in swallowing in Parkinson's disease. *Brain*, *136*(3), 726–738.

Suttrup, I., & Warnecke, T. (2016). Dysphagia in Parkinson's disease. *Dysphagia*, *31*(1), 24–32.

Thomas, M., & Haigh, R. (1995). Dysphagia, a reversible cause not to be forgotten. *Postgraduate Medical Journal*, *71*(832), 94–95.

Tison, F., Wiart, L., Guatterie, M., Fouillet, N., Lozano, V., Henry, P., et al. (1996). Effects of central dopaminergic stimulation by apomorphine on swallowing disorders in Parkinson's disease. *Movement Disorders*, *11*(6), 729–732. http://dx.doi.org/10.1002/mds.870110622.

Tohara, H., Nakane, A., Murata, S., Mikushi, S., Ouchi, Y., Wakasugi, Y., et al. (2010). Inter- and intra-rater reliability in fibroptic endoscopic evaluation of swallowing. *Journal of Oral Rehabilitation*, *37*(12), 884–891. http://dx.doi.org/10.1111/j.1365-2842.2010.02116.x.

Troche, M. S., Brandimore, A. E., Foote, K. D., & Okun, M. S. (2013). Swallowing and deep brain stimulation in Parkinson's disease: A systematic review. *Parkinsonism & Related Disorders*, *19*(9), 783–788. http://dx.doi.org/10.1016/j.parkreldis.2013.05.001.

Troche, M., Brandimore, A., Okun, M., Davenprot, P., & Hegland, K. (2014). Decreased cough sensitivity and aspiration in Parkinson disease. *Chest*, *146*, 294–299.

Troche, M. S., Huebner, I., Rosenbek, J. C., et al. (2011). Respiratory-swallowing coordination and swallowing safety in patients with Parkinson's disease. (Publication no.

http://dx.doi.org/10.1007/s00455-010-9289-x). from Published online 11.07.2010 *Dysphagia*, *26*, 218–224. http://dx.doi.org/10.1007/s00455-010-9289-x.

Troche, M. S., Okun, M. S., Rosenbek, J. C., Musson, N., Fernandez, H. H., Rodriguez, R., et al. (2010). Aspiration and swallowing in Parkinson disease and rehabilitation with EMST: A randomized trial. *Neurology*, *75*, 1912.

Tumilasci, O. R., Cerosimo, M. G., Belforte, J. E., Micheli, F. E., Benarroch, E. E., & Pazo, J. H. (2006). Quantitative study of salivary secretion in Parkinson's disease. *Movement Disorders*, *21*(5), 660–667.

Umemoto, G., Tsuboi, Y., Kitashima, A., et al. (2011). Impaired food transportion in Parkinson's disease related to Lingual Bradykinesia. *Dysphagia*, *26*(3), 250–255.

van Hooren, M. R., Baijens, L. W., Vos, R., Pilz, W., Kuijpers, L. M., Kremer, B., et al. (2015). Voice- and swallow-related quality of life in idiopathic Parkinson's disease. *The Laryngoscope*, *126*(2), 408–414. http://dx.doi.org/10.1002/lary.25481.

van Hooren, M. R., Baijens, L. W., Voskuilen, S., Oosterloo, M., & Kremer, B. (2014). Treatment effects for dysphagia in Parkinson's disease: A systematic review. *Parkinsonism & Related Disorders*, *20*(8), 800–807. http://dx.doi.org/10.1016/j.parkreldis.2014.03.026.

Volonte, M. A., Porta, M., & Comi, G. (2002). Clinical assessment of dysphagia in early phases of Parkinson's disease. *Neurological Sciences: Official Journal of the Italian Neurological Society and of the Italian Society of Clinical Neurophysiology*, *23*(2), 121–122.

Wang, X., You, G., Chen, H., & Cai, X. (2002). Clinical course and cause of death in elderly patients with idiopathic Parkinson's disease. *Chinese Medical Journal*, *115*(9), 1409–1411.

Warnecke, T., Hamacher, C., Oelenberg, S., & Dziewas, R. (2014). Off and on state assessment of swallowing function in Parkinson's disease. *Parkinsonism & Related Disorders*, *20*(9), 1033–1034. http://dx.doi.org/10.1016/j.parkreldis.2014.06.016.

Warnecke, T., Oelenberg, S., Teismann, I., Hamacher, C., Lohmann, H., Ringelstein, E., et al. (2010a). Endoscopic characteristics and levodopa responsiveness of swallowing function in progressive supranuclear palsy. *Movement Disorders*, *25*(9), 1239–1245.

Warnecke, T., Suttrup, I., Schroder, J. B., Osada, N., Oelenberg, S., Hamacher, C., et al. (2016). Levodopa responsiveness of dysphagia in advanced Parkinson's disease and reliability testing of the FEES-Levodopa-test. *Parkinsonism & Related Disorders*, *28*, 100–106. http://dx.doi.org/10.1016/j.parkreldis.2016.04.034.

Williams-Gray, C. H., Mason, S. L., Evans, J. R., Foltynie, T., Brayne, C., Robbins, T. W., et al. (2013). The CamPaIGN study of Parkinson's disease: 10-year outlook in an incident population-based cohort. *Journal of Neurology, Neurosurgery, and Psychiatry*, *84*(11), 1258–1264. http://dx.doi.org/10.1136/jnnp-2013-305277.

Xie, T., Vigil, J., MacCracken, E., Gasparaitis, A., Young, J., Kang, W., et al. (2015). Low-frequency stimulation of STN-DBS reduces aspiration and freezing of gait in patients with PD. *Neurology*, *84*(4), 415–420. http://dx.doi.org/10.1212/wnl.0000000000001184.

CHAPTER FORTY-FOUR

Palliative Care and Nonmotor Symptoms in Parkinson's Disease and Parkinsonism

Nataliya Titova*, K Ray Chaudhuri†,1

*Federal State Budgetary Educational Institution of Higher Education "N.I. Pirogov Russian National Research Medical University" of the Ministry of Healthcare of the Russian Federation, Moscow, Russia
†The Maurice Wohl Clinical Neuroscience Institute, King's College London and National Parkinson Foundation Centre of Excellence, Kings College Hospital, London, United Kingdom
[1]Corresponding author: e-mail address: ray.chaudhuri@nhs.net

Contents

Abstract

The term palliative care (PC) is defined as a collection of interventions and strategies that helps to improve and sustain the quality of life of patients and caregivers in situations and scenarios associated with life-threatening illness. This is usually implemented by means of early identification and treatment of relevant motor and nonmotor issues such as pain, sleep, and autonomic dysfunction, dementia, and depression. In addition, a holistic PC program also includes delivery of physical, psychosocial, and spiritual support. PC as a specific discipline, as well as a treatment strategy for long-term neurological conditions such as Parkinson's disease (PD), is relatively new, but very important as neurodegenerative disorders in the United Kingdom alone affects approximately 10 million people and there are over 130,000 people with PD. With longer life expectancy, the burden of long duration and late stage PD is even more evident, bringing in focus the need for PC. However, the concept of PC in PD is still poorly defined and although there are pockets of excellence, the strategy is poorly implemented into routine clinical practice.

International Review of Neurobiology, Volume 134
ISSN 0074-7742
http://dx.doi.org/10.1016/bs.irn.2017.05.014

The variable progressive nature of the disease, the heterogeneity of clinical subtypes, and also the burden of nonmotor symptoms create challenges for effective PC delivery in PD, but recent clinical trials have started addressing PC in PD and are to be welcomed.

1. INTRODUCTION: WHAT IS PALLIATIVE CARE?

Generally, the term palliative care (PC) is defined as a collection of interventions and strategies that helps to improve and sustain the quality of life (QoL) of patients and caregivers in situations and scenarios associated with life-threatening illness (Sepulveda, Marlin, Yoshida, & Ullrich, 2002). This is usually implemented by means of early identification of relevant motor and nonmotor issues such as dementia and pain, as well as delivery of physical, psychosocial, and spiritual support (Sepulveda et al., 2002). The issue of PC as a specific discipline, as well as a treatment strategy, is relatively new and particularly relevant to long-term neurological conditions (LTNCs) which make up a group of progressive often neurodegenerative disorders, affecting in the United Kingdom alone approximately 10 million people (DoH, 2008; Higginson et al., 2006). People with LTNCs often complain of and experience a range of physical disabilities of the limbs as well as nonmotor symptoms (NMS) such as pain, fatigue, somnolence, depression in addition to cognitive, neuropsychological, autonomic symptoms, and distress (Fig. 1). Some are related to the disease process, while others are related to comorbidities.

Over 25% of people with LTNCs need daily help with activities of daily living complicated by a range of NMS (Bernard, Aspinal, Gridley, & Parker, 2008; Fitzpatrick et al., 2010). In the United Kingdom, nearly one million informal carers are caring for someone with a neurological condition (Bernard et al., 2008; Fitzpatrick et al., 2010). In addition to the burden of the advancing conditions, NMS, in PD is a key determinant of QoL (Martinez-Martin, Rodriguez-Blazquez, Kurtis, & Chaudhuri, 2011).

The world health organization has suggested that PC should be an active strategy involving holistic care of patients with any advanced progressive illness where management of pain and various other symptoms, motor and nonmotor, should be addressed. There should also be provision of psychological, social, and spiritual support, so that the best QoL for patients and their families is achieved. Although in terms of cancer, PC is usually seen as terminal care, application of PC in LTNC is unclear, and in PD PC does

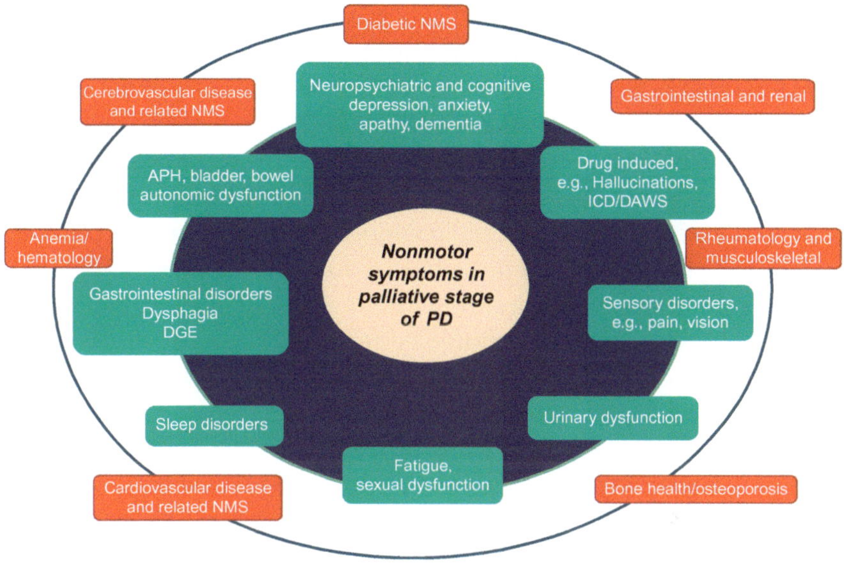

Fig. 1 The circles of palliative care. The *inner circle* represents nonmotor symptoms (NMS) directly related to PD. The *outer circle* relates to NMS related to relevant comorbidity.

not mean terminal care, but palliation of medical, emotional, and social problems. Evidence that PC is effective arises from the field of cancer where PC has been shown to have a proven effect in improving clinical outcomes as well as reducing the caregiver burden and stress and be cost effective (Bakitas et al., 2009; Smith, Brick, O'Hara, & Normand, 2014; Temel et al., 2010). However, delivery of PC remains suboptimal and studies suggest that PC is only offered to less than 50% of subjects who actually need it (Morley et al., 2012; Peters, Fitzpatrick, Doll, Playford, & Jenkinson, 2012). The issue is compounded by the fact that research involving the PC population is particularly difficult, as both patients and caregivers face challenging domestic and emotional circumstances and research participation is not a priority. There is high death rate in this cohort of subjects and therefore, studies are subject to loss of patients at follow-up visits which may affect the credibility of power calculation of related studies (Davies, 2004).

2. PALLIATIVE CARE IN PARKINSON'S DISEASE

The long-term outlook of PD has changed considerably and now many patients live beyond 15–20 years with PD. With the increasing long-term survival of PD patients, PC is thus going to play a major role in the therapeutic strategy for PD. This is further augmented by the fact that management of NMS has emerged as one of the key challenges in the treatment of PD and new and effective therapies have been developed for many NMS evident in the palliative phase of PD, such as pain, constipation, somnolence, and mood disorders (Fig. 1) (Schrag, Sauerbier, & Chaudhuri, 2015). The NMS in the palliative phase would comprised of those related to the disease process and progression and those related to comorbidities such as type 2 diabetes, cerebrovascular, cardiovascular diseases, as well as bone health. These are shown in Fig. 1. Comorbidity-related NMS are discussed in chapter "Personalised medicine and nonmotor symptoms in Parkinson's disease" by Titova and Chaudhuri. An additional issue to consider is the economic burden of nonmotor aspects of PD in the palliative stage. Institutionalization as well as hospitalization are expensive, and the issue is compounded by the cost of care at home. Loss of employment of caregivers who may become fulltime carers may also have major economic implications.

In the United Kingdom, the concept of PC strategy is fairly advanced with a national initiative called the national service framework for long-term conditions which has declared that the provision of PC for people with LTNCs approaching end of life must be regarded as basic quality requirement (DoH, 2005).

PD patients are now recognized to be on a journey through their condition, in excess of 40 years with a prodromal stage which could date back by 20 years before the onset motor PD (Fig. 2) (Berg, Marek, Ross, & Poewe, 2012). The progression of PD varies, motor progression is not always linear with nonmotor progression. The definition of advanced or complex PD is therefore controversial and there is currently no consensus. Palliative therapy or stage would theoretically supervene once the complex stage is over. The National Parkinson Foundation convened a multidisciplinary group and defined advanced PD as either those with Hoehn and Yahr (HY) stage 3 or higher or the presence of dementia, psychosis, and/or associated with significant caregiver strain (Miyasaki, 2013). However, such a definition has many problems, HY staging is purely motor and does not reflect disease progression, and the palliative phase is also associated with significant burden of NMS such as pain and many other symptoms and this is not mentioned. A more real reflection of palliative stage of PD could be gleaned from a study which reported that 25%–27% of PD patients living in Sydney were institutionalized by 15 years of illness with PD and a subsequent follow-up study if this cohort reported a standardized mortality ratio of 3.1 vs a control

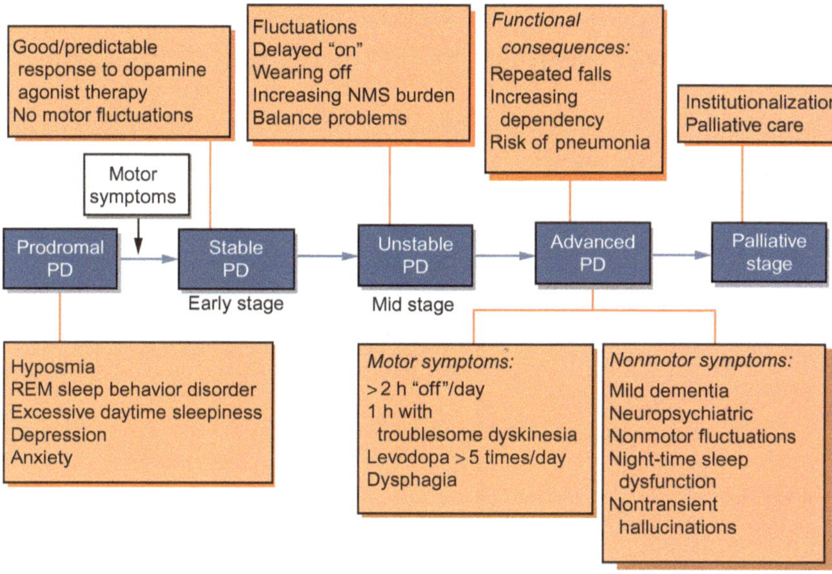

Fig. 2 The "journey" of a person with Parkinson's disease from premotor to palliative. *Adapted from Chaudhuri, K. R., & Fung, V. (2016).* Fast facts: Parkinson's disease *(4th ed.). Oxford: Health Press Ltd.*

population, although this finding may not translate to a larger population (Hely, Morris, Reid, & Trafficante, 2005; Hely, Reid, Adena, Halliday, & Morris, 2008). The authors report that dementia was almost inevitable at 20 years (83% of survivors), while a range of other NMS such as dysarthria, choking, excessive daytime somnolence, and orthostatic hypotension were troublesome (Hely et al., 2008). In a study of Parkinsonism in palliative stage, including PD patients, others have reported that HY stages 3–5 is associated with a range of "practical problems" in 42% which were not met by the healthcare team while 38% wanted more information on the needs and delivery of PC (Higginson et al., 2012; Saleem et al., 2013).

In summary, the major needs and issues at a palliative stage could be described as:

- Multiple symptoms, motor and nonmotor (Fig. 2).
- Deteriorating QoL.
- Progressive disability both in motor and nonmotor domains.
- Emotional and psychological distress.
- High caregiver burden and caregiver distress.
- Refractory/poor response to conventional dopaminergic treatment.

3. NATURAL HISTORY OF PD AND PALLIATIVE CARE

It is generally accepted that after a PD duration of 10 or more years, most patients have a bilateral disease with marked bradykinesia impacting on patients' disability. At this stage, resting tremor is less common while falls and related injuries and fractures are common. The risk of fractures is aggravated by osteoporosis (Fig. 1) a common comorbidity (Coelho et al., 2015). Patients may become wheelchair-bound particularly when in the "off" state. Communication is a problem as dysarthria has a great impact on patient condition and dysphagia is common causing choking and aspiration pneumonia (Salat, Noyce, Schrag, & Tolosa, 2016).

Despite the use of levodopa and other dopamine replacement therapies, PD patients have higher and earlier need for nursing home placements, higher rates of emergency hospital admissions, longer admissions, and higher in-hospital mortality than the general population (Sveinbjornsdottir, 2016). Published results from the Queen Square Brain Bank report that the mean time from diagnosis to death is around 14 years, and the most frequent cause of death is pneumonia, followed by cardiovascular events, stroke, and cancer (Poewe, 2006; Salat et al., 2016; Sveinbjornsdottir, 2016).

4. KEY STRATEGIES FOR MANAGEMENT

4.1 Assessments

Complexity of the palliative stage of PD is underpinned by the range and nature of NMS (Fig. 2) as well as advancing motor disease, patients often being in HY stages 4–5. Scales can be useful to highlight and flag the problems and could include the widely used self-completed nonmotor symptoms questionnaire (NMSQuest), the palliative outcome scale (POS, also used in multiple system atrophy (MSA) and progressive supranuclear palsy (PSP)) or the patient or caregiver completed Edmonton Symptom Assessment Scale for instance, although a whole range of other motor and nonmotor scales could also be applied (Chaudhuri, Healy, & Schapira, 2006; Gao et al., 2016; Gort et al., 2007; Miyasaki et al., 2012). Such scale-based objective assessment is important so that specific needs can be addressed. Caregiver stress and burden can be addressed using the Zarit caregiver burden tool (Gort et al., 2007). A generic tool for PC outcome in people with LTNC has also been validated and is known as IPOS Neuro-S8 (Gao et al., 2016).

This is an eight-item scale for addressing the symptom burden at a palliative stage and can be applied to PD.

4.2 Interventions Related to NMS

NMS are functionally among the most limiting for patients over time, and affect caregiver burden and overall QoL more than motor symptoms. The NMSQuest and NMS scale assessment of advanced HY stage (stages 4 and 5) PD patients from a range of studies also suggest a complex presentation with a range of NMS. A study addressing and grading burden of NMS using the NMSQuest reported that out of 43 patients with HY stages 4 or 5, 32 had severe or very severe NMS burden (Chaudhuri et al., 2015). Out of 125 patients in HY stages 4 and 5, another study using the NMS scale reported 101 subjects either had severe or very severe burden of NMS, thus underpinning the importance of management of NMS at this stage of PD. In a study of 146 consecutive patients' baseline visit, the Edmonton Symptom Assessment Scale was used and the following symptoms were identified as most bothersome (Miyasaki et al., 2012).

- anxiety,
- depression,
- pain,
- constipation,
- stiffness.

In another study, 82 patients with PD, MSA, or PSP (at HY stages 3–5) were followed up in a longitudinal design and outcome was measured using the POS as well as other scales (Higginson et al., 2012) (Table 1).

Seventy-five percentage of patients were severely disabled, with over one-third being wheelchair-bound or bedridden and needing full-time care. During the follow-up of a year, 50% of patients either worsened (24/60) or fluctuated (8/60) in relation to the POS scores, and indicate a severe and complex medley of nonmotor (pain, sleep, cognition, autonomic dysfunction) and motor symptoms in the palliative stage. The symptom prevalence of some key NMS in PD may be as severe as or analogous to patients with advanced cancer. These symptoms are mostly related to pain, fatigue, constipation, shortness of breath, sleep dysfunction, and nausea (Solano, Gomes, & Higginson, 2006). Pain is particularly dominant and while pain was present in 83% of PD patients in a one study, in cancer pain was prevalent in 35%–96%. Another overlapping NMS is fatigue which was rated at 83% in PD and rates in cancer are from 32% to 90%. The management

Table 1 Self-Reported Frequency and Severity of Nonmotor Symptoms (Except Pressure Sores, Communication, Falls, Arms and Hands, and Legs/Walking on the Palliative Outcome Scale (POS) From 82 Patients With Parkinson's Disease (PD), Multiple System Atrophy, and Progressive Supranuclear Palsy)

POS-PD Scores, $N = 82$ (%)

			Not at all (0)		Slight (1)		Moderately (2)		Severely (3)		Overwhelmingly (4)	
	Mean	SD	*n*	%	*n*	%	*N*	%	*n*	%	*n*	%
Pain	1.99	1.15	15	18.3	6	7.3	29	35.4	29	35.4	3	3.7
Spasms/cramps	1.38	1.22	32	39.0	5	6.1	28	34.1	16	19.5	1	1.2
Fatigue	1.84	1.07	13	15.9	11	13.4	38	46.3	16	19.5	4	4.9
Shortness of breath	0.94	1.06	40	48.8	15	18.3	19	23.2	8	9.8	0	0
Nausea	0.46	0.93	62	75.6	8	9.8	7	8.5	4	4.9	1	1.2
Vomiting	0.15	0.61	77	93.9	1	1.2	1	1.2	3	3.7	0	0
Poor appetite	0.41	0.78	61	74.4	10	12.2	9	11.0	2	2.4	0	0
Swallowing	0.94	1.06	40	48.8	15	18.3	19	23.2	8	9.8	0	0
Dribbling/drooling	1.46	1.21	23	28.0	21	25.6	18	22.0	17	20.7	3	3.7
Feeling sleepy	1.52	1.01	14	17.1	27	32.9	26	31.7	14	17.1	1	1.2
Difficulty in sleeping	1.28	1.22	33	40.2	11	13.4	21	25.6	16	19.5	1	1.2
Constipation	1.29	1.24	33	40.2	12	14.6	18	22.0	18	22.0	1	1.2
Bowel control	0.65	1.10	57	69.5	7	8.5	10	12.2	6	7.3	2	2.4

Controlling urine	1.10	1.15	37	45.1	12	14.6	22	26.8	10	12.2	1	1.2
Pressure sores	0.13	0.52	76	92.7	2	2.4	3	3.7	1	1.2	0	0
Arms and hands	1.57	1.33	25	30.5	15	18.3	19	23.2	16	19.5	7	8.5
Legs/walking	2.23	1.35	13	15.9	12	14.6	15	18.3	27	32.9	15	18.3
Communicating	1.44	1.31	28	34.1	17	20.7	14	17.1	19	23.2	4	4.9
Falls	0.95	1.27	48	58.5	8	9.8	9	11.0	16	19.5	1	1.2
Hallucinations	0.34	0.67	61	74.4	16	19.5	3	3.7	2	2.4	0	0
Total score	10.68	3.89										

POS-PD, palliative outcome scale-Parkinson's disease; *SD*, standard deviation. 0 = (best) to 4 = (worst).

Taken from Saleem, T. Z., Higginson, I. J., Chaudhuri, K. R., Martin, A., Burman, R., & Leigh, P. N. (2013). Symptom prevalence, severity and palliative care needs assessment using the Palliative Outcome Scale: A cross-sectional study of patients with Parkinson's disease and related neurological conditions. Palliative Medicine, 27(8), 722–731. http://dx.doi.org/10.1177/0269216312465783. Epub 2012 Dec 3.

strategies therefore need to address these symptoms some of which are addressed in Table 2.

4.3 Multidisciplinary Approach

A multidisciplinary therapy (MDT)-based approach is essential for delivery of good quality PC and currently is an unmet need. The team delivering MDT should include PC specialists and also a liaison nurse. These models have been used in some centers with good success and outcome. A close link with hospice care can also be delivered via the MDT pathway.

Other strategies involved in delivery of PC and improvement in NMS in PD include:

- Early referral to PC team and discussion are placement in hospice if required.
- Pain control.
- Anxiety management.
- Support for the caregiver to prevent caregiver distress.
- Conversation about death and consequences.

The role of MDT is important at this stage so as to also adopt the following practical measures which include:

- Planning for the future.
- Home adaptations to facilitate movement and avoid injury.
- Resuscitation wishes of the sufferer.
- Ascertaining who will complete death certification.

Caregivers of late stage PD patients and parkinsonism such as MSA and PSP play a pivotal role in PC inpsite of often having their own problems or disabilities. These range from exhaustion, anxiety, possible conflict with other family members, and comorbidities. In addition, some may not be able to cope with the increasing care for their partner with PD (Susan, 2003). This issue highlights the role of personalized medicine in PC and the need for individual targeted approach to assessing carer needs at the advanced stages when the patient has greater dependency on support for the activities of daily living.

5. NEW INITIATIVES

The OPTCARE Neuro trial is currently the world's largest PC study in people with LTNCs including PD, recruiting both patients ($N=356$) with and without capacity, as well as their carers (maximum $N=356$). This randomized, double-blind, multicenter study aims to provide key evidence

Table 2 A Summary of Key Management Strategies for Common Nonmotor Symptoms in the Palliative Stage of PD

Key Symptoms	Assessments	Drug Interventions	Multidisciplinary Therapy	Other Approaches
Pain	Use validated PD pain scale such as the Kings Parkinson's Disease Pain Scale	Analgesia: Paracetamol Opiates Oxycodone–naloxone (PANDA study)	Assisted range of motion exercises	Botulinum toxin injection
Severe rigidity and related pain	POS ESAS	Trial of rotigotine transdermal patch or apomorphine subcutaneous injections Liquid levodopa	Physiotherapy as appropriate	Botulinum toxin injection to avoid contractures (may need ultrasound guidance)
Constipation	NMSQuest NMSS ESAS POS	Polyethylene glycol 1–2 capsules/day Senokot 1–4 tablets each day Lactulose Macrogol	Dietary modification and use of whole grains with fibers and adequate hydration	
Urinary dysfunction	POS ESAS	Urinary continence care	Bladder pads, urinals, use of sheaths Catheter placement if required	Avoid and screen for urinary tract infection
Insomnia	NMSQuest PDSS ESS to ascertain cause	Opiates may help insomnia and pain Benzodiazepines for onset insomnia	Pain control at night	Sedative antidepressants

Continued

Table 2 A Summary of Key Management Strategies for Common Nonmotor Symptoms in the Palliative Stage of PD—cont'd

Key Symptoms	Assessments	Drug Interventions	Multidisciplinary Therapy	Other Approaches
EDS		Trial of modafinil for severe EDS		
Confusional Wandering Nocturnal hallucinations		Quetiapine (small dose may be tried)		
RLS		Gabapentin or cautious use of dopamine agonists		
Dementia	POS, MMSE MOCA	Cholinesterase inhibitors ? Memantine Avoid anticholinergics	Home support Social care	Placement in care homes
Psychosis. Delirium	ESAS	? Quetiapine Reduce/downtitrate dopaminergic therapy	Treat underlying infection Often aspiration pneumonia need exclusion	Cholinesterase inhibitors (oral and patch) ? Pimavanserin
Anxiety. Depression	ESAS NMSQuest	Antidepressant use Benzodiazepine	Counseling	
Swallowing problems	ESAS NMSQuest	Assess if off period related. If "yes" then trial of nonoral dopaminergic agents (rotigotine patch, apomorphine injection could be considered)	Referral to speech and language therapy Dietary modification	PEG tube if required

ESAS, Edmonton symptom assessment scale; *ESS*, Epworth sleepiness scale; *MMSE*, mini-mental state examination; *MOCA*, Montreal cognitive assessment scale; *NMSQuest*, nonmotor symptoms questionnaire; *NMSS*, nonmotor symptoms scale; *PANDA study*, prolonged-release oxycodone–naloxone for treatment of severe pain in patients with Parkinson's disease; *PDSS*, Parkinson's disease sleep scale; *PEG*, percutaneous endoscopic gastrostomy; *POS*, palliative outcome scale.

on the shorter term effectiveness of Short-term Integrated Palliative Care Services which is delivered by multidiscipline PC team and will be compared to standard care.

In the United States, through the initiative of the Parkinson's Disease Foundation, a first international meeting on PC and Parkinson's disease (PD) was held and in 2015, the Patient Centered Outcomes Research Institute funded the first comparative effectiveness trial of outpatient PC for PD (Hall, Sumrall, Thelen & Kluger, 2017). This novel approach addresses PC as a strategy to care for PD patients and their families seeking to reduce suffering through a range of spiritual, psychosocial, and medical support which begins at the time of diagnosis and continues through end-of-life care and bereavement. The International Parkinson's and Movement Disorders Society have also set up a new initiative with a task force on PC in PD. The committee will address unmet needs in PC for PD and address its implementation at a global level.

6. PALLIATIVE CARE IN PARKINSONISM: MSA AND PSP

Although MSA and PSP are distinct conditions, but may overlap at presentation and clinically there may be considerable overlap of symptoms. The clinical course of MSA and PSP is faster and more aggressive and PC is therefore important in these conditions. Little work has been done to address PC in MSA and PSP. In an UK study, Higginson et al. reported that MSA and PSP patients experience severe physical and psychological symptoms which were similar to and in some cases even worse than those needing PC with cancer (Higginson et al., 2012). The common symptoms were largely nonmotor and are shown in Table 3 in comparison with PD.

7. CONCLUSIONS

PC is an essential aspect of care of PD, but remains a key unmet need. NMS of PD dominate at the palliative stage and management of these symptoms requires a well-informed multidisciplinary team with PC experts. In addition PC for conditions such as MSA as well as PSP needs to be addressed urgently as there is little research in this area. New research strategies such as OPTcare Neuro study, recent US initiatives as well formation of a PC task force of the Movement Disorders Society are to be welcomed. These strategies are likely to produce evidence-based quality standards for setting up PC services for PD across the world.

Table 3 Percentage of Symptoms, Nonmotor and Motor in Idiopathic Parkinson's Disease (IPD), Progressive Supranuclear Palsy (PSP), and Multiple System Atrophy (MSA)

Item	Symptoms	IPD $n = 50$ (61%)	PSP $n = 15$ (18%)	MSA $n = 17$ (21%)	Total $N = 82$
1	Pain	86.0	60.0	88.2	81.7
2	Spasms	60.0	53.3	70.6	61.0
3	Fatigue/lack of energy	84.0	80.0	88.2	84.2
4	Shortness of breath	54.0	46.7	47.1	51.2
5	Nausea	24.0	6.7	41.2	24.4
6	Vomiting	8.0	0.0	5.9	6.1
7	Poor appetite	22.0	26.7	35.3	25.6
8	Problems swallowing	40.0	60.0	76.5	51.2
9	Mouth problems	70.0	66.7	82.4	72.0
10	Feeling sleepy	86.0	73.3	82.4	82.9
11	Difficulty in sleeping	58.0	40.0	82.4	59.8
12	Constipation	54.0	60.0	76.5	59.8
13	Difficulty with bowel control	24.0	20.0	58.8	30.5
14	Difficulty with controlling urine	52.0	46.7	70.6	54.9
15	Pressure sores	6.0	6.7	11.8	7.3
16	Problems using arms	64.0	73.3	82.4	69.5
17	Problems using legs	80.0	86.7	94.1	84.2
18	Difficulty communicating	58.0	80.0	76.5	65.9
19	Falls	36.0	53.3	47.1	41.5
20	Hallucinations	28.0	13.3	29.4	25.6

Total 82 patients.
Taken from Higginson, I. J., Gao, W., Saleem, T. Z., Chaudhuri, K. R., Burman, R., McCrone, P., et al. (2012). Symptoms and quality of life in late stage Parkinson syndromes: A longitudinal community study of predictive factors. *PloS One,* 7(11), e46327. doi:10.1371/journal.pone.0046327.

ACKNOWLEDGMENTS

Ethical approval and consent to participate: Not applicable.

Consent to publish: Not applicable.

Declaration of competing interests: All authors declare no support from any organization for the submitted work, no financial relationships with any organizations that might have an interest in the submitted work in the previous 3 years, no other relationships or activities that could appear to have influenced the submitted work.

Funding: Nil.

REFERENCES

Bakitas, M., Lyons, K. D., Hegel, M. T., Balan, S., Brokaw, F. C., Seville, J., et al. (2009). Effects of a palliative care intervention on clinical outcomes in patients with advanced cancer: The Project ENABLE II randomized controlled trial. *JAMA*, *302*(7), 741–749. http://dx.doi.org/10.1001/jama.2009.1198.

Berg, D., Marek, K., Ross, G. W., & Poewe, W. (2012). Defining at-risk populations for Parkinson's disease: Lessons from ongoing studies. *Movement Disorders*, *27*(5), 656–665. http://dx.doi.org/10.1002/mds.24985.

Bernard, S., Aspinal, F., Gridley, K., & Parker, G. (2008). Integrated policy making in England for adults with long-term neurological conditions (LTNCs): Some preliminary findings from a scoping study. *International Journal of Integrated Care*, *8*, e60.

Chaudhuri, K. R., Healy, D. G., & Schapira, A. H. (2006). Non-motor symptoms of Parkinson's disease: Diagnosis and management. *Lancet Neurology*, *5*(3), 235–245. http://dx.doi.org/10.1016/s1474-4422(06)70373-8.

Chaudhuri, K. R., Sauerbier, A., Rojo, J. M., Sethi, K., Schapira, A. H., Brown, R. G., et al. (2015). The burden of non-motor symptoms in Parkinson's disease using a self-completed non-motor questionnaire: A simple grading system. *Parkinsonism & Related Disorders*, *21*(3), 287–291. http://dx.doi.org/10.1016/j.parkreldis.2014.12.031.

Coelho, M., Marti, M. J., Sampaio, C., Ferreira, J. J., Valldeoriola, F., Rosa, M. M., et al. (2015). Dementia and severity of parkinsonism determines the handicap of patients in late-stage Parkinson's disease: The Barcelona-Lisbon cohort. *European Journal of Neurology*, *22*(2), 305–312. http://dx.doi.org/10.1111/ene.12567.

Davies, E. (2004). *What are the palliative care needs of older people and how might they be met? (health evidence network report)*. Copenhagen: WHO Regional Office for Europe.

DoH. (2005). *National service framework for long term conditions*. National Health Service, London, UK.

DoH. (2008). *Raising the profile of long term conditions care: A compendium of information*. National Health Service, London, UK.

Fitzpatrick, R., Peters, M., Doll, H., Harris, R., Jenkinson, C., Playford, D., et al. (2010). *The needs and experiences of services by individuals with long-term progressive neurological conditions, and their carers. A benchmarking study*. Oxford: Oxford University Press, HMSO.

Gao, W., Crosby, V., Wilcock, A., Burman, R., Silber, E., Hepgul, N., et al. (2016). Psychometric properties of a generic, patient-centred palliative care outcome measure of symptom burden for people with progressive long term neurological conditions. *PLoS One*, *11*(10), e0165379. http://dx.doi.org/10.1371/journal.pone.0165379.

Gort, A. M., Mingot, M., Gomez, X., Soler, T., Torres, G., Sacristan, O., et al. (2007). Use of the Zarit scale for assessing caregiver burden and collapse in caregiving at home in dementias. *International Journal of Geriatric Psychiatry*, *22*(10), 957–962. http://dx.doi.org/10.1002/gps.1770.

Hall, K., Sumrall, M., Thelen, G., & Kluger, B. (2017). Palliative care for Parkinson's disease: A patient and care partner prescription. NPJ Parkinson's Disease, *3*(16). http://dx.doi.org/10.1038/s41531-017-0016-2.

Hely, M. A., Morris, J. G., Reid, W. G., & Trafficante, R. (2005). Sydney multicenter study of Parkinson's disease: Non-L-dopa-responsive problems dominate at 15 years. *Movement Disorders*, *20*(2), 190–199. http://dx.doi.org/10.1002/mds.20324.

Hely, M. A., Reid, W. G., Adena, M. A., Halliday, G. M., & Morris, J. G. (2008). The Sydney multicenter study of Parkinson's disease: The inevitability of dementia at 20 years. *Movement Disorders*, *23*(6), 837–844. http://dx.doi.org/10.1002/mds.21956.

Higginson, I. J., Gao, W., Saleem, T. Z., Chaudhuri, K. R., Burman, R., McCrone, P., et al. (2012). Symptoms and quality of life in late stage Parkinson syndromes: A longitudinal community study of predictive factors. *PLoS One*, 7(11), e46327. http://dx.doi.org/10.1371/journal.pone.0046327.

Higginson, I. J., Vivat, B., Silber, E., Saleem, T., Burman, R., Hart, S., et al. (2006). Study protocol: Delayed intervention randomised controlled trial within the medical research council (MRC) framework to assess the effectiveness of a new palliative care service. *BMC Palliative Care*, *5*, 7. http://dx.doi.org/10.1186/1472-684x-5-7.

Martinez-Martin, P., Rodriguez-Blazquez, C., Kurtis, M. M., & Chaudhuri, K. R. (2011). The impact of non-motor symptoms on health-related quality of life of patients with Parkinson's disease. *Movement Disorders*, *26*(3), 399–406. http://dx.doi.org/10.1002/mds.23462.

Miyasaki, J. M. (2013). Palliative care in Parkinson's disease. *Current Neurology and Neuroscience Reports*, *13*(8), 367. http://dx.doi.org/10.1007/s11910-013-0367-y.

Miyasaki, J. M., Long, J., Mancini, D., Moro, E., Fox, S. H., Lang, A. E., et al. (2012). Palliative care for advanced Parkinson disease: An interdisciplinary clinic and new scale, the ESAS-PD. *Parkinsonism & Related Disorders*, *18*(Suppl. 3), S6–S9. http://dx.doi.org/10.1016/j.parkreldis.2012.06.013.

Morley, D., Dummett, S., Peters, M., Kelly, L., Hewitson, P., Dawson, J., et al. (2012). Factors influencing quality of life in caregivers of people with Parkinson's disease and implications for clinical guidelines. *Parkinson's Disease, 2012*, 190901. http://dx.doi.org/10.1155/2012/190901.

Peters, M., Fitzpatrick, R., Doll, H., Playford, E. D., & Jenkinson, C. (2012). The impact of perceived lack of support provided by health and social care services to caregivers of people with motor neuron disease. *Amyotrophic Lateral Sclerosis*, *13*(2), 223–228. http://dx.doi.org/10.3109/17482968.2011.649759.

Poewe, W. (2006). The natural history of Parkinson's disease. *Journal of Neurology*, *253*(Suppl. 7), Vii2–Vii6. http://dx.doi.org/10.1007/s00415-006-7002-7.

Salat, D., Noyce, A. J., Schrag, A., & Tolosa, E. (2016). Challenges of modifying disease progression in prediagnostic Parkinson's disease. *Lancet Neurology*, *15*(6), 637–648. http://dx.doi.org/10.1016/s1474-4422(16)00060-0.

Saleem, T. Z., Higginson, I. J., Chaudhuri, K. R., Martin, A., Burman, R., & Leigh, P. N. (2013). Symptom prevalence, severity and palliative care needs assessment using the palliative outcome scale: A cross-sectional study of patients with Parkinson's disease and related neurological conditions. *Palliative Medicine*, *27*(8), 722–731. http://dx.doi.org/10.1177/0269216312465783. Epub 2012 Dec 3.

Schrag, A., Sauerbier, A., & Chaudhuri, K. R. (2015). New clinical trials for nonmotor manifestations of Parkinson's disease. *Movement Disorders*, *30*(11), 1490–1504. http://dx.doi.org/10.1002/mds.26415.

Sepulveda, C., Marlin, A., Yoshida, T., & Ullrich, A. (2002). Palliative care: The World Health Organization's global perspective. *Journal of Pain and Symptom Management*, *24*(2), 91–96.

Smith, S., Brick, A., O'Hara, S., & Normand, C. (2014). Evidence on the cost and cost-effectiveness of palliative care: A literature review. *Palliative Medicine*, *28*(2), 130–150. http://dx.doi.org/10.1177/0269216313493466.

Solano, J. P., Gomes, B., & Higginson, I. J. (2006). A comparison of symptom prevalence in far advanced cancer, AIDS, heart disease, chronic obstructive pulmonary disease and renal disease. *Journal of Pain and Symptom Management*, *31*, 58–69.

Susan, M. C. (2003). The psychosocial impact of late-stage Parkinson's disease. *Journal of Neuroscience Nursing*, *35*(6), 306–313.

Sveinbjornsdottir, S. (2016). The clinical symptoms of Parkinson's disease. *Journal of Neurochemistry*, *139*(Suppl. 1), 318–324. http://dx.doi.org/10.1111/jnc.13691.

Temel, J. S., Greer, J. A., Muzikansky, A., Gallagher, E. R., Admane, S., Jackson, V. A., et al. (2010). Early palliative care for patients with metastatic non-small-cell lung cancer. *The New England Journal of Medicine*, *363*(8), 733–742. http://dx.doi.org/10.1056/NEJMoa1000678.

CHAPTER FORTY-FIVE

Personalized Medicine and Nonmotor Symptoms in Parkinson's Disease

Nataliya Titova*, K Ray Chaudhuri†,[1]
*Federal State Budgetary Educational Institution of Higher Education "N.I. Pirogov Russian National Research Medical University" of the Ministry of Healthcare of the Russian Federation, Moscow, Russia
†National Parkinson Foundation International Centre of Excellence, King's College London and King's College Hospital and The Maurice Wohl Clinical Neuroscience Institute, King's College London, London, United Kingdom
[1]Corresponding author: e-mail address: ray.chaudhuri@nhs.net

Contents

Abstract

Parkinson's disease (PD) is a multineurotransmitter dysfunction related disorder resulting in a range of motor and nonmotor symptoms. Phenotypic heterogeneity is pronounced in PD and nonmotor symptoms dominant subtypes have been described. These endophenotypes may be underpinned by considerable nondopaminergic dysfunction; however, conventional treatment of PD continues to be mostly reliant on dopamine replacement strategy or manipulation of brain dopaminergic pathways. Consequently, treatment of many nondopaminergic nonmotor and some motor symptoms remains a key unmet need. It is also recognized that treatment strategies for PD are influenced by a number of nondrug-related issues. These include factors such as age, personality, and preferences for treatment, cultural beliefs, lifestyle, pharmacoeconomics, pharmacogenetics as well as comorbidity. Therefore, the success

International Review of Neurobiology, Volume 134
ISSN 0074-7742
http://dx.doi.org/10.1016/bs.irn.2017.05.015

of clinical therapy will rest on how much these factors are considered to develop a truly holistic treatment plan.

Personalized medicine is the modern way of delivering this holistic strategy for treatment of PD. Personalized medicine thus encompasses several strands of treatment. From the pharmaceutical point of view, it should involve dopaminergic and non-dopaminergic strategies. In addition, there are substrategies involving precision and tailored medicine to suit the needs and requirements of individual patients. Precision medicine would be relevant for patients who may be at risk of developing the clinical syndrome of Parkinson's as identified by specific gene mutations. Precision medicine in this scenario will attempt to be preventive. Tailored medicine would address the "single multifactorial" complex nature of PD and address symptoms as well subtype-specific strategies. Personalized medicine is now practiced for other conditions such as oncology as well as diabetes. In this chapter, we discuss the rationale and the need to develop strategies for personalized medicine for PD.

1. INTRODUCTION (CORRECTION 1)

Correction 2 In the last 40 years, the treatment of Parkinson's disease (PD) has undergone major advances mainly based on the discovery that the motor syndrome of PD could be reversed by dopamine replacement therapy (DRT). The treatment concept thus remains underpinned by use of levodopa and related compounds and other forms of DRT (Kalia & Lang, 2015). However, while DRT and in particular levodopa therapy has an excellent short-term effect on the motor symptoms of PD, motor complications remain a problem. Newly recognized challenges include the whole spectrum of nonmotor disorders and symptoms (NMS) of PD. PD is a multisystem, multineurotransmitter dysfunction related heterogeneous disorder with considerable phenotypic heterogeneity and studies with biomarkers suggests nondopaminergic syndromes could be part of early or late stage PD and the characteristics of these patients with nonmotor subtypes have been recently described (Jellinger, 2015; Marras & Chaudhuri, 2016; Sauerbier, Jenner, Todorova, & Chaudhuri, 2016; Sauerbier, Qamar, Rajah, & Chaudhuri, 2016; Titova, Padmakumar, Lewis, & Chaudhuri, 2016).

These observations are of and have therapeutic consequences. The variability in underlying pathophysiology (dopaminergic vs nondopaminergic) and clinical presentations call for a revision of the way we prescribe medications in PD. Currently, we have a guideline-driven therapeutic paradigm for the management of PD. Current generic prescribing of DRT in PD does not take into account the variabilities in PD phenotypes. This "one size fits all" treatment concept in PD is thus inadequate, resulting in poor quality of life of patients and carers particularly as nonmotor symptoms (NMS) are usually not addressed. It has also become evident that specific personal needs and choice of patients are key issues to consider before prescribing treatments for PD. A holistic approach which thus takes into account the heterogeneity of PD phenotypes, the genetic basis, pharmacogenomic basis along with personal choice and lifestyle is the basis of the concept of personalized medicine. The definition of personalized medicine is variable and the American Medical Association has defined this as "health care that is informed by each person's unique clinical, genetic, and environmental information." Personalized or individualized medicine strategies are used for metabolic disorders or cancer and are fundamental to the management of "single multifactorial" pathology-driven conditions. These management strategies thus may be different to the conventional strategy of single-symptoms-based treatments and may require the use of "cocktail therapies" (Yeghiazaryan, Peeva, Morelli, Herrera-Marschitz, & Golubnitschaja, 2009).

This concept can be extrapolated to PD and would consist of prescribing based on susceptibility of specific subtypes of PD to DRT side effects (subtype-specific treatment and pharmacogenetics) as well as taking into consideration the lifestyle, genetic framework, personality, and pharmacogenetics. This concept of personalized medicine in PD is relatively new and the defining enablers of this strategy are shown in Fig. 1. Personalized medicine can also comprise of various substrategies ranging from an overarching holistic concept to precision medicine based on genomics. In this chapter, we consider the varying concepts that may underpin development of functionally effective personalized medicine in PD.

2. GENETICS

Prediction of susceptibility for the development of PD in an individual basis maybe possible and is based on genetics of PD. Research has identified "at-risk" individuals through known genetic susceptibility markers in what is regarded as the preprodromal stage of PD (Miller & O'Callaghan, 2015).

Some of these rare forms of monogenic PD can identify specific genes (such as mitochondrial genes DJ-1, PINK1) or gene products (aberrant oligomeric alpha-synuclein aggregates) which may play a part in the pathogenesis of PD. It is possible that this knowledge would identify mechanisms which lead to the cascading pathways such as oxidative stress, mitochondrial dysfunction, and aberrant oligomeric alpha-synuclein aggregate formation.

Fig. 1 The "circle" of the components that enable delivery of true and holistic personalized as well as precision medicine.

The end result is the clinical expression of symptoms of PD, largely nonmotor in the prodromal stage and then motor. The role of personalized medicine here is the development of precision therapies that can then counteract these mechanisms.

Examples of the application of precision medicine linked to the genetics of PD exist. Heterozygote carriers of the glucocerebrocidase gene (*GBA*), which is responsible for Gaucher's disease (GD), have now been shown to impart an increased frequency of PD. Overall, approximately 5%–10% of PD patients have *GBA* mutations and this mutation is currently regarded as the most relevant genetic predisposing risk factor for PD (Beavan & Schapira, 2013). Research has also identified that there is a reciprocal relationship between glucocerebrocidase (GCase) activity and alpha-synuclein activity (Beavan & Schapira, 2013). From a therapeutic angle, this means that enhancing GCase activity may lead to regulation or even attenuation of alpha-synuclein activity which is implicated in the formation of misfolded oligomeric alpha-synuclein. The evidence for this approach has been developed in a mouse model of Gaucher's disease where adeno-associated viral vector delivery (AAV) of the recombinant GCase into the rodent brain caused modulation of alpha-synuclein deposition. Clinically, there was improvement in nonmotor symptoms and specifically memory deficits were improved (Sardi et al., 2013; Schapira & Gegg, 2013). Another target is the heat shock protein 90 (Hsp90) which is known to maintain the functional stability of aberrant neuronal proteins and thus allow and sustain the accumulation of toxic aggregates of alpha-synuclein, for example, (Luo, Rodina, & Chiosis, 2008; Luo, Sun, Taldone, Rodina, & Chiosis, 2010; Waza et al., 2006) (Fig. 2).

In a recent study, Wang et al. (2008) reported that Leucine-rich repeat kinase 2 (*LRRK2*) can form a complex molecule with Hsp90 in vivo. Mutations of *LRRK2* are now recognized as a cause of familial and also some sporadic forms of PD. If the function of Hsp90 is inhibited by a purine-scaffold Hsp90 inhibitor known as PU-H71 then there is a breakdown of the complex formation between Hsp90 and *LRRK2* and subsequent elimination of *LRRK2* by the proteasome.

Other examples include formation of complexes between Hsp90 and PTEN-induced kinase 1 (*PINK1*, an autosomal recessive form of PD

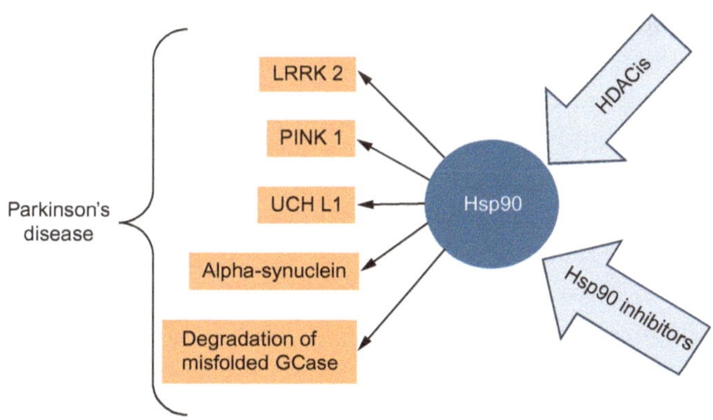

Fig. 2 The role of chaperones and heat shock protein 90 in enabling pathways that may contribute to a Parkinson's disease phenotype and the possibility of precision medicine using HSP90 or histone deacetylase inhibitors. GCase, glucocerebrocidase; HDACis, histone deacetylase inhibitors; Hsp90, heat shock protein 90; *LRRK2*, leucine-rich repeat kinase; *PINK1*, PTEN-induced kinase 1; *UCH-L1*, ubiquitin C-terminal hydrolase L1. *Adapted from Luo et al. (2010).*

associated with a range of NMS), and the links of Hsp90 with alpha-synuclein folding (Fig. 2). In experiments with cells treated with an Hsp90 inhibitor GM, the levels of *PINK1* were substantially diminished via the ubiquitin–proteasome pathway (Moriwaki et al., 2008). Mechanisms via the ubiquitin pathway (*UCH L1*), *LRRK2* and alpha-synuclein pathways have also been suggested (Falsone, Kungl, Rek, Cappai, & Zangger, 2009; Kabuta, Furuta, Aoki, Furuta, & Wada, 2008).

The role of small-molecule chaperones which are competitive inhibitors of GCase is another example. Two oral drugs, ambroxol and isofagomine are small-molecule chaperones which have been shown to increase mutant GCase activity in GD fibroblasts and mouse models related to the two commonest GD mutations, N370S and L444P (Bendikov-Bar, Maor, Filocamo, & Horowitz, 2013; Schapira & Gegg, 2013; Sun et al., 2012, 2011). Histone deacetylase inhibitors are also known to reduce recognition of misfolded GCase by Hsp90β and increase residual enzyme levels (Yang et al., 2013). This strategy has been shown to be effective in GD patient fibroblasts and has potential for treatment in PD with a possibility of affecting a range of nonmotor and motor symptoms. These histone deacetylase inhibitors which act as pharmacological chaperones (ambroxol, isofagomine) are now in clinical trial (Sardi et al., 2013). In terms of personalized medicine, one could adopt a combined chaperone and GCase augmentation based "cocktail" therapy for *GBA*-positive carriers who remain at risk of conversion to clinical PD. These patients, with *GBA* gene mutation variants, are associated with a specific cognitive subtype in PD with rapid cognitive decline progressing to dementia (Mata et al., 2016).

Similarly, identification of the carriers of the leucine-rich repeat kinase 2 gene (*LRRK2*) could be targeted with *LRRK2* inhibitors. In rodent models, the LRRK2 inhibitor GNE-7915 enhanced release of dopamine and also synaptic vesicle mobilization and recycling (Qin et al., 2017).

In addition, recent evidence suggests that some genetic forms of PD may have susceptibility to specific patterns of NMS. For example, triplication of the *SNCA* gene causing early onset autosomal dominant parkinsonism (*PARK1*) can express depression, dysautonomia, and dementia in the majority of cases. In addition, patients are particularly prone to hallucinations (Kasten & Klein, 2014).

In such patients, personalized medicine, taking a tailored approach, should include specific screening for these NMS and also tailoring DRT so as not to aggravate the risk of hallucinations.

Other genetic patterns of PD have a variable frequency of NMS. In some, such as *PINK1* mutation, rates of anxiety are higher. The COMT gene rs6267 allele "T" has been shown to be associated with pain in PD (Li, Chen, Yin, & Zhang, 2014). In another study by Hoogland et al. (2010), the *COMT* LL and HL genotypes were shown to interact with dopaminergic medications negatively influencing cognitive ability. Consequently, although higher levodopa concentration led to better motor performances compared to *COMT* HH carriers, patients with "low" *COMT* genotypes had worse verbal reasoning.

3. PHARMACOGENETICS

Pharmacogenetics is the science of studying the influence of inherited genetic differences in drug metabolic pathways which can affect individual clinical responses to drugs as well as adverse events (Lindpaintner, 2002). Pharmacogenetic approach thus provides the possibility of the development of a personal metabolic passport as the basis of a new and novel pharmacological strategy in PD. In PD, the role of pharmacogenetics is slowly evolving and Fig. 3 shows an example of the most investigated genes and their clinical motor and nonmotor effects. In this section, we focus on the main

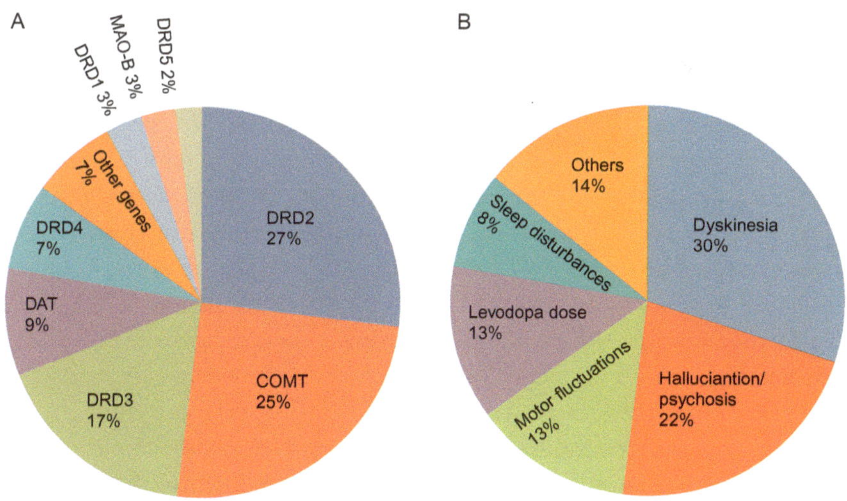

Fig. 3 The most investigated genes and their relative motor and nonmotor contributions in the field of pharmacogenetics. *Taken from Schumacher-Schuh et al. (2014).*

nonmotor outcomes-based pharmacogenetic examples and broadly these are related to drug-induced hallucinations and psychosis, impulse control disorder (ICD) and somnolence including sudden onset of sleep (Table 1). In some pharmacogenetic examples, such as those for excessive daytime sleepiness (*COMT* polymorphism, *DRD2* and *DRD4* (both linked to "sleep attacks"), *HCRT* (prepro-hypocretin)) and psychosis (*DRD4*, *CCK*, *APOE4*, *ACE*), the clinical implications are unclear and the links remain controversial (Džoljic, Novaković, Krajinovic, Grbatinić, & Kostić, 2015; Schumacher-Schuh, Rieder, & Hutz, 2014). Recent evidence suggests that some genetic mutations of PD such as *parkin* and *PINK1* may also impart susceptibility to specific patterns of ICD (Morgante et al., 2016) (Table 1).

Pharmacogenetic principles vary across ethnic groups and studies need to address ethnic variability in drug-related effects and side effects. As an example, susceptibility to drug induced late onset intrusive hallucinations have been described in Caucasian PD patients who carry the C/T SNP (A1) allele of the *TaqIA DRD2* polymorphism (Makoff et al., 2000).

However, in another study by Wang, Zhao, Chen, and Liu (2004), no significant differences were found between Chinese PD subjects in the related genes. There were no reported differences between hallucinators and nonhallucinators in either the genotypic or allelic distributions of polymorphisms of the dopamine receptors *D2 32806 C*>*T* (Taq1A), *D3 Ser9Gly* and Msp1, and *D5 978T*>*C* and dopamine transporter *3′-UTR 40bp VNTR*. This observation suggests an ethnicity based difference in the risk profile of PD patients to develop late onset drug-induced hallucinations between white Caucasians and Chinese subjects. In another example, the *COMT* gene activity has been studied across PD patients of various ethnic groups (Kenyan Africans, white Caucasians and South-west Asians) (McLeod et al., 1998). This study suggests that there is a lower frequency of PD patients carrying the homozygous low activity allele of *COMT* in the Kenyan PD population (9%) when compared to white Caucasians (31%) or South-west Asians (27%). These data further support the notion that there are ethnic variations of *COMT* gene activity, which is a major gene involved in the metabolic pathway of levodopa. Thus, there is molecular evidence of *COMT* activity being less common in East African individuals compared to South Asians as well as a white Caucasian population. In another controlled study, the frequency of low activity *COMT* gene allele in the Japanese controls was 0.29 and significantly lower than rates reported in white Caucasians (0.50) (Kunugi et al., 1997).

Table 1 A List of Investigated Genes Mutations and Polymorphisms and Related Nonmotor Symptoms Playing a Possible Role in Pharmacogenetics

The NMS	Proposed Gene Mutation/ Polymorphism	Clinical Implications
Hallucination	C/T SNP (A1) allele of the TaqIA DRD2 polymorphism	Susceptibility to hallucination
Late-onset hallucinations	DRD2 A1 allele, TaqIA (C/T SNP)	Susceptibility to hallucination
	DRD2 − 141C ins/del	Susceptibility to hallucination
Predisposing to visual hallucinations and psychosis	DAT VNTR (40 bp)	Susceptibility to hallucination and psychosis
Increased risk for visual hallucinations	*CCK,* −45 C/T, T allele	Susceptibility to visual hallucination
Increased risk for visual hallucinations	CCK CT/TT	Susceptibility to visual hallucination
Increased risk for visual hallucinations	CCKAR C allele TC/CC	Susceptibility to visual hallucination
Visual hallucinations	HOMER1 rs4704559 A allele	Increased visual hallucinations
Visual hallucinations	HOMER1 rs4704559 G allele	Decreased visual hallucination
Visual hallucinations	DAT1 839C > T C allele	Increased visual hallucinations
Increased risk for L-DOPA-induced psychosis	ACE ACE–II Ins/ins	L-dopa-induced psychosis
Hallucination and psychosis	APOE*4 allele	Hallucinations and psychosis precipitated
Frequent sleep attacks	HCRT −909 T	Susceptibility to sudden onset of sleep
Sudden onset of sleep	*D2* receptor gene polymorphism *Taq*IA, with predisposing role of *DRD2* allele *A2*	Susceptibility to sudden onset of sleep

Continued

Table 1 A List of Investigated Genes Mutations and Polymorphisms and Related Nonmotor Symptoms Playing a Possible Role in Pharmacogenetics—cont'd

The NMS	Proposed Gene Mutation/ Polymorphism	Clinical Implications
Sudden onset of sleep	48-bp *VNTR* polymorphism in the *DRD4* gene	Susceptibility to sudden onset of sleep
Impulse control disorders	DRD3 (AA genotype), DRD1, DRD4 (48 bp), GRIN2B (CC genotype), HTR2A c.102 T allele)	DRD1, DRD 4: Pathological gambling DRD 3: Violent behavior and impulsiveness DRD 4: Substance abuse
Compulsive shopping Binge eating Punding	Parkin mutation	Increased risk with lower age and smoker status
Compulsive shopping Binge eating Hypersexuality	PINK1 mutation in early onset PD	Susceptibility in rare PINK1 mutation cases

ACE, angiotensin convertin enzyme; APOE*4, apolipoprotein E; C/T SNP (A1) allele of the TaqIA DRD2, single nucleotide polymorphism of the dopamine D2 receptor (DRD2) gene; CCK, Cholecystokinin; CCKAR, Cholecystokinin A Receptor Gene; COMT, catechol o methyl transferase; DA, dopamine agonist; DAT VNTR (40bp), Dopamine transporter DAT1 40 bp variable number tandem repeat; DRD2, dopamine receptor D2; DRD2 (CA_n-STR), dopamine receptor D2 (CA dinucleotide short tandem repeat); DRD3, dopamine receptor D3; DRD4, dopamine receptor D4; GRIN2B (CC genotype), glutamate ionotropic receptor NMDA type subunit 2B; HOMER, Homer protein gene; HTR2A, 5-hydroxytryptamine receptor 2A; Met, methionine; NMS, nonmotor symptoms; PD, Parkinson's disease; PINK1, PTEN-induced kinase 1; Val, valine.

Adapted from Džoljic et al. (2015); Schumacher-Schuh et al. (2014)

4. AGE AND PERSONALIZED MEDICINE

Aging is a complex process and there are differences between chronological vs biological aging. NMS can be evident in the aging PD in the form of natural history, comorbidity as well as side effects of DRT and other interventions such as deep brain stimulation (DBS). Most treatment guidelines for PD do not necessarily take into account the concept of "healthy aging" and patients living with a longer life span, which may influence therapy. Age is considered a clear cut-off for performing DBS in older patients in part related to the risk of development of NMS such as depression and cognitive decline. However, in some well-selected older patients, DBS may still be appropriate and in some neurosurgical centers age is not considered to be an absolute inclusion criteria.

Old age onset of PD has been associated with a distinct phenotype and is likely to be driven by a different natural history as well nonmotor effects of comorbidity (Diederich, Moore, Leurgans, Chmura, & Goetz, 2003). Such patients may have high cognitive morbidity and therapeutic strategies advise caution against the use of DA therapies in particular. However in well-selected "healthy" octogenarians, age alone is not a contraindication to DA therapy and studies have suggested good tolerability (Shulman, Minagar, Rabinstein, & Weiner, 2000). Other studies suggest that transdermal dopamine agonists such as the rotigotine patch are well tolerated in older subjects and thus maybe a suitable and preferable choice in the older PD particularly if there are gastrointestinal issues (Chaudhuri et al., 2016; Farlow & Somogyi, 2011).

Depression appears to be a common NMS in young onset PD (YOPD) with consequent poorer quality of life (Knipe, Wickremaratchi, Wyatt-Haines, Morris, & Ben-Shlomo, 2011; Mehanna, Moore, Hou, Sarwar, & Lai, 2014). In addition, pain related to rigidity as well as painful cramps complicate YOPD. Management of YOPD cases should thus focus on specific treatment of these NMS.

The concept of personalized medicine with aging can be supported by magnetic resonance imaging which may demonstrate focal (medial temporal or global) atrophy or white matter vascular disease. In such cases, there is an increased risk of cognitive impairment and treatment should consider early cognitive testing and possibly earlier use of cholinesterase inhibitors.

On the other hand, younger age and other personality traits could make a PD patient particularly vulnerable to the development of ICD and thus would require a careful, tailored approach to treatment.

5. COMORBIDITIES

A number of comorbidities complicate the phenotypic expression of PD. A range of NMS may be associated with these comorbidities and are summarized in Table 2. Personalized medicine strategies, therefore, need to take into account the NMS associated with these comorbidities. Management in such cases should also focus on active treatment of the underlying comorbidity. For instance, autonomic dysfunction in a PD patient with diabetes may be due to both neurodegeneration related to PD as well as

Table 2 A List of Common or Relatively Common Comorbidities of Parkinson's Disease and Related Nonmotor Symptoms Which Would Drive Delivery of Personalized Medicine

Comorbidities	NMS 1	NMS 2	NMS 3	NMS 4	NMS 5
Diabetes mellitus	Fatigue	Postural hypotension	Erectile and sexual dysfunction	Pain (neuropathic)	Obesity (type 2)
Vascular disease (brain)	MCI	Dementia			
Gastrointestinal	Dysphagia and drooling	Constipation	Nonmotor fluctuations secondary to DGE		
Respiratory/sleep apnoea	EDS	Sleep fragmentation	Anxiety	Memory difficulties	
Thyroid	Anxiety (hyperthyroid)	Depression (hypothyroid)	Weight change		
Testosterone deficiency	Low libido	Depression	Anxiety	Fatigue	
Musculoskeletal disorders	Pain	Fatigue	Depression		
Restless legs syndrome	Insomnia	Fatigue	Anxiety		
Detrusor hyperreflexia/stress incontinence	Urinary incontinence	Sleep disruption due to nocturia			
Osteoporosis	Pain (backpain)				
Low body weight	Depression	Cognitive difficulties	Hyposmia		

DGE, delayed gastric emptying, EDS, excessive daytime sleepiness; MCI, mild cognitive impairment; NMS, nonmotor symptoms.

secondary to diabetes. Management of NMS in such a case should include a multidisciplinary approach including specialist input from a diabetologist.

Relevant strategies include, in those with diabetes, ensuring a good diabetic control and management of autonomic dysfunction and close liaison with a diabetic team. In diabetic patients with PD on advanced therapies involving the skin (apomoporhine infusion) skin care in particular is important to avoid problematic skin nodules.

In patients with vascular risk factors and significant intracerebral vascular disease, holistic management should include lifestyle advise (diet, exercise, sleep hygiene), vascular risk factor management, and active physiotherapy. In some patients, plasma homocysteine levels may need to be monitored particularly in those with high-dose levodopa therapy. In addition, in some patients with severe hyperhomocysteinemia the *MTHFR C677T* polymorphism may need to be checked (Džoljic et al., 2015).

In PD patients requiring antipsychotic medications, cardiac monitoring in relation to QTc interval is necessary. In PD patients with depression, apathy, anxiety, weight loss, or gain, it is important to ensure that a full metabolic screen is performed so as not to miss any comorbid disorders such as diabetes, thyrotoxicosis, or hypothyroidism. In PD patients over the age of 75 years and in older female PD patients, active, primary, and secondary prevention strategies for osteoporosis, using vitamin D2 and bisphosphonates, need to be undertaken (Sato, Honda, & Iwamoto, 2007). Specifically in these patients, physiotherapy to reduce the fear of falling is very important. Low body weight carries the risk of depression and dyskinesias and one would need to consider lower levodopa dosing, nutritional supplements as well as specialist dietary advice (Sharma, Ross, Rascol, & Brooks, 2008). In those with proven GIT disturbances, nonoral therapies such as rotigotine, apomorphine, or intrajejunal levodopa infusion may be indicated (Chaudhuri et al., 2016). Some may have severe hyposmia and special attention may be required related to cooking (smelling gas and decaying food, smoke and fire alarms) and lifestyle at home. Advice from an occupational therapist would be very useful. Some people with a loss of sense of smell also experience loss of appetite, but a normal diet is important for maintaining health. Reminders for eating regular meals are important to maintain weight.

Some practical advice related to hyposmia include:

- Setting fire alarms in all areas of the home, especially the kitchen and around the fireplace, stove
- Replacing gas appliances with electrical ones or fit gas detector

- Accurate monitoring of the shelf life of food, write the use by date on them in big, to know when to throw them away
- Carefully reading the labels on sanitary and hygienic products, such as cleaners, to be aware of the presence of potent chemicals

6. LIFESTYLE

Lifestyle choice and pattern will have an effect on the treatment delivered in PD and is therefore an important aspect of personalized medicine in PD (Fig. 4). Patients with high levels of activity (sports, other professions), sedentary but working lifestyles (people working in open space office areas), and public duties (doctors, politicians, teachers, entertainment workers) may prefer once a day therapies as opposed to DRT being taken several times a day. In these subjects, the mode of delivery of DRT is also important as some would prefer rapidly active nonoral routes such as injections and in future buccal or inhaled routes.

In other lifestyle-related scenarios, such as when a patient relies on high levels of manual work and other dexterity for his/her profession, DRT with rapid and reproducible effects might be chosen. While in some, where high levels of alertness are required, one would have to prescribe avoiding drugs with sedating side effects and in particular in such patients dopamine d3 receptor active agonists are probably best avoided. In summary, therefore,

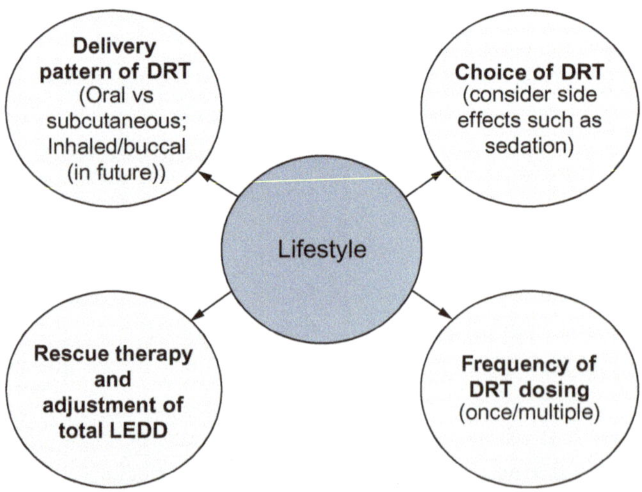

Fig. 4 A "circle" of components of lifestyle that will help complete delivery of personalized medicine. DRT, dopamine replacement therapy; LEDD, levodopa equivalent dose.

consideration of lifestyle should dictate our choice of therapy so as to avoid nonmotor issues such as secondary depression and anxiety related to stress if lifestyle is not considered.

7. BODY WEIGHT

Body weight is determined by environmental, genetic, epigenetic, metabolic, and environmental components and low body weight is common in PD. PD is known to have an association with motor symptoms such as dyskinesias in PD. Dyskinesias are part of the motor fluctuation spectrum in PD and are almost inevitably associated with nonmotor fluctuation. Low body weight has also been linked to lower cognitive scores on testing with MOCA, severe hyposmia as well as a poor quality of life in PD (Akbar et al., 2015).

From a personalized medicine standpoint, therefore, these PD patients with low body weight may particularly benefit from continuous drug delivery strategies to avoid dyskinesias, motor and nonmotor fluctuations in addition to special attention being given to nutritional supplements and diet (Bachmann, Zapf, Brunner, & Trenkwalder, 2009).

8. PERSONALITY AND PERCEPTION OF TREATMENT

Personality and related perception of treatment can play a major role in defining the success of treatment of PD. This issue also underpins the patient choice and informed decision making which is the key to modern delivery of medicine. Several personality-related factors may influence treatment and decisions related to types of treatment chosen. These are shown in Table 3.

ICD, punding, and dopamine dysregulation syndrome are major clinical challenges and personality traits predisposing to these DRT-related side effects should be taken into special consideration before treatment is prescribed (Poletti & Bonuccelli, 2012; Weintraub et al., 2010). Other personality traits can influence the delivery of treatment of PD and require personalized approached. These include inflexible and rigid personality trait. Many authors have reported a premorbid personality profile that is rigid and inflexible which does not change after the onset of motor symptoms of PD (Poewe, 1990; Santangelo, Piscopo, Barone, & Vitale, 2017). The clinical implication is that such patients will have difficulty in accepting the existence of many NMS of PD whereas they have a relatively easier perception of the

Table 3 A Description of Different Trends and Traits Related to Personality That May Affect Delivery of Therapy in Parkinson's Disease and Need for Consideration When Delivering Personalized Medicine

Personality predisposing to reward seeking behavior and risk factors for ICD, dopamine dysregulation, and punding
• Novelty seeking behavior • High alcohol consumption • History of substance abuse and drug addiction • Single status
Personality predisposing to cognitive dysfunction
• Inflexibility • Moral rigidity • Obsessionality
Personality predisposing to depression
• Alexithymia • Negative perception of impending "off" periods
Personality predisposing to anxiety trait (denovo PD)
• Neuroticism
Presence of medical therapy "phobia," particularly levodopa phobia
Belief in alternative or complimentary therapies such as homeopathy
Personal and cultural belief of patient and caregiver
Past and current history of noncompliance to regular prescribing

ICD, impulse control disorder, PD, Parkinson's disease.

motor syndrome of PD. This variable perception by default means that delivery of treatment of some relevant NMS such as depression or anxiety might be difficult. This is because the inflexible and rigid nature of the personality of these patients makes them noncompliant with such therapies. In such patients, specialist input from neuropsychologists is important in order that many NMS which may have an adverse effect on quality of life are treated effectively. Additionally, personality trends such as alexithymia as well as negative perception about impending "off" periods are linked to depression (Poletti et al., 2011). In particular, negative perception of "off period-related distress" may bring on nonmotor fluctuations and call for a tailored or personalized medicine approach (Fernie, Spada, Ray Chaudhuri, Klingelhoefer, & Brown, 2015).

In some PD patients, rigid perceptions may also influence the acceptability of the delivery pattern of DRT and some patients may find nonoral therapies unacceptable. In these patients, therapy needs to be personalized and should include close liaison with primary and secondary care in addition to a detailed explanation for the rationale of the DRT provided. Finally, those with a history of poor compliance with multiple dosing based previous treatment strategies or those who are currently noncompliant for DRT, need to be considered for once a day therapy.

9. PHARMACOECONOMICS

The success and sustenance of DRT in PD depends upon the cost of the drug, the patient's economic status, and the reimbursement system of the local health care. In most countries, health policies allow for management and reimbursement of the motor syndrome of PD while there is poor awareness of the range and impact of NMS. In many situations, expensive anti-PD drugs have to be funded by patients themselves or by private insurance. In such a clinical system, prescribing for management of NMS may be difficult and needs to be personalized based on the overall burden of NMS and prioritizing management of NMS that has the most visible and damaging effect on the patient.

10. BIOMARKER-DRIVEN NONMOTOR SUBTYPE-SPECIFIC TREATMENT OF PD AND PERSONALIZED MEDICINE

The pathophysiological and clinical heterogeneity of PD has been depicted by recent descriptions of nonmotor subtypes of PD (Marras & Chaudhuri, 2016; Sauerbier, Jenner, et al., 2016; Sauerbier, Qamar, et al., 2016). The underlying basis if these endophenotypes is related to the multineurotransmitter dysfunction of PD and resultant clinical phenotype being dominated by specific nonmotor symptoms which ranging from cognitive to sleep dysfunction (Table 4) (Jellinger, 2015; Titova et al., 2016).

Cluster analysis studies of de novo PD cases and individual cohort studies also support the concept of nonmotor subtypes (Erro et al., 2013; Zis, Erro, Walton, Sauerbier, & Chaudhuri, 2015). Nonmotor subtyping is also based on clinical and biomarker-driven identification of phenotypes and cholinergic, noradrenergic, serotonergic, and mixed patterns have been proposed all underpinned by dopamine deficiency (Pavese et al., 2010, 2012; Titova

Table 4 The Link Between Nonmotor Subtypes in Parkinson's Disease and Proposed Personalized Medicine (Subtype Specific) Treatment Paradigm

Neurotransmitter Dysfunction-Based Syndromes in PD	Proposed Clinical NMS Subtypes	Clinical Implications	Biomarkers			Proposed Personalized Medicine Strategy
			Imaging	Genetic	Clinical	
Cholinergic	Park-Cognitive (including MCI, Apathy) Apathy could be cholinergic and dopaminergic	Patients at high risk of dementia, cognitive dysfunction	Brain: PET scans with *N*-[^{11}C]-methyl-4-piperidyl acetate acetylcholinesterase (AChE): ↓ activity (Shimada et al., 2009) Gut: [^{11}*C*]donepezil PET (colon, intestine, pancreas): ↓ activity (Gjerloff et al., 2015)	Microtubule-associated protein Tau (MAPT) gene (H1/H2 genotype) (Miller & O'Callaghan, 2015; Williams-Gray et al., 2009)	↓ semantic fluency (Williams-Gray et al., 2009) Difficulty in pentagon copying (Williams-Gray et al., 2009)	Cognitive decline Counseling regarding lifestyle Forward planning Combining DRT and cholinesterase inhibitor therapy Cognitive training
Noradrenergic	Park-autonomic	Risk of aspiration pneumonia, weight loss, poor oral DRT absorption	Heart: Cardiac Metaiodobenzylguanidine (MIBG) scanning	No robust markers	Upper and lower gastrointestinal dysfunction: Delayed gastric emptying (Chaudhuri et al., 2016)	Consider nonoral therapies as a first option for DRT Nutritional and dietary advice

					Constipation	Probiotic supplementation
					Symptomatic postural hypotension	Supplement DRT with conventional management of postural hypotension Nonmotor fall prevention strategy (extra caution if patient also has osteoporosis)
Serotonergic	Park-Fatigue	Clinical picture similar to chronic fatigue syndrome	Brain: [^{11}C]DASB PET abnormality in limbic striatum and insula (Pavese, Metta, Bose, Chaudhuri, & Brooks, 2010)	No robust marker	Abnormal fatigue scale severity score	Combining DRT with serotonergic agents (receptor agonists)

Continued

Table 4 The Link Between Nonmotor Subtypes in Parkinson's Disease and Proposed Personalized Medicine (Subtype Specific) Treatment Paradigm—cont'd

Neurotransmitter Dysfunction-Based Syndromes in PD	Proposed Clinical NMS Subtypes	Clinical Implications	Biomarkers			Proposed Personalized Medicine Strategy
			Imaging	Genetic	Clinical	
Mixed neurotransmitter dysfunction	Park-Seep	Somnolence, sudden onset of sleep "attacks"	Brain: [^{11}C]DASB PET abnormality in raphe area (Pavese et al., 2012)	DRD2 and DRD4 (both linked to "sleep attacks"), HCRT (preprohypocretin) (Džoljic et al., 2015)	Clinical picture similar to narcolepsy with or without cataplexy Cooccurrence with RBD (Ylikoski, Martikainen, Sarkanen, & Partinen, 2015)	Avoid dopamine D3 receptor active agonists (may precipitate sudden onset of sleep) Consider alerting agents Early advice regarding lifestyle (avoid driving, working with heavy machinery, swimming alone) Probable role of serotonergic agents (receptor agonists) investigational

DASB, 3-amino-4-2-dimethylaminomethylphenylsulfa nyl-benzonitrile; DRD, dopamine receptor; DRT, dopamine replacement therapy; MCI, mild cognitive impairment; NMS, nonmotor symptoms, PET, positron emission tomography, RBD, rapid eye movement behavior disorder.

et al., 2016; Ylikoski et al., 2015). Personalized medicine in these subgroups relates to subtype-specific treatments which take into account the nonmotor clinical pattern as well as susceptibility to nonmotor side effects (such as sudden onset of sleep) of dopaminergic drugs. A multimodal biomarker approach supported by imaging, genetic, pharmacogenetic biomarkers will support a subtype-specific personalized treatment strategy in future and are summarized in Table 4.

11. CONCLUSIONS

Personalized medicine is a poorly understood and multidimensional concept, but perhaps the most important development in relation to treatment delivery of PD. This is because of the current understanding of the complexity and heterogeneity of PD, both from a pathophysiological and clinical standpoint. The general concept of personalized medicine being a genomic only or pharmacogenomic drug delivery strategy is wrong. Personalized medicine, according to our concept, is a multidimensional construct that takes into account not just the genomic and/or phramacogenomic basis, but also lifestyle, personality, individual beliefs, preferences, and choices, as well as comorbidity and pharmacoeconomics. Add to that the crucial and important need to consider the role of aging, body weight, and multimodal biomarker-driven nonmotor endophenotypes of PD. Personalized medicine, therefore, needs to consider all these aspects to make a treatment strategy truly individualized for the PD patient. The multidimensional nature of the concept of personalized medicine allows further subdivisions of the strategy. As such personalized medicine could be delivered in a small number of PD patients in a precise manner (precision medicine) where monogenic parkinsonism is evident. More importantly personalized medicine may play an important role in the growing field of pharmacogenetics once robust genetic associations between antiparkinsonian drugs and their metabolic and side effects are proven. Clinically, personalized medicine can also be directed toward symptoms of PD which may be influenced by the various enabling factors as described in Fig. 1. An additional dimension for the future would be to deliver subtype as well as ethnicity specific packages of care.

In future, personalized medicine could be underpinned by the development of a validated "checklist" of all the components of this strategy as shown in Fig. 1. A holistic personalized strategy of care should have considered all the "enablers" and documented the process. This should ensure that personalized medicine is established as a robust concept for the treatment of PD and serve as a benchmark of excellence in therapy.

Financial Disclosure/Conflict of Interest. The authors declare no conflict of interest.

REFERENCES

Akbar, U., He, Y., Dai, Y., Hack, N., Malaty, I., McFarland, N. R., et al. (2015). Weight loss and impact on quality of life in Parkinson's disease. *PLoS One, 10*(5), e0124541. http://dx.doi.org/10.1371/journal.pone.0124541.

Bachmann, C. G., Zapf, A., Brunner, E., & Trenkwalder, C. (2009). Dopaminergic treatment is associated with decreased body weight in patients with Parkinson's disease and dyskinesias. *European Journal of Neurology, 16*(8), 895–901. http://dx.doi.org/10.1111/j.1468-1331.2009.02617.x.

Beavan, M. S., & Schapira, A. H. (2013). Glucocerebrosidase mutations and the pathogenesis of Parkinson disease. *Annals of Medicine, 45*(8), 511–521. http://dx.doi.org/10.3109/07853890.2013.849003.

Bendikov-Bar, I., Maor, G., Filocamo, M., & Horowitz, M. (2013). Ambroxol as a pharmacological chaperone for mutant glucocerebrosidase. *Blood Cells, Molecules & Diseases, 50*(2), 141–145. http://dx.doi.org/10.1016/j.bcmd.2012.10.007.

Chaudhuri, K., Qamar, M., Rajah, T., Loehrer, P., Sauerbier, A., Odin, P., et al. (2016). Non-oral dopaminergic therapies for Parkinson's disease: Current treatments and the future. *Npj Parkinson's Disease, 2*, 16023. http://dx.doi.org/10.1038/npjparkd.2016.23.

Diederich, N. J., Moore, C. G., Leurgans, S. E., Chmura, T. A., & Goetz, C. G. (2003). Parkinson disease with old-age onset: A comparative study with subjects with middle-age onset. *Archives of Neurology, 60*(4), 529–533. http://dx.doi.org/10.1001/archneur.60.4.529.

Džoljic, E., Novaković, I., Krajinovic, M., Grbatinić, I., & Kostić, V. (2015). Pharmacogenetics of drug response in Parkinson's disease. *The International Journal of Neuroscience, 125*(9), 635–644. http://dx.doi.org/10.3109/00207454.2014.963851.

Erro, R., Vitale, C., Amboni, M., Picillo, M., Moccia, M., Longo, K., et al. (2013). The heterogeneity of early Parkinson's disease: A cluster analysis on newly diagnosed untreated patients. *PloS One, 8*(8), e70244. http://dx.doi.org/10.1371/journal.pone.0070244.

Falsone, S. F., Kungl, A. J., Rek, A., Cappai, R., & Zangger, K. (2009). The molecular chaperone Hsp90 modulates intermediate steps of amyloid assembly of the Parkinson-related protein alpha-synuclein. *The Journal of Biological Chemistry, 284*(45), 31190–31199. http://dx.doi.org/10.1074/jbc.M109.057240.

Farlow, M. R., & Somogyi, M. (2011). Transdermal patches for the treatment of neurologic conditions in elderly patients: A review. *The Primary Care Companion for CNS Disorders, 13*(6). http://dx.doi.org/10.4088/PCC.11r01149.

Fernie, B. A., Spada, M. M., Ray Chaudhuri, K., Klingelhoefer, L., & Brown, R. G. (2015). Thinking about motor fluctuations: An examination of metacognitions in Parkinson's disease. *Journal of Psychosomatic Research, 79*(6), 669–673. http://dx.doi.org/10.1016/j.jpsychores.2015.05.001.

Gjerloff, T., Fedorova, T., Knudsen, K., Munk, O. L., Nahimi, A., Jacobsen, S., et al. (2015). Imaging acetylcholinesterase density in peripheral organs in Parkinson's disease with ^{11}C-donepezil PET. *Brain, 138*(Pt. 3), 653–663. http://dx.doi.org/10.1093/brain/awu369.

Hoogland, J., de Bie, R. M., Williams-Gray, C. H., Muslimovic, D., Schmand, B., & Post, B. (2010). Catechol-O-methyltransferase val158met and cognitive function in Parkinson's disease. *Movement Disorders, 25*(15), 2550–2554. http://dx.doi.org/10.1002/mds.23319.

Jellinger, K. A. (2015). Neuropathobiology of non-motor symptoms in Parkinson disease. *Journal of Neural Transmission (Vienna)*, *122*(10), 1429–1440. http://dx.doi.org/10.1007/s00702-015-1405-5.

Kabuta, T., Furuta, A., Aoki, S., Furuta, K., & Wada, K. (2008). Aberrant interaction between Parkinson disease-associated mutant UCH-L1 and the lysosomal receptor for chaperone-mediated autophagy. *The Journal of Biological Chemistry*, *283*(35), 23731–23738. http://dx.doi.org/10.1074/jbc.M801918200.

Kalia, L. V., & Lang, A. E. (2015). Parkinson's disease. *Lancet*, *386*(9996), 896–912. http://dx.doi.org/10.1016/s0140-6736(14)61393-3.

Kasten, M., & Klein, C. (2014). Non-motor signs in genetic forms of Parkinson's disease. In K. Ray Chaudhuri, E. Tolosa, A. H. V. Schapira, & W. Poewe (Eds.), *Non motor symptoms in Parkinson's disease* (2nd ed., pp. 92–124). Oxford: UK Oxford University Press.

Knipe, M. D., Wickremaratchi, M. M., Wyatt-Haines, E., Morris, H. R., & Ben-Shlomo, Y. (2011). Quality of life in young- compared with late-onset Parkinson's disease. *Movement Disorders*, *26*(11), 2011–2018. http://dx.doi.org/10.1002/mds.23763.

Kunugi, H., Nanko, S., Ueki, A., Otsuka, E., Hattori, M., Hoda, F., et al. (1997). High and low activity alleles of catechol-O-methyltransferase gene: Ethnic difference and possible association with Parkinson's disease. *Neuroscience Letters*, *221*(2–3), 202–204.

Li, W., Chen, Y., Yin, B., & Zhang, L. (2014). Pain in Parkinson's disease associated with COMT gene polymorphisms. *Behavioural Neurology*, *2014*, 304203. http://dx.doi.org/10.1155/2014/304203.

Lindpaintner, K. (2002). Pharmacogenetics and the future of medical practice. *British Journal of Clinical Pharmacology*, *54*(2), 221–230.

Luo, W., Rodina, A., & Chiosis, G. (2008). Heat shock protein 90: Translation from cancer to Alzheimer's disease treatment? *BMC Neuroscience*, *9*(Suppl. 2), S7. http://dx.doi.org/10.1186/1471-2202-9-s2-s7.

Luo, W., Sun, W., Taldone, T., Rodina, A., & Chiosis, G. (2010). Heat shock protein 90 in neurodegenerative diseases. *Molecular Neurodegeneration*, *5*, 24. http://dx.doi.org/10.1186/1750-1326-5-24.

Makoff, A. J., Graham, J. M., Arranz, M. J., Forsyth, J., Li, T., Aitchison, K. J., et al. (2000). Association study of dopamine receptor gene polymorphisms with drug-induced hallucinations in patients with idiopathic Parkinson's disease. *Pharmacogenetics*, *10*(1), 43–48.

Marras, C., & Chaudhuri, K. R. (2016). Nonmotor features of Parkinson's disease subtypes. *Movement Disorders*, *31*(8), 1095–1102. http://dx.doi.org/10.1002/mds.26510.

Mata, I. F., Leverenz, J. B., Weintraub, D., Trojanowski, J. Q., Chen-Plotkin, A., Van Deerlin, V. M., et al. (2016). GBA variants are associated with a distinct pattern of cognitive deficits in Parkinson's disease. *Movement Disorders*, *31*(1), 95–102. http://dx.doi.org/10.1002/mds.26359.

McLeod, H. L., Syvanen, A. C., Githang'a, J., Indalo, A., Ismail, D., Dewar, K., et al. (1998). Ethnic differences in catechol O-methyltransferase pharmacogenetics: Frequency of the codon 108/158 low activity allele is lower in Kenyan than Caucasian or South-west Asian individuals. *Pharmacogenetics*, *8*(3), 195–199.

Mehanna, R., Moore, S., Hou, J. G., Sarwar, A. I., & Lai, E. C. (2014). Comparing clinical features of young onset, middle onset and late onset Parkinson's disease. *Parkinsonism & Related Disorders*, *20*(5), 530–534. http://dx.doi.org/10.1016/j.parkreldis.2014.02.013.

Miller, D. B., & O'Callaghan, J. P. (2015). Biomarkers of Parkinson's disease: Present and future. *Metabolism*, *64*(3 Suppl. 1), S40–S46. http://dx.doi.org/10.1016/j.metabol.2014.10.030.

Morgante, F., Fasano, A., Ginevrino, M., Petrucci, S., Ricciardi, L., Bove, F., et al. (2016). Impulsive-compulsive behaviors in parkin-associated Parkinson disease. *Neurology*, *87*(14), 1436–1441. http://dx.doi.org/10.1212/wnl.0000000000003177.

Moriwaki, Y., Kim, Y. J., Ido, Y., Misawa, H., Kawashima, K., Endo, S., et al. (2008). L347P PINK1 mutant that fails to bind to Hsp90/Cdc37 chaperones is rapidly degraded in a proteasome-dependent manner. *Neuroscience Research*, *61*(1), 43–48. http://dx.doi.org/10.1016/j.neures.2008.01.006.

Pavese, N., Metta, V., Bose, S. K., Chaudhuri, K. R., & Brooks, D. J. (2010). Fatigue in Parkinson's disease is linked to striatal and limbic serotonergic dysfunction. *Brain*, *133*(11), 3434–3443. http://dx.doi.org/10.1093/brain/awq268.

Pavese, N., Simpson, B. S., Metta, V., Ramlackhansingh, A., Chaudhuri, K. R., & Brooks, D. J. (2012). [^{18}F]FDOPA uptake in the raphe nuclei complex reflects serotonin transporter availability. A combined [^{18}F]FDOPA and [^{11}C]DASB PET study in Parkinson's disease. *NeuroImage*, *59*(2), 1080–1084. http://dx.doi.org/10.1016/j.neuroimage.2011.09.034.

Poewe, W. (1990). The premorbid personality of patients with Parkinson's disease. In P. Dostert, P. Riederer, M. Strolin Benedetti, & R. Roncucci (Eds.), *Early markers in Parkinson's and Alzheimer's diseases* (pp. 23–30). Wien: Springer-Verlag.

Poletti, M., & Bonuccelli, U. (2012). Impulse control disorders in Parkinson's disease: The role of personality and cognitive status. *Journal of Neurology*, *259*(11), 2269–2277. http://dx.doi.org/10.1007/s00415-012-6506-6.

Poletti, M., Frosini, D., Pagni, C., Lucetti, C., Del Dotto, P., Ceravolo, R., et al. (2011). Alexithymia is associated with depression in de novo Parkinson's disease. *Psychotherapy and Psychosomatics*, *80*(4), 251–253. http://dx.doi.org/10.1159/000322029.

Qin, Q., Zhi, L. T., Li, X. T., Yue, Z. Y., Li, G. Z., & Zhang, H. (2017). Effects of LRRK2 inhibitors on nigrostriatal dopaminergic neurotransmission. *CNS Neuroscience & Therapeutics*, *23*(2), 162–173. http://dx.doi.org/10.1111/cns.12660.

Santangelo, G., Piscopo, F., Barone, P., & Vitale, C. (2017). Personality in Parkinson's disease: Clinical, behavioural and cognitive correlates. *Journal of the Neurological Sciences*, *374*, 17–25. http://dx.doi.org/10.1016/j.jns.2017.01.013.

Sardi, S. P., Clarke, J., Viel, C., Chan, M., Tamsett, T. J., Treleaven, C. M., et al. (2013). Augmenting CNS glucocerebrosidase activity as a therapeutic strategy for parkinsonism and other Gaucher-related synucleinopathies. *Proceedings of the National Academy of Sciences*, *110*(9), 3537–3542. http://dx.doi.org/10.1073/pnas.1220464110.

Sato, Y., Honda, Y., & Iwamoto, J. (2007). Risedronate and ergocalciferol prevent hip fracture in elderly men with Parkinson disease. *Neurology*, *68*(12), 911–915. http://dx.doi.org/10.1212/01.wnl.0000257089.50476.92.

Sauerbier, A., Jenner, P., Todorova, A., & Chaudhuri, K. R. (2016a). Non motor subtypes and Parkinson's disease. *Parkinsonism & Related Disorders*, *22*(Suppl. 1), S41–S46. http://dx.doi.org/10.1016/j.parkreldis.2015.09.027.

Sauerbier, A., Qamar, M. A., Rajah, T., & Chaudhuri, K. R. (2016b). New concepts in the pathogenesis and presentation of Parkinson's disease. *Clinical Medicine (London, England)*, *16*(4), 365–370. http://dx.doi.org/10.7861/clinmedicine.16-4-365.

Schapira, A. H., & Gegg, M. E. (2013). Glucocerebrosidase in the pathogenesis and treatment of Parkinson disease. *Proceedings of the National Academy of Sciences of the United States of America*, *110*(9), 3214–3215. http://dx.doi.org/10.1073/pnas.1300822110.

Schumacher-Schuh, A. F., Rieder, C. R., & Hutz, M. H. (2014). Parkinson's disease pharmacogenomics: New findings and perspectives. *Pharmacogenomics*, *15*(9), 1253–1271. http://dx.doi.org/10.2217/pgs.14.93.

Sharma, J. C., Ross, I. N., Rascol, O., & Brooks, D. (2008). Relationship between weight, levodopa and dyskinesia: The significance of levodopa dose per kilogram body weight. *European Journal of Neurology*, *15*(5), 493–496. http://dx.doi.org/10.1111/j.1468-1331.2008.02106.x.

Shimada, H., Hirano, S., Shinotoh, H., Aotsuka, A., Sato, K., Tanaka, N., et al. (2009). Mapping of brain acetylcholinesterase alterations in Lewy body disease by PET. *Neurology*, *73*(4), 273–278. http://dx.doi.org/10.1212/WNL.0b013e3181ab2b58.

Shulman, L. M., Minagar, A., Rabinstein, A., & Weiner, W. J. (2000). The use of dopamine agonists in very elderly patients with Parkinson's disease. *Movement Disorders*, *15*(4), 664–668.

Sun, Y., Liou, B., Xu, Y. H., Quinn, B., Zhang, W., Hamler, R., et al. (2012). Ex vivo and in vivo effects of isofagomine on acid beta-glucosidase variants and substrate levels in Gaucher disease. *The Journal of Biological Chemistry*, *287*(6), 4275–4287. http://dx.doi.org/10.1074/jbc.M111.280016.

Sun, Y., Ran, H., Liou, B., Quinn, B., Zamzow, M., Zhang, W., et al. (2011). Isofagomine in vivo effects in a neuronopathic Gaucher disease mouse. *PloS One*, *6*(4), e19037. http://dx.doi.org/10.1371/journal.pone.0019037.

Titova, N., Padmakumar, C., Lewis, S. J., & Chaudhuri, K. R. (2016). Parkinson's: A syndrome rather than a disease? *Journal of Neural Transmission (Vienna)*. http://dx.doi.org/10.1007/s00702-016-1667-6.

Wang, L., Xie, C., Greggio, E., Parisiadou, L., Shim, H., Sun, L., et al. (2008). The chaperone activity of heat shock protein 90 is critical for maintaining the stability of leucine-rich repeat kinase 2. *The Journal of Neuroscience*, *28*(13), 3384–3391. http://dx.doi.org/10.1523/jneurosci.0185-08.2008.

Wang, J., Zhao, C., Chen, B., & Liu, Z. L. (2004). Polymorphisms of dopamine receptor and transporter genes and hallucinations in Parkinson's disease. *Neuroscience Letters*, *355*(3), 193–196.

Waza, M., Adachi, H., Katsuno, M., Minamiyama, M., Tanaka, F., Doyu, M., et al. (2006). Modulation of Hsp90 function in neurodegenerative disorders: A molecular-targeted therapy against disease-causing protein. *Journal of Molecular Medicine (Berl)*, *84*(8), 635–646. http://dx.doi.org/10.1007/s00109-006-0066-0.

Weintraub, D., Koester, J., Potenza, M. N., Siderowf, A. D., Stacy, M., Voon, V., et al. (2010). Impulse control disorders in Parkinson disease: A cross-sectional study of 3090 patients. *Archives of Neurology*, *67*(5), 589–595. http://dx.doi.org/10.1001/archneurol.2010.65.

Williams-Gray, C. H., Evans, J. R., Goris, A., Foltynie, T., Ban, M., Robbins, T. W., et al. (2009). The distinct cognitive syndromes of Parkinson's disease: 5 year follow-up of the CamPaIGN cohort. *Brain*, *132*(Pt. 11), 2958–2969. http://dx.doi.org/10.1093/brain/awp245.

Yang, C., Rahimpour, S., Lu, J., Pacak, K., Ikejiri, B., Brady, R. O., et al. (2013). Histone deacetylase inhibitors increase glucocerebrosidase activity in Gaucher disease by modulation of molecular chaperones. *Proceedings of the National Academy of Sciences of the United States of America*, *110*(3), 966–971. http://dx.doi.org/10.1073/pnas.1221046110.

Yeghiazaryan, K., Peeva, V., Morelli, M., Herrera-Marschitz, M., & Golubnitschaja, O. (2009). Potential targets for early diagnosis and neuroprotection in asphyxiated newborns. In O. Golubnitschaja (Ed.), *Predictive diagnostics and personalized treatment: Dream or reality?* (pp. 509–525). New York: Nova Science.

Ylikoski, A., Martikainen, K., Sarkanen, T., & Partinen, M. (2015). Parkinson's disease and narcolepsy-like symptoms. *Sleep Medicine*, *16*(4), 540–544. http://dx.doi.org/10.1016/j.sleep.2014.12.010.

Zis, P., Erro, R., Walton, C., Sauerbier, A., & Chaudhuri, K. (2015). The range and nature of non-motor symptoms in drug-naive Parkinson's disease patients: A state-of-the-art systematic review. *Npj Parkinson's Disease*, *1*, 15013. http://dx.doi.org/10.1038/npjparkd.2015.13.

SECTION 4

Nonmotor Symptoms in Parkinsonism and Related Disorders

CHAPTER FORTY-SIX

Nonmotor Features in Atypical Parkinsonism

Kailash P. Bhatia*, Maria Stamelou†,‡,§,1
*Institute of Neurology, London, United Kingdom
†HYGEIA Hospital, Athens, Greece
‡Neurology Clinic, Philipps University Marburg, Marburg, Germany
§University of Athens, Athens, Greece
[1]Corresponding author: e-mail address: mariastamelou@gmail.com

Contents

Abstract

Atypical parkinsonism (AP) comprises mainly multiple system atrophy (MSA), progressive supranuclear palsy (PSP), and corticobasal degeneration (CBD), which are distinct pathological entities, presenting with a wide phenotypic spectrum. The classic syndromes are now called MSA-parkinsonism (MSA-P), MSA-cerebellar type (MSA-C), Richardson's syndrome, and corticobasal syndrome. Nonmotor features in AP have been recognized almost since the initial description of these disorders; however, research has been limited. Autonomic dysfunction is the most prominent nonmotor feature of MSA, but also gastrointestinal symptoms, sleep dysfunction, and pain, can be a feature. In PSP and CBD, the most prominent nonmotor symptoms comprise those deriving from the cognitive/neuropsychiatric domain. Apart from assisting the clinician

International Review of Neurobiology, Volume 134
ISSN 0074-7742
http://dx.doi.org/10.1016/bs.irn.2017.06.001

in the differential diagnosis with Parkinson's disease, nonmotor features in AP have a big impact on quality of life and prognosis of AP and their treatment poses a major challenge for clinicians.

1. INTRODUCTION

The term atypical parkinsonism (AP) is used to describe a group of neurodegenerative disorders that are distinct pathological entities, namely multiple system atrophy (MSA), progressive supranuclear palsy (PSP), and corticobasal degeneration (CBD). MSA is defined pathologically by oligodendroglial alpha-synuclein aggregates (Trojanowski, Revesz, Neuropathology Working Group on MSA, 2007), PSP by 4-repeat (4R) tau aggregates in neurons with characteristic tufted astrocytes (Litvan, Agid, et al., 1996), and CBD by 4R tau aggregates with characteristic astrocytic plaques (Armstrong et al., 2013). Advances in the field of AP has led to the recognition of the fact, that these pathological entities may present phenotypically with several syndromes, and conversely, that the typical phenotypes [multiple system atrophy syndrome, Richardson's syndrome (RS), and corticobasal syndrome (CBS)] may be underpinned by diverse pathologies (Ling et al., 2010; Respondek et al., 2014; Stamelou, Quinn, & Bhatia, 2013).

In contrast to Parkinson's disease, for which nonmotor symptoms (NMS) only recently gained attention, these have been recognized in AP almost as early as the description of these disorders (Steele, Richardson, & Olszewski, 1964), being in fact, "red flags" to suspect AP in a patient with parkinsonism such as autonomic or cognitive dysfunction. Historically, therefore, the recognition of NMS in AP precedes that of PD; however, perhaps due to the rarity of these disorders, the quality of studies on NMS in AP is rather limited compared to PD. In this chapter, we discuss nonmotor features of MSA, PSP, and CBD and their relevance for diagnosis, prognosis, quality of life, and treatment.

2. NMS IN MSA

MSA is a sporadic neurodegenerative disorder clinically characterized by a combination of a variable degree of autonomic, parkinsonian, cerebellar, and pyramidal features. MSA-parkinsonism (MSA-P) and cerebellar (MSA-C) subtype are distinct phenotypes according to the predominant syndrome at onset (Wenning, Ben Shlomo, Magalhaes, Daniel, & Quinn, 1994; Wenning et al., 1998). The prevalence is estimated to be between 3.4 and 4.9/100.000. Definite MSA diagnosis is possible only on *postmortem* with evidence of widespread oligodendroglial alpha-synuclein accumulation with pontine and cerebellar atrophy or striatonigral degeneration (Gilman et al., 2008).

Current clinical diagnostic criteria define probable MSA as a sporadic, adult-onset (>30 years), progressive disorder with *autonomic failure* involving urinary incontinence, or an orthostatic decrease of blood pressure within 3 min of standing by at least 30 mm Hg systolic or 15 mm Hg diastolic; *and* poorly levodopa-responsive parkinsonism (bradykinesia with rigidity, tremor, or postural instability) *or* a cerebellar syndrome (gait ataxia with cerebellar dysarthria, limb ataxia, or cerebellar oculomotor dysfunction) (Gilman et al., 2008). Also clinical criteria for possible MSA require at least one autonomic feature combined with parkinsonism or a cerebellar syndrome and one additional sign (Gilman et al., 2008). Autonomic failure, a nonmotor feature, is therefore a sine qua non for clinical diagnosis of MSA.

2.1 Autonomic Dysfunction

Early and severe autonomic failure is a characteristic hallmark feature of MSA. Central and peripheral autonomic networks controlling cardiovascular, respiratory, urogenital, gastrointestinal, and sudomotor functions degenerate causing a variety of symptoms that may prevail the clinical picture. Urogenital and cardiovascular dysfunction are the most frequently affected. Urinary dysfunction presenting with urinary urgency and frequency, urge and stress-incontinence, nocturia, and incomplete bladder emptying that may require intermittent catheterization. Autonomic dysfunction may precede the onset of motor symptoms in up to 50% of the patients in some clinicopathologic studies (Iodice et al., 2012; Schmidt et al., 2008), and urogenital dysfunction precedes orthostatic hypotension in the majority of the patients, which is important for diagnostic and prognostic reasons (Iodice et al., 2012; Jecmenica-Lukic, Poewe, Tolosa, & Wenning, 2012; Kimpinski et al., 2012). Erectile dysfunction occurs in up to 97% of men with definite MSA, while it is reported as the presenting symptom in 48% of male patients (Iodice et al., 2012; Jecmenica-Lukic et al., 2012). Unfortunately, data for genital dysfunction in female patients, perhaps owing to difficulties in assessment, are largely missing (Oertel, Wachter, Quinn, Ulm, & Brandstadter, 2003).

Severe orthostatic hypotension is defined as a blood pressure drop of 30 mmHg systolic or 15 mmHg diastolic within 3 min after standing from the recumbent position (Gilman et al., 2008) and is the main feature of cardiovascular autonomic failure in MSA. Symptoms include syncope, postural dizziness, or pain centered in the neck and shoulders (coat-hanger ache), but it may also be asymptomatic. Up to 50% of the patients may also have postprandial hypotension and supine and nocturnal hypertension (Brisinda et al., 2014; Deguchi et al., 2007; Gilman et al., 2008; Iodice et al., 2012; Jecmenica-Lukic et al., 2012; Wenning et al., 2013).

Respiratory symptoms, such as stridor, sleep-related breathing disorders, and respiratory insufficiency, belong to the MSA phenotype. Although inspiratory sighs and involuntary gasps are common in clinical practice, there are no well-documented published studies in this regard. Inspiratory stridor, which can attributed to laryngeal abductor spasm or result from vocal cords dystonia (Benarroch, Schmeichel, Sandroni, Low, & Parisi, 2006; Magalhaes, Wenning, Daniel, & Quinn, 1995; Wenning, Tison, Ben Shlomo, Daniel, & Quinn, 1997), can be found in 9%–69% of MSA patients. Nocturnal stridor, is highly specific of MSA and is more frequent (42%) than obstructive sleep apnea (37%) (Vetrugno et al., 2007). Central sleep apnea is less common and tends to occur at later stages (Jecmenica-Lukic et al., 2012; Kitae et al., 2001; Munschauer, Loh, Bannister, & Newsom-Davis, 1990; Plazzi et al., 1998; Vetrugno et al., 2007). A loss in the physiological nocturnal dip in blood pressure is an additional suggestive nocturnal feature of MSA (Jecmenica-Lukic et al., 2012; Wenning et al., 1994, 2013).

Not surprisingly, the differential diagnosis of MSA-P with PD and other parkinsonian syndromes is predominantly based on the evaluation of autonomic involvement. Cardiovascular autonomic tests including head up tilt test and Valsalva maneuver have been described to differentiate MSA from PD with 91% sensitivity and 92% specificity. (123)-I-myocardial meta-iodobenzylguanidine (MIBG) scintigraphy has been described to reliably differentiate MSA from PD—the pooled sensitivity and specificity for this differentiation has been found to be 90.2%, respectively 81.9% (Orimo, Suzuki, Inaba, & Mizusawa, 2012). MIBG is abnormal in PD, reflecting a postsynaptic autonomic dysfunction, in contrast to the normal finding in MSA, where there is a presynaptic autonomic dyfunction. However, prospective neuropathologically confirmed studies are largely missing. Also, MIBG does not differentiate between MSA and other atypical parkinsonian disorders, in all of which MIBG is normal.

2.2 Other NMS in MSA

REM-sleep behavior disorder (RBD) is a frequent finding in alpha-synucleinopathies, such as Parkinson's disease (PD), Dementia with Lewy Bodies, and MSA in contrast to other disorders mainly tauopathies such as Alzheimer's disease, CBD, or PSP (3%). The positive predictive value of the presence of RBD, for the presence of alpha-synucleinopathy lies between 91% and 100%. Thus, in a patient with parkinsonism without dementia, the presence of RBD would strongly point to PD or MSA, whereas the absence of RBD is not helpful to differentiate PD from atypical parkinsonian syndromes (Ahmed, Asi, Lees, Revesz, & Holton, 2012; Gilman et al., 2003; Iodice et al., 2012; Iranzo et al., 2005; Jecmenica-Lukic et al., 2012; Plazzi et al., 1997; Tison, Wenning, Quinn, & Smith, 1995). Thus, RBD is very common in MSA and together with autonomic dysfunction, represent a premotor stage in MSA in many patients, which should alert clinicians for this diagnosis.

One of the clinical features that does not support the diagnosis of MSA based on widely accepted clinical criteria is early dementia and hallucinations (Gilman et al., 1998, 2008). However, accumulating evidence suggests that there may be cognitive impairment in MSA mainly of frontal-executive dysfunction. Frontal lobe dysfunction with attention deficits has been reported in one third of cases. Emotional incontinence (i.e., inappropriate laughing or crying in the absence of the appropriate emotional context) can occur, as well as behavioral changes, including depression, anxiety, panic attacks, and suicidal ideation (Balas, Balash, Giladi, & Gurevich, 2010; Brown et al., 2010; Chang et al., 2009; Kao et al., 2009; Kawai et al., 2008; Lee et al., 2013; Lee, Williams, & Storey, 2012; Siri et al., 2013). Despite these data, clinically prominent and early dementia is rather pointing to another disorders mainly tauopathies (Balas et al., 2010; Chang et al., 2009; Kao et al., 2009; Kawai et al., 2008; Lee et al., 2013; Siri et al., 2013).

As many as 50% of patients report disabling pain: advanced disease, dystonia, and female sex are possible risk factors for this symptom. Pain, including musculoskeletal and visceral pain, has been described as fairly common in MSA and PD, even at disease onset, while pain can also be a side effect of dopaminergic treatment in particular causing painful facial dystonia in MSA (Colosimo et al., 2010; Kass-Iliyya, Kobylecki, McDonald, Gerhard, & Silverdale, 2015). Gastrointestinal (GI) symptoms are frequent in MSA, initially mainly dysphagia, which is a prominent characteristic also of PSP (Colosimo et al., 2010).

Depression seems to be very common in MSA (Almeida et al., 2017). Mood and apathy symptoms were found to be more severe in MSA-P patients than MSA-C patients in one study, although the frequency of mood/apathy symptoms was similar between MSA-P and MSA-C patients. Furthermore, drooling, constipation, and pain symptoms were found to be more severe and prevalent in MSA-P patients than in MSA-C patients (Lee et al., 2013).

3. NMS IN PSP AND CBD

3.1 Progressive Supranuclear Palsy

PSP is a progressive neurodegenerative condition characterized by progressive parkinsonism with early falls, supranuclear vertical gaze palsy, frontal subcortical dementia, and axial rigidity (Steele et al., 1964). The prevalence is estimated to be 5–6/100.000. The pathognomonic neuropathological findings in PSP are tau aggregates in astrocytes (tufts), oligodendrocytes (coiled bodies), and neurons (neurofibrillary tangles), which are distinct from other tauopathies (Dickson, 1999). PSP is a sporadic condition in the vast majority of cases; however, a number of familial PSP cases have been reported mainly due to mutations in the *MAPT* gene (Stamelou et al., 2013).

The characteristic syndrome of PSP is now called Richardson's syndrome (PSP-RS) (Williams et al., 2005). Further PSP phenotypes have been described (Williams et al., 2005; Williams & Lees, 2009). Apart from PSP-RS, PSP-Parkinsonism, characterized by a slower disease progression, more asymmetric presentation and initial levodopa-responsiveness, pure akinesia with gait freezing (PAGF), progressive nonfluent aphasia (PNFA), PSP–CBS, and further phenotypes have been described (Respondek et al., 2014). Recently, also a PSP phenotype with cerebellar features has been described (Iwasaki et al., 2013; Kanazawa et al., 2009). The identification of the wide phenotypic spectrum of PSP has led to the revision of the widely used National Institute of Neurological Disorders and Stroke clinical diagnostic criteria for PSP that have been (NINDS)-PSPS widely used since 1996 (Litvan, Agid, et al., 1996). In these criteria, postural instability and supranuclear gaze palsy (two motor signs) were mainly used with specific time frames from onset, to diagnose possible and probable PSP, corresponding mainly to PSP-RS in clinical practice. A consensus has recently been reached for new clinical diagnostic criteria for PSP (Höglinger et al., 2017), that takes into account the various phenotypes of PSP pathology.

3.2 Corticobasal Degeneration

CBD is a rare neurodegenerative disorder typically characterized by progressive asymmetric levodopa-resistant parkinsonism, dystonia, myoclonus, and further cortical signs (e.g., apraxia, alien limb phenomena, cortical sensory loss). The exact prevalence of CBD is unknown, but is thought to be considerably lower than that of PSP or MSA, and males and females are equally affected. Average time from symptom onset to death is 8 years. CBD is a 4R-tauopathy with astrocytic plaques in pathology and almost always a sporadic condition. Patients typically present with strikingly asymmetric rigidity and bradykinesia as well as cortical signs affecting one limb (CBS) (Boeve et al., 1999; Dickson, 1999; Tsuboi et al., 2005; Wenning et al., 1998). The limb may be also affected by dystonia and distal, stimulus-sensitive myoclonus and is often described as "useless," while "alien-limb phenomena" may occur. Dementia is an important part of the phenotype and may be the presenting or predominant clinical feature in CBD. However, similarly to PSP, patients with CBD pathology may present with further phenotypes, such a frontotemporal dementia (FTD), PSP-RS, primary progressive aphasia, or a posterior cortical atrophy syndrome (Stamelou, Alonso-Canovas, & Bhatia, 2012). Conversely, patients with other pathologies such as Alzheimer's disease (AD), PSP and FTD pathology may present with CBS (Ling et al., 2010). The prevalence and natural distribution of these phenotypes are not known.

Clinical criteria for CBD attempt to take into account the different phenotypes, which however, have not been independently validated yet (Armstrong et al., 2013). Definite CBD requires pathological confirmation (Armstrong et al., 2013). There are some suggestions that certain clinical features may partly be helpful to suspect pathology. For example, a symmetric bilateral CBS could imply AD pathology (Josephs et al., 2010) or a patient with RS and earlier age of onset than expected in PSP may have CBD pathology (Ling et al., 2010), however, these suggestions need further confirmation. Moreover, the different phenotypes likely represent the localization (e.g. the brain areas affected) of pathology in each of these disorders (Hassan, Whitwell, & Josephs, 2011; Kouri et al., 2011; Williams & Lees, 2009).

3.3 Cognition and Neuropsychiatric/Behavioral Symptoms in PSP/CBD

Undoubtedly, the most characteristic NMS in PSP and CBD are the ones resulting from cognitive and neuropsychiatric dysfunction. Frontal dysfunction, particularly behavioral and personality changes, is consistently seen in PSP-RS and may appear early in the clinical course, however, in contrast to behavioral variant of frontotemporal dementia (bvFTD), apathy is in PSP the most frequent feature and not disinhibition. Executive deficits, reduced verbal fluency, and a positive applause sign are common frontal features of PSP-RS. Apathy, bradyphrenia (i.e., slowness of thinking), executive dysfunction, reduced phonemic verbal fluency, impulsivity, disinhibition, and perseveration (Aarsland, Litvan, & Larsen, 2001; Borroni, Alberici, Agosti, Cosseddu, & Padovani, 2009; Gerstenecker, Duff, Mast, Litvan, ENGENE-PSP Study Group, 2013; Litvan, Paulsen, Mega, & Cummings, 1998; Rittman et al., 2013) are common clinical features in PSP-RS.

However, PSP pathology has also has been associated with an FTD phenotype, and its frequency in autopsy-confirmed PSP patients has been reported to be up to 5%–20%. A clinical picture of behavioral changes and cognitive dysfunction dominates and resembles that of the bvFTD (Aarsland et al., 2001; Dubois, Slachevsky, Litvan, & Pillon, 2000; Gerstenecker et al., 2013; Kouri et al., 2011; Litvan, 1994; Litvan et al., 1997; Litvan, Mangone, et al., 1996; Litvan et al., 1998; Paviour et al., 2005; Stamelou et al., 2015; Tsuboi et al., 2005; Verny et al., 1996). A recent in-depth review of all studies ever conducted on neuropsychological testing in PSP has shown that the latter are limited by the lack of autopsy confirmation and the inclusion of mostly PSP-RS cases (Respondek et al., 2017). Only a few studies with autopsy-confirmation were available and revealed that PSP may present clinically with features classically attributed to FTD, such as bvFTD (Donker Kaat et al., 2007; Han et al., 2010; Hassan, Parisi, & Josephs, 2012) and nonfluent agrammatic progressive primary aphasia (nfaPPA) (Mochizuki et al., 2003).

The recent largest clinicopathological study reported on PSP, included 206 PSP brains compared to 54 CBD, 51 MS, 53 PD, and 73 bvFTD brains, and has shown that nfaPPA, was present in 19% of definite PSP cases. Frontal dysfunction throughout the course had a sensitivity for PSP of 57% and a specificity of 46%, while sensitivity was 30% and specificity 63%, when frontal dysfunction occurred within the first 3 years (Respondek et al., 2017). Consequently, new PSP clinical criteria have just been proposed (Höglinger et al., 2017), including for the first time cognition, a nonmotor feature, in the core criteria for diagnosing PSP.

Similarly, dementia is an important part of the phenotype and may be the presenting or predominant clinical feature in CBD. Neuropsychometry typically reveals frontal executive defects together with parietal lobe dysfunction. In contrast with Alzheimer's disease, episodic memory is usually well preserved (Borroni et al., 2009; Lee et al., 2012; Rittman et al., 2013). As for PSP, different phenotypes apart from CBS, such as RS, FTD, etc. have been described, and cognition is an important nonmotor features in the clinical criteria for diagnosing the various phenotypes that may be underlined by CBD pathology (Armstrong et al., 2013).

3.4 Other NMS in PSP/CBD

Autonomic dysfunction can also be present in PSP and CBD, but later in the disease course, and rarely severe in the early stages. There is usually no RBD in tauopathies including PSP and CBD, although its presence cannot exclude a tauopathy as seen in some studies (review Respondek et al., 2017). PSP has been associated with sleep architecture changes such as decreased total sleep time and decreased sleep efficiency, while very little is known about sleep changes in CBD (Abbott & Videnovic, 2014). Pain and GI symptoms are common features in both disorders but less frequent than in MSA (Colosimo et al., 2010; Kass-Iliyya et al., 2015; Stamelou, Dohmann, et al., 2012).

4. NMS AND QUALITY OF LIFE IN AP

NMS have a considerable effect on the quality of life (QOL) of MSA patients. A European study on 115 MSA patients found autonomic dysfunction and depression to be associated with poor QOL (Colosimo, 2011). A recent large sample study found that NMS are common both in MSA-C and MSA-P patients, but more severe and prevalent in MSA-P patients, especially mood/apathy and gastrointestinal symptoms. A close correlation between NMS and Parkinson's Disease Questionnaire (PDQ)-39 in MSA-P as well as MSA-C patients has been found. Disease severity, but also cardiovascular and urinary symptoms, sleep problems, fatigue, and GI symptoms determine poor QOL for MSA-P patients, and mood and apathy symptoms in MSA-C patients (Lee et al., 2013).

The causes of death in MSA commonly include bronchopneumonia, urosepsis, or sudden death. Sudden death often occurs at night as a result of either acute bilateral vocal-cord paralysis or acute disruption of the brain-stem cardiorespiratory drive. Older age at onset, a parkinsonian phenotype, and early development of severe autonomic failure are negative prognostic factors, whereas a cerebellar phenotype and later onset of autonomic failure predict slower disease progression (Wenning et al., 2013).

Only a few studies have examined QOL in PSP. One prospective study included 46 PSP patients, 28 of which were followed up after 1 year. Gender, included clinical variables and psychiatric/neuropsychological scales, accounted for 68% of the variance in the Physical Composite Score and 73% of the variance of the Mental Composite Score of the 36-item Short Form Health Survey (SF-36). A significant decline after 1 year in the HR-QOL was detected for physical functioning, vitality, social functioning, and role emotional, as well as in both composite scores (Physical Composite Score and Mental Composite Score). Apathy and depression seem important determinants of QOL in these disorders (Pekmezovic et al., 2015; Winter et al., 2011).

5. TREATMENT OF NMS IN AP

5.1 Autonomic Dysfunction

There are limitations in using antimuscarinic agents like trospium chloride (2–3 × 10 to 15 mg/d) or tolterodine (2 × 2 mg/d) for urinary urge/incontinence, including mainly the anticholinergic side effects, e.g., confusion and cognitive worsening, which pose an important problem in particular in patients that have already cognitive dysfunction, and thus in PSP/CBD more than in MSA. An alternative are botulinum toxin injections in the detrusor muscle (Kollensperger et al., 2010; Novak et al., 2012; Stamelou & Bhatia, 2016). In postvoid residual volumes above 100 mL and urinary retention, clean intermittent self-catheterization should be used, however, with the danger of urethral ulceration, in which case, suprapubic indwelling catheterization may become necessary. Cholinergic agents to enhance vesical detrusor contractility can be added for urinary retention and α1-adrenoreceptor antagonists to promote urethral smooth sphincter relaxation. Sildenafil, a phosphodiesterase type 5 inhibitor, can improve erectile dysfunction; however, this may worsen orthostatic hypotension which can be quite limiting in patients with MSA. Intracavernous injection of alprostadil, a vasodilatory prostaglandin could alternatively be used.

Conservative measures for orthostatic hypotension, such as avoid rapid postural change, heavy meals, exposure to hot temperatures, increasing water and salt intake, keeping the head of the bed raised during sleep, and using compression stockings or abdominal binders should be followed as first line treatment by patients. Midodrine and droxidopa, sympathomimetic agents that increase arteriolar tone, are specifically licensed by the Food and Drug Administration for the symptomatic treatment of neurogenic orthostatic hypotension (Maass, Levin, & Hoglinger, 2016; Stamelou & Bhatia, 2015, 2016).

Continuous positive airway pressure, or, in resistant cases, biphasic positive airway pressure, is the therapy of choice for patients with MSA who have isolated nocturnal inspiratory stridor or sleep apnea. Unilateral botulinum toxin injection in the vocal-cord adductors may be considered for more severely affected patients, but bears the possibility of worsening dysphagia that these patients usually have. Tracheostomy may effectively relieve airway obstruction at the laryngeal level and prevent respiratory crisis due to acute bilateral paralysis of the vocal-cord abductors, but it cannot eliminate the risk of sudden death, because fatal sleep apnea can still occur (Iranzo, 2005, 2007; Iranzo, Santamaria, & Tolosa, 2000; Iranzo et al., 2004).

5.2 Cognitive Impairment

Cognitive impairment in patients with MSA does not usually require treatment, but pharmacologic intervention may be needed for those with severe depression, anxiety, or emotional incontinence. Selective serotonin-reuptake inhibitors are preferable to tricyclic antidepressants because they are more effective and less likely to worsen orthostatic hypotension and urinary retention, and they may help with depression and apathy in some cases. Despite the prominent cognitive dysfunction in PSP and CBD, effective treatments are still not available (Koros & Stamelou, 2016). Donepezil and rivastigmine rather led to a worsening of motor symptoms therefore are not recommended (Fabbrini et al., 2001; Liepelt et al., 2010; Litvan et al., 2001).

5.3 Others

Botulinum toxin injections in the salivary glands can improve sialorrhea, however, with caution not to worsen the dysphagia in advanced MSA, PSP, and CBD. Speech therapy with swallowing training can help with the dysphagia, and a percutaneous endoscopic gastrostomy may reduce the risk of aspiration pneumonia. RBD can be treated with low-dose clonazepam at bedtime, but it may worsen nocturnal stridor or sleep apnea.

6. CONCLUSION

Despite the lack of large and pathology confirmed studies, NMS in AP are well recognized. It is now accepted that a considerable disease burden in AP is caused by NMS. Beyond autonomic dysfunction in MSA and cognitive/neuropsychatric features in PSP/CBD, a wide range of further NMS are present and have an impact on QOL. Management of these symptoms is still difficult and challenging, while they may influence prognosis and survival. Further studies are needed to better describe the prevalence and burden of NMS in AP, understand their pathophysiology, and provide efficacious treatment for these symptoms.

REFERENCES

Aarsland, D., Litvan, I., & Larsen, J. P. (2001). Neuropsychiatric symptoms of patients with progressive supranuclear palsy and Parkinson's disease. *The Journal of Neuropsychiatry and Clinical Neurosciences*, *13*(1), 42–49. http://dx.doi.org/10.1176/jnp.13.1.42.

Abbott, S. M., & Videnovic, A. (2014). Sleep disorders in atypical Parkinsonism. *Movement Disorders Clinical Practice*, *1*(2), 89–96.

Ahmed, Z., Asi, Y. T., Lees, A. J., Revesz, T., & Holton, J. L. (2012). Identification and quantification of oligodendrocyte precursor cells in multiple system atrophy, progressive supranuclear palsy and Parkinson's disease. *Brain Pathology (Zurich, Switzerland)*, *23*, 263–273. http://dx.doi.org/10.1111/j.1750-3639.2012.00637.x.

Almeida, L., Ahmed, B., Walz, R., De Jesus, S., Patterson, A., Martinez-Ramirez, D., et al. (2017). Depressive symptoms are frequent in atypical parkinsonian disorders. *Movement Disorders Clinical Practice*, *4*, 191–197.

Armstrong, M. J., Litvan, I., Lang, A. E., Bak, T. H., Bhatia, K. P., Borroni, B., et al. (2013). Criteria for the diagnosis of corticobasal degeneration. *Neurology*, *80*(5), 496–503. http://dx.doi.org/10.1212/WNL.0b013e31827f0fd1.

Balas, M., Balash, Y., Giladi, N., & Gurevich, T. (2010). Cognition in multiple system atrophy: Neuropsychological profile and interaction with mood. *Journal of Neural Transmission (Vienna, Austria: 1996)*, *117*(3), 369–375. http://dx.doi.org/10.1007/s00702-009-0365-z.

Benarroch, E. E., Schmeichel, A. M., Sandroni, P., Low, P. A., & Parisi, J. E. (2006). Involvement of vagal autonomic nuclei in multiple system atrophy and Lewy body disease. *Neurology*, *66*(3), 378–383. http://dx.doi.org/10.1212/01.wnl.0000196638.98781.bb.

Boeve, B. F., Maraganore, D. M., Parisi, J. E., Ahlskog, J. E., Graff-Radford, N., Caselli, R. J., et al. (1999). Pathologic heterogeneity in clinically diagnosed corticobasal degeneration. *Neurology*, *53*(4), 795–800.

Borroni, B., Alberici, A., Agosti, C., Cosseddu, M., & Padovani, A. (2009). Pattern ofbehavioral disturbances in corticobasal degeneration syndrome and progressive supranuclear palsy. *International Psychogeriatrics*, *21*(3), 463–468. http://dx.doi.org/10.1017/S1041610209008862.

Brisinda, D., Sorbo, A. R., Di Giacopo, R., Venuti, A., Bentivoglio, A. R., & Fenici, R. (2014). Cardiovascular autonomic nervous system evaluation in Parkinson disease and multiple system atrophy. *Journal of the Neurological Sciences*, *336*(1–2), 197–202. http://dx.doi.org/10.1016/j.jns.2013.10.039.

Brown, R. G., Lacomblez, L., Landwehrmeyer, B. G., Bak, T., Uttner, I., Dubois, B., et al. (2010). Cognitive impairment in patients with multiple system atrophy and progressive supranuclear palsy. *Brain*, *133*(Pt. 8), 2382–2393. http://dx.doi.org/10.1093/brain/awq158.

Chang, C. C., Chang, Y. Y., Chang, W. N., Lee, Y. C., Wang, Y. L., Lui, C. C., et al. (2009). Cognitive deficits in multiple system atrophy correlate with frontal atrophy and disease duration. *European Journal of Neurology*, *16*(10), 1144–1150. http://dx.doi.org/10.1111/j.1468-1331.2009.02661.x.

Colosimo, C. (2011). Nonmotor presentations of multiple system atrophy. *Nature Reviews Neurology*, 7(5), 295–298. http://dx.doi.org/10.1038/nrneurol.2011.5.

Colosimo, C., Morgante, L., Antonini, A., Barone, P., Avarello, T. P., Bottacchi, E., et al. (2010). Non-motor symptoms in atypical and secondary parkinsonism: The PRIAMO study. *Journal of Neurology*, *257*(1), 5–14. http://dx.doi.org/10.1007/s00415-009-5255-7.

Deguchi, K., Ikeda, K., Shimamura, M., Urai, Y., Tsukaguchi, M., Touge, T., et al. (2007). Assessment of autonomic dysfunction of multiple system atrophy with laryngeal abductor paralysis as an early manifestation. *Clinical Neurology and Neurosurgery*, *109*(10), 892–895. http://dx.doi.org/10.1016/j.clineuro.2007.07.020.

Dickson, D. W. (1999). Neuropathologic differentiation of progressive supranuclear palsy and corticobasal degeneration. *Journal of Neurology*, *246*(Suppl. 2), II6–II15.

Donker Kaat, L., Boon, A. J., Kamphorst, W., Ravid, R., Duivenvoorden, H. J., & van Swieten, J. C. (2007). Frontal presentation in progressive supranuclear palsy. *Neurology*, *69*(8), 723–729. http://dx.doi.org/10.1212/01.wnl.0000267643.24870.26.

Dubois, B., Slachevsky, A., Litvan, I., & Pillon, B. (2000). The FAB: A frontal assessment battery at bedside. *Neurology*, *55*(11), 1621–1626.

Fabbrini, G., Barbanti, P., Bonifati, V., Colosimo, C., Gasparini, M., Vanacore, N., et al. (2001). Donepezil in the treatment of progressive supranuclear palsy. *Acta Neurologica Scandinavica*, *103*(2), 123–125.

Gerstenecker, A., Duff, K., Mast, B., Litvan, I., & ENGENE-PSP Study Group. (2013). Behavioral abnormalities in progressive supranuclear palsy. *Psychiatry Research*, *210*(3), 1205–1210. http://dx.doi.org/10.1016/j.psychres.2013.08.045.

Gilman, S., Koeppe, R. A., Chervin, R. D., Consens, F. B., Little, R., An, H., et al. (2003). REM sleep behavior disorder is related to striatal monoaminergic deficit in MSA. *Neurology*, *61*(1), 29–34.

Gilman, S., Low, P., Quinn, N., Albanese, A., Ben-Shlomo, Y., Fowler, C., et al. (1998). Consensus statement on the diagnosis of multiple system atrophy. American autonomic society and American Academy of Neurology. *Clinical Autonomic Research*, *8*(6), 359–362.

Gilman, S., Wenning, G. K., Low, P. A., Brooks, D. J., Mathias, C. J., Trojanowski, J. Q., et al. (2008). Second consensus statement on the diagnosis of multiple system atrophy. *Neurology*, *71*(9), 670–676. http://dx.doi.org/10.1212/01.wnl.0000324625.00404.15.

Han, H. J., Kim, H., Park, J. H., Shin, H. W., Kim, G. U., Kim, D. S., et al. (2010). Behavioral changes as the earliest clinical manifestation of progressive supranuclear palsy. *Journal of Clinical Neurology (Seoul, Korea)*, *6*(3), 148–151. http://dx.doi.org/10.3988/jcn.2010.6.3.148.

Hassan, A., Parisi, J. E., & Josephs, K. A. (2012). Autopsy-proven progressive supranuclear palsy presenting as behavioral variant frontotemporal dementia. *Neurocase*, *18*(6), 478–488. http://dx.doi.org/10.1080/13554794.2011.627345.

Hassan, A., Whitwell, J. L., & Josephs, K. A. (2011). The corticobasal syndrome-Alzheimer's disease conundrum. *Expert Review of Neurotherapeutics*, *11*(11), 1569–1578. http://dx.doi.org/10.1586/ern.11.153.

Höglinger, G. U., Respondek, G., Stamelou, M., Kurz, C., Josephs, K. A., Lang, A. E., et al. (2017). Clinical diagnosis of progressive supranuclear palsy: The movement disorder society criteria. *Movement Disorders*, *32*(6), 853–864. http://dx.doi.org/10.1002/mds.26987.

Iodice, V., Lipp, A., Ahlskog, J. E., Sandroni, P., Fealey, R. D., Parisi, J. E., et al. (2012). Autopsy confirmed multiple system atrophy cases: Mayo experience and role of autonomic function tests. *Journal of Neurology, Neurosurgery, and Psychiatry*, *83*(4), 453–459. http://dx.doi.org/10.1136/jnnp-2011-301068.

Iranzo, A. (2005). Management of sleep-disordered breathing in multiple system atrophy. *Sleep Medicine*, *6*(4), 297–300. http://dx.doi.org/10.1016/j.sleep.2005.01.006.

Iranzo, A. (2007). Sleep and breathing in multiple system atrophy. *Current Treatment Options in Neurology, 9*(5), 347–353.

Iranzo, A., Santamaria, J., Rye, D. B., Valldeoriola, F., Marti, M. J., Munoz, E., et al. (2005). Characteristics of idiopathic REM sleep behavior disorder and that associated with MSA and PD. *Neurology, 65*(2), 247–252. http://dx.doi.org/10.1212/01.wnl.0000168864.97813.e0.

Iranzo, A., Santamaria, J., & Tolosa, E. (2000). Continuous positive air pressure eliminates nocturnal stridor in multiple system atrophy. Barcelona multiple system atrophy study group. *Lancet, 356*(9238), 1329–1330.

Iranzo, A., Santamaria, J., Tolosa, E., Vilaseca, I., Valldeoriola, F., Marti, M. J., et al. (2004). Long-term effect of CPAP in the treatment of nocturnal stridor in multiple system atrophy. *Neurology, 63*(5), 930–932.

Iwasaki, Y., Mori, K., Ito, M., Tatsumi, S., Mimuro, M., & Yoshida, M. (2013). An autopsied case of progressive supranuclear palsy presenting with cerebellar ataxia and severe cerebellar involvement. *Neuropathology, 33*(5), 561–567. http://dx.doi.org/10.1111/neup.12012.

Jecmenica-Lukic, M., Poewe, W., Tolosa, E., & Wenning, G. K. (2012). Premotor signs and symptoms of multiple system atrophy. *Lancet Neurology, 11*(4), 361–368. http://dx.doi.org/10.1016/S1474-4422(12)70022-4.

Josephs, K. A., Whitwell, J. L., Boeve, B. F., Knopman, D. S., Petersen, R. C., Hu, W. T., et al. (2010). Anatomical differences between CBS-corticobasal degeneration and CBS-Alzheimer's disease. *Movement Disorders, 25*(9), 1246–1252. http://dx.doi.org/10.1002/mds.23062.

Kanazawa, M., Shimohata, T., Toyoshima, Y., Tada, M., Kakita, A., Morita, T., et al. (2009). Cerebellar involvement in progressive supranuclear palsy: A clinicopathological study. *Movement Disorders, 24*(9), 1312–1318. http://dx.doi.org/10.1002/mds. 22583.

Kao, A. W., Racine, C. A., Quitania, L. C., Kramer, J. H., Christine, C. W., & Miller, B. L. (2009). Cognitive and neuropsychiatric profile of the synucleinopathies: Parkinson disease, dementia with Lewy bodies, and multiple system atrophy. *Alzheimer Disease and Associated Disorders, 23*(4), 365–370. http://dx.doi.org/10.1097/WAD.0b013e3181b5065d.

Kass-Iliyya, L., Kobylecki, C., McDonald, K. R., Gerhard, A., & Silverdale, M. A. (2015). Pain in multiple system atrophy and progressive supranuclear palsy compared to Parkinson's disease. *Brain and Behavior: A Cognitive Neuroscience Perspective, 5*(5), e00320. http://dx.doi.org/10.1002/brb3.320.

Kawai, Y., Suenaga, M., Takeda, A., Ito, M., Watanabe, H., Tanaka, F., et al. (2008). Cognitive impairments in multiple system atrophy: MSA-C vs MSA-P. *Neurology, 70*(16 Pt. 2), 1390–1396. http://dx.doi.org/10.1212/01.wnl.0000310413.04462.6a.

Kimpinski, K., Iodice, V., Burton, D. D., Camilleri, M., Mullan, B. P., Lipp, A., et al. (2012). The role of autonomic testing in the differentiation of Parkinson's disease from multiple system atrophy. *Journal of the Neurological Sciences, 317*(1–2), 92–96. http://dx.doi.org/10.1016/j.jns.2012.02.023.

Kitae, S., Murata, Y., Tachiki, N., Okazaki, M., Harada, T., & Nakamura, S. (2001). Assessment of cardiovascular autonomic dysfunction in multiple system atrophy. *Clinical Autonomic Research, 11*(1), 39–44.

Kollensperger, M., Geser, F., Ndayisaba, J. P., Boesch, S., Seppi, K., Ostergaard, K., et al. (2010). Presentation, diagnosis, and management of multiple system atrophy in Europe: Final analysis of the European multiple system atrophy registry. *Movement Disorders, 25*(15), 2604–2612. http://dx.doi.org/10.1002/mds.23192.

Koros, C., & Stamelou, M. (2016). Interventions in progressive supranuclear palsy. *Parkinsonism & Related Disorders, 22*(Suppl. 1), S93–S95. http://dx.doi.org/10.1016/j.parkreldis.2015.09.033.

Kouri, N., Murray, M. E., Hassan, A., Rademakers, R., Uitti, R. J., Boeve, B. F., et al. (2011). Neuropathological features of corticobasal degeneration presenting as corticobasal syndrome or Richardson syndrome. *Brain, 134*(Pt. 11), 3264–3275. http://dx.doi.org/10.1093/brain/awr234.

Lee, C. N., Kim, M., Lee, H. M., Jang, J. W., Lee, S. M., Kwon, D. Y., et al. (2013). The interrelationship between non-motor symptoms in atypical parkinsonism. *Journal of the Neurological Sciences*, *327*(1–2), 15–21. http://dx.doi.org/10.1016/j.jns.2013.01.034.

Lee, W., Williams, D. R., & Storey, E. (2012). Cognitive testing in the diagnosis of parkinsonian disorders: A critical appraisal of the literature. *Movement Disorders*, *27*(10), 1243–1254. http://dx.doi.org/10.1002/mds.25113.

Liepelt, I., Gaenslen, A., Godau, J., Di Santo, A., Schweitzer, K. J., Gasser, T., et al. (2010). Rivastigmine for the treatment of dementia in patients with progressive supranuclear palsy: Clinical observations as a basis for power calculations and safety analysis. *Alzheimers Dement*, *6*(1), 70–74. http://dx.doi.org/10.1016/j.jalz.2009.04.1231.

Ling, H., O'Sullivan, S. S., Holton, J. L., Revesz, T., Massey, L. A., Williams, D. R., et al. (2010). Does corticobasal degeneration exist? A clinicopathological re-evaluation. *Brain*, *133*(Pt. 7), 2045–2057. http://dx.doi.org/10.1093/brain/awq123.

Litvan, I. (1994). Cognitive disturbances in progressive supranuclear palsy. *Journal of Neural Transmission. Supplementum*, *42*, 69–78.

Litvan, I., Agid, Y., Calne, D., Campbell, G., Dubois, B., Duvoisin, R. C., et al. (1996a). Clinical research criteria for the diagnosis of progressive supranuclear palsy (Steele–Richardson–Olszewski syndrome): Report of the NINDS-SPSP international workshop. *Neurology*, *47*(1), 1–9.

Litvan, I., Campbell, G., Mangone, C. A., Verny, M., McKee, A., Chaudhuri, K. R., et al. (1997). Which clinical features differentiate progressive supranuclear palsy (Steele–Richardson–Olszewski syndrome) from related disorders? A clinicopathological study. *Brain*, *120*(Pt. 1), 65–74.

Litvan, I., Mangone, C. A., McKee, A., Verny, M., Parsa, A., Jellinger, K., et al. (1996b). Natural history of progressive supranuclear palsy (Steele–Richardson–Olszewski syndrome) and clinical predictors of survival: A clinicopathological study. *Journal of Neurology, Neurosurgery, and Psychiatry*, *60*(6), 615–620.

Litvan, I., Paulsen, J. S., Mega, M. S., & Cummings, J. L. (1998). Neuropsychiatric assessment of patients with hyperkinetic and hypokinetic movement disorders. *Archives of Neurology*, *55*(10), 1313–1319.

Litvan, I., Phipps, M., Pharr, V. L., Hallett, M., Grafman, J., & Salazar, A. (2001). Randomized placebo-controlled trial of donepezil in patients with progressive supranuclear palsy. *Neurology*, *57*(3), 467–473.

Maass, S., Levin, J., & Hoglinger, G. (2016). Current treatment of multiple system atrophy. *Current Treatment Options in Neurology*, *18*(12), 51. http://dx.doi.org/10.1007/s11940-016-0435-0.

Magalhaes, M., Wenning, G. K., Daniel, S. E., & Quinn, N. P. (1995). Autonomic dysfunction in pathologically confirmed multiple system atrophy and idiopathic Parkinson's disease—A retrospective comparison. *Acta Neurologica Scandinavica*, *91*(2), 98–102.

Mochizuki, A., Ueda, Y., Komatsuzaki, Y., Tsuchiya, K., Arai, T., & Shoji, S. (2003). Progressive supranuclear palsy presenting with primary progressive aphasia—Clinicopathological report of an autopsy case. *Acta Neuropathologica*, *105*(6), 610–614. http://dx.doi.org/10.1007/s00401-003-0682-5.

Munschauer, F. E., Loh, L., Bannister, R., & Newsom-Davis, J. (1990). Abnormal respiration and sudden death during sleep in multiple system atrophy with autonomic failure. *Neurology*, *40*(4), 677–679.

Novak, P., Williams, A., Ravin, P., Zurkiya, O., Abduljalil, A., & Novak, V. (2012). Treatment of multiple system atrophy using intravenous immunoglobulin. *BMC Neurology*, *12*, 131. http://dx.doi.org/10.1186/1471-2377-12-131.

Oertel, W. H., Wachter, T., Quinn, N. P., Ulm, G., & Brandstadter, D. (2003). Reduced genital sensitivity in female patients with multiple system atrophy of parkinsonian type. *Movement Disorders*, *18*(4), 430–432. http://dx.doi.org/10.1002/mds.10384.

Orimo, S., Suzuki, M., Inaba, A., & Mizusawa, H. (2012). 123I-MIBG myocardial scintigraphy for differentiating Parkinson's disease from other neurodegenerative parkinsonism: A systematic review and meta-analysis. *Parkinsonism & Related Disorders*, *18*(5), 494–500. http://dx.doi.org/10.1016/j.parkreldis.2012.01.009.

Paviour, D. C., Winterburn, D., Simmonds, S., Burgess, G., Wilkinson, L., Fox, N. C., et al. (2005). Can the frontal assessment battery (FAB) differentiate bradykinetic rigid syndromes? Relation of the FAB to formal neuropsychological testing. *Neurocase*, *11*(4), 274–282. http://dx.doi.org/10.1080/13554790590962933.

Pekmezovic, T., Jecmenica-Lukic, M., Petrovic, I., Spica, V., Tomic, A., & Kostic, V. S. (2015). Quality of life in patients with progressive supranuclear palsy: One-year follow-up. *Journal of Neurology*, *262*(9), 2042–2048. http://dx.doi.org/10.1007/s00415-015-7815-3.

Plazzi, G., Corsini, R., Provini, F., Pierangeli, G., Martinelli, P., Montagna, P., et al. (1997). REM sleep behavior disorders in multiple system atrophy. *Neurology*, *48*(4), 1094–1097.

Plazzi, G., Cortelli, P., Montagna, P., De Monte, A., Corsini, R., Contin, M., et al. (1998). REM sleep behaviour disorder differentiates pure autonomic failure from multiple system atrophy with autonomic failure. *Journal of Neurology, Neurosurgery, and Psychiatry*, *64*(5), 683–685.

Respondek, G., Kurz, C., Arzberger, T., Compta, Y., Englund, E., Ferguson, L. W., et al. (2017). Which ante mortem clinical features predict progressive supranuclear palsy pathology? *Movement Disorders*. http://dx.doi.org/10.1002/mds.27034. [Epub ahead of print].

Respondek, G., Stamelou, M., Kurz, C., Ferguson, L. W., Rajput, A., Chiu, W. Z., et al. (2014). The phenotypic spectrum of progressive supranuclear palsy: A retrospective multicenter study of 100 definite cases. *Movement Disorders*, *29*(14), 1758–1766. http://dx.doi.org/10.1002/mds.26054.

Rittman, T., Ghosh, B. C., McColgan, P., Breen, D. P., Evans, J., Williams-Gray, C. H., et al. (2013a). The Addenbrooke's cognitive examination for the differential diagnosis and longitudinal assessment of patients with parkinsonian disorders. *Journal of Neurology, Neurosurgery, and Psychiatry*, *84*, 544–551. http://dx.doi.org/10.1136/jnnp-2012-303618.

Schmidt, C., Herting, B., Prieur, S., Junghanns, S., Schweitzer, K., Globas, C., et al. (2008). Autonomic dysfunction in different subtypes of multiple system atrophy. *Movement Disorders*, *23*(12), 1766–1772. http://dx.doi.org/10.1002/mds.22187.

Siri, C., Duerr, S., Canesi, M., Delazer, M., Esselink, R., Bloem, B. R., et al. (2013). A cross-sectional multicenter study of cognitive and behavioural features in multiple system atrophy patients of the parkinsonian and cerebellar type. *Journal of Neural Transmission (Vienna, Austria: 1996)*, *120*(4), 613–618. http://dx.doi.org/10.1007/s00702-013-0997-x.

Stamelou, M., Alonso-Canovas, A., & Bhatia, K. P. (2012a). Dystonia in corticobasal degeneration: A review of the literature on 404 pathologically proven cases. *Movement Disorders*, *27*(6), 696–702. http://dx.doi.org/10.1002/mds.24992.

Stamelou, M., & Bhatia, K. P. (2015). Atypical parkinsonism: Diagnosis and treatment. *Neurologic Clinics*, *33*(1), 39–56. http://dx.doi.org/10.1016/j.ncl.2014.09.012.

Stamelou, M., & Bhatia, K. P. (2016). Atypical parkinsonism—New advances. *Current Opinion in Neurology*, *29*(4), 480–485. http://dx.doi.org/10.1097/WCO.0000000000000355.

Stamelou, M., Diehl-Schmid, J., Hapfelmeier, A., Kontaxopoulou, D., Stefanis, L., Oertel, W. H., et al. (2015). The frontal assessment battery is not useful to discriminate progressive supranuclear palsy from frontotemporal dementias. *Parkinsonism & Related Disorders*, *21*(10), 1264–1268. http://dx.doi.org/10.1016/j.parkreldis.2015.08.006.

Stamelou, M., Dohmann, H., Brebermann, J., Boura, E., Oertel, W. H., Hoglinger, G., et al. (2012b). Clinical pain and experimental pain sensitivity in progressive supranuclear palsy. *Parkinsonism & Related Disorders*, *18*(5), 606–608. http://dx.doi.org/10.1016/j.parkreldis.2011.11.010.

Stamelou, M., Quinn, N. P., & Bhatia, K. P. (2013). "Atypical" atypical parkinsonism: New genetic conditions presenting with features of progressive supranuclear palsy, corticobasal degeneration, or multiple system atrophy—A diagnostic guide. *Movement Disorders*, *28*(9), 1184–1199. http://dx.doi.org/10.1002/mds.25509.

Steele, J. C., Richardson, J. C., & Olszewski, J. (1964). Progressive supranuclear palsy. A heterogeneous degeneration involving the brain stem, basal ganglia and cerebellum with vertical gaze and Pseudobulbar palsy, nuchal dystonia and dementia. *Archives of Neurology*, *10*, 333–359.

Tison, F., Wenning, G. K., Quinn, N. P., & Smith, S. J. (1995). REM sleep behaviour disorder as the presenting symptom of multiple system atrophy. *Journal of Neurology, Neurosurgery, and Psychiatry*, *58*(3), 379–380.

Trojanowski, J. Q., Revesz, T., & Neuropathology Working Group on MSA. (2007). Proposed neuropathological criteria for the post mortem diagnosis of multiple system atrophy. *Neuropathology and Applied Neurobiology*, *33*(6), 615–620. http://dx.doi.org/10.1111/j.1365-2990.2007.00907.x.

Tsuboi, Y., Josephs, K. A., Boeve, B. F., Litvan, I., Caselli, R. J., Caviness, J. N., et al. (2005). Increased tau burden in the cortices of progressive supranuclear palsy presenting with corticobasal syndrome. *Movement Disorders*, *20*(8), 982–988. http://dx.doi.org/10.1002/mds.20478.

Verny, M., Jellinger, K. A., Hauw, J. J., Bancher, C., Litvan, I., & Agid, Y. (1996). Progressive supranuclear palsy: A clinicopathological study of 21 cases. *Acta Neuropathologica*, *91*(4), 427–431.

Vetrugno, R., D'Angelo, R., Cortelli, P., Plazzi, G., Vignatelli, L., & Montagna, P. (2007). Impaired cortical and autonomic arousal during sleep in multiple system atrophy. *Clinical Neurophysiology*, *118*(11), 2512–2518. http://dx.doi.org/10.1016/j.clinph.2007.08.014.

Wenning, G. K., Ben Shlomo, Y., Magalhaes, M., Daniel, S. E., & Quinn, N. P. (1994). Clinical features and natural history of multiple system atrophy. An analysis of 100 cases. *Brain*, *117*(Pt. 4), 835–845.

Wenning, G. K., Geser, F., Krismer, F., Seppi, K., Duerr, S., Boesch, S., et al. (2013). The natural history of multiple system atrophy: A prospective European cohort study. *Lancet Neurology*, *12*(3), 264–274. http://dx.doi.org/10.1016/S1474-4422(12)70327-7.

Wenning, G. K., Litvan, I., Jankovic, J., Granata, R., Mangone, C. A., McKee, A., et al. (1998). Natural history and survival of 14 patients with corticobasal degeneration confirmed at postmortem examination. *Journal of Neurology, Neurosurgery, and Psychiatry*, *64*(2), 184–189.

Wenning, G. K., Tison, F., Ben Shlomo, Y., Daniel, S. E., & Quinn, N. P. (1997). Multiple system atrophy: A review of 203 pathologically proven cases. *Movement Disorders*, *12*(2), 133–147. http://dx.doi.org/10.1002/mds.870120203.

Williams, D. R., de Silva, R., Paviour, D. C., Pittman, A., Watt, H. C., Kilford, L., et al. (2005). Characteristics of two distinct clinical phenotypes in pathologically proven progressive supranuclear palsy: Richardson's syndrome and PSP-parkinsonism. *Brain*, *128*(Pt. 6), 1247–1258. http://dx.doi.org/10.1093/brain/awh488.

Williams, D. R., & Lees, A. J. (2009). Progressive supranuclear palsy: Clinicopathological concepts and diagnostic challenges. *Lancet Neurology*, *8*(3), 270–279. http://dx.doi.org/10.1016/S1474-4422(09)70042-0.

Winter, Y., Spottke, A. E., Stamelou, M., Cabanel, N., Eggert, K., Hoglinger, G. U., et al. (2011). Health-related quality of life in multiple system atrophy and progressive supranuclear palsy. *Neurodegenerative Diseases*, *8*(6), 438–446. http://dx.doi.org/10.1159/000325829.

CHAPTER FORTY-SEVEN

Nonmotor Symptoms in Vascular and Other Secondary Parkinsonism

Oleg S. Levin[1], Achcha Sh. Chimagomedova, Natalia A. Skripkina, Elena A. Lyashenko, Olga V. Babkina
Russian Medical Academy of Professional Continuous Education, Centre of Extrapyramidal Disorders, Moscow, Russia
[1]Corresponding author: e-mail address: neurolev@gmail.com

Contents

International Review of Neurobiology, Volume 134
ISSN 0074-7742
http://dx.doi.org/10.1016/bs.irn.2017.05.016

Abstract

Vascular parkinsonism (VP) is a relatively frequent variant of secondary parkinsonism caused by ischemic or hemorrhagic lesions of basal ganglia, midbrain, or their links with frontal cortex. According to different investigations, various forms of cerebrovascular disease cause 1%–15% of parkinsonism cases. Nonmotor symptoms are frequently found in VP and may negatively influence on quality of life. However, nonmotor symptoms such as hallucinations, orthostatic hypotension, REM-sleep behavior disorder, and anosmia are rarely revealed in VP, which may be noted to another diagnosis or mixed pathology. Clinical value of nonmotor symptoms in normal pressure hydrocephalus, toxic, and drug-induced parkinsonism is also discussed.

1. VASCULAR PARKINSONISM

Vascular parkinsonism (VP) is a rare variant of parkinsonism caused by ischemic or hemorrhagic lesions of basal ganglia (especially putamen or pallidum), midbrain, or their links with frontal lobes. According to different investigations, various forms of cerebrovascular disease (CVD) may cause 1%–15% of parkinsonism cases.

French neurologist Brissaud (1894) was probably the first person who proposed that vascular damage of substantia nigra may be a substrate of VP. Vogt and Vogt (1920) first linked parkinsonism with lesions of globus pallidus and putamen. In 1929, Critchley published in the "Brain" classical article "Arteriosclerotic parkinsonism" containing thorough clinical description of VP (Critchley, 1929). Critchley divided "arteriosclerotic parkinsonism" into five clinical subtypes:

I. type: rigidity, hypomimia, marche a petit pas,
II. type: pseudobulbar syndrome and variable pyramidal signs,
III. type: dementia and incontinence,
IV. type: pallidopyramidal syndrome,
V. type: pallidocerebellar syndrome.

Later, Critchley modified his conception and proposed the term "vascular pseudoparkinsonism" to highlight principal differences between neurodegenerative Parkinson's disease (PD) and Parkinson-like syndrome in CVD. Other authors proposed to use the term "pseudovascular parkinsonism" in cases of imitation of vascular process (for example, in normal pressure hydrocephalus (NPH)) or "pseudovascular pseudoparkinsonism" in patients with other motor syndrome (for example, primary gait disorder, obsessional slowness, abulia, catatonia, paratonia, or paresis) that mimic some parkinsonian signs but also are caused by nonvascular lesions (Vizcarra, Lang, Sethi, & Espay, 2015).

In recent years, the problem of VP has been studied in several investigations. The results of these works may be presented in next statements. CVD rarely manifests as the syndrome more or less closely imitating classical parkinsonism of idiopathic PD or atypical parkinsonism (parkinsonism-plus) associated with MSA (multiple system atrophy), PSP, and CBD (true VP).

However, in most cases, CVD manifests as motor syndrome principally distinct from idiopathic PD or atypical parkinsonism. This motor syndrome is frequently named as "frontal gait disorder," "frontal dysbasia," "gait apraxia," and sometimes "lower body parkinsonism" (a reference to "Lewy-body parkinsonism"—that is PD). But it is not a variant of parkinsonian syndrome as not being in line with the main formula of parkinsonism: hypokinesia plus resting tremor and/or rigidity. Such patients have no obvious parkinsonian signs in the state of lying and sitting (so the more correct term may be "walking parkinsonism," not "lower body parkinsonism"). Vascular lesions nonselectively involve different functional systems, as a result a complex constellation of different motor syndromes may be in one patient.

Appearance of CT and MRI made the diagnosis of CVD simpler, but formed "diagnostic trap" with the desire to explain any neurological deficit by vascular lesions found. So diagnosis of VP must be only the "diagnosis of exclusion."

Jellinger (2005) found that different variants of vascular damage of the brain (multiinfarct state, leukoencephalopathy, multiple lacunar infarcts in basal ganglia and brainstem) appear to be the cause of parkinsonism in 6% of cases, but epidemiologic data showed higher prevalence of VP—8%–12%. Combination of different forms of CVD and Lewy-body pathology was found in about four times frequently than pure VP.

We found parkinsonian syndrome in 13 from 82 (15%) patients with subcortical vascular encephalopathy. VP is a syndrome that may be caused by different CVDs. Cerebral small vessel disease is the most frequent cause of the VP. Diffuse damage of penetrative long thin noncollateral arteries vascularized deep compartments of the brain (vascular "centrencephalon") associated with long-term arterial hypertension is found in most cases of VP (Table 1). Other CVDs also can lead to VP by causing ischemic or hemorrhagic lesions of area strategically located for parkinsonism.

VP may be divided into five main subtypes according to the main location of motor frontostriatal circuit disruption: (1) nigral (nigrostriatal), (2) striatopallidal, (3) frontostriatal, (4) thalamofrontal, and (5) mixed. Different variants are distinguished by morphologic substrate, pathophysiologic

Table 1 Etiology of Vascular Parkinsonism

Disease	**Ischemic Lesions**	**Hemorrhagic Lesions**
Small cerebral vessel disease (microangiopathy)		
Hypertensive microangiopathy	+	+
Amyloid angiopathy	+	+
Inflammatory angiopathy (vasculitis)	+	+
Heredative angiopathy (CADASIL) and others	+	–
Etat crible	±	–
Large cerebral artery disease		
Atherosclerosis of cerebral arteries	+	–
Meningovascular syphilis	+	–
Cardiogenic damage of brain		
Cardiogenic embolism	+	–
Ischemic–hypoxic encephalopathy	+	–
Others		
Arteriovenous malformation	+	+
Antiphospholipid syndrome	+	–
Coagulopathies	+	+

mechanism, proportionality of the main parkinsonian symptoms, relation of the axial and limbs signs, reaction to the dopaminergic drugs (Levin, 2003).

Proposed criteria of VP diagnosis include (1) parkinsonian syndrome defined as hypokinesia plus rigidity and/or resting tremor; (2) CVD objectively supported by supplemental diagnostic methods such as MRI, MR-angiography, CT-perfusion, CT-angiography, and ultrasound; and (3) causal links between 1 and 2 confirmed by clinical peculiarities or clinical imaging correlations. Clinical peculiarities of VP include absence of resting tremor, predominance of parkinsonian symptoms in lower extremities and axial region, earlier emergence of postural instability and gait disturbances, bilateral onset of the disease, and low efficacy of dopaminergic drugs.

From another point of view, it is noncorrect to associate true VP and "Low body parkinsonism" (Table 2). There are no any pathognomonic signs, which confirmed diagnosis of VP. Striking dissociation between

Table 2 Differential Diagnosis of Parkinsonism in PD and VP and Frontal Gait Disorder

	PD (Lewy Body Parkinsonism)	VP	Frontal Gait Disorder (Lower Body Parkinsonism)
Onset of symptoms	Insidious	Acute, subacute, or insidious	As a rule, insidious
Laterality	Hemi-, then bilateral	Bilateral or (rare) unilateral	Always bilateral
Tremor at rest	Frequently	May be postural, rarely—resting tremor	As a rule, absent
Hypokinesia	Obligate symptom	Obligate symptom	Absent
Hypomimia	Almost always, on early stage may be absent	Almost always	As a rule, absent
Arm movements during gait	Acheirokinesis	Acheirokinesis	Enhanced
Postural instability	May be found	May be found	May be found
Base square	Narrowed	Narrowed	Wided
Step length	Short	Short	Short
Course	Progressive	Progressive or static	Progressive or static
CT/MRI	Normal, may be single lesion of putamen and globus pallidus, thin leukoaraiosis	Spread subcortical leukoaraiosis, focal lesions of putamen and globus pallidus, rarely—substantia nigra and frontal lobes	Spread subcortical leukoaraiosis, multifocal bilateral lacunes in deep frontal lobes

involving of axial region and leg and intact of arm movements has no adequate explanation. It may be proposed that basal ganglia lesions cause dysfunction of subcortical-brainstem projection which via medial brainstem-spinal system control axial muscles provided balance and body-extremity synergism. Positive reaction to L-dopa was found in 10%–48% of patients with VP, but in this case, effect has short duration and moderate or minimal amplitude. Absence of L-dopa effect is explained by damage of striatum or

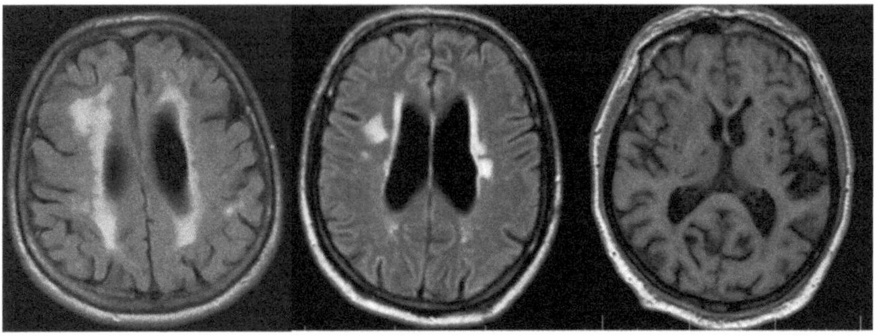

Fig. 1 MRI of patient with VP. FLAIR images: hyperintensive periventricular leukoaraiosis. T1-weighted image: multiple lacunar infarcts in basal ganglia and thalamus.

"exit" ganglia, inner pallidum or reticular substantia nigra, or frontostriatal/thalamofrontal segments of frontostriatal circuit (Fig. 1).

Frequently, VP has a slowly progressive course, but in 20%–25% of cases, symptoms appear acutely or subacutely (usually 1–6 months after stroke). Diagnosis of VP is less possible in next cases: absence of both vascular lesions and lesions in strategic areas on MRI, high persistent L-dopa effect throughout 5 years, downgaze paralysis in the absence of upgaze paralysis, progressive autonomic insufficiency, earlier emergence of visual hallucinations, atrophy of midbrain or putamen, or other signs of atypical parkinsonism.

1.1 Cognitive Impairment and Dementia in Patients With VP

Cognitive impairment is one of the most frequent, almost universal coexisting signs in patients with VP. Colosimo et al. (2010) found attention and memory impairment in 61 of 83 clinically diagnosed VP patients (73.5%). Vale, Caramelli, and Cardoso (2015) reported that all 15 VP patients had cognitive impairment and 80% fulfilled diagnostic criteria for probable vascular dementia. In another study, probable vascular dementia was revealed in 70.5% of VP patients (Vale, Caramelli, & Cardoso, 2013). In the Bambuí study, 8 of 13 (61.5%) patients with VP presented with the concomitant diagnosis of vascular dementia (Barbosa, Caramelli, Maia, et al., 2006). Glass et al. (2012) found concomitant dementia in 39% of the 28 patients with pathologically confirmed VP.

According to some study reports, cognitive impairment is significantly more common in VP than in PD (Benítez-Rivero et al., 2014; Levin, 2003; Santangelo et al., 2010; Zijlmans, Daniel, Hughes, Révész, & Lees, 2004). Cognitive decline was presented or developed early in the course of the disease in about 50% patients with VP (Levin, 1997). Bradvica, Jančuljak, Butković-Soldo, and Vladetić (2011) showed a higher percentage of dementia in patients with VP compared with PD patients.

Dementia in VP is usually subcorticofrontal, manifesting as dysexecutive syndrome with impairment of attention, planning, judgment, goal-directed behavior, abstract thinking, verbal fluency, and early apathy. Santangelo et al. (2010) found that although PD patients had an earlier onset and longer disease duration than VP patients, patients with vascular abnormalities had more severe executive dysfunction. On the other side, Benítez-Rivero et al. (2014) found that VP patients had a global pattern of cognitive impairment, including executive function, verbal memory, and language.

There is no strong correlation between the severity of VP and dementia (Marxreiter & Winkler, 2016). Bombois et al. (2007) found significant correlation between subcortical and periventricular white matter hyperintensities (WMHs) and executive dysfunctions in 170 patients with mild cognitive impairment. Moreover, if cognitive dysfunction correlates with severity of leukoaraiosis, there is no such link according to parkinsonism that shows a lead role of lesion location in parkinsonism emergence.

From another point of view, parkinsonism may be additional criteria of causative link between dementia/MCI and CVD. Thus, Boyle et al. (2005) suggested that parkinsonian symptoms may be more prominent among individuals with nonamnestic MCI as opposed to amnestic MCI. Another study demonstrated that the severity of parkinsonian symptoms correlates with the severity of cognitive impairment, especially in nonmemory domains (Rozzini, Chilovi, & Bertoletti, 2007). Thus, involvement of executive functions is consistent with the dysfunction of frontosubcortical systems and the emergence of parkinsonian signs (Wilson, Schneider, Beckett, Evans, & Bennett, 2002). Leukoencephalopathy and lacunar infarcts in the deep white matter of the frontal lobe, caudate, or putamen may cause dysfunction of frontostriatal circuits and thus induce both executive impairment and parkinsonian symptoms. Moreover, patients with cerebrovascular lesions may have more severe cognitive deficit than patients with PD alone. Dunet et al. (2016) found that cognitive impairment in VP is associated with decreased functional relations between caudate nucleus and cingulate cortex.

1.2 Affective Symptoms

The PRIAMO study found a high prevalence of psychiatric disturbances in patients with VP (77.1%) including apathy (42.2%) (Colosimo et al., 2010).

However, there are a few data about affective symptoms in patients with VP. A lot is written about "vascular depression hypothesis," suggesting that CVD (especially silent cerebral small vessel) affecting the frontolimbic system induces late-onset depression (Alexopoulos, Meyers, Young, Campbell, et al., 1997; Alexopoulos, Meyers, Young, Kakuma, et al., 1997; Baldwin & O'Brien, 2002; Taylor, Aizenstein, & Alexopoulos, 2013; White, McClure, Wallace, et al., 2011; Wu, Li, Tan, Liu, & Huang, 2014). There is not enough convincing evidence supporting this hypothesis. However, LADIS showed that severity of white matter changes is predictive of depressive symptoms, thus supporting hypothesis and strongly implicating white matter lesions in the pathogenesis of late-life depression (The LADIS Study Group, 2011).

According to Glass et al. (2012), mood disorders were presented in 60.9% patients with pathologically confirmed VP. Some authors emphasize the specific links between apathy and severity of movement disorder. In the PRIAMO study, apathy was revealed in 42.2% of VP patients. Diagnostic significance of depression and other affective disorders in VP is low, but nevertheless, it is important as a target for therapeutic management.

1.3 Psychotic Symptoms

The patients with VP have an extremely low frequency of visual hallucinations compared with PD or DLB. So the presence of visual hallucinations makes diagnosis of VP very unlikely (Williams et al., 2008). On the other side, Hasegawa et al. (2013) found that delirium was more frequent in patients with degenerative dementia and coexisting CVD than in those without CVD. But there is no pathological confirmation of dementia etiology in this study. According to the PRIAMO study, psychiatric symptoms, attention, and memory impairment were more frequently observed in DLB patients than in VP.

1.4 Autonomic Insufficiency

Orthostatic hypotension (OH) is one of the commonly occurring nonmotor symptoms (NMS) in elderly patients with parkinsonism. According to Fereshtehnejad and Lökk (2014), symptomatic OH is present in up to 30% patients with PD, 80% patients with MSA, 31% of DLB, and up to 26% patients with VP. Advancing age, male sex, disease severity, and duration and subtype of motor symptoms are predisposing factors. OH increases the risk of postural instability and falls, which affects the quality of life in patients with different parkinsonian types. Postural symptoms due to OH were found in 18% of patients with VP. In several studies, it was shown that OH may be an intermediate link among WMHs and late-life depression (Colloby et al., 2011; Richardson et al., 2009).

Gastrointestinal (GI) dysfunction was found in 65%–80% patients with VP in different studies. According to Park et al. (2015), frequency of GI dysfunction in 62 patients with VP was similar to PD and multisystem degenerations. Constipation was the most common symptom, followed by appetite loss, weight loss, dysphagia, and sialorrhea. Urinary symptoms were found in 65%–90% of patients with VP. In the PRIAMO study, GI symptoms, pain, urinary problems, and postural instability due to OH were most frequent in MSA than in VP, whereas the prevalence of skin and respiratory disorders was rather low in all forms of parkinsonism, ranging between 10% and 30% (for example, respiratory symptoms were found in 25% of patients with VP).

1.5 Sleep Disorders

According to the PRIAMO study, sleep disturbances were found in 70% of patients with VP and other parkinsonism subtypes (except CBD—36%) (Colosimo et al., 2010). Similar to PD, patients with VP identified nocturia and the problem of getting up at night to urinate as being the most frequent and troublesome nocturnal symptom, but nocturnal akinesia is not typical symptom in VP. In total, PD patients in comparison with VP patients reported significantly more troublesome problems with sleep quality (Bhidayasiri et al., 2014). VP patients have more higher score sleep domain compared to PD patients (18.7 ± 8.5 vs 15.8 ± 8.3) (Pan et al., 2012).

It is well known that rapid eye movement sleep behavior disorder (RBD) is strongly associated with synucleinopathies (Iranzo et al., 2014). However, Jiang et al. (2013) found RBD in 8.9% of patients with vascular cognitive impairment and in 4.2% of patients with stroke. Patients with CVD also showed a higher prevalence of restless legs syndrome (RLS)—16.1%, and obstructive sleep apnea syndrome (35.7%). Polysomnography studies revealed reduced proportion of slow-wave sleep and rapid eye movement sleep, prolonged sleep latency, and decreased sleep efficiency index. Subjective sleep disturbances included difficulties in falling asleep, waking early, sleep fragmentation, and RLS. Two basic mechanisms might be hypothesized for the generation of sleep disturbances in these patients (Ahlberg et al., 1988): vascular damage influences directly neural structures involved in sleep regulation, affecting the normal sleep–wake system and then determining sleep structure disruption (Ahn et al., 2015; Schwartz & Roth, 2008) preceding stroke determines a central imbalance of neurotransmitters, such as serotonin, melatonin, and acetylcholine, causing sleep structure abnormalities (Bassetti & Hermann, 2011).

1.6 Sensory Symptoms

It is established that most of patients with PD have impaired olfactory function. This clinical feature corresponds to findings of Lewy bodies in the anterior olfactory nucleus, showing involvement of the olfactory system in the neurodegenerative process. Studies showed that hyposmia occurs in patients with VP, but its prevalence is much less than that in PD. In the PRIAMO study, olfactory dysfunction was found in 39% of patients with VP.

Navarro-Otano et al. found that patients with suspected VP had lower UPSIT score than in controls but higher than in PD patients. Patients with low cardiac (Tullberg et al., 2004) I-MIBG score suggestive of PD were more likely to have a higher UPSIT score. In addition, Katzenschlager et al. (2004) also found that UPSIT scores in VP were significantly better than in PD and did not differ from the healthy controls. Only 4 of the 14 (28.6%) patients with VP and 7 of the 27 (25.9%) controls had scores below published optimal cutoff values for the respective age and sex groups. Katzenschlager et al. (2004) suggested a cutoff value ≤ 23 showing the best balance between sensitivity and specificity in the 65–75 group, with 100% sensitivity and a specificity of 85.7%. In the 76–88 group, an UPSIT score of ≤ 22 yielded a sensitivity of 85.7% and a specificity of 80%.

According to the PRIAMO study, painful sensory symptoms occur in 61% patients with VP—about with the same frequency as in patients with PD (Colosimo et al., 2010).

2. NORMAL PRESSURE HYDROCEPHALUS

NPH is a condition characterized by ventriculomegaly and Hakim-Adams triad of gait apraxia, dementia, and incontinence. There is no nomination of parkinsonism at the original description of this pathology. However, parkinsonian symptoms were frequently described in NPH patients sometimes with good reaction to L-dopa. It is explained by compression of nigrostriatal axons in the periventricular zone (Fig. 2).

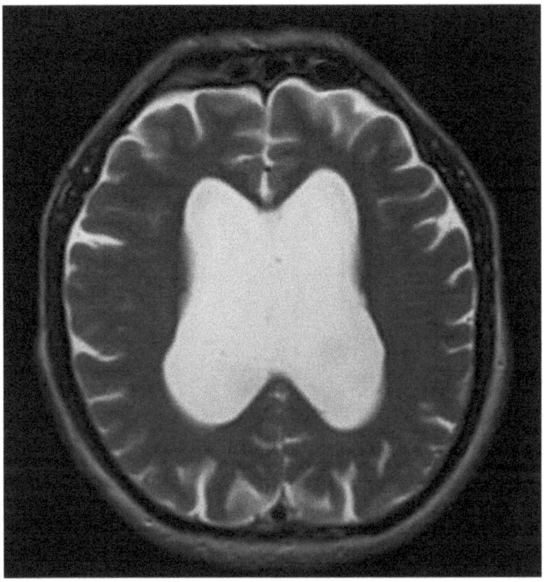

Fig. 2 MRI of patient with NPH. Dilation of lateral ventricles (T2-weighted image).

2.1 Cognitive Impairment

Cognitive impairment is frequent but not obligate feature of idiopathic NPH. Cognitive dysfunction has frontal lobe-like profile and includes psychomotor slowing, impairment of attention, and executive function. But memory especially recognition relatively preserved (Kazui, 2008). This profile is consistent with disruption of the frontal–subcortical circuits. However, a part of patients has cognitive impairment not only subcortical but also cortical types, for example, visuospatial or semantic deficits (Saito et al., 2011).

Dissociation between gait and cognitive dysfunction suggests the existence of different pathophysiological mechanisms. On the other hand, Picascia et al. (2016) founded 4 cognitive profiles in 64 NPH patients: with global cognitive impairment (42%), with frontosubcortical dysfunction (24%), with isolated deficit of a single cognitive domain (17%), and without cognitive impairment (17%).

Neuroimaging studies found that frontal cortical gray matter volume is preserved, but frontal cerebral blood flow (rCBF) is decreased. These findings support the view that frontal lobe-like cognitive dysfunction in NPH is associated with subcortical white matter damage but not frontal lobe damage itself. However, a recent neuroimaging study demonstrated the parietal rCBF reduction associated with visuospatial dysfunctions (Saito et al., 2011).

Some authors think that dementia in NPH is multifactorial. CVD is comorbid in over 60% of patients with NPH and the severity of cognitive deficits appears to correlate with vascular risk factors (Leinonen et al., 2012). AD pathology was found in cortical biopsies taken from subjects with NPH during shunt surgery in 18%–75% of cases. Relationship between NPH and AD pathology has not yet been fully elucidated. One hypothesis is that altered CSF dynamics play a role in impairment of beta-amyloid clearance in the brain.

The only available treatment for NPH is the placement of a CSF shunt; however, identifying patients most likely to have benefit remained a challenge (Cabral et al, 2011). After shunt placement improvement of cognitive functions observes less frequent than gait disorder. The memory, visuospatial, and psychomotor functions appear to respond shunt surgery more likely. Superimposed AD pathology might limit the potential of cognitive improvement (Kazui, 2008). According to Kazui et al. (2013), mild cognitive impairment before shunt surgery, successful visuoconstruction subtest of MMSE performance, and an absence of hypertension were predictors of cognitive impairment disappearance. None of the neuroimaging features predicted the disappearance of symptoms after shunt surgery in this study.

Arterial hypertension is often recognized in patients with NPH (Krauss, Regel, Vach, Droste, Borremans, & Mergner, 1996). It has been suggested that abnormalities in the subcortical white matter may play a part in the development of hydrocephalus. A low cerebral blood flow in the periventricular white matter has been a persistent finding in NPH and several studies have found more extensive white matter abnormalities pointing toward a state of ischemia in the deep white matter (Kristensen et al., 1996).

It has been theorized that the presence of neurological comorbidities in patients with clinical NPH may contribute to shunt responsiveness or lack. If an AD-NPH syndrome exists and is at least partly due to altered CSF dynamics, the restoration of proper CSF circulation would seem a reasonable therapeutic approach. However, studies have shown that the majority of persons with NPH and concomitant AD do not exhibit cognitive improvement after shunt placement. For these reasons, patient selection for shunt surgery has remained a challenge. Surgical treatment is generally discouraged for patients with severe dementia, even in the setting of gait dysfunction and incontinence, regardless of radiographic findings. Klassen and

Ahlskog (2011) reported that after shunting early gait improvement was common (75%), but only one-third of patients had long-lasting improvement by 3 years; baseline postural instability, cognition impairment, and urinary incontinence predicted poor outcome.

2.2 Neuropsychiatric Disturbances

Beyond cognitive impairment NPH patients may present personality changes, anxiety, depression, mania, and psychotic syndromes. Apathy was the most frequent symptom and was present in 70% of the NPH subjects (DeVito et al., 2007; Geda et al., 2008; Kito et al., 2009; Relkin et al., 2005). In NPH, patients with apathy hypoperfusion in the anterior cingulate cortex and thalamus were found (Klinge et al., 2008; Murakami et al., 2007), and the involvement of the orbitofrontal area could induce anxiety, aggression, and stereotyped behavior (Antonucci et al., 2006; Blair, 2001; Miyamoto et al., 2007; Tullberg et al., 2004). Some studies also reported high prevalence of depression in patients with NPH (Behrens et al., 2014; Israelsson et al., 2016; Kanemoto et al., 2016; Markianos et al., 2009; Oliveira et al., 2015).

2.3 Urinary Dysfunction

Urgency and incontinence are cardinal features of NPH. However, urgency with or without incontinence is not specific for NPH. The possibility of urinary problems caused by other neurological or nonneurological disorders (urinary tract infection and a large postvoid residual urine) should be taken into consideration. Patients can be asked to fill out a bladder diary for several days to provide information about how often they urinate, the volume of urine produced, and any episodes of incontinence. Voiding more than 8–12 times in a 24-h period is usually considered compatible with a diagnosis of overactive bladder.

There are five major types of urinary incontinence. Urge incontinence is caused by a sudden, involuntary bladder contraction and is also called "detrusor overactivity." Stress incontinence is when the bladder and urinary sphincter cannot handle the increased compression during exercise, coughing, or sneezing, resulting in loss of urine. This kind of incontinence happens mostly in women with pelvic-organ prolapse (Ghoniem et al., 2008) but also in men after prostate surgery. Mixed incontinence is a combination of both stress and urge incontinence and is most common in older women. Overflow incontinence, in which the bladder becomes distended and cannot be fully emptied, is rarer and is the result of bladder obstruction or injury. In men, it can be the result of an enlarged prostate (Panicker & Fowler, 2010). Functional incontinence is a condition related to dementia and/or gait disturbance interfering with independent toileting skills.

The frequency of urinary problems in NPH is reported to be between 55% and 79% (Governale et al., 2008; Hashimoto et al., 2010; Marmarou et al., 2005; McGirt et al., 2005). Storage symptoms (93%) are more common than voiding symptoms (71%) (Sakakibara et al., 2008). The early urinary symptom was frequently nocturia (Krzastek et al., 2016). In the urodynamic study, it was reported that bladder capacity was small and detrusor overactivity is seen in 95% of NPH patients; thus, detrusor overactivity is postulated to be the basis for most incontinence in NPH. Improvement of bladder hyperactivity after lumbar tap test has been reported (Ahlberg et al., 1988). Recent studies with single-photon emission tomography revealed that a hypoperfusion was present in the right frontal area in NPH patients with moderate or severe incontinence, compared to those with none or mild incontinence (Sakakibara et al., 2008, 2012).

2.4 Sleep Disorders

There are some studies indicated that NPH is associated with sleep disorders, such as obstructive and central apnea. Moreover, it was hypothesized that sleep breathing disorders can be related to the pathophysiology of NPH. Although the mean intracranial pressure is within normal limits, its qualitative analysis is characterized by Lundberg B-waves with an amplitude of up to 50 mmHg, especially at night, which is probably related to different sleep stages, apnea episodes, and rhythmic alterations of cerebral blood flow (Oliveira et al., 2015; Spiegelberg et al., 2016). Thus, this high-resistance and low-complacent environment promotes hypoperfusional deficits and interstitial edema, both of which are associated with apnea and other sleep disorders (Jennum & Børgesen, 1989). Although probably not more common than in the general population of elderly people, the occurrence of sleep-disordered breathing may be a risk factor for further deterioration in a vulnerable setting superimposed on a preexistent CSF hydrodynamic disturbance and small vessel disease (Kristensen et al., 1998). Reduced cerebral perfusion accompanied by apnea-related hypoxemia may further predispose to nocturnal cerebral ischemia and contribute to progression of hydrocephalus (Jennum & Børgesen, 1989). Attier-Zmudka et al. (2016) described a case of 87-year-old man with NPH and sleep apnea syndrome with a central component. In this case, sleep quality and simultaneously symptoms of NPH (gait disorders, progressive loss of autonomy, and memory) improved after continuous positive airway pressure ventilation during 8 months. A marked reduction of apnea was noted in several patients after shunt operations (Kuchiwaki et al., 1985; Oliveira et al., 2015).

2.5 Sensory Symptoms

A few studies examined olfactory function in patients with NPH. Patients with NPH scored significantly higher on the UPSIT than those with AD (26.9 ± 7.1 and 14.1 ± 4.7, respectively), but lower than the matched controls (32.7 ± 6.7). However, test scores of NPH patients were nonetheless within the general normal range. The UPSIT scores did not differ significantly among the three longitudinal time points in the NPH patients. Thus, odor identification test scores of NPH patients are not influenced by the release of intracranial pressure via ELD or 6 months following implantation of a VP shunt.

Podlesek et al. (2012) found that patients with NPH have significantly smaller olfactory bulbs on MRI than controls ($P < 0.04$). Out of 17 NPH patients, only 2 were normosmic (12%), while 9 were hyposmic and 3 had functional anosmia (scores were not present from 3 patients).

3. TOXIN-INDUCED PARKINSONISM

Parkinsonism may be caused by a variety of toxins including manganese, carbon monoxide, cyanide, and many others. Historic clues can suggest toxin exposure, leading to diagnosis. Neuroimaging frequently confirms damage of basal ganglia or deep white matter which may be oversensible to toxic factors. NMSs of different types frequently found in patients with various types of toxic parkinsonism.

3.1 Manganese

The main causes of manganese intoxication (manganism) are inhalation of dust particles or intravenously injection of manganese-contained parenteral nutrition. Over the last years, a few hundred patients with so-called ephedrone encephalopathy, caused with designed "street" drug, have described in the East Europe. Ephedrone (methcathinone) is producing from ephedrine or phenylpropanolamine-containing medicines by means of the special processing including addition of potassium permanganate and acetic acid. The substance turning out as a result possesses an amphetamine-similar action and causes strong mental dependence. In the main part of narcomaniacs parenteral injection of a drug leads to development of an organic damage of a brain with development of serious neurological and psychiatric disturbances.

Initial symptoms of a lesion of CNS were speech illegibility (38%), postural instability and walking difficulties (29%), persistent fatigue (29%), slowness of movements (14%), and affective changes (irritability, emotional lability—19%).

At the advanced stage the majority of patients had the combination of movement disorders (parkinsonism, dystonia, postural kinetic tremor, myoclonus at action) with marked posture instability, pseudobulbar syndrome, subcorticofrontal type of cognitive and affective-behavioral disturbances, and disturbances of sleep.

The dystonia was revealed in 90% of patients involved various regions of a trunk and limbs. The dystonic postures of the low limbs were asymmetric and have been present mainly at walking—a back flexion and twisting feet, turning of toes, the special movement of foot, reminding the "cock gait."

Dysarthria is characterized by a mixed spastiko-hypokinetic features. Due to a spastic component, speech was slowed, rhonchial, intermittent, demanded efforts, the patients are often compelled to stop and clear the throat. The hypokinetic component was present by the monotonous muffled speech and difficulties in the speech initiation. 14% of patients had also the ataxic component of dysarthria and it was present by the tendency to "sticking" speech. The dysarthria was accompanied by primitive reflexes, emotional lability, and pathological laughter, and a dysphagy was mild or moderate.

Both in the case series of welders with neurological symptoms attributed to manganese fumes and in the case series of ephedrone encephalopathy, increased T1 MRI signal in the inner segment of globus pallidus and pars reticularis of the substantia nigra has been reported as a marker of manganese accumulation (Herrero Hernandez et al., 2006; Josephs et al., 2005; Kenangil et al., 2006; Sadek et al., 2003).

Dopamine transporter (DAT) scans confirm an intact nigrostriatal pathway. Although the white matter appears to be normal on T1-weighted MRI scans, diffusion tensor imaging studies showed extensive white matter changes particularly in the frontal and premotor areas and widespread damage to corticopallidal connections (Stepens et al., 2008; Fig. 3).

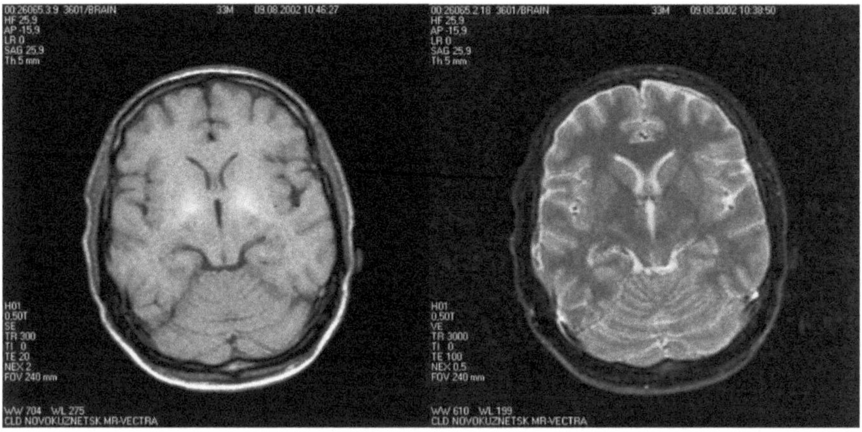

Fig. 3 MRI of patient with ephedrone encephalopathy (6 months from clinical debut). Bilateral homogeneous symmetrical increase of signal from the globus pallidus internus on axial T1-weighted image (A). Pathological signal intensity of basal ganglia on T2-weighted MR-image (B) is not detected.

Among 43 Bay Bridge welders who worked in a confined space up to 2 years, 11 cases of manganism were identified with irreversible neurological syndrome, including cognitive disorder (Bowler et al., 2007). There were identified significant associations between cumulative exposure index and deficits of attention and concentration, working memory, cognitive flexibility, and verbal memory; and between blood Mn level and executive function, cognitive flexibility, visuospatial construction ability, and visual contrast sensitivity. This study suggests that neuropsychological features contribute in a dose-effect-related way to the portrait of manganism. A prospective follow-up study in the same cohort three-and-a-half years after cessation of welding demonstrated improvement of cognitive functions, but other symptoms remained constant or worsened (Bowler et al., 2011; Ellingsen et al., 2008). Neuropsychiatric symptoms in patients with manganism are varied and include emotional lability, depression and anxiety, visual and auditory hallucinations, nervousness, irritability, fatigue, lack of vigor, impulsive/compulsive behavior, and aggression hostility (Bowler et al., 2007). Sleep dysfunctions are commonly reported in patients with manganism (Bouabid et al., 2016). Bowler et al. (2007) revealed sleep disturbances in all 11 patients with parkinsonian signs who were exposed to manganese in welding fumes. Barrington et al. (1998) evaluated cardiovascular autonomic function in eight manganese alloy welders. Seven of the eight workers had symptoms of autonomic dysfunction and significantly decreased heart rate variability, but these did not distinguish the relative exposure. Power spectrum analysis of 24-h ambulatory ECG, indicating a decrease in parasympathetic high-frequency activation of heart rate variability. Autonomic symptoms such as vasomotor disturbances and hyperhidrosis, especially in the field of palm, sexual dysfunction, hypersalivation, and seborrhea were found in majority of patients with ephedrone encephalopathy. Sleep disturbances include insomnia (difficulties of sleep onset, fragmentation) in a half of patients. Hypersomnia (in the form of the extended night sleep, episodes of a daytime drowsiness at the relaxed wakefulness) was found in a third of patients.

Only mild or moderate executive dysfunction has been revealed in patients with ephedrone encephalopathy (Djamshidian et al., 2013; Koksal et al., 2014; Levin, 2005; Sanotsky et al., 2007; Yildirim et al., 2009). Moreover, in the study of Stepens et al. (2008), no one from 23 patients has been reported decline in cognitive function. There was no statistically significant decline in language or visuospatial functions. It has been observed that cognitive dysfunction seen in parkinsonism secondary to chronic manganese intoxication may be long-lasting and may not recover as observed in motor dysfunction. Djamshidian et al. (2013) reported that patients with ephedrone encephalopathy were more impulsive and made more irrational choices on the beads task than controls. However, ephedrone patients had no deficits in working memory or risk taking compared with controls. This may reflect damage to anterior cingulate-basal ganglia loops. There was no correlation reported between motor and cognitive deficit. Hyperintensive MRI changes negatively correlated with the disease duration, but did not depend on the drug abuse duration or its approximate total dosage and also did not correspond to the disease severity, including cognition (Levin, 2005). In the mental status at a half of patients exist the expressed symptoms of apathy, abulia with unspontaneity, limiting of interests, and the tendency to emotional indifference; almost in a third of patients—symptoms of depression in the form of decreased mood, melancholy, uneasiness, feeling of fault and a hopelessness, anhedonia, appetite loss, in serious cases—suicidal intentions. At the same time one-third of patients had euphoria, complacency, and reduction of the awareness to the condition. The disinhibition with impulsiveness, the absence of feeling of a distance, and the tendency to inappropriate jokes were revealed in 21% of patients. Approximately a quarter of patients were considerably expressed irritability with flashes of aggression.

Severity of symptoms is modulated by individual susceptibilities as well as by the length and dose of exposure. Exposed Mn patients can develop motor deficit or manganism without experiencing psychiatric symptoms and neurological function can be deteriorated long after Mn exposure has ceased (Liu et al., 2006). The doses of Mn and exposure time that produce them are not precisely known (Jain & Ferrando, 2011; Sikk et al., 2011).

Repeated courses of calcium and sodium EDTA that accelerates manganese excretion may cause regress of neurological symptoms (Herrero Hernandez et al., 2006; Ono et al., 2002). In patients with ephedrone encephalopathy, $CaNa_2EDTA$ decreases a probability of the further disease progress, though do not contribute significantly to symptoms regress (Levin, 2005; Selikhova et al., 2008).

3.2 Mercury

Clinical manifestations of mercury intoxication vary with the degree and length of exposure. At low levels of exposure, it causes nonspecific symptoms, including fatigue, weakness, depression, sleep disturbance, headache, anorexia, weight loss, and GI symptoms (Taba, 2014). Higher exposure levels are associated with tremor that is most prominent in the hands, but sometimes also involves lips, tongue, eyelids, and lower limbs. It is characteristically a coarse postural and intention-type tremor that in severe cases may affect gait, speech, and handwriting.

Behavioral disorders include excessive timidity, diffidence, loss of self-confidence, anxiety, and excitability. Memory disorders, depression, insomnia, delirium, and hallucinations have also been observed in severe cases. Stomatitis with bleeding gums, loosening of teeth, and hypersalivation may be early signs. Acrodynia with autonomic disorders, erythematous rash, ulceration of oral mucosae, and weakness have been described mostly in children. After cessation of mercury contact, the symptoms may regress, but in many cases, there is no improvement and persistent neurological signs are common (Bernhoft, 2011; Silbernagel et al., 2011).

3.3 Carbon Monoxide

A small concentration of carbon monoxide (CO) can cause hypoxia by reducing oxygen release, especially in brain areas, which require large oxygen consumption such as the basal ganglia (Blumenthal, 2001). Hypoxia may lead to mitochondrial dysfunction, lipid peroxidation, demyelination, and pathological changes in gray matter (Ho et al., 2012).

Neuroimaging changes occur in the bilateral globus pallidus (Ho et al., 2012; Jones, Lagasse, & Zimmerman, 1994; Kao & Nanagas, 2004; Miura, Mitomo, Kawai, et al., 1985; Silver et al., 1996). The cerebellum, caudate nucleus and putamen, deep white matter, amygdala, hippocampus, and parietal and frontal lobes are also involved (Chen et al., 2013; Hopkins and Woon, 2006).

Japanese researchers examined 156 patients 33 years after CO poisoning occurred at the Miike Mikawa Mine, Fukuoka, Japan. Subjective symptoms were observed in 96.8% of the patients (Mimura et al., 1998). Among them, extrapyramidal symptoms were found in 21.8% (parkinsonism in 4.5%, tremor in 10.9%, and muscle rigidity in 16.0%), sensory disturbance—in 25.6%, and peripheral nerve symptoms—in 16.0%.

Above 50% of patients with CO poisoning exhibit some degree of cognitive decline, ranging from mild impairment to dementia (Prockop and Chichkova, 2007; Weaver et al., 2007). Deficits in attention, executive function, visuospatial skills, verbal fluency, and speed of information processing are often observed (Quinn et al., 2009; Weaver et al., 2007). Patients may confabulate in an attempt to hide significant memory problems (Smallwood and Murray, 1999). Symptoms may arise immediately or follow an asymptomatic period. Lucid intervals of up to 240 days have been observed, although the mean latency for cognitive and behavioral symptoms is 3 weeks (Choi, 2002).

Patients who are suspected of having CO poisoning should be given 100% oxygen via a face mask in the accident and emergency department. All patients should undergo baseline and follow-up neuropsychological assessment. Cognitive rehabilitation targeted at maintaining residual cognitive functions and developing compensatory strategies will be beneficial. Donepezil, an acetylcholinesterase inhibitor, has been shown to improve memory in CO poisoning in a case report (Wang et al., 2009).

Yoshiike et al. (2016) revealed near-complete absence of sleep spindles in the first and second PSG recordings, and this significantly increased in the third recording. In contrast, a limited amount of slow wave sleep was observed in the first PSG and did not increase thereafter.

3.4 Carbon Disulfide

Chronic carbon disulfide (CS_2) poisoning (for example, grain storage industry workers, who were exposed to fumigants) may be a cause of parkinsonism and peripheral neuropathy (sensory shading to pin, vibration, and touch to the midcalf of both legs, slowing in sensory conduction velocities, distal muscle wasting) (Peters et al., 1982; Sińczuk-Walczak & Szymczak, 1996).

4. DRUG-INDUCED PARKINSONISM

Drug-induced parkinsonism (DIP) is the side effect of several classes of medications such as antipsychotics, calcium antagonists, and dopamine-depleting agents. DIP is the most common cause of secondary parkinsonism due to epidemiological studies, but rare is diagnosed at autopsy. This paradox may be explained by apparent simplicity of diagnosis that needs only historical data about the use of medication with specific reputation. In practice, DIP frequently misdiagnosed in patients with latent or prodromal neurodegenerative disorders unmasked with medication prescribing.

Morley et al. (2014) performed a retrospective case–control study of 97 DIP subjects and 97 age-matched patients with PD. While total motor scores were similar between groups, postural instability–gait difficulty scores were higher in PD than in DIP.

There are a few data about NMSs of DIP. Moreover, these data may be related to primary disorders that were indication for psychopharmacotherapy such as schizophrenia, bipolar disorder, depression, epilepsy, or prodromal symptoms of neurodegenerative disorders. So the problem of differentiation between NMSs related to DIP and comorbidity remains actual.

Kim, Oh, et al. (2013) and Kim, Youn, Shin, and Cho (2013) have evaluated NMSs in 28 patients with DIP, 35 drug-naïve patients with PD, and 32 healthy controls, using the Nonmotor Symptoms Scale (NMSS). They found that the NMSS total score was higher in PD patients than in subjects with DIP and healthy controls. Moreover, after having controlled for age and gender, NMSs such as urinary and sleep disturbances, attention deficit, and hyposmia were found to be significantly associated with PD. Preliminary evidence exists for sympathetic cardiac scintigraphy to predict dopaminergic pathway abnormalities and to differentiate between DIP and PD. Higher diagnostic accuracy has been demonstrated for DAT imaging.

4.1 Cognitive Impairment

Studies on cognitive changes in DIP are rare. Most of these studies have investigated patients with schizophrenia taking typical or atypical antipsychotics. These studies suggest that cognitive impairment is common among DIP patients, but that there is no specific cognitive profile, which distinguishes DIP from PD. Among 14 patients with DIP caused by antiemetics/prokinetics in study of Ahn et al. (2015), dementia was diagnosed in 4 patients and all other 10 patients had mild cognitive impairment. Masmoudi, Gras-Champel, Masson, et al. (2006) marked cognitive impairment, which was found in 40% of 10 patients with VPA-induced parkinsonism.

According to Kim et al. (2011), cognitive function in DIP was significantly worse than in controls for most domains; there were no significant differences found in the comparisons with the PD patients. However, in other study attention deficit was higher in PD than in DIP patients (Kim, Oh, et al., 2013; Kim, Youn, et al., 2013). After discontinuation of the offending drugs, some patients had transient and reversible cognitive improvement, similar to parkinsonian motor features, and others experienced persistence and eventual worsening of their cognitive dysfunction (Byun et al., 2009; Kim et al., 2011).

The base for cognitive dysfunction is unknown. Because the age of onset of DIP is frequently late in life, it is difficult to determine whether cognitive change in DIP is due to normal aging or to pathological processes underlying or accompanying DIP (Morley et al., 2014). Moreover, some DIP patients are in a prodromal stage of a neurodegenerative disease, which is associated with cognitive dysfunction. However, the results of study by Ahn et al. (2015) suggest that cognitive impairment in patients with DIP reflects the toxic/metabolic process associated with the offending drug in addition to being a risk factor for DIP.

4.2 Autonomic Disturbances

Morley et al. (2014) found that NMSs including constipation and sexual dysfunction were more common in PD than in DIP.

In DIP, OH was associated with the underlying psychiatric diseases and neuroleptics use, whereas prokinetics were not related to OH. The supine blood pressure, nighttime blood pressure, and nocturnal blood pressure dipping did not differ significantly between the DIP and control groups. DIP patients frequently exhibited OH that was associated with the underlying diseases as well as the nature of and exposure time to the offending drugs. Kim and others (2017) found that OH was more frequent in patients with PD (27.1%) or DIP (25%) than in healthy controls.

Cardiac MIBG uptake is reduced in PD patients and a recent meta-analysis suggests that the procedure is 89% sensitive and 95% specific in the diagnosis of PD (Orimo et al., 2012). A study of 20 patients with suspected DIP showed average levels of cardiac MIBG uptake similar to that of controls (Lee et al., 2006). However, two DIP subjects whose uptake fell below normal levels had persistent parkinsonism after drug withdrawal and a robust response to levodopa treatment, suggesting underlying PD.

4.3 Sleep Disorders

Sleep disturbances, according to Kim, Oh, et al. (2013) and Kim, Youn, et al. (2013), assessed with NMSS were more prominent in PD than those in DIP. Scores for daytime sleep, difficulty fall asleep, RLS, and nocturia were especially higher in PD than in DIP.

Morley and Duda (2014) and Morley et al. (2014) found "dream-enactment behavior" in 39% DIP patients. However, this symptom was seen only in 16% patients with reversible (true) DIP and in 86% patients with suspected underlying neurodegeneration, together with hyposmia, constipation, and sexual dysfunction.

4.4 Sensory Symptoms

Presence of smell deficits in DIP patients might be more associated with dopaminergic loss rather than with a drug-mediated dopamine receptor blockade. Thus, olfactory testing may offer a simple and inexpensive method to help predict outcomes in DIP and, potentially, identify a cohort of premotor PD. The notion that hyposmia can accurately discriminate PD from "pure" DIP patients has been confirmed by many studies (Bovi et al., 2010; Brigo et al., 2014; Lee, Yeo, Yong, & Kim, 2007; Morley & Duda, 2014). However, several issues, including smoking history and cognitive functions, can influence smell function assessment. Using the Sniffin' Sticks test in 16 DIP patients, 13 PD patients, and 19 healthy controls, Bovi et al. (2010) found that DIP patients (with normal dopaminergic imaging: FP-CIT SPECT/DaT-Scan) had smell function similar to healthy controls. On the other hand, a subgroup of patients with DIP had impaired smell function and this predicted an abnormal dopaminergic imaging, suggesting that these patients had eventually developed degenerative PD.

Decina et al. (1992) examined the frequency of subjective painful sensory symptoms and their relation to neuroleptic-induced extrapyramidal syndromes (parkinsonism and akathisia) in 107 psychiatric patients. Fourteen (23%) of 60 patients receiving neuroleptics reported experiences of

Table 3 Approximate Frequency of Nonmotor Symptoms in Different Forms of Secondary Parkinsonism in Comparison With PD

	VP	NPH	DIP	Toxic	PD
Cognitive impairment	+++	+++	+	++	++
Depression	++	++	++	++	++
Apathy	++	++	+	++	++
Delusions/hallucinations	–	–	±	+	++
Orthostatic hypotension	+	+	+	+	++
Gastrointestinal disorders	+	+	+	+	++
Urinary symptoms	++	+++	+	+	++
Olfactory dysfunction	–	–	–	–	++
Sleep disorders	++	++	+	++	+++
Sensory abnormalities/pain	+	–	+	+	++

–, <5%, ±5%–10%; +, 10%–25%; ++, 25%–50%; +++, >50%.

spontaneous pain subjectively attributed to pharmacological treatment. Twelve (55%) of the 22 patients with neuroleptic-induced extrapyramidal syndromes reported pain, compared with only 2 (5%) of the 38 patients who received neuroleptics but did not experience extrapyramidal syndromes. Kasantikul and Kanchanatawan (2007) reported pain syndrome in 4 patients with tardive parkinsonism and dystonia from 12 patients with different tardive movement disorders (Table 3).

5. CONCLUSION

We concluded that NMSs are frequent and disable not only in PD but in most patients with secondary parkinsonism (vascular, toxic, drug-induced, etc.). The NMSs are important clues for differential diagnosis with PD and atypical parkinsonism. NMSs in the secondary parkinsonism are the important target for therapy. NMSs seem to be caused by lesions of parallel but segregated frontostriatal circuit at different levels of CNS. Specific peculiarities of different parkinsonian diseases may be explained by three main factors: level of destruction frontosubcortical circuit, secondary dysfunction of related links, and integrity of the contrafunctional links.

REFERENCES

Ahlberg, J., Norlen, L., Blomstrand, C., & Wikkelsö, C. (1988). Outcome of shunt operation on urinary incontinence in normal pressure hydrocephalus predicted by lumbar puncture. *Journal of Neurology, Neurosurgery, and Psychiatry*, *51*(1), 105–108.

Ahn, H. J., Yoo, W. K., Park, J., Ma, H. I., & Kim, Y. J. (2015). Cognitive dysfunction in drug-induced parkinsonism caused by prokinetics and antiemetics. *Journal of Korean Medical Science*, *30*(9), 1328–1333.

Alexopoulos, G. S., Meyers, B. S., Young, R. C., Campbell, S., Silbersweig, D., & Charlson, M. (1997). 'Vascular depression' hypothesis. *Archives of General Psychiatry*, *54*(10), 915–922.

Alexopoulos, G. S., Meyers, B. S., Young, R. C., Kakuma, T., Silbersweig, D., & Charlson, M. (1997). Clinically defined vascular depression. *The American Journal of Psychiatry*, *154*(4), 562–565.

Antonucci, A. S., Gansler, D. A., Tan, S., Bhadelia, R., Patz, S., & Fulwiler, C. (2006). Orbitofrontal correlates of aggression and impulsivity in psychiatric patients. *Psychiatry Research*, *147*, 213–220.

Attier-Zmudka, J., Sérot, J. M., Douadi, Y., Capel, C., Bouzerar, R., & Balédent, O. (2016). Normal-pressure hydrocephalus, sleep apnea syndrome, and quality of life in an octagenarian. *Journal of the American Geriatrics Society*, *64*(9), 1922–1923.

Baldwin, R. C., & O'Brien, J. (2002). Vascular basis of late-onset depressive disorder. *The British Journal of Psychiatry*, *180*, 157–160.

Barbosa, M. T., Caramelli, P., Maia, D. P., et al. (2006). Parkinsonism and Parkinson's disease in the elderly: A community-based survey in Brazil (the Bambui study). *Movement Disorders*, *21*, 800–808.

Barrington, W. W., Angle, C. R., Willcockson, N. K., Padula, M. A., & Korn, T. (1998). Autonomic function in manganese alloy workers. *Environmental Research*, *78*(1), 50–58.

Bassetti, C. L., & Hermann, D. M. (2011). Sleep and stroke. *Handbook of Clinical Neurology*, *99*, 1051–1072.

Behrens, A., Eklund, A., Elgh, E., Smith, C., Williams, M. A., & Malm, J. (2014). A computerized neuropsychological test battery designed for idiopathic normal pressure hydrocephalus. *Fluids and Barriers of the CNS*, *11*, 22.

Benítez-Rivero, S., Lama, M. J., Huertas-Fernández, I., de Toledo, P. Á., Cáceres-Redondo, M. T., Martín-Rodríguez, J. F., et al. (2014). Clinical features and neuropsychological profile in vascular parkinsonism. *Journal of the Neurological Sciences*, *345*(1), 193–197.

Bernhoft, R. A. (2011). Mercury toxicity and treatment: A review of the literature. *Journal of Environmental and Public Health*, *2012*, 460508.

Bhidayasiri, R., Jitkritsadakul, O., Petchrutchatachart, S., Kaewwilai, L., Panyakaew, P., Boonrod, N., et al. (2014). Nocturnal manifestations of atypical and vascular parkinsonism: How do they differ from Parkinson's disease? *Journal of Neural Transmission*, *121*(1), 69–77.

Blair, R. J. (2001). Neurocognitive models of aggression, the antisocial personality disorders, and psychopathy. *Journal of Neurology, Neurosurgery, and Psychiatry*, *71*, 727–731.

Blumenthal, I. (2001). Carbon monoxide poisoning. *Journal of the Royal Society of Medicine*, *94*(6), 270–272.

Bombois, S., Debette, S., Delbeuck, X., Bruandet, A., Lepoittevin, S., Delmaire, C., et al. (2007). Prevalence of subcortical vascular lesions and association with executive function in mild cognitive impairment subtypes. *Stroke*, *38*(9), 2595–2597.

Bouabid, S., Fifel, K., Benazzouz, A., & Lakhdar-Ghazal, N. (2016). Consequences of manganese intoxication on the circadian rest-activity rhythms in the rat. *Neuroscience*, *331*, 13–23.

Bovi, T., Antonini, A., Ottaviani, S., Antonioli, A., Cecchini, M. P., Di Francesco, V., et al. (2010). The status of olfactory function and the striatal dopaminergic system in drug-induced parkinsonism. *Journal of Neurology*, *257*(11), 1882–1889.

Bowler, R. M., Gocheva, V., Harris, M., Ngo, L., Abdelouahab, N., Wilkinson, J., et al. (2011). Prospective study on neurotoxic effects in manganese-exposed bridge construction welders. *Neurotoxicology*, *32*(5), 596–605.

Bowler, R. M., Nakagawa, S., Drezgic, M., Roels, H. A., Park, R. M., Diamond, E., et al. (2007). Sequelae of fume exposure in confined space welding: A neurological and neuropsychological case series. *Neurotoxicology*, *28*, 298–311.

Boyle, P. A., Wilson, R. S., Aggarwal, N. T., Arvanitakis, Z., Kelly, J., Bienias, J. L., et al. (2005). Parkinsonian signs in subjects with mild cognitive impairment. *Neurology*, *65*(12), 1901–1906.

Bradvica, I. Š., Jančuljak, D., Butković-Soldo, S., & Vladetić, M. (2011). Cognitive dysfunction in idiopathic and vascular parkinsonism. *Medicinski Glasnik: Official Publication of the Medical Association of Zenica-Doboj Canton, Bosnia and Herzegovina*, *8*(2), 209–215.

Brigo, F., Erro, R., Marangi, A., Bhatia, K., & Tinazzi, M. (2014). Differentiating drug-induced parkinsonism from Parkinson's disease: An update on non-motor symptoms and investigations. *Parkinsonism & Related Disorders*, *20*(8), 808–814.

Brissaud, É. (1894). *Sur la nature et la pathogénie de la maladie de Parkinson. Coccoz.*

Byun, M. S., Choi, J. S., Yoo, S. Y., Kang, D. H., Choi, C. H., Jang, D. P., et al. (2009). Depressive symptoms and brain metabolite alterations in subjects at ultra-high risk for psychosis: A preliminary study. *Psychiatry Investigation*, *6*(4), 264–271.

Cabral, D., Beach, T. G., Vedders, L., Sue, L. I., Jacobson, S., Myers, K., et al. (2011). Frequency of Alzheimer's disease pathology at autopsy in patients with clinical normal pressure hydrocephalus. *Alzheimer's & Dementia*, 7(5), 509–513.

Chen, H. L., Chen, P. C., Lu, C. H., Hsu, N. W., Chou, K. H., Lin, C. P., et al. (2013). Structural and cognitive deficits in chronic carbon monoxide intoxication: A voxel-based morphometry study. *BMC Neurology*, *13*(1), 129.

Choi, I. S. (2002). Parkinsonism after carbon monoxide poisoning. *European Neurology*, *48*(1), 30–33.

Colloby, S. J., Vasudev, A., O'Brien, J. T., Firbank, M. J., Parry, S. W., & Thomas, A. J. (2011). Relationship of orthostatic blood pressure to white matter hyperintensities and subcortical volumes in late-life depression. *The British Journal of Psychiatry*, *199*(5), 404–410.

Colosimo, C., Morgante, L., Antonini, A., Barone, P., Avarello, T. P., Bottacchi, E., et al. PRIAMO STUDY GROUP, (2010). Non-motor symptoms in atypical and secondary parkinsonism: The PRIAMO study. *Journal of Neurology*, *257*(1), 5.

Critchley, M. (1929). Arteriosclerotic parkinsonism. *Brain*, *52*, 23–83.

Decina, P., Mukherjee, S., Caracci, G., & Harrison, K. (1992). Painful sensory symptoms in neuroleptic-induced extrapyramidal syndromes. *American Journal of Psychiatry*, *149*(8), 1075–1080.

DeVito, E. E., Salmond, C. H., Owler, B. K., Sahakian, B. J., & Pickard, J. D. (2007). Caudate structural abnormalities in idiopathic normal pressure hydrocephalus. *Acta Neurologica Scandinavica*, *116*, 328–332.

Djamshidian, A., Sanotsky, Y., Matviyenko, Y., O'sullivan, S. S., Sharman, S., Selikhova, M., et al. (2013). Increased reflection impulsivity in patients with ephedrone-induced parkinsonism. *Addiction*, *108*(4), 771–779.

Dunet, V., Deverdun, J., Charroud, C., Le Bars, E., Molino, F., de Champfleur, S. M., et al. (2016). Cognitive impairment and basal ganglia functional connectivity in vascular parkinsonism. *American Journal of Neuroradiology*, *37*(12), 2310–2316.

Ellingsen, D. G., Konstantinov, R., Bast-Pettersen, R., Merkurjeva, L., Chashchin, M., Thomassen, Y., et al. (2008). A neurobehavioral study of current and former welders exposed to manganese. *Neurotoxicology*, *29*(1), 48–59.

Fereshtehnejad, S. M., & Lökk, J. (2014). Orthostatic hypotension in patients with Parkinson's disease and atypical parkinsonism. *Parkinson's Disease, 2014*. ID 475854.
Geda, Y. E., Roberts, R. O., Knopman, D. S., Petersen, R. C., Christianson, T. J., Pankratz, V. S., et al. (2008). Prevalence of neuropsychiatric symptoms in mild cognitive impairment and normal cognitive aging: population-based study. *Archives of General Psychiatry, 65*(10), 1193–1198.
Ghoniem, G., Stanford, E., Kenton, K., Achtari, C., Goldberg, R., Mascarenhas, T., et al. (2008). Evaluation and outcome measures in the treatment of female urinary stress incontinence: International urogynecological association (IUGA) guidelines for research and clinical practice. *International Urogynecology Journal and Pelvic Floor Dysfunction, 19*(1), 5–33.
Glass, P. G., Lees, A. J., Bacellar, A., Zijlmans, J., Katzenschlager, R., & Silveira-Moriyama, L. (2012). The clinical features of pathologically confirmed vascular parkinsonism. *Journal of Neurology, Neurosurgery, and Psychiatry, 83*(10), 1027–1029.
Governale, L. S., Fein, N., Logsdon, J., & Black, P. M. (2008). Techniques and complications of external lumbar drainage for normal pressure hydrocephalus. *Neurosurgery, 63*(4), 379–384.
Hasegawa, N., Hashimoto, M., Yuuki, S., Honda, K., Yatabe, Y., Araki, K., et al. (2013). Prevalence of delirium among outpatients with dementia. *International Psychogeriatrics, 25*(11), 1877–1883.
Hashimoto, M., Ishikawa, M., Mori, E., & Kuwana, N. (2010). Diagnosis of idiopathic normal pressure hydrocephalus is supported by MRI-based scheme: A prospective cohort study. *Cerebrospinal Fluid Research*, 7(1), 18.
Herrero Hernandez, E., Discalzi, G., Valentini, C., Venturi, F., Chiò, A., Carmellino, C., et al. (2006). Follow-up of patients affected by manganese-induced parkinsonism after treatment with CaNa2EDTA. *Neurotoxicology, 27*(3), 333–339.
Ho, R. C. M., Cheng, W., Chua, A. N. C., & Mak, A. (2012). Neuropsychiatric aspects of carbon monoxide poisoning: Diagnosis and management. *Advances in Psychiatric Treatment, 18*(2), 94–101.
Hopkins, R. O., & Woon, F. L. M. (2006). Neuroimaging, cognitive, and neurobehavioral outcomes following carbon monoxide poisoning. *Behavioral and Cognitive Neuroscience Reviews, 5*(3), 141–155.
Iranzo, A., Fernández-Arcos, A., Tolosa, E., Serradell, M., Molinuevo, J. L., Valldeoriola, F., et al. (2014). Neurodegenerative disorder risk in idiopathic REM sleep behavior disorder: Study in 174 patients. *PLoS One, 9*(2), e89741.
Israelsson, H., Allard, P., Eklund, A., & Malm, J. (2016). Symptoms of depression are common in patients with idiopathic normal pressure hydrocephalus: The INPH-CRasH study. *Neurosurgery, 78*(2), 161–168.
Jain, S., & Ferrando, S. J. (2011). Manganese neurotoxicity presenting with depression, psychosis and catatonia. *Psychosomatics, 52*(1), 74–77.
Jellinger, K. (2005). Vascular parkinsonism. *Journal of Neurology, 252*(12), 1549.
Jennum, P., & Børgesen, S. E. (1989). Intracranial pressure and obstructive sleep apnea. *Chest, 95*(2), 279–283.
Jiang, B., Ding, C., Yao, G., Yao, C., Zhang, Y., Ge, J., et al. (2013). Polysomnographic abnormalities in patients with vascular cognitive impairment-no dementia. *Sleep Medicine, 14*(11), 1071–1075.
Jones, J. S., Lagasse, J., & Zimmerman, G. (1994). Computerized tomographic findings after acute carbon monoxide poisoning. *American Journal of Emergency Medicine, 12*, 448–451.
Josephs, K. A., Ahlskog, J. E., Klos, K. J., Kumar, N., Fealey, R. D., Trenerry, M. R., et al. (2005). Neurologic manifestations in welders with pallidal MRI T1 hyperintensity. *Neurology, 64*(12), 2033–2039.

Kanemoto, H., Kazui, H., Suzuki, Y., Sato, S., Kishima, H., Yoshimine, T., et al. (2016). Effect of lumbo-peritoneal shunt surgery on neuropsychiatric symptoms in patients with idiopathic normal pressure hydrocephalus. *Journal of the Neurological Sciences*, *361*, 206–212.

Kao, L. W., & Nanagas, K. A. (2004). Carbon monoxide poisoning. *Emergency Medicine Clinics of North America*, *22*, 985.

Kasantikul, D., & Kanchanatawan, B. (2007). Antipsychotic-induced tardive movement disorders: A series of twelve cases. *Journal of the Medical Association of Thailand = Chotmaihet Thangphaet*, *90*(1), 188.

Katzenschlager, R., Zijlmans, J., Evans, A., Watt, H., & Lees, A. J. (2004). Olfactory function distinguishes vascular parkinsonism from Parkinson's disease. *Journal of Neurology, Neurosurgery, and Psychiatry*, *75*(12), 1749–1752.

Kazui, H. (2008). Cognitive impairment in patients with idiopathic normal pressure hydrocephalus. *Brain and Nerve*, *60*(3), 225–231.

Kazui, H., Mori, E., Ohkawa, S., Okada, T., Kondo, T., Sakakibara, R., et al. (2013). Predictors of the disappearance of triad symptoms in patients with idiopathic normal pressure hydrocephalus after shunt surgery. *Journal of the Neurological Sciences*, *328*(1), 64–69.

Kenangil, G., Ertan, S., Sayilir, I., & Ozekmekçi, S. (2006). Progressive motor syndrome in a welder with pallidal T1 hyperintensity on MRI: A two-year follow-up. *Movement Disorders*, *21*(12), 2197–2200.

Kim, Y. D., Kim, J. S., Chung, S. W., Song, I. U., Yang, D. W., Hong, Y. J., et al. (2011). Cognitive dysfunction in drug induced parkinsonism (DIP). *Archives of Gerontology and Geriatrics*, *53*(2), 222–226.

Kim, J. S., Oh, Y. S., Kim, Y. I., Yang, D. W., Chung, Y. A., You, I. R., et al. (2013). Combined use of 123 I-metaiodobenzylguanidine (MIBG) scintigraphy and dopamine transporter (DAT) positron emission tomography (PET) predicts prognosis in drug-induced parkinsonism (DIP): A 2-year follow-up study. *Archives of Gerontology and Geriatrics*, *56*(1), 124–128.

Kim, J. S., Ryu, D. W., Oh, J. H., Lee, Y. H., Park, S. J., Jeon, K., et al. (2017). Cardiovascular autonomic dysfunction in patients with drug-induced parkinsonism. *Journal of Clinical Neurology*, *13*(1), 15–20.

Kim, J. S., Youn, J., Shin, H., & Cho, J. W. (2013). Non-motor symptoms in drug-induced parkinsonism and drug-naïve Parkinson disease. *The Canadian Journal of Neurological Sciences*, *40*(1), 36–41.

Kito, Y., Kazui, H., Kubo, Y., Yoshida, T., Takaya, M., Wada, T., et al. (2009). Neuropsychiatric symptoms in patients with idiopathic normal pressure hydrocephalus. *Behavioural Neurology*, *21*(3), 165–174.

Klassen, B. T., & Ahlskog, J. E. (2011). Normal pressure hydrocephalus how often does the diagnosis hold water? *Neurology*, 77(12), 1119–1125.

Klinge, P. M., Brooks, D. J., Samii, A., Weckesser, E., van den Hoff, J., Fricke, H., et al. (2008). Correlates of local cerebral blood flow (CBF) in normal pressure hydrocephalus patients before and after shunting—A retrospective analysis of [(15)O]H(2) O PET-CBF studies in 65 patients. *Clinical Neurology and Neurosurgery*, *110*, 369–375.

Koksal, A., Keskinkılıc, C., Sozmen, M. V., Dirican, A. C., Aysal, F., Altunkaynak, Y., et al. (2014). Evaluation of cognitive characteristics of patients developing manifestations of parkinsonism secondary to long-term ephedrone use. *European Neurology*, *71*(3–4), 208–212.

Krauss, J. K., Regel, J. P., Vach, W., Droste, D. W., Borremans, J. J., & Mergner, T. (1996). Vascular risk factors and arteriosclerotic disease in idiopathic normal-pressure hydrocephalus of the elderly. *Stroke*, *27*(1), 24–29.

Kristensen, B., Malm, J., Fagerland, M., Hietala, S. O., Johansson, B., Ekstedt, J., et al. (1996). Regional cerebral blood flow, white matter abnormalities, and cerebrospinal fluid hydrodynamics in patients with idiopathic adult hydrocephalus syndrome. *Journal of Neurology, Neurosurgery, and Psychiatry*, *60*(3), 282–288.

Kristensen, B., Malm, J., & Rabben, T. (1998). Effects of transient and persistent cerebrospinal fluid drainage on sleep disordered breathing in patients with idiopathic adult hydrocephalus syndrome. *Journal of Neurology, Neurosurgery, and Psychiatry, 65*(4), 497–501.

Krzastek, S. C., Bruch, W. M., Robinson, S. P., Young, H. F., & Klausner, A. P. (2016). Characterization of lower urinary tract symptoms in patients with idiopathic normal pressure hydrocephalus. *Neurourology and Urodynamics, 36*, 1167–1173.

Kuchiwaki, H., Misu, N., Hirai, N., Inao, S., Ikeda, A., Takada, S., et al. (1985). Pre-and postoperative studies of sleep levels in patients with normal pressure hydrocephalus for an indication of operative treatment. In: H. Dietz, M. Brock, & M. Klinger (Eds.), *Extra-intracranial vascular anastomoses microsurgery at the edge of the tentorium* (pp. 296–303). Berlin, Heidelberg: Springer.

Lee, P. H., Kim, J. S., Shin, D. H., Yoon, S. N., & Huh, K. (2006). Cardiac 123I-MIBG scintigraphy in patients with drug induced parkinsonism. *Journal of Neurology, Neurosurgery, and Psychiatry*, 77(3), 372–374.

Lee, P. H., Yeo, S. H., Yong, S. W., & Kim, Y. J. (2007). Odour identification test and its relation to cardiac 123I-metaiodobenzylguanidine in patients with drug induced parkinsonism. *Journal of Neurology, Neurosurgery & Psychiatry, 78*(11), 1250–1252.

Leinonen, V., Koivisto, A. M., Savolainen, S., Rummukainen, J., Sutela, A., Vanninen, R., et al. (2012). Post-mortem findings in 10 patients with presumed normal-pressure hydrocephalus and review of the literature. *Neuropathology and Applied Neurobiology, 38*(1), 72–86.

Levin, O. S. (1997). Vascular parkinsonism. *Neurologic Journal, 2*(4), 42–51.

Levin, O. S. (2003). Movement disorders in cerebrovascular diseases. In: Extrapyramidal disorders. Diagnosis and treatment by Shtok V.N. et al. (eds). Moscow: MedPress-Inform, 503-519.

Levin, O. S. (2005). Ephedrone encephalopathy. *SS Korsakoff Journal of Neurology & Psychiatry, 105*(7), 12–20.

Liu, X., Sullivan, K. A., Madl, J. E., Legare, M., et al. (2006). Manganese-induced neurotoxicity: The role of astroglial-derived nitric oxide in striatal interneuron degeneration. *Toxicological Sciences, 91*, 521–531.

Markianos, M., Lafazanos, S., Koutsis, G., Sfagos, C., & Seretis, A. (2009). CSF neurotransmitter metabolites and neuropsychiatric symptomatology in patients with normal pressure hydrocephalus. *Clinical Neurology and Neurosurgery, 111*(3), 231–234.

Marmarou, A., Young, H. F., Aygok, G. A., Sawauchi, S., Tsuji, O., Yamamoto, T., et al. (2005). Diagnosis and management of idiopathic normal-pressure hydrocephalus: A prospective study in 151 patients. *Journal of Neurosurgery, 102*(6), 987–997.

Marxreiter, F., & Winkler, J. (2016). Vaskuläres Parkinson-Syndrom (= Parkinsonismus bei vaskulärer Enzephalopathie). *Fortschritte der Neurologie·Psychiatrie, 84*(1), 8–13.

Masmoudi, K., Gras-Champel, V., Masson, H., et al. (2006). Parkinsonism and/or cognitive impairment with valproic acid therapy: A report of ten cases. *Pharmacopsychiatry, 39*, 9–12.

McGirt, M. J., Woodworth, G., Coon, A. L., Thomas, G., Williams, M. A., & Rigamonti, D. (2005). Diagnosis, treatment, and analysis of long-term outcomes in idiopathic normal-pressure hydrocephalus. *Neurosurgery, 57*(4), 699–705.

Mimura, K., Harada, M., Sumiyoshi, S., Tohya, G., Takagi, M., Fujita, E., et al. (1998). Long-term follow-up study on sequelae of carbon monoxide poisoning; serial investigation 33 years after poisoning. *Seishin Shinkeigaku Zasshi = Psychiatria et Neurologia Japonica, 101*(7), 592–618.

Miura, T., Mitomo, M., Kawai, R., et al. (1985). CT of the brain in acute carbon monoxide intoxication: Characteristic features and prognosis. *American Journal of Neuroradiology, 6*, 739–742.

Miyamoto, J., Tatsuzawa, K., Inoue, Y., Imahori, Y., & Mineura, K. (2007). Oxygen metabolism changes in patients with idiopathic normal pressure hydrocephalus before and after shunting operation. *Acta Neurologica Scandinavica, 116*, 137–143.

Morley, J. F., & Duda, J. E. (2014). Use of hyposmia and other non-motor symptoms to distinguish between drug-induced parkinsonism and Parkinson's disease. *Journal of Parkinson's Disease*, *4*(2), 169–173.

Morley, J. F., Pawlowski, S. M., Kesari, A., Maina, I., Pantelyat, A., & Duda, J. E. (2014). Motor and non-motor features of Parkinson's disease that predict persistent drug-induced parkinsonism. *Parkinsonism & Related Disorders*, *20*(7), 738–742.

Murakami, M., Hirata, Y., & Kuratsu, J. I. (2007). Predictive assessment of shunt effectiveness in patients with idiopathic normal pressure hydrocephalus by determining regional cerebral blood flow on 3D stereotactic surface projections. *Acta Neurochirurgica*, *149*, 991–997.

Oliveira, M. F., Nacif, S. R., Castellano, O., Sousa, U. O., Rotta, J. M., & Pinto, F. C. (2015). Improvement of central sleep apneas following ventricular shunt for normal pressure hydrocephalus. *The Journal of Neuropsychiatry and Clinical Neurosciences*, *27*(3), 206–208.

Ono, K., Komai, K., & Yamada, M. (2002). Myoclonic involuntary movement associated with chronic manganese poisoning. *Journal of the Neurological Sciences*, *199*, 93–96.

Orimo, S., Suzuki, M., Inaba, A., & Mizusawa, H. (2012). 123 I-MIBG myocardial scintigraphy for differentiating Parkinson's disease from other neurodegenerative parkinsonism: A systematic review and meta-analysis. *Parkinsonism & Related Disorders*, *18*(5), 494–500.

Pan, M., Gao, H., Long, L., Xu, Y., Liu, M., Zou, J., et al. (2012). Serum uric acid in patients with Parkinson's disease and vascular parkinsonism: A cross-sectional study. *Neuroimmunomodulation*, *20*(1), 19–28.

Panicker, J. N., & Fowler, C. J. (2010). The bare essentials: Uro-neurology. *Practical Neurology*, *10*(3), 178–185.

Park, H., Lee, J. Y., Shin, C. M., Kim, J. M., Kim, T. J., & Kim, J. W. (2015). Characterization of gastrointestinal disorders in patients with parkinsonian syndromes. *Parkinsonism & Related Disorders*, *21*(5), 455–460.

Peters, H. A., Levine, R. L., Matthews, C. G., Sauter, S. L., & Rankin, J. H. (1982). Carbon disulfide-induced neuropsychiatric changes in grain storage workers. *American Journal of Industrial Medicine*, *3*(4), 373–391.

Picascia, M., Minafra, B., Zangaglia, R., Gracardi, L., Pozzi, N. G., Sinforiani, E., et al. (2016). Spectrum of cognitive disorders in idiopathic normal pressure hydrocephalus. *Functional Neurology*, *31*(3), 143.

Podlesek, D., Leimert, M., Schuster, B., Gerber, J., Schackert, G., Kirsch, M., et al. (2012). Olfactory bulb volume in patients with idiopathic normal pressure hydrocephalus. *Neuroradiology*, *54*(11), 1229–1233.

Prockop, L. D., & Chichkova, R. I. (2007). Carbon monoxide intoxication: An updated review. *Journal of the Neurological Sciences*, *262*(1), 122–130.

Quinn, D. K., McGahee, S. M., Politte, L. C., Duncan, G. N., Cusin, C., Hopwood, C. J., et al. (2009). Complications of carbon monoxide poisoning: A case discussion and review of the literature. *Primary Care Companion to the Journal of Clinical Psychiatry*, *11*(2), 74–79.

Relkin, N., Marmarou, A., Klinge, P., Bergsneider, M., & Black, P. M. (2005). Diagnosing idiopathic normal-pressure hydrocephalus. *Neurosurgery*, *57*, 4–16.

Richardson, J., Kerr, S. R., Shaw, F., Kenny, R. A., O'Brien, J. T., & Thomas, A. J. (2009). A study of orthostatic hypotension in late-life depression. *The American Journal of Geriatric Psychiatry*, *17*(11), 996–999.

Rozzini, L., Chilovi, B. V., & Bertoletti, E. (2007). Mild parkinsonian signs and psychobehavioral symptoms in subjects with mild cognitive impairment. *International Psychogeriatrics*, *17*(1), 1–10.

Sadek, A. H., Rauch, R., & Schulz, P. E. (2003). Parkinsonism due to manganism in a welder. *International Journal of Toxicology*, *22*(5), 393–401.

Saito, M., Nishio, Y., Kanno, S., Uchiyama, M., Hayashi, A., Takagi, M., et al. (2011). Cognitive profile of idiopathic normal pressure hydrocephalus. *Dementia and Geriatric Cognitive Disorders Extra*, *1*(1), 202–211.

Sakakibara, R., Kanda, T., Sekido, T., Uchiyama, T., Awa, Y., Ito, T., et al. (2008). Mechanism of bladder dysfunction in idiopathic normal pressure hydrocephalus. *Neurourology and Urodynamics*, *27*(6), 507–510.

Sakakibara, R., Uchida, Y., Ishii, K., Kazui, H., Hashimoto, M., Ishikawa, M., et al. (2012). Correlation of right frontal hypoperfusion and urinary dysfunction in iNPH: A SPECT study. *Neurourology and Urodynamics*, *31*(1), 50–55.

Sanotsky, Y., Lesyk, R., Fedoryshyn, L., Komnatska, I., Matviyenko, Y., & Fahn, S. (2007). Manganic encephalopathy due to "ephedrone" abuse. *Movement Disorders*, *22*(9), 1337–1343.

Santangelo, G., Vitale, C., Trojano, L., De Gaspari, D., Bilo, L., Antonini, A., et al. (2010). Differential neuropsychological profiles in parkinsonian patients with or without vascular lesions. *Movement Disorders*, *25*(1), 50–56.

Schwartz, J. R., & Roth, T. (2008). Neurophysiology of sleep and wakefulness: Basic science and clinical implications. *Current Neuropharmacology*, *6*(4), 367–378.

Selikhova, M., Fedoryshyn, L., Matviyenko, Y., et al. (2008). Parkinsonism and dystonia caused by the illicit use of ephedrone—A longitudinal study. *Movement Disorders*, *23*(15), 2224–2231.

Sikk, K., Haldre, S., Aquilonius, S. M., & Taba, P. (2011). Manganese-induced parkinsonism due to ephedrone abuse. *Parkinson's Disease*, *2011*. ID 865319.

Silbernagel, S. M., Carpenter, D. O., Gilbert, S. G., Gochfeld, M., Groth, E., Hightower, J. M., et al. (2011). Recognizing and preventing overexposure to methylmercury from fish and seafood consumption: Information for physicians. *Journal of Toxicology*, *2011*. ID 983072.

Silver, D. A., Cross, M., Fox, B., et al. (1996). Computerized tomography of the brain in acute carbon monoxide poisoning. *Clinical Radiology*, *51*, 480–483.

Sińczuk-Walczak, H., & Szymczak, M. (1996). Rhythm patterns of basic brain bioelectric activity in workers chronically exposed to carbon disulfide. *International Journal of Occupational Medicine and Environmental Health*, *10*(4), 429–440.

Smallwood, P., & Murray, G. B. (1999). Neuropsychiatric aspects of carbon monoxide poisoning: A review and single case report suggesting a role for amphetamines. *Annals of Clinical Psychiatry*, *11*(1), 21–27.

Spiegelberg, A., Preuß, M., & Kurtcuoglu, V. (2016). B-waves revisited. *Interdisciplinary Neurosurgery*, *6*, 13–17.

Stepens, A., Logina, I., Liguts, V., Aldiņš, P., Ekšteina, I., Platkājis, A., et al. (2008). A parkinsonian syndrome in methcathinone users and the role of manganese. *New England Journal of Medicine*, *358*(10), 1009–1017.

Taba, P. (2014). Movement disorders associated with chronic metal poisoning. In W. Poewe & J. Jankovic (Eds.), *Movement disorders in neurologic and systemic disease* (pp. 220–233). Cambridge: Cambridge University Press.

Taylor, W. D., Aizenstein, H. J., & Alexopoulos, G. S. (2013). The vascular depression hypothesis: Mechanisms linking vascular disease with depression. *Molecular Psychiatry*, *18*(9), 963–974.

The LADIS Study Group, Poggesi, A., Pantoni, L., Inzitari, D., Fazekas, F., Ferro, J., et al. (2011). 2001-2011: A decade of the LADIS (Leukoaraiosis and DISability) study: What have we learned about white matter changes and small-vessel disease? *Cerebrovascular Diseases*, *32*(6), 577–588.

Tullberg, M., Hellstrom, P., Piechnik, S. K., Starmark, J. E., & Wikkelso, C. (2004). Impaired wakefulness is associated with reduced anterior cingulate CBF in patients with normal pressure hydrocephalus. *Acta Neurologica Scandinavica*, *110*, 322–330.

Vale, T. C., Caramelli, P., & Cardoso, F. (2013). Vascular parkinsonism: A case series of 17 patients. *Arquivos de Neuro-Psiquiatria*, *71*(10), 757–762.
Vale, T. C., Caramelli, P., & Cardoso, F. (2015). Clinicoradiological comparison between vascular parkinsonism and Parkinson's disease. *Journal of Neurology, Neurosurgery, and Psychiatry*, *86*(5), 547–553.
Vizcarra, J. A., Lang, A. E., Sethi, K. D., & Espay, A. J. (2015). Vascular parkinsonism: Deconstructing a syndrome. *Movement Disorders*, *30*(7), 886–894.
Vogt, C., & Vogt, O. (1920). *Zur Lehre der Erkrankungen des striären systems. JA Barth.*
Wang, P., Zeng, T., & Chi, Z. (2009). Recovery of cognitive dysfunction in a case of delayed encephalopathy of carbon monoxide poisoning after treatment with donepezil hydrochloride. *Neurology India*, *57*(4), 481–482.
Weaver, L. K., Valentine, K. J., & Hopkins, R. O. (2007). Carbon monoxide poisoning: Risk factors for cognitive sequelae and the role of hyperbaric oxygen. *American Journal of Respiratory and Critical Care Medicine*, *176*(5), 491–497.
White, C. L., McClure, L. A., Wallace, P. M., et al. (2011). The correlates and course of depression in patients with lacunar stroke: Results from the secondary prevention of small subcortical strokes (SPS3) study. *Cerebrovascular Diseases*, *32*, 354–360.
Williams, D. R., Warren, J. D., & Lees, A. J. (2008). Using the presence of visual hallucinations to differentiate Parkinson's disease from atypical parkinsonism. *Journal of Neurology, Neurosurgery, and Psychiatry*, *79*(6), 652–655.
Wilson, R. S., Schneider, J. A., Beckett, L. A., Evans, D. A., & Bennett, D. A. (2002). Progression of gait disorder and rigidity and risk of death in older persons. *Neurology*, *58*(12), 1815–1819.
Wu, R. H., Li, Q., Tan, Y., Liu, X. Y., & Huang, J. (2014). Depression in silent lacunar infarction: A cross-sectional study of its association with location of silent lacunar infarction and vascular risk factors. *Neurological Sciences*, *35*, 1553–1559.
Yildirim, E. A., Eşsizoğlu, A., Köksal, A., Doğu, B., Baybas, S., & Goökalp, P. (2009). Chronic manganese intoxication due to methcathinone (ephedron) abuse: A case report. *Turkish Journal of Psychiatry*, *20*(3), 294–298.
Yoshiike, T., Nishida, M., Yagishita, K., Nariai, T., Ishii, K., & Nishikawa, T. (2016). Altered sleep spindles in delayed encephalopathy after acute carbon monoxide poisoning. *Journal of Clinical Sleep Medicine: JCSM: Official Publication of the American Academy of Sleep Medicine*, *12*(6), 913.
Zijlmans, J. C., Daniel, S. E., Hughes, A. J., Révész, T., & Lees, A. J. (2004). Clinicopathological investigation of vascular parkinsonism, including clinical criteria for diagnosis. *Movement Disorders*, *19*(6), 630–640.

FURTHER READING

Levin, O. S. (2000). Ephedrone parkinsonism. *Neurologic Journal*, *5*(2), 8–15.
Levin, O. S. (2015). *Vascular parkinsonism.* Moscow: MedPress-Inform. 96 pp.

CHAPTER FORTY-EIGHT

Nonmotor Symptoms in Dystonia

Julie Ann Kristy L. Torres, Raymond L. Rosales[1]
University of Santo Tomas Hospital, Manila, Philippines
[1]Corresponding author: e-mail address: rlrosales@ust.edu.ph

Contents

Abstract

With the emergence of quality of life measures as an indicator for the impact of medical and surgical interventions in dystonia, focus has shifted toward unraveling the pathophysiology and neuroanatomical basis of the "nonmotor symptoms" (NMS). To date, the NMS are now recognized as the greater determinant of quality of life in dystonia, going above and beyond the motor symptomatology. This chapter highlights what is currently known in available literature on the NMS among patients with early-onset dystonia, adult-onset focal dystonia, and dystonia-plus syndromes with particular focus on the NMS involving cognition, mood, behavior (even suicide), sleep, and pain.

1. INTRODUCTION

Dystonia is a spectrum of movement disorders characterized by "*sustained or intermittent muscle contractions causing abnormal, often repetitive, movements, postures, or both; typically patterned, twisting, or tremulous. It is often*

International Review of Neurobiology, Volume 134
ISSN 0074-7742
http://dx.doi.org/10.1016/bs.irn.2017.05.003

initiated or worsened by voluntary action and associated with overflow muscle activation" (Albanese et al., 2013). In 2012, an International Consensus Committee brought forth an updated consensus on the phenomenology and classification of dystonia based on its clinical characteristics (i.e., age at onset, body distribution, temporal pattern, associated features) and etiology (i.e., nervous system pathology, inherited, acquired, or idiopathic). Since the early part of the 20th century, when Oppenheim first reported in literature four patients with *dystonia musculorum deformans*, the focus of research has been on the motor phenomenology and the neuroanatomic substrate of this enigmatic disease. It is posited that different "network" connections between the cortical, striatal, thalamic, and cerebellar regions are thought to be involved in the pathophysiology of motor manifestations with changes in synaptic plasticity playing a causative role. Thus, at the turn of the century, advances in the knowledge database and diagnostics have uplifted the lives of these patients as timely therapeutic interventions have been delivered to address the disabling motor problems. But does it all end there? A paradigm shift has occurred recently as health organizations now focus their attention on the impact of these interventions on the patients' health-related quality of life (QoL). Have all our efforts made our patients lives better? The question begs to answer whether there is something missing in the whole picture. What is masked behind the glaringly obvious motor manifestations are a constellation of symptoms that have been increasingly recognized as a prelude perhaps or a harbinger of worse outcomes in patients living with dystonia. Hence, there is a growing interest on the so-called non-motor symptoms (NMS) of dystonia, much akin to that of Parkinson disease (PD) or Huntington disease (HD). These include neuropsychiatric and cognitive symptoms, sleep, sensory abnormalities, autonomic function, and pain. It has been established in various investigations that NMS are more important determinants than motor symptoms with regard to the QoL in movement disorders such as PD and dystonia (Findley et al., 2002; Löhle, Storch, & Reichmann, 2009; Wolters, 2009). This has been an emerging topic in the last decade in the movement disorder field, as the weight of its importance has gradually been acknowledged in numerous case–control studies and review articles.

As a way to expand a previous work from our side (Sunga & Rosales, 2014), we thus present an updated review of the nonmotor phenomenology in dystonia. Segregation between dystonia-plus syndromes, early-onset dystonia, and adult-onset focal dystonia was made if only to strike some differences and similarities between these cohorts.

2. NMS IN DYSTONIA-PLUS SYNDROMES

2.1 DYT 3 (X-Linked Dystonia-Parkinsonism)

X-linked Dystonia-Parkinsonism (OMIM# 314250), otherwise known as "*Lubag*" in the Pilipino vernacular, is an adult-onset, neurodegenerative movement disorder affecting mainly male Filipinos (rarely females) with features of dystonia in the initial stage of the disease and, later, parkinsonism (Lee et al., 2011). Epidemiological studies have reported cases mostly from the Panay islands in the Philippines where the point prevalence is 5.74/10,000 population, though cases of XDP have been reported from different regions in the Philippines (0.31/10,000 population) and from Filipinos living overseas (Lee et al., 2011). Despite these, genetic studies have traced that affected patients have maternal ancestries hailing from the Panay islands and that the genes associated with XDP are DYT3 (chromosome Xq13.1) and TAF1 (TATA-binding protein-associated factor-1) (Makino et al., 2007; Nolte, Niemann, & Müller, 2003). Notwithstanding the X-linked autosomal recessive pattern of inheritance, female cases have been reported who have affected relatives from both paternal and maternal sides of the family, likely due to a codominant disorder or extreme X-mosaicism, and usually their course is milder and dystonia is less disabling (Evidente et al., 2004; Lee et al., 2002).

Lee and colleagues (Lee, Kupke, Caballar-Gonzaga, Hebron-Ortiz & Muller, 1991; Lee, Munoz, Tan, & Reyes, 2001; Lee et al., 2002, 2011; Rosales, 2010) have extensively documented the unique phenomenology and clinical course of XDP in several reports since 1976. XDP starts at around the middle age (usually 39–40 years, range of 12–64 years) as a focal dystonia in 90% of cases. As the disease progresses in 2–7 years in 84% of cases, a segmental (predominantly affecting the axial areas) or generalized form of dystonia is seen. Around the 7th–10th year of illness, dystonia may coexist or is supplanted by features of parkinsonism (resting tremor, rigidity, bradykinesia among others). Nonetheless, by the 15th year of the disease, parkinsonism predominates until the patients succumb to death usually, but not exclusively, from complications of aerodigestive dysfunction at around 55 years. On the average, throughout the duration of the disease, 68% are predominantly dystonic while only 18.6% have parkinsonism. Neuroimaging studies have shown hyperintense putaminal rim (both dystonic and parkinsonian phases) and atrophy of the head of the caudate or putamen (parkinsonian phase) in patients with XDP (Pasco et al., 2011). Autopsy studies have confirmed the presence of atrophy of the caudate nucleus and putamen with mild to severe neuronal loss and gliosis (Evidente et al., 2002). Goto et al. (2005) have identified that changes in the neostriatum of XDP patients are responsible for the hyperkinetic and later hypokinetic state. In the dystonic phase, there is preferential loss of the striosomal pathway and relative sparing of the matrix-based pathway in the substantia nigra leading to disinhibition of nigral dopaminergic neurons hence increasing the action of striatal dopamine on the remaining matrix neurons. Overall, this results to dystonia. With continued dopamine-mediated toxicity, there is greater involvement of the matrix-based pathway inevitably leading to reduction in projections with resultant parkinsonism.

2.1.1 Cognition

Neurocognitive profiles in patients with XDP have been reported by Howe (Howe, Kellison, Fernandez, Okun, & Bowers, 2009), Wadia (Wadia et al., 2010), Kemmotsu (Kemmotsu et al., 2011), Domingo (Domingo, Ledesma, Pasco, Jamora & Lee, 2011), and colleagues. There were two published case reports of Filipino males who underwent neuropsychological screening in preparation for deep brain stimulation (DBS). The first case presented a genetically confirmed 37-year-old Filipino male with XDP who underwent bilateral globus pallidus interna (GPi) DBS. Neuropsychiatric assessment prior to DBS revealed mild executive dysfunction with impairment in visuospatial skills, sequencing, and working memory. In the second case, a postbilateral GPi DBS neuropsychological test was performed in a 69-year-old Filipino male XDP patient 27 months after initial evaluation (Domingo et al., 2011; Howe et al., 2009). Preoperative profile, with and without anticholinergic medication, revealed impairment in domains of processing speed, working memory, verbal memory, mental flexibility, language-related tasks (naming and fluency), and problem solving. Post-DBS, his overall cognitive abilities showed a significant improvement (Dementia Rating Scale 2 score pre = 113 and post = 126/144). However, the total score was still below average for age. Improvements were noted in the domains of verbal memory (retrieval and recognition of information), problem solving, and visuospatial perception and construction. Of note, there was no significant decline in the memory domain. No significant improvement was observed in processing speed, working memory, naming and fluency, and abstract reasoning. Frontal motor control (Luria test) was also impaired. These findings seem promising; nonetheless, Kemmotsu et al. (2011) speculated that decline in the scores in these domains may be seen regardless whether the patient underwent DBS or not, as the integrity of the frontal-subcortical pathway is eventually compromised in the course of XDP. What seems remarkable in this case is that the cognitive function of the patient remained stable post-DBS despite reports that cognitive/mood decline (particularly in frontal lobe functions) has been documented in 5%–33% of patients after DBS (Okun et al., 2009). Targeting the GPi appears to have minimized cognitive decline. From the anatomical perspective, the GPi compared to the subthalamic nucleus, which is another target area in DBS, has distinct advantages, namely, a larger surface area and lesser fiber pathways (Walter & Vitek, 2004; Wichmann & Delong, 2006). A larger, more recent study in 2011 documented the neuropsychological profile of 21 Filipino male XDP patients prior to DBS (Domingo et al., 2011; Jamora, Ledesma, Domingo, Cenina, & Lee, 2014). In these studies, cognitive impairment, as measured using the Mini-mental State Exam (MMSE), Clock Drawing Test (CDT), and Frontal Assessment Battery (FAB), was documented in 76% of the cases involved. Around 24% of the cases were determined to have cognitive impairment based on MMSE score (80% mild and 20% moderately severe impairment; cutoff score of 27). The common errors in MMSE were found in the tests for attention and calculation (76%), orientation (71%), and language (38%). Out of the 21 cases, 67% were unable to follow instructions and complete the CDT accurately, with the most common error identified as placement of the designated time (62%). Nine out of the 21 cases acquired a FAB score of <13%–89% mild impairment, while only one case had severe impairment (FAB score of 9). Impairments were noted in the items of conceptualization and mental flexibility. It is also of note that 67% (14 cases) had below average scores in the test for motor programming (Luria test) despite being able to understand verbal commands in the MMSE, a finding which may be related to deficits in verbal memory, retrieval of information, abstract reasoning, and conceptualization. These findings corroborate with the aforementioned case reports done in other centers.

The etiologic determinant for the cognitive dysfunction has not been fully elucidated in XDP and may not even be attributed to a single critical factor but a variable host of factors, which underlie the disease. Demographic factors such as age and educational level, concurrent medications (i.e., anticholinergic medications), comorbid psychiatric disease, and the neuroanatomical substrate of XDP have been posited as contributory factors as is seen in other related movement disorders such as the primary forms of dystonia, PD, and HD.

2.1.2 Neuropsychiatric Disorders and Autonomic Dysfunction

Morigaki (Morigaki et al., 2013), Jamora (Cenina et al., 2013; Jamora, Cenina, Teleg, & Lee, 2015; Jamora et al., 2014), and colleagues have published three articles on the cooccurring psychiatric disorders among patients with XDP. Depression occurs in 20% of patients with chronic physical health problems including XDP. It occurs two to three times more often compared to age-matched controls in good physical health (Walter & Vitek, 2004). Based on the findings of a field study in 2010 (Morigaki et al., 2013) involving 14 genetically confirmed XDP patients and their age-matched controls, approximately 43% of the participants had cooccurring psychopathologies. These were identified as follows: panic disorder (7.1%), agoraphobia (21.4%), social phobia (28.6%), any anxiety disorder (35.7%), bipolar disorder (7.1%), major depression (14.3%), and any affective disorder (21.4%). Based on the responses on the Zung Self-Rating Depression Scale (SDS), nearly all XDP patients had depressive symptoms (SDS $\geq$ 40). Of these, 50% had minimal to mild, 21.4% moderate to marked, and 21.4% severe to extreme depressive symptoms. However, only two patients met DSM-IV criteria for major depressive disorder. It is significant to note that the mean SDS scores among the XDP patients were higher compared to the age-matched controls. As such, XDP patients were 2.8 times more likely to have symptoms of depression compared to age-matched controls. Furthermore, among the subitems in the SDS, patients with XDP were more likely to report "depressed affect" and "dissatisfaction," and these were correlated to the severity of depression. The mean SDS scores were remarkably also similar to patients with neurodegenerative diseases such as Parkinson disease and Alzheimer disease. A second study conducted in 2013 (Cenina et al., 2013) looked into anxiety and depression in 42 XDP patients using the Hospital Anxiety and Depression Scale—Pilipino (HADS-P) and Hamilton Depression Rating Scale (HAM-D). Out of 42 patients, 16.7% (7/42) had abnormal scores in the anxiety portion of the HADS-P, while 21.4% (9/42) had abnormal scores in the depression portion of the scale. Nearly half of the cases (54.8%) had significant depressive symptoms on HAM-D. Other significant findings in the study were higher mean scores in the Burke–Fahn–Marsden Dystonia Rating Scale (38.78) and lower mean scores in the Schwab and England Activities of Daily Living scale (65.71) among the patients with significant depressive symptoms. It has been described in literature that a growing number of suicides have been reported in the past years among XDP patients (Rosales, 2010). It was previously estimated that the incidence of suicide was as high as 16% (Lee, Kupke, Caballar-Gonzaga, Hebron-Ortiz, & Muller, 1991) in comparison to Parkinson disease (5.3 times higher), amyotrophic lateral sclerosis (ALS, 1.3–5.8 times higher), and Huntington disease (5.7 times higher) (Fang et al., 2008; Farrer, 1986; Kostić et al., 2010). A recent retrospective study by Jamora et al. (2015) has shown that the rate of completed suicide among XDP patients was 10.8%. There were 21 reported deaths attributable to suicide, 1 of which was a female XDP patient and 5 stemming from 2 different families. The patients were mostly in their middle age (around 44 years) with duration of illness of 5–7 years from the time of onset and diagnosis. This is relatively younger compared to the mean age of 55 years at the time of natural death for most XDP patients. Unlike other neurodegenerative diseases such as ALS and HD wherein suicide had been reported in the presymptomatic and early symptomatic phase (Gil, Funalot, & Verschueren, 2008; Kostić et al., 2010), this subset of patients is relatively older and deaths occurred a little later after the individuals became symptomatic. Around the time of suicide, 80% had generalized within 4 years from the onset of dystonia. Although the data are incomplete, nearly half of the patients were married (14/21); all had achieved at least a primary level of formal education but were unemployed, though three were professionals, at the time of suicide. Prior to suicide, depressed mood, irritability, fatigue, insomnia, dependency, and emotional lability were already reported. One case was even admitted to a psychiatric institution, while two cases had episodes of self-harm prior to suicide. The most common method of suicide was hanging (47.6%), and the rest were voluntary starvation (23.8%), self-mutilation, and organophosphate poisoning. Prior to completed suicide, the patients had become dependent on others for assistance in their daily activities (lower ADL score) due to poor functional capacity brought about by worsening of the dystonic features (higher BFM score), which eventually leads to disfigurement (Walter & Vitek, 2004). Interestingly, we accounted for a 44-year-old male XDP patient who went into "status dystonicus" following abrupt discontinuation of on-board medications and refusal to feed with a highly suicidal intention (Rosales, 2010). These data substantiate the findings from previous studies, where depressive symptoms were prevalent among XDP patients and may be a contributory factor to eventual suicide (Cenina et al., 2013; Morigaki et al., 2013). Depression also further translates to incapacity to seek and continue work thus unemployment and loss of financial stability and independence for themselves and their families, as is shown in the aforementioned data. Family separation or disintegration may occur owing to the economic, physical, and emotional burdens

(Rosales, 2010). Some families may go as far as denying the existence of family members with XDP, migration from Panay, or changing surnames (Lee et al., 1991). Stigmatization in XDP leads to social withdrawal and isolation, while catastrophizing the disease leads to severe disability (Rosales, 2010). On another note, a significant association between altered sympathetic skin response (SSR) and XDP has most recently been investigated (Supnet & Rosales, 2017) in a small cohort of eight genetically confirmed XDP patients and their age-matched controls. However, when alteration in SSR was correlated with symptoms suggestive of autonomic dysfunction, no such significant relationship was found.

The nature of the disease as well as the consequent socioeconomic situation it entails have shown that both intrinsic and extrinsic factors are important determinants in the psychiatric comorbidities in these patients. The neuropathologic correlate of mood dysregulation in XDP may in fact be related to the disruption of the frontal-subcortical pathway stemming from the atrophy of the caudate nucleus and putamen, which is a finding consistent not only in XDP but also other neurodegenerative disorders such as HD (Domingo et al., 2011; Limousin et al., 1998).

2.2 DYT 5 (Dopamine-Responsive Dystonia)

GTP cyclohydrolase 1-deficient dopamine-responsive dystonia (GTPCH1-deficient DRD, DYT5a) or Segawa syndrome is an autosomal dominant, childhood-onset dystonia (predominantly in females) with an average age of onset of 6 years, typically presenting as gait disturbance due to foot dystonia which subsequently generalizes though more pronounced in the legs throughout the disease course. In the later stage of the disease, features of parkinsonism are seen. Other symptoms include arm dystonia and postural hand tremor. A diurnal pattern is observed with aggravation of symptoms toward the evening and alleviation of symptoms in the morning. The symptoms have a dramatic and sustained response to low doses of oral levodopa with absence of motor adverse effects associated with chronic therapy. GTPCH1 is the rate-limiting enzyme in the synthesis of tetrahydrobiopterin (BH4), an essential cofactor for the action of tyrosine hydroxylase, tryptophan hydroxylase, phenylalanine hydroxylase, and nitric oxide synthase, which catalyze the synthesis of dopamine, serotonin, tyrosine, and nitric oxide, respectively (Sadahiro et al., 2011). Consequently, deficiency of this essential enzyme and subsequent abnormal metabolism of BH4 has been correlated with neuropsychiatric disorders (Sumi-Ichinose, Ichinose,

Ikemoto, Nomura, & Kondo, 2010). Sadahiro and colleagues assessed the personality traits of GTPCH1-positive Japanese subjects using the Temperament and Character Inventory—a scale which correlates with the neurobiological foundation for personality traits (Sadahiro et al., 2011). Significant correlation was noted between the GTPCH1 genotype and gender. Males with C/C genotype had higher novelty-seeking scores compared to those with the C/T genotype. Among females, no significant difference was noted. Novelty seeking is a personality trait associated with exploratory activity in response to novel stimulation, extravagance in approach to reward cues, impulsive decision-making, avoidance of frustration, and quick to loss of temper (Cloninger, Syrakic, & Przybeck, 1993). A high novelty-seeking score is thought to be associated with dopaminergic pathways (dopamine transporter 1 and dopamine receptor 4) and is significantly correlated with low basal dopaminergic activity (Cloninger et al., 1993). A number of case series by Hahn (Hahn et al., 2001), Van Hove (Van Hove et al., 2005), Tadic (Tadic et al., 2012), López-Laso (López-Laso et al., 2011), Brüggemann (Brüggemann et al., 2014), and colleagues assessed the psychiatric profiles of 83 patients with GTPCH1 deficiency. In a family found to have the GTPCH1 genotype, four had severe depression, six had anxiety symptoms, and one had obsessive–compulsive traits (Hahn et al., 2001). A larger case series ($n = 22$) by Van Hove et al. (2005) showed that 50% of the cases had symptoms of recurrent depression. Selective serotonin and noradrenaline reuptake inhibitors, serotonin agonists, and hydroxytryptophan were used to successfully treat nine patients. Four patients met clinical criteria for obsessive–compulsive disorder (OCD) ($n = 4$). However, the rate of anxiety disorder (22%) was similar to the general population. The onset of psychiatric symptoms appeared to be in adulthood (20–30 years). In another case series ($n = 23$) by Tadic et al. (2012), mild to moderate depressive symptoms were identified in 65% of the female patients while none among the males. In the study by López-Laso et al. (2011), impulsivity ($n = 4$) and anxiety ($n = 1$) were found among the adult cohorts. However, depression or obsessive–compulsive traits were not manifest in the other cases. Nine of the cases were noted to have IQ scores in the borderline intellectual functioning to mild mental retardation range. Sleep disturbance such as chronic sleep insufficiency and sleepwalking was documented in four cases. However, the study by Brüggemann et al. (2014) showed no significant differences in rates of depression, anxiety, or sleep disturbance among 23 cases and their age-matched controls. It is noteworthy that among the 23 cases, 6 underwent full polysomnography and the results showed abnormalities in initiation and maintenance of sleep, increased REM latency ($n = 5$), increased spontaneous arousal ($n = 4$), sleep fragmentation ($n = 3$), increased sleep latency ($n = 2$), and reduced sleep efficiency ($n = 2$). The QoL scores among cases were reduced, though not significantly, compared to controls. Unsurprisingly, physical health domain scores among the 23 cases differed significantly from the controls and were inversely related to sleep quality and depression rating scores.

Tyrosine hydroxylase (TH)-deficient dopamine-responsive dystonia (DYT5b), a mild form of TH deficiency, generally begins between 12 months and 6 years. Lower limb dystonia with gait disturbance is usually the initial presenting complaint. Thereafter, gradual progression to generalized dystonia is observed with some parkinsonian features such as bradykinesia, tremors (postural), and rigidity. Diurnal pattern of symptoms has been reported in 33% of cases. Similar to GTPCH1-deficient DRD, a dramatic and sustained response to low dose levodopa is also reported. A cohort of patients ($n = 6$) was found to have varying degrees of psychomotor (mild $n = 1$, moderate $n = 2$, severe $n = 1$) and mental retardation (IQ score 75–95) (Chi, Lee, & Tsai, 2012). These findings were consistent with the results in a larger group of patients ($n = 36$), wherein mental retardation (IQ score < 70) was found in 32% of the progressive hypokinetic-rigid phenotype.

Autosomal recessive sepiapterin reductase (SR)-deficient dopamine-responsive dystonia is a rare and very mild form of SR deficiency with age of onset at 7 months (Chi et al., 2012). Clinical features are motor and speech delay, axial hypotonia, oculogyric crises, dystonia, and parkinsonism. As with the aforementioned types of DRD, it also has a diurnal pattern and response to levodopa is sustained but may be partial or complete. Friedman et al. (2012) sought to determine the neuropsychiatric profile of 43 patients with varying severity of the disease. In the neurocognitive evaluation, 20 patients (47%) had mental retardation while 15 had mild/moderate learning disability (35%). Observed psychiatric symptoms were inattention (33%), irritability (25%), anxiety (16%), hyperactivity (11%), aggression (9%), obsessive–compulsive (6%), depression (4%), impulsivity (4%), panic disorder (6%), and psychosis (6%). Sleep disturbances were also reported—hypersomnolence (75%), difficulty maintaining sleep (50%), and drowsiness (25%).

Levodopa is a mainstay in the treatment of DRD, though this has the potential for exacerbating psychiatric manifestations; as for example, increased dopaminergic activity in the mesolimbic pathway may produce hallucinations and delusions (Murray, Lappin, & Di Forti, 2017). The same holds true for cannabinoid agonists, which increase dopaminergic activity in the nucleus accumbens ultimately leading to risk-taking behaviors and substance use, as well as the glutamate receptor inhibitors, which increase glutamate activity in the cortical regions leading to anxiety, psychosis, and addiction (Lee, Oleson, Diergaarde, Cheer, & Pattij, 2012; Linden & Schoepp, 2006; Sperlágh, Windisch, Andó, & Vizi, 2009).

2.3 DYT 11 (Myoclonus-Dystonia)

Myoclonus-dystonia (MD) is an autosomal dominant progressive movement disorder equally affecting males and females during the first two decades of life (Asmus et al., 2007). Myoclonus most commonly affects the neck and arms followed by the trunk and bulbar muscles, and occurs in combination with dystonic features (focal or segmental presenting as cervical dystonia or writer's cramp) (Klein et al., 2002). Involvement of the lower limbs is rare. There has been no reports of worsening or generalization of dystonia during the clinical course. There is a reported dramatic reduction in myoclonus after alcohol ingestion, and the clinical course is benign with affected individuals having a normal life span. The disorder is caused by mutation of the *SGCE* gene that encodes for the protein epsilon-sarcoglycan which is mostly associated with cerebellar Purkinje cells (Asmus et al., 2002). Abnormal function of this protein causes altered function of nuclear membranes (blebbing) (Yokoi, Dang, Zhou, & Li, 2012). On functional MRI, subcortical activation (Nitschke et al., 2006) has been reported such as bilateral hypermetabolism in the thalamus and cerebellum (Klein et al., 2002) as well as reduced striatal D2 receptor binding in the fronto-striatal areas (Beukers et al., 2009; Papapetropoulos et al., 2008; Tai et al., 2017). This may explain the occurrence of psychiatric symptoms among affected individuals (Rosales, 2010). A systematic review by Peall, Waite, Blake, Owen, and Morris (2011) reported generalized anxiety disorder and obsessive–compulsive disorder as the most common psychiatric comorbidities in these patients. Twelve case series were undertaken to assess the psychiatric symptomatology in this cohort (Asmus et al., 2005; Beukers et al., 2009; Doheny, Brin, et al., 2002; Doheny, Danisi, et al., 2002; Foncke et al., 2009; Hess et al., 2007; Huang et al., 2010; Maréchal et al., 2003; Peall et al., 2013, 2011; Ritz et al., 2011; Roze et al., 2008; Saunders-Pullman et al., 2002; van Tricht et al., 2012; Weissbach et al., 2013). Depression, anxiety, obsessive–compulsive disorder, alcohol dependence/abuse, and phobic disorder were the most frequently occurring psychiatric comorbidities, corroborating with the findings in the systematic review. The overall depression and anxiety scores among the cohort were significantly higher compared to controls, and these scores also correlated with motor symptom severity (Foncke et al., 2009). Weissbach et al. (2013) also reported a case of heroin abuse and one dependent personality disorder in their cohort. Neurocognitive assessments were also undertaken by Foncke (Foncke et al., 2009), Huang (Huang et al., 2010), van Tricht (van Tricht et al., 2012), and colleagues. In the first two studies, there was absence of marked impairment in the cognitive functions of the cohort. However, in the study by van Tricht, significant differences were found with nonverbal reasoning and executive function (letter-number sequencing and verbal fluency) in the cohort compared to the psychiatric case controls (van Tricht et al., 2012). Beukers and colleagues were able to assess the QoL in 10 cases pre- and post-DBS surgery (Beukers et al., 2009). Postoperative SF-36 quality of life scores showed a significant improvement in the domains of general perception of health and change in health state. Nevertheless, the scores in the physical and mental domains remained unchanged postsurgery.

2.4 DYT 12 (Rapid-Onset Dystonia-Parkinsonism)

Rapid-onset dystonia parkinsonism (RDP) is an autosomal dominant movement disorder caused by missense mutations in the ATP1A3 gene (encodes the alpha 3 subunit of the Na, K-ATPase found in neurons) (Brashear et al., 2007). Affected individuals classically present with an abrupt onset of dystonia with features of parkinsonism during the second and third decades of life, at times associated with triggers such as stress (Brashear, Farlow, Butler, Kasarskis, & Dobyns, 1996). The bulbar muscles are most prominently involved with concomitant dysarthria and dysphagia and involvement of other regions proceeds in a rostrocaudal fashion (Dobyns et al., 1993). This is typically poorly responsive to levodopa. Barbano and colleagues documented the neuropsychiatric profiles of an Italian family (mother and three offsprings) with genetically confirmed ATP1A3 mutation (Barbano et al., 2012). Using the Composite International Diagnostic Interview (CIDI-Auto), all of the mutation-positive offsprings ($n = 3$) demonstrated these psychiatric disorders: obsessive–compulsive disorder ($n = 2$), panic disorder ($n = 2$), posttraumatic stress disorder ($n = 2$), alcohol abuse/dependence ($n = 3$), cannabis abuse/dependence ($n = 1$). One of these two offsprings was previously treated for depression and anxiety. There was also a history of suicidal ideation and conduct disorder among the offsprings. Psychotic symptoms appeared after the onset of motor symptoms. Remarkably, two offsprings reported drinking excessive amounts of alcohol prior to the onset of motor symptoms. A larger longitudinal study in families with RDP has shown that alcohol is one of the triggers in 40% of affected individuals (Brashear et al., 2010). The other known triggers are heat exposure, psychological stress, and childbirth (Brashear et al., 2007; Dobyns et al., 1993). The results of the cognitive assessment showed that the most significant impairments were in verbal fluency and executive functioning tasks. All four affected individuals showed moderate impairment on the linguistic verbal fluency task, but only two exhibited deficits on the semantic fluency task. The unaffected sibling was also demonstrated to have a linguistic fluency deficit. In terms of executive function assessed using the Controlled Oral Word Association (COWA) and Trail Making Tests (both sensitive to frontal lobe dysfunction), the affected individuals also showed significant impairments on the Trail Making B task (mental flexibility) as well as in inductive reasoning. The unaffected sibling also exhibited the same patterns of impairment. Cook et al. (2014) also examined a cohort of 29 ATP1A3 mutation-positive patients, 3 of which did not exhibit any movement disorder as well as their age-matched mutation-negative controls all from the same families. The results demonstrated significant differences in the following cognitive domains: memory, design memory, learning, psychomotor speed and attention, executive functioning, as well as semantic verbal fluency. Brashear et al. (2010) conducted a familial case–control comparison of 26 mutation-positive motor manifesting (MC) individuals with 3 mutation-positive non-motor manifesting (NMC) and 27 mutation-negative individuals (NC). The most significant differences were observed between MC and NC cases in terms of mood disorders (50% vs 20.8%) and psychosis (19% vs 0%). Psychotic disorders were identified in five MC cases, with the mean age of onset of psychosis at 12.5 years. There were also elevated rates of anxiety, depression but no significant differences in OCD scores. Of note 10 affected individuals presented with psychiatric symptoms first prior to the motor manifestations and these include mood disorder, anxiety disorder, psychosis, and substance use. None of the NC individuals have psychiatric symptoms.

In experimental models, mutations in ATP1A3 have been associated with cognitive impairment (Moseley et al., 2007). Alpha 3-deficient mice had shown difficulty remembering locations of platforms in the Morris water maze (test for spatial learning). On histopathology, a 40% reduction in the expression of NMDA receptors, which are essential for synaptic plasticity, learning, and memory, were documented in the hippocampi of the alpha 3 knocked-out mice. These findings may offer an explanation for the deficit in spatial learning.

3. NMS IN EARLY-ONSET DYSTONIA

3.1 DYT 1 (Early-Onset Torsion Dystonia, Oppenheim Dystonia)

DYT 1 is an early-onset autosomal dominant form of dystonia with age of onset in childhood or adolescence, rarely in adulthood (Ozelius & Bressman, 2011). Genetic studies have shown specific deletion of three base pairs (GAG) in the TOR1A gene. Isolated dystonia apparent with specific actions is the usual presentation but over time this affects other body regions (60%–70%) or remains focal (20%). It has not been associated with other systemic abnormalities.

Heiman et al. (2004, 2007) documented the psychiatric symptoms in 96 mutation-positive individuals (MC) and compared these to 125 controls (60 NMC and 65 NC). The study reported a higher relative risk of recurrent major depressive disorder and an earlier age at onset of symptoms independent of motor symptom severity among MC individuals. In the cohort, only seven individuals met criteria for obsessive–compulsive disorder but the scores were not significantly different between the groups. A study by Arkadir et al. (2016) showed that individuals with DYT 1 dystonia were more likely to display risk-taking behavior in an operant-learning paradigm with probabilistic reward. This was also correlated positively with severity of motor symptoms (Ghilardi et al., 2003). The investigators surmised that people with DYT 1 dystonia learn excessively from unexpected wins but show weakened learning after losses thereby resulting to overestimation of the value of risky choices. This event mirrors an inappropriate strengthening of connections in the corticostriatal pathway thus increased long-term potentiation and decreased long-term depression in corticostriatal synapses which translates to abnormal risk taking. This may also translate to overlearning movements that had a desired effect in the past and being unable to suppress movements that are no longer useful. These results correspond to a dominant model of reinforcement learning in the basal ganglia wherein prediction-error modulated long-term potentiation and long-term depression in corticostriatal synapses change the propensity to repeat actions that previously led to positive or negative prediction errors, respectively.

The neurocognitive profiles of DYT 1 affected individuals are variable. In a study by Ghilardi et al. (2003), NMC cases had increased difficulty with sequence learning tasks compared to NC controls. Among MC cases, there was reduced recall after distraction and semantic fluency as well as lower executive function scores compared to controls (Balas, Peretz, Badarny, Scott, & Giladi, 2006). Both MC and NMC cases were noted to have reduced motor and visual sequence learning. Cognitive profiles were also compared between DYT 1 MC and DYT 1-negative generalized dystonia patients post-DBS. There was no significant difference in the cognitive profile or psychiatric symptoms (Jahanshahi et al., 2014). Please refer to Table 1 for a summary of the NMS in dystonia-plus syndromes and early-onset dystonia.

4. NMS IN FOCAL DYSTONIA

4.1 Blepharospasm

Blepharospasm (BPS) is a primary focal cranial dystonia characterized by involuntary contractions of the orbicularis oculi muscle resulting to increased blinking and involuntary closing of the eyes. Similar to the secondary forms of dystonia, patients with BPS have shown abnormalities in cognition apart from psychiatric manifestations. The putative cause of which has also been linked to abnormalities in the cortico-striatal-thalamic-cortical circuits.

In a study by Yang and colleagues in China (Yang et al., 2016), 68 BPS patients from a Movement Disorder clinic and their matched controls (age, sex, and education level) underwent neuropsychological testing using the Chinese version of the Addenbrook's Cognitive Examination-Revised (ACE-R) and Hamilton Rating Scales for both Anxiety and Depression (HAM-A and HAM-D). In terms of cognition, the BPS cohort had significantly lower total ACE-R scores, ACE-R subdomain scores, as well as MMSE score. Impairments in the following ACE-R subdomains were reported: visuospatial (30.9%), language (30.9%), memory (27.9%), orientation/attention (26.4%), and verbal fluency (22.0%). The prevalence of cognitive impairment was pegged at 32.3% based on ACE-R and 22.0% based on MMSE. The presence of cognitive impairment was correlated significantly with the female gender and lower educational level, with the latter being a potential independent determinant of cognitive impairment in the logistic regression analysis. Poor QoL scores were also observed among patients with cognitive impairment, most especially in the domains of

Table 1 Nonmotor Symptoms in Early Dystonia and Dystonia-Plus Syndromes

Disease	Gene	Inheritance	Age of Onset	Motor Phenotype	Nonmotor Symptoms Psychiatric	Nonmotor Symptoms Cognitive
Early-onset torsion dystonia (dystonia musculorum deformans 1)	DYT 1 GAG deletion of *Torsin A*	AD	~12 years; median 9–11 years	Focal (leg or arm) → generalized (60%–70%); remains focal in 20%	**MC**: recurrent major depression, obsessive–compulsive, risk-taking behavior	**MC**: memory deficits, executive dysfunction, impaired semantic fluency **NMC**: motor and visual sequence learning
X-linked dystonia-parkinsonism	DYT 3 TAF 1	AR	39–40 years; range 12–64 years	Focal → segmental/ generalized → parkinsonism 68% predominantly dystonic 18% parkinsonism	Depressive symptoms (20%–50%) Bipolar disorder (7%) Anxiety symptoms/ disorder (16%–35%) Panic disorder (7%) Agoraphobia (21.4%) Social phobia (29%) Suicide (10%–16%)	**MMSE**: 80% mild, 20% mod severe **FAB**: 89% mild, 1% severe **CDT**: 67% incomplete **Pre- and post-DBS**: Impairment in working memory, sequencing, visuospatial skills, processing speed, problem solving

Dopamine-responsive dystonia	DYT 5a *GCH1* DYT 5b *TH* DYT 5b *SPR*	DYT 5a: AD DYT 5b: AR	DYT 5a: 1–12 years; mean 6 years F > M DYT 5b TH deficiency (mild): 1–6 years DYT 5b SPR deficiency (very mild): 7 months	**DYT 5a**: focal (foot) ➔ generalized (more pronounced lower limbs) ➔ parkinsonism (+) caudal to rostral progression (+) diurnal fluctuation (+) dystonic tremor **DYT 5b TH mild def**: focal (lower limb) ➔ generalized (+) diurnal fluctuation **DYT 5b SPR def**: axial hypotonia, dystonia, oculogyric crisis, parkinsonism (+) diurnal fluctuation	**DYT 5a**: Higher novelty seeking in males Depressive symptoms (65%) Anxiety disorder (22%) Obsessive–compulsive disorder Sleep disorders **DYT 5b:** Depressive symptoms (4%) Anxiety symptoms (16%) Obsessive–compulsive disorder (6%) Panic disorder (6%)	**DYT 5b**: Psychomotor and mental retardation Learning disability
Myoclonus-dystonia	DYT 11 *SGCE*	AD	6 months–80 years M = F	Myoclonic jerks of the neck, arms, trunk, bulbar muscles; Focal/segmental dystonia (+) rostrocaudal progression (+) alcohol responsive	Depressive disorder Anxiety disorder Obsessive–compulsive disorder Alcohol dependence/abuse	Impaired nonverbal reasoning Executive dysfunction

Continued

Table 1 Nonmotor Symptoms in Early Dystonia and Dystonia-Plus Syndromes—cont'd

Disease	Gene	Inheritance	Age of Onset	Motor Phenotype	Nonmotor Symptoms Psychiatric	Nonmotor Symptoms Cognitive
Rapid onset dystonia-parkinsonism	DYT 12 *ATP1A3*	AD	4–55 years	Abrupt onset of dystonia with parkinsonism (bradykinesia and postural instability) (+) rostrocaudal progression: face ➔ arm ➔ leg (+) prominent bulbar findings (+) no response to L-dopa (+) triggering events	Psychotic disorder Obsessive–compulsive disorder Panic disorder Posttraumatic stress disorder Alcohol abuse/dependence Cannabis use Conduct disorder Suicidal ideation	Impaired verbal and semantic fluency, memory, processing speed Executive dysfunction

AD, autosomal dominant; *AR*, autosomal recessive; *ATP1A3*, ATPase Na+/K+ transporting subunit alpha 3; *CDT*, clock drawing test; *F*, female; *FAB*, frontal assessment battery; *GCH1*, GTP cyclohydrolase 1; *M*, male; *MC*, mutation-positive motor manifesting; *MMSE*, mini-mental status exam; *NC*, mutation-negative; *NMC*, mutation-positive nonmotor manifesting; *SGCE*, epsilon sarcoglycan; *SPR*, sepiapterin reductase; *TAF1*, TATA-binding protein-associated factor-1; *TH*, tyrosine hydroxylase.

physical functioning, role physical, and social functioning. The mean HAM-D score in the cohort with BPS was significantly higher (9.24) compared to their healthy counterparts, more so among patients with cognitive impairment who had higher mean scores (11.77) indicating mild depression. Nonetheless, this finding was not statistically significant, and the HAM-D score was not a significant determinant of cognitive impairment in the logistic regression analysis. In another study by Dias et al. (2009), screening for executive dysfunction and cognitive impairment was carried out using the FAB and MMSE among 22 patients with BPS. FAB scores among patients with BPS correlated negatively with age and positively with educational level as well as MMSE score. Other studies have shown sustained attention deficits prior to and after botulinum toxin injection as well as impaired complex movement planning, motor dexterity, visuospatial working memory, and tactile object recognition (Aleman, de Erausquin, & Micheli, 2009; Allam, Frank, Pereira, & Tomaz, 2007).

Sleep disturbance has also been documented in BPS, as shown in studies by Silvestri (Silvestri et al., 1990), Sforza (Sforza, Montagna, Defazio, & Lugaresi, 1991), and colleagues. Reduced REM sleep and increased awakening were positively correlated with disease severity. In another study by Paus et al. (2011), impaired sleep quality was documented in 46% of BPS patients compared to 20% of healthy controls using the Pittsburgh Sleep Quality Index (PSQI). Furthermore, this was associated with symptoms of depression (26%) and restless legs syndrome (19%). More female patients reported impaired sleep quality compared to their male counterparts. Excessive daytime sleepiness measured by the Epworth Sleepiness Scale (ESS) was rarely noted among patients with BPS. In contrast to the aforementioned studies by Silvestri and Sfoza, no significant difference was documented between sleep quality and severity of dystonic symptoms as assessed by the Jankovic Rating Scale. Furthermore, botulinum toxin treatment did not ameliorate poor sleep. It is important to note that around 28% of the participants in each cohort were already on antidepressants and sedatives prior to the onset of the study.

Fontenelle et al. (2011) evaluated 22 patients with BPS and hemifacial spasm (HFS) for obsessive–compulsive and other psychiatric symptoms using the Mini International Neuropsychiatric Interview, Obsessive–Compulsive Inventory-Revised (OCI-R), Beck Depression Inventory (BDI), Beck Anxiety Inventory (BAI), and the MMSE. The severity of OCD symptom dimensions was positively correlated with the duration of abnormal movements and negatively correlated with the severity of depression. In the OCI-R subset, checking was significantly associated with shorter duration of BPS compared to HFS. Though, in terms of the BDI, BAI, and MMSE scores, no correlation was found with the duration of BPS. A similar study used the Hamburg Obsession/Compulsion Inventory-Short Form (HOCI-S) and reported a greater severity of obsessive–compulsive symptoms in a small sample of patients with BPS, though none of the cases met the diagnostic criteria for obsessive–compulsive disorder (Paus et al., 2011). Similar findings were not replicated by other authors who used instruments such as the Yale-Brown Obsessive–Compulsive Scale (Broocks, Thiel, Angerstein, & Dressler, 1998; Dias et al., 2010; Fontenelle et al., 2011; Hall et al., 2005; Munhoz et al., 2005; Scheidt, Schuller, Rayki, Kommerell, & Deuschl, 1996).

Sensory abnormalities are also observed prior to and at the onset of BPS, in particular burning sensation, grittiness or dryness of the eyes, and photophobia (Martino et al., 2005). Relatives of patients with focal dystonia who eventually develop BPS are more likely to have eye symptoms before the onset of BPS than relatives who develop other forms of focal dystonia (Defazio et al., 1998).

4.2 Cervical Dystonia

Cervical dystonia (CD) is another form of primary focal dystonia characterized by involuntary muscle contractions causing abnormal postures and/or twisting movements of the head and neck (Defazio, Jankovic, Giel, & Papapetropoulos, 2013). It represents the most common form of adult-onset focal dystonia with a prevalence estimated at 28–183 cases/million (Lehéricy, Tijssen, Vidailhet, Kaji, & Meunier, 2013). Women are more commonly affected than men with a ratio of 2:1. As with blepharospasm, NMS have also been extensively studied in this group of patients. As postulated, the cortico-striato-thalamo-cortical and cortico-cerebellar circuitry may modulate NMS such as sensory, emotional, autonomic, cognitive, and sleep/wake cycle-related domains (Jahanshahi, 1991).

Klingelhoefer et al. (2014) described the prevalence of NMS in 102 patients with CD (mean UDRS score 5.09) from a single dystonia clinic, all of them receiving botulinum toxin injection. Several items from the NMSQuest, a self-reported questionnaire developed and validated originally for PD, were used as a tool for screening of NMS in the sample. Nearly all (95%) patients reported NMS—36% had at least seven symptoms. Loss of self-confidence (61.8%), insomnia of sleep onset and maintenance

(59.8%), and fatigue (51%) were the most frequently reported NMS in the study. Self-deprecation, as a consequence of loss of self-confidence due to the stigma of CD, has been established as the most salient predictor of depression in CD (Lehéricy et al., 2013). These findings are in line with Camfield's (Camfield, Ben-Shlomo, & Warner, 2002) findings, wherein 289 CD patients had a disproportionate negative impact on physical role limitation compared with chronic diseases such as stroke and PD as well as Papathanasiou's (Papathanasiou, MacDonald, Whurr, & Jahanshahi, 2001) findings in which a large proportion of 79 CD patients ranked self-consciousness highest. In the latter study, perceived stigma due to visible neck dystonia was negatively correlated with age; hence, they found that younger patients had significant mood disturbances compared to older patients (Jahanshahi & Marsden, 1990a; Jankovic, Leder, Warner, & Schwartz, 1991). No significant correlation was found between NMS, age, sex, and disease duration. Conversely, a weak correlation was noted between NMS and motor severity (UDRS) perhaps indicating that patients with severe CD may have more NMS (Papathanasiou et al., 2001).

Pain is one of the most common and disabling complaints in dystonia, with a prevalence of 67%–75% in patients with cervical dystonia (Duane, 1988; Kutvonen, Dastidar, & Nurmikko, 1997; Lowenstein & Aminoff, 1988). Around 43% of patients in Klingelhoefer's study reported pain as a NMS, this figure is lower compared to previous studies. The most frequent muscles involved are the trapezius and sternocleidomastoid in the neck region, often also involving the ipsilateral arm on the side to which the head is rotated (Defazio et al., 1998; Lobbezoo, Tanguay, Thu Thon, & Lavigne, 1996). Changes in pain-pressure thresholds and alteration in pain processing might help explain the presence of pain in CD. Pain-pressure thresholds, even in nonaffected muscles, were about two times lower in CD compared to age- and gender-matched controls (Tinazzi, Rosso, & Fiaschi, 2003). Neurophysiologic studies have shown changes in excitability, and somatosensory cortical activity during movement while dystonic body parts were shown to have abnormal representation in S1 (Avanzino et al., 2010).

The quality of sleep is also impaired in CD as with BPS, with a prevalence of 44%–70% (Eichenseer, Stebbins, & Comella, 2014; Paus et al., 2011). Sleep onset and maintenance insomnia (60%) and lack of sleep refreshment (40%) were common sleep-related problems (Trotti, Esper, Feustel, Bliwise, & Factor, 2009). In a prospective study by Eichenseer et al. (2014), 65% of CD patients were identified as poor sleepers (PSQI >5). Significant worse quality of sleep and impairment in sleep efficiency, daytime dysfunction, sleep latency, and habitual sleep efficiency were significantly greater in comparison to control participants. The study concluded that impaired sleep quality was twice as common in CD compared to age- and sex-matched controls and sleep disturbances did not improve after botulinum toxin injection. Conflicting data exists, however, regarding excessive daytime sleepiness and its association with CD. A higher percentage of excessive daytime sleepiness (ESS score >11) compared to other focal movement disorders and an age-matched control was observed in one study (Lobbezoo, Thu Thon, Remillard, Montplaisir, & Lavigne, 1996), but the same finding was not replicated in another. Excessive daytime sleepiness was, however, not correlated with scores on the Toronto Western Spasmodic Torticollis Rating Scale (TWSTRS) or its subscores (Lobbezoo, Thu Thon, et al., 1996). Similar to BPS, reduced REM sleep and impaired sleep efficiency were also positively correlated with severity of motor symptoms (Eichenseer et al., 2014; Sforza et al., 1991). Apart from sleep disturbances, sleep may also mitigate cervical spinal pain intensity and unpleasantness overnight by approximately 50% (Stamelou, Edwards, Hallett, & Bhatia, 2012).

Patients with CD also complain of sensory abnormalities that develop prior to or concomitant with dystonia (Conte et al., 2016). Unpleasant sensation such as numbness, tingling, or pins and needles (30%), balance problem (49%), and visual disturbance (21.6%) has been reported among the CD patients in Klingelhoefer's study (Klingelhoefer et al., 2014). Errors in the spatial representation of body part location and failure in the integration of proprioceptive information with motor output have been postulated as the causes for the alteration in proprioceptive afferent-related functions including kinesthesia and vibration-induced illusion of movement in patients with focal dystonia such as CD (Rome & Grünewald, 1999). Sensory abnormalities involve both affected and unaffected body regions in CD patients and their asymptomatic first-degree relatives (Frima, Nasir, & Grünewald, 2008; Molloy, Carr, Zeuner, Dambrosia, & Hallett, 2003; Putzki et al., 2006). Afferent processing of sensory stimuli has been extensively discussed in dystonia. The spatial discrimination threshold (defined as the shortest distance interval at which two stimuli applied to the same part of the body can be recognized as spatially separate) and the temporal discrimination threshold (defined as the minimum interstimulus interval at which participants recognize a pair of tactile stimuli as temporally separate) are abnormally higher in affected and unaffected body parts of dystonic patients than in healthy controls (Bradley et al., 2009). These findings have also been reported in patients' unaffected relatives (Bradley et al., 2012; Conte et al., 2014; Tamura et al., 2008; Walsh et al., 2009). Alteration in temporal discrimination is thought to be due to abnormal basal ganglia processing of sensory information, as documented by the significant correlation between putamen enlargement and temporal discrimination threshold values in dystonic patients and unaffected relatives (Bradley et al., 2012; Conte et al., 2014; Tamura et al., 2008; Walsh et al., 2009). Synaptic plasticity in cortical sensorimotor area is also thought to play a role in the involvement of the sensory system in dystonia. Abnormal inhibitory interneuron activity and sensory motor integration were documented using transcranial magnetic stimulation (Belvisi et al., 2013; Gündel et al., 2003). Using a plasticity-inducing protocol (paired associative stimulation) that couples sensory input with primary motor cortex activation, Belvisi and colleagues were able to document increased TMS-motor-evoked potentials in target as well as nontarget muscles (Belvisi et al., 2013).

Psychiatric comorbidity is significantly higher in patients with CD in comparison to the general population and among patients with other medical conditions (Heiman et al., 2004; Jankovic et al., 1991; Kuyper, Parra, Aerts, Okun, & Kluger, 2011). Up to 90% of CD patients have a current or lifetime diagnosis of a psychiatric illness. As such, the lifetime risk of depression (15%–53%) and anxiety (26%–83%) disorders is increased among patients with CD (Conte et al., 2016; Fabbrini et al., 2010; Gündel, Wolf, Xidara, Busch, & Gündel, 2001; Kuyper et al., 2011; Miller et al., 2007). However, the association of dystonia severity with depression and anxiety is not clear-cut as data are conflicting in the available studies (Lencer et al., 2009; Lewis, Butler, & Jahanshahi, 2008). Personality disorders may also occur concurrently in CD. Using the Five Factor Personality Inventory, the lifetime prevalence for any personality disorder was 70.9%, in particular anxious personality (32.6%), obsessive–compulsive (22.1%), and avoidant personality disorders (16.3%) (Lauterbach, Freeman, & Vogel, 2004). The likelihood of social phobia, agoraphobia, and panic disorder was 4.5-fold higher in CD (Stamelou et al., 2012). In other studies, the estimated lifetime prevalence of social phobia using the Social Phobia and Social Interaction Anxiety Scales was 71% or about 10-fold compared to the general adult population (Lewis et al., 2008). Social phobia correlated with body image and a "maladaptive attitude" toward their illness. It was not associated with objective severity of the dystonia. Alcohol use was also found to be significantly higher in a small cohort of CD patients compared to healthy controls (Wenzel et al., 1998). A study by Wenzel and colleagues has shown that these psychiatric manifestations often precede the onset of motor manifestations (Moraru et al., 2002). On long-term follow-up in CD patients, psychiatric symptoms remained stable despite reduction in the severity of motor symptoms (Tiple et al., 2008).

Patients with cervical dystonia were also observed to have mild subclinical changes in some measures of heart rate variability and baroreflex sensitivity even before botulinum toxin injections. Despite this, no report of erectile, sexual, urinary, or bowel dysfunction has yet been published (Lewis et al., 2008).

NMS may play a role in determining the QoL among patients with CD. As a matter of fact, a psychiatric comorbidity is the most important predictor of a decreased HR-QoL as was supported by studies (Pekmezovic et al., 2009; Slawek et al., 2007; Soeder et al., 2009). A large retrospective study among patients with dystonia showed that both physical and mental aspects of QoL were strongly associated with depression and anxiety (Jahanshahi & Marsden, 1990a, 1990b). Several factors also come into play such as longstanding disability, pain, deformity, and lifestyle changes, which also contribute to depression and poor QoL. Despite treatment, improvement in depression and health-related QoL remains uncertain (Cano et al., 2004; Muller et al., 2002; Whurr, Nye, & Lorch, 1998). Using the Cervical Dystonia Impact Scale, a large proportion of CD patients has reported that sleep, fatigue, and tiredness have an impact on their QoL (Romano et al., 2014) and out of these the tired subscore of the Visual Analogue Mood Scale was found to be the foremost factor affecting overall QoL (Jahanshahi & Marsden, 1990a, 1990b). In another study by Smit et al. (2017), the first eight domains of health-related QoL (using the RAND-36 item Health Survey) were significantly lower in CD due to the presence and severity of depressive as well as anxiety symptoms, pain, and disability.

Cognitive impairment has also been documented among patients with CD. In a study by Romano et al. (2014), patients with CD showed deficits in the following cognitive domains: working memory (n-Back test), mental control and visual production (Wechsler Memory Scale), information processing speed and set-shifting capacity (Trail Making Test A and B). These deficits might reflect the changes occurring in the network circuitry between the cortex and subcortical regions. In another study by Loetscher, McDonnell, and Bradnam (2015), random number generation (RNG), which taps into the working memory and executive function, was used to assess 10 CD patients and health controls. The results revealed that patients with CD have enhanced response stereotypy and difficulties in suppressing habitual responses in their RNG performance. These impairments were not significantly correlated to their disability on TWSTRS.

4.3 Focal Hand Dystonia

Focal hand dystonia (FHD) is characterized by task-specific sustained muscle contraction that leads to impaired use (i.e., writer's cramp, musician's dystonia). In a study by Voon et al. (2010) using a semistructured interview for the diagnosis of DSM-IV Axis I disorders in 40 patients with FHD, obsessive–compulsive disorder and subsyndromal OCD was diagnosed in 12.82% vs 2.3% in general population. Lifetime, current, and subsyndromal OCD prevalence rates were reported in 10%, 7.6%, and 2.5% of cases, respectively. The other psychiatric comorbidities identified were generalized anxiety disorder (15%), simple phobia (5%), social phobia (8%), and panic disorder (2.5%). A lifetime history of major depressive disorder was found in 25% of cases (17% in the general population) while 18% had recurrent major depression, 7.6% had current major depression, and 5% had current dysthymia. Significantly, all patients with OCD and subsyndromal OCD had comorbid depression. Family history of psychiatric disorder was also reported in the cohort—depression, "nervous breakdown," as well as alcohol use disorder. The total score on Yale-Brown Obsessive–Compulsive Scale (YBOCS) was greater in patients with OCD. Anxiety and depression rating scales (BDI and BAI) also showed a trend toward higher scores in patients with OCD.

FHD, like OCD, is characterized by frontostriatal dysfunction with mechanisms of decreased inhibition, greater plasticity, and sensory impairments (Smit et al., 2016). Functional neuroanatomical correlates include disorganized putaminal somatotopy, larger putaminal volume, and hyperechogenicity in the lenticular nucleus (Black, Ongur, & Perlmutter, 1998; Breakefield et al., 2008; Delmaire et al., 2005; Naumann, Becker, Toyka, Supprian, & Reiners, 1996). OCD is associated with abnormalities in tasks associated with striatal function such as implicit sequence learning, procedural learning, and response reversal (Menzies et al., 2008). Please refer to Table 2 for a summary of the NMS in adult-onset primary focal dystonia.

5. CONCLUSION

In the advent of QoL-driven therapeutic interventions, the emergence of the "nonmotor" symptoms of dystonia has proven to be indeed a challenge for the movement disorder specialist. Hence, the wide array of NMS in genetically defined, adult-onset, or dystonia-plus syndromes—from

Table 2 Nonmotor Symptoms in Adult-Onset Primary Focal Dystonia

Disease	Motor Phenomenology	Nonmotor Symptoms Psychiatric	Nonmotor Symptoms Cognitive and Others
Blepharospasm	Involuntary contractions of the orbicularis oculi muscle with increased blinking and involuntary closure of the eyes	Depressive symptoms Obsessive–compulsive symptoms	**Cognitive impairment** (22%–32%): impaired visuospatial, language, memory, orientation/attention, and verbal fluency **Sleep disturbance**: reduced REM sleep and sleep efficiency **Sensory abnormalitie**s: dry eyes, burning sensation, photophobia
Cervical dystonia	Involuntary contractions of the muscles of the neck with abnormal posture and/or twisting of the neck	Depressive disorder (15%–53%) Anxiety disorder (26%–83%) Personality disorder (71%) Social phobia (71%)	**Cognitive impairment**: deficits in working memory, mental control and visual production, information processing speed and set-shifting capacity, enhance response stereotypy, difficulty suppressing habitual responses **Pain** (43%): decreased pain-pressure threshold, altered pain processing **Sleep disturbance** (44%–70%): impaired sleep efficiency with daytime dysfunction, sleep latency, significantly worse quality of sleep **Sensory abnormalities**: both affected and unaffected body regions; numbness, tingling, pins and needles, imbalance, visual disturbance **Autonomic symptoms**: mild subclinical changes in heart rate variability and baroreflex sensitivity
Hand dystonia	Task-specific sustained contractions of the hand and arm muscles (i.e., writing, playing a musical instrument)	Depressive disorder (25%) Obsessive–compulsive disorder (13%) Anxiety disorder (15%) Social phobia (8%) Panic disorder (3%)	—

neurocognitive, psychiatric, pain, sensory abnormalities, and finally sleep, is an open topic for research among our colleagues. This review has highlighted many similarities between the cohorts in terms of cognitive and neuropsychiatric symptoms. What seems glaringly obvious from the data presented is no matter how the dystonia presented or how it progresses, nearly the same gamut of NMS is observed in the affected individuals. Thus, alluding to dystonia as a network disorder with various cortical and subcortical connections converging at the basal ganglia, with resultant failure of inhibitory mechanisms and sensory integration to create a motor activity and modulate mood and behavior possibly stemming from maladaptive plasticity (Sunga & Rosales, 2014). Given the impact of NMS on the patient's QoL, further studies on its prevalence (given a dystonia NMS questionnaire was developed in 2015 by Chaudhuri et al., *personal communication*), effects, and subsequent treatment are needed if we are to make the picture whole once and for all.

REFERENCES

Albanese, A., Bhatia, K., Bressman, S. B., Delong, M. R., Fahn, S., Fung, V. S. C., et al. (2013). Phenomenology and classification of dystonia: A consensus update. *Movement Disorders*, *28*(7), 863–873. https://doi.org/10.1002/mds.25475.

Aleman, G. G., de Erausquin, G. A., & Micheli, F. (2009). Cognitive disturbances in primary blepharospasm. *Movement Disorders*, *24*(14), 2112–2120. https://doi.org/10.1002/mds.22736.

Allam, N., Frank, J. E., Pereira, C., & Tomaz, C. (2007). Sustained attention in cranial dystonia patients treated with botulinum toxin. *Acta Neurologica Scandinavica*, *116*(3), 196–200. https://doi.org/10.1111/j.1600-0404.2007.00862.x.

Arkadir, D., Radulescu, A., Raymond, D., Lubarr, N., Bressman, S. B., Mazzoni, P., et al. (2016). DYT1 dystonia increases risk taking in humans. *eLife*, 5, 1–13. https://doi.org/10.7554/eLife.14155.

Asmus, F., Hjermind, L. E., Dupont, E., Wagenstaller, J., Haberlandt, E., Munz, M., et al. (2007). Genomic deletion size at the epsilon-sarcoglycan locus determines the clinical phenotype. *Brain*, *130*(10), 2736–2745. https://doi.org/10.1093/brain/awm209.

Asmus, F., Salih, F., Hjermind, L. E., Ostergaard, K., Munz, M., Kühn, A. A., et al. (2005). Myoclonus-dystonia due to genomic deletions in the epsilon-sarcoglycan gene. *Annals of Neurology*, *58*(5), 792–797. https://doi.org/10.1002/ana.20661.

Asmus, F., Zimprich, A., Du Montcel, S. T., Kabus, C., Deuschl, G., Kupsch, A., et al. (2002). Myoclonus-dystonia syndrome: ε-Sarcoglycan mutations and phenotype. *Annals of Neurology*, *52*(4), 489–492. https://doi.org/10.1002/ana.10325.

Avanzino, L., Martino, D., Marchese, R., Aniello, M. S., Minafra, B., Superbo, M., et al. (2010). Quality of sleep in primary focal dystonia: A case–control study. *European Journal of Neurology*, *17*(4), 576–581. https://doi.org/10.1111/j.1468-1331.2009.02884.x.

Balas, M., Peretz, C., Badarny, S., Scott, R. B., & Giladi, N. (2006). Neuropsychological profile of DYT1 dystonia. *Movement Disorders*, *21*(12), 2073–2077. https://doi.org/10.1002/mds.21070.

Barbano, R. L., Hill, D. F., Snively, B. M., Light, L. S., Boggs, N., McCall, W. V., et al. (2012). New triggers and non-motor findings in a family with rapid-onset dystonia-parkinsonism. *Parkinsonism & Related Disorders*, *18*(6), 737–741. https://doi.org/10.1016/j.parkreldis.2012.03.020.

Belvisi, D., Suppa, A., Marsili, L., Di Stasio, F., Parvez, A. K., Agostino, R., et al. (2013). Abnormal experimentally- and behaviorally-induced LTP-like plasticity in focal hand dystonia. *Experimental Neurology, 240*, 64–74. https://doi.org/10.1016/j.expneurol.2012.11.003.

Beukers, R. J., Booij, J., Weisscher, N., Zijlstra, F., Van Amelsvoort, T. A. M. J., & Tijssen, M. A. J. (2009). Reduced striatal D2 receptor binding in myoclonus-dystonia. *European Journal of Nuclear Medicine and Molecular Imaging, 36*(2), 269–274. https://doi.org/10.1007/s00259-008-0924-9.

Black, K. J., Ongur, D., & Perlmutter, J. S. (1998). Putamen volume in idiopathic focal dystonia. *Neurology, 51*(3), 819–824.

Bradley, D., Whelan, R., Kimmich, O., O'Riordan, S., Mulrooney, N., Brady, P., et al. (2012). Temporal discrimination thresholds in adult-onset primary torsion dystonia: An analysis by task type and by dystonia phenotype. *Journal of Neurology, 259*(1), 77–82. https://doi.org/10.1007/s00415-011-6125-7.

Bradley, D., Whelan, R., Walsh, R., Reilly, R. B., Hutchinson, S., Molloy, F., et al. (2009). Temporal discrimination threshold: VBM evidence for an endophenotype in adult onset primary torsion dystonia. *Brain, 132*(9), 2327–2335. https://doi.org/10.1093/brain/awp156.

Brashear, A., Cook, J. F., Hill, D. F., Amponsah, A., Snively, B. M., & Light, L. S. (2010). New triggers in rapid-onset dystonia-parkinsonism (RDP). *Neurology, 74*(Suppl), A2015.

Brashear, A., Dobyns, W. B., De Carvalho Aguiar, P., Borg, M., Frijns, C. J. M., Gollamudi, S., et al. (2007). The phenotypic spectrum of rapid-onset dystonia-parkinsonism (RDP) and mutations in the ATP1A3 gene. *Brain, 130*(3), 828–835. https://doi.org/10.1093/brain/awl340.

Brashear, A., Farlow, M. R., Butler, I. J., Kasarskis, E. J., & Dobyns, W. B. (1996). Variable phenotype of rapid-onset dystonia-parkinsonism. *Movement Disorders, 11*(2), 151–156. https://doi.org/10.1002/mds.870110206.

Breakefield, X. O., Blood, A. J., Li, Y., Hallett, M., Hanson, P. I., & Standaert, D. G. (2008). The pathophysiological basis of dystonias. *Nature Reviews Neuroscience, 9*(3), 222–234. https://doi.org/10.1038/nrn2337.

Broocks, A., Thiel, A., Angerstein, D., & Dressler, D. (1998). Higher prevalence of obsessive-compulsive symptoms in patients with blepharospasm than in patients with hemifacial spasm. *American Journal of Psychiatry, 155*(4), 555–557.

Brüggemann, N., Stiller, S., Tadic, V., Kasten, M., Münchau, A., Graf, J., et al. (2014). Non-motor phenotype of dopa-responsive dystonia and quality of life assessment. *Parkinsonism & Related Disorders, 20*(4), 428–431. https://doi.org/10.1016/j.parkreldis.2013.12.014.

Camfield, L., Ben-Shlomo, Y., & Warner, T. T. (2002). Impact of cervical dystonia on quality of life. *Movement Disorders, 17*(4), 838–841. https://doi.org/10.1002/mds.10127.

Cano, S. J., Warner, T. T., Linacre, J. M., Bhatia, K. P., Thompson, A. J., Fitzpatrick, R., et al. (2004). Capturing the true burden of dystonia on patients: The cervical dystonia impact profile (CDIP-58). *Neurology, 63*(9), 1629–1633.

Cenina, A. R. F., dela Llana, P., Domingo, A., Jamora, R. D. G., Lee, P., & Lee, L. V. (2013). Depression among patients with X-linked dystonia parkinsonism [abstract]. *Movement Disorders*, 28(Suppl. 1), 3.

Chi, C.-S., Lee, H.-F., & Tsai, C.-R. (2012). Tyrosine hydroxylase deficiency in Taiwanese infants. *Pediatric Neurology, 46*(2), 77–82. https://doi.org/10.1016/j.pediatrneurol.2011.11.012.

Cloninger, C., Syrakic, D., & Przybeck, T. (1993). A psychobiological model of temperament and character. *Archives of General Psychiatry, 50*(12), 975–990.

Conte, A., Berardelli, I., Ferrazzano, G., Pasquini, M., Berardelli, A., & Fabbrini, G. (2016). Non-motor symptoms in patients with adult-onset focal dystonia: Sensory and psychiatric disturbances. *Parkinsonism & Related Disorders, 22*, S111–S114. https://doi.org/10.1016/j.parkreldis.2015.09.001.

Conte, A., Rocchi, L., Ferrazzano, G., Leodori, G., Bologna, M., Li Voti, P., et al. (2014). Primary somatosensory cortical plasticity and tactile temporal discrimination in focal hand dystonia. *Clinical Neurophysiology: Official Journal of the International Federation of Clinical Neurophysiology, 125*(3), 537–543. https://doi.org/10.1016/j.clinph.2013.08.006.

Cook, J. F., Hill, D. F., Snively, B. M., Boggs, N., Suerken, C. K., Haq, I., et al. (2014). Cognitive impairment in rapid-onset dystonia-parkinsonism. *Movement Disorders, 29*(3), 344–350. https://doi.org/10.1002/mds.25790.

Defazio, G., Berardelli, A., Abbruzzese, G., Lepore, V., Coviello, V., Acquistapace, D., et al. (1998). Possible risk factors for primary adult onset dystonia: A case-control investigation by the Italian movement disorders study group. *Journal of Neurology, Neurosurgery, and Psychiatry, 64*(1), 25–32. https://doi.org/10.1136/jnnp.64.1.25.

Defazio, G., Jankovic, J., Giel, J. L., & Papapetropoulos, S. (2013). Descriptive epidemiology of cervical dystonia. *Tremor and Other Hyperkinetic Movements (New York, N.Y.)*, 3. Retrieved from http://tremorjournal.org/article/view/193.

Delmaire, C., Krainik, A., Tezenas du Montcel, S., Gerardin, E., Meunier, S., Mangin, J.-F., et al. (2005). Disorganized somatotopy in the putamen of patients with focal hand dystonia. *Neurology, 64*(8), 1391–1396. https://doi.org/10.1212/01.WNL.0000158424.01299.76.

Dias, F. M. V., Doyle, F. C. P., Kummer, A., Cardoso, F., Caramelli, P., & Teixeira, A. L. (2009). Executive functioning in patients with blepharospasm in comparison with patients with hemifacial spasm. *Arquivos de Neuro-Psiquiatria, 67*(1), 12–15. https://doi.org/10.1590/S0004-282X2009000100004.

Dias, F. M., Doyle, F., Kummer, A., Cardoso, F., Fontenelle, L. F., & Teixeira, A. L. (2010). Frequency of psychiatric disorders in blepharospasm does not differ from hemifacial spasm. *Acta Neuropsychiatrica, 22*(5), 223–227. https://doi.org/10.1111/j.1601-5215.2009.00436.x.

Dobyns, W. B., Ozelius, L. J., Kramer, P. L., Brashear, A., Farlow, M. R., Perry, T. R., et al. (1993). Rapid-onset dystonia-parkinsonism. *Neurology, 43*(12), 2596–2602.

Doheny, D. O., Brin, M. F., Morrison, C. E., Smith, C. J., Walker, R. H., Abbasi, S., et al. (2002). Phenotypic features of myoclonus-dystonia in three kindreds. *Neurology, 59*(8), 1187–1196. https://doi.org/10.1212/WNL.59.8.1187.

Doheny, D., Danisi, F., Smith, C., Morrison, C., Velickovic, M., de Leon, D., et al. (2002). Clinical findings of a myoclonus-dystonia family with two distinct mutations. *Neurology, 59*(8), 1244–1246. https://doi.org/10.1212/WNL.59.8.1244.

Domingo, A., Ledesma, L. K., Pasco, P., Jamora, R. D. G., & Lee, L. V. (2011). Neuropsychological profile of X-linked dystonia parkinsonism (XDP, DYT3). *Movement Disorders*, 26(Suppl), S264.

Duane, D. D. (1988). Spasmodic torticollis. *Advances in neurology: Vol. 49* (pp. 135–150). United States: Raven Press.

Eichenseer, S. R., Stebbins, G. T., & Comella, C. L. (2014). Beyond a motor disorder: A prospective evaluation of sleep quality in cervical dystonia. *Parkinsonism & Related Disorders, 20*(4), 405–408. https://doi.org/10.1016/j.parkreldis.2014.01.004.

Evidente, V. G. H., Nolte, D., Niemann, S., Advincula, J., Mayo, M. C., Natividad, F. F., et al. (2004). Phenotypic and molecular analyses of X-linked dystonia-parkinsonism ("lubag") in women. *Archives of Neurology, 61*(12), 1956–1959. https://doi.org/10.1001/archneur.61.12.1956.

Evidente, V. G., et al. (2002). Novel neuropathological findings in Lubag or x-linked dystonia-parkinsonism. *Movement Disorders*, *17*, 294.

Fabbrini, G., Berardelli, I., Moretti, G., Pasquini, M., Bloise, M., Colosimo, C., et al. (2010). Psychiatric disorders in adult-onset focal dystonia: A case-control study. *Movement Disorders*, *25*(4), 459–465. https://doi.org/10.1002/mds.22983.

Fang, F., Valdimarsdóttir, U., Fürst, C. J., Hultman, C., Fall, K., Sparén, P., et al. (2008). Suicide among patients with amyotrophic lateral sclerosis. *Brain*, *131*(10), 2729–2733. https://doi.org/10.1093/brain/awn161.

Farrer, L. (1986). Suicide and attempted suicide in Huntington disease: Implications for preclinical testing of persons at risk. *American Journal of Medical Genetics*, *24*(2), 305–311.

Findley, L., Eichhorn, T., Janca, A., Kazenwadel, J., Baker, M., Currie-Gnjesda, D., et al. (2002). Factors impacting on quality of life in Parkinson's disease: Results from an international survey. *Movement Disorders*, *17*(1), 60–67. https://doi.org/10.1002/mds.10010.

Foncke, E. M. J., Cath, D., Zwinderman, K., Smit, J., Schmand, B., & Tijssen, M. (2009). Is psychopathology part of the phenotypic spectrum of myoclonus-dystonia?: A study of a large Dutch M-D family. *Cognitive and Behavioral Neurology*, *22*(2), 127–133. Retrieved from http://journals.lww.com/cogbehavneurol/Fulltext/2009/06000/Is_Psychopathology_Part_of_the_Phenotypic_Spectrum.9.aspx.

Fontenelle, L. F., Pacheco, P. G., Nascimento, P. M., de Freitas, A. R., Rosso, A. L., Teixeira, A. L., et al. (2011). Obsessive-compulsive symptoms among patients with blepharospasm and hemifacial spasm. *General Hospital Psychiatry*, *33*(5), 476–481. https://doi.org/10.1016/j.genhosppsych.2011.05.016.

Friedman, J., Roze, E., Abdenur, J. E., Chang, R., Gasperini, S., Saletti, V., et al. (2012). Sepiapterin reductase deficiency: A treatable mimic of cerebral palsy. *Annals of Neurology*, *71*(4), 520–530. https://doi.org/10.1002/ana.22685.

Frima, N., Nasir, J., & Grünewald, R. A. (2008). Abnormal vibration-induced illusion of movement in idiopathic focal dystonia: An endophenotypic marker? *Movement Disorders*, *23*(3), 373–377. https://doi.org/10.1002/mds.21838.

Ghilardi, M. F., Carbon, M., Silvestri, G., Dhawan, V., Tagliati, M., Bressman, S., et al. (2003). Impaired sequence learning in carriers of the DYT1 dystonia mutation. *Annals of Neurology*, *54*(1), 102–109. https://doi.org/10.1002/ana.10610.

Gil, J., Funalot, B., & Verschueren, A. (2008). Causes of death amongst French patients with amyotrophic lateral sclerosis: A prospective study. *European Journal of Neurology*, *5*(11), 1245–1251.

Goto, S., Lee, L. V., Munoz, E. L., Tooyama, I., Tamiya, G., Makino, S., et al. (2005). Functional anatomy of the basal ganglia in X-linked recessive dystonia-parkinsonism. *Annals of Neurology*, *58*(1), 7–17. https://doi.org/10.1002/ana.20513.

Gündel, H., Wolf, A., Xidara, V., Busch, R., & Gündel, H. (2001). Social phobia in spasmodic torticollis. *Journal of Neurology, Neurosurgery, and Psychiatry*, *71*(4), 499–504.

Gündel, H., Wolf, A., Xidara, V., Busch, R., Ladwig, K.-H., Jacobi, F., et al. (2003). High psychiatric comorbidity in spasmodic torticollis: A controlled study. *The Journal of Nervous and Mental Disease*, *191*(7), 465–473. https://doi.org/10.1097/01.NMD.0000081667.02656.21.

Hahn, H., Trant, M., Brownstein, M., Harper, R., Milstein, S., & Butlrer, I. (2001). Neurologic and psychiatric manifestations in a family with a mutation in exon 2 of the guanosine triphosphate-cyclohydrolase gene. *Archives of Neurology*, *5*, 749–755.

Hall, T. A., McGwin, G., Searcey, K., Xie, A., Hupp, S. L., Owsley, C., et al. (2005). Benign essential blepharospasm: Risk factors with reference to hemifacial spasm. *Journal of Neuro-Ophthalmology: The Official Journal of the North American Neuro-Ophthalmology Society*, *25*(4), 280–285. https://doi.org/10.1097/01.wno.0000189063.43423.04.

Heiman, G. A., Ottman, R., Saunders-Pullman, R. J., Ozelius, L. J., Risch, N. J., & Bressman, S. B. (2004). Increased risk for recurrent major depression in DYT1 dystonia mutation carriers. *Neurology*, *63*, 631–637. https://doi.org/10.1212/WNL.64.10.1821-a.

Heiman, G. A., Ottman, R., Saunders-Pullman, R. J., Ozelius, L. J., Risch, N. J., & Bressman, S. B. (2007). Obsessive-compulsive disorder is not a clinical manifestation of the DYT1 dystonia gene. *American Journal of Medical Genetics. Part B: Neuropsychiatric Genetics*, *144*(3), 361–364. https://doi.org/10.1002/ajmg.b.30431.

Hess, C. W., Raymond, D., Aguiar, P. D. C., Frucht, S., Shriberg, J., Heiman, G. A., et al. (2007). Myoclonus-dystonia, obsessive-compulsive disorder, and alcohol dependence in SGCE mutation carriers. *Neurology*, *68*(7), 522–524. https://doi.org/10.1212/01.wnl.0000253188.76092.06.

Howe, L. L. S., Kellison, I. L., Fernandez, H. H., Okun, M. S., & Bowers, D. (2009). Neuropsychological profile of a Filipino gentleman with X-linked dystonia-parkinsonism: A case report of Lubag disease. *The Clinical Neuropsychologist*, *23*, 100–117. https://doi.org/10.1080/13854040801894714.

Huang, C. L., Lan, M. Y., Chang, Y. Y., Hsu, C. Y., Lai, S. C., Chen, R. S., et al. (2010). Large SGCE deletion contributes to Taiwanese myoclonus-dystonia syndrome. *Parkinsonism & Related Disorders*, *16*(9), 585–589. https://doi.org/10.1016/j.parkreldis.2010.06.016.

Jahanshahi, M. (1991). Psychosocial factors and depression in torticollis. *Journal of Psychosomatic Research*, *35*(4–5), 493–507.

Jahanshahi, M., & Marsden, C. D. (1990a). A longitudinal follow-up study of depression, disability, and body concept in torticollis. In *Behavioural Neurology: Vol. 3* (pp. 233–246). Netherlands: IOS Press. https://doi.org/10.3233/BEN-1990-3403.

Jahanshahi, M., & Marsden, C. D. (1990b). Body concept, disability, and depression in patients with spasmodic torticollis. *Behavioural Neurology*, *3*(2), 117–131. https://doi.org/10.3233/BEN-1990-3206.

Jahanshahi, M., Torkamani, M., Beigi, M., Wilkinson, L., Page, D., Madeley, L., et al. (2014). Pallidal stimulation for primary generalised dystonia: Effect on cognition, mood and quality of life. *Journal of Neurology*, *261*(1), 164–173. https://doi.org/10.1007/s00415-013-7161-2.

Jamora, R. D. G., Cenina, A. R. F., Teleg, R. A., & Lee, L. V. (2015). Suicidality among patients with sex-linked dystonia-Parkinsonism (XDP). *Acta Medica Philippina*, *49*(1), 20–23.

Jamora, R. D. G., Ledesma, L. K., Domingo, A., Cenina, A. R. F., & Lee, L. V. (2014). Nonmotor features in sex-linked dystonia parkinsonism. *Neurodegenerative Disease Management*, *4*, 283–289. https://doi.org/10.2217/nmt.14.16.

Jankovic, J., Leder, S., Warner, D., & Schwartz, K. (1991). Cervical dystonia: Clinical findings and associated movement disorders. *Neurology*, *41*(7), 1088–1091.

Kemmotsu, N., Price, C. C., Oyama, G., Okun, M. S., Kelly, D., Howe, L. L. S., et al. (2011). Pre- and post-GPi DBS neuropsychologic profiles in a case of X-linked dystonia parkinsonism. *The Clinical Neuropsychologist*, *25*(1), 141–159. https://doi.org/10.1080/13854046.2010.532812.

Klein, C., Liu, L., Doheny, D., Kock, N., Muller, B., de Carvalho Aguiar, P., et al. (2002). Epsilon-sarcoglycan mutations found in combination with other dystonia gene mutations. *Annals of Neurology*, *52*(5), 675–679. https://doi.org/10.1002/ana.10358.

Klingelhoefer, L., Martino, D., Martinez-Martin, P., Sauerbier, A., Rizos, A., Jost, W., et al. (2014). Nonmotor symptoms and focal cervical dystonia: Observations from 102 patients. *Basal Ganglia*, *4*(3–4), 117–120. https://doi.org/10.1016/j.baga.2014.10.002.

Kostić, V. S., Pekmezović, T., Tomić, A., Ječmenica-Lukić, M., Stojković, T., Špica, V., et al. (2010). Suicide and suicidal ideation in Parkinson's disease. *Journal of the Neurological Sciences*, *289*(1), 40–43. https://doi.org/10.1016/j.jns.2009.08.016.

Kutvonen, O., Dastidar, P., & Nurmikko, T. (1997). Pain in spasmodic torticollis. *Pain, 69*(3), 279–286. https://doi.org/10.1016/S0304-3959(96)03296-4.

Kuyper, D. J., Parra, V., Aerts, S., Okun, M. S., & Kluger, B. M. (2011). The non-motor manifestations of dystonia: A systematic review. *Movement Disorders, 26*(7), 1206–1217. https://doi.org/10.1002/mds.23709.

Lauterbach, E. C., Freeman, A., & Vogel, R. L. (2004). Differential DSM-III psychiatric disorder prevalence profiles in dystonia and Parkinson's disease. *The Journal of Neuropsychiatry and Clinical Neurosciences, 16*(1), 29–36. https://doi.org/10.1176/jnp.16.1.29.

Lee, L. V., Kupke, K. G., Caballar-Gonzaga, F., Hebron-Ortiz, M., & Muller, U. (1991). The phenotype of the X-linked dystonia-parkinsonism syndrome. An assessment of 42 cases in the Philippines. *Medicine, 70*(3), 179–187.

Lee, L. V., Maranon, E., Demaisip, C., Peralta, O., Borres-Icasiano, R., Arancillo, J., et al. (2002). The natural history of sex-linked recessive dystonia parkinsonism of Panay, Philippines (XDP). *Parkinsonism & Related Disorders, 9*(1), 29–38. https://doi.org/10.1016/S1353-8020(02)00042-1.

Lee, L. V., Munoz, E. L., Tan, K. T., & Reyes, M. T. (2001). Sex linked recessive dystonia parkinsonism of Panay, Philippines (XDP). *Molecular Pathology: MP, 54*(6), 362–368.

Lee, A. M., Oleson, E. B., Diergaarde, L., Cheer, J. F., & Pattij, T. (2012). Cannabinoids and value-based decision making: Implications for neurodegenerative disorders. *Basal Ganglia, 2*, 131–138. https://doi.org/10.1016/j.baga.2012.06.005.

Lee, L. V., Rivera, C., Teleg, R. A., Dantes, M. B., Pasco, P. M. D., Jamora, R. D. G., et al. (2011). The unique phenomenology of sex-linked dystonia parkinsonism (XDP, DYT3, "Lubag"). *International Journal of Neuroscience, 121*(Suppl. 1), 3–11. https://doi.org/10.3109/00207454.2010.526728.

Lehéricy, S., Tijssen, M. A. J., Vidailhet, M., Kaji, R., & Meunier, S. (2013). The anatomical basis of dystonia: Current view using neuroimaging. *Movement Disorders, 28*(7), 944–957. https://doi.org/10.1002/mds.25527.

Lencer, R., Steinlechner, S., Stahlberg, J., Rehling, H., Orth, M., Baeumer, T., et al. (2009). Primary focal dystonia: Evidence for distinct neuropsychiatric and personality profiles. *Journal of Neurology, Neurosurgery, and Psychiatry, 80*(10), 1176–1179. https://doi.org/10.1136/jnnp.2008.170191.

Lewis, L., Butler, A., & Jahanshahi, M. (2008). Depression in focal, segmental and generalized dystonia. *Journal of Neurology, 255*(11), 1750–1755. https://doi.org/10.1007/s00415-008-0020-x.

Limousin, P., Krack, P., Pollak, P., Benazzouz, A., Ardouin, C., & Hoffman, D. (1998). Electrical stimulation of the subthalamic nucleus in advanced Parkinson's disease. *New England Journal of Medicine, 339*(16), 1105–1111.

Linden, A.-M., & Schoepp, D. D. (2006). Metabotropic glutamate receptor targets for neuropsychiatric disorders. *Drug Discovery Today: Therapeutic Strategies, 3*, 507–517. https://doi.org/10.1016/j.ddstr.2006.10.018.

Lobbezoo, F., Tanguay, R., Thu Thon, M., & Lavigne, G. J. (1996). Pain perception in idiopathic cervical dystonia (spasmodic torticollis). *Pain, 67*(2), 483–491. 10.1016/0304-3959(96)03153-3.

Lobbezoo, F., Thu Thon, M., Remillard, G., Montplaisir, J. Y., & Lavigne, G. J. (1996). Relationship between sleep, neck muscle activity, and pain in cervical dystonia. *The Canadian Journal of Neurological Sciences, 23*(4), 285–290.

Loetscher, T., McDonnell, M., & Bradnam, L. (2015). Impaired cognitive functioning in cervical dystonia. *Frontiers in Human Neuroscience, 271*. https://doi.org/10.3389/conf.fnhum.2015.217.00271.

Löhle, M., Storch, A., & Reichmann, H. (2009). Beyond tremor and rigidity: Non-motor features of Parkinson's disease. *Journal of Neural Transmission*, *116*(11), 1483. https://doi.org/10.1007/s00702-009-0274-1.

López-Laso, E., Sánchez-Raya, A., Moriana, J. A., Martínez-Gual, E., Camino-León, R., Mateos-González, M. E., et al. (2011). Neuropsychiatric symptoms and intelligence quotient in autosomal dominant Segawa disease. *Journal of Neurology*, *258*(12), 2155–2162. https://doi.org/10.1007/s00415-011-6079-9.

Lowenstein, D. H., & Aminoff, M. J. (1988). The clinical course of spasmodic torticollis. *Neurology*, *38*(4), 530. https://doi.org/10.1212/WNL.38.4.530.

Makino, S., Kaji, R., Ando, S., Tomizawa, M., Yasuno, K., Goto, S., et al. (2007). Reduced neuron-specific expression of the TAF1 gene is associated with X-linked dystonia-parkinsonism. *American Journal of Human Genetics*, *80*, 393–406.

Maréchal, L., Raux, G., Dumanchin, C., Lefebvre, G., Deslandre, E., Girard, C., et al. (2003). Severe myoclonus-dystonia syndrome associated with a novel epsilon-sarcoglycan gene truncating mutation. *American Journal of Medical Genetics. Part B, Neuropsychiatric Genetics: The Official Publication of the International Society of Psychiatric Genetics*, *119B*(1), 114–117. https://doi.org/10.1002/ajmg.b.10062.

Martino, D., Defazio, G., Alessio, G., Abbruzzese, G., Girlanda, P., Tinazzi, M., et al. (2005). Relationship between eye symptoms and blepharospasm: A multicenter case-control study. *Movement Disorders*, *20*(12), 1564–1570. https://doi.org/10.1002/mds.20635.

Menzies, L., Chamberlain, S. R., Laird, A. R., Thelen, S. M., Sahakian, B. J., & Bullmore, E. T. (2008). Integrating evidence from neuroimaging and neuropsychological studies of obsessive-compulsive disorder: The orbitofronto-striatal model revisited. *Neuroscience and Biobehavioral Reviews*, *32*(3), 525–549. https://doi.org/10.1016/j.neubiorev.2007.09.005.

Miller, K. M., Okun, M. S., Fernandez, H. F., Jacobson, C. E., IV, Rodriguez, R. L., & Bowers, D. (2007). Depression symptoms in movement disorders: Comparing Parkinson's disease, dystonia, and essential tremor. *Movement Disorders*, *22*, 666–672. https://doi.org/10.1002/mds.21376.

Molloy, F. M., Carr, T. D., Zeuner, K. E., Dambrosia, J. M., & Hallett, M. (2003). Abnormalities of spatial discrimination in focal and generalized dystonia. *Brain: A Journal of Neurology*, *126*(Pt. 10), 2175–2182. https://doi.org/10.1093/brain/awg219.

Moraru, E., Schnider, P., Wimmer, A., Wenzel, T., Birner, P., Griengl, H., et al. (2002). Relation between depression and anxiety in dystonic patients: Implications for clinical management. *Depression and Anxiety*, *16*(3), 100–103. https://doi.org/10.1002/da.10039.

Morigaki, R., Nakataki, M., Kawarai, T., Lee, L. V., Teleg, R. A., Tabuena, M. D. P., et al. (2013). Depression in X-linked dystonia-parkinsonism: A case-control study. *Parkinsonism & Related Disorders*, *19*(9), 844–846. https://doi.org/10.1016/j.parkreldis.2013.04.027.

Moseley, A. E., Williams, M. T., Schaefer, T. L., Bohanan, C. S., Neumann, J. C., Behbehani, M. M., et al. (2007). Deficiency in Na,K-ATPase alpha isoform genes alters spatial learning, motor activity, and anxiety in mice. *The Journal of Neuroscience: The Official Journal of the Society for Neuroscience*, *27*(3), 616–626. https://doi.org/10.1523/JNEUROSCI.4464-06.2007.

Muller, J., Kemmler, G., Wissel, J., Schneider, A., Voller, B., Grossmann, J., et al. (2002). The impact of blepharospasm and cervical dystonia on health-related quality of life and depression. *Journal of Neurology*, *249*(7), 842–846. https://doi.org/10.1007/s00415-002-0733-1.

Munhoz, R. P., Teive, H. A. G., Coletta, M. V. D., Germiniani, F. M. B., Iwamoto, F. M., Camargo, C. H. F., et al. (2005). Frequency of obsessive and compulsive symptoms in

patients with blepharospasm and hemifacial spasm. *Arquivos de Neuro-Psiquiatria, 63*(2), 213–216. https://doi.org/10.1590/S0004-282X2005000300037.
Murray, R. M., Lappin, J., & Di Forti, M. (2017). Schizophrenia: From developmental deviance to dopamine dysregulation. *European Neuropsychopharmacology, 18*, S129–S134. https://doi.org/10.1016/j.euroneuro.2008.04.002.
Naumann, M., Becker, G., Toyka, K. V., Supprian, T., & Reiners, K. (1996). Lenticular nucleus lesion in idiopathic dystonia detected by transcranial sonography. *Neurology, 47*(5), 1284–1290.
Nitschke, M. F., Erdmann, C., Trillenberg, P., Sprenger, A., Kock, N., Sperner, J., et al. (2006). Functional MRI reveals activation of a subcortical network in a 5-year-old girl with genetically confirmed myoclonus-dystonia. *Neuropediatrics, 37*(2), 79–82. https://doi.org/10.1055/s-2006-924109.
Nolte, D., Niemann, S., & Müller, U. (2003). Specific sequence changes in multiple transcript system DYT3 are associated with X-linked dystonia parkinsonism. *Proceedings of the National Academy of Sciences of the United States of America, 100*(18), 10347–10352. https://doi.org/10.1073/pnas.1831949100.
Okun, M. S., Fernandez, H. H., Wu, S. S., Kirsch-Darrow, L., Bowers, D., Bova, F., et al. (2009). Cognition and mood in Parkinson's disease in subthalamic nucleus versus globus pallidus interna deep brain stimulation: The COMPARE trial. *Annals of Neurology, 65*(5), 586–595. https://doi.org/10.1002/ana.21596.
Ozelius, L. J., & Bressman, S. B. (2011). Genetic and clinical features of primary torsion dystonia. *Neurobiology of Disease, 42*(2), 127–135. https://doi.org/10.1016/j.nbd.2010.12.012.
Papapetropoulos, S., Argyriou, A. A., Polychronopoulos, P., Spyridonidis, T., Gourzis, P., & Chroni, E. (2008). Frontotemporal and striatal SPECT abnormalities in myoclonus-dystonia: Phenotypic and pathogenetic considerations. *Neurodegenerative Diseases, 5*(6), 355–358. https://doi.org/10.1159/000119458.
Papathanasiou, I., MacDonald, L., Whurr, R., & Jahanshahi, M. (2001). Perceived stigma in spasmodic torticollis. *Movement Disorders, 16*(2), 280–285. Retrieved from http://www.ncbi.nlm.nih.gov/pubmed/11295781.
Pasco, P. M. D., Ison, C. V., Muñoz, E. L., Magpusao, N. S., Cheng, A. E., Tan, K. T., et al. (2011). Understanding XDP through imaging, pathology, and genetics. *The International Journal of Neuroscience, 121*(Suppl. 1), 12–17.
Paus, S., Gross, J., Moll-Müller, M., Hentschel, F., Spottke, A., Wabbels, B., et al. (2011). Impaired sleep quality and restless legs syndrome in idiopathic focal dystonia: A controlled study. *Journal of Neurology, 258*(10), 1835–1840. https://doi.org/10.1007/s00415-011-6029-6.
Peall, K. J., Smith, D. J., Kurian, M. A., Wardle, M., Waite, A. J., Hedderly, T., et al. (2013). SGCE mutations cause psychiatric disorders: Clinical and genetic characterization. *Brain: A Journal of Neurology, 136*(Pt 1), 294–303. https://doi.org/10.1093/brain/aws308.
Peall, K. J., Waite, A. J., Blake, D. J., Owen, M. J., & Morris, H. R. (2011). Psychiatric disorders, myoclonus dystonia, and the epsilon-sarcoglycan gene: A systematic review. *Movement Disorders, 26*(10), 1939–1942. https://doi.org/10.1002/mds.23791.
Pekmezovic, T., Svetel, M., Ivanovic, N., Dragasevic, N., Petrovic, I., Tepavcevic, D. K., et al. (2009). Quality of life in patients with focal dystonia. *Clinical Neurology and Neurosurgery, 111*(2), 161–164. https://doi.org/10.1016/j.clineuro.2008.09.023.
Putzki, N., Stude, P., Konczak, J., Graf, K., Diener, H.-C., & Maschke, M. (2006). Kinesthesia is impaired in focal dystonia. *Movement Disorders, 21*(6), 754–760. https://doi.org/10.1002/mds.20799.
Ritz, K., van Schaik, B. D., Jakobs, M. E., van Kampen, A. H., Aronica, E., Tijssen, M. a., et al. (2011). SGCE isoform characterization and expression in human brain: Implications for myoclonus-dystonia pathogenesis? *European Journal of Human Genetics: EJHG, 19*(4), 438–444. https://doi.org/10.1038/ejhg.2010.206.

Romano, R., Bertolino, A., Gigante, A., Martino, D., Livrea, P., & Defazio, G. (2014). Impaired cognitive functions in adult-onset primary cranial cervical dystonia. *Parkinsonism & Related Disorders*, *20*(2), 162–165. https://doi.org/10.1016/j.parkreldis.2013.10.008.

Rome, S., & Grünewald, R. a. (1999). Abnormal perception of vibration-induced illusion of movement in dystonia. *Neurology*, *53*, 1794–1800. https://doi.org/10.1212/WNL.53.8.1794.

Rosales, R. L. (2010). X-linked dystonia parkinsonism: Clinical phenotype, genetics and therapeutics. *Journal of Movement Disorders*, *3*(2), 32–38. https://doi.org/10.14802/jmd.10009.

Roze, E., Apartis, E., Clot, F., Dorison, N., Thobois, S., Guyant-Marechal, L., et al. (2008). Myoclonus-dystonia: Clinical and electrophysiologic pattern related to SGCE mutations. *Neurology*, 70(13 Pt. 1), 1010–1016. https://doi.org/10.1212/01.wnl.0000297516.98574.c0.

Sadahiro, R., Suzuki, A., Matsumoto, Y., Shibuya, N., Enokido, M., Kamata, M., et al. (2011). Functional polymorphism of the GTP cyclohydrolase 1 gene affects the personality trait of novelty seeking in healthy subjects. *Neuroscience Letters*, *503*, 220–223.

Saunders-Pullman, R., Shriberg, J., Heiman, G., Raymond, D., Wendt, K., Kramer, P., et al. (2002). Myoclonus dystonia: Possible association with obsessive-compulsive disorder and alcohol dependence. *Neurology*, *58*(2), 242–245. https://doi.org/10.1212/WNL.58.2.242.

Scheidt, C. E., Schuller, B., Rayki, O., Kommerell, G., & Deuschl, G. (1996). Relative absence of psychopathology in benign essential blepharospasm and hemifacial spasm. *Neurology*, *47*(1), 43–45.

Sforza, E., Montagna, P., Defazio, G., & Lugaresi, E. (1991). Sleep and cranial dystonia. *Electroencephalography and Clinical Neurophysiology*, *79*(3), 166–169. https://doi.org/10.1016/0013-4694(91)90135-Q.

Silvestri, R., De Domenico, P., Di Rosa, A. E., Bramanti, P., Serra, S., & Di Perri, R. (1990). The effect of nocturnal physiological sleep on various movement disorders. *Movement Disorders*, *5*(1), 8–14. https://doi.org/10.1002/mds.870050104.

Slawek, J., Friedman, A., Potulska, A., Krystkowiak, P., Gervais, C., Banach, M., et al. (2007). Factors affecting the health-related quality of life of patients with cervical dystonia and the impact of botulinum toxin type A injections. *Functional Neurology*, *22*(2), 95–100.

Smit, M., Kuiper, A., Han, V., Jiawan, V. C. R., Douma, G., van Harten, B., et al. (2016). Psychiatric co-morbidity is highly prevalent in idiopathic cervical dystonia and significantly influences health-related quality of life: Results of a controlled study. *Parkinsonism & Related Disorders*, *30*, 7–12. https://doi.org/10.1016/j.parkreldis.2016.06.004.

Smit, M., Kuiper, A., Han, V., Jiawan, V. C. R., Douma, G., van Harten, B., et al. (2017). Psychiatric co-morbidity is highly prevalent in idiopathic cervical dystonia and significantly influences health-related quality of life: Results of a controlled study. *Parkinsonism & Related Disorders*, *30*, 7–12. https://doi.org/10.1016/j.parkreldis.2016.06.004.

Soeder, A., Kluger, B. M., Okun, M. S., Garvan, C. W., Soeder, T., Jacobson, C. E., et al. (2009). Mood and energy determinants of quality of life in dystonia. *Journal of Neurology*, *256*(6), 996–1001. https://doi.org/10.1007/s00415-009-5060-3.

Sperlágh, B., Windisch, K., Andó, R. D., & Vizi, E. S. (2009). Neurochemical evidence that stimulation of {CB1} cannabinoid receptors on {GABAergic} nerve terminals activates the dopaminergic reward system by increasing dopamine release in the rat nucleus accumbens. *Neurochemistry International*, *54*(7), 452–457. https://doi.org/http://doi.org/10.1016/j.neuint.2009.01.017.

Stamelou, M., Edwards, M. J., Hallett, M., & Bhatia, K. P. (2012). The non-motor syndrome of primary dystonia: Clinical and pathophysiological implications. *Brain*, *135*(6), 1668–1681. https://doi.org/10.1093/brain/awr224.

Sumi-Ichinose, C., Ichinose, H., Ikemoto, K., Nomura, T., & Kondo, K. (2010). Advanced research on dopamine signaling to develop drugs for the treatment of mental disorders: Regulation of dopaminergic neural transmission by tyrosine hydroxylase protein at nerve terminals. *Journal of Pharmacological Sciences*, *114*, 17–24.

Sunga, M. A. P., & Rosales, R. L. (2014). Mental dysfunctions in dystonia-plus syndromes. *Journal of Parkinson's Disease*, *4*(2), 161–167. https://doi.org/10.3233/JPD-130283.

Supnet, M., & Rosales, R. L. (2017). Assessment of autonomic nervous system dysfunction in X-linked dystonia-parkinsonism using sympathetic skin response (SSR) and symptoms questionnaire: A pilot study [abstract]. *Neurology*, *88*(16 Suppl), P3.025.

Tadic, V., Kasten, M., Brüggemann, N., Stiller, S., Hagenah, J., & Klein, C. (2012). Dopa-responsive dystonia revisited: Diagnostic delay, residual signs, and nonmotor signs. *Archives of Neurology*, *69*(12), 1558–1562. Retrieved from http://dx.doi.org/10.1001/archneurol.2012.574.

Tai, C.-H., Yen, R.-F., Lin, C.-H., Yen, K.-Y., Yip, P.-K., Wu, R.-M., et al. (2017). Focal brain glucose hypermetabolism in myoclonus-dystonia syndrome caused by an epsilon-sarcoglycan gene mutation. *Parkinsonism & Related Disorders*, *15*(8), 614–616. https://doi.org/10.1016/j.parkreldis.2008.11.007.

Tamura, Y., Matsuhashi, M., Lin, P., Ou, B., Vorbach, S., Kakigi, R., et al. (2008). Impaired intracortical inhibition in the primary somatosensory cortex in focal hand dystonia. *Movement Disorders*, *23*(4), 558–565. https://doi.org/10.1002/mds.21870.

Tinazzi, M., Rosso, T., & Fiaschi, A. (2003). Role of the somatosensory system in primary dystonia. *Movement Disorders*, *18*, 605–622. https://doi.org/10.1002/mds.10398.

Tiple, D., Strano, S., Colosimo, C., Fabbrini, G., Calcagnini, G., Prencipe, M., et al. (2008). Autonomic cardiovascular function and baroreflex sensitivity in patients with cervical dystonia receiving treatment with botulinum toxin type a. *Journal of Neurology*, *255*(6), 843–847. https://doi.org/10.1007/s00415-008-0753-6.

Trotti, L. M., Esper, C. D., Feustel, P. J., Bliwise, D. L., & Factor, S. A. (2009). Excessive daytime sleepiness in cervical dystonia. *Parkinsonism & Related Disorders*, *15*(10), 784–786. https://doi.org/10.1016/j.parkreldis.2009.04.007.

Van Hove, J. L. K., Steyaert, J., Matthijs, G., Legius, E., Theys, P., Wevers, R., et al. (2005). Expanded motor and psychiatric phenotype in autosomal dominant Segawa syndrome due to GTP cyclohydrolase deficiency. *Journal of Neurology, Neurosurgery & Psychiatry*, 77(1), 18–23. Retrieved from http://jnnp.bmj.com/content/77/1/18.abstract.

van Tricht, M. J., Dreissen, Y. E. M., Cath, D., Dijk, J. M., Contarino, M. F., van der Salm, S. M., et al. (2012). Cognition and psychopathology in myoclonus-dystonia. *Journal of Neurology, Neurosurgery, and Psychiatry*, *83*(8), 814–820. https://doi.org/10.1136/jnnp-2011-301386.

Voon, V., Butler, T. R., Ekanayake, V., Gallea, C., Ameli, R., Murphy, D. L., et al. (2010). Psychiatric symptoms associated with focal hand dystonia. *Movement Disorders*, *25*(13), 2249–2252. https://doi.org/10.1002/mds.23250.

Wadia, P. M., Lim, S.-Y., Lozano, A. M., Adams, J. R., Poon, Y.-Y., Torres Diaz, C. V., et al. (2010). Bilateral pallidal stimulation for X-linked dystonia parkinsonism. *Archives of Neurology*, *67*(8), 1012–1015. https://doi.org/10.1001/archneurol.2010.187.

Walsh, R. A., Whelan, R., O'Dwyer, J., O'Riordan, S., Hutchinson, S., O'Laoide, R., et al. (2009). Striatal morphology correlates with sensory abnormalities in unaffected relatives of cervical dystonia patients. *Journal of Neurology*, *256*(8), 1307–1313. https://doi.org/10.1007/s00415-009-5119-1.

Walter, B. L., & Vitek, J. L. (2004). Surgical treatment for Parkinson's disease. *The Lancet. Neurology*, *3*(12), 719–728. https://doi.org/10.1016/S1474-4422(04)00934-2.

Weissbach, A., Kasten, M., Grünewald, A., Brüggemann, N., Trillenberg, P., Klein, C., et al. (2013). Prominent psychiatric comorbidity in the dominantly inherited movement disorder myoclonus-dystonia. *Parkinsonism & Related Disorders, 19*(4), 422–425. https://doi.org/10.1016/j.parkreldis.2012.12.004.

Wenzel, T., Schnider, P., Wimmer, A., Steinhoff, N., Moraru, E., & Auff, E. (1998). Psychiatric comorbidity in patients with spasmodic torticollis. *Journal of Psychosomatic Research, 44*(6), 687–690. https://doi.org/10.1016/S0022-3999(97)00229-8.

Whurr, R., Nye, C., & Lorch, M. (1998). Meta-analysis of botulinum toxin treatment of spasmodic dysphonia: A review of 22 studies. *International Journal of Language & Communication Disorders, 33*(Suppl), 327–329.

Wichmann, T., & Delong, M. R. (2006). Deep brain stimulation for neurologic and neuropsychiatric disorders. *Neuron, 52*(1), 197–204. https://doi.org/10.1016/j.neuron.2006.09.022.

Wolters, E. C. (2009). Non-motor extranigral signs and symptoms in Parkinson's disease. *Parkinsonism & Related Disorders, 15*, S6–S12. https://doi.org/10.1016/S1353-8020(09)70770-9.

Yang, J., Song, W., Wei, Q., Ou, R., Cao, B., Liu, W., et al. (2016). Screening for cognitive impairments in primary blepharospasm. *PLoS One, 11*(8), e0160867. https://doi.org/10.1371/journal.pone.0160867.

Yokoi, F., Dang, M. T., Zhou, T., & Li, Y. (2012). Abnormal nuclear envelopes in the striatum and motor deficits in DYT11 myoclonus-dystonia mouse models. *Human Molecular Genetics, 21*(4), 916–925. https://doi.org/10.1093/hmg/ddr528.

CHAPTER FORTY-NINE

Nonmotor Symptoms in Essential Tremor and Other Tremor Disorders

Alessandro F. Fois*[,2], Hugo M. Briceño*[,2], Victor S.C. Fung*[,†,1]
*Movement Disorders Unit, Westmead Hospital, Sydney, NSW, Australia
†Sydney Medical School, University of Sydney, Sydney, NSW, Australia
[1]Corresponding author: e-mail address: vscfung@ozemail.com.au

Contents

Abstract

Tremor, like dystonia, is a term used at the phenomenological, syndromic, and aetio-pathological level. Parkinsonian, essential, and dystonic tremor are the three most common tremor diagnoses encountered in clinical practice. Investigation of nonmotor

[2] Both authors contributed equally.

International Review of Neurobiology, Volume 134
ISSN 0074-7742
http://dx.doi.org/10.1016/bs.irn.2017.05.010

symptoms in essential tremor and dystonic tremor syndromes is significantly hampered by the lack of clear clinical diagnostic criteria for these groups at a syndromic level, and the absence of biomarkers which allow definitive diagnosis at an aetiopathological level. Much work is needed in clarifying the motor features of these disorders in order to allow delineation of the nonmotor features of the most common tremor syndromes. With this limitation in mind, this chapter reviews what is known about nonmotor symptoms in these two tremor types. The final sections deal with nonmotor symptoms observed in patients with lesional tremor, thankfully a much more clearly defined albeit less common group of patients.

1. INTRODUCTION

Tremor, like dystonia, is a term used at the phenomenological, syndromic, and aetiopathological level. Parkinsonian, essential, and dystonic tremor are the three most common tremor diagnoses encountered in clinical practice. The nonmotor symptoms (NMS) of Parkinson's disease (PD) will not be covered in this chapter. Investigation of NMS in essential tremor (ET) and dystonic tremor syndromes is significantly hampered by the lack of clear clinical diagnostic criteria for these groups at a syndromic level, and the absence of biomarkers which allow definitive diagnosis at an aetiopathological level. Much work is needed in clarifying the motor features of these disorders in order to allow delineation of the nonmotor features of the most common tremor syndromes. Much of the data that are reviewed in this section need to be interpreted with these limitations in mind. The final sections deal with NMS observed in patients with lesional tremor, thankfully a much more clearly defined albeit less common group of patients.

2. ESSENTIAL TREMOR

ET derives its name from the absence of other neurological features in this condition. This putatively monosymptomatic, often familial, syndrome of tremor has been recognized as disease for centuries. In one of the earliest systematic reports entitled "Hereditary tremor," Dana describes "the affection in question consists of a fine tremor, constantly present in waking hours in typical cases during waking hours, voluntarily controlled for a brief time, affecting nearly all the voluntary muscles, chronic, beginning in very early life, not progressive, not shortening life, *not accompanied with paralysis or any other disturbance of nervous function*" (Dana, 1887). In a later description of "Essential (Heredofamilial) Tremor," Critchley also notes the lack of any other accompanying neurological dysfunction, going so far as to declare "in its nature as a constitutional monosymptomatic peculiarity it can scarcely be regarded as a 'disease'" (Critchley, 1949). By the 1990s the monosymptomatic nature of ET was codified in a consensus statement of the Movement Disorder Society on Tremor by the definition of "other abnormal neurological signs" as an exclusion criteria for the diagnosis (Deuschl, Bain, & Brin, 1998).

This seeming lack of nonmotor features and the absence of uniform pathological hallmarks led some to consider ET a functional syndrome rather than a disease with a single well-defined etiology. However, just as the nonmotor phenomena of PD were initially overlooked ("the senses and intellect being uninjured"; Parkinson, 1817) only to later emerge as a major source of morbidity and disability, so too has close observation of ET unmasked a number of neurological features extrinsic to the tremor itself. The observation of other motor phenomena (such as gait impairment Singer, Sanchez-Ramos, & Weiner, 1994, dysmetria Deuschl, Wenzelburger, Loffler, Raethjen, & Stolze, 2000, impaired conditioning of the blink reflex Kronenbuerger, Gerwig, Brol, Block, & Timmann, 2007, and eye movement abnormalities Helmchen et al., 2003) was closely followed by an increasing appreciation of nonmotor manifestations including cognitive, psychiatric, sleep, and sensory abnormalities.

The lack of a diagnostic gold standard for ET and the ongoing controversy surrounding the status of ET as a functional (i.e., caused by abnormal neuronal oscillation; Deuschl & Elble, 2009) vs a neurodegenerative disease (as suggested by some postmortem data; Louis, 2010) make the interpretation of nonmotor features difficult. Are they evidence that ET is a single neurodegenerative disease or do they reflect that tremor can be a common feature of a number of heterogeneous disease processes which are labeled ET if tremor is the salient feature? Herein we discuss the reported nonmotor features of ET as well as the pathological and radiological data that give insight into the mechanisms that underpin them.

2.1 Cognitive Deficit

With the growing appreciation of nonmotor features in PD, and the advent of deep brain stimulation (DBS) for the treatment of ET in the late 1990s (Hubble et al., 1996) prompting preoperative neuropsychiatric assessment of ET patients, a number of systematic studies of cognition in ET began to emerge in the early 21st century. Gasparini et al. compared cohorts of patients with ET (and a family history of either ET or PD), PD, and controls using a neuropsychiatric panel of tests assessing frontal lobe function and demonstrated significantly worse performance in the Stroop test, Wisconsin Card Sorting Test (WCST), and verbal fluency test in the patients with hereditary ET compared to controls (Gasparini et al., 2001). In a study of ET and PD patients being assessed for DBS, Lombardi et al. performed a more comprehensive neuropsychological battery that assessed a number of cognitive domains as well as mood (Lombardi, Woolston, Roberts, & Gross, 2001). This study indicated a number of domains of underperformance of ET patients relative to controls, including verbal fluency, naming, mental set-shifting, and verbal memory (as well as a few domains of outperformance in ET), but the only relationships to survive the Bonferroni correction were decreased verbal fluency (for letters and animals) and increased depression. Interestingly, the authors also found no correlation between tremor scores and medication use with performance on the neuropsychiatric battery, suggesting that poor performance was mediated by cognition rather than tremor. Medeiros et al. also reported worse verbal fluency (animals) and memory (Brown–Peterson task) in ET, as well as impaired manual dexterity (Pegboard test) (Medeiros et al., 2016).

The results of these small, neuropsychology studies have been borne out in larger epidemiological studies that have confirmed an association between ET and dementia. The Neurological Disorders in Central Spain (NEDICES) Study has made a substantial contribution to this field. In NEDICES, a population-based study of older people in central Spain, 7.8% of 206 ET cases (clinically confirmed by the investigators) developed incident dementia over a mean follow-up of 1.7 years, compared with 3.9% of controls (RR 2.08) (Bermejo-Pareja, Louis, & Benito-Leon, 2007). Interestingly, when the cases were stratified by age of onset of ET, the elevated risk of dementia was confined to the subgroup with older onset (at age >65 years old) (Benito-Leon, Louis, & Bermejo-Pareja, 2006). A similar but weaker association has been demonstrated between older-onset ET and mild cognitive impairment in the NEDICES cohort (OR 1.57) (Benito-Leon, Louis, Mitchell, & Bermejo-Pareja, 2011). Moreover, through extended follow-up cognition was shown to decline more quickly in ET patients compared to controls, including those in whom the motor aspects of ET were not yet manifest (Benito-Leon, Louis, Sanchez-Ferro, & Bermejo-Pareja, 2013). The association between ET and dementia has been validated in a separate cohort of elders in Manhattan, New York (Thawani, Schupf, & Louis, 2009), though the methodology for diagnosis of ET was largely dependent on analysis of writing tasks (with clinical validation of a small number of subjects) and thus not as rigorous as in the

NEDICES cohort. While most data for impaired cognition in ET have been collected in older North American or European subjects, impaired performance on cognitive testing has also been demonstrated in a young ET subjects (Sengul et al., 2015) and in a cohort of Chinese ET subjects (Wu et al., 2016).

The pathophysiology underpinning cognitive impairment in ET remains unclear, particularly given the controversies surrounding the pathophysiology of ET more generally. Given the mild and very slowly progressive nature of most cases of ET, postmortem data are limited; however, some themes have emerged from recent studies. Some series have identified either Lewy bodies, especially in the locus ceruleus with sparing of the substantia nigra pars compacta, or cerebellar changes (with torpedoes and loss of Purkinje cells) as two pathological subtypes of ET (Louis, 2010; Louis et al., 2007), though other series have challenged these findings (Rajput, Adler, Shill, & Rajput, 2012; Symanski et al., 2014). The possibility of cerebellar degeneration representing the cause of ET has led to comparisons between the ET cognitive profile and the cerebellar cognitive affective syndrome described particularly in posterior cerebellar or vermis lesions (Schmahmann & Sherman, 1998; Troster et al., 2002). Another more recent hypothesis, prompted by the finding of PSP postmortem in 12% of cases of an ET pathological series (Louis, Babij, Ma, Cortes, & Vonsattel, 2013), is that tau accumulation may be the mechanism for cognitive decline in ET. Pan et al. examined this hypothesis through a comparison of neuropathology from 40 ET subjects free of dementia clinically and of Alzheimer's findings at postmortem with 32 controls (Pan et al., 2014). The investigators found an increased burden of neurofibrillary tangles in the ET cases as evidenced by significantly increased Braak stages. Imaging studies also support neuronal dysfunction in ET that goes beyond the cerebellum and involves cerebral cortex (Fang et al., 2013; Ha et al., 2015).

In summary, impaired performance on cognitive testing and dementia are overrepresented in ET, with loss of verbal fluency an affected domain identified in several studies. Cognitive impairment in ET is particularly apparent in older patients. The heterogeneity of the clinical and pathological data leaves open the question as to whether ET is a single, neurodegenerative disease in which cognitive impairment may develop with time (especially in older age), or is composed of several entities including a relatively benign; "classical" ET; and a late-onset, dementia-, and mortality-associated ET (age-related tremor/"senile tremor") (Deuschl, Petersen, Lorenz, & Christensen, 2015).

2.2 Psychiatric Features

2.2.1 Depression

In their early study of cognitive deficits in ET, Lombardi et al. also screened their ET patients for depression using the Geriatric Depression Scale (GDS) (Lombardi et al., 2001). They noted that there was no correlation between GDS scores and performance on the other cognitive tests in the battery and also that the ET group exhibited significantly higher GDS scores compared to normative data (and similar scores to the PD comparison cohort). The appreciation of depression as a common feature of PD also prompted further investigation in other movement disorders, including ET. Miller et al. explored this association in a study of PD, ET, and dystonia patients at a single movement disorders centre in Florida who were assessed for symptoms of depression with the Beck Depression Inventory (BDI) (Miller et al., 2007). 34% of ET patients were categorized as depressed (BDI score > 9), compared with 37% of dystonia patients and 48% of PD patients (*P* value 0.09). The profiles of BDI symptoms for each group differed significantly for work difficult and somatic preoccupation, both of which were significantly lower in the ET cohort compared to PD. However, unlike cognition, BDI score was positively associated with tremor severity (Clinical Rating Scale for Tremor-motor subscale score), suggesting that depression may be attributable to chronic illness or disability rather than being an intrinsic part of ET neurobiology. NEDICES has yielded similar results, demonstrating lower morale on the Philadelphia Geriatric Centre Morale Scale (PCGMS) in ET patients (including those undiagnosed prior to the study) compared with controls (Louis, Benito-Leon, Bermejo-Pareja, & Neurological Disorders in Central Spain Study Group, 2008). Wu et al., however, did not find a significant relationship between ET and depression in their Chinese cohort (Wu et al., 2016). Depressed ET patients also appear to have a different mood phenotype from patients with primary affective disorders, with more lassitude and concentration issues and less reported sadness, pessimistic thoughts, and inability to feel (Li et al., 2011).

Depression is a major cause of disability and reduced quality of life (QoL) in ET (Chandran & Pal, 2013; Lorenz, Poremba, Papengut, Schreiber, & Deuschl, 2011; Louis, Huey, Gerbin, & Viner, 2012b; Woods, Scott, Fields, Poquette, & Troster, 2008). The complexity of its pathophysiological relationship with ET is further illustrated by data from treatment studies. Depression appears not respond to Vim DBS despite improvements in tremor severity and ADLs (Graff-Radford et al., 2010). Interestingly, a case of ET receiving electroconvulsive therapy has been reported with temporary improvement of both tremor and depressive symptoms (Kushner, Prudic, & Louis, 2007).

2.2.2 Anxiety

Anxiety has been observed in ET in a number of studies. In the Columbia University Assessment of Disability in ED (CADET) study, Louis et al. assessed cognition, anxiety, and depression in a cohort of 43 ET patients from Manhattan, New York (Louis et al., 2001). The investigators found that anxiety (assessed with the Hamilton Anxiety Rating Scale—HARS) was elevated relative to controls and (along with depression) was independently associated with greater functional disability. Tan et al. studied the psychological symptoms of an Asian cohort of 84 ET patients using the Symptom Checklist-90R (SCL-90R) and found higher scores in the ET cohort for anxiety, phobic anxiety, and psychoticism relative to controls (Tan et al., 2005). Tremor severity was correlated with anxiety and depression scores. Sengul et al. demonstrated the association between anxiety and ET in a cohort of younger patients (average age 25 years old), with 71% of their cohort having moderate to severe anxiety levels (Sengul et al., 2015). Like depression, anxiety has been shown to correlate with QoL in ET (Chandran & Pal, 2013).

2.2.3 Personality

In addition to the psychological manifestations described earlier, a few studies have attempted to identify other personality changes or traits overrepresented in ET. Louis et al. used the tridimensional personality questionnaire to assess the personality traits of 55 ET patients in three broad dimensions (harm avoidance, novelty seeking, and reward dependence) (Chatterjee, Jurewicz, Applegate, & Louis, 2004). The authors found higher harm avoidance scores in the ET cohort compared with controls, and harm avoidance scores did not correlate with tremor severity or disability. Another study, also based in New York, compared ET, PD, and dystonia patients using the Apathy Evaluation Scale and found higher apathy scores in all three groups relative to controls (Louis, Huey, Gerbin, & Viner, 2012a).

As is the case for PD and a number of other chronic diseases, the psychosocial burden of ET, depression and anxiety in particular, is a key determinant of QoL. Screening for and managing these symptoms is therefore an important facet in the holistic care of these patients.

2.3 Sleep

Like cognition and mood, sleep has been well recognized to be affected in a number of movement disorders. The association between the REM sleep behavior disorder (RBD) and the synucleinopathies has been particularly well studied and is of special interest since RBD appears to be a premotor phenomenon in PD (Adler et al., 2011; Boeve, Silber, Ferman, Lucas, & Parisi, 2001; Gagnon et al., 2002; Schenck, Bundlie, & Mahowald, 1996). In ET, the data regarding sleep abnormalities are heterogeneous. Adler et al. compared patients with PD, RLS, and ET using the Mayo Sleep Questionnaire (MSQ) and Epworth Sleepiness Scale (ESS) to ascertain rates of probably RBD, though no polysomnography was performed (Adler et al., 2011). They found probably RBD in 69% of PD patients but only 13% of ET patients (and 13% of controls). Moreover, the mean ESS was comparable in ET and controls. By contrast, Gerbin et al. found a significantly greater proportion of ET patients with an ESS > 10 (22.5%) compared to controls (9.2%), and comparable to the proportion in PD (25%) (Gerbin, Viner, & Louis, 2012). The NEDICES data have also been used to address this question through comparing self-reported daily sleep duration at baseline with incident ET at 3-year follow-up (Benito-Leon, Louis, & Bermejo-Pareja, 2013). Interestingly, even after adjustment for confounders (including age), short sleep duration conferred a twofold elevation in risk of incident ET, suggesting that short sleep may represent a premotor biomarker.

Polysomnography data have clarified the nature of sleep disturbances in ET and shown clear differences with PD, but less clear differences from controls. Barut et al. compared cohorts of ET, PD patients, and controls and found no cases of clinical RBD in the ET cohort, but 18% of patients had subclinical RBD (compared with 52% of the PD cohort) (Barut, Tascilar, & Varo, 2015). Other polysomnographic parameters of sleep did not vary significantly between the groups apart from nocturnal mean SpO_2, which was significantly lower in ET than in PD and controls.

2.4 Sensory Abnormalities

2.4.1 Olfaction

There have been conflicting reports regarding whether olfaction is normal or abnormal in ET. Louis et al. have reported a mild olfactory deficit in ET patients relative to controls using the University of Pennsylvania Smell Identification Test score (UPSIT) (Applegate & Louis, 2005; Louis, Bromley, Jurewicz, & Watner, 2002). There was a trend toward older age and longer smoking history in the ET group relative to controls in these studies, though the authors adjusted for these confounders. In a separate study, this group has found higher levels of harmane, a neurotoxin used to induce an animal model of ET, correlate with lower UPSIT scores in the ET cohort but not in controls (Louis, Rios, et al., 2008).

However, the finding of olfactory dysfunction ET has not been replicated by other researchers. In order to reduce potential cognitive bias, Shah et al. studied olfaction in PD, ET, and controls using the UPSIT as well as olfactory event-related potentials, an objective neurophysiological measure of olfaction (Shah, Muhammed, Findley, & Hawkes, 2008). The investigators found no significant decrement in smell in the ET cohort relative to controls, unlike the decrement found for the PD group, though in their study fewer controls were smokers relative to the patient groups. Quagliato et al. found no significant difference in UPSIT in ET relative to controls (Quagliato, Viana, Quagliato, & Simis, 2009).

Given the heterogeneity of these results and the fact that olfactory dysfunction in ET is, at worst, very mild compared to controls (and certainly not as prominent as in PD), an obvious explanation is that the ET cohorts of these studies are contaminated by small numbers of patients who have PD pathology. Indeed, this would be compatible with the aforementioned postmortem findings of Lewy bodies in some ET patients. Nuclear medicine imaging may be a useful tool to clarify whether ET patients with olfactory deficits also have radiological evidence of dopaminergic neuronal loss to suggest these patients in fact have PD. Djaldetti et al. partly addressed this question through performing [^{123}I]-FP-CIT SPECT scans and UPSIT of patients with rest and postural tremor and comparing them with ET, PD, and control cohorts; however, their numbers were too small to show a significant correlation between the SPECT result and the olfaction (Djaldetti et al., 2008). Moreover, these authors also found significant olfactory loss only in the PD (not ET) cohort relative to controls.

2.4.2 Hearing

Unlike the other NMS heretofore discussed, hearing dysfunction appears to be a feature of ET that is not prominent in PD. In an early report, Ondo et al. assessed hearing in cohorts of ET, PD, and normal controls using the Nursing Home Handicap Index (NHHI) survey, audiology, and hearing aid use (Ondo, Sutton, Dat Vuong, Lai, & Jankovic, 2003). ET patients had significantly higher survey scores (indicating worse hearing) and also higher hearing aid use (16.8%) compared to PD (1.6%) and controls (0.8%).

Interestingly, these differences were driven largely by a subset of "pure ET" patients without dystonic features. Louis et al. also demonstrated a high prevalence of symptoms of deafness (21%) and hearing aid use (19%) in a cohort of 94 ET patients, though with a mean age of 74 years old and no control group, it was difficult to correct for the potential confounders of age and presbycusis (Louis, Borden, & Moskowitz, 2005). The NEDICES data have confirmed the association between hearing dysfunction and ET most convincingly through analysis of a large dataset (including 248 ET cases) and over 4000 controls and correcting for multiple confounders including age (Benito-Leon, Louis, Bermejo-Pareja, & Neurological Disorders in Central Spain Study Group, 2007). The mechanism of this association remains unclear, but given the implication of cerebello-thalamo-cortical pathways in ET, the ventral thalamus has been suggested as an anatomical site whose dysfunction could give rise to both tremor and deafness (and is also a recognized DBS target for the treatment of ET) (Ondo & Lai, 2005). Hearing therefore represents an important nonmotor feature distinguishing ET from PD (and normal controls) with potential implications for the underlying pathological mechanisms of this disease.

3. DYSTONIC TREMOR

The presence of tremor is a common feature of adult-onset isolated focal or segmental dystonia. Tremor in a dystonic limb is mainly postural and kinetic in nature; however, associated rest tremor may be observed in 40% of patients (Erro et al., 2014). The prevalence of this type of tremor has been investigated in several cohorts and estimated to be 17%–30% (Defazio et al., 2013; Erro et al., 2014; Williams et al., 2017); however, a large cohort study reported that 63% of all patients with isolated focal or segmental dystonia have dystonic tremor (Norris et al., 2016). Although current observational studies have shed light on the frequency and phenomenology of dystonic tremor, the lack of specific biomarkers has made it challenging to investigate NMS in this group of patients. In this section we will not review studies of NMS in dystonia where tremor was not the dominant inclusion criterion, considering them beyond the remit of this chapter.

Given the scarcity of studies evaluating NMS in isolated dystonia syndromes, it is instructive to consider another group of tremor patients with features of dystonia, namely patients with scans without evidence of dopaminergic deficits (SWEDDs). An important caveat in interpreting the results of studies of cohorts with SWEDDs is that they almost certainly represent a heterogeneous group of patients with different underlying diseases, including early-stage PD and depression with psychomotor slowing as two examples. However, some patients with SWEDDs, in addition to Parkinsonian features such as asymmetric rest tremor and reduced arm swing, also have evidence of neck and position-specific arm tremor, laryngeal dystonia, thumb hyperextension, and lack of akinesia and reemergent tremor (Erro et al., 2014; Schneider et al., 2007). It has been proposed that after careful clinical examination and follow-up these patients, who represent approximately 70% in some series, should be reclassified as having a dystonic tremor syndrome (De Rosa et al., 2014; Schwingenschuh et al., 2010). This is also supported by the PRECEPT study that demonstrated normal DaT scans at 22-month follow-up in 91% of SWEDDs (Marek et al., 2014). Therefore, studies of SWEDDs patients may also provide insight into the prevalence of NMS in dystonic tremor syndromes.

3.1 Olfaction

Despite the lack of clarity surrounding the diagnosis of SWEDDs there are important and significant differences in nonmotor features when compared with controls or PD patients. In the study of Silveira-Moriyama et al., olfactory function (assessed using UPSIT) in SWEDDs not only seemed to be unaffected in comparison to PD but also similar to dystonia and ET patients (Silveira-Moriyama et al., 2009). Likewise, other studies have found similar UPSIT scores in SWEDDs and healthy controls; however, hyposmia and anosmia were detected in SWEDDs in 17% and 4%, respectively (Taylor et al., 2016). In this same reported cohort when evaluated by different authors, hyposmia was reported in 19%, 67%, and 6.5% of SWEDDs, PD, and healthy controls, respectively (Sprenger et al., 2015). These differences between disease groups are also detected using the Nonmotor Symptoms Scale (NMSS)·(Yang et al., 2014). Altogether, the evidence suggests that SWEDDs have better olfaction than PD but a higher frequency of olfactory dysfunction when compared to healthy controls.

3.2 Mood Disorders

There is conflicting evidence with regards to the prevalence of affective symptoms in SWEDDs. In the Parkinson's Progression Markers Initiative (PPMI) study, comparing SWEDDs with controls and patients with PD using the GDS, SWEDDs patients had higher rates of depression compared to PD and controls; however, the severity of depression was worse in PD

(Taylor et al., 2016). Nonetheless, even with the PPMI cohort Sprenger et al. reported a high depression rate in the three groups (Sprenger et al., 2015), with the difference in results most likely explained by the use of different methodological analyses. In addition, in this same study, depression, anxiety, and apathy scores in the MDS-UPDRS Part I were higher in SWEDDs compared to healthy control but not compared to PD. Others have found similar depression scores in PD and SWEDDs patients using the Geriatric Depression Scale-Short Form (GDS-SF)·(Wyman-Chick, Martin, Minar, & Schroeder, 2016).

Early reports observed normal scores in anxiety scales in SWEDDs, whereas more recent studies have found some differences from healthy controls (Schwingenschuh et al., 2010). In the PPMI study, true SWEDDs patients had more severe state and trait anxiety symptoms than healthy controls, but with similar severity to those with PD (Taylor et al., 2016). However, no differences between the three groups were found in this same cohort reported by others (Sprenger et al., 2015). The available evidence suggests that anxiety and depression are more frequent in SWEDDs when compared to healthy controls but similar to PD.

3.3 Sleep Disorders

RBD is a well-recognized feature of PD and is now being increasingly recognized to be a common feature of SWEDDs. Using RBD screening questionnaires the prevalence in the SWEDDs population is estimated to be 18%–30% (Taylor et al., 2016; Yang et al., 2014). Some studies have found the prevalence of RBD to be marginally higher in SWEDDs than in PD; however, the severity was greater in PD (Taylor et al., 2016). Importantly, this is not a consistent finding among other cohorts (Yang et al., 2014).

Excessive daytime sleepiness can be a result of poor quality sleep or medication side effects. Here also there is conflicting evidence as to whether its prevalence is higher in SWEDDs or PD. One study found that excessive daytime sleepiness evaluated by the ESS was more frequent in SWEDDs than controls (31% vs 11%) and with higher scores (ESS $>$ 10 points) in comparison to PD patients (31% vs 14%) (Sprenger et al., 2015). In addition, daytime sleepiness as measured in item 8 of MDS-UPDRS Part I was more frequent than in controls. These findings were not confirmed by others in the same cohort (Taylor et al., 2016). What is agreed is that RBD is more common in SWEDDs as compared with the normal population. As RBD is a common associated finding in neurodegenerative disorders, especially synucleinopathies, this finding raises the possibility that SWEDDs cohorts may include patients with neurodegenerative disorders.

3.4 Autonomic Symptoms

Autonomic symptoms have been found in SWEDDs. In the PPMI study SWEDDs were associated with cardiovascular, thermoregulatory, gastrointestinal, and urinary autonomic symptoms. Cardiovascular and thermoregulatory symptoms were significantly worse in SWEDDs compared with PD in the SCOPA-AUT questionnaire (Sprenger et al., 2015). Gastrointestinal and urinary symptoms were also worse than in healthy controls. Others have found that postural hypotension is more common in the NMSS in SWEDDs compared with PD (Taylor et al., 2016). However, this has not been a reliable finding (Yang et al., 2014). One study that used objective measures of blood pressure found that patients with SWEDDs had nocturnal hypertension and were less likely to demonstrate the normal nocturnal dip in blood pressure (Kang et al., 2016). This pattern differentiated SWEDDs from healthy controls, however, was similar to what is found in PD. Cardiac sympathetic innervation has also been found to be abnormal in SWEDDs patients when compared to healthy controls, but notably not equivalent to the denervation seen in PD (Jang et al., 2013). Taken together, these results suggest that patients with SWEDDs have more autonomic and particularly the cardiovascular and thermoregulatory dysfunction when compared to those with PD.

3.5 Cognition

Cognitive decline, indicated by Montreal Cognitive Assessment (MoCA) scores below 26 after a 2-year follow-up, has been observed in 30% of SWEDDs·(Wyman-Chick et al., 2016). Interestingly, in the same study only 15% of PD patients had cognitive decline on follow-up, suggesting and alternative diagnosis in SWEDDs. Other studies have found variable results of cognitive symptoms. One study showed similar cognitive symptoms to PD (Yang et al., 2014), while other found more cognitive symptoms compared to PD and controls using only NMSS·(Sprenger et al., 2015).

3.6 Quality of Life

QoL does not appear to be different from PD, but comparison with other disease groups has not been performed. In one study, health-related QoL scores using EQ-5D instrument were not different between SWEDDs and PD including in the mobility, self-care, activity, pain, anxiety, and depression dimensions (Yang et al., 2014).

4. ORTHOSTATIC TREMOR

The distinctive syndrome of lower limb tremor after standing was initially recognized in 1970 by Pazzaglia, Sabattini, and Lugaresi (1970). In 1984 Heilman highlighted the phenomenology of three cases and coined the term orthostatic tremor (OT) to denote the consistent appearance of tremor after this specific task (Heilman, 1984). Patients with OT describe unsteadiness or leg tremor seconds or minutes after adopting a standing position. Sitting or walking gives temporary relief to their symptoms, which tend to worsen over time despite treatment (Fung, Sauner, & Day, 2001). The diagnosis of OT requires the electrophysiological demonstration of a 16 Hz EMG tremor in the lower limbs while in the standing position (Deuschl et al., 1998; Koster et al., 1999; Thompson et al., 1986). Furthermore, in view of its association with other neurodegenerative disease such as PD or other movements disorders, OT has been subclassified as OT-plus (when combined with other features) or as primary OT (when seen in isolation) (Gerschlager et al., 2004). However, this classification has been called into question due to the possibility of referral bias and the finding of a small number of patients with OT-plus in a large study of OT (Benito-Leon, Domingo-Santos, et al., 2016). Symptomatic OT due to structural, metabolic, and autoimmune disorders has also been described (Gerschlager & Brown, 2011). Compared with other, more common causes of tremor, the recognition and study of NMS in OT are lacking.

4.1 Cognition and Affective Features

The main neuropsychological features of OT that have been reported are depression and a dysexecutive syndrome with visuospatial deficits. In a case–control study where the average age of participants was 57 years, those with OT showed impairments in a number of cognitive domains, specifically executive function, verbal memory, visual memory, visuospatial ability, and language (Benito-Leon, Louis, et al., 2016). Among patients with OT, older age (over 60 years) was associated with relatively greater impairment of measures of neuropsychologic testing. The same study also found those with OT had comparatively greater impairments in affective and personality domains compared to healthy controls. These included increased somatic concerns, anxiety-related disorders, depression, and antisocial personality traits. Mild to severe depression has also been reported in other studies (Schöberl et al., 2017).

4.2 Quality of Life

While the evidence is sparse, those studies that have examined the effect of OT on QoL have shown significant impairment in this area. The most comprehensive study found that disease severity, defined as patients' ability to remain standing, had a moderate correlation with physical functioning in the SF-36 survey. However, there was no correlation between depression scores on BDI and disease severity. Furthermore, mean scores of physical function and social functioning were markedly reduced (Gerschlager et al., 2003). Half of those with OT were depressed, ranging from mild to severe depression. Other studies have also corroborated low QoL scores in OT (Benito-Leon, Louis, et al., 2016).

5. LESIONAL TREMOR: THALAMIC, CEREBELLAR, AND BRAIN STEM LESIONS

Tremor is well recognized to be a manifestation of isolated cerebral lesions. Specifically, lesions involving the cerebellum, brain stem, and thalamus have also been associated with a spectrum of NMS and tremor. Specifically, cognitive and behavioral manifestations are often a consequence of disrupted functional connectivity of specific subcortical brain regions linked to associative and limbic cortices (Boes et al., 2015). Our current knowledge of NMS arising from lesions of these circuits mostly derives from studies of patients with isolated vascular lesions.

5.1 Thalamic Tremor

Thalamic tremor is characterized by a combination of postural and irregular kinetic tremor. An abnormal hand posture with flexion of the metacarpophalangeal and extension of the interphalangeal joints (thalamic hand) has a localizing value (Lee & Marsden, 1994; Lehericy et al., 2001). Most of the lesions that generate tremor are localized at or adjacent to the ventrolateral and posterolateral nuclei. These regions receive ascending projection from the cerebellum but also have reciprocal connections to prefrontal regions and involved in cortico-subcortical circuits (Catani et al., 2012). Therefore, lesions along the cerebellar–thalamic pathway not only generate tremor but also cognitive and behavioral deficits.

The frequency and severity of cognitive and behavioral deficits depend on the location and lateralization of the lesion within the thalamus. Anteromedian thalamic infarcts result in executive dysfunction in almost 90% of cases, in contrast posterolateral lesions cause this impairment in approximately 40% of patients (Carrera & Bogousslavsky, 2006; De Witte et al., 2011). Isolated lesions of the mediodorsal thalamic nucleus and mammillothalamic tract (MTT) impair memory; however, MTT lesions cause more severe deficits in both verbal recall and recognition memory (Danet et al., 2015; Van der Werf, Witter, Uylings, & Jolles, 2000). Eighty percent of patients with bilateral thalamic lesions causing tremor are accompanied by psychiatric and behavioral changes. Apathy and abulia are commonly observed following damage to the medial thalamus, which may also result in grasping and clonic perseveration (Krause et al., 2012).

5.2 Cerebellar Tremor

Tremor is a common clinical finding in patients with cerebellar disorders. Cerebellar tremor is mainly a 2–7-Hz kinetic tremor that appears during the execution of movement and is usually maximal as the limb approaches the target (Deuschl et al., 1998). A postural component of cerebellar tremor is often observed in lesions involving the outflow tracts of the cerebellar nuclei to the brain stem. In addition to its function in the coordination of movement, the cerebellum contributes to the regulation of cognitive, language, and affective functions. The cognitive region of the cerebellum is located in the posterior lobe, whereas the limbic cerebellum is represented in the posterior vermis (Buckner, Krienen, Castellanos, Diaz, & Yeo, 2011; Stoodley, MacMore, Makris, Sherman, & Schmahmann, 2016). Cerebellar motor syndromes predominantly result from lesions of the anterior lobe (Schmahmann, Macmore, & Vangel, 2009).

The cerebellar contribution to cognition and emotion was unveiled by Schmahmann and Sherman, in their seminal neuropsychological work in patients with cerebellar lesions. They found that most patients had impairment in executive function, working memory, motor, or ideational set-shifting and perseveration (Schmahmann & Sherman, 1998). These deficits are more noticeable in patients with large-volume lesions with acute onset (Hokkanen, Kauranen, Roine, Salonen, & Kotila, 2006; Malm et al., 1998). Combined cognitive and linguistic impairment is observed up to 80% of the cases (Baillieux, De Smet, Paquier, De Deyn, & Marien, 2008) with lateralization of damage influencing the specific cognitive deficit. Right-sided lesions cause a language processing disorder while those on the disrupt attention and visuospatial skills (Baillieux et al., 2008; Tedesco et al., 2011).

The behavioral component of the syndrome relates to the disturbance of behavior and personality modulation, producing a spectrum ranging from exaggeration to diminution of facets of normal behavior. Manifested deficits range from hyperactivity, distractibility, and compulsive and ritualistic behavior to negative symptoms such as inattention and perseveration. Patients may show a lack of emotional control manifested as anxiety, panic, disinhibition, and impulsiveness or display depression and dysphoria (Schmahmann, Weilburg, & Sherman, 2007). Pathological crying and laughing can occur in patients with chronic cerebellar lesions as well as in anywhere within the pontocerebellar circuit, as observed in neurodegenerative atypical parkinsonisms such as multiple system atrophy. Psychosis with paranoia and hallucinations has also been described in patients with cerebellar lesions. An orbitofrontal and medial frontal syndrome can also be the consequence of cerebellar lesions (Baillieux et al., 2010). A well-recognized syndrome of mutism is seen in children after cerebellar tumor resection involving the midline cerebellar structures.

5.3 Brain Stem Lesional Tremor

Brain stem lesions associated with tremor are located in regions within the cerebello-thalamic-red nucleus connections. According to current consensus (Deuschl et al., 1998), Holmes tremor is defined as a 3–4-Hz rest and intention tremor with irregular amplitude; however, in clinical practice, it is mainly brought by posture (Raina et al., 2016). Lesions involving the midbrain, posterior thalamus, and outflow cerebellar tracts are associated with Holmes tremor.

Other clinical accompaniments help to identify this type of tremor. Myorhythmia, a slow, 2–3-Hz rest tremor, can develop after damage of the superior cerebellar peduncle, dentate nucleus, or central tegmental tract (Baizabal-Carvallo, Cardoso, & Jankovic, 2015; Vidailhet, Jedynak, Pollak, & Agid, 1998). Myorhythmia can be evident in the palate, cranial muscles, or limbs. Symptomatic palatal tremor has been classically associated with lesions involving the Guillain–Mollaret triangle with or without olivary hypertrophy. Occasionally, pendular nystagmus accompanies palatal movements (oculopalatal tremor). Current studies in Holmes tremor and myorhythmia have focused on the motor manifestations, while NMS studies have been scarce.

Cognitive and affective disorders after isolated brain stem lesions may resemble those observed in cerebellar disorders. Executive dysfunction, and disturbance of memory and attention are observed in more than two-thirds of the cases, whereas deficits of language, visuospatial skills, and praxis occur in 15%. Furthermore, affective symptoms or behavioral changes can be observed in nearly half of the patients. These symptoms include delusions, irritability, aggression, apathy, emotional liability, disinhibition, paranoia, and mania. A diaschisis mechanism between the brain stem and the cerebral cortex is suggested by the presence of hypoperfusion affecting predominantly the frontal, parietal, and temporal cortices in SPECT (D'Aes & Marien, 2015).

6. CONCLUSIONS

NMS are common in patients with disorders where tremor is the dominant motor feature. Despite limitations in research thus far, it seems clear that, as for many motor disorders, the nonmotor symptoms can significantly impact on QoL and should be a focus of further research, but also contemporary treatment.

REFERENCES

Adler, C. H., Hentz, J. G., Shill, H. A., Sabbagh, M. N., Driver-Dunckley, E., Evidente, V. G., et al. (2011). Probable RBD is increased in Parkinson's disease but not in essential tremor or restless legs syndrome. *Parkinsonism & Related Disorders*, *17*(6), 456–458.

Applegate, L. M., & Louis, E. D. (2005). Essential tremor: Mild olfactory dysfunction in a cerebellar disorder. *Parkinsonism & Related Disorders*, *11*(6), 399–402.

Baillieux, H., De Smet, H. J., Dobbeleir, A., Paquier, P. F., De Deyn, P. P., & Marien, P. (2010). Cognitive and affective disturbances following focal cerebellar damage in adults: A neuropsychological and SPECT study. *Cortex*, *46*(7), 869–879.

Baillieux, H., De Smet, H. J., Paquier, P. F., De Deyn, P. P., & Marien, P. (2008). Cerebellar neurocognition: Insights into the bottom of the brain. *Clinical Neurology and Neurosurgery*, *110*(8), 763–773.

Baizabal-Carvallo, J. F., Cardoso, F., & Jankovic, J. (2015). Myorhythmia: Phenomenology, etiology, and treatment. *Movement Disorders*, *30*(2), 171–179.

Barut, B. O., Tascilar, N., & Varo, A. (2015). Sleep disturbances in essential tremor and Parkinson disease: A polysomnographic study. *Journal of Clinical Sleep Medicine*, *11*(6), 655–662.

Benito-Leon, J., Domingo-Santos, A., Hassan, A., Ahlskog, J. E., Matsumoto, J. Y., & Bower, J. H. (2016a). Orthostatic tremor: Clinical, electrophysiologic, and treatment findings in 184 patients. *Neurology*, *87*(3), 341.

Benito-Leon, J., Louis, E. D., & Bermejo-Pareja, F. (2006). Elderly-onset essential tremor is associated with dementia. *Neurology*, *66*(10), 1500–1505.

Benito-Leon, J., Louis, E. D., & Bermejo-Pareja, F. (2013a). Short sleep duration heralds essential tremor: A prospective, population-based study. *Movement Disorders*, *28*(12), 1700–1707.

Benito-Leon, J., Louis, E. D., Bermejo-Pareja, F., & Neurological Disorders in Central Spain Study Group. (2007). Reported hearing impairment in essential tremor: A population-based case-control study. *Neuroepidemiology*, *29*(3–4), 213–217.

Benito-Leon, J., Louis, E. D., Mitchell, A. J., & Bermejo-Pareja, F. (2011). Elderly-onset essential tremor and mild cognitive impairment: A population-based study (NEDICES). *Journal of Alzheimer's Disease*, *23*(4), 727–735.

Benito-Leon, J., Louis, E. D., Puertas-Martin, V., Romero, J. P., Matarazzo, M., Molina-Arjona, J. A., et al. (2016b). Cognitive and neuropsychiatric features of orthostatic tremor: A case-control comparison. *Journal of the Neurological Sciences*, *361*, 137–143.

Benito-Leon, J., Louis, E. D., Sanchez-Ferro, A., & Bermejo-Pareja, F. (2013b). Rate of cognitive decline during the premotor phase of essential tremor: A prospective study. *Neurology*, *81*(1), 60–66.

Bermejo-Pareja, F., Louis, E. D., & Benito-Leon, J. (2007). Risk of incident dementia in essential tremor: A population-based study. *Movement Disorders*, *22*(11), 1573–1580.

Boes, A. D., Prasad, S., Liu, H., Liu, Q., Pascual-Leone, A., Caviness, V. S., Jr., et al. (2015). Network localization of neurological symptoms from focal brain lesions. *Brain*, *138*(Pt. 10), 3061–3075.

Boeve, B. F., Silber, M. H., Ferman, T. J., Lucas, J. A., & Parisi, J. E. (2001). Association of REM sleep behavior disorder and neurodegenerative disease may reflect an underlying synucleinopathy. *Movement Disorders*, *16*(4), 622–630.

Buckner, R. L., Krienen, F. M., Castellanos, A., Diaz, J. C., & Yeo, B. T. (2011). The organization of the human cerebellum estimated by intrinsic functional connectivity. *Journal of Neurophysiology*, *106*(5), 2322–2345.

Carrera, E., & Bogousslavsky, J. (2006). The thalamus and behavior: Effects of anatomically distinct strokes. *Neurology*, *66*(12), 1817–1823.

Catani, M., Dell'acqua, F., Bizzi, A., Forkel, S. J., Williams, S. C., Simmons, A., et al. (2012). Beyond cortical localization in clinico-anatomical correlation. *Cortex*, *48*(10), 1262–1287.

Chandran, V., & Pal, P. K. (2013). Quality of life and its determinants in essential tremor. *Parkinsonism & Related Disorders*, *19*(1), 62–65.

Chatterjee, A., Jurewicz, E. C., Applegate, L. M., & Louis, E. D. (2004). Personality in essential tremor: Further evidence of non-motor manifestations of the disease. *Journal of Neurology, Neurosurgery, and Psychiatry*, *75*(7), 958–961.

Critchley, M. (1949). Observations on essential (heredofamial) tremor. *Brain*, *72*(Pt. 2), 113–139.

D'Aes, T., & Marien, P. (2015). Cognitive and affective disturbances following focal brainstem lesions: A review and report of three cases. *Cerebellum*, *14*(3), 317–340.

Dana, C. (1887). Hereditary tremor, a hitherto undescribed form of motor neurosis. *American Journal of the Medical Sciences*, *188*, 386–393.

Danet, L., Barbeau, E. J., Eustache, P., Planton, M., Raposo, N., Sibon, I., et al. (2015). Thalamic amnesia after infarct: The role of the mammillothalamic tract and mediodorsal nucleus. *Neurology*, *85*(24), 2107–2115.

De Rosa, A., Carducci, C., Carducci, C., Peluso, S., Lieto, M., Mazzella, A., et al. (2014). Screening for dopa-responsive dystonia in patients with scans without evidence of dopaminergic deficiency (SWEDD). *Journal of Neurology*, *261*(11), 2204–2208.

De Witte, L., Brouns, R., Kavadias, D., Engelborghs, S., De Deyn, P. P., & Marien, P. (2011). Cognitive, affective and behavioural disturbances following vascular thalamic lesions: A review. *Cortex*, *47*(3), 273–319.

Defazio, G., Gigante, A. F., Abbruzzese, G., Bentivoglio, A. R., Colosimo, C., Esposito, M., et al. (2013). Tremor in primary adult-onset dystonia: Prevalence and associated clinical features. *Journal of Neurology, Neurosurgery, and Psychiatry*, *84*(4), 404–408.

Deuschl, G., Bain, P., & Brin, M. (1998a). Consensus statement of the Movement Disorder Society on tremor. Ad hoc scientific committee. *Movement Disorders*, *13*(Suppl. 3), 2–23.

Deuschl, G., & Elble, R. (2009). Essential tremor—Neurodegenerative or nondegenerative disease towards a working definition of ET. *Movement Disorders*, *24*(14), 2033–2041.

Deuschl, G., Petersen, I., Lorenz, D., & Christensen, K. (2015). Tremor in the elderly: Essential and aging-related tremor. *Movement Disorders*, *30*(10), 1327–1334.

Deuschl, G., Wenzelburger, R., Loffler, K., Raethjen, J., & Stolze, H. (2000). Essential tremor and cerebellar dysfunction clinical and kinematic analysis of intention tremor. *Brain*, *123*(Pt. 8), 1568–1580.

Djaldetti, R., Nageris, B. I., Lorberboym, M., Treves, T. A., Melamed, E., & Yaniv, E. (2008). [(123)I]-FP-CIT SPECT and olfaction test in patients with combined postural and rest tremor. *Journal of Neural Transmission*, *115*(3), 469–472.

Erro, R., Rubio-Agusti, I., Saifee, T. A., Cordivari, C., Ganos, C., Batla, A., et al. (2014). Rest and other types of tremor in adult-onset primary dystonia. *Journal of Neurology, Neurosurgery, and Psychiatry*, *85*(9), 965–968.

Fang, W., Lv, F., Luo, T., Cheng, O., Liao, W., Sheng, K., et al. (2013). Abnormal regional homogeneity in patients with essential tremor revealed by resting-state functional MRI. *PLoS One*, *8*(7), e69199.

Fung, V. S., Sauner, D., & Day, B. L. (2001). A dissociation between subjective and objective unsteadiness in primary orthostatic tremor. *Brain*, *124*(Pt. 2), 322–330.

Gagnon, J. F., Bedard, M. A., Fantini, M. L., Petit, D., Panisset, M., Rompre, S., et al. (2002). REM sleep behavior disorder and REM sleep without atonia in Parkinson's disease. *Neurology*, *59*(4), 585–589.

Gasparini, M., Bonifati, V., Fabrizio, E., Fabbrini, G., Brusa, L., Lenzi, G. L., et al. (2001). Frontal lobe dysfunction in essential tremor: A preliminary study. *Journal of Neurology*, *248*(5), 399–402.

Gerbin, M., Viner, A. S., & Louis, E. D. (2012). Sleep in essential tremor: A comparison with normal controls and Parkinson's disease patients. *Parkinsonism & Related Disorders*, *18*(3), 279–284.

Gerschlager, W., & Brown, P. (2011). Orthostatic tremor—A review. *Handbook of Clinical Neurology*, *100*, 457–462.

Gerschlager, W., Katzenschlager, R., Schrag, A., Lees, A. J., Brown, P., Quinn, N., et al. (2003). Quality of life in patients with orthostatic tremor. *Journal of Neurology*, *250*(2), 212–215.

Gerschlager, W., Munchau, A., Katzenschlager, R., Brown, P., Rothwell, J. C., Quinn, N., et al. (2004). Natural history and syndromic associations of orthostatic tremor: A review of 41 patients. *Movement Disorders*, *19*(7), 788–795.

Graff-Radford, J., Foote, K. D., Mikos, A. E., Bowers, D., Fernandez, H. H., Rosado, C. A., et al. (2010). Mood and motor effects of thalamic deep brain stimulation surgery for essential tremor. *European Journal of Neurology*, *17*(8), 1040–1046.

Ha, S. W., Yang, Y. S., Song, I. U., Chung, Y. A., Oh, J. K., & Chung, S. W. (2015). Changes in regional brain glucose metabolism measured with F-18-FDG-PET in essential tremor. *Acta Radiologica*, *56*(4), 482–486.

Heilman, K. M. (1984). Orthostatic tremor. *Archives of Neurology*, *41*(8), 880–881.

Helmchen, C., Hagenow, A., Miesner, J., Sprenger, A., Rambold, H., Wenzelburger, R., et al. (2003). Eye movement abnormalities in essential tremor may indicate cerebellar dysfunction. *Brain*, *126*(Pt. 6), 1319–1332.

Hokkanen, L. S., Kauranen, V., Roine, R. O., Salonen, O., & Kotila, M. (2006). Subtle cognitive deficits after cerebellar infarcts. *European Journal of Neurology*, *13*(2), 161–170.

Hubble, J. P., Busenbark, K. L., Wilkinson, S., Penn, R. D., Lyons, K., & Koller, W. C. (1996). Deep brain stimulation for essential tremor. *Neurology, 46*(4), 1150–1153.

Jang, W., Kim, J. S., Cho, J. W., Kim, Y. H., Kim, J. Y., Choi, Y. Y., et al. (2013). Cardiac sympathetic denervation in Parkinson's disease patients with SWEDDs. *Neurological Sciences, 34*(8), 1375–1382.

Kang, S. J., Ahn, J. Y., Kim, J. S., Cho, J. W., Kim, J. Y., Choi, Y. Y., et al. (2016). 24-Hour ambulatory blood pressure monitoring in SWEDDs patients with parkinsonism. *The Canadian Journal of Neurological Sciences, 43*(3), 390–397.

Koster, B., Lauk, M., Timmer, J., Poersch, M., Guschlbauer, B., Deuschl, G., et al. (1999). Involvement of cranial muscles and high intermuscular coherence in orthostatic tremor. *Annals of Neurology, 45*(3), 384–388.

Krause, M., Mahant, N., Kotschet, K., Fung, V. S., Vagg, D., Wong, C. H., et al. (2012). Dysexecutive behaviour following deep brain lesions—A different type of disconnection syndrome? *Cortex, 48*(1), 97–119.

Kronenbuerger, M., Gerwig, M., Brol, B., Block, F., & Timmann, D. (2007). Eyeblink conditioning is impaired in subjects with essential tremor. *Brain, 130*(Pt. 6), 1538–1551.

Kushner, S. A., Prudic, J., & Louis, E. D. (2007). Transient improvement of essential tremor during electroconvulsive therapy. *The Journal of ECT, 23*(2), 99–102.

Lee, M. S., & Marsden, C. D. (1994). Movement disorders following lesions of the thalamus or subthalamic region. *Movement Disorders, 9*(5), 493–507.

Lehericy, S., Grand, S., Pollak, P., Poupon, F., Le Bas, J. F., Limousin, P., et al. (2001). Clinical characteristics and topography of lesions in movement disorders due to thalamic lesions. *Neurology, 57*(6), 1055–1066.

Li, Z. W., Xie, M. J., Tian, D. S., Li, J. J., Zhang, J. P., Jiao, L., et al. (2011). Characteristics of depressive symptoms in essential tremor. *Journal of Clinical Neuroscience, 18*(1), 52–56.

Lombardi, W. J., Woolston, D. J., Roberts, J. W., & Gross, R. E. (2001). Cognitive deficits in patients with essential tremor. *Neurology, 57*(5), 785–790.

Lorenz, D., Poremba, C., Papengut, F., Schreiber, S., & Deuschl, G. (2011). The psychosocial burden of essential tremor in an outpatient- and a community-based cohort. *European Journal of Neurology, 18*(7), 972–979.

Louis, E. D. (2010). Essential tremor: Evolving clinicopathological concepts in an era of intensive post-mortem enquiry. *Lancet Neurology, 9*(6), 613–622.

Louis, E. D., Babij, R., Ma, K., Cortes, E., & Vonsattel, J. P. (2013). Essential tremor followed by progressive supranuclear palsy: Postmortem reports of 11 patients. *Journal of Neuropathology and Experimental Neurology, 72*(1), 8–17.

Louis, E. D., Barnes, L., Albert, S. M., Cote, L., Schneier, F. R., Pullman, S. L., et al. (2001). Correlates of functional disability in essential tremor. *Movement Disorders, 16*(5), 914–920.

Louis, E. D., Benito-Leon, J., Bermejo-Pareja, F., & Neurological Disorders in Central Spain Study Group. (2008a). Philadelphia geriatric morale scale in essential tremor: A population-based study in three Spanish communities. *Movement Disorders, 23*(10), 1435–1440.

Louis, E. D., Borden, S., & Moskowitz, C. B. (2005). Essential tremor centralized brain repository: Diagnostic validity and clinical characteristics of a highly selected group of essential tremor cases. *Movement Disorders, 20*(10), 1361–1365.

Louis, E. D., Bromley, S. M., Jurewicz, E. C., & Watner, D. (2002). Olfactory dysfunction in essential tremor: A deficit unrelated to disease duration or severity. *Neurology, 59*(10), 1631–1633.

Louis, E. D., Faust, P. L., Vonsattel, J. P., Honig, L. S., Rajput, A., Robinson, C. A., et al. (2007). Neuropathological changes in essential tremor: 33 Cases compared with 21 controls. *Brain, 130*(Pt. 12), 3297–3307.

Louis, E. D., Huey, E. D., Gerbin, M., & Viner, A. S. (2012a). Apathy in essential tremor, dystonia, and Parkinson's disease: A comparison with normal controls. *Movement Disorders*, *27*(3), 432–434.

Louis, E. D., Huey, E. D., Gerbin, M., & Viner, A. S. (2012b). Depressive traits in essential tremor: Impact on disability, quality of life, and medication adherence. *European Journal of Neurology*, *19*(10), 1349–1354.

Louis, E. D., Rios, E., Pellegrino, K. M., Jiang, W., Factor-Litvak, P., & Zheng, W. (2008b). Higher blood harmane (1-methyl-9H-pyrido[3,4-b]indole) concentrations correlate with lower olfactory scores in essential tremor. *Neurotoxicology*, *29*(3), 460–465.

Malm, J., Kristensen, B., Karlsson, T., Carlberg, B., Fagerlund, M., & Olsson, T. (1998). Cognitive impairment in young adults with infratentorial infarcts. *Neurology*, *51*(2), 433–440.

Marek, K., Seibyl, J., Eberly, S., Oakes, D., Shoulson, I., Lang, A. E., et al. (2014). Longitudinal follow-up of SWEDD subjects in the PRECEPT Study. *Neurology*, *82*(20), 1791–1797.

Medeiros, L. M., de Castro, P. C., Felicio, A. C., Queiros, B. B., Silva, S. M., Ferraz, H. B., et al. (2016). Patients with essential tremor can have manual dexterity and attention deficits with no impairments in other cognitive functions. *Arquivos de Neuro-Psiquiatria*, *74*(2), 122–127.

Miller, K. M., Okun, M. S., Fernandez, H. F., Jacobson, C. E., 4th, Rodriguez, R. L., & Bowers, D. (2007). Depression symptoms in movement disorders: Comparing Parkinson's disease, dystonia, and essential tremor. *Movement Disorders*, *22*(5), 666–672.

Norris, S. A., Jinnah, H. A., Espay, A. J., Klein, C., Bruggemann, N., Barbano, R. L., et al. (2016). Clinical and demographic characteristics related to onset site and spread of cervical dystonia. *Movement Disorders*, *31*(12), 1874–1882.

Ondo, W. G., & Lai, D. (2005). Olfaction testing in patients with tremor-dominant Parkinson's disease: Is this a distinct condition? *Movement Disorders*, *20*(4), 471–475.

Ondo, W. G., Sutton, L., Dat Vuong, K., Lai, D., & Jankovic, J. (2003). Hearing impairment in essential tremor. *Neurology*, *61*(8), 1093–1097.

Pan, J. J., Lee, M., Honig, L. S., Vonsattel, J. P., Faust, P. L., & Louis, E. D. (2014). Alzheimer's-related changes in non-demented essential tremor patients vs. controls: Links between tau and tremor? *Parkinsonism & Related Disorders*, *20*(6), 655–658.

Parkinson, J. (1817). *An essay on the shaking palsy*. London: Sherwood, Neely, and Jones.

Pazzaglia, P., Sabattini, L., & Lugaresi, E. (1970). On an unusual disorder of erect standing position (observation of 3 cases). *Rivista Sperimentale di Freniatria e Medicina Legale delle Alienazioni Mentali*, *94*(2), 450–457.

Quagliato, L. B., Viana, M. A., Quagliato, E. M., & Simis, S. (2009). Olfaction and essential tremor. *Arquivos de Neuro-Psiquiatria*, *67*(1), 21–24.

Raina, G. B., Cersosimo, M. G., Folgar, S. S., Giugni, J. C., Calandra, C., Paviolo, J. P., et al. (2016). Holmes tremor: Clinical description, lesion localization, and treatment in a series of 29 cases. *Neurology*, *86*(10), 931–938.

Rajput, A. H., Adler, C. H., Shill, H. A., & Rajput, A. (2012). Essential tremor is not a neurodegenerative disease. *Neurodegenerative Disease Management*, *2*(3), 259–268.

Schenck, C. H., Bundlie, S. R., & Mahowald, M. W. (1996). Delayed emergence of a parkinsonian disorder in 38% of 29 older men initially diagnosed with idiopathic rapid eye movement sleep behaviour disorder. *Neurology*, *46*(2), 388–393.

Schmahmann, J. D., Macmore, J., & Vangel, M. (2009). Cerebellar stroke without motor deficit: Clinical evidence for motor and non-motor domains within the human cerebellum. *Neuroscience*, *162*(3), 852–861.

Schmahmann, J. D., & Sherman, J. C. (1998). The cerebellar cognitive affective syndrome. *Brain*, *121*(Pt. 4), 561–579.

Schmahmann, J. D., Weilburg, J. B., & Sherman, J. C. (2007). The neuropsychiatry of the cerebellum—Insights from the clinic. *Cerebellum*, *6*(3), 254–267.

Schneider, S. A., Edwards, M. J., Mir, P., Cordivari, C., Hooker, J., Dickson, J., et al. (2007). Patients with adult-onset dystonic tremor resembling parkinsonian tremor have scans without evidence of dopaminergic deficit (SWEDDs). *Movement Disorders*, *22*(15), 2210–2215.

Schöberl, F., Feil, K., Xiong, G., Bartenstein, P., la Fougère, C., Jahn, K., et al. (2017). Pathological ponto-cerebello-thalamo-cortical activations in primary orthostatic tremor during lying and stance. *Brain*, *140*(1), 83–97.

Schwingenschuh, P., Ruge, D., Edwards, M. J., Terranova, C., Katschnig, P., Carrillo, F., et al. (2010). Distinguishing SWEDDs patients with asymmetric resting tremor from Parkinson's disease: A clinical and electrophysiological study. *Movement Disorders*, *25*(5), 560–569.

Sengul, Y., Sengul, H. S., Yucekaya, S. K., Yucel, S., Bakim, B., Pazarci, N. K., et al. (2015). Cognitive functions, fatigue, depression, anxiety, and sleep disturbances: Assessment of nonmotor features in young patients with essential tremor. *Acta Neurologica Belgica*, *115*(3), 281–287.

Shah, M., Muhammed, N., Findley, L. J., & Hawkes, C. H. (2008). Olfactory tests in the diagnosis of essential tremor. *Parkinsonism & Related Disorders*, *14*(7), 563–568.

Silveira-Moriyama, L., Schwingenschuh, P., O'Donnell, A., Schneider, S. A., Mir, P., Carrillo, F., et al. (2009). Olfaction in patients with suspected parkinsonism and scans without evidence of dopaminergic deficit (SWEDDs). *Journal of Neurology, Neurosurgery, and Psychiatry*, *80*(7), 744–748.

Singer, C., Sanchez-Ramos, J., & Weiner, W. J. (1994). Gait abnormality in essential tremor. *Movement Disorders*, *9*(2), 193–196.

Sprenger, F. S., Seppi, K., Djamshidian, A., Reiter, E., Nocker, M., Mair, K., et al. (2015). Nonmotor symptoms in subjects without evidence of dopaminergic deficits. *Movement Disorders*, *30*(7), 976–981.

Stoodley, C. J., MacMore, J. P., Makris, N., Sherman, J. C., & Schmahmann, J. D. (2016). Location of lesion determines motor vs. cognitive consequences in patients with cerebellar stroke. *NeuroImage. Clinical*, *12*, 765–775.

Symanski, C., Shill, H. A., Dugger, B., Hentz, J. G., Adler, C. H., Jacobson, S. A., et al. (2014). Essential tremor is not associated with cerebellar Purkinje cell loss. *Movement Disorders*, *29*(4), 496–500.

Tan, E. K., Fook-Chong, S., Lum, S. Y., Gabriel, C., Koh, K. K., Prakash, K. M., et al. (2005). Non-motor manifestations in essential tremor: Use of a validated instrument to evaluate a wide spectrum of symptoms. *Parkinsonism & Related Disorders*, *11*(6), 375–380.

Taylor, S., Gafton, J., Shah, B., Pagano, G., Chaudhuri, K. R., Brooks, D. J., et al. (2016). Progression of nonmotor symptoms in subgroups of patients with non-dopamine-deficient parkinsonism. *Movement Disorders*, *31*(3), 344–351.

Tedesco, A. M., Chiricozzi, F. R., Clausi, S., Lupo, M., Molinari, M., & Leggio, M. G. (2011). The cerebellar cognitive profile. *Brain*, *134*(Pt. 12), 3672–3686.

Thawani, S. P., Schupf, N., & Louis, E. D. (2009). Essential tremor is associated with dementia: Prospective population-based study in New York. *Neurology*, *73*(8), 621–625.

Thompson, P. D., Rothwell, J. C., Day, B. L., Berardelli, A., Dick, J. P., Kachi, T., et al. (1986). The physiology of orthostatic tremor. *Archives of Neurology*, *43*(6), 584–587.

Troster, A. I., Woods, S. P., Fields, J. A., Lyons, K. E., Pahwa, R., Higginson, C. I., et al. (2002). Neuropsychological deficits in essential tremor: An expression of cerebello-thalamo-cortical pathophysiology? *European Journal of Neurology*, *9*(2), 143–151.

Van der Werf, Y. D., Witter, M. P., Uylings, H. B., & Jolles, J. (2000). Neuropsychology of infarctions in the thalamus: A review. *Neuropsychologia*, *38*(5), 613–627.

Vidailhet, M., Jedynak, C. P., Pollak, P., & Agid, Y. (1998). Pathology of symptomatic tremors. *Movement Disorders*, *13*(Suppl. 3), 49–54.

Williams, L., McGovern, E., Kimmich, O., Molloy, A., Beiser, I., Butler, J. S., et al. (2017). Epidemiological, clinical and genetic aspects of adult onset isolated focal dystonia in Ireland. *European Journal of Neurology*, *24*(1), 73–81.

Woods, S. P., Scott, J. C., Fields, J. A., Poquette, A., & Troster, A. I. (2008). Executive dysfunction and neuropsychiatric symptoms predict lower health status in essential tremor. *Cognitive and Behavioral Neurology*, *21*(1), 28–33.

Wu, Y., Wang, X., Wang, C., Sun, Q., Song, N., Zhou, Y., et al. (2016). Prevalence and clinical features of non-motor symptoms of essential tremor in Shanghai rural area. *Parkinsonism & Related Disorders*, *22*, 15–20.

Wyman-Chick, K. A., Martin, P. K., Minar, M., & Schroeder, R. W. (2016). Cognition in patients with a clinical diagnosis of Parkinson disease and scans without evidence of dopaminergic deficit (SWEDD): 2-Year follow-up. *Cognitive and Behavioral Neurology*, *29*(4), 190–196.

Yang, H. J., Kim, Y. E., Yun, J. Y., Ehm, G., Kim, H. J., & Jeon, B. S. (2014). Comparison of sleep and other non-motor symptoms between SWEDDs patients and de novo Parkinson's disease patients. *Parkinsonism & Related Disorders*, *20*(12), 1419–1422.

CHAPTER FIFTY

Nonmotor Symptoms in Huntington Disease

Francisco Cardoso[1]
UFMG, Belo Horizonte, Brazil
[1]Corresponding author: e-mail address: cardosofe@terra.com.br

Contents

Abstract

Huntington disease (HD) is characterized by the triad of motor abnormalities, cognitive dysfunction, and behavioral changes. The aim of this chapter is to describe the frequency, clinical features, and management of behavioral and cognitive dysfunction in HD. Depression, suicidal ideation, apathy, irritability, aggressiveness, obsessions, and compulsions are the most common behavioral abnormalities in HD. All HD patients develop cognitive decline. Recent data suggest that these nonmotor changes are found in premanifest HD gene carriers. Finally, we review the therapeutic options available for behavioral and cognitive abnormalities in HD. The management of these conditions is based on experts' agreement since there is a lack of empirical data.

1. INTRODUCTION

Huntington disease (HD) is an autosomal dominant fully penetrant condition associated to an unstable CAG expansion on the short arm of the chromosome 4. The American physician George Huntington is credited with the original description of the disease published in 1872 (Huntington, 1872). There are, however, previously published papers referring to the condition now known to be HD (Vale & Cardoso, 2015).

The classical clinical features of HD are the triad of a movement disorder, cognitive decline, and behavioral abnormalities with average onset at age

International Review of Neurobiology, Volume 134
ISSN 0074-7742
http://dx.doi.org/10.1016/bs.irn.2017.05.004

42 years. Initially called Huntington chorea, the nomenclature was changed to disease after the realization that chorea is just one of the motor phenomena seen in patients with HD. Most adult patients initially display a combination of chorea and dystonia, although tics and myoclonus are also often present. With progression of the disease, the phenotype becomes increasingly hypokinetic, with prominent fixed dystonia, rigidity and parkinsonism. Less than 10% of patients develop HD before age 20 years. The latter corresponds to the juvenile or Westphal variant whose phenotype is characterized by rigid-akinetic syndrome in the context of an encephalopathy with severe cognitive as well as behavioral problems and occasional seizures (Cardoso, 2009).

Classically, the prevalence of HD is estimated to be 5/100,000 people. However, this is quite variable depending on ethnicity and geography. For instance, in Japan and Asia in general HD is less common (Cardoso, 2009). There are also data suggesting that in Sub-Saharan Africa HD-Like 2 disease, a cause of phenocopy of HD, is at least as common as HD (Cardoso, 2014; Krause et al., 2015). A recent study suggests that the prevalence in British Columbia, Canada, can be as high as 14/100,000 (Fisher & Hayden, 2014). In Latin America, there are clusters of HD throughout the continent. The best known is in Maracaibo, Venezuela but there are others in Peru and Brazil. This is a reflection of the colonization of the continent with development of remote and isolated communities. Interestingly, the HD gene, now known as IT15 gene, was identified in patients from Maracaibo as a result of the collaboration between local professionals and investigators from the United States (The Huntington's Disease Collaborative Research Group, 1993).

The original concept established that people with 39 or more copies of the IT15 will be always symptomatic if live long enough. Those individuals with 36–39 repetitions were in a "gray zone" with a variable expression and penetrance (Cardoso, 2009). However, careful observations show that subjects with intermediate alleles (repetitions between 27 and 35) develop motor and cognitive manifestations although at a greater age than individuals with more than 35 CAG repeats (Cubo et al., 2016). The number of CAG repeats is directly correlated with the severity of phenotype and inverse correlated with the age at onset of the disease. Another genetic phenomenon of HD is the so-called anticipation, i.e., the decrease of age at onset in subsequent generations in a given family as a result of the increase in the number of CAG repeats (Cardoso, 2009). Current criteria establish that individuals must present motor features in association with a positive genetic test to meet diagnosis for HD. According to the Unified Huntington Disease Rating Scale (UHDRS) Diagnostic Criteria, depending on the severity of the motor features, the diagnostic confidence may range from less than 50% up to ≥99%. The first problem of these criteria is the excessive emphasis on motor findings disregarding the importance of nonmotor abnormalities. Another issue is not to take into consideration the uncertainty as to when the disease actually starts. From a biological point of view, it is reasonable to accept that pathological changes may take place already intra-utero. Moreover, careful prospective clinical and radiological observations of at risk or nonmanifest individuals, i.e., subjects with a positive genetic test who do meet currently diagnostic criteria, also support the notion that detectable abnormalities are present long before these individuals can currently be diagnosed with HD (Huntington Study Group, 1996, 2016; Tabrizi et al., 2012, 2013).

From a historical point of view, George Huntington had clearly described nonmotor features in his original study when he described "a tendency to insanity and suicide":

> *"The tendency to insanity, and sometimes that form of insanity which leads to suicide, is marked. I know of several instances of suicide of people suffering from this form of chorea, or who belonged to families in which the disease existed. As the disease progresses the mind becomes more or less impaired, in many amounting to insanity, while in others mind and body both gradually fail until death relieves them of their sufferings"*
>
> **Huntington (1872).**

Many nonmotor domains are abnormal in HD. Regarding autonomic features, the most commonly reported is weight loss. It can happen as an early feature of the disease. In this case, it is thought to represent a consequence of an energy production dysfunction at the mitochondrial level (van der Burgh, Björkqvist, & Brundin, 2009). This finding tends to worsen with progression of HD as a result of dysphagia. In fact many studies identify aspiration pneumonia and other pulmonary abnormalities as the main cause of death in HD (Lanska, Lanska, Lavine, & Schoenberg, 1988). Urinary incontinence is also a common finding in end stage disease likely representing a nonspecific consequence of widespread central nervous system lesion. More recently, there are studies showing that several hitherto unrecognized nonmotor problems are in fact present in HD subjects. An analysis of the database of the REGISTRY study showed that 41% of a cohort of 1474 subjects experienced pain. This symptom was associated with depression, anxiety, and irritability. Although the pathogenesis of the pain in HD is not understood, its correlation with psychological factors, particularly depression, is similar to what has been described in many different conditions. It is possible that pain may be cause and consequence of depression (Underwood, Bonas, Dale, & REGISTRY Investigators of the European Huntington's Disease Network, 2017). There are also reports describing sleep disorders in HD subjects. Although the published studies have involved a relatively small number of subjects, the available data suggest that sleep abnormalities are present in up to 70% of HD individuals, becoming more frequent with progression of the disease although lacking correlation with the number of CAG repeats. The findings include reduced sleep efficiency, insomnia, excessive daytime somnolence, REM sleep behavioral disorder, and periodic leg movements among others (Goodman & Barker, 2010; Videnovic, Leurgans, Fan, Jaglin, & Shannon, 2009). Of note a more recent study demonstrated that nonmanifest HD gene carriers have a higher frequency of disrupted sleep, suggesting that sleep abnormalities are early signs of the disease (Lazar, Panin, Goodman, et al., 2015). Undoubtedly, though, behavioral and cognitive abnormalities are the most common nonmotor features of HD and the major contributors to functional disability in these individuals (Ross, Pantelyat, Kogan, & Brandt, 2014). The aim of this chapter is to provide an overview of behavioral and cognitive abnormalities in HD, addressing their epidemiology, clinical features, and management.

2. BEHAVIORAL DISTURBANCES

Behavioral disturbances are often the most disabling features of HD. Several well-designed prospective studies demonstrate that they occur before the motor diagnosis is done (Biglan et al., 2009; Harrington et al., 2012; Huntington Study Group, 2016; Stout et al., 2011; Tabrizi et al., 2012, 2013). The recent publication of the results of the Prospective Huntington at Risk Observational Study show that subjects with CAG $\geq$ 37 repeats had statistically significant greater scores on the behavioral items of the UHDRS at baseline (Huntington Study Group, 2016). This is similar to the PREDICT and TRACK-HD studies which had a similar design although, unlike in PHAROS, the raters were not blinded to the genetic status of the participants (Huntington Study Group, 2016; Paulsen et al., 2008; Tabrizi et al., 2012, 2013). With passing of time, the differences in behavior between at risk individuals and control subjects become more marked (Epping et al., 2016), although this was not the case of the PHAROS cohort (Huntington Study Group, 2016). Despite their frequency and clinical importance, there are few valid rating scales for behavioral symptoms in HD. A recent study of the International Parkinson's Disease and Movement Disorders Society (MDS) concluded that there were just three rating scales classified as recommended exclusively for screening purposes in HD: the Irritability Scale for irritability, the Beck Depression Inventory-II, and the Hospital Anxiety and Depression Scale for depression. Of note the authors reviewed 27 scales for several behavioral domains: irritability, anxiety, depression, apathy, obsessive–compulsive behaviors, psychosis, and suicidal ideation (Mestre et al., 2016). It is possible that the lack of standardized and validated assessment tools accounts for some of the discrepancies between studies of behavioral abnormalities in HD.

The REGISTRY study is a large, prospective study observing the natural course, clinical spectrum, and management of HD in European countries (Handley, van Walsem, Juni, et al., 2011). It has been replaced by the multicontinental ENROLL-HD (Landwehrmeyer et al., 2016). Both studies include gene carriers regardless of the clinical status. Data reported on 1993 carriers of HD mutation of the REGISTRY cohort show that 42.1% of individuals had depression with 12.7% reporting as severe forms. There is an association of this finding with lower total functional capacity, obsessive compulsive behaviors, and suicide attempts (van Duijn et al., 2014). In general, these numbers are in line with other previous studies, although there are reports describing up to 76% of HD individuals with depression (van Duijn, Kingma, & van der Mast, 2007). Although other studies demonstrate that depression is more common at the early stages of HD and its frequency diminishes in late stages (Paulsen, Nehl, Hoth, et al., 2005), in the REGISTRY cohort depression the opposite happened.

George Huntington had already recognized suicide as a manifestation of HD (Huntington, 1872). Several contemporary studies confirm that the rate of suicides in individuals with HD is greater than in controls. In REGISTRY study, nearly 7% of HD individuals commit suicide, whereas in the general population, the figure is approximately 1%. There is an association between suicide attempts not only with depression but also with irritability/aggression, obsessive compulsive behaviors, and apathy (van Duijn et al., 2014). The proportion of gene carries that have suicidal ideation is actually higher, 9.9% (Hubers, van Duijn, Roos, et al., 2013). When the analysis is restricted to individuals at risk, suicidal ideation is present in 6.3% of individuals and associated with depression, hopelessness, irritability, aggression, anxiety, impulsivity, and elevated motor score (Anderson et al., 2016).

Irritability and aggression are observed in 38.2% of individuals of the REGISTRY study. This was rated as moderate to severe in 13.9% of the individuals and there was a direct correlation between stage of HD and frequency of this behavioral abnormality (Epping et al., 2016). A previous analysis of the same cohort reported irritability in 23% of gene carriers (Bouwens, van Duijn, van der Mast, Roos, & Giltay, 2015). This symptom is more common in males, younger patients, and individuals with psychosis and previous suicide attempt (van Duijn et al., 2014). In clinical practice, severe irritability and aggressiveness are among the most disabling features of HD, leading to intense distress for relatives and care givers in general.

Apathy is common in HD. In the survey of participants of the REGISTRY study, this neuropsychiatric finding was observed in 47.4% of individuals with increase of the frequency as disease progressed. In stages 4–5, 54.6% of individuals reported to have apathy (van Duijn et al., 2014). Another survey showed that 32% of mutation carriers far from the diagnosis had apathy, whereas this symptom was reported by 62% of early-HD patients (Martinez-Horta et al., 2016). In the REGISTRY database, apathy was associated with older age, previous suicide attempt, depression, obsessive compulsive behaviors, and lower total functional capacity (van Duijn et al., 2014).

Given the fact that obsessive compulsive behaviors are related to basal ganglia disorders, it is not surprising that they are found in 25.8% of HD individuals (van Duijn et al., 2014). Once more, the longer duration of the disease, the more frequent obsessions and compulsions are: in the REGISTRY cohort, the prevalence of moderate to severe obsessive compulsive behaviors increased from 4.5% in stage 1 to 25.8% in stages 4–5. A study of the PREDICT-HD cohort demonstrated that checking, cleanliness, compulsive rituals, and pathologic impulses were more common in premanifest gene carriers (Beglinger et al., 2008).

Psychotic disorders rarely happen in HD. Among 1993 mutation carries of the REGISTRY cohort, psychosis was recorded just in 4.1% of individuals. There was an association with intermediate stages of the disease, longer duration of HD, and lower total functional capacity (van Duijn et al., 2014).

3. COGNITIVE DYSFUNCTION

Cognitive impairment, one of the elements of the triad that characterizes the phenotype of HD, is always present in the course of the disease (Cardoso, 2009). There is a controversy if patients may present with isolated cognitive dysfunction characterizing a purely cognitive form of HD (Loy &

McCusker, 2013). Currently available data from prospective follow-up suggest that even when cognitive impairment is the major feature, patients also have a significant degree of motor dysfunction (Dorsey et al., 2013; Huntington Study Group, 2016; Tabrizi et al., 2013). HD is always mentioned as the prototype cause of the so-called subcortical dementia. In fact, in the early stages of the disease, most of the affected cognitive domains are related to basal ganglia dysfunction. In contrast, cortex-dependent domains such as semantic memory, language comprehension, and spatial awareness and orientation are usually preserved at this point (Dumas, van den Bogaard, Middelkoop, & Roos, 2013). The practical implication of these findings is that tools to assess cognition that are centered on cortical functions, such as the mini-mental status examination (MMSE), are not particularly well suited to study cognition in HD. Recent data show, however, that there is cortical involvement even in early stages of the disease. They may account, for instance, for visuomotor problems found in early stages of HD (Say, Jones, Scahill, et al., 2011). As the disease progresses, the pathological process disseminates involving cortical areas leading to a change the profile of the cognitive impairment. At advanced stages of the disease, patients invariably have dementia with global impairment of cognition (Papoutsi, Labuschagne, Tabrizi, & Stout, 2014).

As previously discussed in this chapter, current diagnosis of HD is solely centered on motor findings. Gene mutation carriers that are more than 10 years from diagnosis do not usually have cognitive abnormalities (Papoutsi et al., 2014). In contrast those who are 10 or fewer years from diagnosis display a wide range of impairment of cognitive functions: visuomotor integration, psychomotor speed, executive function, and emotion recognition (Harrington et al., 2012; Huntington Study Group, 2016; Papoutsi et al., 2014; Paulsen et al., 2008; Stout et al., 2011; Tabrizi et al., 2012, 2013). At the time of the diagnosis cognitive, abnormalities become more severe and more widespread in terms of domains: psychomotor speed, executive function, attention, episodic memory, working memory, learning, and emotion (Papoutsi et al., 2014). There is a relentless progression of the cognitive impairment over the course of the disease. Even using the MMSE, prospective follow up of HD patients show an annual decline of 0.7 points (95% CI, 0.6–0.8) (Dorsey et al., 2013).

More recently, there has been growing interest in the presence of anosognosia, lack of awareness of deficits. Patients with HD consistently under-report the presence of involuntary movements and under-estimate or deny cognitive and behavioral abnormalities (McCusker & Loy, 2014;

Sitek, Thompson, Craufurd, & Snowden, 2014). There is an inverse correlation between insight and severity of disease. This suggests that anosognsosia is at least partly related to nondominant prefrontal impairment (McCusker & Loy, 2014). There also suggestions that it may reflect a denial mechanism (Sitek et al., 2014). Regardless of the underlying causal mechanism, anosognosia in HD has important practical implications. First, individuals may place them at risk if engage in activities for which they are not fit. Second, health professionals must bear in mind this lack of awareness during the management of HD patients.

There are also recent studies in social cognition in HD. A meta-analysis of studies involving 2226 mutation carriers showed that patients with HD have severe impairment in recognizing negative emotions, particularly anger, disgust, and fear. These individuals also have deficits in theory of mind. Although less severe, there is also impairment of social cognition in premanifest HD (Bora, Velakoulis, & Walterfang, 2016).

4. MANAGEMENT

In the past few years, there has been an exponential growth in the number of publications related to HD. Unfortunately, to date, very little empirical evidence supports the therapeutic choices facing health professionals caring for HD patients. Management is almost exclusively based on personal choices, recommendations by experts, and off label medications (Killoran & Biglan, 2014; Shannon & Fraint, 2015).

Despite the high frequency of depression in HD, there are few and of low quality studies of antidepressants in these patients. Agents so far studied have been citalopram, venlafaxine, fluoxetine, atomoxetine, modafinil, lithium, and five different antipsychotics. Unfortunately, no studies were of adequate duration, size, or outcome, and no controlled trial resulted in improvement of depression (Moulton, Hopkins, & Bevan-Jones, 2014). Nevertheless, there is agreement among experts that antidepressant treatment should be started whenever depression is identified (Désaméricq et al., 2016). Many physicians choose selective serotonin reuptake inhibitors (SSRIs), particularly citalopram, since there is a trial studying cognitive effect of this agent that suggested antidepressant efficacy in HD patients (Beglinger, Adams, Langbehn, et al., 2014). Suicidal ideation, an issue related to depression and other domains of behavioral dysfunction in HD, is increased in mutation carriers. Unfortunately, there are no published studies of therapeutic intervention in suicidal ideation in HD. Practitioners usually provide aggressive treatment of depression and impulsivity in these patients as well as multidisciplinary approach including psychiatrist, psychologist, and social worker (Désaméricq et al., 2016).

As previously discussed, irritability and aggression are among the most disabling features of HD. Unfortunately, there are no controlled trials of interventions for these behavioral disorders. Current recommendations are based on experts' agreement (Groves et al., 2011). Management is based on the use of dopamine receptor blocking agents, particularly olanzapine and risperidone, if there is an association with psychosis. In the event of relationship to depression, SSRIs can be used (Désaméricq et al., 2016). Mood stabilizers, such as lithium and valproic acid, are occasionally reported to provide some benefit to these patients (Killoran & Biglan, 2014).

The opposite of irritability and aggression is apathy, another common behavioral problem in HD. In the few subjects where this is caused by depression, antidepressants can result in improvement of this symptom. However, in the majority of individuals, it is related to cognitive decline, a nonmotor finding for which there are few, if any, options (Désaméricq et al., 2016).

Psychosis in HD is often managed by the use of atypical antipsychotics, although this is based on experts' agreement rather than empirical data (Désaméricq et al., 2016; Shannon & Fraint, 2015). Interestingly, Europeans are more prone to use these agents than North Americans (Groves et al., 2011). Similarly, there are no guidelines based on empiric data on how to treat obsessions and compulsions in HD. Based on other conditions, most physicians use SSRIs in the rare instances where these behavioral problems are disabling (Désaméricq et al., 2016). Finally, a few trials of acetyl cholinesterase inhibitors have failed to show cognitive benefit to HD individuals (Li, Hai, Zhou, & Dong, 2015; Shannon & Fraint, 2015).

5. CONCLUSIONS

Nonmotor, particularly behavioral and cognitive, features are present in essentially all carriers of HD mutation. They are found before individuals meet current diagnostic criteria for the disease and become more frequent and severe over the course of the disease progression. These nonmotor findings account for a significant proportion of the disability of in HD. Given the lack of empirical data, the management of these nonmotor problems is based on experts' agreement.

REFERENCES

Anderson, K. E., Eberly, S., Groves, M., Kayson, E., Marder, K., Young, A. B., et al. (2016). Risk factors for suicidal ideation in people at risk for Huntington's disease. *Journal of Huntington's Disease*, *5*, 389–394.

Beglinger, L. J., Adams, W. H., Langbehn, D., et al. (2014). Results of the citalopram to enhance cognition in Huntington disease trial. *Movement Disorders*, *29*, 401–405.

Beglinger, L. J., Paulsen, J. S., Watson, D. B., Wang, C., Duff, K., Langbehn, D. R., et al. (2008). Obsessive and compulsive symptoms in prediagnosed Huntington's disease. *The Journal of Clinical Psychiatry*, *69*, 1758–1765.

Biglan, K. M., Ross, C. A., Langbehn, D. R., Aylward, E. H., Stout, J. C., Queller, S., et al. (2009). Motor abnormalities in premanifest persons with Huntington's disease: The PREDICT-HD study. *Movement Disorders*, *24*, 1763–1772.

Bora, E., Velakoulis, D., & Walterfang, M. (2016). Social cognition in Huntington's disease: A meta-analysis. *Behavioural Brain Research*, *297*, 131–140.

Bouwens, J. A., van Duijn, E., van der Mast, R. C., Roos, R. A., & Giltay, E. J. (2015). Irritability in a prospective cohort of Huntington's disease mutation carriers. *The Journal of Neuropsychiatry and Clinical Neurosciences*, *27*, 206–212.

Cardoso, F. (2009). Huntington disease and other choreas. *Neurologic Clinics*, *27*, 719–723.

Cardoso, F. (2014). Differential diagnosis of Huntington's disease: What the clinician should know. *Neurodegenerative Disease Management*, *4*, 67–72.

Cubo, E., Ramos-Arroyo, M. A., Martinez-Horta, S., Martínez-Descalls, A., Calvo, S., Gil-Polo, C., et al. (2016). Clinical manifestations of intermediate allele carriers in Huntington disease. *Neurology*, *87*, 571–578.

Désamérícq, G., Youssov, K., Charles, P., Saleh, N., Olivier, A., Sherer-Gagou, C., et al. (2016). Guidelines for clinical pharmacological practices in Huntington's disease. *Revue Neurologique*, *172*, 423–432.

Dorsey, E. R., Beck, C. A., Darwin, K., Nichols, P., Brocht, A. F., Biglan, K. M., et al. (2013). Natural history of Huntington disease. *JAMA Neurology*, *70*, 1520–1530.

Dumas, E. M., van den Bogaard, S. J., Middelkoop, H. A., & Roos, R. A. (2013). A review of cognition in Huntington's disease. *Frontiers in Bioscience (Scholar Edition)*, *5*, 1–18.

Epping, E. A., Kim, J. I., Craufurd, D., Brashers-Krug, T. M., Anderson, K. E., McCusker, E., et al. (2016). Longitudinal psychiatric symptoms in prodromal Huntington's disease: A decade of data. *American Journal of Psychiatry*, *173*, 184–192.

Fisher, E. R., & Hayden, M. R. (2014). Multisource ascertainment of Huntington disease in Canada: Prevalence and population at risk. *Movement Disorders*, *29*, 105–114.

Goodman, A. O., & Barker, R. A. (2010). How vital is sleep in Huntington's disease? *Journal of Neurology*, *257*, 882–897.

Groves, M., van Duijn, E., Anderson, K., Craufurd, D., Edmondson, M. C., Goodman, N., et al. (2011). An international survey-based algorithm for the pharmacologic treatment of irritability in Huntington's disease. *PLoS Currents Huntington Disease, Edition, 1*. http://dx.doi.org/10.1371/currents.RRN1259.

Handley, O., van Walsem, M., Juni, P., et al. (2011). Study protocol of Registry version 2.0 European Huntington's Disease Network (EHDN). *Hygeia Public Health*, *46*, 115–182.

Harrington, D. L., Smith, M. M., Zhang, Y., Carlozzi, N. E., Paulsen, J. S., & PREDICT-HD Investigators of the Huntington Study Group. (2012). Cognitive domains that predict time to diagnosis in prodromal Huntington disease. *Journal of Neurology, Neurosurgery, and Psychiatry*, *83*, 612–619.

Hubers, A. A., van Duijn, E., Roos, R. A., et al. (2013). Suicidal ideation in a European Huntington's disease population. *Journal of Affective Disorders*, *151*, 248–258.

Huntington, G. (1872). On chorea. *The Medical and Surgical Reporter: A Weekly Journal*, *26*, 317–321.

Huntington Study Group. (1996). Unified Huntington's disease rating scale: Reliability and consistency. *Movement Disorders, 11*, 136–142.
Huntington Study Group PHAROS Investigators. (2016). Clinical-genetic associations in the prospective Huntington at risk observational study (PHAROS): Implications for clinical trials. *JAMA Neurology, 73*, 102–110.
Killoran, A., & Biglan, K. M. (2014). Current therapeutic options for Huntington's disease: Good clinical practice versus evidence-based approaches? *Movement Disorders, 29*, 1404–1413.
Krause, A., Mitchell, C., Essop, F., Tager, S., Temlett, J., Stevanin, G., et al. (2015). Junctophilin 3 (JPH3) expansion mutations causing Huntington disease like 2 (HDL2) are common in South African patients with African ancestry and a Huntington disease phenotype. *American Journal of Medical Genetics. Part B, Neuropsychiatric Genetics, 168*, 573–585.
Landwehrmeyer, G. B., Fitzer-Attas, C. J., Giuliano, J. D., Gonçalves, N., Anderson, K. E., Cardoso, F., et al. (2016). Data analytics from Enroll-HD, a global clinical research platform for Huntington's disease. *Movement Disorders Clinical Practice*. http://dx.doi.org/10.1002/mdc3.12388.
Lanska, D. J., Lanska, M. J., Lavine, L., & Schoenberg, B. S. (1988). Conditions associated with Huntington's disease at death. A case-control study. *Archives of Neurology, 45*, 878–880.
Lazar, A. S., Panin, F., Goodman, A. O., et al. (2015). Sleep deficits but no metabolic deficits in premanifest Huntington's disease. *Annals of Neurology, 78*, 630–648.
Li, Y., Hai, S., Zhou, Y., & Dong, B. R. (2015). Cholinesterase inhibitors for rarer dementias associated with neurological conditions. *Cochrane Database of Systematic Reviews, 3*, CD009444.
Loy, C. T., & McCusker, E. A. (2013). Is a motor criterion essential for the diagnosis of clinical Huntington disease? Version 2. *PLoS Currents*, 5. pii: ecurrents.hd.f4c66bd51e8db11f55e1701af937a419.
Martinez-Horta, S., Perez-Perez, J., van Duijn, E., Fernandez-Bobadilla, R., Carceller, M., Pagonabarraga, J., et al. (2016). Neuropsychiatric symptoms are very common in premanifest and early stage Huntington's disease. *Parkinsonism & Related Disorders, 25*, 58–64.
McCusker, E., & Loy, C. T. (2014). The many facets of unawareness in Huntington disease. *Tremor and Other Hyperkinetic Movements, 4*, 257.
Mestre, T. A., van Duijn, E., Davis, A. M., Bachoud-Lévi, A. C., Busse, M., Anderson, K. E., et al. (2016). Rating scales for behavioral symptoms in Huntington's disease: Critique and recommendations. *Movement Disorders, 31*, 1466–1478.
Moulton, C. D., Hopkins, C. W., & Bevan-Jones, W. R. (2014). Systematic review of pharmacological treatments for depressive symptoms in Huntington's disease. *Movement Disorders, 29*, 1556–1561.
Papoutsi, M., Labuschagne, I., Tabrizi, S. J., & Stout, J. C. (2014). The cognitive burden in Huntington's disease: Pathology, phenotype, and mechanisms of compensation. *Movement Disorders, 29*, 673–683.
Paulsen, J. S., Langbehn, D. R., Stout, J. C., Aylward, E., Ross, C. A., Nance, M., et al. (2008). Detection of Huntington's disease decades before diagnosis: The PREDICT-HD study. *Journal of Neurology, Neurosurgery, and Psychiatry, 79*, 874–880.
Paulsen, J. S., Nehl, C., Hoth, K. F., et al. (2005). Depression and stages of Huntington's disease. *The Journal of Neuropsychiatry and Clinical Neurosciences, 17*, 496–502.
Ross, C. A., Pantelyat, A., Kogan, J., & Brandt, J. (2014). Determinants of functional disability in Huntington's disease: Role of cognitive and motor dysfunction. *Movement Disorders, 29*, 1351–1358.
Say, M. J., Jones, R., Scahill, R. I., et al. (2011). Visuomotor integration deficits precede clinical onset in Huntington's disease. *Neuropsychologia, 49*, 264–270.

Shannon, K. M., & Fraint, A. (2015). Therapeutic advances in Huntington's disease. *Movement Disorders, 30*, 1539–1546.
Sitek, E. J., Thompson, J. C., Craufurd, D., & Snowden, J. S. (2014). Unawareness of deficits in Huntington's disease. *Journal of Huntington's Disease, 2014*(3), 125–135.
Stout, J. C., Paulsen, J. S., Queller, S., Solomon, A. C., Whitlock, K. B., Campbell, J. C., et al. (2011). Neurocognitive signs in prodromal Huntington disease. *Neuropsychology, 25*, 1–14.
Tabrizi, S. J., Reilmann, R., Roos, R. A., Durr, A., Leavitt, B., Owen, G., et al. (2012). Potential endpoints for clinical trials in premanifest and early Huntington's disease in the TRACK-HD study: Analysis of 24 month observational data. *Lancet Neurology, 11*, 42–53.
Tabrizi, S. J., Scahill, R. I., Owen, G., Durr, A., Leavitt, B. R., Roos, R. A., et al. (2013). Predictors of phenotypic progression and disease onset in premanifest and early-stage Huntington's disease in the TRACK-HD study: Analysis of 36-month observational data. *Lancet Neurology, 12*, 637–649.
The Huntington's Disease Collaborative Research Group. (1993). A novel gene containing a trinucleotide repeat that is expanded and unstable on Huntington's disease chromosomes. *Cell, 72*, 971–983.
Underwood, M., Bonas, S., Dale, M., & REGISTRY Investigators of the European Huntington's Disease Network. (2017). Huntington's disease: Prevalence and psychological indicators of pain. *Movement Disorders Clinical Practice, 4*, 198–204. http://dx.doi.org/10.1002/mdc3.12376.
Vale, T. C., & Cardoso, F. (2015). Chorea: A journey through history. *Tremor and Other Hyperkinetic Movements, 5*. pii: tre-5-296.
van der Burgh, J. M., Björkqvist, M., & Brundin, P. (2009). Beyond the brain: Widespread pathology in Huntington's disease. *Lancet Neurology, 8*, 765–774.
van Duijn, E., Craufurd, D., Hubers, A. A., Giltay, E. J., Bonelli, R., Rickards, H., et al. (2014). Neuropsychiatric symptoms in a European Huntington's disease cohort (REGISTRY). *Journal of Neurology, Neurosurgery, and Psychiatry, 85*, 1411–1418.
van Duijn, E., Kingma, E. M., & van der Mast, R. C. (2007). Psychopathology in verified Huntington's disease gene carriers. *The Journal of Neuropsychiatry and Clinical Neurosciences, 19*, 441–448.
Videnovic, A., Leurgans, S., Fan, W., Jaglin, J., & Shannon, K. M. (2009). Daytime somnolence and nocturnal sleep disturbances in Huntington disease. *Parkinsonism & Related Disorders, 15*, 471–474.

CHAPTER FIFTY-ONE

Nonmotor Symptoms in Amyotrophic Lateral Sclerosis: A Systematic Review

Ton Fang, Felix Jozsa, Ammar Al-Chalabi[1]
Maurice Wohl Clinical Neuroscience Institute, King's College London, London, United Kingdom
[1]Corresponding author: e-mail address: ammar.al-chalabi@kcl.ac.uk

Contents

Abstract

Background: ALS is a progressive neurodegenerative disease with no curative treatment. Nonmotor symptoms presenting in ALS may cause significant distress, worsen prognosis, and affect survival. *Objective*: To systematically review evidence for the prevalence of nonmotor ALS symptoms, and treatment options. *Methods*: Multiple medical literature databases were searched and studies screened using predefined inclusion criteria. Of 4580 studies, 44 were eligible for inclusion with 25 relating to treatment and 19 to the prevalence of nonmotor symptoms in ALS. *Results*: Nonmotor symptoms involve neuropsychiatric, autonomic, gastrointestinal, and vascular systems, and affect between 5% and 80% of people with ALS. Screening tools for individual nonmotor symptoms are useful in classifying symptom severity and to compare between treatment options.

ISSN 0074-7742
http://dx.doi.org/10.1016/bs.irn.2017.04.009

Several methods to relieve nonmotor symptoms have been trialed with varying success rates. *Conclusions*: Many of the current studies of nonmotor symptoms in ALS have small sample sizes, requiring more evidence to increase precision in prevalence estimates. Further research is needed to assess the efficacy of current treatments and to find new therapies. Symptom relief or treatment of these nonmotor symptoms should therefore be considered during the clinical management of ALS.

1. INTRODUCTION

Amyotrophic lateral sclerosis (ALS) is a neurodegenerative disease of motor neurons in the brain and spinal cord. ALS is clinically characterized by the progressive loss of motor neurons, leading to increasing weakness and eventually death from neuromuscular respiratory failure. Age is a risk factor for ALS with peak onset around 60 years old (Knibb et al., 2016; Wijesekera & Leigh, 2009). Involvement of motor neurons is usually initiated in a single functional region (bulbar, cervical, lumbosacral, and thoracic) and progressively spreads to involve the entire voluntary motor system (Cedarbaum et al., 1999). Functional impairment is initially seen in individuals with difficulties performing activities of daily living (Cedarbaum et al., 1999), progressively deteriorating to involve respiratory muscle weakness and dysphagia requiring medical intervention. Many studies have focused on the motor symptoms and survival of ALS, while the prevalence and treatment of nonmotor symptoms have been comparatively less considered.

ALS is a multisystem disorder, and unlike Parkinson's disease (Chaudhuri, Healy, Schapira, & National Institute for Clinical, 2006) has nonmotor symptoms that are poorly defined with unexplained mechanisms. Like other neurodegenerative diseases, nonmotor symptoms in ALS can arise from pathological disease spread of motor areas into neighboring nonmotor regions of the brain (Brundin, Melki, & Kopito, 2010; Gunther et al., 2016) or arising from other unknown causes. Common symptoms such as dementia, fatigue, pain, and sialorrhea frequently contribute to depression experienced by people with ALS (Pizzimenti, Aragona, Onesti, & Inghilleri, 2013). Many of these nonmotor symptoms are under diagnosed and left unreported during clinical consultation, where patients are either embarrassed or unsuspecting that these symptoms relate to their underlying pathological condition (Chaudhuri, Healy, et al., 2006). Without a clear definition of what nonmotor symptoms of ALS include, it is impossible to specifically target, screen, and treat them.

In ALS, nonmotor symptoms can be broadly categorized into neuropsychiatric, autonomic, vascular, and gastrointestinal disturbances, and are symptoms that do not have an underlying motor pathology in the form of hypertonia, hyperreflexia, weakness, or spasticity (Table 1). The distinction between motor and nonmotor causes is not always clear-cut, because the direct manifestation of motor weakness is most easily recognized in limb weakness rather than nonlimb weakness. For those, with some motor component, such as sialorrhea from dysphagia, treatment of the motor cause may not always be possible, or may not resolve the symptom. Symptoms that are not a result of limb weakness, but are nevertheless a consequence of motor neuron loss, such as sialorrhea, dyspnoea, and dysphagia, should therefore be considered and treated as nonmotor symptoms.

For common symptoms, prevalence ranges from 10% for major depression (Patten, Svenson, White, Khaled, & Metz, 2007; Rabkin et al., 2005) to 50% for sialorrhea (Stone & O'Leary, 2009) and are considered to diminish

Table 1 Nonmotor Symptoms in Amyotrophic Lateral Sclerosis

Nonmotor Symptoms	
Neuropsychiatric	Depression
	Anxiety
	Suicidal ideation
	Cognitive-behavioral impairment
	Fatigue
	Pseudobulbar affect
	Sleep disturbance
Autonomic	Pain
	Dyspnoea
	Urinary incontinence
Gastrointestinal	Sialorrhea
	Dysphagia (weight loss)
	Constipation
Vascular	Dyslipidemia
Others	Itching
	Pressure sores

health outcomes and reduce quality of life (QoL) for patients (Pizzimenti et al., 2013). Screening tools for nonmotor symptoms have been suggested, such as the nonmotor symptom questionnaire (NMSQuest) (Gunther et al., 2016), which has been validated for use in Parkinson's disease (Chaudhuri, Martinez-Martin, et al., 2006), to enable classification of symptom severity and identification for treatment. Other tools similar to those used for disease staging (Balendra et al., 2014; Fang, Al Khleifat, et al., 2017; Roche et al., 2012; Tramacere et al., 2015), measure the severity of nonmotor symptoms, enabling comparisons between treatment groups (Table 2). Treatments may be pharmacological or non-pharmacological and include invasive options.

Given that in neurodegenerative diseases, nonmotor symptoms may cause significant distress to patients, and worsen prognosis (Dupuis et al., 2008; Koerner et al., 2013; Moreau et al., 2012), an overview of the frequency and treatment options is much needed. To date, no systematic review has been established to determine the prevalence and available therapies for nonmotor symptoms in ALS. This systematic review aims to investigate the prevalence of and various treatments for nonmotor symptoms in ALS; and to evaluate the efficacy, side effects, and effectiveness of these treatments, with the goal to allow healthcare professionals to make informed clinical decisions in improving patient health outcomes.

2. METHODS

2.1 Search Strategy

The Preferred Reporting Items for Systematic Reviews and Meta-Analysis (PRISMA) guidelines (Liberati et al., 2009) were followed. A combination of online databases, Ovid Medline, Embase, PubMed, PsychInfo, and Web of Science, were searched using combinations of the terms "nonmotor symptoms," "prevalence," "treatment," "clinical studies," and "ALS". Relevant articles published since the start of the database until the last search conducted on 21 January 2017 were screened. The full protocol is listed in the PROSPERO register (CRD: 42017057137) (Fang, Jozsa, & Al-Chalabi, 2017).

2.2 Inclusion/Exclusion Criteria

The Participants, Interventions, Comparators, Outcomes, and Study design (PICOS) checklist was used to describe the inclusion criteria used during eligibility screening (Liberati et al., 2009). Participants were patients of

Table 2 Nonmotor Symptom Rating Scales

Nonmotor Rating Scales	
Depression	Hospital Anxiety and Depression Scale (HADS-D)
	ALS Depression Index (ADI)
	Beck's Depression Index (BDI)
	Centre for Epidemiological Studies Depression scale (CESD)
	Major Depression Inventory (MDI)
	Patient Health Questionnaire (PHQ)
	Zung self-rating depression score (SDS)
Anxiety	Hospital Anxiety and Depression Scale (HADS-A)
Suicidal ideation	Beck Hopelessness Scale (BHS)
Cognitive-behavioral impairment	ALS Cognitive Behavioral Screen (ALS-CBS)
	Mini-Mental State Exam (MMSE)
	Frontal Systems Behavior Scale (FrSBe)
Fatigue	Fatigue Severity Scale (FSS)
Pseudobulbar affect	Centre for Neurologic Study Liability Scale (CNS-LS)
Sleep disturbance	Apnoea index
	Epworth Sleepiness Scale (ESS)
Pain	Brief Pain Inventory (BPI)
	Ashworth Spasticity Score (ASS)
	Sullivan Pain Catastrophizing Scale (PCS)
Urinary incontinence	International Consultation on Incontinence Modular Questionnaire (ICIQ-SF)
	Urinary Distress Inventory (UDI-6)
Sialorrhea	Sialorrhea Severity Scale (SSS)
	Drooling Severity and Frequency Scale (DSFS)
Dysphagia (weight loss)	Body Mass Index (BMI)
Constipation	Cleveland Clinic Incontinence Score (CCIS)
Quality of life	The Schedule for the Evaluation of Individual Quality of Life (SEI-QoL)
	ALS specific Quality of Life (ALSSQoL)
	McGill Quality of Life (McGill-QoL)
	Short Form-36 (SF36)

any age reported as diagnosed with ALS and experiencing any of the nonmotor symptoms in the search criteria. Interventions included treatment for nonmotor symptoms, studies of the prevalence of nonmotor symptoms, or screening tools for nonmotor symptoms in ALS. Comparators comprised articles mentioning treatments or placebo-controlled trials for nonmotor symptoms. For studies establishing prevalence or describing screening tools, comparators were not analyzed because the study designs did not allow sufficient evidence. Outcomes measuring treatment efficacy, adverse reactions, treatment duration, and treatment dosage were all accepted. Study design included all clinical studies that were experimental (randomized-controlled trials, controlled clinical trials) or observational (cohort, case-controlled, cross-sectional studies) in nature. Searches were limited to English language and human studies, with reviews, case reports, and nonfull-text articles excluded. Articles were assessed for eligibility initially by title, abstract and then full text, with studies being excluded if they did not meet our PICOS criteria. Eligibility was assessed by two authors (T.F. and A.A.) and any disagreements were resolved after discussion.

2.3 Quality Assessment

The "Consolidated Standards of Reporting Trials" (CONSORT) (Moher et al., 2010) was used to assess the risk of reporting bias of experimental studies and the "Strengthening the Reporting of Observational Studies in Epidemiology" (STROBE) statement (von Elm et al., 2007) was used for observational studies. Articles were assigned a bias grade of low, medium, or high depending on the number of items checked using the relevant list. A low possibility of bias was assigned to studies achieving more than 66% of the items, medium was given to studies scoring between 33% and 66% and high for studies scoring less than 33%. The CONSORT list contained 25 items and STROBE list 22 and studies were allocated a low (C: 17–25, S: 15–22), medium (C: 9–16, S: 8–21), or high risk (C: 0–8, S: 0–7) depending on the nature of the study.

Data quality was scored using the GRADE quality assessment tool (Atkins et al., 2004) to highlight the quality and strength of the evidence. Studies were rated in accordance with the grading criteria of high, moderate, low, and very low. Evidence derived from randomized-controlled trials was assumed to be of high quality and observational studies assumed to be of low quality. Studies could be upgraded or downgraded based on the GRADE criteria (Atkins et al., 2004). Upgrades occurred when articles had a large magnitude of effect size, adjusted for all plausible confounders, or used a dose–response gradient. Downgrades occurred when articles had limitations in the study design, indirectness of evidence, unexplained inconsistencies in results, low precision in results, or a high probability of publication bias. Any studies which were upgraded or downgraded had the reason for this decision recorded. Data extraction was conducted on Covidence using Cochrane's standardized extraction template (Veritas-Health-Innovation, 2017). Quality assessments and data extraction were conducted independently by two review authors (T.F. and F.J.), with any discrepancies later resolved through discussion.

2.4 Data Analysis

Due to the heterogeneity between studies involving different nonmotor symptoms which use an assortment of interventions and comparisons, there was not enough homogeneous data for a meta-analysis. Data analysis was therefore descriptive.

3. RESULTS

3.1 Studies

Of 4580 records identified, after removal of duplicates, animal trials, and non-English language texts, 4205 articles were screened and eligibility tested using the PICOS criteria. 44 studies were left, of which 25 were related to treatment (Table 3) and 19 to prevalence (Table 4) of nonmotor symptoms in ALS (Fig. 1). One study determining the prevalence of depression in ALS showed a high risk of bias on the STROBE grading criteria and was excluded from this review on that basis. Apart from the excluded study, no other studies had a high risk of bias on STROBE/CONSORT checklists or a very low quality of evidence using the GRADE criteria. There were no instances where both an upgrade and downgrade occurred simultaneously for a single study using the GRADE scoring.

3.2 Neuropsychiatric

Depression is one of the major cognitive and psychiatric disorders encountered by people with ALS, affecting between 6% and 29% of patients (Atassi et al., 2011; McElhiney et al., 2009; Rabkin et al., 2005), depending on the criteria used and timing of the survey. This is likely to be a significant under

Table 3 Study Characteristics (Treatment)

Author/s (Year)	Objective (NMS Subgroup)	Participants and Study Design (n = Group/ Comparator)	Intervention	Summary of Outcome Measures	Summary of Results	Risk of Bias (CONSORT/ STROBE)	Evidence Quality (GRADE)
Aksu and Citak-karakaya (2002)	Pain	Retrospective cohort study, $n = 13/13$	Supervised exercise program Home exercise program	• VAS (pain intensity) • Muscle stiffness • Muscle strength • Muscle shortening	Supervised exercise program reduced pain and improved muscle function	STROBE: Low	Low
Assouline et al. (2014)	Sialorrhea	Prospective cohort study, $n = 30/20$	10 Gy 20 Gy	• SSS	Radiotherapy reduces sialorrhea	STROBE: Low	Medium
Beck et al. (2005)	Autonomic dysfunction	Prospective cohort study, $n = 13$	Intrathecal BDNF 25/150 μg/day	• Sympathetic function test • Parasympathetic function test • Serum noradrenaline	Intrathecal BDNF has no effect on autonomic dysfunction	STROBE: Medium	Low
Bourry et al. (2013)	Sialorrhea	Retrospective cohort study, $n = 21$	Electron or photon fraction radiotherapy	• Personal interview • ALSFRS	Radiotherapy reduces sialorrhea	STROBE: Medium	Low
Brooks et al. (2004)	Pseudobulbar affect	Randomized-controlled trial, $n = 65/30/34$	AVP-923 or Dextromethorphan or Quinidine	• CNS-LS • Crying or laughing • VAS (QoL) • VAS (QoR)	AVP-923 reduces pseudobulbar events	CONSORT: Low	High

Costa et al. (2008)	Sialorrhea	Open label prospective study, $n = 16$	Botox B	• VAS (severity/disability) • Drooling severity and frequency scale • VAS (QoL) • Cotton wool weight	Botox B reduces sialorrhea	CONSORT: Medium	Medium
Dupuis et al. (2008)	Dyslipidemia	Case–control study, $n = 368/286$	Survival	• Serum lipids • Liver autopsy	Dyslipidemia protects against mortality	STROBE: Low	Medium
Gilio et al. (2010)	Sialorrhea	Open-label controlled trial, $n = 26$	Botox A	• VAS (severity/disability) • Cotton wool weight	Botox A reduces sialorrhea	CONSORT: Medium	Medium
Guidubaldi et al. (2011)	Sialorrhea	Randomized-controlled trial, $n = 7/7$	Botox A or Botox B, crossover	• Cotton wool weight	Both Botox reduces sialorrhea, latency shorter for Botox B	CONSORT: Low	High
Jackson et al. (2009)	Sialorrhea	Randomized-controlled trial, $n = 11/9$	Botox B or saline	• Global impression score • VAS (severity/thickness) • VAS (caregiver) • ALSFRS • SEIQoL-DW	Botox B reduces sialorrhea	CONSORT: Low	High

Continued

Table 3 Study Characteristics (Treatment)—cont'd

Author/s (Year)	Objective (NMS Subgroup)	Participants and Study Design (n = Group/ Comparator)	Intervention	Summary of Outcome Measures	Summary of Results	Risk of Bias (CONSORT/ STROBE)	Evidence Quality (GRADE)
Katzberg et al. (2013)	Sleep	Prospective cohort study, $n = 12$	Home NIV	• Apnoea index • Total sleep period • Total sleep time • Sleep efficiency • REM duration	NIV improved maximum overnight oxygen saturation and reduced time spent below 90% oxygen saturation	STROBE: Low	Low
Kleinbub et al. (2015)	Depression	Prospective cohort study, $n = 15/15$	Hypnosis or control	• DSQ • HSS • PCI • HADS • ALSSQoL • ALSFRS	Hypnosis reduced patient's anxiety and reduced caregiver's anxiety and depression	STROBE: Low	Low
Koerner et al. (2013)	Weight loss	Prospective cohort study, $n = 121$		• BDI • SF36 • ALSFRS	Patients taking high-calorie nutritional supplements and had PEGs had a better weight stabilization and QoL	STROBE: Medium	Low

Korner et al. (2013)	Depression	Prospective cohort study, $n = 38$	Speech therapy and/or communication device	• BDI • SF36 • ALSFRS	Speech device improves depression and QoL	STROBE: Medium	Low
Lyall et al. (2001)	Sleep	Prospective cohort study, $n = 16$	NIV	• SF36 • ALSFRS	NIV improves QoL	STROBE: Low	Low
Maggiani et al. (2016)	Pain	Randomized-controlled trial, $n = 7/7$	OMT or normal physiotherapy	• Somatic dysfunction • BPI • McGill-QoL	OMT does not improve pain or QoL	STROBE: Low	Medium
McClelland et al. (2008)	Pain	Retrospective cohort study, $n = 8$	Intrathecal baclofen	• VAS (Pain) • Ashworth spasticity score	Intrathecal baclofen reduced pain	STROBE: Low	Low
Norris, Tan, Fallat, and Elias (1993)	Weight loss	Randomized-controlled trial, $n = 5/8$	Physostigmine or placebo, crossover	• Weight loss • ALS score • FVC	Physostigmine has no effect on weight loss	CONSORT: Medium	Medium
Rabkin et al. (2009)	Fatigue	Randomized-controlled trial, $n = 21/7$	Modafinil or placebo	• CGI • FSS • ESS • BDI • RFS • VAS (stamina/energy/depression)	Modafinil did not improve fatigue, but improved CGI, VAS (energy/stamina)	CONSORT: Low	High

Continued

Table 3 Study Characteristics (Treatment)—cont'd

Author/s (Year)	Objective (NMS Subgroup)	Participants and Study Design (n = Group/ Comparator)	Intervention	Summary of Outcome Measures	Summary of Results	Risk of Bias (CONSORT/ STROBE)	Evidence Quality (GRADE)
Raglio et al. (2016)	Depression and anxiety	Randomized-controlled trial, $n = 15/15$	Active music therapy or standard care	• ALSFRS • HADS • McGill-QoL • Music Therapy Rating Scale	Active music therapy had no effect on depression or anxiety	CONSORT: Medium	High
Van Groenestijn et al. (2015)	Depression	Randomized-controlled trial, $n = 10/5$	CBT or standard care	• SF36 • ALSAQ-40 • HADS • CSI	CBT had no effect on depression	CONSORT: Low	Medium
Verschueren, Monnier, Attarian, Lardillier, and Pouget (2009)	Weight loss	Prospective cohort study, $n = 30/35$	PN or PEG	• BMI • ALSFRS • SVC	PN is equally effective to PEG with respiratory insufficiency, but PEG without respiratory insufficiency is more effective than other groups	STROBE: Low	Low
Vrijsen et al. (2015)	Sleep	Prospective cohort study, $n = 24$	NIV	• PSQI • ESS • REM duration • QoL	NIV increased PSQI, ESS, and REM duration	STROBE: Low	Low

Weber, Goldman, and Truniger (2010)	Pain	Randomized-controlled trial, $n = 14/13$	THC or placebo, crossover	• VAS (cramp/fasciculation) • Number of cramps • ALSAQ-40 • SDQ • FAACT • HADS	THC has no effect on pain	CONSORT: Low	High
Weikamp et al. (2016)	Sialorrhea	Randomized-controlled trial, $n = 10/10$	Radiotherapy or Botox A	• DRS • Degree of salivation • Global change of drooling • Level of satisfaction	Radiotherapy is as effective as Botox A at reducing sialorrhea at 4 weeks after treatment. Radiotherapy have a higher reduction than Botox A at 12 weeks	CONSORT: Low	High

ALS, amyotrophic lateral sclerosis; *ALSAQ-40*, ALS Assessment Questionnaire; *ALSFRS*, ALS Functional Rating Scale; *ALSSQoL*, ALS-Specific QoL; *AVP 923*, dextromethorphan and quinidine; *BDI*, Beck's Depression Index; *Botox*, botulinum toxin; *BPI*, Brief Pain Inventory; *CBT*, Cognitive Behavioral Therapy; *CGI*, Clinical Global Impressions-Improvement Scale; *CNS-LS*, Centre for Neurologic Study Lability Scale; *CONSORT*, Consolidated Standards of Reporting Trials; *CSI*, Caregiver Strain Index; *DRS*, Drooling Rating Scale; *DSQ*, Defense Style Questionnaire; *ESS*, Epworth Sleepiness Scale; *FAACT*, the Functional Assessment of Anorexia/Cachexia Therapy; *FSS*, Fatigue Severity Scale; *FVC*, Forced Vital Capacity; *GRADE*, Grading Quality of Evidence and Strength of Recommendations; *Gy*, Gray unit; *HADS*, Hospital Anxiety and Depression Scale; *HSS*, Hypnoidal State Score; *McGill-QoL*, McGill QoL; *NIV*, noninvasive ventilation; *OMT*, osteopathic manual treatment; *PCI*, Phenomenology of Consciousness Inventory; *PEG*, percutaneous endoscopic gastrostomy; *PN*, parenteral nutrition; *PSQI*, Pittsburgh Sleep Quality Assessment; *QoL*, quality of life; *QoR*, quality of relationship; *REM*, rapid eye movement; *RFS*, role function scale; *SDQ*, sleep Disorders Questionnaire; *SEIQoL-DW*, the Schedule for Evaluation of Individual QoL; *SF36*, Short Form Survey; *SSS*, Sialorrhea Scale Score; *STROBE*, Strengthening the Reporting of Observational Studies in Epidemiology; *THC*, tetrahydrocannabinol; *VAS*, Visual Analogue Scale.

Table 4 Study Characteristics (Prevalence)

Author (Year)	Objective (NMS Subgroup)	Study Design, Population and Country (n = Group/ Comparator)	Mean Age (SD)	Prevalence	Risk of Bias (STROBE)	Evidence Quality (GRADE)
Atassi et al. (2011)	Depression	Cross-sectional study, ALS outpatient clinic, USA ($n = 127$)	61	• 29% Mild depression • 6% Severe depression	Low	Medium
Bock, Duong, Kim, Murphy, and Lomen-Hoerth (2015)	Cognitive-behavioral impairment	Prospective cohort study, ALS outpatient clinic, USA ($n = 86$)	64.1 (9.5)	• 62% Cognitive impairment • 37% Behavioral impairment	Low	Medium
Chio et al. (2012)	Pain	Case-controlled study, population, Italy ($n = 160/160$)	62.4 (10.6)	• 57% Pain in ALS vs 33% Pain in controls • Pain more frequent in extremities • Pain correlated with worsened functional score	Low	Low
Fang et al. (2008)	Suicide	Retrospective cohort study, Swedish inpatient register, Sweden ($n = 6642$)	67.6 (12.1)	• Sixfold increased risk of suicide in people with ALS • People who committed suicide were on average 7 years younger at the time of hospitalization • Highest relative risk of suicide within the first year of hospitalization	Low	Low
Ferentinos et al. (2011)	Depression	Prospective cohort study, tertiary referral centre, Greece ($n = 37$)	62 (10.7)	• 13% Major depression following onset of ALS	Medium	Low

Hanisch, Skudlarek, Berndt, and Kornhuber (2015)	Pain	Cross-sectional study, outpatient clinic, Germany ($n = 46/46$)	64.0 (10.8)	• 78% Pain in ALS • 54% Pain in control	Low	Low
Ho, Ruthazer, and Russell (2011)	Pain	Retrospective cohort study, medical records, USA ($n = 193$)	63	• 23% Shoulder pain	Low	Low
Massman et al. (1996)	Cognition	Prospective cohort study, ALS outpatient clinic, USA ($n = 146$)	59.1 (12.9)	• 36% Neuropsychologically impaired	Low	Low
McElhiney, Rabkin, Gordon, Goetz, and Mitsumoto (2009)	Fatigue and depression	Prospective cohort study, ALS outpatient clinic, USA ($n = 223$)	61 (12)	• 44% Fatigue • 7% Major depression	Low	Low
Moisset et al. (2015)	Pain	Cross-sectional study, ALS outpatient clinic, France ($n = 93$)	62 (12.5)	• 66% Pain • 9% Neuropathic pain	Low	Low
Moreau et al. (2012)	Hypertension	Case-controlled study, population, France ($n = 102/408$)	62.7 (12.5)	• 57% Hypertension in ALS vs 41% in controls • 26% Smoking in ALS vs 15% in controls	Low	Low
Nuebling et al. (2014)	Bladder and bowel dysfunction	Prospective cohort study, ALS outpatient clinic, Germany ($n = 43$)	66.5 (9.6)	• Urinary incontinence higher in 60+ ALS than baseline population • 46% Constipation • 9% Stool incontinence	Low	Low

Continued

Table 4 Study Characteristics (Prevalence)—cont'd

Author (Year)	Objective (NMS Subgroup)	Study Design, Population and Country (*n* = Group/Comparator)	Mean Age (SD)	Prevalence	Risk of Bias (STROBE)	Evidence Quality (GRADE)
Ogawa, Tanaka, and Hirata (2009)	Cognition	Case-controlled study, Japan ($n = 19/19$)	67.7 (7.4)	• No global cognitive impairment difference	Medium	Low
Patten et al. (2007)	Affective disorders	Retrospective cohort study, population, Canada ($n = 336$)	—	• 9% Affective disorder	Low	Low
Pizzimenti et al. (2013)	Depression and pain	Cross-sectional study, ALS outpatient clinic, Italy ($n = 36$)	63.7 (10.9)	• 72% Pain	Low	Low
Rabkin et al. (2005)	Depression	Prospective cohort study, ALS outpatient clinic, USA ($n = 80$)	Range: 27–85	• 9% Major depression	Low	Low
Tremolizzo et al. (2014)	Primitive reflexes	Prospective cohort study, ALS outpatient clinic, Italy ($n = 50$)	63.9 (10.4)	• Palmomental, corneomandibular, and glabella tap reflexes were more prevalent in ALS patients	Medium	Low
Wallace et al. (2014)	Pain	Case-controlled study, ALS outpatient clinic, UK ($n = 42/41$)	63	• 85% Pain in ALS • 35% Pain in control	Low	Low
Witgert et al. (2010)	Cognition and Behavior	Prospective cohort study, ALS outpatient clinic, USA ($n = 225$)	57.7 (14.0)	• 24% Behavioral dysfunction	Low	Low

ALS, amyotrophic lateral sclerosis.

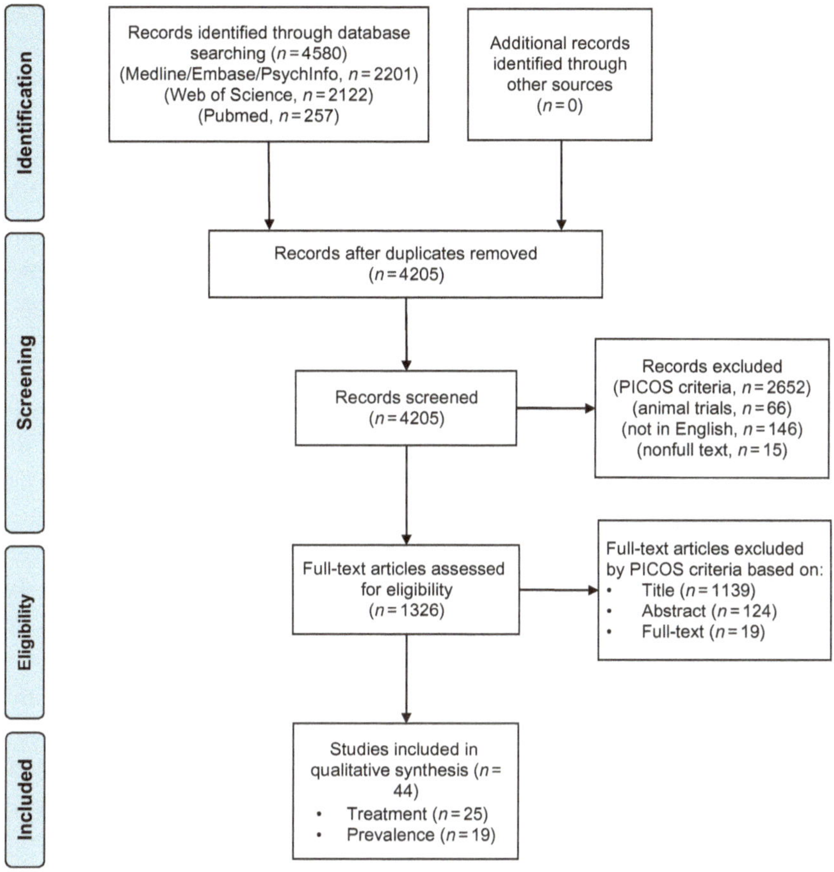

Fig. 1 PRISMA (preferred reporting items for systematic reviews and meta-analysis) flow diagram.

estimate. Criteria for diagnosing depression used by clinicians, such as the DSM-4 or ICD-9, are useful in indicating clinical depression (Patten et al., 2007). However, the severity of depression is difficult to stratify using these methods and therefore multiple severity scales have been designed for this purpose. Most frequently the ALS Depression Inventory (ADI-12) (Hammer, Hacker, Hautzinger, Meyer, & Kubler, 2008), Beck's Depression Inventory (Hautzinger, Keller, & Kühner, 2006), Hospital Anxiety and Depression Scale (Zigmond & Snaith, 1983), Zung Self-Rating Depression Scale (Zung, 1965), and Patient Health Questionnaire depression scales are used to indicate the extent of depression (Spitzer, Kroenke, Williams, & Group, 1999), with ADI-12 specific to ALS.

Mild depression, associated with low mood, anhedonia, apathy, and increased irritability while being able to function socially, was reported by 29% of patients using the ADI-12 questionnaire (Atassi et al., 2011). Major depression where social functioning is also impaired was seen in only 6% of this patient cohort (Atassi et al., 2011), with other studies reporting similar levels at 7% (McElhiney et al., 2009), 9% (Patten et al., 2007; Rabkin et al., 2005), and 13% (Ferentinos et al., 2011). In healthy populations, a 12-month prevalence of major depression affected between 5% and 6% of people (Bromet et al., 2011; Hasin, Goodwin, Stinson, & Grant, 2005; Kessler et al., 2003), suggesting that the risk of major depression in people with ALS is slightly higher than that of the normal population. However, this depends on the population of the ALS study being compared with and therefore further studies using larger sample sizes preferably matched with controls are needed to confirm this. The duration of disease from onset does not seem to correlate with depression (Atassi et al., 2011; McElhiney et al., 2009; Rabkin et al., 2005), however, when the severity staging of disease is taken into consideration, disease progression correlates with worsening of depression and anxiety (Jones et al., 2014). Many ALS patients with depression are treated with antidepressants, the most frequently used being selective serotonin reuptake inhibitors followed by dopamine reuptake inhibitors and serotonin–noradrenaline reuptake inhibitors (Atassi et al., 2011). Hypnosis and active music therapy have also been trialed as methods for improving psychological disorders in ALS, showing no effects on depression (Kleinbub et al., 2015; Raglio et al., 2016) but hypnosis is effective at reducing anxiety (Kleinbub et al., 2015). Mindfulness is also being considered a method for improving depression symptoms, where meditation techniques may be an effective intervention for treatment of depression (Gould et al., 2015; Pagnini, Phillips, Bosma, Reece, & Langer, 2015).

Depression may also stem from other nonmotor symptoms such as poor sleep, pain, fatigue, weight loss, and psychological distress. A study examining the use of tetrahydrocannabinol, the psychoactive compound found in cannabis, in treating pain in ALS showed no effect on depression (Weber et al., 2010). Cognitive behavioral therapy was also ineffective at reducing depression in psychologically distressed patients (Van Groenestijn et al., 2015). A randomized-controlled trial into the effectiveness of modafinil, a drug promoting wakefulness, in treating fatigue showed no effect on depression compared with placebo (Rabkin et al., 2009). Another study showed no differences for depression in patients with or without weight loss (Koerner et al., 2013).

Suicidal ideation is more frequent in ALS patients with depression. A population study in Sweden found a near sixfold increase in suicide over a 40-year period for people with ALS. Highest risk of suicide was seen during the first year after diagnosis with an 11.2 risk ratio. Length of time since diagnosis reduced this risk to 5.5 for 1–3 years and to 2.4 after having the diagnosis for 3 years (Fang et al., 2008).

Severe fatigue is described by 30%–40% of people with ALS (McElhiney et al., 2009; Rabkin et al., 2009) and patients with fatigue are more likely to be depressed (McElhiney et al., 2009). Fatigue is characterized by lack of energy, loss of stamina, sleepiness, and diminished alertness, and can be rated based on the Fatigue Severity Scale (Krupp, LaRocca, Muir-Nash, & Steinberg, 1989). Modafinil is a eugeroic drug used to promote wakefulness but has not shown a significant effect in reducing severe fatigue for people with ALS, although this may be because the study was underpowered (Rabkin et al., 2009). The randomized-controlled trial showed that people with ALS taking modafinil had a mean reduction of 21 points on the Fatigue Severity Scale, when compared to the 10 points for placebo $P = 0.066$, with increased patient-reported energy and stamina. Side effects encountered were agitation, headaches, nausea, and dizziness (Rabkin et al., 2009).

Emotional lability, also known as pseudobulbar affect, is a nonmotor symptom characterized by the loss of control for emotional expressions, with episodes of disinhibited laughing or crying. Up to 50% of people with ALS may have emotional lability and it is more common in those with pseudobulbar palsy. Dextromethorphan, an *N*-methyl-D-aspartate (NMDA) receptor antagonist, has been shown to be ineffective at treating pseudobulbar affect in ALS (Brooks et al., 2004). However, quinidine alters the pharmacokinetics of dextromethorphan by reducing first pass hepatic metabolism and in turn increases the bioavailability of the NMDA antagonist. A randomized control trial using a combined dose of dextromethorphan and quinidine showed a reduction in the frequency of episodes of uncontrolled laughing or crying (Panitch et al., 2006).

Cognitive impairment in ALS is thought to overlap with frontotemporal dementia and includes apathy, executive dysfunction and disinhibition. Up to 60% of people with ALS exhibit some form of measurable cognitive impairment (Bock et al., 2015; Massman et al., 1996; Ogawa et al., 2009; Witgert et al., 2010), with 25% of people experiencing frontal lobe-mediated dysfunction (Witgert et al., 2010). Frequently used neuropsychological tests to determine cognitive impairments are the Wisconsin Card Sorting Test (Heaton, Chelune, Curtiss, Kay, & Talley, 1993), Stroop Test

(Golden & Freshwater, 1978) and Visual Search and Attention Test (Trenerry, Crosson, DeBoe, & Leber, 1990), but the most sensitive measure is reduced verbal fluency (Abrahams et al., 1996). Other methods involving evoked receptor potentials have also been used (Ogawa et al., 2009). An effective, validated screening tool for detecting cognitive impairment is the Edinburgh Cognitive Assessment Screen, which allows for a useful assessment in about 20 min by a trained operator (Abrahams, Newton, Niven, Foley, & Bak, 2014).

Sleep disturbance is commonly reported in ALS. Disease spread into neighboring brain regions controlling sleep–wake cycles could be contributing factors to sleep disorders in ALS. Other factors such as pain, limited mobility, and skeletal muscle weakness leading to respiratory distress may also contribute to sleep difficulties. Sleep disorders can be measured by the Epworth Sleepiness Scale (Johns, 1991), apnoea index, total sleep time, and sleep efficiency. Respiratory weakness causing daytime sleepiness is assessed by overnight oximetry, looking for frequent oxygen desaturations during REM sleep. As people with ALS eventually reach respiratory failure and require noninvasive ventilation (NIV), prospective studies have considered the effects of NIV on sleep. Two studies found an improvement in night-time oxygen saturation after 1 month of nocturnal NIV (Katzberg et al., 2013; Vrijsen et al., 2015) but this was not seen in a third study (Lyall et al., 2001). Sleep quality was also improved, with increased slow wave sleep and rapid eye movement sleep after intervention with NIV (Vrijsen et al., 2015).

3.3 Autonomic

Musculoskeletal pain often arise as a result of secondary complications from motor weakness in ALS and is reported to affect 20%–80% of patients depending on the type and location of pain and disease duration (Chio et al., 2012; Hanisch et al., 2015; Ho et al., 2011; Moisset et al., 2015; Pizzimenti et al., 2013; Wallace et al., 2014). The most frequent location of musculoskeletal pain is in the neck and shoulder regions, affecting nearly 40% of those with pain (Moisset et al., 2015; Wallace et al., 2014). Pain with neuropathic tendencies has a much lower prevalence at less than 10% (Moisset et al., 2015) with other studies reporting no association between ALS and neuropathic pain (Wallace et al., 2014). Under managed in the early stages of disease, pain usually contributes to increased depression and worsened QoL. The Brief Pain Inventory (Cleeland & Ryan, 1994), Pain Severity Index, and Pain Interference Index are used to assess the severity of pain and sometimes accompanied with McGill QoL questionnaire (Cohen, Mount, Strobel, & Bui, 1995) and Beck's Depression Inventory (Hautzinger et al., 2006) to assess depression and health-related QoL. Pain in most patients is managed with nonsteroidal antiinflammatories (Chio et al., 2012; Moisset et al., 2015), but it is not uncommon for patients also to be prescribed opiates or chronic pain treatments such as anticonvulsants. Other therapies for pain include supervised exercise therapy, which show a decrease in pain intensity and muscle stiffness when compared to home exercise therapy (Aksu & Citak-karakaya, 2002). Pain arising from spasticity is treated with muscle relaxants, and intrathecal baclofen has also been seen to improve pain in ALS, with an average reduction of 54% and complete remission of pain in some (McClelland et al., 2008). Pain arising from cramp may respond to quinine, and early studies suggest a possible benefit from Mexiletine, a nonselective sodium channel blocker (Weiss et al., 2016). Tetrahydrocannabinol and osteopathic manual treatment for pain is ineffective at reducing cramp intensities, number of cramps, or fasciculation in patients with ALS (Maggiani et al., 2016; Weber et al., 2010).

Breathlessness in ALS occurs once muscles involved in respiration are affected. People with ALS requiring NIV are classed at stage 4 in the King's ALS staging system, with a median time to death of less than 14 weeks (Balendra et al., 2014; Fang, Al Khleifat, et al., 2017). Apart from NIV, benzodiazepines and opiates may also be used for symptom relief (Borasio et al., 2001; Clemens & Klaschik, 2008), particularly in the later stages. Anxiety correlates strongly with dyspnoea intensity (Clemens & Klaschik, 2008) and therefore anxiolytics are useful, especially during the end stages of the disease when withdrawing from ventilation.

Urinary incontinence in people with ALS aged over 60 is higher than in the baseline population, with 73% of those affected having urge incontinence (Nuebling et al., 2014). This may be related to difficulties accessing facilities due to poor mobility and anxiety resulting from this. There are no curative treatments for urinary incontinence, which may be iatrogenic, as a side effect in muscle relaxant and anticholinergic medication used in ALS.

Constipation is found in 46% of patients and is higher than in the baseline population, but incontinence of feces occurs in 8%, the same as the baseline prevalence (Nuebling et al., 2014). This suggests that poor mobility resulting from ALS have less of an impact on defecation than urination.

Brain-derived neurotrophic factor (BDNF) is a nerve growth agent with a role to support, grow, and differentiate existing neurons of the central and peripheral nervous system. This nerve growth factor may be important to help sustain and increase motor nerve growth in the nervous system of people experiencing ALS. Intrathecal BDNF was tested to determine any positive effect on the autonomic system for people with ALS and at 12 months no positive effects were found when compared to placebo (Beck et al., 2005).

3.4 Gastrointestinal

Sialorrhea is common and affects in the region of 20%–50% of people with ALS (Bourry et al., 2013; Jackson et al., 2009). Treatments can be categorized into medical and invasive forms. Medical treatments include anticholinergics such as glycopyrrolate and scopolamine as well as the tricyclic antidepressant amitriptyline, which is effective in about 70% of patients with mild to moderate sialorrhea (Weikamp et al., 2016). However, those requiring higher doses often exhibit severe side effects like blurred vision and constipation that limits the usefulness of medical treatments (Weikamp et al., 2016). For these patients, more invasive therapy is needed with botulinum toxin (Botox) A, B, and radiotherapy being the three most extensive invasive methods studied.

Most frequently, invasive treatments target the parotid and submandibular glands as these produce most of the saliva for the stimulated and fasting states, respectively. The sublingual glands, which produce less fasting saliva than the submandibular glands, are often spared to prevent complete dehydration of mucosal membranes, which could lead to dry mouth. Saliva reduction is usually measured using cotton wool weight, but sometimes by subjective scoring. A validated method for staging of severity of sialorrhea is the Sialorrhea Severity Score (Weikamp et al., 2016). Botox A injections result in a reduction in cotton wool weight and improvement in overall saliva production (Gilio et al., 2010; Guidubaldi et al., 2011). Botox B injections showed an improvement to saliva reduction by 60% (Costa et al., 2008) and continue to show more improvement over placebo receivers for 12 weeks (Guidubaldi et al., 2011; Jackson et al., 2009). Botox B has a shorter latency to effect initiation than Botox A, but both treatments are equally effective at treating sialorrhea in ALS (Guidubaldi et al., 2011). Side effects for Botox treatment include dry mouth and saliva thickening.

Radiotherapy is successful at reducing saliva for 60%–100% of patients with ALS (Assouline et al., 2014; Bourry et al., 2013). A bigger target dose of >16 Gy has a larger treatment effect than radiotherapy with a smaller target dose <16 Gy (Bourry et al., 2013). There is a significant lasting effect of radiotherapy, as patients continue to have beneficial effects 6 months after treatment (Assouline et al., 2014). This is a key benefit of radiotherapy over Botox, shown by a study reporting that the effects of radiotherapy lasted longer than Botox treatment at the 12-week follow-up (Weikamp et al., 2016). Apart from the lasting effect of radiotherapy, Botox A was equally effective at reducing saliva in the ALS group after a 4-week follow-up check (Weikamp et al., 2016). However, radiotherapy can cause side effects at the point of intervention such as jaw pain (Weikamp et al., 2016).

Dysphagia can lead to weight loss, and in people with ALS severe dysphagia may require intervention in the form of percutaneous endoscopic gastrostomy (PEG). Weight loss may also arise from less understood causes, possibly from increased energy expenditure from heavy breathing, muscle fasciculation, hypermetabolism, or because of lower energy intake. One study has shown that people with ALS on average consume about 84% of the calories that they need (Kasarskis, Berryman, Vanderleest, Schneider, & McClain, 1996). Weight loss is also a negative prognostic factor for survival, where studies show that patients who had weight loss had a shorter survival time than those who had stable weight (Koerner et al., 2013; Rio et al., 2010). Parenteral nutrition has also been compared to PEG for effectiveness as a feeding mechanism, and no differences were found for post procedural survival for people with respiratory insufficiency. However, patients without respiratory insufficiency using PEG had a longer post procedural survival when compared to both PEG with respiratory insufficiency and parenteral nutrition groups (Verschueren et al., 2009). Physostigmine, a cholinesterase inhibitor, has not been shown to reduce weight loss when compared to pretrial periods (Norris et al., 1993), leaving high-calorie nutritional supplements as the only current option for weight stabilization (Koerner et al., 2013).

3.5 Vascular

Dyslipidemia in healthy populations can lead to atherosclerosis and liver steatosis, resulting in long-term vascular, liver, and cardiac dysfunction. In ALS, the stability of weight can lead to longer survival times (Koerner et al., 2013) and patients are sometimes encouraged to consume high-calorie nutritional supplements. One study has shown that dyslipidemia, where LDL:HDL ratio is high, is twice as common in people with ALS as in a control group, but people with ALS also had a higher incidence of liver steatosis than controls. Survival for patients who had an abnormally elevated LDL: HDL ratio was as much as 12 months longer than in patients with normal blood lipids (Dupuis et al., 2008).

Although not a symptom, hypertension is an important nonmotor change that may be detected. It has been found to decrease survival in people with ALS. In a case–control study, hypertension was seen in 57% of the ALS group compared with 41% of controls. The study showed that hypertensive patients were at a 4% yearly increased risk of death. Median survival time for patients was also decreased depending on the duration of clinical hypertension with long-term hypertensive patients having significantly shorter disease duration (Moreau et al., 2012). These effects on survival may result from reduced neural perfusion, leading to motor neuron death. Treatments for hypertension includes antihypertensives such as angiotensin-converting enzyme inhibitors, angiotensin receptor blockers, calcium channel blockers, β-blockers, and diuretics, as well as lifestyle changes in the form of diet, exercise, and smoking cessation, although the role of exercise is more limited or challenging in ALS.

3.6 Others

Itchiness and pressure sores are other nonmotor symptoms that have been reported by people with ALS. In one study pressure sores were reported by 5% of the ALS group (Hayashi et al., 2007) and itchiness was shown to have a higher prevalence in ALS than in a healthy control group (Wallace et al., 2014). Both symptoms are likely to result from reduced mobility rather than some systemic change.

3.7 Nonmotor Symptoms and Quality of Life

Health-related QoL is an important marker that contributes to a person's overall sense of wellbeing and is often related to the presenting disease. These are determined by impacts of the person's health on their wellbeing and are measured by standardized fixed systems such as the Short Form-36 (SF36) questionnaire. Individual QoL monitors the person's wellbeing through the patient's perspective by using patient-generated assessments, where varying weight is placed on categories that patients perceive to affect them the most. One method for measuring individual QoL is by using the Schedule for the Evaluation of Individual QoL-Direct Weighting scale (SEIQoL-DW). In ALS, most studies show that while health-related QoL decreases as expected, individual QoL that are unrelated to health, remains remarkably high throughout the disease course (Neudert, Wasner, & Borasio, 2004).

Studies of depression show that health-related QoL is closely related to symptoms, with lower QoL associated with more depressive symptoms (Bock et al., 2015; Pizzimenti et al., 2013). Active music therapy has been studied as a possible remedy for depression. A randomized-controlled trial found a significant improvement to health-related QoL in people with ALS given 4 weeks of active music therapy on top of their usual standard care (Raglio et al., 2016). However, this may not always be possible depending on the financial resources of the countries' healthcare system or individuals themselves.

Weight loss also reduces health-related QoL (Korner et al., 2013). People with ALS on a diet of high-calorie nutritional supplements had better weight stabilization when compared to those who were not. These patients also experienced a significant improvement to their QoL when measured using the SF-36 scale (Koerner et al., 2013).

Pain has a negative impact on health-related QoL (Hanisch et al., 2015). Other treatments suggested to relieve pain such as tetrahydrocannabinol (Weber et al., 2010) and osteopathic manual treatments (Maggiani et al., 2016) are not effective at treating pain and therefore as expected have no impact on health-related QoL in ALS.

Urinary incontinence and emotional lability decrease health-related QoL (Brooks et al., 2004; Nuebling et al., 2014). A trial showing successful treatment of emotional lability using combined dextromethorphan and quinidine improved both health-related QoL and quality of relationships, which is a determinant for individual QoL (Brooks et al., 2004).

Sleep disorders also reduce health-related QoL (Lyall et al., 2001; Vrijsen et al., 2015). NIV is important for improving nocturnal oxygen saturation, sleep efficiency, and health-related QoL in ALS individuals with sleep-disordered breathing (Lyall et al., 2001; Vrijsen et al., 2015).

4. DISCUSSION

We identified 44 studies for this systematic review, 25 related to the treatment of nonmotor symptoms in ALS and 19 describing the prevalence of these symptoms. Most studies had a small sample size of less than 100 patients and had a follow-up period of up to 12 weeks. Validated clinical severity scales established for nonmotor symptoms were used in many of the studies but different scales were used between studies, creating heterogeneity. This also made estimating the prevalence of nonmotor symptoms in different ALS populations difficult, as prevalence estimates often had large ranges. All studies included were of an appropriate evidence quality and had an acceptable risk of bias in study quality.

As nonmotor symptoms in ALS are relatively under studied, variation in the aims and interventions for nonmotor symptoms were high between studies and the use of different outcome measures and comparators made meta-analysis impossible. On the other hand, the number of studies found and the variety of interventions described, suggest that there is increasing awareness of the importance of nonmotor symptoms experienced by people with ALS.

Nonmotor ALS symptoms have been incompletely studied however, with the focus placed on common nonmotor symptoms such as depression and sialorrhea, with other rarer nonmotor symptoms inadequately researched. This bias toward coverage of frequent symptoms is understandable, as treatments and symptom relief would be targeted at a wider patient population and be beneficial for more people with ALS. In addition, studying uncommon nonmotor symptoms is difficult because of the poor prognosis and relatively low numbers of people affected at any one time (Chio et al., 2009; Fang, Al Khleifat, et al., 2017; Roche et al., 2012). Combining this with the infrequent nature of rare nonmotor symptoms means that any prospective study will encounter difficulties in recruiting the needed sample size to meaningfully study the prevalence or treatment of such symptoms in ALS.

Other neurodegenerative diseases therefore have more literature in nonmotor symptoms due to the longer prognosis. For example, Parkinson's disease has an estimated prognosis of 16 years (Forsaa, Larsen, Wentzel-Larsen, & Alves, 2010) with nonmotor symptoms having being researched in more depth and first appear with signs of constipation, rapid eye movement sleep disorders and anosmia, before motor involvement (Poewe, 2008). Corresponding with the Braak hypothesis of disease progression, these preclinical indicators of disease may prove useful for early identification and treatment in slowing the disease. In ALS, less is understood about when nonmotor symptoms arise, because of the fast progressing nature of the disease. Nevertheless, nonmotor symptoms from both diseases cause distress, discomfort (Chaudhuri, Healy, et al., 2006), and worsen prognosis for patients (Dupuis et al., 2008; Koerner et al., 2013; Moreau et al., 2012; Stubendorff, Aarsland, Minthon, & Londos, 2012). A study using the NMSQuest tool to examine the prevalence of nonmotor symptoms in 525 patients with Parkinson's disease found them comparable to those experienced in ALS and with similar frequencies (Martinez-Martin et al., 2007).

Because treatments are mostly targeting symptom relief, related therapies could be given to patients regardless of the cause being Parkinson's or ALS.

The need for a formal definition, and a robust and clear outline of nonmotor symptoms associated with ALS is necessary for clinicians to provide the best care and improvement to patient reported outcomes. This systematic review covers the major nonmotor symptoms encountered in ALS and the prevalence, screening methods, and treatment options for such symptoms. Further research is needed for nonmotor symptoms with no available treatments. Evidence derived from larger sample sizes is needed for nonmotor symptoms with existing treatments, to allow for comparisons of benefits and harms when supporting clinical decisions. As ALS is a disease with no curative treatment, and rapidly progresses ending in a lethal outcome, management of nonmotor symptoms is a critical aspect of the care for patients with this disease.

ACKNOWLEDGMENTS

This work is part of an EU Joint Programme—Neurodegenerative Disease Research (JPND) project. The project is supported through the following funding organizations under the aegis of JPND—www.jpnd.eu (*United Kingdom, Medical Research Council* (MR/L501529/1) *and Economic and Social Research Council* (ES/L008238/1)). A.A.C. receives salary support from the National Institute for Health Research (NIHR) Dementia Biomedical Research Unit at South London and Maudsley NHS Foundation Trust and King's College London. The work leading up to this publication was funded by the European Community's Health Seventh Framework Programme (FP7/2007–2013; grant agreement number 259867) and Horizon 2020 Programme (H2020-PHC-2014-two-stage; grant agreement number 633413).

Conflicts of interest. The authors declare no conflict of interest.

REFERENCES

Abrahams, S., Goldstein, L. H., Kew, J. J., Brooks, D. J., Lloyd, C. M., Frith, C. D., et al. (1996). Frontal lobe dysfunction in amyotrophic lateral sclerosis. A PET study. *Brain, 119*(Pt. 6), 2105–2120.

Abrahams, S., Newton, J., Niven, E., Foley, J., & Bak, T. H. (2014). Screening for cognition and behaviour changes in ALS. *Amyotrophic Lateral Sclerosis and Frontotemporal Degeneration, 15*(1–2), 9–14. http://dx.doi.org/10.3109/21678421.2013.805784.

Aksu, S., & Citak-karakaya, I. (2002). Effect of exercise therapy on pain complaints in patients with amyotrophic lateral sclerosis. *The Pain Clinic, 14*(4), 353–359.

Assouline, A., Levy, A., Abdelnour-Mallet, M., Gonzalez-Bermejo, J., Lenglet, T., Le Forestier, N., et al. (2014). Radiation therapy for hypersalivation: A prospective study in 50 amyotrophic lateral sclerosis patients. *International Journal of Radiation Oncology Biology Physics, 88*(3), 589–595.

Atassi, N., Cook, A., Pineda, C. M. E., Yerramilli-Rao, P., Pulley, D., & Cudkowicz, M. (2011). Depression in amyotrophic lateral sclerosis. *Amyotrophic Lateral Sclerosis, 12*(2), 109–112.

Atkins, D., Best, D., Briss, P. A., Eccles, M., Falck-Ytter, Y., Flottorp, S., et al. (2004). Grading quality of evidence and strength of recommendations. *BMJ, 328*(7454), 1490. http://dx.doi.org/10.1136/bmj.328.7454.1490.

Balendra, R., Jones, A., Jivraj, N., Knights, C., Ellis, C. M., Burman, R., et al. (2014). Estimating clinical stage of amyotrophic lateral sclerosis from the ALS functional rating scale. *Amyotrophic Lateral Sclerosis and Frontotemporal Degeneration, 15*(3–4), 279–284. http://dx.doi.org/10.3109/21678421.2014.897357.

Beck, M., Flachenecker, P., Magnus, T., Giess, R., Reiners, K., Toyka, K. V., et al. (2005). Autonomic dysfunction in ALS: A preliminary study on the effects of intrathecal BDNF. *Amyotrophic Lateral Sclerosis and Other Motor Neuron Disorders, 6*(2), 100–103. http://dx.doi.org/10.1080/14660820510028412.

Bock, M., Duong, Y. N., Kim, A., Murphy, J., & Lomen-Hoerth, C. (2015). In *Cognitive-behavioral changes in amyotrophic lateral sclerosis: Prevalence, progression, and impact on patients and caregivers Neurology. Conference: 67th American Academy of Neurology Annual Meeting, AAN, 84(no pagination).*

Borasio, G. D., Shaw, P. J., Hardiman, O., Ludolph, A. C., Sales Luis, M. L., Silani, V., et al. (2001). Standards of palliative care for patients with amyotrophic lateral sclerosis: Results of a European survey. *Amyotrophic Lateral Sclerosis and Other Motor Neuron Disorders, 2*(3), 159–164.

Bourry, N., Guy, N., Achard, J. L., Verrelle, P., Clavelou, P., & Lapeyre, M. (2013). Salivary glands radiotherapy to reduce sialorrhea in amyotrophic lateral sclerosis: Dose and energy. *Cancer/Radiotherapie, 17*(3), 191–195.

Bromet, E., Andrade, L. H., Hwang, I., Sampson, N. A., Alonso, J., de Girolamo, G., et al. (2011). Cross-national epidemiology of DSM-IV major depressive episode. *BMC Medicine, 9*, 90. http://dx.doi.org/10.1186/1741-7015-9-90.

Brooks, B. R., Thisted, R. A., Appel, S. H., Bradley, W. G., Olney, R. K., Berg, J. E., et al. (2004). Treatment of pseudobulbar affect in ALS with dextromethorphan/quinidine: A randomized trial. *Neurology, 63*(8), 1364–1370.

Brundin, P., Melki, R., & Kopito, R. (2010). Prion-like transmission of protein aggregates in neurodegenerative diseases. *Nature Reviews. Molecular Cell Biology, 11*(4), 301–307. http://dx.doi.org/10.1038/nrm2873.

Cedarbaum, J. M., Stambler, N., Malta, E., Fuller, C., Hilt, D., Thurmond, B., et al. (1999). The ALSFRS-R: A revised ALS functional rating scale that incorporates assessments of respiratory function. BDNF ALS Study Group (Phase III). *Journal of the Neurological Sciences, 169*(1–2), 13–21.

Chaudhuri, K. R., Healy, D. G., Schapira, A. H., & National Institute for Clinical Excellence (2006a). Non-motor symptoms of Parkinson's disease: Diagnosis and management. *Lancet Neurology, 5*(3), 235–245. http://dx.doi.org/10.1016/S1474-4422(06)70373-8.

Chaudhuri, K. R., Martinez-Martin, P., Schapira, A. H., Stocchi, F., Sethi, K., Odin, P., et al. (2006b). International multicenter pilot study of the first comprehensive self-completed nonmotor symptoms questionnaire for Parkinson's disease: The NMSQuest study. *Movement Disorders, 21*(7), 916–923. http://dx.doi.org/10.1002/mds.20844.

Chio, A., Canosa, A., Gallo, S., Moglia, C., Ilardi, A., Cammarosano, S., et al. (2012). Pain in amyotrophic lateral sclerosis: A population-based controlled study. *European Journal of Neurology, 19*(4), 551–555. http://dx.doi.org/10.1111/j.1468-1331.2011.03540.x.

Chio, A., Logroscino, G., Hardiman, O., Swingler, R., Mitchell, D., Beghi, E., et al. (2009). Prognostic factors in ALS: A critical review. *Amyotrophic Lateral Sclerosis, 10*(5–6), 310–323. http://dx.doi.org/10.3109/17482960802566824.

Cleeland, C. S., & Ryan, K. (1994). *The brief pain inventory.* Available online https://www.mdanderson.org/education-andresearch/departments-programs-and-labs/departments-and-divisions/symptom-research/symptomassessment-tools/BPI_UserGuide.pdf [accessed on 1 March 2016].

Clemens, K. E., & Klaschik, E. (2008). Morphine in the management of dyspnoea in ALS. A pilot study. *European Journal of Neurology*, *15*(5), 445–450. http://dx.doi.org/10.1111/j.1468-1331.2008.02102.x.

Cohen, S. R., Mount, B. M., Strobel, M. G., & Bui, F. (1995). The McGill quality of life questionnaire: A measure of quality of life appropriate for people with advanced disease. A preliminary study of validity and acceptability. *Palliative Medicine*, *9*(3), 207–219.

Costa, J., Rocha, M. L., Ferreira, J., Evangelista, T., Coelho, M., & de Carvalho, M. (2008). Botulinum toxin type-B improves sialorrhea and quality of life in bulbaronset amyotrophic lateral sclerosis. *Journal of Neurology*, *255*(4), 545–550. http://dx.doi.org/10.1007/s00415-008-0738-5.

Dupuis, L., Corcia, P., Fergani, A., De Aguilar, J. L. G., Bonnefont-Rousselot, D., Bittar, R., et al. (2008). Dyslipidemia is a protective factor in amyotrophic lateral sclerosis. *Neurology*, *70*(13), 1004–1009. http://dx.doi.org/10.1212/01.wnl.0000285080.70324.27.

Fang, T., Al Khleifat, A., Stahl, D. R., Lazo La Torre, C., Murphy, C., Uk-Mnd Lical, S., et al. (2017). Comparison of the King's and MiToS staging systems for ALS. *Amyotrophic Lateral Sclerosis and Frontotemporal Degeneration*, *18*, 227–232.

Fang, T., Jozsa, F., & Al-Chalabi, A. (2017). *Non-motor symptoms in ALS: A systematic review*. Retrieved 21/01/17, 2017, from https://www.crd.york.ac.uk/PROSPERO/display_record.asp?ID=CRD42017057137.

Fang, F., Valdimarsdottir, U., Furst, C. J., Hultman, C., Fall, K., Sparen, P., et al. (2008). Suicide among patients with amyotrophic lateral sclerosis. *Brain*, *131*(Pt. 10), 2729–2733.

Ferentinos, P., Paparrigopoulos, T., Rentzos, M., Zouvelou, V., Alexakis, T., & Evdokimidis, I. (2011). Prevalence of major depression in ALS: Comparison of a semi-structured interview and four self-report measures. *Amyotrophic Lateral Sclerosis*, *12*(4), 297–302. http://dx.doi.org/10.3109/17482968.2011.556744.

Forsaa, E. B., Larsen, J. P., Wentzel-Larsen, T., & Alves, G. (2010). What predicts mortality in Parkinson disease? A prospective population-based long-term study. *Neurology*, *75*(14), 1270–1276. http://dx.doi.org/10.1212/WNL.0b013e3181f61311.

Gilio, F., Iacovelli, E., Frasca, V., Gabriele, M., Giacomelli, E., Picchiori, F., et al. (2010). Botulinum toxin type a for the treatment of sialorrhoea in amyotrophic lateral sclerosis: A clinical and neurophysiological study. *Amyotrophic Lateral Sclerosis*, *11*(4), 359–363. http://dx.doi.org/10.3109/17482960903264998.

Golden, C. J., & Freshwater, S. M. (1978). Stroop color and word test. *Age*, *15*, 90.

Gould, R. L., Coulson, M. C., Brown, R. G., Goldstein, L. H., Al-Chalabi, A., & Howard, R. J. (2015). Psychotherapy and pharmacotherapy interventions to reduce distress or improve well-being in people with amyotrophic lateral sclerosis: A systematic review. *Amyotrophic Lateral Sclerosis and Frontotemporal Degeneration*, *16*(5–6), 293–302.

Guidubaldi, A., Fasano, A., Ialongo, T., Piano, C., Pompili, M., Masciana, R., et al. (2011). Botulinum toxin a versus B in Sialorrhea: A prospective, randomized, double-blind, crossover pilot study in patients with amyotrophic lateral sclerosis or Parkinson's disease. *Movement Disorders*, *26*(2), 313–319. http://dx.doi.org/10.1002/mds.23473.

Gunther, R., Richter, N., Sauerbier, A., Chaudhuri, K. R., Martinez-Martin, P., Storch, A., et al. (2016). Non-motor symptoms in patients suffering from motor neuron diseases. *Frontiers in Neurology*, *7*, 117. http://dx.doi.org/10.3389/fneur.2016.00117.

Hammer, E. M., Hacker, S., Hautzinger, M., Meyer, T. D., & Kubler, A. (2008). Validity of the ALS-Depression-Inventory (ADI-12)—A new screening instrument for depressive disorders in patients with amyotrophic lateral sclerosis. *Journal of Affective Disorders*, *109*(1–2), 213–219. http://dx.doi.org/10.1016/j.jad.2007.11.012.

Hanisch, F., Skudlarek, A., Berndt, J., & Kornhuber, M. E. (2015). Characteristics of pain in amyotrophic lateral sclerosis. *Brain and Behavior*, *5*(3), 8.

Hasin, D. S., Goodwin, R. D., Stinson, F. S., & Grant, B. F. (2005). Epidemiology of major depressive disorder: Results from the National Epidemiologic Survey on Alcoholism and Related Conditions. *Archives of General Psychiatry*, *62*(10), 1097–1106. http://dx.doi.org/10.1001/archpsyc.62.10.1097.

Hautzinger, M., Keller, F., & Kühner, C. (2006). *Beck depressions-inventar (BDI-II)*. Harcourt Test Services Frankfurt.

Hayashi, T., Narita, Y., Okugawa, N., Hamaguchi, E., Shibahara, M., & Kuzuhara, S. (2007). Pressure ulcers in ALS patients on admission at a university hospital in Japan. *Amyotrophic Lateral Sclerosis*, *8*(5), 310–313. http://dx.doi.org/10.1080/17482960701538825.

Heaton, R. K., Chelune, G. J., Curtiss, G., Kay, G. G., & Talley, J. L. (1993). *Wisconsin card sorting test: Psychological assessment resources*.

Ho, D. T., Ruthazer, R., & Russell, J. A. (2011). Shoulder pain in amyotrophic lateral sclerosis. *Journal of Clinical Neuromuscular Disease*, *13*(1), 53–55.

Jackson, C. E., Gronseth, G., Rosenfeld, J., Barohn, R. J., Dubinsky, R., Simpson, C. B., et al. (2009). Randomized double-blind study of botulinum toxin type B for sialorrhea in ALS patients. *Muscle and Nerve*, *39*(2), 137–143.

Johns, M. W. (1991). A new method for measuring daytime sleepiness: The Epworth sleepiness scale. *Sleep*, *14*(6), 540–545.

Jones, A. R., Jivraj, N., Balendra, R., Murphy, C., Kelly, J., Thornhill, M., et al. (2014). Health utility decreases with increasing clinical stage in amyotrophic lateral sclerosis. *Amyotrophic Lateral Sclerosis and Frontotemporal Degeneration*, *15*(3–4), 285–291. http://dx.doi.org/10.3109/21678421.2013.872149.

Kasarskis, E. J., Berryman, S., Vanderleest, J. G., Schneider, A. R., & McClain, C. J. (1996). Nutritional status of patients with amyotrophic lateral sclerosis: Relation to the proximity of death. *The American Journal of Clinical Nutrition*, *63*(1), 130–137.

Katzberg, H. D., Selegiman, A., Guion, L., Yuan, N., Cho, S. C., Katz, J. S., et al. (2013). Effects of noninvasive ventilation on sleep outcomes in amyotrophic lateral sclerosis. *Journal of Clinical Sleep Medicine*, *9*(4), 345–351.

Kessler, R. C., Berglund, P., Demler, O., Jin, R., Koretz, D., Merikangas, K. R., et al. (2003). The epidemiology of major depressive disorder: Results from the National Comorbidity Survey Replication (NCS-R). *JAMA*, *289*(23), 3095–3105. http://dx.doi.org/10.1001/jama.289.23.3095.

Kleinbub, J. R., Palmieri, A., Broggio, A., Pagnini, F., Benelli, E., Sambin, M., et al. (2015). Hypnosis-based psychodynamic treatment in ALS: A longitudinal study on patients and their caregivers. *Frontiers in Psychology*, *6*, 822. http://dx.doi.org/10.3389/fpsyg.2015.00822.

Knibb, J. A., Keren, N., Kulka, A., Leigh, P. N., Martin, S., Shaw, C. E., et al. (2016). A clinical tool for predicting survival in ALS. *Journal of Neurology, Neurosurgery, and Psychiatry*, *87*(12), 1361–1367. http://dx.doi.org/10.1136/jnnp-2015-312908.

Koerner, S., Hendricks, M., Kollewe, K., Zapf, A., Dengler, R., Silani, V., et al. (2013). Weight loss, dysphagia and supplement intake in patients with amyotrophic lateral sclerosis (ALS): Impact on quality of life and therapeutic options. *BMC Neurology*, *13*, 84. http://dx.doi.org/10.1186/1471-2377-13-84.

Korner, S., Sieniawski, M., Kollewe, K., Rath, K. J., Krampfl, K., Zapf, A., et al. (2013). Speech therapy and communication device: Impact on quality of life and mood in patients with amyotrophic lateral sclerosis. *Amyotrophic Lateral sclerosis & Frontotemporal Degeneration*, *14*(1), 20–25.

Krupp, L. B., LaRocca, N. G., Muir-Nash, J., & Steinberg, A. D. (1989). The fatigue severity scale: Application to patients with multiple sclerosis and systemic lupus erythematosus. *Archives of Neurology*, *46*(10), 1121–1123.

Liberati, A., Altman, D. G., Tetzlaff, J., Mulrow, C., Gotzsche, P. C., Ioannidis, J. P., et al. (2009). The PRISMA statement for reporting systematic reviews and meta-analyses of studies that evaluate healthcare interventions: Explanation and elaboration. *BMJ, 339*, b2700. http://dx.doi.org/10.1136/bmj.b2700.

Lyall, R. A., Donaldson, N., Fleming, T., Wood, C., Newsom-Davis, I., Polkey, M. I., et al. (2001). A prospective study of quality of life in ALS patients treated with noninvasive ventilation. *Neurology, 57*(1), 153–156.

Maggiani, A., Tremolizzo, L., Valentina, A. D., Mapelli, L., Sosio, S., Milano, V., et al. (2016). Osteopathic manual treatment for amyotrophic lateral sclerosis: A feasibility pilot study. *The Open Neurology Journal, 10*, 59–66. http://dx.doi.org/10.2174/1874205x01610010059.

Martinez-Martin, P., Schapira, A. H., Stocchi, F., Sethi, K., Odin, P., MacPhee, G., et al. (2007). Prevalence of nonmotor symptoms in Parkinson's disease in an international setting; study using nonmotor symptoms questionnaire in 545 patients. *Movement Disorders, 22*(11), 1623–1629. http://dx.doi.org/10.1002/mds.21586.

Massman, P. J., Sims, J., Cooke, N., Haverkamp, L. J., Appel, V., & Appel, S. H. (1996). Prevalence and correlates of neuropsychological deficits in amyotrophic lateral sclerosis. *Journal of Neurology, Neurosurgery & Psychiatry, 61*(5), 450–455.

McClelland, S., 3rd, Bethoux, F. A., Boulis, N. M., Sutliff, M. H., Stough, D. K., Schwetz, K. M., et al. (2008). Intrathecal baclofen for spasticity-related pain in amyotrophic lateral sclerosis: Efficacy and factors associated with pain relief. *Muscle & Nerve, 37*(3), 396–398.

McElhiney, M. C., Rabkin, J. G., Gordon, P. H., Goetz, R., & Mitsumoto, H. (2009). Prevalence of fatigue and depression in ALS patients and change over time. *Journal of Neurology, Neurosurgery and Psychiatry, 80*(10), 1146–1149.

Moher, D., Hopewell, S., Schulz, K. F., Montori, V., Gotzsche, P. C., Devereaux, P. J., et al. (2010). CONSORT 2010 explanation and elaboration: Updated guidelines for reporting parallel group randomised trials. *BMJ, 340*, c869. http://dx.doi.org/10.1136/bmj.c869.

Moisset, X., Cornut-Chauvinc, C., Clavelou, P., Pereira, B., Dallel, R., & Guy, N. (2015). Is there pain with neuropathic characteristics in patients with amyotrophic lateral sclerosis? A cross-sectional study. *Palliative Medicine, 30*(5), 486–494.

Moreau, C., Brunaud-Danel, V., Dallongeville, J., Duhamel, A., Laurier-Grymonprez, L., De Reuck, J., et al. (2012). Modifying effect of arterial hypertension on amyotrophic lateral sclerosis. *Amyotrophic Lateral Sclerosis, 13*(2), 194–201.

Neudert, C., Wasner, M., & Borasio, G. D. (2004). Individual quality of life is not correlated with health-related quality of life or physical function in patients with amyotrophic lateral sclerosis. *Journal of Palliative Medicine*, 7(4), 551–557. http://dx.doi.org/10.1089/jpm.2004.7.551.

Norris, F. H., Tan, Y., Fallat, R. J., & Elias, L. (1993). Trial of oral physostigmine in amyotrophic lateral sclerosis. *Clinical Pharmacology & Therapeutics, 54*(6), 680–682.

Nuebling, G. S., Mie, E., Bauer, R. M., Hensler, M., Lorenzl, S., Hapfelmeier, A., et al. (2014). Increased prevalence of bladder and intestinal dysfunction in amyotrophic lateral sclerosis. *Amyotrophic Lateral Sclerosis and Frontotemporal Degeneration, 15*(3–4), 174–179. http://dx.doi.org/10.3109/21678421.2013.868001.

Ogawa, T., Tanaka, H., & Hirata, K. (2009). Cognitive deficits in amyotrophic lateral sclerosis evaluated by event-related potentials. *Clinical Neurophysiology, 120*(4), 659–664.

Pagnini, F., Phillips, D., Bosma, C. M., Reece, A., & Langer, E. (2015). Mindfulness, physical impairment and psychological well-being in people with amyotrophic lateral sclerosis. *Psychology & Health, 30*(5), 503–517. http://dx.doi.org/10.1080/08870446.2014.982652.

Panitch, H. S., Thisted, R. A., Smith, R. A., Wynn, D. R., Wymer, J. P., Achiron, A., et al. (2006). Randomized, controlled trial of dextromethorphan/quinidine for pseudobulbar affect in multiple sclerosis. *Annals of Neurology*, *59*(5), 780–787. http://dx.doi.org/10.1002/ana.20828.

Patten, S. B., Svenson, L. W., White, C. M., Khaled, S. M., & Metz, L. M. (2007). Affective disorders in motor neuron disease: A population-based study. *Neuroepidemiology*, *28*(1), 1–7. http://dx.doi.org/10.1159/000097849.

Pizzimenti, A., Aragona, M., Onesti, E., & Inghilleri, M. (2013). Depression, pain and quality of life in patients with amyotrophic lateral sclerosis: A cross-sectional study. *Functional Neurology*, *28*(2), 115–119.

Poewe, W. (2008). Non-motor symptoms in Parkinson's disease. *European Journal of Neurology*, *15*(Suppl. 1), 14–20. http://dx.doi.org/10.1111/j.1468-1331.2008.02056.x.

Rabkin, J. G., Albert, S. M., Del Bene, M. L., O'Sullivan, I., Tider, T., Rowland, L. P., et al. (2005). Prevalence of depressive disorders and change over time in late-stage ALS. *Neurology*, *65*(1), 62–67. http://dx.doi.org/10.1212/01.wnl.0000167187.14501.0c.

Rabkin, J. G., Gordon, P. H., McElhiney, M., Rabkin, R., Chew, S., & Mitsumoto, H. (2009). Modafinil treatment of fatigue in patients with ALS: A placebo-controlled study. *Muscle & Nerve*, *39*(3), 297–303. http://dx.doi.org/10.1002/mus.21245.

Raglio, A., Giovanazzi, E., Pain, D., Baiardi, P., Imbriani, C., Imbriani, M., et al. (2016). Active music therapy approach in amyotrophic lateral sclerosis: A randomized-controlled trial. *International Journal of Rehabilitation Research*, *39*(4), 365–367. http://dx.doi.org/10.1097/mrr.0000000000000187.

Rio, A., Ellis, C., Shaw, C., Willey, E., Ampong, M. A., Wijesekera, L., et al. (2010). Nutritional factors associated with survival following enteral tube feeding in patients with motor neurone disease. *Journal of Human Nutrition and Dietetics*, *23*(4), 408–415. http://dx.doi.org/10.1111/j.1365-277X.2010.01057.x.

Roche, J. C., Rojas-Garcia, R., Scott, K. M., Scotton, W., Ellis, C. E., Burman, R., et al. (2012). A proposed staging system for amyotrophic lateral sclerosis. *Brain*, *135*(Pt. 3), 847–852. http://dx.doi.org/10.1093/brain/awr351.

Spitzer, R. L., Kroenke, K., Williams, J. B., & Group, P. H. Q. P. C. S (1999). Validation and utility of a self-report version of PRIME-MD: The PHQ primary care study. *JAMA*, *282*(18), 1737–1744.

Stone, C. A., & O'Leary, N. (2009). Systematic review of the effectiveness of botulinum toxin or radiotherapy for sialorrhea in patients with amyotrophic lateral sclerosis. *Journal of Pain and Symptom Management*, *37*(2), 246–258. http://dx.doi.org/10.1016/j.jpainsymman.2008.02.006.

Stubendorff, K., Aarsland, D., Minthon, L., & Londos, E. (2012). The impact of autonomic dysfunction on survival in patients with dementia with Lewy bodies and Parkinson's disease with dementia. *PLoS One*, 7(10), e45451. http://dx.doi.org/10.1371/journal.pone.0045451

Tramacere, I., Dalla Bella, E., Chiò, A., Mora, G., Filippini, G., Lauria, G., et al. (2015). The MITOS system predicts long-term survival in amyotrophic lateral sclerosis. *Journal of Neurology, Neurosurgery & Psychiatry*, *86*(11), 1180–1185. http://dx.doi.org/10.1136/jnnp-2014-310176.

Tremolizzo, L., Susani, E., Lunetta, C., Corbo, M., Ferrarese, C., & Appollonio, I. (2014). Primitive reflexes in amyotrophic lateral sclerosis: Prevalence and correlates. *Journal of Neurology*, *261*(6), 1196–1202.

Trenerry, M., Crosson, B., DeBoe, J., & Leber, W. (1990). *Visual search and attention test.* Odessa, FL: Psychological Assessment Resources.

Van Groenestijn, A. C., Schroder, C. D., Visser-Meily, J. M. A., Reenen, E. T. K. V., Veldink, J. H., & Van Den Berg, L. H. (2015). Cognitive behavioural therapy and quality of life in psychologically distressed patients with amyotrophic lateral sclerosis and their caregivers: Results of a prematurely stopped randomized controlled trial. *Amyotrophic Lateral Sclerosis and Frontotemporal Degeneration*, *16*(5–6), 309–315.

Veritas-Health-Innovation. (2017). *Covidence systematic review software. Retrieved 21/01, 2017, from www.covidence.org*.

Verschueren, A., Monnier, A., Attarian, S., Lardillier, D., & Pouget, J. (2009). Enteral and parenteral nutrition in the later stages of ALS: An observational study. *Amyotrophic Lateral Sclerosis*, *10*(1), 42–46.

von Elm, E., Altman, D. G., Egger, M., Pocock, S. J., Gotzsche, P. C., Vandenbroucke, J. P., et al. (2007). The strengthening the reporting of observational studies in epidemiology (STROBE) statement: Guidelines for reporting observational studies. *Annals of Internal Medicine*, *147*(8), 573–577.

Vrijsen, B., Buyse, B., Belge, C., Robberecht, W., Van Damme, P., Decramer, M., et al. (2015). Noninvasive ventilation improves sleep in amyotrophic lateral sclerosis: A prospective polysomnographic study. *Journal of Clinical Sleep Medicine*, *11*(5), 559–566.

Wallace, V. C., Ellis, C. M., Burman, R., Knights, C., Shaw, C. E., & Al-Chalabi, A. (2014). The evaluation of pain in amyotrophic lateral sclerosis: A case controlled observational study. *Amyotrophic Lateral sclerosis & Frontotemporal Degeneration*, *15*(7–8), 520–527.

Weber, M., Goldman, B., & Truniger, S. (2010). Tetrahydrocannabinol (THC) for cramps in amyotrophic lateral sclerosis: A randomised, double-blind crossover trial. *Journal of Neurology, Neurosurgery and Psychiatry*, *81*(10), 1135–1140.

Weikamp, J. G., Schinagl, D. A., Verstappen, C. C., Schelhaas, H. J., de Swart, B. J., & Kalf, J. G. (2016). Botulinum toxin-A injections vs radiotherapy for drooling in ALS. *Acta Neurologica Scandinavica*, *134*(3), 224–231. http://dx.doi.org/10.1111/ane.12559.

Weiss, M. D., Macklin, E. A., Simmons, Z., Knox, A. S., Greenblatt, D. J., Atassi, N., et al. (2016). A randomized trial of mexiletine in ALS: Safety and effects on muscle cramps and progression. *Neurology*, *86*(16), 1474–1481. http://dx.doi.org/10.1212/wnl.0000000000002507.

Wijesekera, L. C., & Leigh, P. N. (2009). Amyotrophic lateral sclerosis. *Orphanet Journal of Rare Diseases*, *4*, 3. http://dx.doi.org/10.1186/1750-1172-4-3.

Witgert, M., Salamone, A. R., Strutt, A. M., Jawaid, A., Massman, P. J., Bradshaw, M., et al. (2010). Frontal-lobe mediated behavioral dysfunction in amyotrophic lateral sclerosis. *European Journal of Neurology*, *17*(1), 103–110. http://dx.doi.org/10.1111/j.1468-1331.2009.02801.x.

Zigmond, A. S., & Snaith, R. P. (1983). The hospital anxiety and depression scale. *Acta Psychiatrica Scandinavica*, *67*(6), 361–370.

Zung, W. W. (1965). A self-rating depression scale. *Archives of General Psychiatry*, *12*(1), 63–70.

CHAPTER FIFTY-TWO

Nonmotor Manifestations of Wilson's Disease

Samar Biswas*, Neelanjana Paul†, Shyamal K. Das*,[1]
*Bangur Institute of Neurosciences, Institute of Post graduate Medical Education and Research, Kolkata, India
†ICARE Institute of Medical Sciences and Research, Haldia, India
[1]Corresponding author: e-mail address: das_sk70@hotmail.com

Contents

Abstract

Wilson disease (WD) is an autosomal genetic disorder characterized by excessive copper deposition initially in liver (hepatic variant) followed by brain (neuropsychiatric variant) and other organs such as cornea and kidney due to defect in biliary copper excretion. Predominant presentations of neuropsychiatric variant are extrapyramidal motor dysfunctions such as dystonias, Parkinsonism, choreoathetosis, tremor, and ataxias. Nonmotor symptoms (NMS) can appear before clinical disease expression and during ongoing disease process. NMS may cause confusion and delay in clinical diagnosis. In the early stage, presence of asymptomatic or symptomatic evidence of acute or chronic liver disease with or without KF ring in young subjects against the background of family history of liver disease may be indicative of underlying WD. In WD, common NMS are personality disorders, mood changes, psychosis, cognitive abnormalities, sleep disorders, and autonomic disturbances besides few systemic dysfunctions. Cognitive

International Review of Neurobiology, Volume 134
ISSN 0074-7742
http://dx.doi.org/10.1016/bs.irn.2017.04.010

changes can be diagnosed by neuropsychological assessment, MRI, and SPECT study of brain. Nonmotor manifestations can be managed by metal chelator, antipsychotic agents, mood stabilizers, rarely electroconvulsive therapy, and occasional hepatic transplantation.

Wilson's disease (WD) is an autosomal recessively inherited disorder of copper metabolism due to dysfunction of the copper transporter *ATP7B* gene located on chromosome 13. Till now, over 600 mutations of the genes have been described (www.Wilsondisease.med.ualberta.ca accessed on 12th March, 2017) and can cause impaired biliary copper excretion leading to its accumulation in the liver followed by spillage into brain and other organs. WD has a prevalence of approximately 1 in 30,000 live births and carrier frequency of 1 in 90 in United States. This is more frequent in communities isolated and characterized by high frequency of consanguinity.

Commonly, the affected subjects may present at the end of the first decade up to 50 years of age and rarely up to eighth decade (Ala, Borjigin, Rochwarger, & Schilsky, 2005). Hepatic variant of Wilson disease usually present in the first decade as early as 4 years (Rukunuzzaman, 2015) and neuropsychiatric variety present in second to fourth decade (Bagchi et al., 2016). Copper accumulation leads to gradual cellular damage by way of apoptosis and free radical production, thereby producing clinical manifestations of hepatic and neuropsychiatric disturbances. Diagnosis of this disorder is made by supportive clinical features of either hepatic or basal ganglia affection with evidence of lower level of serum ceruloplasmin (below 20 mg per 100 mL), high urinary copper excretion (>100 μg in 24 h), and hepatic copper accumulation (more than 250 μg per 100 g of dry tissue). Kayser–Fleischer (KF) ring in cornea is an important clinical finding and may be positive in 95% of cases in neuropsychiatric form and 50% in hepatic form. Neuroimaging in neuropsychiatric variety may be helpful in supplementing the diagnosis (Bagchi et al., 2016; Sinha, Taly, Ravishankar, et al., 2006). MRI imaging of liver may be beneficial also in evaluation of hepatic form of WD (Akhon et al., 2009; Cheon et al., 2010).

The commonest presentation of neuropsychiatric variety of WD is motor manifestations pertaining to extrapyramidal disorders such as Parkinsonism, tremor, coordination difficulty and hyperkinetic disorders such as dystonia, chorea–athetosis, myoclonus, and tics (Bagchi et al., 2016; Fahn & Jankovic, 2007). Apart from motor features, patient with Parkinsonism may present with various nonmotor manifestations which may cause initial diagnostic confusion and management (Srinivas et al., 2008). Similarly, hepatic presentation may be in the form of acute hepatitis or fulminant hepatic failure or chronic active hepatitis, hepatic carcinoma, and cirrhosis of liver (Rukunuzzaman, 2015). Asymptomatic hepatomegaly or unexplained rise of liver enzymes may occur (Giannini, Testa, & Savarino, 2005). Consideration of hepatic presentation in early age, with family history of liver disease and premature death in siblings emphasizes search for WD. Associated hemolytic anemia with liver disease is another pointer for suspicion of WD. Since KF ring may be absent in 50% of hepatic cases, final diagnosis has to be based on liver biopsy.

Nonmotor symptoms (NMS) in WD are not well characterized and may precede many years prior to expression of typical clinical symptoms. Patients with neuropsychiatric-WD predominantly present with psychiatric, cognitive, behavioral changes, sleep disorders, and autonomic disturbances. NMS are often difficult to identify before clinical motor expression and by second decade motor features are so overwhelmed that NMS may be missed. Table 1 shows a spectrum of NMS manifestation in WD (Das & Ray, 2006). In addition to diverse clinical features, we have planned to focus on common

Table 1 Showing the Spectrum of Nonmotor Manifestations in Wilson Disease

System	Presentation
Hepatological	Asymptomatic hepatomegaly, fulminant hepatitis, chronic active hepatitis, cirrhosis of liver
Central nervous system	Psychiatric manifestation, cognitive changes, sleep and autonomic disturbances
Rare nonmotor manifestations	
Hematological	Epistaxis Acute hemolytic anemia (nonimmunological)
Bone and orthopedic	Metabolic bone disease (rickets and osteomalacia), juvenile polyarthritis, chondrocalcinosis
Renal	Renal tubular acidosis, hypercalcuria, microscopic hematuria, and/or minimal proteinuria
Cardiovascular	Arrhymias, rheumatic fever like manifestation
Dermatological	Hyperpigmentation
Ocular	Kayser–Fleischer ring (asymptomatic) Sunflower cataract
Gynecological	Primary or secondary amenorrhea, repeated and unexplained spontaneous abortion

nonmotor manifestations of WD in this communication. Apart from literature support, we have also added our own experience (unpublished) in this message.

Half of the patients with neurological manifestations have a history of behavioral abnormality within 5 years of diagnosis (Carta, Mura, Sorbello, Farina, & Demelia, 2012). The psychiatric symptoms can be the effect of brain tissue damage due to underlying pathology caused by copper accumulation. It might also be the consequence of comorbidity with bipolar disorder which might cosegregate with the mutations causing WD. Additionally, metabolic disturbances such as hepatic encephalopathy also contribute to cognitive behavioral changes (Ferenci, Litwin, Seniow, & Czlonkowska, 2015). Copper chelator may cause exacerbation of underlying neuropsychiatric disturbances possibly due to mobilization of excess copper from periphery to brain.

1. NMS IN WD

Usually, NMS include mood disorders such as depression, mania, anxiety, irritability, schizophrenia like behavioral states (Bidaki et al., 2012; Brewer, 2005), hallucination and delusion (Akil & Brewer, 1995; Araujo-de-Freitas, Rocha, Gondik, Quarinitini, & Miranda-cippa, 2014; Benhamla, Tirouche, Abaoub-Germaine, & Theodore, 2007; Carta et al., 2012; Dening & Berrios, 1989; Saint-Laurent, 1992; Shanmugiah et al., 2008; Srinivas et al., 2008; Zimbrean & Schilsky, 2014), cognitive changes, slowing of thought, language, memory difficulties (Chung, Kim, Kim, Ji, & Seo, 2016; Forta, Caramelli, & Barbosa, 2009; Hegde, Sinha, Rao, Taly, & Vasudev, 2010), personality changes (Genc, Genc, & Guveneroglu, 2013), mental retardation (Lingam, Wilson, Nazer, & Mowat, 1987), impairment of sexual functions (Akil & Brewer, 1995), alcohol abuse (Spyridi et al., 2008), sleep disturbances (Netto, Sinha, Tally, Panda, & Rao, 2010; Nevsimalova et al., 2011), and autonomic dysfunctions (Meenakshi-Sundaram et al., 2001; Soni, Shukla, Singh, Goyal, & Behari, 2009). We shall elaborate these aspects in subsequent paragraphs.

1.1 Psychiatric and Behavioral Abnormality

Psychiatric symptoms appear either before or at some point in most individuals with WD during the course of the disease (Saint-Laurent, 1992; Zimbrean & Schilsky, 2014). Often they are associated with neurological dysfunctions, suggesting that both originate from central nervous system pathology. Most common psychiatric features include emotional lability or "emotionalism," impulsiveness, disinhibition, and self injurious behavior. About one-third of the patients may present with psychiatric symptoms in early stage and 20% of them precede all other manifestations of WD (Schwartz, Feics, Polak, & Sharf, 1993). About 66% eventually develop psychiatric disturbances during the course of the illness (Akil & Brewer, 1995). In children, progressive scholastic deterioration can be one of the important presentations and early diagnosis may be of extremely beneficial for the affected children (Lingam et al., 1987). A study has tried to establish relationship between component of basal ganglia nuclei and clinical neuropsychiatric abnormalities such as depression with parkinsonian states; personality changes with caudate or putaminal involvement; psychosis, impulsivity, and sexual disorders with caudate lesions; dementia and mania with caudate and pallidal lesions; and compulsions with pallidal disease (Lauterbach et al., 1998).

In original paper, Wilson described psychiatric problem in 8 out of 12 subjects and 2 of them had schizophrenia-like psychosis (Wilson, 1912). According to case series, one-third of patients with WD may initially present behavioral abnormality and may lead to misdiagnosis and delay in starting specific treatment (Srinivas et al., 2008).

1.2 Pattern of Psychiatric and Cognitive Manifestation in WD

1. Personality disorder
2. Affective disorder
3. Psychosis
4. Other psychiatric abnormalities
5. Cognitive impairment

Personality disorder is the most common among psychiatric problems. Life time prevalence is about 46%–71% as typical symptoms include irritability or aggressiveness and importantly these correlated with the motor symptoms (Brewer & Yuzbasiyan-Gurkan, 1992). Often violence, socially disinhibited behaviors, obsession (Duggal & Nizamie, 2000; Kumawat, Sharma, Tripathi, Ralot, & Dixit, 2007; Sahu, Singhi, & Malhotra, 2013), catatonic signs (Bhattacharya, Debnath, & Sarkar, 2015; Davis & Borde, 1993), and paranoid features may be observed (Genc et al., 2013).

Affective disorder is the next common psychiatric dysfunction. Depression has a frequency between 30% and 60% (Akil & Brewer, 1995). It is still unsettled whether depression is a psychological reaction to WD or is a biological impairment. In a study of 23 patients with WD, it is observed that depression is related to alteration of presynaptic serotonin transporter (SERT) availability as measured by SPECT (Eggers et al., 2003). Overall, the prevalence of depression in extrapyramidal disorders such as Parkinsonism may be higher than in patients with chronic disabling diseases (Ehman, Beninger, Gawel, & Riopelle, 1990; Lauterbach et al., 1998). Because of consequences of depression, higher rate of suicide (4%–16%) in these patients has also been reported (Oder et al., 1991).

Other affective disorders like hypomania, mania, and bipolar disorders have been reported. However, the presence of these disturbances does not correlate with degree of motor disability. A wide range of psychiatric symptoms ranging from catatonia, agitation, aggression, delusional thought, and mania have been documented (Akil & Brewer, 1995; Benhamla et al., 2007; Dening & Berrios, 1989; Machado et al., 2006).

The association between WD and psychiatric disturbances appears more frequently than believed in the past, due to increasing studies over the last decade. Higher prevalence of major depressive disorder and bipolar disorder has been associated with higher brain damage on SPECT study (Giagheddu et al., 2001; Piga et al., 2008). WD should be considered if any young person develops unexplained or treatment resistant psychiatric dysfunction especially in the presence of florid neurological manifestations with family history of liver disease (Srinivas et al., 2008).

Psychosis: Although schizophrenia and other psychosis are as frequent in WD as in general population but this is more frequent (8%) in patients with neurological involvement (Dening & Berrios, 1989; Grover, Sarkar, Jhanda, & Chawla, 2014). These subjects also have the propensity to develop side effects to neuroleptic agents. Thus, presence of neuroleptic adverse events in subjects with schizophreniform behavior points to the need of search for WD (Srinivas et al., 2008).

1.3 Other Psychiatric Disorder

Anxiety disorder, substance abuse, catatonia, anorexia, etc. have been reported to be associated with WD but their prevalence in WD is difficult to determine from available literature (Carta et al., 2015; Saint-Laurent, 1992).

1.4 Imaging Abnormalities

Various imaging modalities such as MRI, SPECT study have been used for evaluation in WD (Bagchi et al., 2016; Das & Ray, 2006; Giagheddu et al., 2001; Piga et al., 2008). Out of these modalities, MRI of brain has been studied widely. The common findings in MRI are basal ganglionic and brainstem signal intensity and diffuse atrophy. SPECT has shown an abnormality of either focal or diffuse change in 86% of WD patients (Giagheddu et al., 2001). However, imaging abnormalities have also been described in patients with comorbid bipolar disorders and WD associated with predominant basal ganglia affection (Carta et al., 2015). Imaging abnormalities also suggest the possible role of fronto-limbic loops involved in the expression of emotion as the pathogenetic mechanism of bipolar disorder. Relation of copper/other trace metal accumulation in the brain as a cause of bipolar disorder in WD has been hypothesized (Pfeil & Lynn, 1999).

2. TREATMENT OF PSYCHIATRIC MANIFESTATION IN WD

The most important step in treatment of these patients is the copper chelating agent which can improve psychiatric symptoms and behavioral manifestations (Brewer et al., 1996; Modai, Karep, Liberman, & Munitz, 1985). If chelating agents are not effective antipsychiatric agents should be used (Srinivas et al., 2008). However, sometimes D-penicillamine can lead to neurological exacerbation or the occurrence of neuropsychiatric manifestations particularly seizure and psychosis. Many studies have shown mood stabilizers are effective. Some reports have shown the efficacy of antipsychotics such as clozapine, quetiapine, and olanzapine (Krim & Barroso, 2001). However, neuroleptics are better avoided routinely as their use may lead to parkinsonian symptom or tardive dyskinesia which may create confusion and make error in diagnosis and management of these patients (Chroni, Lekka, Tsibri, Econmou, & Paschalis, 2001; Srinivas et al., 2008; Tu, 1981). Sometimes lithium may be effective in bipolar disorder with WD when antipsychotics fail to respond (Rybakowski, Litwin, Chlopocka-Wozniak, & Czlonkowska, 2013). However, worsening of tremor may be a problem. One of the important advantages of lithium use that this is not metabolized by liver.

Electroconvulsive therapy may be useful at times (Sahoo, Avasthi, Sahoo, Modi, & Biswas, 2010). Favorable report have not been documented for role of orthotropic liver transplantation with neuropsychiatric features of WD except in mild cases (Medici et al., 2005; Pabón et al., 2008) and occasional anecdotal report showing posttransplant resolution of psychosis (Sorbello, Riccio, Sini, Carta, & Deemlia, 2011). Additional therapy in the form of interpersonal psychotherapy and neuro-rehabilitation will be extremely helpful (Keller, Tarta, Lagget, Crasto, & Bergamasco, 1999). Further study with large number of patients is needed to resolve this issue.

2.1 Cognitive Impairment

Cognitive impairment has been documented in both asymptomatic and symptomatic cases of WD (Forta et al., 2009). Cognitive impairment can reduce independency, cause mood fluctuations, and induce lack of adherence to treatment. Verbal intelligence and episodic memory abilities may remain preserved in WD patients. Patients showing neurological motor symptoms usually have behavior and cognitive decline in 25% of patients (Lauterbach et al., 1998). Clinically, WD dementia is considered as a subcortical dementia, but activities of daily living are often preserved. So in the true sense it should better be called a type of minor cognitive dysfunction. Executive functioning such as working memory and abstract reasoning involving frontostriatal circuits are impaired. Neuropsychological examination is useful in initial evaluation and diagnosis for WD patients before the development of frank neurological signs. MRI has good correlation with cognitive deficits. A study had shown, neuropsychological assessment of cognitive deficits over multiple domain in all subjects in a series of 12 established WD patients (Hegde et al., 2010). Neuropsychological evaluation revealed affection of deficits in various domains such as mental speed, verbal working memory, visual memory, attention, verbal learning and fluency, visuoconstructional ability, and verbal recognition.

3. IMAGING

MRI in WD has shown cortico subcortical white matter lesions with frontal predilection in 83.3% of cases (Hegde et al., 2010). Since cerebral cortex is relatively spared in WD, cognitive and behavioral abnormalities are probably explained by the dysfunction of the subcortical–frontal circuits. Two patients showed neuropsychological abnormalities in spite of normal MRI. Follow-up of serial magnetic resonance analysis and neuropsychological tests showed improvements following treatment with metal chelators (Chung et al., 2016).

Several studies have called the dementia syndrome as reversible syndrome, since cognitive and motor symptoms may improve after administration of proper treatment or hepatic transplantation (Rosseli, Lorenzana, Rosseli, & Vergara, 1987).

3.1 Sleep in WD

There are very few studies which highlighted sleep-related aspects of WD. WD patients woke up later than general population, suffer from daytime somnolence, cataplexy-like episodes, and poor nocturnal sleep (Nevsimalova et al., 2011). Nocturnal awakening is common in WD patients and more common among females (36%) than males (23%). They often take naps and suffer from fatigue. Patients with WD had lower latency of stage 1 and stage 2 of nonrapid eye movement (NREM) sleep and less amount of NREM sleep stage 2. WD patients usually have altered REM sleep function due to disturbance in monoaminergic neurons present in brainstem, an important site of involvement in WD. Another study based on polysomnogram on WD patients has documented sleep abnormalities. Patients had significantly reduced total sleep-time, sleep-efficiency, percentage of deep sleep, and REM-sleep with prolonged sleep-onset latency (Netto et al., 2010). Men had significantly more bradycardia in both awake and sleep states. Younger patients (<20 years) had frequent tachycardia, higher periodic limb movement Index, and lesser frequency of REM sleep. After chelation, patients had prolonged REM-sleep-onset latency. The isolated limb movements in sleep were more in the severe form of disease.

3.2 Autonomic Disturbance in WD

Involvement of autonomic nervous system in WD has received scanty attention (Chu, Chu, Huang, 1997). Rarely, WD can present with only autonomic symptoms (Kumar, 2005). A study based on clinical and laboratory investigation has suggested that at least 50% patients of WD had some form of dysautonomia detected clinically and or electrophysiologically (Meenakshi-Sundaram et al., 2001). Mostly, the disturbance is subclinical in WD and probably central in origin. It is more frequent among those with neurological presentation. Sympathetic and parasympathetic systems are equally affected. Peripheral nerve conduction was normal, though a study by von Giesen et al. based on quantitative testing indicated damage to unmyelinated warm specific C fibers suggesting peripheral dysautonomia (Von Giesen, Weiss, Arendt, & Hefter, 2003). An another study evaluated central cardiovascular and sudomotor responses in patients of neurological WD and documented abnormality predominantly in parasympathetic system. It has no correlation with severity of the diseases. In a study of autonomic dysfunction in 4 WD patients, two had sensory complaint 2 and another two had autonomic complaints, including syncopal spells. Skin sympathetic response was normal in only two of them and all had normal peripheral nerve conduction. This is ascribed due to as small fiber neuropathy (Francisco de Assis, Davi, Ítalo, & Otoni, 2014).

3.3 Influence of Nonmotor Manifestations in Quality of Life in WD

Quality of life (QOL) is a subjective concept of well being depending on number of factors, such as duration and severity of illness, medication use, and stress events. There are few studies which have taken the issue of QOL in WD (Carta et al., 2012). All these studies have shown lower scores on different domain of QOL with neurological and psychiatric symptoms compared to WD with predominant hepatic forms. Among mood disorders, bipolar disorder is associated with lower QOL. However with successful liver transplantation, long-term QOL may be comparable to age and sex matched control for general population.

3.4 Our Experience of NMS in Subjects With WD in Bangur Institutes of Neurosciences, Kolkata, India (Tables 2 and 3)

We have studied a cohort of 90 subjects in our institute and preliminary findings are given below:

- *Total number of subjects*: 90, including 4 (4.4%) KF negative symptomatic WD and 2 (2.2%) asymptomatic WD individuals identified by screening of index WD cases

Gender: Males 56 (62.2%) and females 34 (48.8%)

Residence: Urban 24 (26.7%), Suburban 15 (16.7%), Rural background 51 (56.7%)

Religion: Hindus 67 (74.45) and Muslims 23 (25.6%)

Consanguineous parentage present in 10 (11.1%)

Family history of WD positive in 22 (24.4%)

Family history of psychiatric disorders:

Absent	59 (65.6%)
Affective disorders	21 (23.3%)
Psychosis	One (1.1%)
Suicide	5 (5.6%)
No information in 4 (4.4%)	

Table 2 Showing Frequency and Age of Onset of Motor, Behavioral, and Cognitive Symptoms

Variable	n[a]	Range	Mean ± SD
Age (years)	90	8–50	18.39 ± 7.21
Age of onset of motor symptoms (years)	84	5–48	14.94 ± 7.56
Age of onset of behavioral abnormalities (years)	84	6–49	15.53 ± 6.98
Age of onset of cognitive symptoms (years)	44	6–43	14.38 ± 6.4
Duration of motor symptoms (months)	84	2–168	42.43 ± 40.41
Duration of behavioral abnormality (months)	83	30 days to 14 years	30.51 ± 34.0
Duration of cognitive symptoms (months)	43	2–120	33.23 ± 30.82
Serum ceruloplasmin levels (in mg/dL)	70	0.08–48.50	12.36 ± 12.32
Urinary copper levels (mg/24 h)	59	48–2144	324.54 ± 334.03

[a] *n* refers to the number of subjects where data are available.

Table 3 Showing Frequency of Various Neuropsychiatric Symptoms in Our WD Cohort

Symptom	Frequency	%
Aberrant motor behavior	7	7.8
Aggression	51	56.7
Anxiety	8	8.9
Apathy	5	5.6
Appetite changes	35	38.9
Authoritarian manner[a]	27	30.0
Delusions	11	12.2
Demanding behavior[a]	45	50.0
Depression/dysphoria	10	11.1
Disinhibition	24	26.7
Euphoria	35	38.9
Hallucinations—Auditory[a]	6	6.7
Hallucinations—Visual[a]	10	11.1
Hyperactivity	36	40
Hypersexuality[a]	11	12.2
Irritability	74	82.2
Lability of emotions[a]	15	16.7
Obsessions & compulsions[a]	2	2.2
Oppositional & conduct problems[a]	34	37.8
Personality change[a]	15	16.7
Sleep disturbances	44	48.9
Increased socialization[a]	16	17.8
Stubbornness[a]	50	55.6
Suicidal behaviors[a]	20	22.2

[a]Items added to the checklist based on Neuropsychiatric Inventory (NPI).

4. DISCUSSION

This is a naturalistic study of a group of WD patients from a neurology specialty clinic, where they were directly examined, observed prospectively as per clinic protocol, and retrospective data collected from the patients' guardians regarding previous psychiatric aspects.

Almost all patients were found to have behavioral problems atleast sometime during the span of WD, reflecting the fact that they are commoner in neurological WD, compared to hepatic type (Oder et al., 1991). However, the behavioral abnormality was often not a cause for concern, nor was the chief complaint among many of the guardians, being attributed to the distress of the chronic disease process and prolonged treatment itself (Table 4).

We found externalizing behaviors to be the commonest type of abnormality, similar to most case series (Dening & Berrios, 1989; Shanmugiah et al., 2008). However, contrary to studies from developed and Western countries, which found more of depressed or anxious patients (Svetel et al., 2009), our cohort had a higher proportion of individuals meeting a diagnosis of mania and bipolar illness. This could represent a separate genetic pool from the other ethnic groups, especially as another study from India (Shanmugiah et al., 2008), had comparable results. Higher frequency of manic symptoms may also be due to (a) different stage or severity of the WD cohort (i.e., stable-treated vs untreated or inadequately treated); (b) presence of a genetic predisposition to develop affective disorders, as shown by positive family history of mood disorders; (c) a different pattern

Table 4 Showing Syndromal Psychiatric Diagnoses That Could Be Made After Analysis of Behavioral Abnormalities ($N = 90$)

Diagnoses	Frequency (%)
No diagnosis	45 (50%)
Manic episode/bipolar affective disorder	27 (30%)
Depressive episode	3 (3.3%)
Schizophrenia-like psychosis	2 (2.2%)
Other psychotic conditions	7 (7.8%)
Personality change	4 (4.4%)
Obsessive compulsive disorder	2 (2.2%)

of involvement of WD with predilection for the basal ganglia, whose involvement leads to psychiatric symptoms by their own merit; (d) socio-familial differences allowing greater leniency and acceptance of abnormal behaviors so that they are unaddressed in initial stages causing them to worsen; and (e) presence of a psychiatrist in the core team who actually probed for and unearthed the behavioral symptoms.

5. CONCLUSION

Thus, WD has an abundance of NMS which are in the domain of neuropsychiatric and cognitive disorders besides hepatic and multisystem abnormalities. These have not been adequately explored in a systemic manner. These NMS can be present before the full expression and also during the clinical progression of the disease. These psychiatric–cognitive disturbances are best picked by neuropsychological assessment and neuroimaging such as MRI and SPECT study when the basic diagnosis has already been made by clinical and biochemical investigation. Various modes of treatment are available and application of these modalities in early stage will be very beneficial for alleviation of the suffering subjects and also improving their QOL. Well-conducted research is needed to find out the detail spectrum of NMS, early diagnosis of these abnormalities, and treatment need in future. Genetic study should also be undertaken to know if there is any particular risk factor for development of these NMS.

REFERENCES

Akhon, O., Akpinar, E., Karcaaltincaba, M., Haliloglu, M., Akata, D., Karaosmanoglu, A. D., et al. (2009). Imaging findings of liver involvement of Wilson's disease. *European Journal of Radiology*, *69*, 147–155.

Akil, M., & Brewer, G. J. (1995). Psychiatric and behavioural abnormalities in Wilson's disease. *Advances in Neurology*, *65*, 171–178.

Ala, A., Borjigin, J., Rochwarger, A., & Schilsky, M. (2005). Wilson disease in septuagenarian siblings: Raising the bar for diagnosis. *Hepatology*, *41*, 668–670.

Araujo-de-Freitas, L., Rocha, M., Gondik, V., Quarinitini, L., & Miranda-cippa, A. (2014). Major depression caused by Wilson's disease. *Revista Brasileira de Psiquiatria*, *36*, 184. http://dx.doi.org/10.1590/1516-4446-2013-1188.

Bagchi, M., Paul, N., Ghosh, S., Majumdar, S., Saha, S., Roy, A., et al. (2016). Spectrum of abnormal movements in a cohort of neuro-Wilson patients with radiological associations. *Journal of Neurological Disorders*, *4*, 259–266. http://dx.doi.org/10.4172/2329-6895.1000259.

Benhamla, T., Tirouche, Y. D., Abaoub-Germaine, A., & Theodore, F. (2007). The onset of psychiatric disorders and Wilson's disease. *Encephale*, *33*, 924–932.

Bhattacharya, T., Debnath, A., & Sarkar, S. (2015). Catatonic disorder due to Wilson's disease—A rare presentation. *Bengal Journal of Psychiatry*, *20*, 17–20.

Bidaki, R., Zarei, M., Mirhosseini, S. M. M., Moghadami, S., Hejrati, M., Kohnavard, M., et al. (2012). Mismanagement of Wilson's disease as psychotic disorder. *Advanced Biomedical Research, 1*, 61.

Brewer, G. J. (2005). Behavioral abnormalities in Wilson's disease. *Advances in Neurology, 96*, 262–274.

Brewer, G. J., Johnson, V., Dick, R. D., Klin, K. J., Fink, J. K., & Brunberg, J. A. (1996). Treatment of Wilson disease with ammonium tetrathiomolybdate. II. Initial therapy in 33 neurologically affected patients and follow with zinc therapy. *Archives of Neurology, 53*, 1017–1025.

Brewer, G. J., & Yuzbasiyan-Gurkan, V. (1992). Wilson disease. *Medicine (Baltimore), 71*, 139–164.

Carta, M. G., Mura, G., Sorbello, O., Farina, G., & Demelia, L. (2012). Quality of life and psychiatric symptoms in Wilson's disease: The relevance of bipolar disorders. *Clinical Practice and Epidemiology in Mental Health, 8*, 102–109.

Carta, M. G., Saba, L., Moro, M. F., Demelia, E., Sorbello, O., Pintus, M., et al. (2015). Homogeneous magnetic resonance imaging of brain abnormalities in bipolar spectrum disorders comorbid with Wilson disease. *General Hospital Psychiatry, 37*, 134–138.

Cheon, J.-E., Kim, I.-O., Seo, J. K., KO, J. S., Lee, J. M., Shin, C.-I., et al. (2010). Clinical application of liver MR imaging in Wilson's disease. *Korean Journal of Radiology, 11*, 665–672.

Chroni, E., Lekka, N. P., Tsibri, E., Econmou, A., & Paschalis, C. (2001). Acute progressive akinetic-rigid syndrome induced by neuroleptics in a case of Wilson's disease. *Journal of Neuropsychiatry and Clinical Neurosciences, 13*, 531–532.

Chu, E. C., Chu, N. S., & Huang, C. C. (1997). Autonomic involvement in Wilson's disease: A study of sympathetic skin response and RR interval variation. *Journal of the Neurological Sciences, 149*, 131–137.

Chung, E. J., Kim, E. G., Kim, S. J., Ji, K. H., & Seo, J. H. (2016). Wilson disease with cognitive impairment and without extrapyramidal signs: Improvement of neuropsychological performance and reduction of MRI abnormalities with trientine treatment. *Neurocase, 22*, 40–44.

Das, S. K., & Ray, K. (2006). Wilson's disease: An update. *Nature Clinical Practice Neurology, 2*, 482–493.

Davis, E. J., & Borde, M. (1993). Wilson's disease and catatonia. *The British Journal of Psychiatry, 162*, 256–259.

Dening, T. R., & Berrios, G. E. (1989). Wilson's disease: Psychiatric symptoms in 195 cases. *Archives of General Psychiatry, 46*, 1126–1134.

Duggal, H. S., & Nizamie, S. H. (2000). Wilson's disease presenting with obsessive-compulsive disorder. *Indian Journal of Psychiatry, 42*, 312–316.

Eggers, B., Hermann, W., Barthel, H., Sabri, O., Wagner, A., & Hesse, S. (2003). The degree of depression in Hamilton rating scale is correlated with the density of presynaptic serotonin transporters 23 patients with Wilson's disease. *Journal of Neurology, 250*, 576–580.

Ehman, T. S., Beninger, R. J., Gawel, M. J., & Riopelle, R. J. (1990). Depressive symptoms in Parkinson's disease: A comparison with disabled control subjects. *Journal of Geriatric Psychiatry and Neurology, 3*, 3–9.

Fahn, S., & Jankovic, J. (2007). *Wilson's disease. Principles and Practice of movement disorders.* (pp. 589–612). Philadelphia: Churchill Living Stone, Elsevier.

Ferenci, P., Litwin, T., Seniow, J., & Czlonkowska, A. (2015). Encephalopathy in Wilson disease: Copper toxicity or liver failure? *Journal of Clinical and Experimental Hepatology, 5*(Suppl. 1), S88–S95.

Forta, N. A. F., Caramelli, P., & Barbosa, E. R. (2009). Cognitive impairment in Wilson's disease. *Dementia and Neuropsychologia, 3*, 16–21.

Francisco de Assis, A. G., Davi, F. A., Ítalo, S. O., & Otoni, C. D. V. (2014). Small fiber dysfunction in patients with Wilson's disease. *Arquivos de Neuro-psiquiatria, 72*, 592–595.

Genc, E. S., Genc, A., & Guveneroglu, N. U. (2013). Personality changes in a patient with Wilson's disease. *Düşünen Adam: The Journal of Psychiatry and Neurological Sciences*, *26*, 299–302.

Giagheddu, M., Tamburini, G., Piga, M., Tacconi, P., Giagheddu, A., Serra, A., et al. (2001). Comparison of MRI, EEG, EPs and ED-SPECT in Wilson's disease. *Acta Neurologica Scandinavica*, *103*, 71–81.

Giannini, E. G., Testa, R., & Savarino, V. (2005). Liver enzyme alteration: A guide for clinicians. *Canadian Medical Association Journal*, *172*, 367–379.

Grover, S., Sarkar, S., Jhanda, S., & Chawla, Y. (2014). Psychosis in an adolescent with Wilson's disease: A case report and review of the literature. *Indian Journal of Psychiatry*, *56*(4), 395–398. http://dx.doi.org/10.4103/00195545.146530.

Hegde, S., Sinha, S., Rao, S. L., Taly, A. B., & Vasudev, M. K. (2010). Cognitive profile and structural findings in Wilson's disease: A neuropsychological and MRI based study. *Neurology India*, *58*, 708–713.

Keller, R., Tarta, R., Lagget, M., Crasto, S., & Bergamasco, B. (1999). Psychiatric symptoms as late onset of Wilson's disease: Neuroradiological findings, clinical features and treatment. *Italian Journal of Neurological Sciences*, *20*, 49–54.

Krim, E., & Barroso, B. (2001). Psychiatric disorders treated with clozapine in a patient with Wilson's disease. *Presse Médicale*, *30*, 738.

Kumar, S. (2005). Severe autonomic dysfunction as a presenting feature of Wilson's disease. *Journal of Postgraduate Medicine*, *51*, 75–76.

Kumawat, B. L., Sharma, C. M., Tripathi, G., Ralot, T., & Dixit, S. (2007). Wilson's disease presenting as isolated obsessive-compulsive disorder. *Indian Journal of Medical Sciences*, *61*, 607–610.

Lauterbach, E. C., Cummings, J. L., Duffy, J., Coffey, C. E., Kaufer, D., Lovell, M., et al. (1998). Neuropsychiatric correlates and treatment of lenticulostriatal disease: A review of the literature and overview of research opportunities in Huntington's, Wilson's and Fahr's disease. *The Journal of Neuropsychiatry and Clinical Neuroscience*, *10*, 249–266.

Lingam, S., Wilson, J., Nazer, H., & Mowat, A. P. (1987). Neurological abnormalities in Wilson's disease are reversible. *Neuropediatrics*, *18*, 11–12.

Machado, A., Chien, H. F., Deguti, M. M., Cançado, E., Azevedo, R. S., Scaff, M., et al. (2006). Neurological manifestations in Wilson's disease. Report of 119 cases. *Movement Disorders*, *21*, 2192–2196.

Medici, V., Mirante, V. G., Fassati, L. R., Pompili, M., Forti, D., & Gaudio, M. D. (2005). Liver transplantation for Wilson's disease: The burden of neurological and psychiatric disorders. *Liver Transplantation*, *11*, 1056–1063.

Meenakshi-Sundaram, S., Taly, A. B., Kamath, V., Arunoday, G. R., Rao, S., & Swamy, H. S. (2001). Autonomic dysfunction in Wilson's disease—A clinical and electrophysiological study. *Clinical Autonomic Research*, *12*, 190–192.

Modai, I., Karep, I., Liberman, U. A., & Munitz, H. (1985). Penicillaimine therapy for schizophreniform psychosis in Wilson's disease. *Journal of Nervous and Mental Disorders*, *173*, 689–701.

Netto, A. B., Sinha, S., Tally, A. B., Panda, S., & Rao, S. (2010). Sleep in Wilson's disease: A polysomnography-based study. *Neurology India*, *58*, 933–938.

Nevsimalova, S., Buskova, J., Bruha, R., Kemlink, D., Sonka, K., Vitek, L., et al. (2011). Sleep disorders in Wilson disease. *European Journal of Neurology*, *8*, 184–190.

Oder, W., Grimm, G., Kollegger, H., Ferenci, P., Schneider, B., & Deecke, L. (1991). Neurological and neuropsychological spectrum of Wilson's disease: A prospective study of 45 cases. *Journal of Neurology*, *238*, 281–287.

Pabón, V., Dumortier, J., Gincul, R., Baulieux, J. C., Trépo, C., & Souquet, J.-C. (2008). Long term results of liver transplantation for Wilson disease. *Gastroentérologie Clinique et Biologique*, *32*, 378–381.

Pfeil, S. A., & Lynn, D. J. (1999). Wilson's disease: Copper unfettered. *Journal of Clinical Gastroenterology*, *29*, 22–31.

Piga, M., Murru, A., Satta, L., Serra, A., Sias, A., Loi, G., et al. (2008). Brain MRI and SPECT in the diagnosis of early neurological involvement in Wilson's disease. *European Journal of Nuclear Medicine and Molecular Imaging*, *35*, 716–724.

Rosseli, M., Lorenzana, P., Rosseli, A., & Vergara, I. (1987). Wilson disease, a reversible dementia: Case report. *Journal of Clinical and Experimental Neuropsychology*, *9*, 99–406.

Rukunuzzaman, M. (2015). Wilson disease in Bangladeshi children: Analysis of 100 cases. *Pediatric Gastroenterology, Hepatology, & Nutrition*, *18*, 121–127.

Rybakowski, J. K., Litwin, T., Chlopocka-Wozniak, M., & Czlonkowska, A. (2013). Lithium treatment of a bipolar patient with Wilson's disease: A case report. *Pharmacopsychiatry*, *46*, 1201.

Sahoo, M. K., Avasthi, A., Sahoo, M., Modi, M., & Biswas, P. (2010). Psychiatric manifestations of Wilson disease and treatment with electroconvulsive therapy. *Indian Journal of Psychiatry*, *52*, 66–68.

Sahu, J. K., Singhi, P., & Malhotra, S. (2013). Late occurrence of isolated obsessive-compulsive behavior in a boy with Wilson's disease on treatment. *Journal of Child Neurology*, *28*, 277.

Saint-Laurent, M. (1992). Schizophrenia and Wilson's disease. *Canadian Journal of Psychiatry*, *37*, 358–360.

Schwartz, M., Feics, S., Polak, H., & Sharf, B. (1993). Psychiatric manifestations in Wilson's disease. *Harefuah*, *124*, 75–77.

Shanmugiah, A., Sinha, S., Taly, A. B., Prashanth, L. K., Tomar, M., Arunodaya, G. R., et al. (2008). Psychiatric manifestations in Wilson's disease: A cross-sectional analysis. *The Journal of Neuropsychiatry and Clinical Neurosciences*, *20*, 81–85.

Sinha, S., Taly, A. B., Ravishankar, S., et al. (2006). Wilson's disease: Cranial MRI observations and clinical correlation. *Neuroradiology*, *48*, 613–621.

Soni, D., Shukla, G., Singh, S., Goyal, V., & Behari, M. (2009). Cardiovascular and sudomotor autonomic dysfunction in Wilson's disease—Limited correlation with clinical severity. *Autonomic Neuroscience*, *151*, 154–158.

Sorbello, O., Riccio, D., Sini, M., Carta, M., & Deemlia, L. (2011). Resolved psychosis after liver transplantation in a patient with Wilsons's disease. *Clinical Practice Epidemiology in Mental Health*, *7*, 182–184.

Spyridi, S., Diakoglannis, I., Michaelides, M., Sokolakis, S., Icovides, A., & Kaprinis, G. (2008). Delusional disorder and alcohol abuse in a patient with Wilson's disease. *General Hospital Psychiatry*, *30*, 585–586.

Srinivas, K., Sinha, S., Taly, A. B., Prashanth, L. K., Arunodaya, G. R., Janardhana Reddy, Y. C., et al. (2008). Dominant psychiatric manifestations in Wilson's disease: A diagnostic and therapeutic challenge! *Journal of the Neurological Sciences*, *266*, 104–108.

Svetel, M., Potrebić, A., Pekmezović, T., Tomić, A., Kresojević, N., Jesić, R., et al. (2009). Neuropsychiatric aspects of treated Wilson disease. *Parkinsonism & Related Disorders*, *15*, 772–775.

Tu, J. (1981). The inadvisability of neuroleptic medication in Wilson's disease. *Biological Psychiatry*, *16*, 963–968.

Von Giesen, H. J., Weiss, P., Arendt, G., & Hefter, H. (2003). Potential c-fibre damage in Wilson's disease. *Acta Neurologica Scandinavica*, *108*(3), 257–261.

Wilson, S. A. K. (1912). Progressive lenticular degeneration: A familial nervous disease associated with cirrhosis of the liver. *Brain*, *34*, 295–507.

Zimbrean, P. C., & Schilsky, M. L. (2014). Psychiatric aspects of Wilson disease: A review. *General Hospital Psychiatry*, *36*, 53–62.

CHAPTER FIFTY-THREE

Tourette Syndrome and Chronic Tic Disorders: The Clinical Spectrum Beyond Tics

Davide Martino*[,1], Christos Ganos[†], Tamara M. Pringsheim*
*University of Calgary, Calgary, AB, Canada
[†]University Medical Center Hamburg-Eppendorf, Hamburg, Germany
[1]Corresponding author: e-mail address: davide.martino@ucalgary.ca

Contents

Abstract

The clinical surveillance and active management of Tourette syndrome (TS) and other primary chronic tic disorders cannot be limited to tics, as these patients manifest a spectrum of sensory-, behavioral-, cognitive-, and sleep-related problems that have a major impact on their functioning and quality of life, influencing enormously clinical decision making on a routine basis. The sensory phenomena of primary tic disorders consist of premonitory urges and heightened sensitivity to external somatosensory and interoceptive stimuli. Recent evidence suggests that raised interoceptive awareness may be related to the classical premonitory urges associated with tics. The burden of behavioral comorbidities is very important in determining the degree of disability of patients with primary tic disorders. Only 10%–15% of these patients presents exclusively with a tic disorder. Obsessive–compulsive disorder (OCD) are common in TS, and the clinical distinction between compulsions and complex tics may be difficult in some cases. "Tic-related OCD" represents a phenomenologically characteristic subtype of OCD, also associated with "just right" phenomena. Probably the presence of comorbid attention

International Review of Neurobiology, Volume 134
ISSN 0074-7742
http://dx.doi.org/10.1016/bs.irn.2017.05.006

deficit/hyperactivity disorder is the main determinant of cognitive dysfunction in TS patients and influences heavily also the risk of developing disruptive behaviors. Mood and anxiety disorders, impulse control disorders, rage attacks, "impulsive" tic-like behaviors (e.g., nonobscene socially inappropriate behaviors, and self-injurious behaviors), and autism spectrum disorders complete the wide psychopathological spectrum of primary chronic tic disorders. Moreover, specific sleep abnormalities have been reported in TS patients, although more research is needed on this specific clinical problem. As in other areas of clinical neuroscience, a comprehensive approach to both motor and nonmotor aspects of this group of disorders will help personalizing treatment interventions and, ultimately, improve quality of care.

1. INTRODUCTION

Primary tic disorders are neurodevelopmental conditions beginning in the first two decades, which may or may not persist as clinically relevant during adulthood. The Diagnostic and Statistical Manual of Mental Disorders (5th ed.) from the American Psychiatric Association (2013) lists three main types of chronic primary tic disorders. Of these, Tourette syndrome (TS) reaches most frequently medical attention and surveillance, and is diagnosed when multiple motor and vocal tics of onset in childhood and adolescence coexist for more than 1 year. It is differentiated from chronic motor or chronic vocal tic disorders by the coexistence of these two types of tics. Tics, which represent their core defining feature, can be defined as recurrent, patterned, usually rapid, nonrhythmic movements, and vocalizations that can be suppressed voluntarily to a variable degree. The clinical surveillance and active management of TS and other chronic tic disorders, however, cannot be limited to tics, as these patients manifest a spectrum of sensory, behavioral, cognitive, and sleep-related problems that have a major impact on their functioning and quality of life, influencing enormously clinical decision making on a routine basis (see Fig. 1 for a summary of proposed clinical assessment procedures). This chapter is an overview of the phenomenology, core pathophysiological aspects, and basic management principles of the nonmotor features of chronic tic disorders.

2. SENSORY ASPECTS OF TIC DISORDERS

Different from other hyperkinesias, tics bear the greatest resemblance to voluntary actions with regard to their phenomenology, neurophysiology, and associated features, as, for example, *premonitory urges* (Ganos & Martino,

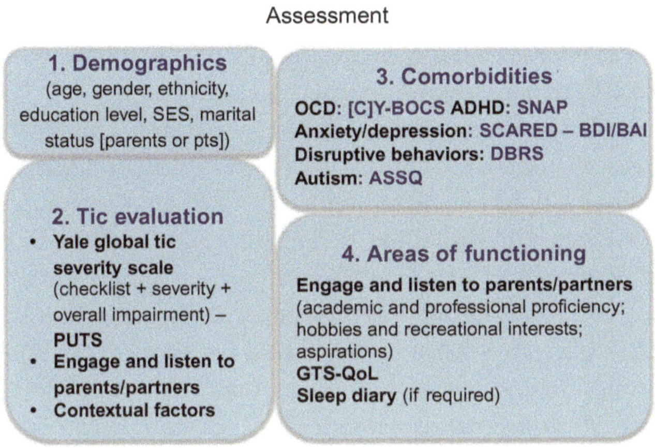

Fig. 1 Assessment plan in Tourette syndrome. *ASSQ*: Autism Spectrum Screening Questionnaire (Ehlers, Gillberg, & Wing, 1999). *BAI*, Beck Anxiety Inventory (Steer, Beck, Riskind, & Brown, 1986); *BDI*, Beck Depression Inventory (Joe, Woolley, Brown, Ghahramanlou-Holloway, & Beck, 2008); *[C]Y-BOCS*, [Children]Yale-Brown Obsessive–Compulsive Scale (Scahill, Riddle, McSwiggin-Hardin, et al., 1997); *DBRS*, Disruptive Behavior Rating Scale (Silva et al., 2005); *OCD*, obsessive–compulsive disorder; *PUTS*, Premonitory Urge for Tics Scale (Woods, Piacentini, Himle, & Chang, 2005); *SCARED*, Screen for Child Anxiety and Related Emotional Disorders (Birmaher et al., 1997); *SNAP*, Swanson, Nolan, and Pelham questionnaire (Bussing et al., 2008).

2015). The latter describe a group of sensory phenomena that commonly precede tics and are often viewed as the driving force behind their manifestation. Patients often find it difficult to uniformly label the sensory premonitions and, thereby, use descriptive terms such as urge to move/tic, urge to apply pressure, increased tension, pressure/fullness, or an urge to stretch (Kwak, Dat Vuong, & Jankovic, 2003). Other less common descriptions include ache, itching, or burning (Kwak et al., 2003).

The clinical relevance of premonitory urges in the pathophysiology of tic disorders was first highlighted by Bliss, a physician who also had TS (Bliss, 1980). Bliss suggested that tics were mere responses to the unpleasant premonitory urges and, therefore, viewed the latter as an essential, inseparable part of tic disorders. Bliss's view was reinforced by subsequent reports (Cohen & Leckman, 1992; Kane, 1994; Shapiro, Shapiro, Young, & Feinberg, 1988), and in a simple and elegant clinical study in 1991 Lang demonstrated that, unlike patients with other hyperkinetic movement disorders such as chorea or myoclonus, adults with tic disorders often viewed their tics as *voluntary* responses to *involuntary* premonitory urges (Lang, 1991). Soon thereafter, it became clear that the majority of patients with tics and TS reported premonitory urges (Leckman, Walker, & Cohen, 1993), which, though, might be less well consciously perceived in patients younger than 10 years (Banaschewski, Woerner, & Rothenberger, 2003; Leckman et al., 1993). Importantly, the distribution of premonitory urges across the different body parts of patients with TS is not random, but covaries with the distribution of their tics (Ganos et al., 2012; Leckman et al., 1993). This indicates a possibly common underlying neural factor that mediates both the presence of premonitory urges and the emergence of tics, and reinforces the view of tics being the mere consequence of abnormal somatosensory perception. The subjective sense of relief that patients report upon releasing urge-driven tics further lends clinical support to this notion.

However, several lines of evidence do not conform to this simplistic view, and might, in fact, suggest a different pathophysiological interplay between the emergence of tics and the presence of premonitory urges. First, as highlighted earlier, premonitory urges may, indeed, be reported for the majority, but not for every single tic. Hence, some tics can emerge without the presence of premonitory urges. Indeed, patients with TS and other primary tic disorders are often not consciously aware of some of their most common tics, as for example eye tics, and, therefore, also do not perceive associated premonitory urges (Leckman et al., 1993; Martino, Cavanna, Robertson, & Orth, 2012). A study that investigated the long-term course of tic disorders could demonstrate, for example, the presence of tics on video recordings during the follow-up examination, despite the belief of the recorded individuals, that they were "tic free" (Pappert, Goetz, Louis, Blasucci, & Leurgans, 2003). Similarly, patients with autism spectrum disorders, who often also exhibit tics, appear to be less aware of their hyperkinetic movements, and thereby, their tic-related urges (Kahl et al., 2015). Second, although preliminary experimental evidence has shown a correlation between the fluctuations of the awareness of premonitory urges with tic emergence (Brandt et al., 2016), this has not been demonstrated for all tics in a single individual. Indeed, particularly in cases with severe, frequent, and multifocal tics, it is very difficult to conceive two parallel streams of shared conscious awareness, one related to physiological signals associated with voluntary actions and one to premonitory urges and tic movements. In the case where one attentional resource would be required for both ongoing voluntary motor behavior and tics, this would potentially lead to frequent tic-related motor blocks, but these are in fact very rare phenomena (Ganos, Mueller-Vahl, & Bhatia, 2015). Third, patients with the same tic burden perceive different intensities of premonitory urges (Crossley & Cavanna,

2013; Eddy & Cavanna, 2013a; Muller-Vahl, Riemann, & Bokemeyer, 2014; Reese et al., 2014; Steinberg et al., 2010; Sutherland Owens, Miguel, & Swerdlow, 2011; Woods et al., 2005). This highlights that the perceptual source of premonitory urges is distinct from a putative tic generator and could rely on different factors, other than the sole presence of hyperkinetic movements. Taken together, these data-driven observations demonstrate that the relation between premonitory urges and tics is more complex than previously considered and that tic and urge generation could rely on different neurocognitive correlates. Moreover, they suggest that there might be distinct tic behaviors in patients, as, for example, such which are more voluntary and habitual (tics related to premonitory urges) and others that are more automatic or reflexive (tics without urges). The presence of tics preceded by premotor potentials and tics without (Duggal & Nizamie, 2002; Karp, Porter, Toro, & Hallett, 1996; Obeso, Rothwell, & Marsden, 1982; van der Salm, Tijssen, Koelman, & van Rootselaar, 2012) also supports this suggestion.

Several studies have attempted to elucidate the neurocognitive correlates of premonitory urges. On one hand, the clinical knowledge that patients with tic disorders and TS experience *heightened sensitivity to external somatosensory stimuli* (e.g., clothing tags, tight sock bands, etc.) (Belluscio, Jin, Watters, Lee, & Hallett, 2011; Cohen & Leckman, 1992; Taylor, Conelea, McKay, Crowe, & Abramowitz, 2014) generated the hypothesis that premonitory urges are a result of altered exteroceptive perception. Neuroimaging findings, which associated alterations within the primary somatosensory cortex of adult patients with TS with tic severity, supported this suggestion (reviewed in Ganos, Roessner, & Munchau, 2013; Hashemiyoon, Kuhn, & Visser-Vandewalle, 2017). However, two studies, which examined thresholds of exteroceptive stimuli in adults with TS, did not find differences between patients and age-matched healthy controls (Belluscio et al., 2011; Schunke et al., 2016). Importantly, no associations between either the reported perception of external stimuli or the intensity of premonitory urges and actual threshold detection levels within different somatosensory modalities were observed in patients (Belluscio et al., 2011; Schunke et al., 2016). Although these results demonstrate that basic exteroceptive perception and premonitory urges are not tightly linked, they do not provide further insights in higher-order aspects of somatosensory processing in patients with TS. However, this was assessed by a third study, which demonstrated that children with TS, 48% of whom with comorbid attention deficit/hyperactivity disorder (ADHD) and 28% with comorbid obsessive–compulsive disorder, showed impaired performance in tactile perception and adaptation tasks compared to 67 controls (Puts et al., 2015). The same study also employed magnetic resonance spectroscopy and found reduced concentrations of the neurotransmitter GABA for the corresponding primary sensorimotor cortex of TS patients providing a pathophysiological correlate of behavioral (under) performance (Puts et al., 2015).

A different hypothesis associates the perception of premonitory urges with the awareness of other bodily urges, such as yawning or micturition (Jackson, Parkinson, Kim, Schuermann, & Eickhoff, 2011). This view considers premonitory urges as a class of somatic (interoceptive) signals attributed to tics. Accordingly, perception of premonitory urges would depend on the capacity to perceive bodily signals (e.g., heartbeats), also referred to as *interoceptive sensitivity*. To date, one study examined interoceptive sensitivity using an established heartbeat perception task (Schandry, 1981) in a sample of 19 adults with TS and 25 healthy controls and showed reduced interoception in patients (Ganos, Garrido, et al., 2015). Importantly, a strong positive correlation was demonstrated between the capacity to perceive bodily signals, such as heartbeats, and premonitory urges. The same study also showed that greater tic severity was correlated to more urges. These behavioral results were further corroborated by two neuromaging studies, which associated premonitory urges with a key brain structure of interoceptive sensitivity, the insular cortex (Draper, Jackson, Morgan, & Jackson, 2016; Tinaz, Malone, Hallett, & Horovitz, 2015). Draper et al. demonstrated structural changes within the insular cortex of 29 children and adolescents with TS compared to controls (Draper et al., 2016). Importantly, an inverse correlation between thickness of the insular cortex and the perceived intensity of premonitory urges was demonstrated in patients, linking imaging findings to behavior (Draper et al., 2016). In the study by Tinaz et al., 13 adult patients were compared to 13 control subjects (Tinaz et al., 2015): changes in functional connectivity were demonstrated for the patient group between the right dorsal anterior insula and predetermined frontostriatal nodes, including the supplementary motor area, as part of a putative urge–tic network (Tinaz et al., 2015). Here, premonitory urge ratings correlated with connectivity between the right dorsal anterior insula and the supplementary motor area bilaterally, again implicating the insular cortex with the clinical symptom of premonitory urges. Taken together, these results highlight the role of interoceptive sensitivity in the perception of premonitory urges. Patients who are more able to perceive their interoceptive signals may also more frequently misinterpret them as urges to tic (Ganos, Garrido, et al., 2015).

3. BEHAVIORAL PROBLEMS IN TIC DISORDERS: COMORBID ADHD

ADHD is a complex neurodevelopmental disorder and the most common comorbidity observed in children and youth with tic disorders seeking medical attention. The frequency of this comorbidity ranges from 38% of TS patients in community-based studies to more than 60% in service-based studies (Cavanna & Rickards, 2013). In a recent large service-based study, a lifetime diagnosis of ADHD was given in 54.3% of patients, with males being at higher risk (Hirschtritt et al., 2015). This high comorbidity rate justifies the need to screen each patient with a tic disorder for the presence of ADHD.

The spectrum of ADHD comprises three major clinical subtypes: *inattentive*, *hyperactive/impulsive*, and *combined*. DSM-5 diagnostic criteria for the combined subtype of ADHD require children to present with six or more symptoms of inattention and six or more symptoms of hyperactivity-impulsivity (Box 1) that have been present for at least 6 months and are inappropriate for their developmental level. Inattentive or hyperactive/impulsive subtypes are diagnosed if children manifest at least six symptoms of either of the two subtypes, but not of both. In addition, DSM-5 criteria require symptoms to be present before age 12 years and to occur in two or more settings (i.e., home and school). In the majority of patients, though, ADHD symptoms typically begin between 3 and 5 years of age, and therefore predate the onset of tics. The ascertainment of ADHD comorbidity in youth with a chronic tic disorder is a crucial aspect of the assessment of these patients, given its major impact on psychosocial development, externalizing and internalizing symptoms, cognitive functioning, and quality of life.

A body of evidence indicates that ADHD is probably the most relevant determinant of *problems in executive functions* in patients with TS across different age groups (Rizzo, Gulisano, Cali, & Curatolo, 2013). Several clinical studies highlight that, in these patients, comorbid ADHD symptoms are more strongly related to deficits in planning ability, inhibitory function, working memory, and visual attention than core TS symptoms, i.e., tics and premonitory urges (Termine et al., 2016). In addition, the risk of *aggressive and delinquent behavior* in children with tic disorders is largely due to the presence of ADHD and is a major source of individual patient disability and

BOX 1 DSM-5 Criteria for ADHD

People with ADHD show a persistent pattern of inattention and/or hyperactivity–impulsivity that interferes with functioning or development:

1. *Inattention*: Six or more symptoms of inattention for children up to age 16, or 5 or more for adolescents 17 and older and adults; symptoms of inattention have been present for at least 6 months, and they are inappropriate for developmental level:
   - o Often fails to give close attention to details or makes careless mistakes in schoolwork, at work, or with other activities.
   - o Often has trouble holding attention on tasks or play activities.
   - o Often does not seem to listen when spoken to directly.
   - o Often does not follow through on instructions and fails to finish schoolwork, chores, or duties in the workplace (e.g., loses focus, side-tracked).
   - o Often have trouble organizing tasks and activities.
   - o Often avoids, dislikes, or is reluctant to do tasks that require mental effort over a long period of time (such as schoolwork or homework).
   - o Often loses things necessary for tasks and activities (e.g., school materials, pencils, books, tools, wallets, keys, paperwork, eyeglasses, and mobile telephones).
   - o Is often easily distracted.
   - o Is often forgetful in daily activities.
2. *Hyperactivity and impulsivity*: Six or more symptoms of hyperactivity–impulsivity for children up to age 16, or 5 or more for adolescents 17 and older and adults; symptoms of hyperactivity–impulsivity have been present for at least 6 months to an extent that is disruptive and inappropriate for the person's developmental level:
   - o Often fidgets with or taps hands or feet, or squirms in seat.
   - o Often leaves seat in situations when remaining seated is expected.
   - o Often runs about or climbs in situations where it is not appropriate (adolescents or adults may be limited to feeling restless).
   - o Often unable to play or take part in leisure activities quietly.
   - o Is often "on the go" acting as if "driven by a motor."
   - o Often talks excessively.
   - o Often blurts out an answer before a question has been completed.
   - o Often has trouble waiting his/her turn.
   - o Often interrupts or intrudes on others (e.g., butts into conversations or games).

In addition, the following conditions must be met:

- Several inattentive or hyperactive–impulsive symptoms were present before age 12 years.

BOX 1 DSM-5 Criteria for ADHD—cont'd

- Several symptoms are present in two or more setting (such as at home, school, or work; with friends or relatives; in other activities).
- There is clear evidence that the symptoms interfere with, or reduce the quality of, social, school, or work functioning.
- The symptoms are not better explained by another mental disorder (such as a mood disorder, an anxiety disorder, a dissociative disorder, or a personality disorder). The symptoms do not happen only during the course of schizophrenia or another psychotic disorder.

Based on the types of symptoms, three kinds (presentations) of ADHD can occur:

Combined presentation: if enough symptoms of both criteria inattention and hyperactivity–impulsivity were present for the past 6 months.

Predominantly inattentive presentation: if enough symptoms of inattention, but not hyperactivity–impulsivity, were present for the past 6 months.

Predominantly hyperactive–impulsive presentation: if enough symptoms of hyperactivity–impulsivity but not inattention were present for the past 6 months.

Because symptoms can change over time, the presentation may change over time as well.

familial distress (Rizzo et al., 2007). A recent observation study reported that *disruptive behavior disorders* were present in 14.5% of a clinic-based cohort of 578 patients, and that this was related to the presence of ADHD and to a family history of aggressive and violent behavior and forensic encounters (Robertson, Cavanna, & Eapen, 2015).

Screening for ADHD can be performed using one of the available validated standardized rating scales, which can be completed by parents and teachers (Chang, Wang, & Tsai, 2016); many of these scales are publically available without a user fee.

Common neurodevelopmental abnormalities underpin the occurrence of TS and ADHD, and justify their high comorbidity rate. These two disorders may share a similar etiology, potentially subdued by pathological mechanisms including shared genetic substrate, exposure to neurotoxic substances, excitotoxicity, and neuroinflammation associated with sustained microglial activation. Ultimately, this complex etiology may lead to abnormal connectivity within the sensory motor and associative loops of the corticobasal ganglia circuitry (Cortese, 2012). Both conditions may be associated with alterations in noradrenergic and dopaminergic transmission, resulting in inadequate modulation of corticostriatal circuits and subsequent failure to inhibit sensory–motor input as well as fine-tune the execution of complex, adaptive behavioral sequences (Fig. 2).

In a substantial proportion of youth with tic disorders, comorbid ADHD requires active intervention, in particular when it impacts academic performance, impairs social functioning, and is associated with severe oppositional or aggressive behavior. The most effective treatments for ADHD and related oppositional and aggressive behaviors are psychoactive stimulants. This class of medications includes methylphenidate, which blocks the reuptake of dopamine and norepinephrine into the presynaptic neuron, and

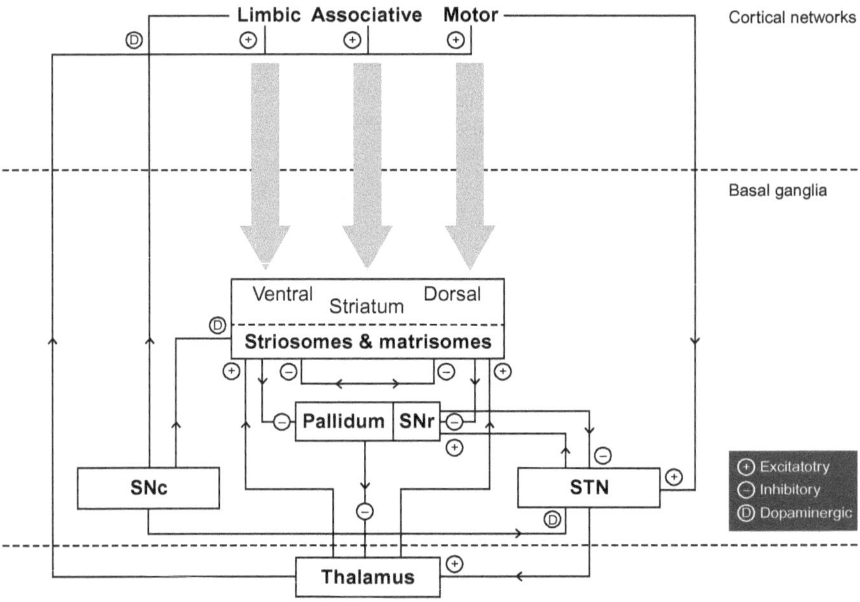

Fig. 2 Simplified schematic illustration of the major functional cortico-striato-thalamo-cortical loops. The three main cortical networks (limbic/associative/motor) convey converging information to the striatum (ventral and dorsal) comprised of matrisomal and striosomal compartments. The ventral striatum and in particular the striosomes receive their main input from limbic and associative areas, while the dorsal striatum is predominantly reached by motor efferents. Depicted are also basal ganglia nuclei and their functional cross interactions, with two main subcortical–cortical efferents. The thalamus relays processed information from the basal ganglia back to cortical networks. The substantia nigra pars compacta (SNc) communicates through dopaminergic efferents with the limbic network modulating motivation. SNr = substantia nigra pars reticulata; STN = subthalamic nucleus. *From Ganos, C., Roessner, V., & Munchau, A. (2013). The functional anatomy of Gilles de la Tourette syndrome.* Neuroscience and Biobehavioral Reviews, 37, *1050–1062.*

amphetamines, which increase the release of these monoamines into the extraneuronal space. Available evidence shows that the effect sizes related to improvement of ADHD symptom severity are much larger for psychoactive stimulants (methylphenidate in particular) than for other medication classes, suggesting that psychoactive stimulants should represent the first-line pharmacological treatment (Pringsheim & Steeves, 2009). A randomized, controlled trial by The Tourette Syndrome Study Group (2002) compared the effect on ADHD symptoms, measured using the ADHD Conner's Abbreviated Symptoms Questionnaire for Teachers (ASQ), of a flexible dose of methylphenidate, clonidine, clonidine plus methylphenidate, or placebo in 136 children with TS for 16 weeks. Methylphenidate alone and clonidine plus methylphenidate were both significantly more effective than placebo in ameliorating ADHD symptoms. Importantly, the frequency of tic worsening did not differ significantly between active treatments and placebo, and subsequent trials showed improvements in both ADHD and tic severity with methylphenidate in children with coexisting conditions.

A single, small trial of an amphetamine in children with comorbid tics and ADHD showed improvement of ADHD symptoms but worsening of tics when the amphetamine daily dose used was higher than 25 mg per day (Castellanos et al., 1997). Considering the whole class of psychoactive stimulants, the observed risk of tic worsening is about 20% and should be considered before and during treatment (Ganos, Martino, & Pringsheim, 2016). Long-acting formulations allow single daily dosing and are associated with less risk of abuse. Overall, family history of sudden cardiac death or arrhythmias, blood pressure measurement, and cardiac examination plus ECG should be evaluated before initiating psychoactive stimulant treatment.

If psychoactive stimulants are not well tolerated or are associated with tic worsening, α2A-agonists like clonidine or guanfacine should be considered. These compounds showed efficacy against tics, ADHD symptoms and oppositional behavior, but their effect size is smaller than that of psychoactive stimulants and the latency of effect is usually of several weeks (Gorman, Gardner, Murphy, et al., 2015). Psychoactive stimulants and α2A-agonists may also be combined in order to achieve a similarly satisfactory control of tics and ADHD.

Atomoxetine, a selective norepinephrine reuptake inhibitor with limited abuse potential, is the most commonly prescribed nonstimulant medication for the treatment of ADHD. It is recommended as monotherapy for the treatment of ADHD not only in youth unresponsive to psychoactive stimulants but also in adult populations. Only two trials showed effective response to atomoxetine on ADHD symptoms in patients with tics; the risk of tic worsening with this drug seems lower than with psychoactive stimulants, but available data are limited (Hutchison, Ghuman, Ghuman, Karpov, & Schuster, 2016).

Psychosocial interventions, comprising parent training, family therapy addressing parent–child relationships and communication, and cognitive behavioral therapy (CBT), are safe and effective treatment strategies and, if available, represent valid treatment alternatives, despite lower quality of evidence (Scotto Rosato et al., 2012).

4. BEHAVIORAL PROBLEMS IN TIC DISORDERS: COMORBID OBSESSIVE–COMPULSIVE SYMPTOMS AND OBSESSIVE–COMPULSIVE DISORDER

Obsessive–compulsive disorder (OCD) is the second most common comorbidity of TS. Natural history studies indicate that OCD symptoms typically peak later than tics, toward the end of the first decade. It may be often challenging to differentiate complex tics from *compulsive behaviors* (Worbe, Mallet, Golmard, et al., 2010). These have similar aggravating (e.g., stress and excitement) and attenuating factors (e.g., focusing attention on specific physical or mental activities). However, they notably differ in the fact that most tics, including the most complex, are typically released in response to an unpleasant sensory experience (the premonitory urge), whereas compulsions are elaborate voluntary actions or "mental acts," i.e., mental operations or mentally rehearsed or imagined actions, that typically prevent or relieve a state of anxiety, are performed according to a rigid pattern, and may or may not be responding to obsessions. *Obsessions* are defined as recurrent and persistent thoughts that are unwanted, intrusive, and invariably distressing. Sometimes the content of obsessions is related to the possibility that negative and dreaded events might unfold if compulsive rituals are not properly and timely deployed.

The DSM-5 diagnostic criteria for OCD require that obsessions, compulsions, or both should be time consuming (occupying at least 1 h on average each day) and/or cause significant clinical distress or impairment in functioning (American Psychiatric Association, 2013). In addition, the DSM-5 recognizes the existence of *tic-related OCD* as a separate subtype of OCD in which there is a current or past history of a tic disorder. Tic-related OCD seems to have an earlier age at onset compared to OCD without tics and follows the male predominance that is typical of TS and other chronic tic disorders (Holzer et al., 1994). There is evidence that the obsessive–compulsive symptom phenomenology might differ between OCD with and without tics. In tic-related OCD there seems to be a higher frequency of aggressive, sexual, religious, and symmetry-related obsessions, as well as of counting, ordering, touching, and hoarding compulsions (George, Trimble, Ring, Sallee, & Robertson, 1993; Leckman, Grice, Barr, et al., 1994). Tic-related OCD is also associated with "*just right*" *phenomena*, e.g., having to cross a door threshold in a certain manner, touching a surface, or performing a routine action in a certain way until it conveys a "*just right*" feeling to the subject. This "*just right*" element could even extend to the execution of tics that should happen in a particular way, thus leading to the repetition of certain tics in rapid succession over time (Neal & Cavanna, 2013). Their negative counterpart, *not just right phenomena*, are probably less tightly associated with OCD, are probably much more common than *just right* phenomena, and may phenomenologically overlap premonitory urges. In OCD without tics, typically obsessive–compulsive symptoms are related to fear of dirt and germ contamination, and manifest overtly as cleaning or washing compulsions.

The majority of studies evaluating the rates of OCD comorbidity included less than 200 participants and varied in study design and population characteristics, leading to estimates ranging from 11% to 66% (Gadow, Nolan, Sprafkin, & Schwartz, 2002; Khalifa & Von Knorring, 2006). A recent, cross-sectional study used structured diagnostic interviews to assess comorbidity rate in 1374 TS patients and 1142 unaffected family members (Hirschtritt et al., 2015). A lifetime comorbid diagnosis of OCD was present in 50% of TS patients, with the age of greatest risk for its onset between 4 and 10 years and a higher likelihood of OCD comorbidity in females. Observational studies on large service-based cohorts of TS patients suggest that obsessive–compulsive behaviors that do not fulfill criteria for a diagnosis of OCD may be up to 6 times more frequent than a comorbid diagnosis of OCD (Bloch & Leckman, 2009). Likewise, a significantly higher rate of OCD tics has been repeatedly reported in first-degree relatives of TS patients, in both tic-related OCD and TS without OCD (Hirschtritt et al., 2015).

These findings support a complex genetic relationship between TS and OCD, indicated by a high degree of heritability of both conditions. Susceptibility genes playing a major causative role remain unidientified or there may be exclusively a polygenic contribution involving small effect genetic or epigenetic factors. Interestingly, the most recent genome-wide association studies support a significant polygenic component in OCD, which seems less significant in TS (Yu, Mathews, Scharf, et al., 2015). Indeed, the coexistence of TS in OCD patients attenuates the significance of a polygenic signal, suggesting a difference in genetic architecture between OCD alone and tic-related OCD or OCD + TS. More recently, symptom-level and latent class analyses conducted in families of 3494 individuals with TS, OCD, and ADHD (Darrow, Hirschtritt, Davis, et al., 2016) showed two cross-disorder symptom-based phenotypes characterized by high heritability, consisting of symmetry (symmetry, evening up, and checking obsessions; ordering, arranging, counting, and writing–rewriting compulsions; and repetitive writing tics), and disinhibition (uttering syllables/words, echopalilalia, coprophenomena, and obsessive urges to offend/mutilate/be destructive).

Obsessive–compulsive symptoms should be evaluated in each patient with a diagnosis of chronic tic disorder. The Children's Yale-Brown Obsessive Compulsive Scale (CY-BOCS, or Y-BOCS in adults) is the most widespread rating scale, which allows the self-completion of a symptom checklist before clinic visits and the subsequent severity rating by the clinician (Scahill et al., 1997).

The available evidence suggests CBT as first-line treatment for tic-related OCD. Subanalyses in randomized controlled trials (POTS and POTS II studies; Pediatric OCD Treatment Study (POTS) Team, 2004; March et al., 2007) on children/adolescents with both OCD (with or without tics) suggested that tic-related OCD is more likely than OCD alone to respond partially to selective serotonin reuptake inhibitors (SSRIs) alone, despite similar OCD severity at baseline between the two subgroups. General experience on pharmacotherapy of tic-related OCD is that SSRIs often need to be administered at doses at the higher end of the recommended range. At difference, in these trials, CBT combined to SSRIs (sertraline) was significantly more effective than CBT alone, and CBT alone was significantly more effective than placebo in reducing OCD severity in both children with tic-related OCD and OCD alone. Limited evidence supports antipsychotic augmentation with risperidone or aripiprazole as adjuncts to SSRIs in adults with treatment-resistant OCD (Komossa, Depping, Gaudchau, Kissling, & Leucht, 2010; Selvi et al., 2011). Adults with tic-related OCD could benefit from this augmentation strategy more than adults with OCD without tics. One naturalistic comparative study showed equal efficacy of risperidone and aripiprazole in reducing OCD severity in 69 children with tic-related OCD who did not respond adequately to either sertraline or fluvoxamine alone. In this study, 39 of the 69 children improved with antipsychotic augmentation with respect to OCD symptoms, and 47 of the 69 improved also in tic severity (Masi, Pfanner, & Brovedani, 2013). There is uncontrolled evidence that thalamic deep brain stimulation in severe, pharmacorefractory TS might improve obsessive–compulsive symptoms alongside tics (Porta et al., 2009), but confirmation from randomized controlled trials is required to support this very preliminary observation.

5. BEHAVIORAL DISTURBANCES IN TIC DISORDERS: COMORBID ANXIETY AND DEPRESSION

In recent years, there has been increasing research on comorbid *anxiety disorders* and *depression* in individuals with TS. The reported prevalence of comorbid anxiety and depression in individuals with TS has varied considerably between studies, depending on the population studied and study methodology. Population-based studies of depression in children and youth suggest that individuals with tics do not have higher rates of depression (Kurlan et al., 2002; Robertson, 2006), while clinically ascertained samples demonstrate depressive symptoms which are often correlated with tic severity, comorbid OCD diagnoses, and older age (Chou, Lin, Lin, et al., 2013; Mol Debes, Hjalgrim, & Skov, 2008; Robertson, Williamson, & Eapen, 2006). In contrast, both population-based (Kurlan et al., 2002) and clinically ascertained samples (Coffey, Biederman, Smoller, et al., 2000) show higher rates of anxiety disorders in children and youth with TS, as well as correlations between tic severity and anxiety symptoms or diagnoses (Coffey et al., 2000; Johnco, McGuire, McBride, et al., 2016). Some studies have focused on self-reported symptoms of anxiety or depression without confirmation of clinical diagnoses of these conditions (Mol Debes et al., 2008; Robertson et al., 2006).

In a school-based community study in New York state assessing comorbid psychopathology in children with tics (Kurlan et al., 2002), overanxious disorder (diagnosis now subsumed under generalized anxiety disorder) was present in 21.2% of children with tics, which was more common than OCD (10.9%) and major depressive disorder (9.7%). The authors of this study state that the low rate of OCD observed in their study may have been due to fact that their sample was restricted to children, and symptoms meeting OCD diagnostic criteria may not have developed yet. Longitudinal studies have shown that OCD symptoms tend to peak later than tic symptoms (Bloch, Peterson, Scahill, et al., 2006), and that the prevalence of OCD in the general population is significantly higher in 13–15-year olds compared to younger children (Ford, Goodman, & Meltzer, 2003). Children younger than 12-year old with TS are, therefore, likely to meet DSM criteria for *generalized anxiety disorder* than OCD, though nondisabling obsessive–compulsive behaviors are frequently present.

Studies have also reported a high prevalence of *separation anxiety* in children with TS, including the earlier mentioned community-based study in which separation anxiety symptoms where present in 14.8% of the children with tics and only 7.2% of children without tics (Kurlan et al., 2002). A clinic-based study in children with TS reported a higher prevalence of both *panic disorder* and separation anxiety in children with severe tics than in children with mild to moderate tics. They also found that separation anxiety disorder was the disorder that most robustly predicted tic severity (Coffey et al., 2000). The association between separation anxiety symptoms and tic disorders may be the result of embarrassment due to tics in social situations and efforts to suppress tics, resulting in feelings of stress or mental fatigue.

A qualitative study with youth with TS found that young people feel anxious and worried about their tics (which worsens the tics) and that the anxiety-related emotions associated with tics are distressing (Cuenca, Glazebrook, Kendall, et al., 2015). Previous studies (Coffey et al., 2000; Johnco et al., 2016; Kurlan et al., 2002) suggest that tic severity is worse in individuals with comorbid anxiety disorders. Disentangling the cause effect relationship between tics and anxiety is challenging and ideally, treatments aimed at targeting both tic and anxiety symptoms may result in more favorable responses. The success of the Comprehensive Behavioral Intervention for Tics, which includes not only habit reversal training but also relaxation training and a functional intervention to address situations that sustain or worsen tics, suggests this may be true (Piacentini, Woods, Scahill, et al., 2010). Furthermore, other behavioral interventions used in the treatment of anxiety disorders have encouraging pilot data in individuals with tics, including mindfulness-based stress reduction (Reese, Vallejo, Rasmussen, et al., 2015) and acceptance and commitment therapy (Franklin, Best, Wilson, et al., 2011).

Population-based prevalence studies of mental health disorders in children and adolescents consistently demonstrate that while non-OCD anxiety disorders tend to peak between the age of 8 and 12, depressive disorders peak later in adolescence (from age 15) and continue increasing into adulthood (Costello, Mustillo, Erkanli, et al., 2003; Ford et al., 2003). Studies of adults with Tourette have been able to demonstrate a more consistent association between TS, depression diagnoses, and related functional impairment (Jalenques, Galland, Malet, et al., 2012; Lewin, Storch, Conelea, et al., 2011; Piedad, 2016). In contrast, studies in children and youth have reported inconsistent associations between TS and depression, likely due to differences between studies in the mean age of the included sample, symptom severity of the included sample, and the screening instruments used. A study of comorbid disorders in youth with TS using the Tourette Syndrome International Database Consortium found that the prevalence of mood disorders increased in each successive age group studied, with the highest rates reported in 14–17-year olds (Wanderer, Freeman, Bock, Rothenberger, & Becker, 2012). A population-based cohort study in Taiwan evaluating comorbid depression in youth with TS found that the mean age at depression diagnosis was 14.4 years (Chou et al., 2013). In one of the largest studies of comorbid DSM-IV-TR disorders using a clinic-based sample of individuals recruited for genetic studies of TS, the lifetime prevalence of major depressive disorder, dysthymia, and bipolar I and II disorder was 29.8% (Hirschtritt et al., 2015). Prevalence was significantly higher in females than in males, and individuals with comorbid OCD were more likely to have a mood disorder. Depression, therefore, should be screened for in individuals seeking specialist care for TS, especially in adolescents and adults.

6. BEHAVIORAL DISTURBANCES IN TIC DISORDERS: COMORBID IMPULSE CONTROL DISORDERS, IMPULSIVE BEHAVIORS, AND "IMPULSIVE" TICS

Impulse control disorders (ICDs) are a group of conditions in which patients have significant difficulties in controlling their impulses to perform rewarding, often risk-seeking, behaviors. An exploratory study evaluated the frequency and clinical characteristics of ICDs in 31 adult patients with TS using the Minnesota Impulsive Disorders Interview (Frank, Piedad, Rickards, & Cavanna, 2011). Seventy-four percent of them presented with at least one ICD: the most common two were *intermittent explosive disorder* (52%) and *compulsive buying disorder* (52%), and kleptomania, pathological gambling, and pyromania were also observed. Importantly, the number of ICDs correlated negatively with health-related quality of life in this clinical sample. This preliminary observation needs confirmation from a larger sample of TS patients. In other observational studies, the prevalence of ICDs in a major comorbid disorder-like OCD ranged between 16%–17% (Grant, Mancebo, Eisen, & Rasmussen, 2010; Grant, Mancebo, Pinto, Eisen, & Rasmussen, 2006) and 35% (Fontenelle, Mendlowicz, & Versiani, 2005), which are lower estimates than those reported in TS patients by Frank et al. (2011).

The most common type of impulsive behavior not specifically associated with a coexisting ICD in TS is represented by sudden, explosive episodes of rage, or "*rage attacks*." TS patients referred to specialist clinics experience episodes of aggressive impulses that are neither proportionate to the intensity of precipitants nor to the severity of a coexisting intermittent explosive disorder (Budman, Bruun, Park, & Olson, 1998). Precipitating factors for rage attacks include patients being told that they are wrong or mistaken about a topic, deviation from patients' own wishes, or alteration in their schedule (Kano, Ohta, Nagai, Spector, & Budman, 2008). Not surprisingly, Mol Debes et al. (2008) showed that the frequency of rage attacks tends to increase with the comorbidity burden, especially with ADHD and OCD. A more recent, single-center, clinical sample of 1032 German TS patients yielded a frequency of rage attacks close to 58%. Although this study found a weak association between ADHD and rage attacks, interestingly 44% of the patients who suffered from rage attacks did not suffer from comorbid ADHD (Sambrani, Jakubovski, & Muller-Vahl, 2016). Another study has found a positive correlation between anger symptoms and TS patients' attachment styles as well as with anxiety associated with close relationships (Dehning, Burger, Krause, et al., 2015).

Regardless of the comorbidity with ICDs, TS patients may exhibit complex behaviors that phenomenologically overlap tics and impulsive actions. These "impulsive," tic-like behaviors are often preceded by an urge to release them and associated with the ability to suppress them actively, but are also charged by a high risk of physical injury or social rejection. A first example of this type of behavior is represented by *nonobscene socially inappropriate behaviors* (NOSIBs), also referred to as "*social disinhibition*," which share the social inappropriateness of coprophenomena but lacking an obscene content (Kurlan et al., 1996). Typically, these behaviors consist of inappropriate personal comments like making explicit and potentially insulting remarks on someone's physical aspect, height, weight, intelligence, and comments that are triggered by a specific context but socially disruptive, e.g., shouting "bomb" at an airport. Urges associated with NOSIBs were reported by up to two-thirds of TS patients attending a specialist outpatient clinic, but not all of them led to actions (Eddy & Cavanna, 2013b). Interestingly, patients with NOSIBs presented with greater tic, urge and obsessive–compulsive symptom severity, higher rate of attention problems, conduct problems and coprolalia, and poorer quality of life. Coprolalia and severity of obsessive–compulsive symptoms and urges predicted NOSIBs in this patient population. Recent fMRI research highlighted a link between the temporoparietal junction dysfunction and impulse control problems and NOSIBs in TS patients (Eddy, 2016). Finally, NOSIBs were recently confirmed to be a highly heritable phenomenological feature within the TS spectrum (Hirschtritt et al., 2016).

Another example of "impulsive" tic-like behavior is represented by *self-injurious behaviors* (SIBs), which consist of head banging; body, face, or head slapping; banging into hard objects with their bodies; and self-poking with sharp objects. Like NOSIBs, SIBs are often associated with premonitory urges. The main feature that distinguishes SIBs from other forms of complex tic is the invariably present element of self-harm as consequence of the motor action, which suggests a degree of goal-directedness that is absent in "typical" tics, but it is present in impulsive actions. Freeman et al. (2000) reported a 15% frequency of SIBs from an international registry of 6805 TS patients around the world, with a higher rate in patients with ADHD comorbidity. Mathews et al. (2004) observed a correlation between a milder form of SIBs and obsessive–compulsive symptom severity, whereas a more severe form of SIBs correlated with tic severity, thereby suggesting the existence of two distinct forms of SIBs, phenomenologically different in the severity of the physical harm caused by the behaviors. Sambrani et al. (2016) reported a lifetime prevalence of SIBs of about 39%. In this study, there was a strong association between SIBs and coprophenomena, as well as with presence of complex motor tics, with weaker associations with obsessive–compulsive behaviors and anxiety. This finding suggests that the pathogenesis of SIBs is much closer to that of complex tics rather than to that of OCD. Indeed, these clinical observations support the indication to treat NOSIBs and SIBs with behavioral therapy or antitic medications (particularly antipsychotics). In a restricted proportion of cases with severe SIBs, deep brain stimulation surgery of the centromedian/parafascicularis nuclei of the thalamus or the globus palllidus pars interna has reported to be efficacious (Servello, Porta, Sassi, Brambilla, & Robertson, 2008).

7. BEHAVIORAL DISTURBANCES IN TIC DISORDERS: COMORBID AUTISTIC SPECTRUM DISORDERS

Autistic spectrum disorders (ASD) encompass a spectrum of developmental syndromes manifesting with a deficit in social communication, language development, and stereotyped behaviors. Autism in its classical form requires deficits in all three domains, while the presence of one or two domains results in disorders along the disease spectrum. ASD may coexist with TS in 6.5%–50% of cases. Such a wide variability of prevalence estimates could be due to the clinical characteristics, disease severity, cognitive abilities, and other coexisting brain abnormalities in the populations studied (Baron-Cohen, O'Riordan, Stone, Jones, & Plaisted, 1999; Burd, Kerbeshian, Wikenheiser, & Fisher, 1986; Simonoff et al., 2008).

A significant overlap and enrichment of genes implicated in ASD (e.g., *NRXN1*, *OFCC1*) in patients with TS when compared to control subjects in studies on copy number variants indicates important pathogenic commonalities between the two groups of disorders (Cukier et al., 2014; Fernandez et al., 2012; Nag et al., 2013). Similarly, common environmental factors, e.g., lower socioeconomic status (Miller, Scharf, Mathews, & Ben-Shlomo, 2014) and prenatal stress (Fine, Zhang, & Stevens, 2014), and dysregulated mechanisms in the cross talk between immune system, e.g., microglial dysregulation (Frick & Pittenger, 2016), endocrine system (in particular, the hypothalamic–pituitary–adrenal axis), and the CNS might be involved in the developmental abnormalities underlying TS and ASD (Hornig & Lipkin, 2013; Martino, Zis, & Buttiglione, 2015). TS and ASD share similar maturational connectivity problems to those underlined earlier with respect to TS–ADHD comorbidity, e.g., long-range underconnectivity and short-range overconnectivity. Factor analyses of TS phenotypes have shown that autistic symptoms might represent a specific factor which could reflect specific underlying causative factors for TS patients with a high burden of comorbid autistic features (Huisman-van Dijk, Schoot, Rijkeboer, Mathews, & Cath, 2016).

On the basis of the high prevalence of neurodevelopmental disorders such as ADHD and ASD in TS, Burd, Freeman, Klug, and Kerbeshian (2005) explored the rate of *learning disabilities* (LD) in TS patients using the Tourette Syndrome International Consortium database. Of 5450 patients with TS across an international scale, 1235 patients were found to have comorbid LD, with a diagnosis of either specific LD or learning disorders not otherwise specified in the DSM-IV. Logistic regression analysis was used to produce a five-variable model, which predicted the presence of LD in 65.2% of patients with GTS; according to this analysis, male gender, younger age at diagnosis, complications during labor and delivery, fewer family members affected, and increased comorbid disease were associated with LD cooccurrence. Cubo et al. (2013) performed an association study looking at the relationship between tic disorders and poor academic performance in 1867 school-aged children and found that the rate of LD and the need for academic support were similar in children with and without tics. More recently, distinct patterns of daily executive function problems in TS and ASD children were reported; in particular, children with TS manifest more problems with emotional control than cognitive flexibility compared to ASD children (Hovik, Egeland, Isquith, et al., 2014).

8. SLEEP DISORDERS

The high ADHD comorbidity rate prompted the exploration of sleep problems also in TS patients, given the significant occurrence of sleep disorders in ADHD (among which increasing sleep onset latency, reduced sleep duration, parasomnias, periodic limb movements during sleep, and sleep-disordered breathing). The large international database from Freeman et al. (2000) reported clinical complaints of sleep problems in about 25% of children with TS, with rates increasing to 65% in the case of comorbid ADHD. Another study that investigated sleep disorders ad hoc using a prospective questionnaire in 123 TS patients aged ≤21 years (61% with comorbid ADHD) surprisingly detected an almost identical percentage (65% and 64%) of DSM-5, coded sleep disorders in patients without and with ADHD, respectively (Ghosh et al., 2014). ADHD comorbidity contributed relevantly to the cooccurrence of *insomnia*, which was more common among TS + ADHD patients (75%) than in those with TS alone (32%).

Only a handful of studies investigated a potential relationship between sleep architecture and neurobehavioral functioning in TS. Kirov, Kinkelbur, Banaschewski, and Rothenberger (2007) reported *shorter REM latency*, *increased percentage of REM sleep*, and *increased movements during sleep stages 1 and 2* in TS + ADHD patients. In a subsequent polysomnographic study (Kirov, Brand, Banaschewski, & Rothenberger, 2017), the same group reported *increased overall, slow-wave and REM sleep onset latencies*, and *lower REM sleep duration and percentage of REM sleep of total sleep time* in both children with TS patients alone and children with TS + ADHD (25%–30% of whom on medications) compared to typically developing children. In the overall patient population from the same study (ADHD alone, TS alone, and

TS + ADHD), the percentage of REM sleep correlated positively with scores of inattention and negatively with performance IQ. Another study conducted a 2-night consecutive polysomnographic sleep study in 90 children with TS alone, ADHD alone, or both disorders in comorbidity, aged 6–16 years (Stephens, Chung, Jovanovic, et al., 2013). The study showed few differences between groups, specifically a higher number of total arousals and of arousals from slow wave sleep in ADHD-only. Interestingly, this report also showed that children with more arousals from sleep were more likely to manifest conduct problems, hyperactivity/immaturity, and restless/disorganized behavior. The observation of an adverse impact of REM sleep overdrive and sleep arousals on daytime neurobehavioral functions should be investigated further to tease out the confounding role of medications and understand the pathophysiological mechanisms involved.

9. CONCLUSIONS

The assessment and clinical management of primary chronic tic disorders cannot be restricted to tics. It appears evident that primary tic disorders are not unitary conditions, as originally believed. At least 85% of these patients present with a comorbid behavioral disorder which in many cases may be their main reason to seek medical attention. Fig. 1 summarizes the complexity of the comprehensive assessment of patients with TS and chronic tic disorders in general within a specialized outpatient setting. The increased focus of attention from clinicians and clinical researchers on nonmotor aspects of tic disorders, in particular behavioral comorbidities, is contributing to a better understanding of the clinical classification of patients into different subgroups, as well as of those pathomechanisms and genetic/environmental causative factors that are shared between coexisting disorders and those which are, rather, specific for each of them. As in other areas of clinical neuroscience, a comprehensive approach to both motor and nonmotor aspects of this group of disorders will help personalizing treatment interventions and, ultimately, improve quality of care.

REFERENCES

American Psychiatric Association. (2013). *Diagnostic and statistical manual for mental disorders (DSM-5)* (5th ed.). Washington, DC: American Psychiatric Association.

Banaschewski, T., Woerner, W., & Rothenberger, A. (2003). Premonitory sensory phenomena and suppressibility of tics in Tourette syndrome: Developmental aspects in children and adolescents. *Developmental Medicine and Child Neurology*, *45*(10), 700–703.

Baron-Cohen, S., O'Riordan, M., Stone, V., Jones, R., & Plaisted, K. (1999). Recognition of faux pas by normally developing children and children with Asperger syndrome or high-functioning autism. *Journal of Autism and Developmental Disorders*, *29*, 407–418.

Belluscio, B. A., Jin, L., Watters, V., Lee, T. H., & Hallett, M. (2011). Sensory sensitivity to external stimuli in Tourette syndrome patients. *Movement Disorders*, *26*(14), 2538–2543.

Birmaher, B., Khetarpal, S., Brent, D., Cully, M., Balach, L., Kaufman, J., et al. (1997). The Screen for Child Anxiety Related Emotional Disorders (SCARED): Scale construction and psychometric characteristics. *Journal of the American Academy of Child and Adolescent Psychiatry*, *36*(4), 545–553.

Bliss, J. (1980). Sensory experiences of Gilles de la Tourette syndrome. *Archives of General Psychiatry*, *37*(12), 1343–1347.

Bloch, M. H., & Leckman, J. F. (2009). Clinical course of Tourette syndrome. *Journal of Psychosomatic Research*, *67*, 497–501.

Bloch, M. H., Peterson, B. S., Scahill, L., et al. (2006). Adult outcome of tic and obsessive-compulsive symptom severity in children with Tourette syndrome. *Archives of Pediatric and Adolescent Medicine*, *2006*, 65–69.

Brandt, V. C., Beck, C., Sajin, V., Baaske, M. K., Baumer, T., Beste, C., et al. (2016). Temporal relationship between premonitory urges and tics in Gilles de la Tourette syndrome. *Cortex*, *77*, 24–37.

Budman, C. L., Bruun, R. D., Park, K. S., & Olson, M. E. (1998). Rage attacks in children and adolescents with Tourette's disorder: A pilot study. *The Journal of Clinical Psychiatry*, *59*, 576–580.

Burd, L., Freeman, R. D., Klug, M. G., & Kerbeshian, J. (2005). Tourette syndrome and learning diabilites. *BMC Pediatrics*, *5*, 34.

Burd, L., Kerbeshian, J., Wikenheiser, M., & Fisher, W. (1986). A prevalence study of Gilles de la Tourette syndrome in North Dakota school-age children. *Journal of the American Academy of Child and Adolescent Psychiatry*, *25*, 552–553.

Bussing, R., Fernandez, M., Harwood, M., Wei Hou, Garvan, C. W., Eyberg, S. M., et al. (2008). Parent and teacher SNAP-IV ratings of attention deficit hyperactivity disorder symptoms: Psychometric properties and normative ratings from a school district sample. *Assessment*, *15*(3), 317–328. http://dx.doi.org/10.1177/1073191107313888. Epub 2008 Feb 29.

Castellanos, F. X., Giedd, J. N., Elia, J., Marsh, W. L., Ritchie, G. F., Hamburger, S. D., et al. (1997). Controlled stimulant treatment of ADHD and comorbid Tourette's syndrome: Effects of stimulant and dose. *Journal of the American Academy of Child and Adolescent Psychiatry*, *36*, 589–596.

Cavanna, A. E., & Rickards, H. (2013). The psychopathological spectrum of Gilles de la Tourette syndrome. *Neuroscience and Biobehavioral Reviews*, *37*, 1008–1015.

Chang, L. Y., Wang, M. Y., & Tsai, P. S. (2016). Diagnostic accuracy of rating scales for attention-deficit/hyperactivity disorder: A meta-analysis. *Pediatrics*, *137*, 1–13.

Chou, I., Lin, H., Lin, C., et al. (2013). Tourette syndrome and risk of depression: A population-based cohort study in Taiwan. *Journal of Developmental & Behavioral Pediatrics*, *34*, 181–185.

Coffey, B. J., Biederman, M. D., Smoller, J. W., et al. (2000). Anxiety disorders and tic severity in juveniles with Tourette's disorder. *Journal of the American Academy of Child and Adolescent Psychiatry*, *39*, 562–568.

Cohen, A. J., & Leckman, J. F. (1992). Sensory phenomena associated with Gilles de la Tourette's syndrome. *The Journal of Clinical Psychiatry*, *53*(9), 319–323.

Cortese, S. (2012). The neurobiology and genetics of attention-deficit/hyperactivity disorder (ADHD): What every clinician should know. *European Journal of Paediatric Neurology*, *16*, 422–433.

Costello, E. J., Mustillo, S., Erkanli, A., et al. (2003). Prevalence and development of psychiatric disorders in childhood and adolescence. *Archives of General Psychiatry*, *60*, 837–844.

Crossley, E., & Cavanna, A. E. (2013). Sensory phenomena: Clinical correlates and impact on quality of life in adult patients with Tourette syndrome. *Psychiatry Research*, *209*(3), 705–710.

Cubo, E., Trejo, J., Ausin, V., Saez, S., Delgado, V., Macarron, J., et al. (2013). Association of tic disorders with poor academic performance in central Spain: A population-based study. *The Journal of Pediatrics*, *163*, 217–223.

Cuenca, J., Glazebrook, C., Kendall, T., et al. (2015). Perceptions of treatment for tics among young people with Tourette syndrome and their parents: A mixed methods study. *BMC Psychiatry*, *15*, 46.

Cukier, H. N., Dueker, N. D., Slifer, S. H., Lee, J. M., Whitehead, P. L., Lalanne, E., et al. (2014). Exome sequencing of extended families with autism reveals genes shared across neurodevelopmental and neuropsychiatric disorders. *Molecular Autism*, *5*, 1.

Darrow, S. M., Hirschtritt, M. E., Davis, L. K., et al. (2016). Identification of two heritable cross-disorder endophenotypes for Tourette syndrome. *American Journal of Psychiatry*, *174*, 387–396.

Dehning, S., Burger, M. B., Krause, D., et al. (2015). Tourette syndrome is associated with insecure attachment and higher aggression. *International Journal of Neuroscience*, *125*, 521–525.

Draper, A., Jackson, G. M., Morgan, P. S., & Jackson, S. R. (2016). Premonitory urges are associated with decreased grey matter thickness within the insula and sensorimotor cortex in young people with Tourette syndrome. *Journal of Neuropsychology*, *10*(1), 143–153.

Duggal, H. S., & Nizamie, S. H. (2002). Bereitschaftspotential in tic disorders: A preliminary observation. *Neurology India*, *50*(4), 487–489.

Eddy, C. M. (2016). The junction between self and other? Temporo-parietal dysfunction in neuropsychiatry. *Neuropsychologia*, *89*, 465–477.

Eddy, C. M., & Cavanna, A. E. (2013a). Premonitory urges in adults with complicated and uncomplicated Tourette syndrome. *Behavior Modification*, *38*, 264–275.

Eddy, C. M., & Cavanna, A. E. (2013b). "It's a curse!": Coprolalia in Tourette syndrome. *European Journal of Neurology*, *20*, 1467–1470.

Ehlers, S., Gillberg, C., & Wing, L. (1999). A screening questionnaire for Asperger syndrome and other high-functioning autism spectrum disorders in school age children. *Journal of Autism and Developmental Disorders*, *29*(2), 129–141.

Fernandez, T. V., Sanders, S. J., Yurkiewicz, I. R., Ercan-Sencicet, A. G., Kim, Y. S., Fishman, D. O., et al. (2012). Rare copy number variants in tourette syndrome disrupt genes in histaminergic pathways and overlap with autism. *Biological Psychiatry*, *71*, 392–402.

Fine, R., Zhang, J., & Stevens, H. E. (2014). Prenatal stress and inhibitory neuron systems: Implications for neuropsychiatric disorders. *Molecular Psychiatry*, *19*, 641–651.

Fontenelle, L. F., Mendlowicz, M. V., & Versiani, M. (2005). Impulse control disorders in patients with obsessive-compulsive disorder. *Psychiatry and Clinical Neurosciences*, *59*, 30–37.

Ford, T., Goodman, R., & Meltzer, H. (2003). The British child and adolescent mental health survey 1999: The prevalence of DSM-IV disorders. *Journal of the American Academy of Child and Adolescent Psychiatry*, *42*, 1203–1211.

Frank, M., Piedad, J., Rickards, H., & Cavanna, A. E. (2011). The role of impulse control disorders in Tourette syndrome: An exploratory study. *Journal of the Neurological Sciences*, *310*, 276–278.

Franklin, M. E., Best, S. H., Wilson, M. A., et al. (2011). Habit reversal training and acceptance and commitment therapy for Tourette syndrome: A pilot project. *Journal of Developmental and Physical Disabilities*, *23*, 49–60.

Freeman, R. D., Fast, D. K., Burd, L., Kerbeshian, J., Robertson, M. M., & Sandor, P. (2000). An international perspective on Tourette syndrome: Selected findings, from 3,500 individuals in 22 countries. *Developmental Medicine and Child Neurology*, *42*, 436–447.

Frick, L., & Pittenger, C. (2016). Microglial dysregulation in OCD, Tourette syndrome, and PANDAS. *Journal of Immunology Research*, *2016*, 8606057.

Gadow, K. D., Nolan, E. E., Sprafkin, J., & Schwartz, J. (2002). Tics and psychiatric comorbidity in children and adolescents. *Developmental Medicine and Child Neurology*, *44*, 330–338.

Ganos, C., Garrido, A., Navalpotro-Gomez, I., Ricciardi, L., Martino, D., Edwards, M. J., et al. (2015). Premonitory urge to tic in Tourette's is associated with interoceptive awareness. *Movement Disorders*, *30*(9), 1198–1202.

Ganos, C., Kahl, U., Schunke, O., Kuhn, S., Haggard, P., Gerloff, C., et al. (2012). Are premonitory urges a prerequisite of tic inhibition in Gilles de la Tourette syndrome? *Journal of Neurology, Neurosurgery, and Psychiatry*, *83*(10), 975–978.

Ganos, C., & Martino, D. (2015). Tics and tourette syndrome. *Neurologic Clinics*, *33*(1), 115–136.

Ganos, C., Martino, D., & Pringsheim, T. M. (2016). Tics in the pediatric population: Pragmatic management. *Movement Disorders Clinical Practice*, *4*, 160–172.

Ganos, C., Mueller-Vahl, K., & Bhatia, K. (2015). Blocking phenomena in Gilles de la Tourette syndrome. *Movement Disorders Clinical Practice*, *2*, 438–439.

Ganos, C., Roessner, V., & Munchau, A. (2013). The functional anatomy of Gilles de la Tourette syndrome. *Neuroscience and Biobehavioral Reviews*, *37*(6), 1050–1062.

George, M. S., Trimble, M., Ring, H. A., Sallee, F. R., & Robertson, M. (1993). Obsessions in obsessive compulsive disorder with and without Gilles de la Tourette's syndrome. *American Journal of Psychiatry*, *150*, 93–97.

Ghosh, D., Rajan, P. V., Das, D., Datta, P., Rothner, A. D., & Erenberg, G. (2014). Sleep disorders in children with Tourette syndrome. *Pediatric Neurology*, *51*, 31–35.

Gorman, D. A., Gardner, D., Murphy, A., et al. (2015). Canadian guidelines on pharmacotherapy for disruptive and aggressive behaviour in children and adolescents with attention-deficit hyperactivity disorder, oppositional defiant disorder or conduct disorder. *Canadian Journal of Psychiatry*, *60*, 62–76.

Grant, J. E., Mancebo, M. C., Eisen, J. L., & Rasmussen, S. A. (2010). Impulse-control disorders in children and adolescents with obsessive compulsive disorder. *Psychiatry Research*, *175*, 109–113.

Grant, J. E., Mancebo, M. C., Pinto, A., Eisen, J. L., & Rasmussen, S. A. (2006). Impulse control disorders in adults with obsessive compulsive disorder. *Journal of Psychiatric Research*, *40*, 494–501.

Hashemiyoon, R., Kuhn, J., & Visser-Vandewalle, V. (2017). Putting the pieces together in Gilles de la Tourette syndrome: Exploring the link between clinical observations and the biological basis of dysfunction. *Brain Topography*, *30*(1), 3–29.

Hirschtritt, M. E., Darrow, S. M., Illmann, C., Osiecki, L., Grados, M., Sandor, P., et al. (2016). Social disinhibition is a heritable subphenotype of tics in Tourette syndrome. *Neurology*, *87*, 497–504.

Hirschtritt, M. E., Lee, P. C., Pauls, D. L., Dion, Y., Grados, M. A., Illmann, C., et al. (2015). Lifetime prevalence, age of risk, and genetic relationships of comorbid psychiatric disorders in Tourette syndrome. *JAMA Psychiatry*, *72*, 325–333.

Holzer, J. C., Goodman, W. K., McDougle, C. J., Baer, L., Boyarsky, B. K., Leckman, J. F., et al. (1994). Obsessive-compulsive disorder with and without a chronic tic disorder. A comparison of symptoms in 70 patients. *British Journal of Psychiatry*, *164*, 469–473.

Hornig, M., & Lipkin, W. I. (2013). Immune-mediated animal models of Tourette syndrome. *Neuroscience and Biobehavioral Reviews*, *37*, 1120–1138.

Hovik, K. T., Egeland, J., Isquith, P. K., et al. (2014). Distinct patterns of everyday executive function problems distinguish children with Tourette syndrome from children with ADHD or autism spectrum disorders. *Journal of Attention Disorders*. pii 1087054714550336 [Epub ahead of print].

Huisman-van Dijk, H. M., Schoot, R. V., Rijkeboer, M. M., Mathews, C. A., & Cath, D. C. (2016). The relationship between tics, OC, ADHD, and autism symptoms: A cross-disorder symptom analysis in Gilles de la Tourette syndrome patients and family-members. *Psychiatry Research*, *237*, 138–146.

Hutchison, S. L., Ghuman, J. K., Ghuman, H. S., Karpov, I., & Schuster, J. M. (2016). Efficacy of atomoxetine in the treatment of attention-deficit hyperactivity disorder in patients with common comorbidities in children, adolescents and adults: A review. *Therapeutic Advances in Psychopharmacology*, *6*, 317–334.

Jackson, S. R., Parkinson, A., Kim, S. Y., Schuermann, M., & Eickhoff, S. B. (2011). On the functional anatomy of the urge-for-action. *Cognitive Neuroscience*, *2*(3–4), 227–243.

Jalenques, I., Galland, F., Malet, L., et al. (2012). Quality of life in adults with Gilles de la Tourette syndrome. *BMC Psychiatry*, *12*, 109.

Joe, S., Woolley, M. E., Brown, G. K., Ghahramanlou-Holloway, M., & Beck, AT. (2008). Psychometric properties of the Beck Depression Inventory-II in low-income, African American suicide attempters. *Journal of Personality Assessment*, *90*(5), 521–523. http://dx.doi.org/10.1080/00223890802248919.

Johnco, C., McGuire, J. F., McBride, N. M., et al. (2016). Suicidal ideation in youth with tic disorders. *Journal of Affective Disorders*, *200*, 204–211.

Kahl, U., Schunke, O., Schöttle, D., David, N., Brandt, V., Bäumer, T., et al. (2015). Tic phenomenology and tic awareness in adults with autism. *Movement Disorders Clinical Practice*, *2*, 237–242.

Kane, M. J. (1994). Premonitory urges as "attentional tics" in Tourette's syndrome. *Journal of the American Academy of Child and Adolescent Psychiatry*, *33*(6), 805–808.

Kano, Y., Ohta, M., Nagai, Y., Spector, I., & Budman, C. (2008). Rage attacks and aggressive symptoms in Japanese adolescents with Tourette syndrome. *CNS Spectrums*, *13*, 325–332.

Karp, B. I., Porter, S., Toro, C., & Hallett, M. (1996). Simple motor tics may be preceded by a premotor potential. *Journal of Neurology, Neurosurgery, and Psychiatry*, *61*(1), 103–106.

Khalifa, N., & Von Knorring, A.-L. (2006). Psychopathology in a Swedish population of school children with tic disorders. *Journal of the American Academy of Child and Adolescent Psychiatry*, *45*, 1346–1353.

Kirov, R., Brand, S., Banaschewski, T., & Rothenberger, A. (2017). Opposite impact of REM sleep on neurobehavioral functioning in children with common psychiatric disorders compared to typically developing children. *Frontiers in Psychology*, 7, 2059.

Kirov, R., Kinkelbur, J., Banaschewski, T., & Rothenberger, A. (2007). Sleep patterns in children with attention-deficit/hyperactivity disorder, tic disorder, and comorbidity. *Journal of Child Psychology and Psychiatry*, *48*, 561–570.

Komossa, K., Depping, A. M., Gaudchau, A., Kissling, W., & Leucht, S. (2010). Second-generation antipsychotics for major depressive disorder and dysthymia. *The Cochrane Database of Systematic Reviews*, (12), CD008121. http://dx.doi.org/10.1002/14651858.CD008121.pub2. Review.

Kurlan, R. C., Como, P. G., Miller, B., Palumbo, D., Deeley, C., Andresen, E. M., et al. (2002). The behavioral spectrum of tic disorders. A community-based study. *Neurology*, *59*, 414–420.

Kurlan, R., Daragjati, C., Como, P. G., McDermott, M. P., Trinidad, K. S., Roddy, S., et al. (1996). Non-obscene complex socially inappropriate behavior in Tourette's syndrome. *The Journal of Neuropsychiatry and Clinical Neurosciences*, *8*, 311–317.

Kwak, C., Dat Vuong, K., & Jankovic, J. (2003). Premonitory sensory phenomenon in Tourette's syndrome. *Movement Disorders*, *18*(12), 1530–1533.
Lang, A. (1991). Patient perception of tics and other movement disorders. *Neurology*, *41*(2 (Pt. 1)), 223–228.
Leckman, J. F., Grice, D. E., Barr, L. C., et al. (1994). Tic-related vs non-tic-related obsessive compulsive disorder. *Anxiety*, *1*, 208–215.
Leckman, J. F., Walker, D. E., & Cohen, D. J. (1993). Premonitory urges in Tourette's syndrome. *The American Journal of Psychiatry*, *150*(1), 98–102.
Lewin, A. B., Storch, E. A., Conelea, C. A., et al. (2011). The roles of anxiety and depression in connecting tic severity and functional impairment. *Journal of Anxiety Disorders*, *25*, 164–168.
March, J. S., Franklin, M. E., Leonard, H., Garcia, A., Moore, P., Freeman, J., et al. (2007). Tics moderate treatment outcome with sertraline but not cognitive-behavior therapy in pediatric obsessive-compulsive disorder. *Biological Psychiatry*, *61*, 344–347.
Martino, D., Cavanna, A. E., Robertson, M. M., & Orth, M. (2012). Prevalence and phenomenology of eye tics in Gilles de la Tourette syndrome. *Journal of Neurology*, *259*(10), 2137–2140.
Martino, D., Zis, P., & Buttiglione, M. (2015). The role of immune mechanisms in Tourette syndrome. *Brain Research*, *1617*, 126–143.
Masi, G., Pfanner, C., & Brovedani, P. (2013). Antipsychotic augmentation of selective serotonin reuptake inhibitors in resistant tic-related obsessive-compulsive disorder in children and adolescents: A naturalistic comparative study. *Journal of Psychiatric Research*, *47*, 1007–1012.
Mathews, C. A., Waller, J., Glidden, D., Lowe, T. L., Herrera, L. D., Budman, C. L., et al. (2004). Self injurious behaviour in Tourette syndrome: Correlates with impulsivity and impulse control. *Journal of Neurology, Neurosurgery and Psychiatry*, *75*, 1149–1155.
Miller, L. L., Scharf, J. M., Mathews, C. A., & Ben-Shlomo, Y. (2014). Tourette syndrome and chronic tic disorder are associated with lower socio-economic status: Findings from the Avon longitudinal study of parents and children cohort. *Developmental Medicine and Child Neurology*, *56*, 157–163.
Mol Debes, N. M. M., Hjalgrim, H., & Skov, L. (2008). Validation of the presence of comorbidities in a Danish clinical cohort of children with Tourette syndrome. *Journal of Child Neurology*, *23*, 1017–1027.
Muller-Vahl, K. R., Riemann, L., & Bokemeyer, S. (2014). Tourette patients' misbelief of a tic rebound is due to overall difficulties in reliable tic rating. *Journal of Psychosomatic Research*, *76*(6), 472–476.
Nag, A., Bochukova, E. G., Kremeyer, B., Campbell, D. D., Muller, H., Valencia-Duarte,-A. V., et al. (2013). CNV analysis in Tourette syndrome implicates large genomic rearrangements in COL8A1 and NRXN1. *PloS One*, *8*, e59061.
Neal, M., & Cavanna, A. E. (2013). "Not just right experiences" in patients with Tourette syndrome: Complex motor tics or compulsions? *Psychiatry Research*, *210*, 559–563.
Obeso, J. A., Rothwell, J. C., & Marsden, C. D. (1982). The neurophysiology of Tourette syndrome. *Advances in Neurology*, *35*, 105–114.
Pappert, E. J., Goetz, C. G., Louis, E. D., Blasucci, L., & Leurgans, S. (2003). Objective assessments of longitudinal outcome in Gilles de la Tourette's syndrome. *Neurology*, *61*(7), 936–940.
Pediatric OCD Treatment Study (POTS) Team. (2004). Cognitive-behavior therapy, sertraline, and their combination for children and adolescents with obsessive-compulsive disorder: The pediatric OCD treatment study (POTS) randomized controlled trial. *JAMA*, *292*, 1969–1976.
Piacentini, J., Woods, D. W., Scahill, L., et al. (2010). Behaviour therapy for children with Tourette disorder: A randomized controlled trial. *JAMA*, *303*, 1929–1937.

Piedad, J. C., & Cavanna, A. E. (2016). Depression in Tourette syndrome: A controlled and comparison study. *Journal of Neurological Sciences, 364*, 128–132.

Porta, M., Brambilla, A., Cavanna, A. E., Servello, D., Sassi, M., Rickards, H., et al. (2009). Thalamic deep brain stimulation for treatment-refractory Tourette syndrome: Two-year outcome. *Neurology, 73*, 1375–1380.

Pringsheim, T., & Steeves, T. (2009). Pharmacological treatment for attention deficit hyperactivity disorder in children with co-morbid tic disorders [serial online]. *Cochrane Database of Systematic Reviews, 4*, CD007990.

Puts, N. A., Harris, A. D., Crocetti, D., Nettles, C., Singer, H. S., Tommerdahl, M., et al. (2015). Reduced GABAergic inhibition and abnormal sensory symptoms in children with Tourette syndrome. *Journal of Neurophysiology, 114*(2), 808–817.

Reese, H. E., Scahill, L., Peterson, A. L., Crowe, K., Woods, D. W., Piacentini, J., et al. (2014). The premonitory urge to tic: Measurement, characteristics, and correlates in older adolescents and adults. *Behavior Therapy, 45*(2), 177–186.

Reese, H. E., Vallejo, Z., Rasmussen, J., et al. (2015). Mindfulness-based stress reduction for Tourette syndrome and chronic tic disorder: A pilot study. *Journal of Psychosomatic Research, 78*, 293–298.

RIzzo, R., Curatolo, P., Gulisano, M., Virzì, M., Arpino, C., & Robertson, M. M. (2007). Disentangling the effects of Tourette syndrome and attention deficit hyperactivity disorder on cognitive and behavioral phenotypes. *Brain Development, 29*, 413–420.

Rizzo, R., Gulisano, M., Cali, P. V., & Curatolo, P. (2013). Tourette syndrome and comorbid ADHD: Current pharmacological treatment options. *European Journal of Paediatric Neurology, 17*, 421–428.

Robertson, M. (2006). Mood disorders and Gilles de la Tourette's syndrome: An update on prevalence, etiology, comorbidity, clinical associations, and implications. *Journal of Psychosomatic Research, 61*, 349–358.

Robertson, M. M., Cavanna, A. E., & Eapen, V. (2015). Gilles de la Tourette syndrome and disruptive behavior disorders: Prevalence, associations, and explanation of the relationships. *The Journal of Neuropsychiatry and Clinical Neurosciences, 27*, 33–41.

Robertson, M., Williamson, F., & Eapen, V. (2006). Depressive symptomatology in young people with Gilles de la Tourette syndrome—A comparison of self-report scales. *Journal of Affective Disorders, 91*, 265–268.

Sambrani, T., Jakubovski, E., & Muller-Vahl, K. R. (2016). New insights into clinical characteristics of Gilles de la Tourette syndrome: Findings in 1032 patients from a single German center. *Frontiers in Neuroscience, 10*, 415.

Scahill, L., Riddle, M. A., McSwiggin-Hardin, M., et al. (1997). Children's Yale-Brown obsessive compulsive scale: Reliability and validity. *Journal of the American Academy of Child and Adolescent Psychiatry, 36*, 844–852.

Schandry, R. (1981). Heart beat perception and emotional experience. *Psychophysiology, 18*(4), 483–488.

Schunke, O., Grashorn, W., Kahl, U., Schottle, D., Haggard, P., Munchau, A., et al. (2016). Quantitative sensory testing in adults with Tourette syndrome. *Parkinsonism & Related Disorders, 24*, 132–136.

Scotto Rosato, N., Correll, C. U., Pappadopulos, E., Chait, A., Crystal, S., & Jensen, P. S. (2012). Treatment of maladaptive aggression in youth steering committee. Treatment of maladaptive aggression in youth: CERT guidelines II. Treatments and ongoing management. *Pediatrics, 129*, e1577–e1586.

Selvi, Y., Atli, A., Aydin, A., Besiroglu, L., Ozdemir, P., & Ozdemir, O. (2011). The comparison of aripiprazole and risperidone augmentation in selective serotonin reuptake inhibitor-refractory obsessive-compulsive disorder: A single-blind, randomised study. *Human Psychopharmacology, 26*, 51–57.

Servello, D., Porta, M., Sassi, M., Brambilla, A., & Robertson, M. M. (2008). Deep brain stimulation in 18 patients with severe Gilles de la Tourette syndrome refractory to treatment: The surgery and stimulation. *Journal of Neurology, Neurosurgery and Psychiatry, 79*, 136–142.

Shapiro, A. K., Shapiro, E. S., Young, J. G., & Feinberg, T. E. (1988). *Gilles de la Tourette syndrome*. New York, NY: Raven Press.

Silva, R. R., Alpert, M., Pouget, E., Silva, V., Trosper, S., Reyes, K., et al. (2005). A rating scale for disruptive behavior disorders, based on the DSM-IV item pool. *The Psychiatric Quarterly, 76*(4), 327–339.

Simonoff, E., Pickles, A., Charman, T., Chandler, S., Loucas, T., & Baird, G. (2008). Psychiatric disorders in children with autism spectrum disorders: Prevalence, comorbidity, and associated factors in a population-derived sample. *Journal of the American Academy of Child and Adolescent Psychiatry, 47*, 921–929.

Steer, R. A., Beck, A. T., Riskind, J. H., & Brown, G. (1986). Differentiation of depressive disorders from generalized anxiety by the Beck Depression Inventory. *Journal of Clinical Psychology, 42*(3), 475–478.

Steinberg, T., Shmuel Baruch, S., Harush, A., Dar, R., Woods, D., Piacentini, J., et al. (2010). Tic disorders and the premonitory urge. *Journal of Neural Transmission, 117*(2), 277–284.

Stephens, R. J., Chung, S. A., Jovanovic, D., et al. (2013). Relationship between polysomnographic sleep architecture and behavior in medication-free children with TS, ADHD, TS and ADHD, and controls. *Journal of Developmental Behavioral Pediatrics, 34*, 688–696.

Sutherland Owens, A. N., Miguel, E. C., & Swerdlow, N. R. (2011). Sensory gating scales and premonitory urges in Tourette syndrome. *TheScientificWorldJournal, 11*, 736–741.

Taylor, S., Conelea, C. A., McKay, D., Crowe, K. B., & Abramowitz, J. S. (2014). Sensory intolerance: Latent structure and psychopathologic correlates. *Comprehensive Psychiatry, 55*(5), 1279–1284.

Termine, C., Luoni, C., Fontolan, S., Selvini, C., Perego, L., Pavone, F., et al. (2016). Impact of co-morbid attention-deficit and hyperactivity disorder on cognitive function in male children with Tourette syndrome: A controlled study. *Psychiatry Research, 243*, 263–267.

Tinaz, S., Malone, P., Hallett, M., & Horovitz, S. G. (2015). Role of the right dorsal anterior insula in the urge to tic in Tourette syndrome. *Movement Disorders, 30*(9), 1190–1197.

Tourette's Syndrome Study Group. (2002). Treatment of ADHD in children with tics: A randomized controlled trial. *Neurology, 58*, 527–536.

van der Salm, S. M., Tijssen, M. A., Koelman, J. H., & van Rootselaar, A. F. (2012). The bereitschaftspotential in jerky movement disorders. *Journal of Neurology, Neurosurgery, and Psychiatry, 83*(12), 1162–1167.

Wanderer, S., Roessner, V., Freeman, R., Bock, N., Rothenberger, A., & Becker, A. (2012). Relationship of obsessive-Comulsive disorder to age-related comorbidity in children and adolescents with Tourette syndrome. *Journal of Developmental & Behavioral Pediatrics, 33*, 124–133.

Woods, D. W., Piacentini, J., Himle, M. B., & Chang, S. (2005). Premonitory Urge for Tics Scale (PUTS): Initial psychometric results and examination of the premonitory urge phenomenon in youths with tic disorders. *Journal of Developmental and Behavioral Pediatrics, 26*(6), 397–403.

Worbe, Y., Mallet, L., Golmard, J. L., et al. (2010). Repetitive behaviours in patients with Gilles de la Tourette syndrome: Tics, compulsions, or both [serial online]? *PloS One, 5*, e12959.

Yu, D., Mathews, C. A., Scharf, J. M., et al. (2015). Cross-disorder genome-wide analyses suggest a complex genetic relationship between Tourette's syndrome and OCD. *American Journal of Psychiatry*, *172*, 82–93.

FURTHER READING

Muscatello, M. R. A., Bruno, A., Pandolfo, G., et al. (2011). Effect of aripiprazole augmentation of serotonin reuptake inhibitors or clomipramine in treatment-resistant obsessive-compulsive disorder. *The Journal of Clinical Psychiatry*, *31*, 174–179.

SECTION 5

Nonmotor Parkinson's and Future Directions

CHAPTER FIFTY-FOUR

Nonmotor Parkinson's and Future Directions

Nataliya Titova*, K Ray Chaudhuri†,‡,1

*Federal State Budgetary Educational Institution of Higher Education "N.I. Pirogov Russian National Research Medical University" of the Ministry of Healthcare of the Russian Federation, Moscow, Russia
†National Parkinson Foundation International Centre of Excellence, King's College London and King's College Hospital, London, United Kingdom
‡The Maurice Wohl Clinical Neuroscience Institute, King's College London, London, United Kingdom
[1]Corresponding author: e-mail address: ray.chaudhuri@nhs.net

Contents

Abstract

Nonmotor symptoms (NMS) of Parkinson's disease (PD) are integral to the condition largely regarded as a motor syndrome. A range of NMS underpin the prodromal stage of Parkinson's and are present with variable frequency, range, and nature across the whole journey of a patient with Parkinson's from the onset of the motor disease to palliative stage. These symptoms also are key determinants of quality of life of the patient as well as the carer. Despite this, recognition management and focused treatment of NMS of PD remain poor. Future would, therefore, need to focus on better definition and management of NMS of PD. This would include development of robust animal models of specific NMS such as cognitive, sleep, and autonomic dysfunctions as well as pain to understand the mechanistic pathways of these symptoms. In turn this will lead to better drug development using a bench to bedside model. Nonmotor clinical subtypes of PD have also been described and, in future, proper biomarkers will consolidate these findings in addition to defining the natural history of the subtypes. Revised versions of established scales and questionnaires will enable the adoption of good clinical practice

International Review of Neurobiology, Volume 134
ISSN 0074-7742
http://dx.doi.org/10.1016/bs.irn.2017.05.017

with recognition of these subtypes in clinic. This will enhance the delivery of true subtype-specific therapies. Drug development should also include nondopaminergic and cell replacement restorative therapies with a nonmotor focus. An additional key area of future research would be the formalizing of true personalized medicine for PD. Personalized medicine pathways should concentrate on the role of exercise, complementary medicine as well as age, body weight, ethnicity on various NMS of PD. Genetics and pharmacogenetic developments in PD will add to the precision of the individualized approach.

1. INTRODUCTION

James Parkinson was an astute observer of clinical symptoms and signs and in his essay of 1817 described the classic motor features of what became known as Parkinson's disease (PD). He however also described a myriad of nonmotor symptoms (NMS) ranging from delirium to sleep dysfunction. For many years, comprehensive assessment and recognition of these NMS of PD remained poorly defined, relatively unknown and underresearched. The 21st century saw a recognition of the impact of global burden of NMS in PD addressed by validated tools such as the self-completed NMS questionnaire and the physician completed NMS scale. Quality-of-life-based studies showed the major impact of the NMS burden on the patient as well as caregiver. Some NMS emerged as key clinical biomarkers for the prodromal phase of PD while nonmotor fluctuation, in response to dopaminergic therapy, became a well-established concept. By 2017, several clinical trials are addressing some key NMS of PD such as dementia, depression, pain, sleep dysfunction, and constipation using NMS tools as outcome measures. A revised criteria for the diagnosis of PD are now proposed including NMS (Schrag, Sauerbier, & Chaudhuri, 2015). Therefore, there has been much progress and to some extent, the circle is leading back to the observations of James Parkinson that PD is a complex disorder. We have also recognized that the clinical phenotypic heterogeneity of PD is not simply related to motor dysfunction, but a range of nonmotor dysfunctions related to underlying neurotransmitter dysfunction (Titova, Padmakumar, Lewis, & Chaudhuri, 2016). The consequent clinical scenario is the emergence of specific nonmotor subtypes which would need to be better defined for prognostic purposes. Such strategies would also help to enrich patient groups selected for specific NMS targeted clinical trials. In our view, therefore, the future of nonmotor PD would comprise of (Fig. 1):

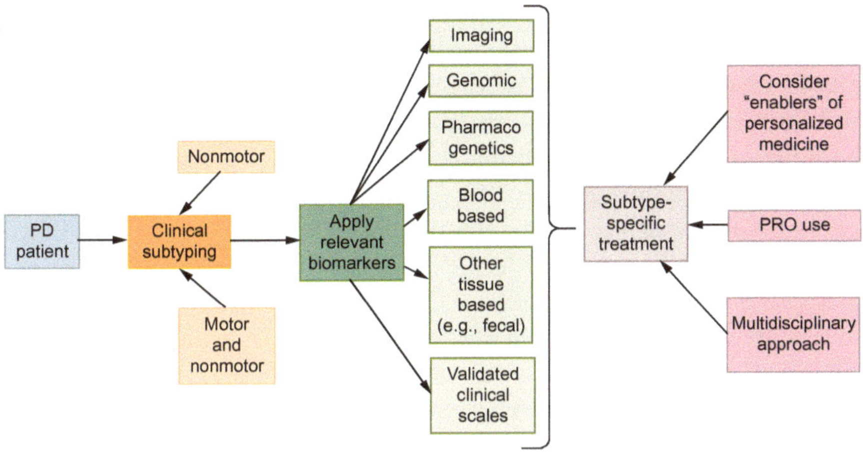

Fig. 1 A proposed strategy for clinical subtyping of PD patients in clinic. Imaging could include magnetic resonance imaging, ultrasound, and functional imaging. PRO = patient-related outcome. "Enablers" of personalized medicine include consideration of genetics and pharmacogenetics, age, lifestyle, personality, comorbidity, body weight, ethnicity, and pharmacoeconomics.

- Development of robust animal models for specific NMS in PD
- Remove barriers toward the understanding of NMS in PD
- Validation of the nonmotor subtypes of PD (using multimodal biomarkers)
- Better evidence base for the treatment of specific NMS in PD based on mechanistic data (bench to bedside translation)
- Make NMS assessment (including burden grading) an obligatory assessment in the clinic
- Personalized medicine (combining precision, with subtype-specific and pharmacogenetic approach)
- Exploration of the role of exercise and complementary medicine in the management of NMS in PD.

2. ANIMAL MODELS

Animal models which can be pathfinders to study-specific NMS in PD are an unmet need in spite of the extensive use of experimental models of PD. Such models have been extensively used for studying the dopaminergic system and motor response to dopamine replacement therapies. The methodologies have included lesioning the nigrostriatal pathway using toxins and thus not reflecting the widespread pathological changes involving multiple neurotransmitters that underpin the NMS of PD. Most NMS in PD are only partially responsive to dopaminergic medication. These NMS are largely related to the degeneration of the nondopaminergic nuclei in the brain and little research has been carried out to develop animal models based on functional abnormalities of these central and peripheral nervous system area. A list of animal models that have been used to delineate aspects of NMS are shown in Table 1. Future should see robust animal models of specific NMS such as dementia, sleep dysfunction such as rapid eye movement behavior disorder (RBD), excessive daytime somnolence and dysautonomia so that focused and tailored therapy can occur. Unraveling NMS related to common genetic mutations in animal models of PD for instance, glucocerebrocidase mutation may also help precision medicine-based treatment in future. Another example would be the validation of the prion hypothesis using relevant animal models. This hypothesis suggests that neuronal α-synuclein aggregates can be transported intraaxonally to different brain regions and taken up by surrounding neurons. Animal model-based experiments may indicate whether a prion like propagation could be possible from the gut to the brain.

3. OVERCOMING BARRIERS TOWARD THE UNDERSTANDING OF NMS IN PD

A key problem underlying the nonrecognition of NMS in clinical practice has been the lack of understanding of the impact of NMS on quality of life (Martinez-Martin, Rodriguez-Blazquez, Kurtis, & Chaudhuri, 2011). NMS thus truly represent the hidden face of PD. A recent study by Hurt et al. (2017) explored probable barriers by studying PD patients using two theoretical frameworks: the Common Sense Model of illness perception and Theoretical Domains Framework (Leventhal, Nerenz, & Steele, 1984). The notion of uncertainty about the relationship of NMS to PD, as well as the lack of clear guidance regarding treatments, were commonly cited problems. In addition, embarrassment and communication difficulties related to potentially sensitive symptoms such as sexual dysfunction also led to barriers. This has also been shown in a previous European multicenter study (Chaudhuri et al., 2010). A list of potential barriers, unearthed in this work, are shown in Table 2.

Future strategies should address these barriers so as not to compromise management of PD related to undeclared and unexplored NMS.

Table 1 Parkinsonian Animal Models Which Could Address Arrange of Nonmotor Symptoms and Underlying Pathophysiology

6-OHDA lesioned rodents	• Olfaction • Sensory/pain threshold • Sleep/wakefulness • Circadian rhythms • Cognitive function • Altered cardiovascular function • Bladder hyperactivity • Altered motility of gastrointestinal tract	Duty and Jenner (2011)
α-Synuclein overexpression (ASO = Thy1-aSYN) mice	• Olfaction • Autonomic • Constipation • Sleep (circadian dysfunction) • Cognition	Chesselet et al. (2012)
MPTP-treated primates	• Bladder hyperreflexia • Constipation • Drooling • Altered cardiovascular function • Sleep disturbance • Cognitive disturbance	Duty and Jenner (2011)
Mice model of intragastric rotenone administration	• α-Synuclein accumulation in dorsal vagal nucleus • Potential for investigating autonomic symptoms such as constipation • Sleep disturbances	Pan-Montojo et al. (2010)
Göttingen minipigs (Ellegaard Minipigs ApS)	• Cognition • Sleep disturbances	Lind et al. (2007)
GBA deficiency mouse models	• Memory problems • Cognition problems	Sardi et al. (2011)
Transgenic mouse model overexpressing LRRK2 gene mutation	• Gastrointestinal dysfunction • Olfactory dysfunction	Bichler, Lim, Zeng, and Tan (2013)
DJ-1 knocked out	• Cognition disturbances	Aron et al. (2010)

Continued

Table 1 Parkinsonian Animal Models Which Could Address Arrange of Nonmotor Symptoms and Underlying Pathophysiology—cont'd

Parkin knockout model	• Anxiety • Cognition	Zhu et al. (2007)
VMAT2-deficient mice	• Olfactory discrimination • Delayed gastric emptying • Sleep disturbances • Anxiety-like behavior • Depression	Taylor et al. (2009)
Cycad-fed rats	• Excessive daytime somnolence • RBD	McDowell and Chesselet (2012)

OHDA: hydroxydopamine; *MPTP*: 1-methyl-4-phenyl-1,2,3,6-tetrahydropyridine; *RBD*: rapid eye movement behavior sleep disorder; *VMAT*: vesicular monoamine transporter; *LRRK2*: leucine-rich repeat kinase 2; *GBA*: glucocerebrosidase.

Adapted from Todorova, A., Jenner, P., & Ray Chaudhuri, K. (2014). Non-motor Parkinson's: Integral to motor Parkinson's, yet often neglected. *Practical Neurology, 14*(5), 310–322. http://dx.doi.org/10.1136/practneurol-2013-000741; Titova, N., Padmakumar, C., Lewis, S. J., & Chaudhuri, K. R. (2016). Parkinson's: A syndrome rather than a disease? *Journal of Neural Transmission*. http://dx.doi.org/10.1007/s00702-016-1667-6.

4. VALIDATION OF THE NONMOTOR SUBTYPES OF PD (USING MULTIMODAL BIOMARKERS)

Studies suggest that commonly described motor subtypes of PD may be inconstant and variable in follow up, while nonmotor subtypes, based on underlying neurotransmitter dysfunctions, may be more robust although the matter remains somewhat controversial (Marras & Chaudhuri, 2016; Titova et al., 2016). Identification of specific subtypes underpinned by potential neurochemical and neurotransmitter deficits are now possible in some cases using functional imaging or other biomarker-based studies. A future strategy should be to better validate these specific nonmotor subtypes (cholinergic, serotonergic, noradrenergic, and all underpinned by dopaminergic deficiency) so that subtype-specific treatment strategy, as a strand of personalized medicine for PD, can be applied (Fig. 1). Whether specific genetic mutations can be markers for certain NMS (dementia, sleep dysfunction, and pain) will need to be researched in future. Although a genetic basis for side effects of dopamine replacement therapy has been described, the evidence base is not strong. Future research should focus on large-scale pharmacogenetic studies to clearly define the role of genetic mutations in determining the susceptibility of PD patient to dopamine replacement therapy-related adverse events. In most cases, these side effects are nonmotor in nature

Table 2 Common Sense Model of Illness Perception in PD Which May Act as Barriers to Seeking Help for NMS in PD

CSM Domains	Symptom Belief	Number of Interviews (%)	Example Quote
Identity	Uncertainty whether the symptom is part of Parkinson's disease	16 (67)	I just wonder if any other … is this normal, is it Parkinson's related or is it age-related?
Consequences	Belief that the symptom is not serious	13 (54)	It's slightly tedious—it's not going to kill me
Consequences	Acceptance of the symptom as "part of life"	7 (29)	The trouble is it sort of becomes sort of part of your life to deal with it, and you do not see … eventually you do not see it as a problem
Personal control	Belief that self-help strategies are a preferable or acceptable option	12 (50)	So I have tried to sort of keep a diary on occasions just to sort of try and work out for myself what, what triggers it to be worse and when my better days are, and you know, what I can do to you know try and ease … like I say, to try and help myself
Treatment control	Uncertainty that effective treatment is available	19 (79)	You sometimes think there's nothing they can do so it's not worth saying anything
Treatment control	Concern the treatment of symptom will require change of PD medication or taking extramedication	18 (75)	I mean I'm a little bit worried um, you know about changing the medication, so as long as, the longer I can stay on the low, lowest dose
Emotional representations	A perception of the symptom as embarrassing	11 (46)	It's an embarrassment to me I think is the best way to describe it
Timeline	Fluctuating presentation	12 (50)	It's worse when I get tired and sometimes when I go to bed and I'm shaking, it's difficult to get … to relax, to get to sleep, but um, there are other times it's not too bad and I can get to sleep straight away

From Hurt, C. S., Rixon, L., Chaudhuri, K. R., Moss-Morris, R., Samuel, M., & Brown, R. G. (2017). Identifying barriers to help seeking for non-motor symptoms in people with Parkinson's disease. *Journal of Health Psychology*. http://dx.doi.org/10.1177/1359105316683239.

(psychosis, impulse control disorders, and excessive daytime sleepiness). Development of a strong multimodal and practical biomarker approach is therefore crucial and required. Clear and objective biomarker-driven definition of clinical subtypes will also enable enrich populations for clinical trials of specific NMS of PD as well as define natural history (Schrag et al., 2015).

5. BETTER EVIDENCE BASE FOR TREATMENT OF SPECIFIC NMS IN PD

While major advances have been made for treating motor symptoms of PD and the motor complications of dopaminergic therapy, evidence base for treatment of NMS in PD remains poor and considered an unmet need (Schrag et al., 2015). Many disabling NMS such as dribbling of saliva, dementia, psychosis, sleep dysfunction, and dysautonomia have no strong level 1 evidence to support treatment. Treatment is thus empirical and usually driven by expert opinions which may be variable and in some situations no treatment is offered. Most clinical trials still do not use validated NMS tools as outcome measures and efficacy of a drug is usually based on motor response in PD. In future, we should see a reversal of this key drawback in our management strategy of PD. Better awareness of the importance and impact of NMS in PD of prescribing committees as well as policymakers is key to the implementation of this development. Management and awareness of nonmotor fluctuation is also poorly researched, and future strategies should focus on specific treatment paradigms where nonmotor fluctuations of PD may be more disabling than motor fluctuations. New drugs (dopaminergic and nondopaminergic) are being developed, focusing on novel targets, but largely for motor symptoms of PD. Examples are the glucagon-like peptide 1 agonist, urate precursor agents, or calcium channel antagonists. Nonmotor effects of these drugs are unknown and research needs to address nonmotor effects of these agents if there are any. Only such measures would make clinical application of these new drugs relevant to the patient in future. Similarly, future research should also focus on unraveling the nonmotor effects of conventional drugs as well as advanced therapies including neurosurgical approaches. NMS measures as surrogate markers or outcome measures in cell transplantation-based therapies is also a key unmet need and needs to be explored in the future. Mesenchymal stem cell therapy-related management of NMS in PD has been described and could become an important development of cell therapy in PD (Pantcheva, Reyes, Hoover, Kaelber, & Borlongan, 2015).

Mechanistic knowledge of NMS is gradually unfolding. This may help define better treatments for specific NMS in PD, currently a key unmet need. No robust treatment strategies exist for dementia, aspects of sleep dysfunction, autonomic failure, fatigue, apathy, and gastrointestinal problems in PD. A multitude of studies have shown the impact of these NMS on quality of life of patients with PD and carers. Animal model-based unraveling of the pathways underpinning, these NMS may open up new drug discovery via the bench to bedside translational route (Fig. 1).

Currently, cell-based therapies only address the dopaminergic system and not the nondopaminergic pathways which are dysfunctional in PD. As such management of NMS using cell-based therapies remains suboptimal. Mesenchymal stem cells which may be capable of multiple lineage differentiation and be able to secrete growth factors and therapeutic molecules may rescue the dopaminergic system as well as other neurotransmitter systems affected in PD. In theory, this will have effects on a range of NMS of PD and thus needs to be studied. An approach for the future would, therefore, be the development of stem cells that are capable of differentiating into nondopaminergic neurotransmitter phenotypes or be able to secrete growth factors to rescue the nondopaminergic system. A multimodal approach combining activity on dopaminergic as well as nondopaminergic neurons will help. It is known, for instance, that many different nondopaminergic receptors that regulate behavior and other NMS of PD such as GABA, cannabinoid, purinergic, and opioid also modify striatal dopamine levels (Cachope & Cheer, 2014). This strategy, which is likely to have a major effect on NMS of PD, has not been explored properly.

Gene therapy, including viral vector-mediated growth factor or neurotransmitter-synthesizing enzyme expression, active (using short peptides homologous to α-synuclein attached to a carrier) and passive (using monoclonal antibodies against α-synuclein) immunizations are being developed for managing PD. The effect of these strategies on NMS of PD is unknown and such information will enhance the role of these therapies in the clinical arena.

6. MAKE NMS ASSESSMENT (INCLUDING BURDEN GRADING) AN OBLIGATORY ASSESSMENT IN THE CLINIC

Studies suggest that even in expert centers for PD management, NMS are often undeclared and therefore untreated (Chaudhuri et al., 2010). Parkinson's UK, the foremost patient charity of the UK, runs a yearly audit of Parkinson services and the 2010 report showed that NMS are not recorded in 28% of elderly care and up to 54% of neurology services (Parkinson's UK, 2010). This lack of good clinical practice relates to demands on time during clinic consultations as well as the lack of use of self-declaration tools which allow patients to "flag" symptoms. Future paradigms of clinical care should thus have obligatory use of self-completed NMS tools in the clinic so that key NMS are not missed (Fig. 2).

7. PERSONALIZED MEDICINE

As discussed previously a true paradigm shift in PD would occur when we adopt holistic delivery of personalized medicine usually considered to be related to genomic precision medicine. Personalized medicine is however a much broader construct including many facets as discussed in the relevant chapter in this book (Fig. 1). Future management strategies for PD should adopt a personalized medicine strategy whereby therapy could be directed in a precision manner for a preventative approach related to gene-defined "at risk" cases for PD. In other symptomatic cases, treatment could be either tailored based on drug-related side effects and underlying pharmacogenetic factors or more clinically dictated subtype-specific approach. Individual personalized medicine "checklists" may help direct proper delivery of therapy so that issues such as lifestyle, age, ethnicity, and personality are all taken into account.

8. EXPLORATION OF THE ROLE OF EXERCISE AND COMPLEMENTARY MEDICINE IN THE MANAGEMENT OF NMS IN PD

Exercise and complementary therapies have remained of anecdotal value until recently when several studies have shown the valuable role of exercise on nonmotor aspects of PD in animal models and humans. The role of complementary therapies, such as acupuncture and yoga, are also being developed. Nutrition is emerging as an important aspect for management of some NMS of PD such as constipation, and probiotic use may, for instance, be beneficial. The future should bring about a large-scale gathering of evidence for the precise role of exercise and other complementary therapies including diet in PD and its role in alleviating aspects of a range of NMS.

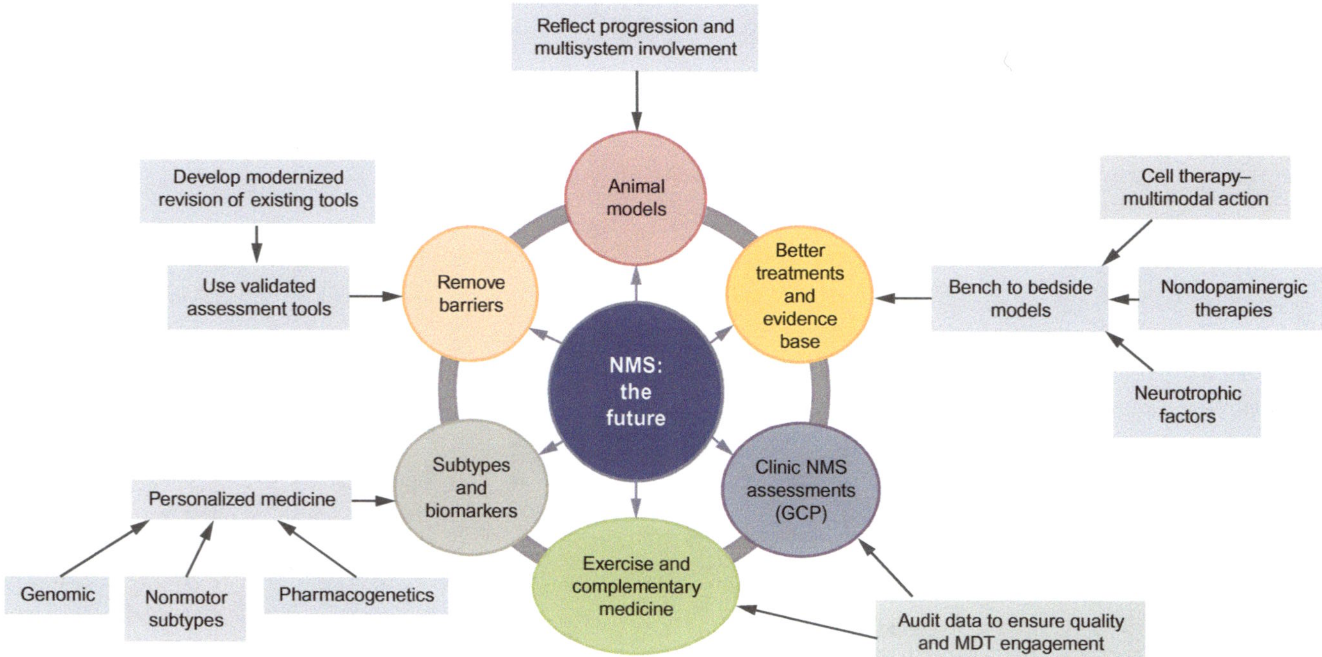

Fig. 2 A diagram outlining unmet needs that have to be addressed so as to consolidate the role of nonmotor symptoms in PD. MDT = multidisciplinary therapy, GCP = good clinical practice.

9. CONCLUSIONS

The nonmotor aspect of PD is complex and wide ranging in pathophysiology and clinical phenotypic expression. Treatment of these NMS, a key determinant of quality of life in PD, remains underresearched and, as such, poor in quality. Future strategies should have a concerted effort toward development of good and robust animal models for specific NMS of PD to provide better understanding of the pathophysiology, which will enable development of focused and targeted therapeutic strategies. In part, the relative nonrecognition of the burden of NMS and its effects are related to perceived barriers and these need to be removed by education and communication (Table 2). Only then NMS will cease to be the hidden painful face of PD.

Financial disclosure/conflict of interest: The authors declare no conflict of interest.

REFERENCES

Aron, L., Klein, P., Pham, T. T., Kramer, E. R., Wurst, W., & Klein, R. (2010). Pro-survival role for Parkinson's associated gene DJ-1 revealed in trophically impaired dopaminergic neurons. *PLoS Biology*, *8*(4), e1000349. http://dx.doi.org/10.1371/journal.pbio.1000349.

Bichler, Z., Lim, H. C., Zeng, L., & Tan, E. K. (2013). Non-motor and motor features in LRRK2 transgenic mice. *PLoS One*, *8*(7), e70249. http://dx.doi.org/10.1371/journal.pone.0070249.

Cachope, R., & Cheer, J. F. (2014). Local control of striatal dopamine release. *Frontiers in Behavioral Neuroscience*, *8*, 188. http://dx.doi.org/10.3389/fnbeh.2014.00188.

Chaudhuri, K. R., Prieto-Jurcynska, C., Naidu, Y., Mitra, T., Frades-Payo, B., Tluk, S., et al. (2010). The nondeclaration of nonmotor symptoms of Parkinson's disease to health care professionals: An international study using the nonmotor symptoms questionnaire. *Movement Disorders*, *25*(6), 704–709. http://dx.doi.org/10.1002/mds.22868.

Chesselet, M. F., Richter, F., Zhu, C., Magen, I., Watson, M. B., & Subramaniam, S. R. (2012). A progressive mouse model of Parkinson's disease: The Thy1-aSyn ("line 61") mice. *Neurotherapeutics*, *9*(2), 297–314. http://dx.doi.org/10.1007/s13311-012-0104-2.

Duty, S., & Jenner, P. (2011). Animal models of Parkinson's disease: A source of novel treatments and clues to the cause of the disease. *British Journal of Pharmacology*, *164*(4), 1357–1391. http://dx.doi.org/10.1111/j.1476-5381.2011.01426.x.

Hurt, C. S., Rixon, L., Chaudhuri, K. R., Moss-Morris, R., Samuel, M., & Brown, R. G. (2017). Identifying barriers to help seeking for non-motor symptoms in people with Parkinson's disease. *Journal of Health Psychology*. http://dx.doi.org/10.1177/1359105316683239.

Leventhal, H., Nerenz, D., & Steele, C. (1984). Illness representations and coping with health threats. In A. Baum & J. Singer (Eds.), *A handbook of psychology and health* (pp. 219–252). Hillsdale, NJ: Erlbaum.

Lind, N. M., Moustgaard, A., Jelsing, J., Vajta, G., Cumming, P., & Hansen, A. K. (2007). The use of pigs in neuroscience: Modeling brain disorders. *Neuroscience and Biobehavioral Reviews*, *31*(5), 728–751. http://dx.doi.org/10.1016/j.neubiorev.2007.02.003.

Marras, C., & Chaudhuri, K. R. (2016). Nonmotor features of Parkinson's disease subtypes. *Movement Disorders*, *31*(8), 1095–1102. http://dx.doi.org/10.1002/mds.26510.

Martinez-Martin, P., Rodriguez-Blazquez, C., Kurtis, M. M., & Chaudhuri, K. R. (2011). The impact of non-motor symptoms on health-related quality of life of patients with Parkinson's disease. *Movement Disorders*, *26*(3), 399–406. http://dx.doi.org/10.1002/mds.23462.

McDowell, K., & Chesselet, M. F. (2012). Animal models of the non-motor features of Parkinson's disease. *Neurobiology of Disease*, *46*(3), 597–606. http://dx.doi.org/10.1016/j.nbd.2011.12.040.

Pan-Montojo, F., Anichtchik, O., Dening, Y., Knels, L., Pursche, S., Jung, R., et al. (2010). Progression of Parkinson's disease pathology is reproduced by intragastric administration of rotenone in mice. *PLoS One*, *5*(1), e8762. http://dx.doi.org/10.1371/journal.pone.0008762.

Pantcheva, P., Reyes, S., Hoover, J., Kaelber, S., & Borlongan, C. V. (2015). Treating non-motor symptoms of Parkinson's disease with transplantation of stem cells. *Expert Review of Neurotherapeutics*, *15*(10), 1231–1240. http://dx.doi.org/10.1586/14737175.2015.1091727.

Parkinson's, U. K. (2010). *Parkinson's audit 2010 report*. Retrieved from London, UK.

Sardi, S. P., Clarke, J., Kinnecom, C., Tamsett, T. J., Li, L., Stanek, L. M., et al. (2011). CNS expression of glucocerebrosidase corrects alpha-synuclein pathology and memory in a mouse model of Gaucher-related synucleinopathy. *Proceedings of the National Academy of Sciences of the United States of America*, *108*(29), 12101–12106. http://dx.doi.org/10.1073/pnas.1108197108.

Schrag, A., Sauerbier, A., & Chaudhuri, K. R. (2015). New clinical trials for nonmotor manifestations of Parkinson's disease. *Movement Disorders*, *30*(11), 1490–1504. http://dx.doi.org/10.1002/mds.26415.

Taylor, T. N., Caudle, W. M., Shepherd, K. R., Noorian, A., Jackson, C. R., Iuvone, P. M., et al. (2009). Nonmotor symptoms of Parkinson's disease revealed in an animal model with reduced monoamine storage capacity. *The Journal of Neuroscience*, *29*(25), 8103–8113. http://dx.doi.org/10.1523/jneurosci.1495-09.2009.

Titova, N., Padmakumar, C., Lewis, S. J., & Chaudhuri, K. R. (2016). Parkinson's: A syndrome rather than a disease? *Journal of Neural Transmission*. http://dx.doi.org/10.1007/s00702-016-1667-6. [Epub ahead of print].

Zhu, X. R., Maskri, L., Herold, C., Bader, V., Stichel, C. C., Gunturkun, O., et al. (2007). Non-motor behavioural impairments in Parkin-deficient mice. *The European Journal of Neuroscience*, *26*(7), 1902–1911. http://dx.doi.org/10.1111/j.1460-9568.2007.05812.x.

FSC
www.fsc.org
MIX
Papier | Fördert gute Waldnutzung
FSC® C083411

Kolibri 360 GmbH
Ferdinand-Jühlke-Straße 7
99095 Erfurt, Deutschland
produktsicherheit@kolibri360.de